ALGEBRA 1

Third Edition

Mark Wetzel
Gene Bucholtz
Tamera Knisely
Larry Hall

ALGEBRA 1
Third Edition

Mark Wetzel
Gene Bucholtz, MEd
Tamera Knisely
Larry Hall, MS

First Edition Authors
Kathy Pilger, EdD
Ron Tagliapietra, EdD

Contributing Author
Steve McKisic, MEd

Editor
Mary Schleifer

Bible Integration
Brian Collins, PhD
Bryan Smith, PhD

Project Manager
Kevin Neat

Consultants
Kathy Kohler, MEd
Steve McKisic, MEd
Kathy Pilger, EdD

Composition
Dzign Associates,
Patricia Tirado

Cover Design
Drew Fields

Illustrators
John Cunningham
Kathy Pflug
Del Thompson

Permissions
Sylvia Gass
Ashley Hobbs
Rita Mitchell

Photograph credits appear on pages 613–14.

Greenville, South Carolina 29609

ISBN 978-1-62856-562-1

15 14 13 12 11 10 9 8 7 6 5

Probing the Unknown

The Great Pyramid of Giza is a place of ancient innovation and modern mystery. We still don't know how it was even built! For 4,500 years people have probed this pyramid, the oldest and largest in Egypt, to uncover its secrets.

But how do you probe a pyramid without drilling holes in it? That's generally discouraged when you're dealing with one of the Seven Ancient Wonders of the World! Science and algebra have come to the rescue to gather data and piece it together in surprising ways that can give us breakthroughs.

A team of engineers and scientists from around the world have joined in a project called ScanPyramids. This project takes infrared images of the pyramid from different angles. These images can be plotted with algebra to form 3D models of the pyramid's innards, helping them understand what might be hiding just under the surface. What they've found is astounding.

There seems to be another corridor and room yet to be discovered, and they're big. Archaeologists analyzing some of the oldest Egyptian texts suggest that this room at the apex of the pyramid may contain a great iron throne made of meteorite.

Thousands of years later, the algebra that allowed Egyptians to innovate still helps us to probe the unknown. With just a few symbols, we can probe the unknown to make models and solve real-world problems. Algebra can help us model events like hurricanes and lightning. It can also help us solve problems like making a profit, improving the internet, or conserving energy to help people who bear God's image. When we innovate like this, we can celebrate not just the glory of exercising good and wise dominion over God's creation (Gen. 1:28), but also the glory of His creativity and character. The monuments of our creative work like the Great Pyramid are faint echoes of God's creative genius.

Take a Peek Inside!

We've designed this textbook with you in mind. We hope it will help you appreciate what it means to manage God's world even more. Flip through the following pages to see the features that we've included to help you succeed in *Algebra 1*. In the back of the book you'll find answers to selected odd-numbered problems, a glossary, and an index. Inside the back cover are handy quick references you'll need this year as you do your own innovating in *Algebra 1*.

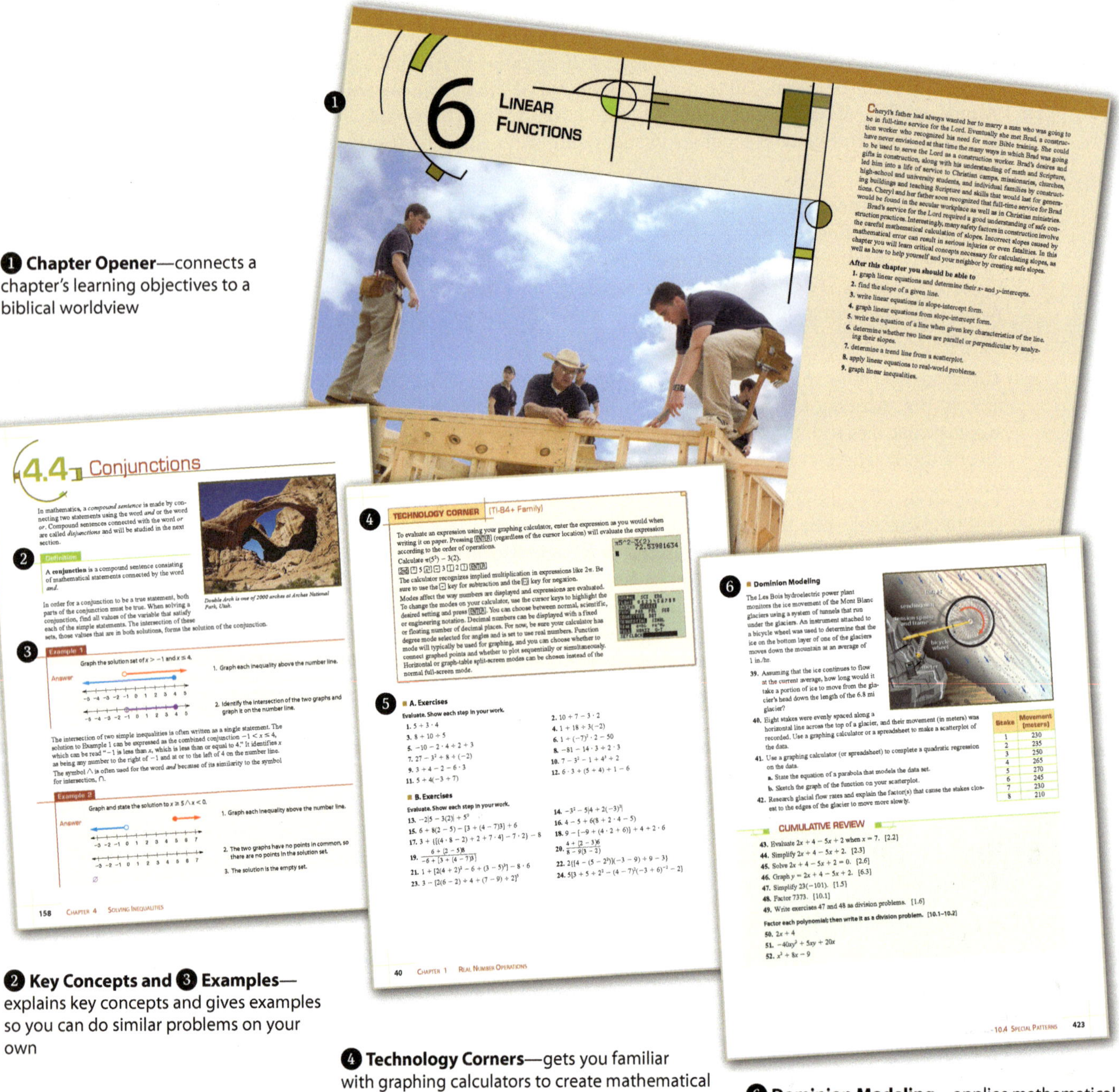

❶ **Chapter Opener**—connects a chapter's learning objectives to a biblical worldview

❷ **Key Concepts and** ❸ **Examples**—explains key concepts and gives examples so you can do similar problems on your own

❹ **Technology Corners**—gets you familiar with graphing calculators to create mathematical models and solve problems

❺ **Practice Exercises**—builds, maintains, and extends your skills with carefully sequenced practice exercises and reviews

❻ **Dominion Modeling**—applies mathematical concepts to solve real-world problems from a Christian worldview

CONTENTS

INTRODUCTION

What Is Algebra?

The Arabic word *al-jabr*, from which we get our word *algebra*, carries the idea of the reunion or restoration of broken parts. Algebra allows us to take the fragmented information of a problem and organize it into a useful mathematical form. In the ninth century AD one of the best known of the ancient mathematicians, al-Khwarizmi, wrote a very important book called *Kitab al-Jabr w'al-Muqabala*, roughly meaning "the book of integration and equation." From this book, algebra took shape as a separate mathematical discipline. Algebra is a structured system of analyzing and expressing both concrete and abstract conditions with known and unknown values.

The power of algebra is seen in determining unknown quantities by simply manipulating symbols according to established principles. The value of algebra is realized not only in its use in finding solutions to problems but also in its ability to develop our analytical thinking skills—skills that allow us to analyze and solve problems beyond the mathematics classroom.

Our Calling in God's World

The world declares the glory of God.

All creation exists to declare the glory of God—to reveal His greatness and His goodness (Ps. 19:1; Rom. 11:36). Humans, however, play a special role. Shortly after the creation of the first man and woman, God commanded them, "Be fruitful, and multiply, and replenish the earth, and subdue it: and have dominion over the fish of the sea, and over the fowl of the air, and over every living thing that moveth upon the earth" (Gen. 1:28). This dominion is a mandate to control and manage what God has provided. Proper management of His creation requires ever-increasing knowledge, wisdom, and skills.

The Christian student learns the same algebraic concepts as the nonbeliever, but the Christian applies these concepts with a different view of life and of the world. Viewing our world as the creation of God and ourselves as created in the image of God (Gen. 1:26–27) provides value and purpose for life as well as unique insight into the world around us. This Christian worldview affects how we apply the math skills we learn through the study of algebra.

Why Is Algebra Important?

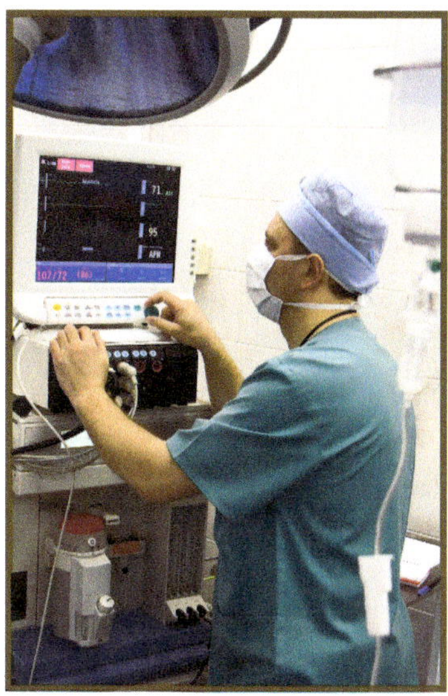

Algebra helps us create models and technology that enhance our lives.

The language of algebra expresses complicated relationships using a few simple symbols. It is used to model relationships in nature and science. Understanding the complex dynamics of our physical world requires a thorough understanding of mathematics. Scientific inquiry, from physics to business, depends on algebra's power of symbolic abstraction.

Our entire technological culture depends on algebra's ability to encapsulate the concepts and principles used in the study of related quantities. Algebra, as a tool of analysis, comprehension, and deduction, helps us solve problems in many areas of life, including business, science, engineering, building, medicine, technology, and manufacturing. In this way, algebra enables us to better fulfill God's command to manage His creation.

But Is It Worth the Effort?

Algebra provides powerful tools for understanding our world.

Algebra is a worthwhile study for all students. It develops our thinking skills, training our minds to analyze, organize, rationalize, evaluate, solve, and express results with detailed precision. These skills are valuable in every facet of life. Many careers demand the attention to detail and the analytical thinking skills that are developed in the mathematics classroom.

But the development of thinking skills is often a long and sometimes even a painful process. So is it worth it? Algebra provides the skills and experiences required for future education, careers, and challenges in life. The tremendous benefits of proper preparation make the difficulties and challenges well worth the effort. Without an understanding of algebra, many of life's challenges become much more difficult. A thorough understanding of the foundations of mathematics is the key that unlocks many opportunities.

Fulfilling our responsibilities at work, home, and church requires the attainment of much knowledge and many skills through diligence and hard work. For some, algebra hides inside the instruments of their trade as an unseen partner, but for others it is never any farther away than the end of their pencil or their calculator's display. If Christians are to exercise dominion as God commands, algebra remains an important tool in completing His will.

The hand of the diligent shall bear rule: but the slothful shall be under tribute. Proverbs 12:24

1 Real Number Operations

"Clouds are not spheres, mountains are not cones, coastlines are not circles, and bark is not smooth, nor does lightning travel in a straight line," explains Benoit Mandelbrot in his 1982 book, *The Fractal Geometry of Nature*. Traditional geometric shapes are rapidly being replaced by fractals as the scientist's model of the seemingly random processes of the real world. By using fractals, scientists are now just beginning to understand some of the amazing complexity of God's created universe.

Fractals are rough or fragmented geometric shapes that are repeated at an ever smaller scale. They are generated by repeated applications of a simple procedure or *iteration*. In true mathematical fractals, this procedure is applied infinitely. Some of the first fractals were created by Helge von Koch in 1904 and Waclaw Sierpinski in 1915. Though fractals were initially developed to test new definitions, some mathematicians avoided studying these "monsters" due to their unusual and puzzling properties. With advances in computer graphics and Mandelbrot's investigations in the 1960s, the beauty of fractals became apparent and scientists began to harness their power to explain chaotic phenomena.

Meteorologists now use fractals to model lightning strikes and hurricanes. Fractals generate realistic landscapes for video games and simulate coastlines and watersheds in topographical studies. From the distribution of galaxies to the growth of crystals, fractals model the "ordered randomness" of reality at any scale. Biologists use fractals to simulate arteries and veins, the lungs, the brain, and DNA. Fractals have even been applied to the study of languages and financial markets. While much of the research being done today is beyond the scope of this course, many fractals can be easily generated and their foundational concepts explained using the basic operations of arithmetic.

After this chapter you should be able to

1. identify the standard subsets of the real numbers, their relationships to each other, and examples of each subset.
2. determine the union and intersection of sets and illustrate these operations using Venn diagrams.
3. graph numbers on a number line.
4. determine the absolute value of numbers and numerical expressions.
5. find the length and midpoint of a segment on a number line.
6. add, subtract, multiply, and divide rational numbers.
7. identify and apply properties of real numbers.
8. use unit multipliers to determine equivalent quantities with different units.
9. evaluate numerical expressions containing integral exponents.
10. evaluate numerical expressions using the order of operations.
11. produce simple fractals and generate expressions modeling their characteristics.

1.1 Sets of Numbers

In football gains and losses are represented by positive and negative integers.

Classifying objects into groups is an essential ability in fulfilling God's command to be wise managers of His creation. Scientists group living organisms into categories so they can efficiently study the many forms of life that surround us. In algebra you will usually deal with groups of numbers.

Definitions

A **set** is a collection of objects.

Each object in the set is called an **element** or **member** of the set.

A set with no elements is called the **empty set** or **null set** and is denoted { } or $\varnothing$.

Consider the following sets:

$A = \{1, 3, 7, 8, 10\}$ and $B = \{1, 4, 6, 10\}$.

Sets are named using capital letters and denoted using braces, { }, which are read "the set whose members are." The fact that 7 is an element of set A but not of set B can be stated symbolically as $7 \in A$ and $7 \notin B$.

Union and intersection are the two main set operations. To find the union of sets A and B, put the elements from A and B together into one set.

$A \cup B = \{1, 3, 4, 6, 7, 8, 10\}$

To find the intersection of sets A and B, list the elements that A and B have in common.

$A \cap B = \{1, 10\}$

Definitions

The **union** of sets ($\cup$) is the set of elements that appear in any of the sets.

The **intersection** of sets ($\cap$) is the set of elements common to all of the sets.

Sets are often represented in picture form using a Venn diagram. From it you can easily determine the elements in the union and in the intersection of the sets.

Hopefully there will not be a union of vehicles at this intersection.

Example 1

Using the following sets, complete each set operation and then illustrate the operation using a Venn diagram: $C = \{1, 2, 3, 4, 5, 6\}$ and $D = \{2, 4, 6, 8, 10\}$.

a. $C \cup D$ **b.** $C \cap D$

Answer

a. $C \cup D = \{1, 2, 3, 4, 5, 6, 8, 10\}$

Combine all the elements into one set.

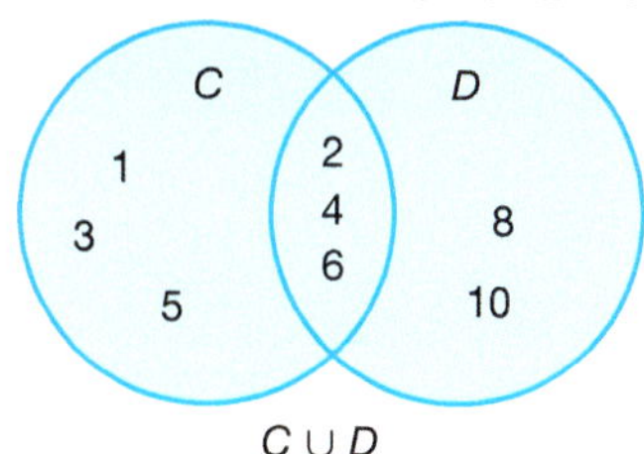

$C \cup D$

Shade all regions of both circles.

b. $C \cap D = \{2, 4, 6\}$

List the elements common to both sets.

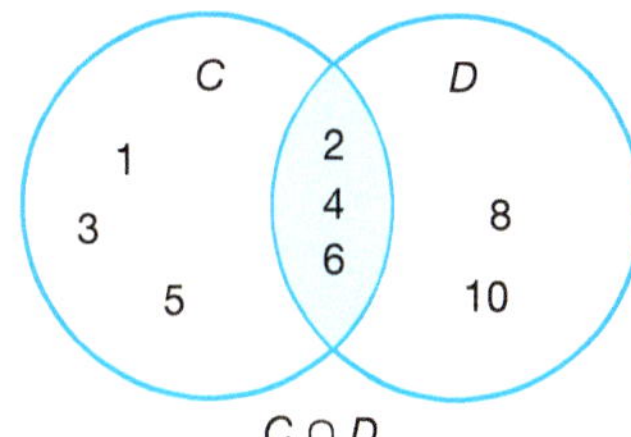

$C \cap D$

Shade the overlapping area.

Definition

One set is a **subset** of another set if every element of the first is contained in the second. $A \subseteq B$ means set A is a subset of set B.

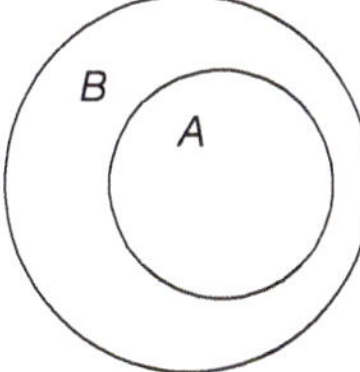

Example 2

Using sets C and D from the previous example, state whether each statement is true or false and explain why.

a. $\{5, 6\} \subseteq C$ **b.** $C \cup D \subseteq C$ **c.** $\varnothing \subseteq C$

d. $C \nsubseteq C$ **e.** $6 \subseteq D$ **f.** $C \cap D \subseteq D$

Answer

a. true Both 5 and 6 are in C.

b. false $C \cup D = \{1, 2, 3, 4, 5, 6, 8, 10\}$, which includes 8 and 10, neither of which are in C.

c. true There is no element in $\varnothing$ that is not in C. The empty set is a subset of every set.

d. false Every set is a subset of itself since every element in the set is in itself.

e. false $6 \in D$; it is not a subset of D. Instead, $\{6\} \subseteq D$.

f. true $C \cap D = \{2, 4, 6\}$, and all of these numbers are members of D.

Two of the first number sets that students encounter are the *natural* (or counting) numbers and the *whole* numbers.

$$\mathbb{N} = \{1, 2, 3, \ldots\} \text{ and } \mathbb{W} = \{0, 1, 2, 3, \ldots\}$$

The *ellipses* (...) mean that the numbers continue on forever in the same pattern. Therefore, the following statements are true:

$$4 \in \mathbb{N} \text{ and } -2 \notin \mathbb{W}.$$

If the number of elements in a set is a whole number, then the set is called a *finite set*. An *infinite set* is a set that is not finite. The set of natural numbers is an example of an infinite set since there is no largest natural number.

The set of *integers* is written as

$$\mathbb{Z} = \{\ldots, -3, -2, -1, 0, 1, 2, 3, \ldots\}.$$

The set of integers can be represented on a number line. Positive numbers are typically to the right of zero, while negative numbers are to the left of zero. Zero is neither positive nor negative.

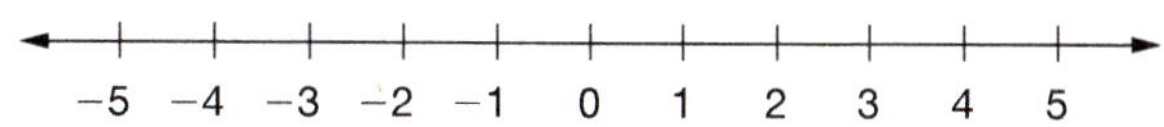

Since every counting number is a whole number, $\mathbb{N} \subseteq \mathbb{W}$. The integers are composed of the natural numbers, their opposites, and zero. Therefore, $\mathbb{N} \subseteq \mathbb{W} \subseteq \mathbb{Z}$. The integers are a subset of an even larger set of numbers, the rational numbers.

Rational numbers are numbers that can be written as a ratio of integers. Listing the set of rational numbers is not possible because there is not a regular pattern, but the set can be described as follows:

$$\mathbb{Q} = \left\{\frac{a}{b} \,\middle|\, a \in \mathbb{Z}, b \in \mathbb{Z}, \text{and } b \neq 0\right\},$$

which is read "the set of all ratios a over b such that a and b are integers and b is not equal to zero." The restriction $b \neq 0$ must be included because division by zero is not defined.

Definitions

The set of **rational numbers**, $\mathbb{Q}$, consists of numbers that can be expressed as a ratio of two integers when the denominator is not equal to zero.

The set of **irrational numbers**, $\mathbb{Q}'$ (read "Q prime"), consists of numbers that cannot be expressed as a ratio of integers.

Because any integer can be written over 1, integers are rational numbers. Decimals that either terminate or repeat are also rational numbers because they can be written as a ratio of integers.

The decimal form of irrational numbers never terminates and never repeats. The numbers π, $\sqrt{2}$, e, and $\sqrt{3}$ frequently appear in mathematical models of God's creation. The rational approximations 3.14159, 1.4142, 2.718, and 1.732 can be used in computations involving these numbers.

Example 3

State whether each number is rational or irrational.

a. 5 **b.** $\sqrt{9}$ **c.** $\sqrt{6}$ **d.** 3.14159

e. $\frac{\pi}{4}$ **f.** $0.1\overline{6}$ **g.** 0.121122111222...

Answer

a. rational $5 = \frac{5}{1}$

b. rational $\sqrt{9} = 3 = \frac{3}{1}$

c. irrational $\sqrt{6} \approx 2.449489743\ldots$; The decimal does not terminate and does not repeat.

d. rational The decimal terminates.

e. irrational This fraction cannot be expressed as a ratio of integers.

f. rational The decimal repeats.

g. irrational The decimal does not terminate and does not repeat.

The set of *real numbers*, $\mathbb{R}$, is the union of the sets of rational numbers, $\mathbb{Q}$, and irrational numbers, $\mathbb{Q}'$. The real numbers are a subset of an even larger set called the complex numbers, but this course will deal only with numbers in the real number system. A Venn diagram can be used to illustrate the relationship of these different sets of numbers. Notice that $\mathbb{N} \subseteq \mathbb{W} \subseteq \mathbb{Z} \subseteq \mathbb{Q}$.

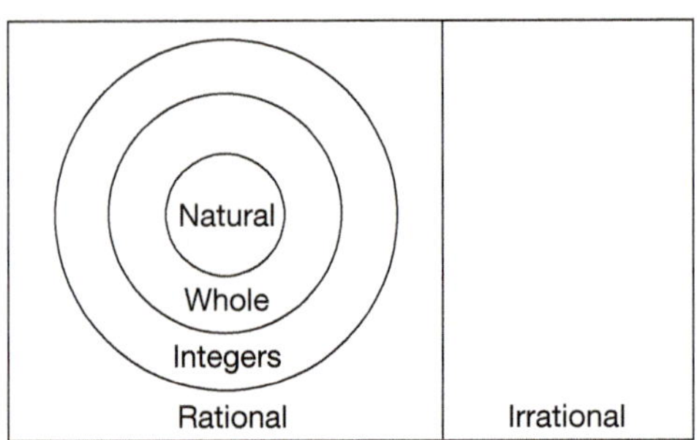

Real numbers

A. Exercises

Draw a copy of the Venn diagram representing the real number system; then place each number within the innermost set to which it belongs.

1. -6
2. $4.868668666\ldots$
3. 0.5
4. $31.313131\ldots$
5. $-\frac{4}{11}$
6. $-7.21721172111\ldots$
7. 0
8. $\sqrt{5}$
9. $\frac{3}{1}$
10. $\sqrt{4}$

True or false

11. $\mathbb{Q} \cup \mathbb{Q}' = \mathbb{R}$
12. $\mathbb{Q} \cap \mathbb{Q}' = \varnothing$
13. $\mathbb{Q} \subseteq \mathbb{Z}$
14. $\mathbb{N} \subseteq \mathbb{Z}$
15. $\mathbb{W} \nsubseteq \mathbb{Z}$
16. $\mathbb{Z} \nsubseteq \mathbb{Q}'$
17. $\varnothing \subseteq \mathbb{R}$
18. $\mathbb{N} \in \mathbb{R}$
19. $0 \in \varnothing$

B. Exercises

Use the following sets for exercises 20–34.
$A = \{1, 3, 5, \ldots\}$; $B = \{-2, -1, 1, 3\}$; $C = \{-1, 2, 5, 6\}$; $D =$ {odd integers}; $E =$ {even integers}

True or false

20. $3 \in B$
21. $\varnothing \in B$
22. $\{3, 5, 7\} \subseteq A$
23. $\{2, 3\} \subseteq C$
24. $5 \notin C$
25. $D \cup E = \varnothing$
26. $D \cap E = \mathbb{Z}$
27. C is an infinite set.
28. E is an infinite set.

Identify each set using set notation.

29. $A \cap B$
30. $C \cup \mathbb{N}$
31. $B \cap C$
32. $A \cap E$
33. $(A \cup B) \cap C$
34. $(A \cap B) \cup (A \cap C)$

Use the following sets for exercises 35–42.
$D =$ {US states bordering the Mississippi} = {MN, WI, IA, IL, MO, KY, TN, AR, MS, LA}
$E =$ {US states beginning with the letter M} = {ME, MA, MD, MI, MN, MO, MS, MT}
$F =$ {US states bordering the Gulf of Mexico} = {TX, LA, MS, AL, FL}
$G =$ {US states with two-word names} = {NM, ND, SD, WV, NC, SC, NJ, NY, RI, NH}
$H =$ {US states sharing a land border with Canada} = {WA, ID, MT, ND, MN, NY, VT, NH, ME}

35. $D \cup E$
36. $E \cup F$
37. $F \cup G$
38. $E \cap F$
39. $D \cap G$
40. $E \cap H$
41. $(F \cup H) \cap G$
42. $(D \cap E) \cup F$

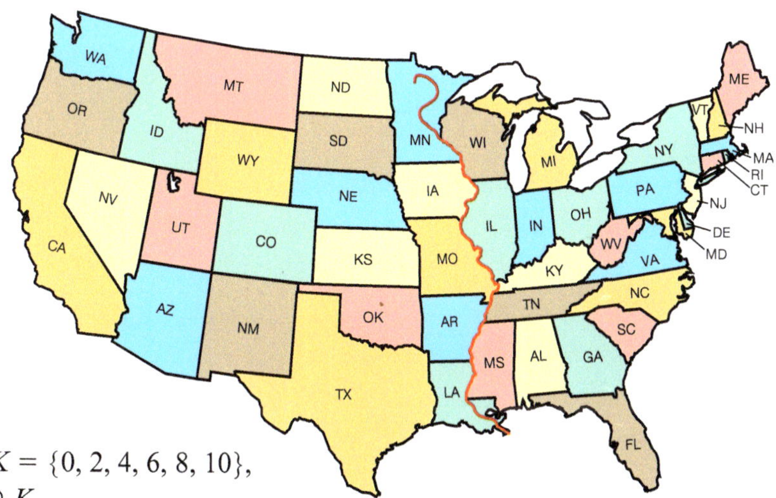

43. List the set of whole numbers.
44. List the set of integers.
45. If $J = \{-5, -3, -1, 0, 1, 2, 3\}$ and $K = \{0, 2, 4, 6, 8, 10\}$, draw a Venn diagram to represent $J \cap K$.

C. Exercises

For exercises 46–48, translate each sentence into symbols.

46. Nine is not an element of the union of sets R and S.

47. The set whose members are 5 and 7 is a subset of the set containing the positive odd integers.

48. The union of sets A and B is commutative.

49. Identify the finite and infinite sets in exercise 47.

50. Given any set A, list set I such that $A \cup I = A$. This set I is the identity element for the set operation of union.

Dominion Modeling

The Koch curve is one of the earliest fractals studied. It begins with a line segment having a length of 1 unit.

Drawing an equilateral triangle based on the middle third of the segment and then removing the middle third of the segment produces the first iteration.

Performing the same steps to each segment produces the next iteration: For each segment, draw an equilateral triangle on the middle third of the segment and then remove the middle third of the segment.

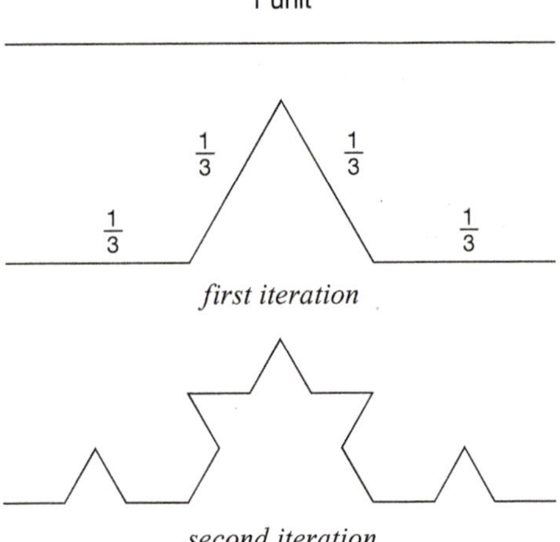

51. What is the total length of the first iteration of the curve?

52. Describe the second iteration of the Koch curve.

- **a.** How many segments do we now have?
- **b.** How long is each segment?
- **c.** What is the total length of the curve?

53. Draw the next iteration of the fractal and use your sketch to answer the following questions.

- **a.** How many segments do we now have?
- **b.** How long is each segment?
- **c.** What is the total length of the curve?

CUMULATIVE REVIEW

Find the greatest common factor (GCF) and the least common multiple (LCM) of each group of numbers.

54. 20 and 45

55. 48, 72, and 80

Perform the indicated operations.

56. $3\frac{2}{5} + 1\frac{4}{5}$

57. $\frac{7}{8} + \frac{2}{3}$

58. $3\frac{2}{5} - 1\frac{4}{5}$

59. $\frac{7}{8} - \frac{2}{3}$

60. $3\frac{2}{5} \times 1\frac{4}{5}$

61. $\frac{7}{8} \cdot \frac{2}{3}$

62. $3\frac{2}{5} \div 1\frac{4}{5}$

63. $\frac{7}{8} \div \frac{2}{3}$

1.2 Number Lines, Opposites, and Absolute Values

Many times in mathematics, particularly algebra, graphs and number lines will aid your understanding of algebraic ideas. Later in this chapter you will review how to use number lines to perform basic operations on integers. In other chapters you will learn about graphs and how to use them. Numbers can be represented in picture or symbolic form; the picture form of a number is shown on a number line.

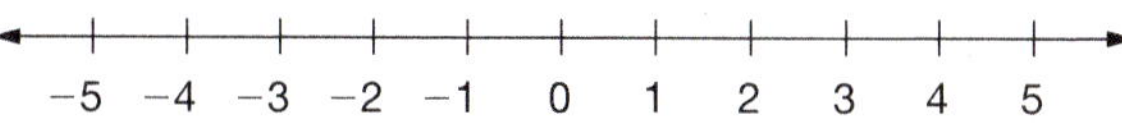

The arrows at the ends of the number line indicate that it extends without limit in both directions. Can you think of anything else that has no end? God has always existed and will always exist. The illustration of a number line can help our finite minds visualize the concept of infinity.

Measuring temperatures requires both positive and negative integers.

The point at zero is called the *origin*, and all numbers to the right of the origin are *positive numbers*. The numbers to the left of zero are *negative numbers*. A point on the number line is called the *graph* of a number and can be named with a capital letter. The number associated with the point is called the *coordinate*. The coordinate of a point can be thought of as the graph's "address."

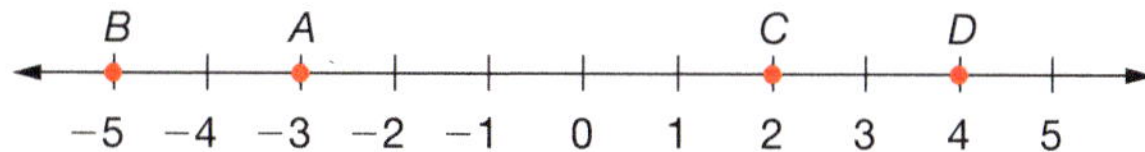

Note that the numbers increase in value as they go from left to right. Since A (the graph of -3) is farther to the right on the number line than B (the graph of -5), then -3 is greater than -5, which is written $-3 > -5$. Likewise, since C is to the left of D, $2 < 4$.

The arrow on the left end of the number line, indicating the continuation of the number line in that direction, includes numbers less than the last number on the line. Is -6 less than -5? Is 6° below 0 colder than 5° below 0?

Graphing a Number on a Number Line

1. Draw a line, choose an arbitrary point as the origin, and label it 0.
2. Mark the line with segments of equal length.
3. Number the marks consistently with the numbers to be graphed.
4. Place a point at the indicated number on the number line.

Example 1

Graph the following points on a number line.

A: 3 B: -2 C: 1.5 D: $-3\frac{2}{3}$

Answer

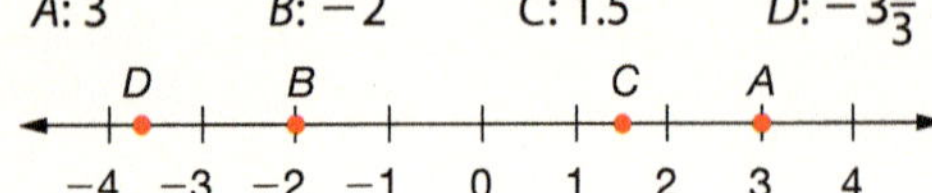

The number line does not always have to be divided into units of one. It can be divided into units greater than one or smaller than one. Number lines are typically drawn horizontally; but you can also graph on a vertical number line, where the positive numbers are typically located above zero and the negative numbers are below zero.

Example 2

Graph the following points on a vertical number line.

X: -0.5 Y: $1\frac{1}{4}$ Z: -1.4

Answer Notice that the number line is divided into tenths.

For point Y, $1\frac{1}{4} = 1.25$, so it is located halfway between 1.2 and 1.3.

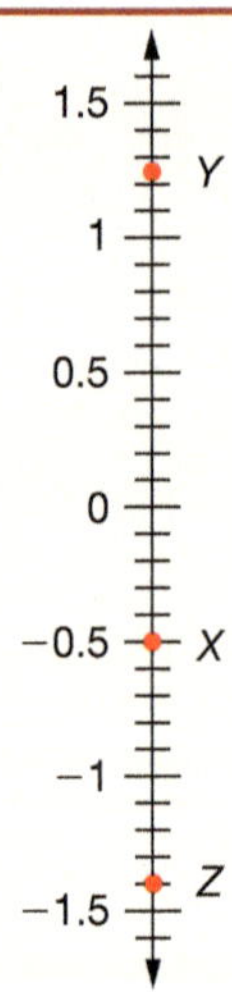

Definition

Opposite numbers are numbers that are the same distance from zero but are on opposite sides of zero.

Opposite times of day at the Arc de Triomphe in Paris

Example 3

State the opposite of each number.

a. 12 **b.** -37 **c.** -4.7 **d.** $\frac{\pi}{2}$

Answer **a.** -12 **b.** 37 **c.** 4.7 **d.** $-\frac{\pi}{2}$

How should $-(-5)$ be interpreted? The negative sign outside the parentheses indicates that you should take the opposite of -5. The result is 5.

Before you can become proficient in algebra, you must become very skilled in performing the basic operations of addition, subtraction, multiplication, and division with both positive and negative numbers. If you can perform these calculations quickly and accurately, you will have an advantage later in this course. A thorough knowledge of number lines, opposites, and absolute values is foundational for successfully performing operations with integers.

Definition

The **absolute value** of a number, denoted $|x|$, is the number of units (distance) between zero and the number's graph on a number line.

Example 4

Use a number line to find each absolute value.

a. $|4|$ **b.** $|-3|$

Answer

a. $|4| = 4$ 4 is 4 units from 0.

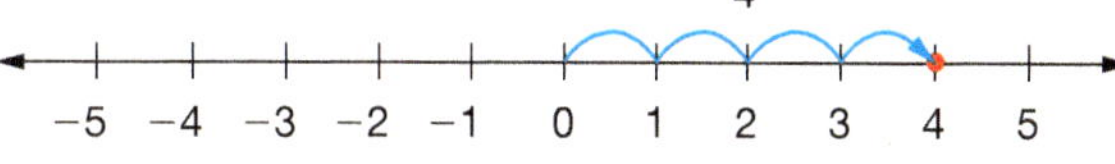

b. $|-3| = 3$ −3 is 3 units from 0.

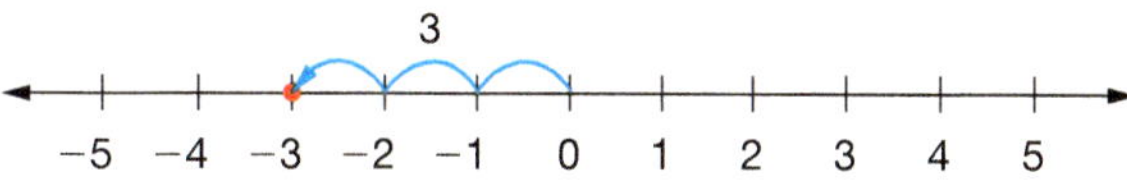

Since distances are never negative, an absolute value is always positive or zero. You must be careful when applying operations to absolute values of numbers. The first expression in the following example indicates the difference of two absolute values, while the second indicates the absolute value of a difference.

Example 5

Simplify.

a. $|74| - |-24|$ **b.** $|82 - 31|$

Answer

a. $|74| - |-24| = 74 - 24$ 1. Find the absolute values.

$= 50$ 2. Subtract.

b. $|82 - 31| = |51|$ 1. Subtract within the grouping symbols first.

$= 51$ 2. Find the absolute value.

Plotting two points on a number line defines a segment. The length of segment AB on the number line below can be found by taking the absolute value of the difference of the coordinates. The segment's length is $|10 - 4| = |6| = 6$. Notice how choosing a different order for the endpoints produces that same length: $|4 - 10| = |-6| = 6$.

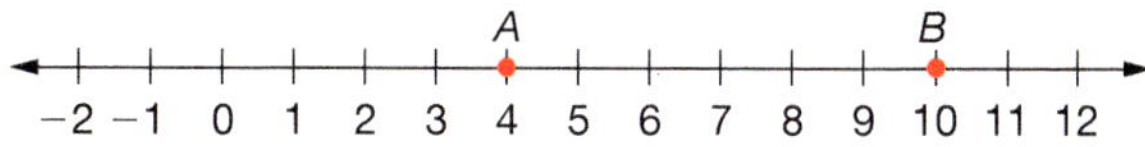

Formula for the Length of a Segment on a Number Line	
$l = \|b - a\|$	The length, l, of a segment is the absolute value of the difference of the endpoints' coordinates.

The coordinate of the midpoint of a segment can be found by averaging the coordinates of the endpoints. For example, the coordinate of M, the midpoint of segment AB, is found by averaging 4 and 10.

$$\frac{4 + 10}{2} = \frac{14}{2} = 7$$

Formula for the Midpoint of a Segment on a Number Line	
$m = \frac{a + b}{2}$	The coordinate of the midpoint, m, of a segment is the average of the endpoints' coordinates.

A. Exercises

Graph each point on a horizontal number line.

1. A: 4; B: -3
2. C: -24; D: 32
3. E: -1.3; F: 0.4
4. G: $1\frac{3}{8}$; H: $-\frac{3}{4}$

Graph each point on a vertical number line.

5. I: 40; J: -25
6. K: -4; L: 1
7. M: $-\frac{3}{2}$; N: $\frac{3}{4}$
8. P: -0.7; Q: 1.2

State the coordinate of each point.

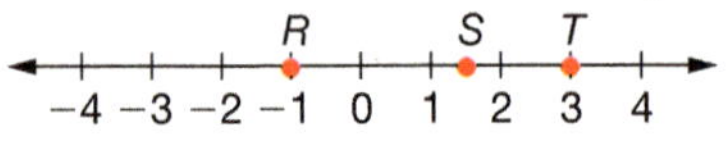

V W U
−20 −10 0 10 20

Z X Y
1 0 −1

9. R
10. S
11. T
12. U
13. V
14. W
15. X
16. Y
17. Z

State the opposite of each number.

18. 17
19. -9
20. -1.3
21. $\frac{1}{2}$

Find each absolute value.

22. $|-2|$
23. $|17|$
24. $\left|-\frac{1}{2}\right|$
25. $|2.8|$

B. Exercises

Perform the indicated operations.

26. $|18 + 27|$
27. $|341 - 26|$
28. $|12.46 - 3.89|$
29. $\left|\frac{5}{8} - \frac{1}{2}\right|$
30. $|-9| + |-4|$
31. $|18| - |-6|$
32. $|-86| - |10|$
33. $|3.0| - |-1.4|$
34. $\left|-\frac{2}{7}\right| - \left|\frac{2}{7}\right|$

A segment has endpoints with the given coordinates. Find the length of the segment and the midpoint of the segment.

35. 12 and 22
36. 137 and 259
37. 25 and 12
38. 102 and 15
39. What is true about the absolute value of a number and the absolute value of its opposite?
40. Find the difference of the absolute value of a number and the absolute value of its opposite.
41. Tim knows that his family is staying at a hotel right off of exit 49. If they just passed mile marker 182, how many more miles do they need to travel before they stop for the night?

C. Exercises

Perform the indicated operations.

42. $|18 - 7| - |-4| + |2 - 5|$

43. $|-12 - 5| - |-12| - |5|$

44. The coordinate of a segment's midpoint is 19. If the coordinate of one of its endpoints is 7, find the coordinate of the other endpoint.

45. A segment with a length of 4 units has a midpoint whose coordinate is -1. State the coordinates of the segment's endpoints.

Dominion Modeling

46. Make a table similar to the one below and summarize the data for the first three iterations of the Koch curve that you studied in Section 1.1, Dominion Modeling. Round each decimal equivalent to the nearest hundredth.

Iteration	Number of Segments	Length of Each Segment	Total Length of the Curve	Decimal Equivalent of the Total Length
0	1	1	1	1
1	4	$\frac{1}{3}$	$\frac{4}{3}$	1.33
2				
3				

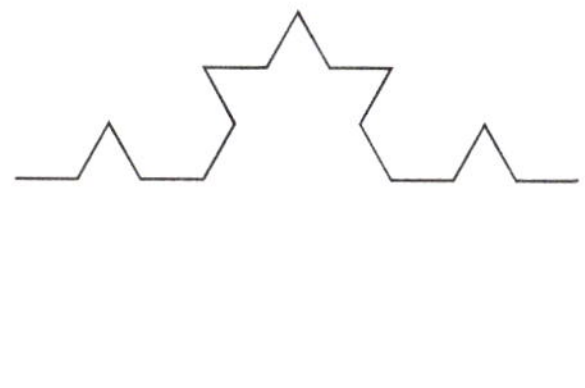

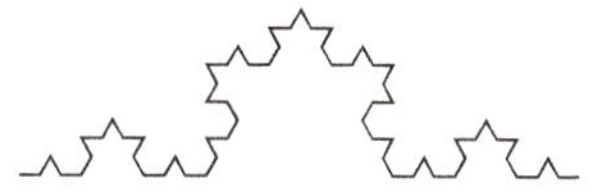

47. Analyze the following patterns in the table and state the relationships in successive iterations of the Koch curve.

a. the number of segments in successive iterations

b. the length of each segment in successive iterations

c. the total length of the curve in successive iterations

48. Extend the table from exercise 46 to predict the number of segments, the length of each segment, and the total length of the curve for the fourth and fifth iterations of the Koch curve.

CUMULATIVE REVIEW

Use the following sets for exercises 49–52. [1.1]
$V = \{a, e, i, o, u\}$; $C = \{h, j, k, l, m, n\}$; $M = \{m, a, t, h\}$

49. Find $C \cap M$.

50. Find $V \cup C$.

True or false [1.1]

51. $a \in M$

52. $M \subseteq C$

53. Draw a Venn diagram illustrating the relationship of the following subsets of the real number system: integers, irrational numbers, natural numbers, rational numbers, and whole numbers. [1.1]

SEQUENCES

Evens, Odds, and Multiples

Challenge **Can you supply the next two numbers in each of the following sequences?**

a. 2, 4, 6, 8, 10, …

b. −5, −10, −15, −20, −25, …

c. 8, 16, 24, 32, 40, …

Recognizing patterns is one of the fundamental skills needed by anyone who wishes to make use of God's creation, mathematicians in particular. This skill of recognizing patterns, illustrated on a simple level by the Challenge above, is an example of *inductive reasoning*. Logical arrangements of numbers, called *sequences*, will be the topic of a series of features in this textbook. Mathematicians seek to provide a general formula that describes a sequence whenever possible. Inductive reasoning is often described as using specific examples to lead to a general conclusion. Care must be taken, however, to keep from jumping to a false conclusion from a few specific examples. For instance, since $0^2 = 0$ and $1^2 = 1$, is it acceptable to conclude that $n^2 = n$?

The numbers in a sequence are represented algebraically with subscripts, which indicate the position of each number in the list. For example, A_5 is used to designate the fifth number, or fifth term, in a sequence. In the third sequence of the Challenge above, $A_5 = 40$. The *general term* is represented by A_n and is called the *n*th term. The even counting numbers listed in the first sequence can be described as 2, 4, 6, 8, 10, …, $2n$. The general-term formula used to represent this sequence is $A_n = 2n$, where n is the number of the term. Similarly, the odd counting numbers can be represented as $A_n = 2n - 1$. One advantage to such general-term formulas is their capability of producing a specific term in the sequence. For example, the one-hundredth odd counting number would be $A_{100} = 2(100) - 1 = 199$.

On the elementary-school level, learning the sequences of the multiples of the counting numbers is crucial to developing a knowledge of the multiplication tables. Mastering the sequence $A_n = 9n$, which is the list of the numbers 9, 18, 27, 36, 45, 54, 63, 72, 81, …, is an important part of mastering the facts for multiplication by 9.

Exercises

List the first five terms of each of the following sequences.

1. $A_n = 4n$
2. $A_n = -6n$
3. $A_n = 11n$
4. $A_n = 12n$

Find A_{20} for each of the following sequences.

5. $A_n = n$
6. $A_n = -5n$
7. $A_n = 2n - 1$
8. Give two descriptions of inductive reasoning.

1.3 Adding Rational Numbers

The first operation you performed on numbers was probably the addition of natural numbers. Over the years you have learned how to add decimals and fractions as well. Each number being added is an *addend*, and the answer is called the *sum*.

The ability to add both positive and negative numbers is essential to your success in algebra. Examining the addition of integers on a number line illustrates the rules for addition of signed numbers.

This machine adds thousands of wood chips to a pile.

Example 1

Add the following numbers with the same sign.

a. $3 + 2$

b. $-1 + (-3)$

Answer

a. From the origin, move 3 units to the right. From 3, move 2 more units to the right. The answer is 5.

b. From the origin, move 1 unit to the left. From -1, move 3 more units to the left. The answer is -4.

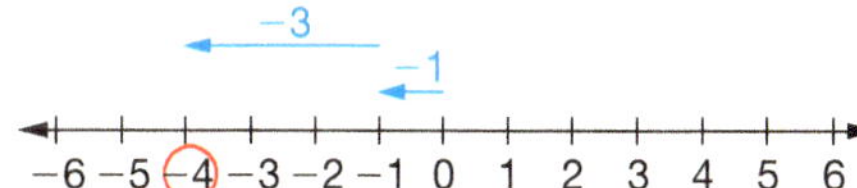

You can see that when the numbers have the same sign, the absolute values of the numbers are added together and the answer must have the sign of the original numbers.

Example 2

Add the following combinations of a positive and a negative number.

a. $3 + (-2)$

b. $-7 + 5$

Answer

a. From the origin, move 3 units to the right. From 3, move 2 units to the left. The answer is 1.

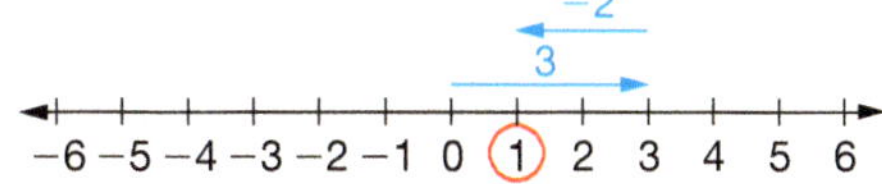

b. From the origin, move 7 units to the left. From -7, move 5 units to the right. The answer is -2.

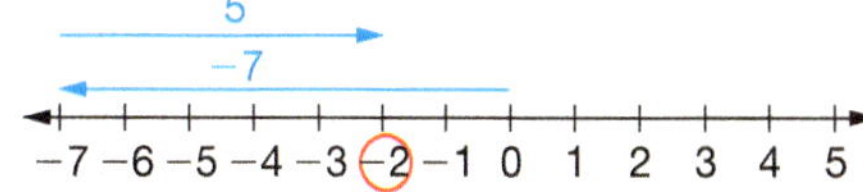

Since the arrows are in different directions, the final distance from the origin is the difference of the absolute values of the addends. The longer arrow (greater absolute value) determines which side of the origin the final answer is on.

Using a number line to add numbers would be very time consuming. Instead, use the following procedures.

ADDING POSITIVE AND NEGATIVE NUMBERS	
Addends with the Same Sign	**Addends with Different Signs**
1. Add the absolute values of the addends.	1. Subtract the smaller absolute value from the larger absolute value.
2. Give the sum the sign of the two addends.	2. Give the difference the sign of the addend with the larger absolute value.

Example 3

Add.

a. $-1 + (-3)$ **b.** $8 + (-3)$ **c.** $4 + (-10)$

Answer

a. $|-1| + |-3| = 1 + 3 = 4$
1. When addends have the same sign, add their absolute values.

$-1 + (-3) = -4$
2. The sum of two negatives must be negative.

b. $|8| - |-3| = 8 - 3 = 5$
1. When addends have different signs, subtract the smaller absolute value from the larger. This is the absolute value of the answer.

$8 + (-3) = 5$
2. The sum is positive because $|8| > |-3|$.

c. $|-10| - |4| = 10 - 4 = 6$
1. When addends have different signs, subtract the smaller absolute value from the larger. This is the absolute value of the answer.

$4 + (-10) = -6$
2. The sum is negative because $|-10| > |4|$.

These principles apply to adding all real numbers, not just integers. When you are finding the sum or difference of decimal numbers, be sure to align the decimal points in your calculations. Recall that using common denominators is necessary for finding sums or differences of fractions.

Example 4

Add.

a. $-112.6 + 39.27$ **b.** $-2.39 + (-14.682)$ **c.** $\frac{3}{4} + \frac{1}{12}$ **d.** $\frac{3}{32} + \left(-\frac{5}{8}\right)$

Answer

a.

$$\begin{array}{r} 112.6 \\ -\ 39.27 \\ \hline 73.33 \end{array}$$

1. When addends have different signs, find the difference of their absolute values.

$-112.6 + 39.27 = -73.33$

2. The sum is negative because $|-112.6| > |39.27|$.

b.

$$\begin{array}{r} 2.39 \\ +\ 14.682 \\ \hline 17.072 \end{array}$$

1. When addends have the same sign, find the sum of their absolute values.

$-2.39 + (-14.682) = -17.072$

2. The sum of two negatives must be negative.

c. $\frac{3}{4} + \frac{1}{12} = \frac{3}{4}\left(\frac{3}{3}\right) + \frac{1}{12} = \frac{9}{12} + \frac{1}{12}$

1. Rename so the fractions have a common denominator.

$\frac{9}{12} + \frac{1}{12} = \frac{10}{12} = \frac{5}{6}$

2. When addends have the same sign, find the sum of their absolute values.

$\frac{3}{4} + \frac{1}{12} = \frac{5}{6}$

3. The sum of two positives must be positive.

d. $\frac{3}{32} + \left[-\frac{5}{8}\left(\frac{4}{4}\right)\right] = \frac{3}{32} + \left(-\frac{20}{32}\right)$

1. Rename so the fractions have a common denominator.

$\frac{20}{32} - \frac{3}{32} = \frac{17}{32}$

2. When addends have different signs, find the difference of their absolute values.

$\frac{3}{32} + \left(-\frac{5}{8}\right) = -\frac{17}{32}$

3. The sum is negative because $\left|-\frac{5}{8}\right| > \left|\frac{3}{32}\right|$.

Mathematicians have given special names to characteristics of operations on numbers that occur frequently.

Definition

A **mathematical property** or *identity* is an equation or statement that is true for any value of the variable.

Properties of Addition	
1. **Commutative Property of Addition** $a + b = b + a$	If you add two numbers together in different orders, the sum is the same. $3 + 5 = 5 + 3$
2. **Associative Property of Addition** $(a + b) + c = a + (b + c)$	If you group numbers together differently when adding, the sum is the same. $(2 + 9) + 4 = 2 + (9 + 4)$
3. **Additive Identity Property** $a + 0 = 0 + a = a$	The sum of any number and zero is that number. $3 + 0 = 0 + 3 = 3$
4. **Additive Inverse Property** $a + (-a) = 0$	The sum of any number and its additive inverse (opposite) is zero. $3 + (-3) = 0$

Applying these properties can often simplify addition in more complicated problems. The Commutative and Associative Properties allow numbers to be added in any order. Rearranging the order of the addends can produce a simpler expression than you would get by simply adding from left to right.

Example 5

Add.

a. $-1 + 5 + (-3) + (-7) + 4 + (-6)$

b. $8 + (-3) + 15 + (-8)$

Answer **a.** $-1 + 5 + (-3) + (-7) + 4 + (-6)$
$= 5 + 4 + (-1) + (-3) + (-7) + (-6)$
$= 9 + (-17)$
$= -8$

1. Using the Commutative Property of Addition, add all the positive numbers and add all the negative numbers.
2. Add the results.

b. $8 + (-3) + 15 + (-8)$
$= 8 + (-8) + (-3) + 15$
$= 0 + 12$
$= 12$

1. Use the Commutative Property of Addition.
2. Use the Additive Inverse Property.
3. Use the Additive Identity Property.

A. Exercises

Use a number line to add.

1. $4 + (-7)$ **2.** $2 + 4$ **3.** $-2 + (-2)$ **4.** $-2 + 5$

Add.

5. $3 + 21$ **6.** $-13 + 32$ **7.** $26 + (-28)$
8. $-7 + (-8)$ **9.** $-126 + 35$ **10.** $7 + (-3)$
11. $-14 + (-7)$ **12.** $384 + (-24)$ **13.** $6 + (-6)$
14. $4 + 0$

B. Exercises

Add.

15. $-527.69 + (-89.41)$
16. $0.23 + (-624.4)$
17. $\frac{1}{7} + \frac{14}{5}$
18. $\frac{4}{7} + \left(-\frac{11}{7}\right)$
19. $-\frac{2}{15} + \frac{3}{5}$
20. $-\frac{12}{21} + \left(-\frac{10}{3}\right)$
21. $-\frac{3}{8} + \frac{2}{11}$
22. $-\frac{4}{27} + \frac{8}{21}$
23. $-3 + 2 + 7 + (-6)$
24. $8 + (-2) + (-9) + 3$
25. $-7 + (-8) + 18 + (-5)$
26. $14 + (-9) + 6 + (-28)$
27. $-8 + |-10| + (-9) + |17|$
28. $6 + (-8) + |-4| + 7$
29. $-8 + |-10 + (-9)| + 17$
30. $6 + (-8) + |3|$

Find the midpoint of a segment having endpoints with the given coordinates.

31. -8 and -16 **32.** -15 and 3 **33.** 0 and -8 **34.** -5 and 5

Translate each phrase into a numerical expression. Then simplify your expression.

35. five plus two
36. nine added to four
37. the sum of two and seven
38. the opposite of eight increased by three
39. five more than the opposite of six

Solve.

40. If the temperature on a January day is 0°F at 7:00 AM, increases 14° by noon, and then decreases 20° from noon to 10:00 PM, what is the temperature at 10:00 PM?

41. A motorboat travels at an average speed of 17 knots (nautical miles per hour) in still water. Determine the actual speed of the boat when it is traveling against a current of 5 knots.

C. Exercises

State the property of addition that justifies each numbered step in the following simplification.

$(-3 + 10) + (-7)$

42. $= [10 + (-3)] + (-7)$

43. $= 10 + [-3 + (-7)]$

$= 10 + (-10)$ [addition]

44. $= 0$

Dominion Modeling

45. Mathematicians model patterns using expressions for the general, or nth, case. For example, in the nth iteration there are 4^n segments. Continue your table from Section 1.2, Dominion Modeling, by filling in expressions for the nth iteration of the Koch curve.

a. the length of each segment

b. the total length of the curve

46. Use these general expressions and a calculator to complete the following table describing the tenth and the one-hundredth iterations of the Koch curve.

Iteration	Number of Segments	Length of Each Segment	Total Length of the Curve	Decimal Equivalent of the Total Length
nth	4^n			NA
10				
100				

CUMULATIVE REVIEW

True or false [1.1]

47. $\mathbb{N} \subseteq \mathbb{Z}$

48. $-5 \in \mathbb{Z}$

49. $\mathbb{Q} \cup \mathbb{Z} = \mathbb{R}$

50. $\mathbb{W} \cap \mathbb{Q}' = \varnothing$

51. $\pi \notin \mathbb{Q}'$

52. $\mathbb{W} \subseteq \mathbb{N}$

Simplify. [1.2]

53. $-(-5)$

54. $-|-5|$

55. $-[-(-7)]$

56. $-\{-[-(-|6|)]\}$

1.4 Subtracting Rational Numbers

In the last section you were asked to find the difference of the absolute values of two numbers, which assumes the ability to subtract a smaller positive number from a larger positive number. We must define what is meant by subtraction in other cases.

Definition

Subtraction is defined as adding the opposite:
For any real numbers a and b, $a - b = a + (-b)$.

This equation is read "a minus b equals a plus the opposite of b." The number being subtracted from, a, is called the *minuend*; and the number being subtracted, b, is called the *subtrahend*. The answer is called the *difference*.

The definition implies that subtracting $8 - 3$ is done by adding $8 + (-3) = 5$. We can see that our previous method of subtracting a smaller positive number from a larger positive number has not really changed and can be performed as before. More complicated subtractions such as $4 - (-2)$ and $-3 - 2$ are performed by adding the opposite.

Example 1

Subtract.

a. $4 - (-2)$ **b.** $-3 - 2$ **c.** $2 - 5$

Answer

a. $4 - (-2) = 4 + 2 = 6$ Subtract -2 by adding its opposite, 2.

b. $-3 - 2 = -3 + (-2) = -5$ Subtract 2 by adding its opposite, -2.

c. $2 - 5 = 2 + (-5) = -3$ Subtract 5 by adding its opposite, -5.

The definition of subtraction applies to all real numbers, not just integers. Note how this definition is applied to rational numbers in the following example.

Example 2

Subtract.

a. $39.27 - 112.6$ **b.** $2.39 - (-14.682)$ **c.** $-\frac{3}{4} - \frac{2}{3}$ **d.** $-\frac{3}{32} - \left(-\frac{5}{8}\right)$

Answer

a. $39.27 - 112.6 = 39.27 + (-112.6)$
$= -73.33$

1. To subtract, add the opposite.
2. Find the difference of the absolute values and choose the sign of the number with the larger absolute value.

b. $2.39 - (-14.682) = 2.39 + 14.682$
$= 17.072$

1. To subtract, add the opposite.
2. Add the absolute values and keep the sign of the numbers.

c. $-\frac{3}{4} - \frac{2}{3} = -\frac{3}{4} + \left(-\frac{2}{3}\right)$
$= -\frac{3}{4}\left(\frac{3}{3}\right) + \left(-\frac{2}{3}\right)\left(\frac{4}{4}\right)$
$= -\frac{9}{12} + \left(-\frac{8}{12}\right) = -\frac{17}{12}$

1. To subtract, add the opposite.
2. Add the absolute values and keep the sign of the numbers.

d. $-\frac{3}{32} - \left(-\frac{5}{8}\right) = -\frac{3}{32} + \frac{5}{8}$
$= -\frac{3}{32} + \frac{5}{8}\left(\frac{4}{4}\right)$
$= -\frac{3}{32} + \frac{20}{32} = \frac{17}{32}$

1. To subtract, add the opposite.
2. Find the difference of the absolute values and choose the sign of the number with the larger absolute value.

In algebra, it is generally preferable to think in terms of adding positive and negative numbers instead of subtracting. In this way, complicated expressions are simplified to addition problems. The expression

$$3 + (-5) - (-9) - 4 - [-(-5)]$$

should be simplified to

$$3 - 5 + 9 - 4 - 5$$

and thought of as

$$3 + (-5) + 9 + (-4) + (-5).$$

The sum of the positive numbers can then be added to the sum of the negative numbers.

$$12 - 14 = -2, \text{ thought of as } 12 + (-14) = -2.$$

By "adding opposites" instead of "subtracting," we can use the Commutative and Associative Properties of Addition to perform the simplification in any order that we prefer.

Example 3

Simplify $|-3| - (-4) + (-2) + 9 - [-(-7)]$.

Answer

$|-3| - (-4) + (-2) + 9 - [-(-7)]$
$= 3 + 4 - 2 + 9 - 7$
$= -2 + 9 = 7$

1. Simplify each term.
2. The Additive Inverse Property implies that $(3 + 4) + (-7) = 0$.

A. Exercises

Subtract.

1. $3 - (-2)$
2. $4 - 5$
3. $-12 - 5$
4. $-8 - (-2)$
5. $-29 - (-5)$
6. $7 - (-3)$
7. $6 - 8$
8. $-16 - (-31)$
9. $-83 - 47$
10. $4 - (-9)$
11. $-27 - (-41)$
12. $-12 - (-3)$

B. Exercises

Subtract.

13. $621.4 - 0.23$
14. $3.896 - (-11.42)$
15. $\frac{2}{9} - \left(-\frac{5}{9}\right)$
16. $4 - \frac{14}{9}$
17. $\frac{3}{5} - \frac{7}{2}$
18. $\frac{3}{8} - \frac{11}{16}$
19. $-\frac{3}{4} - \left(-\frac{8}{9}\right)$
20. $5\frac{3}{7} - 9\frac{1}{5}$

Perform the indicated operations.

21. $-1 + 2 - (-4) - 3$
22. $5 - 3 - [-(-9)] + 13$
23. $|-5| + (-4) - 3 + (-6) - (-8)$
24. $14 - 6 - (-15) - |18| + (-7)$
25. $-\{-[-(-2)]\} - |-8| - 6 - (-3)$
26. $-29 + 13 - (-26) - 4 + 7 - (-|-12|)$

Translate each phrase into a numerical expression. Then simplify your expression.

27. the difference of six and two
28. seven minus four
29. three subtracted from eight
30. sixteen less than five
31. sixteen less five

Solve.

32. If the temperature on January 5 increased from -3°F to 14°F, how much did it increase?

33. On Monday the price of stock in an electronics company opened at \$15.75, fell \$0.19 to its daily low before climbing \$0.56 to its daily high, and then fell \$0.20 to its daily closing price. What was the value of the stock at the end of the day on Monday?

34. Emily is sewing a top treatment for her bedroom window that requires a special self-enclosed French seam. The first part of the seam measures $\frac{1}{4}$ in., and the second part enclosing the first seam measures $\frac{3}{8}$ in. What is the total amount of fabric that Emily should allow for the top treatment seam?

Fractions describe the lengths of cloth.

C. Exercises

Simplify.

35. $7 - |-6 + 18| - (-9)$
36. $|-8| - |7| - |-8 - 7|$
37. $4 - (-5) + |2 - 19| - (-6) - 4$
38. $2.9 + 13.05 - 26.472$
39. $\frac{3}{7} + \frac{5}{14} - \frac{7}{21} + \frac{9}{28}$

Dominion Modeling

40. The Koch snowflake is the result of performing the iterations of the Koch curve on an equilateral triangle. The figures below illustrate the fourth and fifth iterations of the Koch snowflake. If each side of the original triangle has a length of one, what is the perimeter of the fifth iteration of the Koch snowflake?

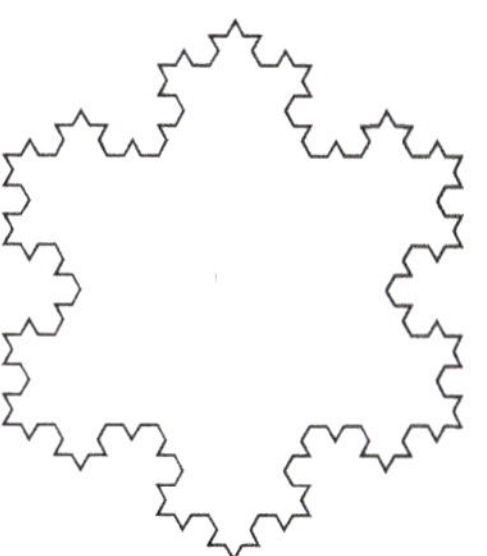

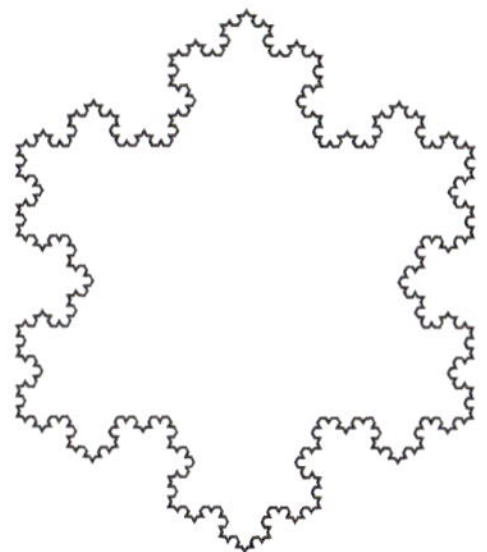

41. If the iterations of the Koch curve continued forever, describe what would happen to each of the following.

a. the number of sides

b. the length of each segment

c. the length of the curve

Mathematically, it can be shown that the lengths of both the Koch curve and the Koch snowflake are infinite when there are an infinite number of iterations. Our finite human minds have difficulty with this concept, but thankfully that does not prevent mathematicians from using fractals as models in the solution of a variety of problems.

CUMULATIVE REVIEW

42. What is the most specific subset of the real numbers to which the number 12.36333 belongs? [1.1]

43. Simplify | 12   7|. [1.2]

44. Simplify $|-12| - |7|$. [1.2]

45. Is the absolute value of a difference always equal to the difference of the absolute values? Explain. [1.2]

46. Add 11.25 and $0.\overline{3}$. [1.3]

47. Name the property illustrated by the following statement: $(2 + 8) + 5 = 2 + (8 + 5)$. [1.3]

48. Is subtraction commutative? Explain by giving an example. [1.3]

Find the length of a segment having endpoints with the given coordinates. [1.2]

49. -8 and -16

50. -15 and 3

51. -5 and 5

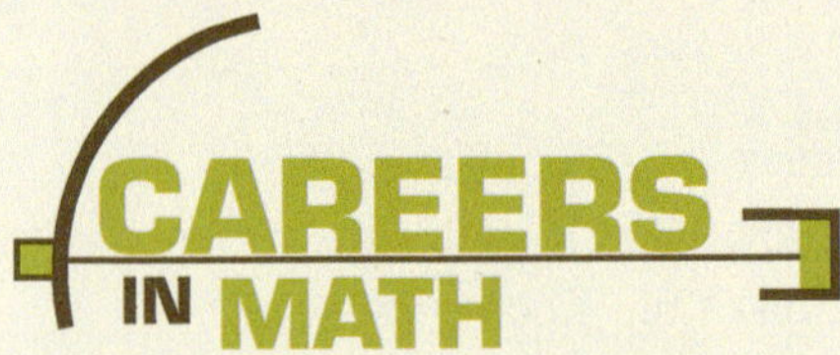

Mathematician

What is the top job in the United States? Perhaps a professional athlete or a media star? According to an evaluation by CareerCast.com published in *The Wall Street Journal* (Jan. 26, 2009), the top job based on environment, income, employment outlook, physical demands, and stress was being a mathematician! You probably are asking, "What does a mathematician do?" According to the US Department of Labor, "Mathematicians use mathematical theory, computational techniques, algorithms, and the latest computer technology to solve economic, scientific, engineering, physics, and business problems." In other words, they are problem solvers.

Theoretical mathematicians focus on creating pure, abstract mathematical knowledge by studying unsolved problems and formulating new relationships from currently known mathematical ideas. On the other hand, *applied mathematicians* focus on applying known theories and techniques to solve practical problems using mathematical modeling and computational methods. Without the discovery of calculus by Isaac Newton and Gottfried Leibniz, the physics of classical mechanics would not be possible. The development of fractals and chaos theory by Benoit Mandelbrot and others initiated current research of apparently random events within the physical, natural, and even social sciences.

If God has given you ability and interest in mathematics, prayerfully consider mathematics as a college major and future career choice. Whether your skills are used to benefit others directly or indirectly by increasing mathematical knowledge, you can use your work to glorify God and show your love for others.

1.5 Multiplying Rational Numbers

To estimate his harvest, a farmer estimates the number of plants in each row and multiplies by the number of rows and the average number of ears per plant.

In elementary school you learned that multiplication is repeated addition.

$$3 \cdot 5 = 5 + 5 + 5 = 15$$

To prevent confusion between the variable x and the multiplication sign $\times$, a small dot in the middle of the line is often used to indicate multiplication. Multiplication is also intended when no sign is shown, as in $3x$ or $3(2 + 5)$. Therefore, $3x$ means $3 \cdot x$ and $3(2 + 5)$ means $3 \cdot (2 + 5)$.

Properties of Multiplication	
1. **Commutative Property of Multiplication** $ab = ba$	If you multiply two numbers in different orders, the product is the same. $3 \cdot 5 = 5 \cdot 3$
2. **Associative Property of Multiplication** $a(bc) = (ab)c$	If you group numbers differently when multiplying, the product is the same. $2(5 \cdot 3) = (2 \cdot 5)3$
3. **Multiplicative Identity Property** $a \cdot 1 = 1 \cdot a = a$	The product of any number and one is that number. $5 \cdot 1 = 1 \cdot 5 = 5$
4. **Distributive Property** $a(b + c) = ab + ac$	The product of a number and a sum is equal to the sum of the products of the number and each addend. $3(6 + 7) = (3 \cdot 6) + (3 \cdot 7)$
5. **Zero Property of Multiplication** $a \cdot 0 = 0$	The product of any number and zero is zero. $4 \cdot 0 = 0$

A *product* is the answer to a multiplication problem, and a *factor* is any number that is being multiplied in a problem. Factors, like addends, can be positive or negative. Look at the following illustrations of multiplication as repeated addition. Watch carefully the signs of the factors and the resulting products.

$3 \cdot 5$ $\quad$ $3(-5)$

$5 + 5 + 5 = 15$ $\quad$ $-5 + (-5) + (-5) = -15$

In the first illustration, both factors are positive and the product is also positive. In the second, one factor is positive and the other negative; the product is negative. By applying the Commutative Property to the second example, we see that it does not matter whether the first or second factor carries the negative sign: $3 \cdot (-5) = -5 \cdot 3$. If exactly one factor of a multiplication problem is negative, the resulting product is negative.

What happens if both factors are negative? Observe the pattern in the following products.

$3(-5) = -15$ found using the illustration above

$2(-5) = -10$

$1(-5) = -5$

$0(-5) = 0$ by the Zero Property of Multiplication

$-1(-5) = 5$ continuing the pattern

$-2(-5) = 10$

$-3(-5) = 15$

As the first factor decreases by one, the product increases by five. Notice that the product of a negative factor and a positive factor is negative but that the product of two negatives is positive.

Multiplying Positive and Negative Numbers
1. If the factors are both positive or both negative, the product is positive.
2. If one factor is positive and the other is negative, the product is negative.

Example 1

Multiply.

a. $-2(9)$ **b.** $-4(-6)$

Answer **a.** $-2(9) = -18$ Two factors with different signs produce a negative product.

b. $-4(-6) = 24$ Two factors with the same sign produce a positive product.

What if there are more than two factors? What is the sign of the product? Observe the pattern in the following products.

$1(-1) = -1$	one negative → negative product
$-1(-1) = 1$	two negatives → positive product
$-1(-1)(-1) = -1$	three negatives → negative product
$-1(-1)(-1)(-1) = 1$	four negatives → positive product
$-1(-1)(-1)(-1)(-1) = -1$	five negatives → negative product
$-1(-1)(-1)(-1)(-1)(-1) = 1$	six negatives → positive product

Since the product of every pair of negative factors is positive and the product of any number of positive factors is positive, the following generalizations can be made about the products of nonzero factors.

Multiplication Rule

1. If there is an even number of negative factors, the product is positive.
2. If there is an odd number of negative factors, the product is negative.

Example 2

Multiply.

a. $-2(-3)(-5)$ **b.** $4(-3)(-1)$

Answer **a.** $|-2| \cdot |-3| \cdot |-5| = 2 \cdot 3 \cdot 5 = 30$ 1. Multiply the absolute values.

$-2(-3)(-5) = -30$ 2. The product of an odd number of negative factors is negative.

b. $|4| \cdot |-3| \cdot |-1| = 4 \cdot 3 \cdot 1 = 12$ 1. Multiply the absolute values.

$4(-3)(-1) = 12$ 2. The product of an even number of negative factors is positive.

These principles apply to all real numbers, not just integers. Do you remember how to multiply decimals? This example may remind you. Be sure you know how to perform this operation.

Example 3

Multiply $-4.28 \cdot 5.31$.

Answer

```
    4.28
  × 5.31
     428
   1284
  2140
 22.7268
```

1. Find the product of the absolute values of the factors.

$-4.28 \cdot 5.31 = -22.7268$ 2. The product of a positive factor and a negative factor is negative.

Do you need a common denominator for multiplying fractions? No, you can multiply the numerators and then multiply the denominators.

Example 4

Multiply $-\frac{3}{4}\left(-\frac{7}{8}\right)$.

Answer $\frac{3}{4} \times \frac{7}{8} = \frac{3 \times 7}{4 \times 8} = \frac{21}{32}$ 1. Find the product of the absolute values of the factors by multiplying the numerators and the denominators.

$-\frac{3}{4}\left(-\frac{7}{8}\right) = \frac{21}{32}$ 2. The product of two negative factors is positive.

When you multiply rational numbers in fractional form, you can reduce the terms before or after you multiply.

Cancel first: $\frac{12}{7} \times \frac{14}{3} = \frac{\overset{4}{\cancel{12}}}{\cancel{7}} \times \frac{\overset{2}{\cancel{14}}}{\cancel{3}} = 8$

Multiply first: $\frac{12}{7} \times \frac{14}{3} = \frac{168}{21} = \frac{2^3 \times \cancel{3} \times \cancel{7}}{\cancel{3} \times \cancel{7}} = 2^3 = 8$

Canceling first is usually easier and can often be done mentally. Multiplying first generally produces large numbers that must be factored while you are reducing the result. Furthermore, if you leave $\frac{168}{21}$ as your answer, it will be unacceptable since it is not reduced.

Therefore, you should always reduce and cancel first to save time and to avoid possible errors. Reducing first will also prepare you for reducing fractions with variables later. Remember to convert mixed numbers to improper fractions before multiplying. Answers can be left as improper fractions except in word problems, where mixed numbers are preferred.

Example 5

Multiply.

a. $2\frac{1}{5} \times 4\frac{2}{7}$ **b.** $-\frac{3}{4} \times \frac{18}{7}$

Answer **a.** $2\frac{1}{5} \times 4\frac{2}{7} = \frac{11}{5} \times \frac{30}{7}$ 1. Rewrite the mixed numbers as improper fractions.

$= \frac{11}{\cancel{5}} \times \frac{\overset{6}{\cancel{30}}}{7} = \frac{66}{7}$ 2. Cancel before multiplying.

b. $\frac{3}{\underset{2}{\cancel{4}}} \times \frac{\overset{9}{\cancel{18}}}{7} = \frac{27}{14}$ 1. Find the product of the absolute values of the factors by multiplying the numerators and the denominators after canceling.

$-\frac{3}{4} \times \frac{18}{7} = -\frac{27}{14}$ 2. The product of a positive factor and a negative factor is negative.

Definition

A **unit multiplier** is a fraction with two different expressions for the same quantity in the numerator and the denominator.

Unit multipliers are used to change the units of a number. Any fraction with the same quantity in the numerator and the denominator is equal to one. Since 12 in. = 1 ft, the fractions $\frac{12\text{ in.}}{1\text{ ft}}$ and $\frac{1\text{ ft}}{12\text{ in.}}$ are both equal to one. The Multiplicative Identity Property guarantees that multiplying a quantity by a unit multiplier results in the same quantity.

$\overset{5}{\cancel{60}}\ \cancel{\text{in.}}\left(\frac{1\text{ ft}}{\cancel{12}\ \cancel{\text{in.}}}\right) = 5\text{ ft}$ and $5\ \cancel{\text{ft}}\left(\frac{12\text{ in.}}{1\ \cancel{\text{ft}}}\right) = 60\text{ in.}$

Notice that the same quantity is represented using a different unit. Use the unit multiplier that cancels the original unit, leaving the desired unit in the result.

Example 6

Convert the following lengths.

a. 53.25 in. to centimeters **b.** 51 in. to feet **c.** 2 mi to inches

Answer **a.** $53.25 \text{ in.}\left(\frac{2.54 \text{ cm}}{1 \text{ in.}}\right) = 135.255 \text{ cm}$

Placing 1 in. in the denominator allows the inches to cancel, leaving centimeters.

b. $\overset{17}{51} \text{ in.}\left(\frac{1 \text{ ft}}{\underset{4}{12} \text{ in.}}\right) = \frac{17}{4} \text{ ft} = 4\frac{1}{4} \text{ ft}$

Placing 12 in. in the denominator allows the inches to cancel, leaving feet.

c. $2 \text{ mi}\left(\frac{5280 \text{ ft}}{1 \text{ mi}}\right)\left(\frac{12 \text{ in.}}{1 \text{ ft}}\right) = 126{,}720 \text{ in.}$

Notice how unit multipliers can be used consecutively.

A. Exercises

Multiply.

1. $3(-2)$ **2.** $7(8)$ **3.** $-4(-6)$ **4.** $-8(6)$

5. $-3(-1)$ **6.** $3(-7)$ **7.** $4(9)$ **8.** $-32(3)$

9. $-41(-7)$ **10.** $79(-8)$ **11.** $-182(3)$ **12.** $372(-24)$

B. Exercises

Multiply.

13. $-14.1 \cdot 5.6$ **14.** $-3.9(-8.42)$ **15.** $3(-7)(6)$

16. $-10(-2)(-4)(-8)$ **17.** $98(4)(0)(-5)$ **18.** $\frac{4}{7} \cdot \frac{5}{12}$

19. $-\frac{4}{11} \cdot \frac{7}{12}$ **20.** $-\frac{1}{6}\left(-\frac{18}{19}\right)$ **21.** $-\frac{1}{10}\left(-\frac{15}{4}\right)$

22. $-\frac{18}{35} \cdot \frac{7}{16}$ **23.** $3\frac{7}{9}\left(-\frac{3}{17}\right)$ **24.** $-\frac{3}{7} \cdot 5\frac{3}{5}$

25. $-2\frac{5}{8}\left(-2\frac{1}{3}\right)$ **26.** $3\frac{5}{6} \cdot 2\frac{2}{5}$

Translate each phrase into a numerical expression. Then simplify your expression.

27. five times four **28.** seven multiplied by three

29. the product of the opposite of two and nine **30.** twice the opposite of eleven

31. six sets of eight objects each

Solve.

32. Kevin's bucket contains $2\frac{1}{2}$ gal of water. If he gives his horse $\frac{2}{3}$ of the water, how much water does the horse get?

33. Mike, Melissa, and Matt each picked $\frac{2}{5}$ gal of strawberries. How many gallons of strawberries did they pick altogether?

34. If one side of a square is $\frac{3}{8}$ in., what is the perimeter of the square?

35. Convert 88.2 ft to yards.

36. Convert 24,816 ft to miles (1 mi = 5280 ft).

37. Convert 13 in. to centimeters (1 in. = 2.54 cm).

38. Convert 100 cm to yards.

C. Exercises

Find the pattern in each series of statements and express it as a general property using a variable.

39. $-3 = -1 \cdot 3$
$-4 = -1 \cdot 4$
$-7 = -1 \cdot 7$

40. $-(-1) = 1$
$-(-3) = 3$
$-(-6) = 6$

41. Write $8(-3)$ using the definition of multiplication as repeated addition.

Dominion Modeling

The figures below illustrate the first three iterations of the Sierpinski triangle, another famous fractal. Each iteration is performed by connecting the midpoints of any upward-pointing triangles.

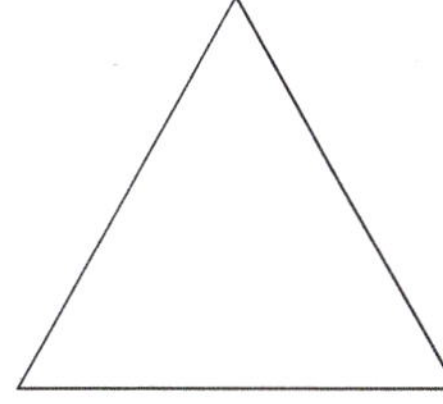
initial triangle

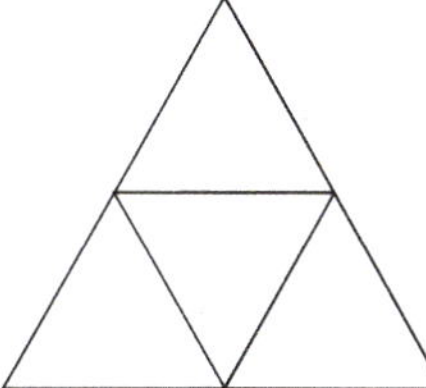
1st iteration

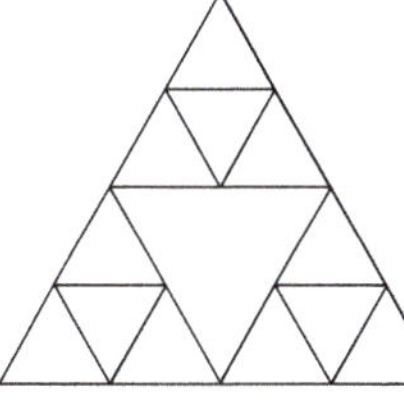
2nd iteration

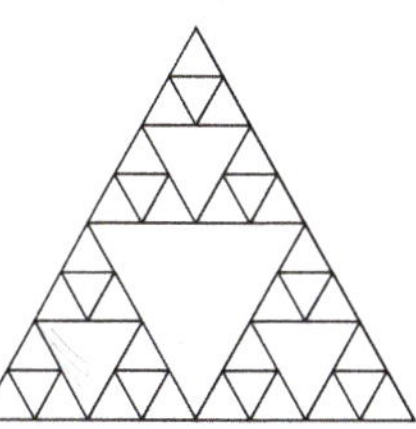
3rd iteration

42. Beginning with your own equilateral triangle, complete the first four iterations of the Sierpinski triangle. Shading the downward-pointing triangles created in each iteration with a different color will produce "fractal art."

A Sierpinski triangle can also be generated from the three vertices of a triangle and any randomly graphed point using many iterations of the following procedure: Plot the midpoint of the segment connecting a randomly chosen vertex to the previously plotted point. Initially the result appears to be a random plotting of points, but after sufficient iterations, a pattern can be observed. When our observations are limited, it is often difficult to generalize the underlying patterns in complex events that appear random or chaotic.

43. Download the Chaos Game from the Wolfram Demonstrations Project website (a link is provided on the Algebra 1 Resources page on the BJU Press website) and explore the different fractals that can be drawn. Use the polygon sides setting of 3 and the point size of tiny to answer the following questions.

a. What happens to the fractal as distances less than 0.5 are chosen?

b. What happens to the fractal as distances greater than 0.5 are chosen?

c. Set the distance for 0.65 and describe the resulting pattern in the fractal. What is the smallest number of points for which the pattern is evident?

CUMULATIVE REVIEW

Find the length of a segment having endpoints with the given coordinates. [1.2]

44. 26 and -13 **45.** -42 and -86

Find the midpoint of a segment having endpoints with the given coordinates. [1.2]

46. 26 and -13 **47.** -42 and -86

Add. [1.3]

48. $\frac{7}{24} + \left(-\frac{5}{8}\right)$ **49.** $-\frac{1}{6} + \left(-\frac{3}{4}\right)$ **50.** $-9.07 + (-5.96)$

Subtract. [1.4]

51. $\frac{19}{24} - \left(-\frac{3}{8}\right)$ **52.** $-\frac{1}{6} - \left(-\frac{3}{4}\right)$ **53.** $-9.07 - (-5.96)$

Find the digit that will properly fill in each box in the following multiplication problems.

1.

```
  □□□
×  39
 47□3
1□□1
20,□□3
```

2.

```
   7□□□
×    □8
  □81□2
 14□□8
20□,□92
```

3.

```
     □2□
   × □81
     □□6
   □4□8
 □278
16□,□06
```

4.

```
   5□4□
 ×   □3
  17□□8
□□□□2
43□,0□8
```

1.6 Dividing Rational Numbers

Recall that the answer to a division problem is called a *quotient*. The *dividend* is the number being divided into equal parts by the *divisor*.

Just as subtraction is the inverse operation of addition, division is the inverse operation of multiplication. Every multiplication problem has two corresponding division problems.

A section of this stadium has 16 rows and seats 480. If you are distributing flyers to everyone in this section, how many should you pass down each row?

$9 \cdot 2 = 18$ so $18 \div 2 = 9$ and $18 \div 9 = 2$

$-6 \cdot (-4) = 24$ so $24 \div (-4) = -6$ and $24 \div (-6) = -4$

$-5 \cdot 3 = -15$ so $-15 \div 3 = -5$ and $-15 \div (-5) = 3$

Because division and multiplication are inverse operations, the sign rules for division correspond to the sign rules for multiplication.

Dividing Positive and Negative Numbers

1. If the divisor and the dividend have like signs, the quotient is positive.
2. If the divisor and the dividend have unlike signs, the quotient is negative.

Example 1

Divide.

a. $14 \div 2$ **b.** $\frac{-72}{9}$ **c.** $105 \div (-21)$ **d.** $\frac{-864}{-36}$

Answer

a. $14 \div 2 = 7$ The quotient is positive when the divisor and dividend have the same sign.

b. $\frac{-72}{9} = -8$ The quotient is negative when the divisor and dividend have different signs.

c. $105 \div (-21) = -5$ The quotient is negative when the divisor and dividend have different signs.

d. $\frac{-864}{-36} = 24$ The quotient is positive when the divisor and dividend have the same sign.

A good way to check your division is to multiply the quotient by the divisor. The answer should be the same as the dividend.

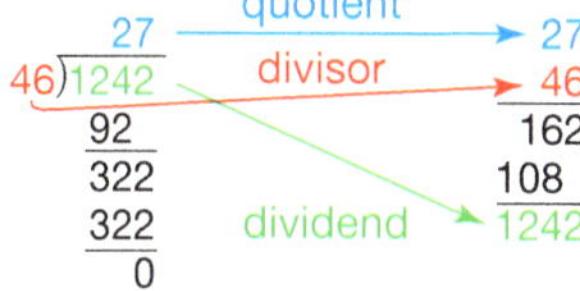

What is $18 \div 0$ or $\frac{18}{0}$? Since multiplication is the inverse operation of division, this is the same as asking what number can be multiplied by zero to get 18. According to the Zero Property of Multiplication, the product of any number and zero is always zero. Therefore, there is no number that when multiplied by zero gives a product of 18. Consequently, division by zero is not possible and is, therefore, undefined.

$18 \div 0$ is undefined, but what about $0 \div 18$? There is a definite answer to the problem $18 \cdot ? = 0$; the missing number is zero.

In conclusion, $\frac{0}{18} = 0$, but $\frac{18}{0}$ is undefined.

Since division is the inverse operation of multiplication, division problems are often restated as multiplication by the reciprocal of the divisor.

Definition

Reciprocals (multiplicative inverses) are two numbers whose product is 1.

The reciprocal of $\frac{2}{3}$ is $\frac{3}{2}$ because $\frac{2}{3} \cdot \frac{3}{2} = 1$. -5 and $-\frac{1}{5}$ are reciprocals since $-5\left(-\frac{1}{5}\right) = 1$.

The *Multiplicative Inverse Property* states that the product of a number and its reciprocal is always one (the multiplicative identity). In symbols, $a \cdot \frac{1}{a} = 1$ and $\frac{1}{a} \cdot a = 1$.

Definition

Division is defined as multiplication by the reciprocal of the divisor. For any real numbers a and b, $b \neq 0$, $a \div b = \frac{a}{b} = a \cdot \frac{1}{b}$.

Example 2

Divide.

a. $\frac{4}{5} \div \frac{9}{10}$ **b.** $-\frac{5}{7} \div (-5)$ **c.** $-1\frac{7}{12} \div 7\frac{3}{5}$

Answer

a. $\frac{4}{5} \div \frac{9}{10} = \frac{4}{5} \times \frac{10}{9}$ — 1. Rewrite the division as multiplication by the reciprocal.

$= \frac{8}{9}$ — 2. The product of two factors with the same sign is positive.

b. $-\frac{5}{7} \div (-5) = -\frac{5}{7}\left(-\frac{1}{5}\right)$ — 1. Rewrite the division as multiplication by the reciprocal.

$= \frac{1}{7}$ — 2. The product of two factors with the same sign is positive.

c. $-1\frac{7}{12} \div 7\frac{3}{5} = -\frac{19}{12} \div \frac{38}{5}$ — 1. Convert the mixed numbers to improper fractions before dividing.

$= -\frac{19}{12} \times \frac{5}{38}$ — 2. Rewrite the division as multiplication by the reciprocal.

$= -\frac{5}{24}$ — 3. The product of two factors with different signs is negative.

The division problem $\frac{1}{2} \div \frac{3}{4}$ can also be written as the fraction $\frac{\frac{1}{2}}{\frac{3}{4}}$. A fraction whose numerator or denominator contains a fraction is called a *complex fraction*. When you are simplifying a complex fraction, remember to rewrite the division problem as multiplication by the reciprocal of the divisor.

Example 3

Divide.

a. $\frac{\frac{4}{5}}{-\frac{1}{2}}$ **b.** $\frac{\frac{2}{3}}{4}$

Answer

a. $\frac{4}{5} \div \left(-\frac{1}{2}\right) = \frac{4}{5}\left(-\frac{2}{1}\right)$ — 1. Rewrite the division as multiplication by the reciprocal.

$= -\frac{8}{5}$ — 2. The product of two factors with different signs is negative.

b. $\frac{2}{3} \div 4 = \frac{2}{3} \times \frac{1}{4}$ — 1. Rewrite the division as multiplication by the reciprocal.

$= \frac{1}{6}$ — 2. The product of two factors with the same sign is positive.

Do you remember how to divide decimals? The next example may remind you. Be sure you know how to perform this operation.

Example 4

Divide 1.26 by 0.3.

Answer

$$\begin{array}{r} 4.2 \\ 0.3\overline{)1.26} \\ \underline{12} \\ 6 \\ \underline{6} \\ 0 \end{array}$$

Now that you have reviewed all four operations, you should be able to identify and use all of the following properties.

Property	of Addition	of Multiplication
Commutative	$a + b = b + a$ $2 + 3 = 3 + 2$	$a \cdot b = b \cdot a$ $2 \cdot 3 = 3 \cdot 2$
Associative	$a + (b + c) = (a + b) + c$ $2 + (3 + 4) = (2 + 3) + 4$	$a(b \cdot c) = (a \cdot b)c$ $2(3 \cdot 4) = (2 \cdot 3)4$
Identity	$a + 0 = 0 + a = a$ $2 + 0 = 0 + 2 = 2$	$a \cdot 1 = 1 \cdot a = a$ $2 \cdot 1 = 1 \cdot 2 = 2$
Inverse	$a + (-a) = 0$ $2 + (-2) = 0$	$a \cdot \frac{1}{a} = \frac{1}{a} \cdot a = 1$ $3 \cdot \frac{1}{3} = \frac{1}{3} \cdot 3 = 1$
Zero		$a \cdot 0 = 0 \cdot a = 0$ $2 \cdot 0 = 0 \cdot 2 = 0$
Distributive		$a(b + c) = ab + ac$ $2(3 + 4) = 2(3) + 2(4)$

A. Exercises

Divide.

1. $27 \div 3$

2. $-56 \div 8$

3. $36 \div 0$

4. $-20 \div (-4)$

5. $54 \div (-6)$

6. $-48 \div 6$

7. $588 \div (-84)$

8. $-2352 \div (-56)$

9. $\frac{198}{-22}$

10. $\frac{0}{-37}$

11. $\frac{-124}{0}$

12. $\frac{-516}{86}$

13. $\frac{-6584}{-823}$

B. Exercises

Divide.

14. $-56.4 \div 0.2$

15. $61 \div 12.2$

16. $-6.09 \div (-7)$

17. $-\frac{4}{9} \div \frac{8}{11}$

18. $-\frac{1}{4} \div \left(-\frac{9}{16}\right)$

19. $\frac{5}{11} \div \left(-\frac{3}{22}\right)$

20. $\frac{2}{17} \div 8$

21. $\frac{54}{19} \div \frac{18}{7}$

22. $-\frac{4}{9} \div \frac{2}{3}$

23. $-\frac{18}{35} \div \frac{-16}{-7}$

24. $-2\frac{5}{6} \div \left(-3\frac{2}{5}\right)$

25. $\frac{\frac{2}{3}}{\frac{5}{4}}$

26. $\frac{-\frac{6}{11}}{\frac{2}{7}}$

27. $\frac{-6}{\frac{5}{7}}$

28. $\frac{-\frac{6}{5}}{7}$

Translate each phrase into a numerical expression, using the fraction bar to denote division. Then simplify your expression.

29. eighty-one divided by the opposite of three
30. divide four into twelve
31. the quotient of nine and two
32. divide the opposite of fifty by six
33. the ratio of eleven to four

Solve.

34. Clark and Cris picked $1\frac{1}{2}$ gal of blackberries and divided them evenly between themselves. How many gallons of blackberries did each get?
35. If an equilateral triangle has a perimeter of $2\frac{2}{5}$ ft, what is the length of each side of the triangle?
36. Explain why division by zero is undefined.
37. Write the Zero Property of Multiplication as a division principle.

C. Exercises

38. Simplify $[14{,}175 \div (-5)] \div [-2916 \div (-36)]$.
39. Simplify $-\frac{13}{42} \div \left(-\frac{5}{7}\right) + \frac{1}{15} - 0.427$.
40. The product of two numbers is 7904, and the absolute value of one of the numbers is 247. Find two possible solutions for the other number.
41. Is division commutative? Explain and give an example.

Dominion Modeling

42. In Section 1.5, Dominion Modeling, notice that each iteration of the Sierpinski triangle produces identical smaller upward-pointing triangles. Use a table like the one below to record the following data for each iteration. Assume the original triangle has an area of one square unit.
    a. the number of upward-pointing triangles
    b. the area of each upward-pointing triangle
43. Determine the total area of all the upward-pointing triangles for each iteration and record the answers in your table.

Is this tree bark or fractal art?

Iteration	0	1	2	3	4
Number of upward-pointing triangles	1				
Area of each upward-pointing triangle	1				
Total area of upward-pointing triangles	1				

CUMULATIVE REVIEW

44. Sketch Venn diagrams illustrating the union and the intersection of sets $A = \{1, 3, 5, 7, 9\}$ and $B = \{0, 1, 2, 3, 4\}$. [1.1]

State the property of addition that justifies each numbered step in the following simplification. [1.3]

$(56 + 0) + (79 + 44)$

45. $= 56 + (79 + 44)$

46. $= 56 + (44 + 79)$

47. $= (56 + 44) + 79$

$= 100 + 79 = 179$ [addition]

Simplify. [1.2–1.5]

48. $|-21 - 5| - |-9| + |3 - 8|$

49. $|-4.7 + 1.2| + |-6| - |5.3 - 9.1|$

50. $\frac{3}{8} + \left(-\frac{1}{4}\right) - \frac{11}{16}$

51. $-\frac{3}{4} + \frac{5}{6} - \left(-\frac{2}{3}\right)$

52. $9.1 \cdot 3.07 \cdot 5.64$

53. $\frac{9}{16}\left(-\frac{2}{3}\right) \div \frac{7}{4}$

1.7 Exponents

Do you know a short way to write $6 \cdot 6 \cdot 6$ or $5 \cdot 5$? To save time and space, you can use exponential form. It is a shorter way to write repeated multiplication, just as multiplication is a shorter way to write repeated addition.

Definitions

For $x \in \mathbb{N}$, b^x is the product of x factors of b.

The repeated factor, b, is the **base**, and x, which indicates the number of times b is used as a factor, is the **exponent**. When numbers are written in this way, they are said to be in **exponential form** or **exponential notation**.

In exponential form, $6 \cdot 6 \cdot 6 = 6^3$ and $5 \cdot 5 = 5^2$. In 6^3 the 3 is the exponent and the 6 is the base. The expression 5^2 is read "five to the second power" or "five squared," and 6^3 is read "six to the third power" or "six cubed."

The reasons for this terminology are illustrated by the area of a square and the volume of a cube.

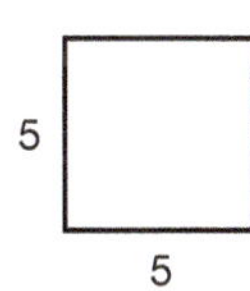

Area $= 5 \cdot 5 = 5^2$
5 squared

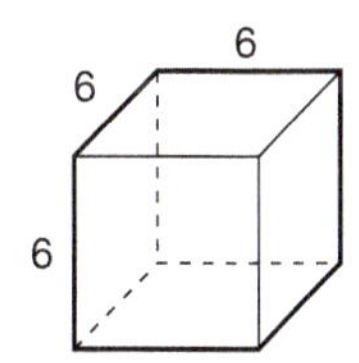

Volume $= 6 \cdot 6 \cdot 6 = 6^3$
6 cubed

Example 1

Write each of the following in exponential form.

a. $2 \cdot 2 \cdot 2 \cdot 2 \cdot 2 \cdot 2 \cdot 2$ **b.** $-3 \cdot (-3) \cdot (-3) \cdot (-3)$

Answer **a.** $2 \cdot 2 \cdot 2 \cdot 2 \cdot 2 \cdot 2 \cdot 2 = 2^7$ The base is 2, which is used as a factor seven times.

b. $-3 \cdot (-3) \cdot (-3) \cdot (-3) = (-3)^4$ The base is −3, which is used as a factor four times.

This example illustrates an important distinction. The expression -3^4 represents the opposite of three to the fourth power:

$$-3^4 = -1 \cdot 3^4 = -1(3 \cdot 3 \cdot 3 \cdot 3) = -81.$$

However,

$$(-3)^4 = -3 \cdot (-3) \cdot (-3) \cdot (-3) = 81.$$

When you are asked to evaluate an expression, you need to find the number that is represented by the expression. In numerical expressions, this is the same as simplifying the expression.

Example 2

Evaluate the following exponential forms.

a. 9^3 **b.** $(-7)^4$ **c.** -4^2

Answer **a.** $9^3 = 9 \cdot 9 \cdot 9 = 729$

b. $(-7)^4 = -7 \cdot (-7) \cdot (-7) \cdot (-7) = 2401$

c. $-4^2 = -(4 \cdot 4) = -16$

What happens when you multiply with like bases?

$$5^2 \cdot 5^3 = (5 \cdot 5)(5 \cdot 5 \cdot 5) = 5^{2+3} = 5^5$$

When the bases are the same and the operation is multiplication, add the exponents.

What happens when you divide with like bases?

$$8^5 \div 8^2 = \frac{8^5}{8^2} = \frac{8 \cdot 8 \cdot 8 \cdot \cancel{8} \cdot \cancel{8}}{\cancel{8} \cdot \cancel{8}} = 8^{5-2} = 8^3$$

When the bases are the same and the operation is division, subtract the exponents.

What happens when you raise an exponential expression to a power?

$$(3^2)^4 = 3^2 \cdot 3^2 \cdot 3^2 \cdot 3^2 = 3^{2+2+2+2} = 3^{2(4)} = 3^8$$

When an exponential expression is raised to a power, multiply the exponents.

Properties of Exponents		
Product Property	$x^a \cdot x^b = x^{a+b}$	To multiply like bases, add the exponents.
Quotient Property	$\frac{x^a}{x^b} = x^{a-b}$ for $x \neq 0$	To divide like bases, subtract the exponents.
Power Property	$(x^a)^b = x^{ab}$	To raise a power to a power, multiply the exponents.

Example 3

Simplify, leaving each answer in exponential form.

a. $5^2 \cdot 5^{10}$ **b.** $3^2 \cdot 3^4 \cdot 3^9$ **c.** $2^4 \div 2$
d. $10^5 \div 10^3$ **e.** $(5^3)^6$

Answer

a. $5^2 \cdot 5^{10} = 5^{2+10} = 5^{12}$ Add the exponents to multiply like powers.
b. $3^2 \cdot 3^4 \cdot 3^9 = 3^{2+4+9} = 3^{15}$ Add the exponents to multiply like powers.
c. $2^4 \div 2 = 2^{4-1} = 2^3$ Subtract the exponents to divide like powers.
d. $10^5 \div 10^3 = 10^{5-3} = 10^2$ Subtract the exponents to divide like powers.
e. $(5^3)^6 = 5^{3(6)} = 5^{18}$ Multiply the exponents to find powers of powers.

The Quotient Property gives meaning to the expression x^0. Since $\frac{x^2}{x^2} = x^{2-2} = x^0$ and anything divided by itself equals one, $x^0 = 1$.

The Quotient Property also gives meaning to exponential expressions with negative exponents. $\frac{x^2}{x^3} = \frac{\cancel{x} \cdot \cancel{x}}{x \cdot \cancel{x} \cdot \cancel{x}} = \frac{1}{x}$ and $\frac{x^2}{x^3} = x^{2-3} = x^{-1}$ by the Quotient Property. Therefore, $\frac{1}{x} = x^{-1}$. In general, $x^{-a} = \frac{1}{x^a}$ since $x^{-a} = \frac{x^0}{x^a} = x^{0-a} = x^{-a}$.

Definitions

Any nonzero number to the **zero power** equals 1. That is, when $x \neq 0$, $x^0 = 1$.

Negative exponent: For any nonzero number x and integer a, $x^{-a} = \frac{1}{x^a}$.

How do we deal with a negative exponent in the denominator? Applying the definition of a negative exponent, $\frac{1}{x^{-a}} = \frac{1}{\left(\frac{1}{x^a}\right)} = 1 \div \frac{1}{x^a} = 1 \times \frac{x^a}{1} = x^a$, so $\frac{1}{x^{-a}} = x^a$.

Example 4

Evaluate the following exponential forms.

a. $(-2)^{-4}$ **b.** -3^{-2} **c.** $\frac{1}{6^{-3}}$

Answer

a. $(-2)^{-4} = \frac{1}{(-2)^4} = \frac{1}{-2(-2)(-2)(-2)} = \frac{1}{16}$

b. $-3^{-2} = -\frac{1}{3^2} = -\frac{1}{3 \times 3} = -\frac{1}{9}$

c. $\frac{1}{6^{-3}} = 6^3 = 6 \cdot 6 \cdot 6 = 216$

The properties of exponents apply to expressions with negative exponents as well.

Example 5

Simplify, leaving each answer in positive exponential form.

a. $3^{-9} \cdot 3^7$ **b.** $2^{-3} \cdot 2^{-2}$ **c.** $4^{-2} \div 4^{-3}$ **d.** $(5^{-3})^4$

Answer
a. $3^{-9} \cdot 3^7 = 3^{-9+7} = 3^{-2} = \frac{1}{3^2}$
b. $2^{-3} \cdot 2^{-2} = 2^{-3+(-2)} = 2^{-5} = \frac{1}{2^5}$
c. $4^{-2} \div 4^{-3} = 4^{-2-(-3)} = 4^{-2+3} = 4^1 = 4$
d. $(5^{-3})^4 = 5^{-3(4)} = 5^{-12} = \frac{1}{5^{12}}$

A. Exercises

Write using repeated factors; then evaluate.

1. 4^3
2. $(-2)^4$
3. $(-3)^5$
4. 1^4
5. 2314^0
6. $\left(\frac{3}{4}\right)^2$
7. 4^{-2}
8. $(-3)^{-3}$
9. -2^{-4}
10. $\frac{1}{8^{-2}}$
11. $\frac{1}{(-3)^{-4}}$
12. $\left(\frac{2}{3}\right)^{-3}$

B. Exercises

Simplify, leaving each answer in exponential form.

13. $3^2 \cdot 3^3$
14. $4^{16} \div 4^{-2}$
15. $(8^3)^5$
16. $(3^{11})^0$
17. $12^2 \cdot 12^{-7}$
18. $(10^{10})^5$
19. $10^{12} \cdot 10^3$
20. $\frac{4^{-2}}{4^{-9}}$
21. $\frac{1}{3^2} \cdot \frac{1}{3^3}$
22. $\left(\frac{1}{7^3}\right)^4$
23. $\frac{1}{9^2} \div \frac{1}{9^4}$
24. $2^2 \cdot 2^3 \cdot 3^4 \cdot 5^2 \cdot 5$

Simplify, leaving each answer in positive exponential form.

25. $7^{-2} \div 7^9$
26. $8^7 \div 8^3$
27. $17^{81} \div 17^{27}$
28. $(7^{-2})^3$
29. $(28^{-4})^{-2}$
30. $5^3 \cdot 5^{-2} \cdot 5^{10}$
31. $2^3 \div 2$
32. $(3^{-6})^0$
33. $4^{-5} \cdot \frac{1}{4^2}$
34. $8^{-3} \cdot \frac{1}{8^5}$
35. $\frac{1}{3^2} \cdot \frac{1}{3^{-2}}$
36. $10^2 \cdot 10^4 \cdot 3^3 \cdot 3^{-6}$
37. Which of the following expressions are positive?
    a. $(-2)^3$ **b.** $(-2)^4$ **c.** $(-3)^{-2}$ **d.** 3^{-2}
38. When a negative number is raised to an odd power, is the result positive or negative?
39. When a negative number is raised to an even power, is the result positive or negative?

Translate each phrase into a numerical expression. Then evaluate your expression.

40. two to the seventh power
41. nine to the fourth power
42. the sixth power of ten
43. the square of eight
44. five cubed

C. Exercises

Simplify, leaving each answer in positive exponential form.

45. $\dfrac{2^3 \cdot 3^{10} \cdot 2^7 \cdot 5^3}{2^4 \cdot 5^4(3^5)^2}$

46. $\dfrac{2^4 \cdot 7^8(11^0)^2}{2^3(3^2)^3 \cdot 3^5}$

47. $[3x^{-2}(x^5)]^{-4}$

48. $(-2x^{-3}y^4)^{-2}$

Dominion Modeling

49. Use your Sierpinski triangles from Section 1.5, Dominion Modeling, to extend your table from Section 1.6 by determining the following expressions for the nth iteration.

a. the number of upward-pointing triangles

b. the area of each upward-pointing triangle

c. the total area of all the upward-pointing triangles

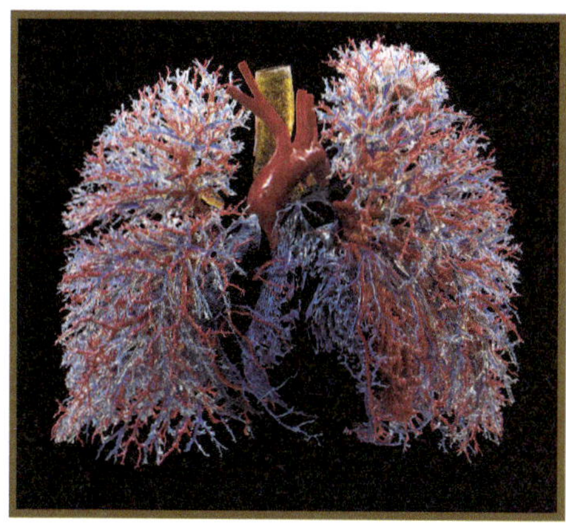

Structures within the human body exhibit fractal characteristics.

50. Use the expressions from above to complete the following table for the tenth and the one-hundredth iterations of Sierpinski's triangle.

Iteration	n	10	100
Number of upward-pointing triangles			
Area of each upward-pointing triangle			
Total area of upward-pointing triangles			

CUMULATIVE REVIEW

Simplify. [1.3–1.6]

51. 37 − 85

52. 126 + 395

53. 27(−82)

54. $\dfrac{-135}{5}$

55. −389 + (−27)

56. 23 + (−82)

57. −83(−24)

58. −146 − 29 − (−35)

59. Find the length of a segment having endpoints at 2.5 and −3.75. [1.2]

60. Find the coordinate of the midpoint of a segment having endpoints at 2.5 and −3.75. [1.2]

1.8 Order of Operations

Evaluate this expression: $3 + 2 \cdot 4 - 6 \div 3$.

Did you get $\frac{14}{3}$? $\frac{5}{3}$? $-\frac{10}{3}$? 9? 18? confused? If you are confused, you have good reason to be; all of those numbers are possible answers, but only one of them is correct. In mathematics we are not left to wonder which answer is correct. Mathematicians have agreed that numerical expressions without grouping symbols are evaluated in the following order.

1. Evaluate all exponential expressions.
2. Perform all multiplication and division from left to right.
3. Perform all addition and subtraction from left to right.

We now know exactly how to simplify the introductory expression. Since this problem has no grouping symbols and no exponential expressions, you should perform the multiplication and division first and then perform the addition and subtraction.

$$3 + 2 \cdot 4 - 6 \div 3$$
$$= 3 + 8 - 2$$
$$= 11 - 2$$
$$= 9$$

Example 1

Evaluate.

a. $2 + 3^2 - 4 \div 2 \cdot 3$

b. $2^3 - 1^2 + 5 \cdot 4 - 3^2 + 9 \div 3$

Answer

a. $2 + 3^2 - 4 \div 2 \cdot 3$

$2 + 9 - 4 \div 2 \cdot 3$	1. Evaluate the exponential expression first.
$2 + 9 - 2 \cdot 3$	2. Evaluate all multiplication and division from left to right.
$2 + 9 - 6$	
5	3. Evaluate all addition and subtraction from left to right.

b. $2^3 - 1^2 + 5 \cdot 4 - 3^2 + 9 \div 3$

$8 - 1 + 5 \cdot 4 - 9 + 9 \div 3$	1. Evaluate the exponential expressions first.
$8 - 1 + 20 - 9 + 3$	2. Evaluate all multiplication and division from left to right.
21	3. Evaluate all addition and subtraction from left to right.

Sometimes you may want to perform a sequence of operations in an order that is different from the one prescribed by the order of operations. For instance, you may want to add two numbers before you divide. If so, you must have some way to indicate the revised order. *Grouping symbols* let you indicate which operation to perform first.

Parentheses ( ), *brackets* [ ], and *braces* { } are used to indicate that the operations enclosed by them are to be performed prior to any other operations. Absolute value symbols | |, the fraction bar, and the radical sign are also used as grouping symbols.

Order of Operations

Evaluate within the innermost grouping symbols first.
Evaluate all exponential expressions.
Perform all multiplication and division from left to right.
Perform all addition and subtraction from left to right.

Example 2

Evaluate.

a. $2 + 3(5 + 7) \div 2$ **b.** $\dfrac{2(-3 + 7) - (7 - 5)}{5 - 3(1 - 4)}$ **c.** $9 - 3|2 + 7 \cdot 2^3(3 - 5)| - 4$

Answer

a. $2 + 3(5 + 7) \div 2$

$2 + 3(12) \div 2$ — 1. Evaluate within the grouping symbols first.

$2 + 36 \div 2$ — 2. Evaluate all multiplication and division from left to right.

$2 + 18$

20 — 3. Evaluate the addition.

b. $\dfrac{2(-3 + 7) - (7 - 5)}{5 - 3(1 - 4)}$

$\dfrac{2(4) - 2}{5 - 3(-3)}$ — 1. Evaluate within the grouping symbols first.

$\dfrac{8 - 2}{5 + 9}$ — 2. Evaluate the multiplication indicated in both the numerator and the denominator.

$\dfrac{6}{14}$ — 3. Evaluate the addition and subtraction.

$\dfrac{3}{7}$ — 4. Reduce the resulting fraction.

c. $9 - 3|2 + 7 \cdot 2^3(3 - 5)| - 4$

$9 - 3|2 + 7 \cdot 2^3(-2)| - 4$ — 1. Evaluate within the innermost grouping symbols first.

$9 - 3|2 + 7 \cdot 8(-2)| - 4$ — 2. Evaluate within the next innermost grouping symbols by evaluating the exponential expression, multiplication, and then addition.

$9 - 3|2 - 112| - 4$

$9 - 3|-110| - 4$

$9 - 3(110) - 4$ — 3. Evaluate the absolute value, which functions as a grouping symbol.

$9 - 330 - 4$ — 4. Evaluate the resulting expression by performing the multiplication and then the addition and subtraction.

-325

TECHNOLOGY CORNER (TI-84+ Family)

To evaluate an expression using your graphing calculator, enter the expression as you would when writing it on paper. Pressing ENTER (regardless of the cursor location) will evaluate the expression according to the order of operations.

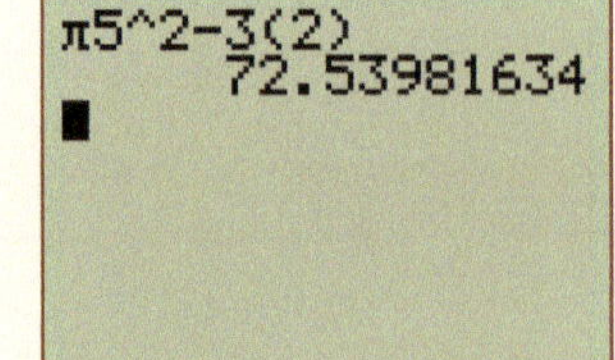

Calculate $\pi(5^2) - 3(2)$.

2nd ^ 5 x^2 − 3 (2) ENTER

The calculator recognizes implied multiplication in expressions like 2π. Be sure to use the − key for subtraction and the (-) key for negation.

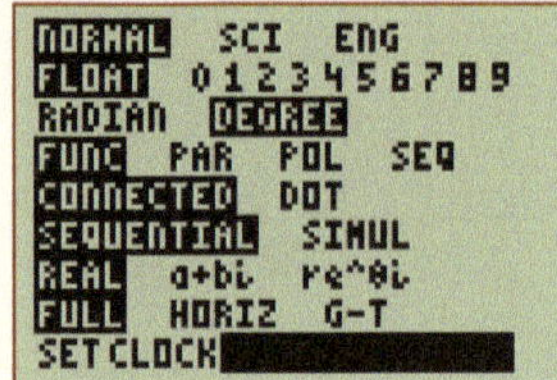

Modes affect the way numbers are displayed and expressions are evaluated. To change the modes on your calculator, use the cursor keys to highlight the desired setting and press ENTER. You can choose between normal, scientific, or engineering notation. Decimal numbers can be displayed with a fixed or floating number of decimal places. For now, be sure your calculator has degree mode selected for angles and is set to use real numbers. Function mode will typically be used for graphing, and you can choose whether to connect graphed points and whether to plot sequentially or simultaneously. Horizontal or graph-table split-screen modes can be chosen instead of the normal full-screen mode.

A. Exercises

Evaluate. Show each step in your work.

1. $5 + 3 \cdot 4$

2. $10 + 7 - 3 \cdot 2$

3. $8 + 10 \div 5$

4. $1 + 18 \div 3(-2)$

5. $-10 - 2 \cdot 4 \div 2 + 3$

6. $1 + (-7)^2 \cdot 2 - 50$

7. $27 - 3^2 + 8 \div (-2)$

8. $-81 - 14 \cdot 3 \div 2 \cdot 3$

9. $3 + 4 - 2 - 6 \cdot 3$

10. $7 - 3^2 - 1 + 4^3 \div 2$

11. $5 + 4(-3 + 7)$

12. $6 \cdot 3 \div (5 + 4) + 1 - 6$

B. Exercises

Evaluate. Show each step in your work.

13. $-2|5 - 3(2)| + 5^0$

14. $-3^2 - 5|4 + 2(-3)^2|$

15. $6 + 8(2 - 5) - [3 + (4 - 7)3] + 6$

16. $4 - 5 + 6(8 + 2 \cdot 4 - 5)$

17. $3 + \{[(4 \cdot 8 - 2) + 2 + 7 \cdot 4] - 7 \cdot 2\} - 8$

18. $9 - [-9 + (4 \cdot 2 + 6)] + 4 + 2 \cdot 6$

19. $\dfrac{6 + (2 - 5)8}{-6 + [3 + (4 - 7)3]}$

20. $\dfrac{4 + (2 - 3)6}{8 - 9(3 - 2)}$

21. $1 + [2(4 + 2)^2 - 6 + (3 - 5)^3] - 8 \cdot 6$

22. $2\{[4 - (5 - 2^2)](-3 - 9) \div 9 - 3\}$

23. $3 - [2(6 - 2) \div 4 + (7 - 9) \div 2]^5$

24. $5[3 + 5 \div 2^2 - (4 - 7)^2(-3 + 6)^{-2} - 2]$

Translate each phrase into a numerical expression. Then evaluate your expression.

25. five more than twice the square root of four

26. the product of five and the sum of six and seven

27. the square of the difference of twenty and thirteen

28. the difference of the squares of twenty and thirteen

29. the sum of two and five, times six

30. the sum of five times six and two

31. the product of five and six increased by the quotient of twenty and five

32. seven less than the product of eleven and the reciprocal of nine

33. nine reduced by the absolute value of the sum of two and negative fifteen

C. Exercises

Evaluate. Show each step in your work.

34. $\frac{5}{6} \div \frac{4}{3} + \frac{1}{3} \cdot \frac{2}{5}$

35. $-\frac{2}{3} + \left(\frac{1}{2} - \frac{3}{4} \cdot \frac{5}{6}\right)$

36. $0.8 - 3 \cdot 2.2 + 1.8 \div 2$

37. $-1.7 + 2(1.2^2 \cdot 2.5)$

Dominion Modeling

Fractals are used to generate lifelike landscapes in computer games as well as a variety of other phenomena in training simulators.

38. Research and briefly describe how the midpoint-displacement algorithm is used to model landscapes.

Whether teaching firefighters the characteristics of wildfires, training surgeons virtually, or generating special effects, the mathematics of fractals enhances our lives. The opportunities provided by this new branch of mathematics to better understand God's creation and to implement solutions to problems troubling our society are as breathtaking as the art that it produces.

CUMULATIVE REVIEW

State the coordinate of each point. [1.2]

39. J

40. K

41. L

42. Graph the following points on a number line: R: 3; S: $1\frac{1}{4}$; T: -1.75. [1.2]

43. Find the length of a segment having endpoints at 17 and -4. [1.2]

44. Find the coordinate of the midpoint of a segment having endpoints at 17 and -4. [1.2]

Classify each number as natural, whole, integer, rational, or irrational. Give the most specific set. [1.1]

45. 3

46. $\frac{1}{3}$

47. $\sqrt{3}$

48. -3

CHAPTER 1 REVIEW

Draw a copy of the Venn diagram representing the real number system; then place each number within the innermost set to which it belongs.

1. $0.\overline{3}$ **2.** -4 **3.** $\sqrt{7}$

True or false

4. $0 \in \mathbb{W}$ **5.** $\mathbb{Q} \cup \mathbb{Q}' = \mathbb{R}$ **6.** $\mathbb{Z} \subseteq \mathbb{N}$

Identify each set using set notation.
$A = \{2, 4, 6, 8\}$; $B = \{-2, -1, 0, 1, 2\}$; $C = \{-5, -3, -1\}$

7. $A \cap B$

8. $B \cup C$

9. $B \cap (A \cup C)$

10. Graph the following points on a number line: A: -4; B: 2.5; C: $-\frac{4}{3}$.

State the coordinate of each point.

D E F
−4 −3 −2 −1 0 1 2 3 4

11. D **12.** E **13.** F

Find each absolute value.

14. $|-1.3|$ **15.** $|7|$

A segment has endpoints with the given coordinates. Find the length of the segment and the midpoint of the segment.

16. -12 and 22 **17.** 8 and 17

Evaluate.

18. $|-9| - |-4|$ **19.** $-18 + 14$ **20.** $2(-14)$

21. $12 - 32$ **22.** $(-2)^3$ **23.** $\frac{-32}{-8}$

24. $-5 + 3$ **25.** 15^0 **26.** $-7 + 1.8$

27. $-3.7(-1.2)$ **28.** $\frac{1}{8} - \left(-\frac{3}{8}\right)$ **29.** $\frac{4}{7} \div \frac{3}{8}$

30. $-\frac{1}{12}\left(-\frac{4}{5}\right)$ **31.** $\frac{5}{8} - \frac{7}{9}$ **32.** $-\frac{2}{3} \div \frac{1}{3}$

Simplify, leaving each answer in exponential form.

33. $8^9 \cdot 8^6$ **34.** $4^8 \cdot 4^{-3} \div 4^2$ **35.** $(3^4)^6$

Evaluate.

36. $4 + 6 \cdot 8 \div 2 - 14$ **37.** $3 + 8 - 5 \cdot 4 + 6$

38. $5 + 3(2 - 6 \cdot 2) - 4 + 6$ **39.** $3 + 6^2 - 7 - 2 \cdot 4 - 9$

40. $6 - 2(9 \div 3 + 8) - 6 \cdot 4$ **41.** $2^3 + 2[4(3 + 9) - 7] - 15 + 3 \cdot 6$

42. $4^2 - 1 \cdot 8 + 3 - 7 + 10 \div 2$ **43.** $\sqrt{9} + |-9|(2 - 6) + 4 \cdot 5^0$

Translate each phrase into a numerical expression. Then evaluate your expression.

44. the opposite of negative six

45. eight less than five

46. the product of fifteen and negative three decreased by twelve

47. the sum of three cubed and four

Match each statement to the property it illustrates.

48. $8 \cdot 0 = 0$

49. $-8 + 0 = -8$

50. $(3 + 5) + 4 = 3 + (5 + 4)$

51. $2 + (-2) = 0$

52. $6(-7) = -7(6)$

53. $-9 \cdot 1 = -9$

54. $3(2 - 7) = 6 - 21$

55. $-3 + 4 = 4 - 3$

56. $\frac{1}{5}(20a) = 4a$

a. Associative Property of Addition

b. Associative Property of Multiplication

c. Commutative Property of Addition

d. Commutative Property of Multiplication

e. Distributive Property

f. Additive Identity Property

g. Multiplicative Identity Property

h. Additive Inverse Property

i. Multiplicative Inverse Property

j. Zero Property of Multiplication

2 Variables and Equations

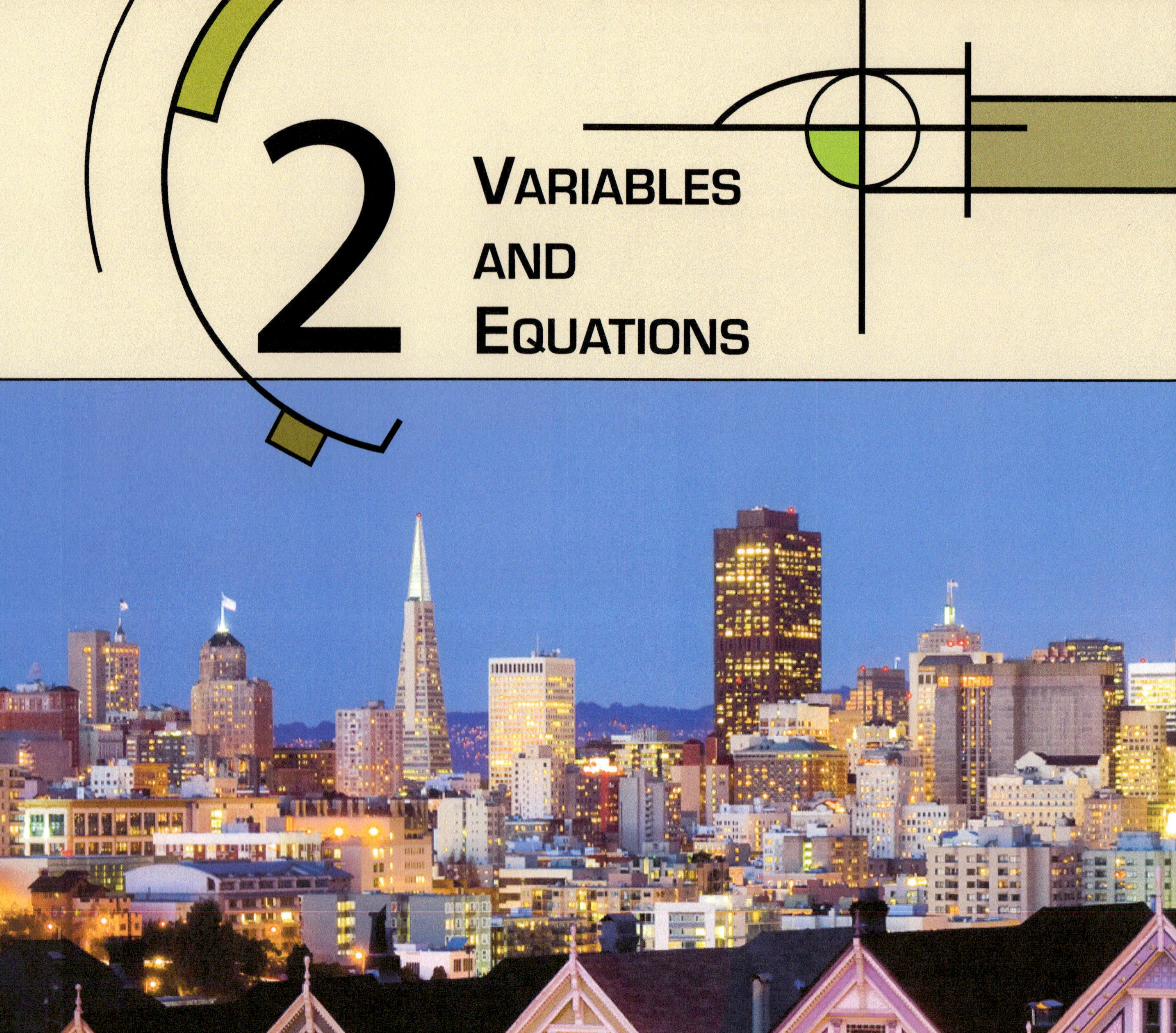

"Turn the lights off!" is a phrase you have probably heard over and over again. The theme of energy conservation resounds through our lives on a daily basis. The Dominion Mandate found in Genesis 1:28 demands that we not waste the resources God has provided for our use. Christ's example of gathering what was left after feeding the crowd in John 6:12–13 illustrates wise management of God's resources. In everyday life the following types of questions arise: How much does it cost to leave that light on? Should I purchase the more expensive, energy-efficient appliance? When bombarded by the media's cries of impending crisis, you must carefully analyze marketers' claims to discern the best solutions to everyday problems.

When challenging His listeners to true discipleship, Christ commends those who count the costs prior to beginning a new enterprise (Luke 14:28–30). A business venture must be carefully planned to avoid failure. First Timothy 6:17–19 warns us not to trust in the assets God provides but rather to trust in the God Who provides them for us to enjoy and to use for the benefit of others. As a Christian attempts to fulfill the command to be a wise manager of his God-given time and resources, mathematics will play a leading role in his decision-making processes.

After this chapter you should be able to

1. identify variables and constants in an algebraic expression.
2. translate word phrases and sentences into algebraic expressions and equations.
3. evaluate a formula or algebraic expression when given values for its variables.
4. use the Distributive Property to simplify algebraic expressions.
5. identify and combine like terms.
6. solve one-step, two-step, and multi-step equations.
7. identify the properties used to solve equations.
8. apply translation and equation-solving techniques to solve word problems.
9. clear equations of fractions and decimals when solving equations.
10. use expressions and equations to manage God-given resources.

2.1 Variables and Algebraic Expressions

A church is planning a fall cookout. The food committee decided to shop for the best buy in canned soft drinks. Here is what they found.

Soft Drinks	Price per Can
Ryans Root Beer	75¢
Cross Cola	70¢
Kwik Zip	65¢

Brand	Price per Can p	Total Price $62p$
Ryans	0.75	\$46.50
Cross	0.70	\$43.40
Kwik	0.65	\$40.30

The committee chairman asked for the total cost of each brand based on one can for each of the 62 members. If p is used to represent the price of each drink, the committee can represent this cost with the expression $62p$ and determine the total cost of beverages by multiplying each p value by 62.

If symbols are written next to each other with no operational sign between them, multiply them.

On the sheets above, p represents three different values: {\$0.75, \$0.70, \$0.65}. We say p is a variable because its value may change. A variable is denoted by a letter that stands for any member of a set. This set of possible replacement values is called the *domain* of the variable. Variables are used extensively in algebra to indicate unknown quantities.

Frequently, a mathematical expression will contain a *constant*, a number or letter whose value does not change. The Greek letter pi (π) represents a constant approximately equal to 3.14. The constant in the illustration above is 62.

Definitions

A **variable** is a symbol used to represent any element of a given set of numbers. The set of numbers is the domain of the variable.

A **constant** is a fixed number.

An **algebraic expression** is a mathematical expression containing numbers, variables, and operational signs.

You already know how valuable variables are for summarizing properties. It is much easier to write $a + b = b + a$ than to try to list all number statements separately ($2 + 3 = 3 + 2, 2 + 4 = 4 + 2$, etc.).

To *evaluate* an algebraic expression means to substitute values for the variable or variables and to simplify the result.

Example 1

Evaluate the expression $x + 9$ for each member of the domain $\{-5, \frac{1}{2}, 2, 3\}$.

Answer If $x = -5$, then $x + 9 = -5 + 9 = 4$.

If $x = \frac{1}{2}$, then $x + 9 = \frac{1}{2} + 9 = 9\frac{1}{2} = \frac{19}{2}$.

If $x = 2$, then $x + 9 = 2 + 9 = 11$.

If $x = 3$, then $x + 9 = 3 + 9 = 12$.

The corresponding set of values, called the *range*, is $\{4, \frac{19}{2}, 11, 12\}$.

Example 2

If c represents the cost of one pencil in cents, the algebraic expression $10c$ represents the cost of ten pencils. If the domain of c is {5, 10, 17}, evaluate the expression for each member of the domain.

Answer If $c = 5$, then $10c = 10(5) = 50$.

If $c = 10$, then $10c = 10(10) = 100$.

If $c = 17$, then $10c = 10(17) = 170$.

Thus, the cost of the pencils is 50 cents, or \$0.50; 100 cents, or \$1.00; and 170 cents, or \$1.70.

As a student of algebra, you must be able to take information from our God-created universe and translate it into algebraic language. After you have symbolized the problem, you can solve it and relate it back to the physical world. Look for words that indicate certain operations. The following table lists words typically associated with each operation.

Addition	Subtraction	Multiplication	Division
sum add more increased	difference subtract less decreased diminished	product times twice of doubled	quotient divided by half per average

The ability to translate phrases into algebraic expressions is foundational to success in algebra.

Example 3

Translate each word phrase into an algebraic expression.

a. the sum of a number and four

b. the difference between a number and five

c. one-third of a number

d. a number divided by six

e. the sum of three and eight times a number

f. eight less than three times a number

g. the quotient of the difference of a number and eight, and twice the number

h. the opposite of the reciprocal of the square of a number

Answer

a. $n + 4$

b. $n - 5$

c. $\frac{1}{3}n$ or $\frac{n}{3}$

d. $\frac{n}{6}$ or $n \div 6$

e. $3 + 8x$

f. $3n - 8$

g. $\frac{n - 8}{2n}$

h. $-\frac{1}{s^2}$

Knowing general expressions for several frequently encountered phrases will assist you in translating them into algebraic expressions. The numbers 7, 8, 9, ... are called consecutive integers. If the first integer in a sequence is symbolized by n, then the second number is symbolized by $n + 1$, the third number by $n + 2$, and so on. Consecutive even integers are two units apart. If n represents the first even number, then $n + 2$ represents the next even number and $n + 4$ represents the third even number. Consecutive odd integers are also two units apart and can also be represented by n, $n + 2$, $n + 4$, and so on, where n represents the first odd integer.

Example 4

Write an expression for the sum of three consecutive integers.

Answer

n = the first integer

$n + 1$ = the second integer

$n + 2$ = the third integer

1. Represent each of the numbers with an expression.

$n + (n + 1) + (n + 2)$

2. The sum is the result of adding the numbers.

General expressions for the supplement and complement of an angle are also helpful. Recall that supplementary angles are those whose sum is 180° and complementary angles are those whose sum is 90°. If an angle has a measure of a degrees, then its supplement measures $(180 - a)$ degrees and its complement measures $(90 - a)$ degrees.

Example 5

Write an expression for the difference of twice an angle and its complement.

Answer

a = the measure of the angle

$90 - a$ = the measure of its complement

1. Represent the angle and its complement with expressions.

$2a - (90 - a)$

2. Translate the word phrase into an algebraic expression.

A. Exercises

List the variables and constants in each expression.

1. $3x + 6$
2. $\frac{ab}{3}$
3. $\frac{t}{4} + 6$
4. πr^2

Evaluate each expression for each member of the domain {−4, 1, 3, 8}.

5. $x - 4$
6. $y + 6$
7. $3z$
8. $2a + 4$
9. $\frac{1}{2}n$
10. $\frac{a}{4} + 5$
11. $x^2 + 4x$
12. $|b^2 - 10|$

Translate each word phrase into an algebraic expression.

13. the product of eight and a number
14. a number decreased by seventeen
15. a number n less ten
16. twenty-four greater than a number
17. four more than five times a number
18. the sum of twice a number and six
19. the difference between a number and twice the same number

20. the square root of a number

21. a number q to the fifth power

22. the cube of the number s

B. Exercises

Translate each word phrase into an algebraic expression.

23. the calories Joe consumes eating n sweet rolls when one sweet roll has 325 calories

24. Joy's keyboarding speed if Joy types twice as fast as Monique

25. the number of history books in a store if there are four more than three times the number of math books

26. the perimeter of a rectangle 24 units long and w units wide

27. the amount Evelyn makes the week she gives x haircuts at \$25 each and y perms at \$90 each

28. the average of the three numbers p, q, and r

29. a number cubed plus the same number squared

30. the sum of the squares of a and b

31. the quotient of twice a number, and the sum of the number and 7

32. the reciprocal of the absolute value of a number

33. the sum of three consecutive odd numbers

34. 126 diminished by the product of two consecutive integers

35. three less than the product of two consecutive even integers

36. three times the complement of an angle

37. the sum of an angle's complement and its supplement

38. the difference of an angle's supplement and the measure of the angle itself

39. the amount spent if a customer buys six large drinks at d dollars each and twelve hot dogs at h dollars each at a baseball game

40. the number of animal legs on a farm containing x horses, y cows, and z chickens

41. the number of points a basketball player gets in one game when he makes x three-point shots, y two-point shots, and z free throws (at one point each)

42. Write separate expressions for Sue's score and Jim's score if Sue got six more points on the test than Mark and Jim got ten fewer points than Mark. Mark's score was m.

C. Exercises

Translate each word phrase into an algebraic expression.

43. the amount of water in a pool if three pipes deliver water at the rates x, y, and z gal per hour respectively (The first and third are used to fill the pool, while the second is a drainpipe. The fill pipes are opened and run for 4 hr, and the drainpipe is inadvertently left open for 2 hr and then shut off.)

44. the amount of sales tax paid at 6% for x shirts at \$24 each and y pairs of socks at \$4 each

Write two word phrases to describe each algebraic expression.

45. $2a + 5$

46. $3(x + 2)$

Dominion Modeling

An electric device is often labeled with the amount of power (measured in watts, W) required to operate it. The appliance's wattage indicates the rate at which energy is consumed. Power companies typically charge consumers for the electric energy consumed measured in kilowatt-hours (kWh). A wise manager will be aware of the cost of operating an electric device.

Example: Determine the typical daily, monthly, and yearly cost of operating a 65 W incandescent light bulb turned on for 8 hr a day if the electric company charges $0.12/kWh.

$$65 \text{ W} \times 8 \text{ hr} = 520 \text{ WH} \times \frac{1 \text{ kW}}{1000 \text{ W}} = 0.520 \text{ kWh each day}$$

$$0.520 \times \frac{\$0.12}{\text{kWh}} = \$0.0624 \approx \$0.06/\text{day}$$

$$\$0.0624 \times 30 \text{ days} = \$1.872 \approx \$1.87/\text{mo}$$

$$\$0.0624 \times 365 \text{ days} = \$22.776 \approx \$22.78/\text{yr}$$

47. Using the same assumptions as above, determine the daily, monthly, and yearly cost of operating a 15 W compact fluorescent light bulb (CFL) that produces the same amount of light as a 65 W incandescent bulb for 8 hr a day.

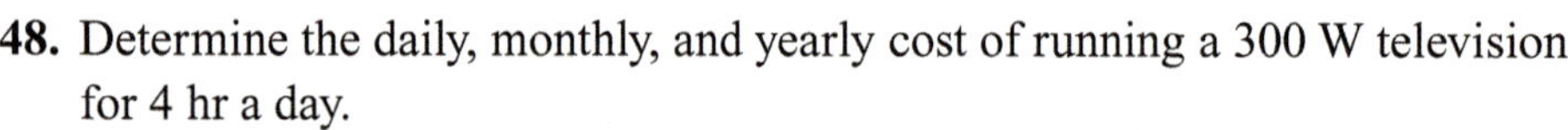

48. Determine the daily, monthly, and yearly cost of running a 300 W television for 4 hr a day.

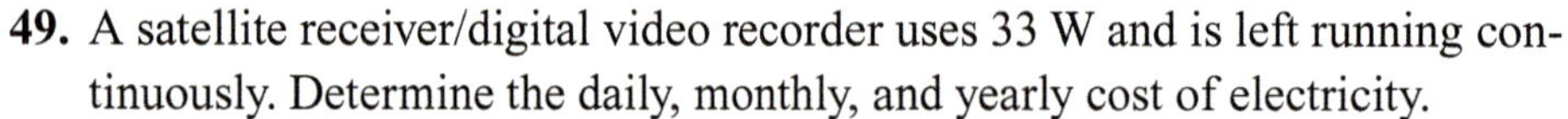

49. A satellite receiver/digital video recorder uses 33 W and is left running continuously. Determine the daily, monthly, and yearly cost of electricity.

50. Many electronic devices consume electricity to maintain a standby mode while turned off but still plugged in. Determine the cost of keeping an appliance plugged in for 1 yr if its standby mode uses 1 W.

CUMULATIVE REVIEW

Evaluate. [1.3–1.6]

51. $-18 + (-6)$

52. $-18 - (-6)$

53. $-18(-6)$

54. $\frac{-18}{-6}$

Use algebraic equations to express the following properties of exponents. [1.7]

55. Product Property

56. Quotient Property

57. Power Property

Use the properties of exponents to simplify the following algebraic expressions. [1.7]

58. x^7x^{-3}

59. $\frac{x^7}{x^{-3}}$

60. $(x^7)^{-3}$

2.2 Evaluating Algebraic Expressions

Recall that evaluating an expression means to substitute values for the variable or variables and then simplify the result. To evaluate $x + 10$ when $x = 5$, replace x with its given value, and the expression becomes $5 + 10 = 15$.

Evaluating an Expression

1. Substitute the given values for the variables.
2. Evaluate by following the order of operations.

Example 1

Evaluate $3a^2 - b + c^3$ when $a = 1$, $b = 4$, and $c = 3$.

Answer $3a^2 - b + c^3 = 3(1)^2 - 4 + 3^3 = 3 - 4 + 27 = 26$

Example 2

Evaluate x^2yz^2 when $x = -\frac{1}{2}$, $y = -2$, and $z = \frac{1}{4}$.

Answer $x^2yz^2 = \left(-\frac{1}{2}\right)^2(-2)\left(\frac{1}{4}\right)^2 = \frac{1}{4}\left(-\frac{2}{1}\right)\left(\frac{1}{16}\right) = -\frac{1}{32}$

When using formulas, you must evaluate the expression given in the formula.

When Ahmed dives, he feels additional pressure on his ears above normal atmospheric air pressure. Near the surface he feels less pressure than when he dives for pearls. The amount of additional pressure, p, in pounds per square inch (psi) depends upon the height, h, of the water above the diver, or his depth. The relationship is shown in the formula $p = 0.433h$. Diving to a depth of 20 ft, he feels $0.433(20) = 8.66$ psi of additional pressure. At a depth of 3 ft he feels only $0.433(3) = 1.299$ psi of additional pressure.

Definition

A **formula** is an equation that expresses the relationship between related quantities.

When preparing for her youth group's mission trip, Sara found that the average daily high temperature in Spain during the month of her visit is 25°C. She uses the formula $F = \frac{9}{5}C + 32$ to convert from Celsius to Fahrenheit in order to plan which clothes to pack.

$$F = \frac{9}{5}C + 32 = \frac{9}{5}(25) + 32 = 45 + 32 = 77°F$$

Formulas are mathematical sentences that are used extensively in science, business, engineering, carpentry, and many other professions. You can distinguish between mathematical phrases and sentences by identifying a verb—similar to how you distinguish between phrases and sentences in English grammar. Can you find the verb in the sentence $p = 2l + 2w$? This is read, "the perimeter is twice the length plus twice the width." In formulas, the verb "is," or "equals," is symbolized by "=." Here are some other mathematical verbs you should know.

Math Verb	Meaning	Math Verb	Meaning
$=$	is, equals	$\notin$	is not an element of
$\neq$	is not equal to	$>$	is greater than
$\approx$	is approximately equal to	$<$	is less than
$\subseteq$	is a subset of	$\geq$	is at least, is greater than or equal to
$\in$	is an element of	$\leq$	is at most, is less than or equal to

You should also know these formulas.

Perimeter and Area Formulas		
rectangle	$p = 2l + 2w$	$A = lw$
square	$p = 4s$	$A = s^2$
circle	$c = 2\pi r$	$A = \pi r^2$
triangle	$p = a + b + c$	$A = \frac{1}{2}bh$

Example 3

Find the perimeter of a rectangle whose length is 8 cm and whose width is 3 cm.

Answer

$p = 2l + 2w$ — 1. Write the formula for the perimeter of a rectangle.

$= 2(8) + 2(3)$ — 2. Substitute the given values into the formula.

$= 16 + 6$ — 3. Evaluate.

$= 22$ cm

Example 4

Find the area of a circle with a radius of 9 ft.

Answer

$A = \pi r^2$ — 1. Write the formula for the area of a circle.

$= \pi(9)^2$ — 2. Substitute the length of the radius into the formula.

$= 81\pi \approx 81(3.14)$ — 3. Evaluate, using 3.14 to estimate π.

$= 254.34 \text{ ft}^2$

Using Formulas
1. Write the correct formula.
2. Substitute the given values for the variables in the formula.
3. Evaluate the expression.

A. Exercises

Evaluate when $x = 4$, $y = -2$, $z = \frac{4}{5}$, and $h = -1$.

1. x^2y
2. $x + yz$
3. $\frac{x + y}{h}$
4. $x^2 + h - 3$
5. $\frac{x}{y} + \frac{z}{h}$
6. $(x + y)(x - y)$
7. $x^3 - z^3$
8. $h(x + z)$
9. $h^2 + y^2 - x^3$
10. $\frac{3x^2 - y}{5}$

Write the appropriate formula, substitute the given values into it, and then evaluate.

11. Using the formula $p = 0.433h$, determine how much additional pressure you will feel from the water when you dive to a depth of 32 ft.
12. Convert 8°C to Fahrenheit.
13. Mr. Denton wants to put a fence around his rectangular garden, but first he needs to know the perimeter. If his garden is 60 ft long and 20 ft wide, how many feet of fencing does he need?
14. A picture that is 11 in. by 14 in. is to be framed using wood that is 1 in. wide. Find the perimeter of the picture and then explain why the length of the perimeter would not be enough to frame the picture.
15. A carpenter plans to cut Formica to go around the outer edge of a circular table with a 2 ft radius. Find the circumference of the table (using $\pi \approx 3.14$) to determine the length of the piece of Formica.

The volume of a rectangular box can be found using the formula $V = lwh$. Find the volume of the boxes with the following dimensions.

16. $l = 4$ in., $w = 2$ in., and $h = 5$ in.
17. $l = 10$ in., $w = 3$ in., and $h = 2$ in.
18. $l = 8$ in., $w = 7$ in., and $h = 6$ in.

B. Exercises

Evaluate xy^2z for the following values.

19. $x = 2$, $y = 3.8$, and $z = -2$
20. $x = 10$, $y = \frac{2}{7}$, and $z = \frac{1}{5}$
21. $x = a$, $y = a$, and $z = a$
22. $x = c$, $y = c^3$, and $z = c^{-2}$

Evaluate when $p = 12$, $q = 196$, and $r = -13$.

23. $r^2 + \sqrt{q}$
24. $\frac{|p + r|}{p + r}$

Evaluate when $x = \frac{1}{2}$, $y = -\frac{1}{4}$, $z = 1$, and $h = \frac{1}{3}$.

25. $x^2 + h - 3$
26. $(x + y)(x - y)$
27. $h^2 + y^2 - x^3$
28. $3x^2y - \frac{z}{5}$

Use the given formula to answer each question.

29. The formula $d = rt$ calculates distance as the rate multiplied by the time. If a bus carrying spectators to a soccer game travels 47 mi/hr for 2 hr, how far does the bus travel?
30. A simplified formula for normal weight is $w = \frac{11(h - 40)}{2}$. The variable h represents height in inches. If Ben is 5 ft 2 in. tall and weights 154 lb, what is his normal weight? Should he lose or gain weight to be at his normal weight? How much?

31. The amount of simple interest Shanda earns on the money she saves from her baby-sitting jobs is expressed by $I = Prt$. In this formula, I is simple interest, P is the principal (or amount invested), r is the percentage rate of the investment per year in decimal form, and t is the time or length of the investment in years. If Shanda invests \$450 at 5% interest, how much interest will she earn in one year?

Solve using the formula $R = xp$, where R is revenue (business income), x is the number of items sold, and p is the price per item.

32. If 48 cans of Ryan's soda are sold at \$0.79 per can, how much revenue is collected?
33. The senior class washed 110 cars and charged \$6.50 per car. How much money did they raise?
34. A company sold 33 bicycles for \$315 each. How much revenue did it receive?

Hourly employees working more than 40 hr a week are often paid "time and a half" for *overtime,* any hours beyond 40. In the formula $p = wr + 1.5we$, gross pay (p) is computed by multiplying the hourly wage (w) by the number of regular hours worked ($r \leq 40$) and adding 1.5 times the product of the hourly wage (w) and the number of extra hours (e) beyond 40. Find your gross pay if you work the given number of hours at the given rate.

35. 20 hr at \$7.75/hr
36. 40 hr at \$8.12/hr
37. 45 hr at \$8.50/hr
38. 47 hr at \$10.92/hr

Write each of the following mathematical formulas in symbols.

39. Distance (d) divided by rate (r) equals time (t).
40. The perimeter (p) of an isosceles right triangle is the sum of the length of its hypotenuse (c) and twice the length of a leg of the triangle (l).
41. The diameter (d) of a circle is twice the radius (r).

Translate each mathematical sentence into words.

42. $5 < 7$
43. $\pi \neq 3.14$

C. Exercises

Evaluate when $a = -\frac{2}{3}$, $b = \frac{4}{7}$, and $c = -3$.

44. $|a - b| - \frac{c^2}{b}$
45. $\sqrt{14abc} + \left(\frac{1}{a}\right)^2$

Solve.

46. Chris plans to paint a cylindrical gasoline tank. After he finds the number of square feet of surface area on the cylinder using the formula $S = 2\pi rH + 2\pi r^2$, he can compute the amount of paint needed. If the height of the cylinder is 5 ft and its radius is 2 ft, what is its total surface area? Use $\pi \approx 3.14$.
47. If a circle has radius r, find the effect on the circumference and the area when the radius is increased by 50%.

Dominion Modeling

How much does it cost to run a computer? Manufacturer labels can be misleading since they must list the theoretical maximum amount of electricity used, not the typical amount used. Typical desktop computers can use anywhere from 65 to 250 W. Peripheral devices such as monitors, printers, modems, routers, and speakers also add to the electric usage. Assume a rate of \$0.12/kWh for the following exercises.

48. Calculate the annual cost of continuously running a 250 W large computer with a gaming-level graphics card and an old 80 W CRT monitor.
49. Calculate the annual cost of running a more efficient 65 W computer with a 35 W LCD monitor for 3 hr a day if the computer and monitor are in sleep mode (using 5 W each) for the rest of the day.

CUMULATIVE REVIEW

Translate each mathematical sentence into words. [1.1]

50. $\varnothing \subseteq \mathbb{N}$

51. $-1 \notin \mathbb{N}$

Name the property each statement illustrates. [1.6]

52. $a(b + c) = ab + ac$

53. $t(sq) = (ts)q$

54. $n\left(\frac{1}{n}\right) = 1$

55. List the order of operations. [1.8]

Simplify. [1.7]

56. $\frac{(x^3)^5x^{-2}}{x^4}$

57. $\frac{y^5y^{-2}}{(y^{-1})^4}$

Translate each word phrase into an algebraic expression. [2.1]

58. the product of a number and four

59. the midpoint between m and n

Here are five incorrect equations. To correct each equation, reposition just one match.

VI − II = II

VII − II = VIII

IX + IV = XV

VII + V = III

VI − VII = I

TECHNOLOGY CORNER (TI-84+ Family)

Any numerical value, including the result of an evaluated expression, can be stored as a variable A–Z.

Assign $-\frac{1}{2}$ to the variable A and $\frac{4}{5}$ to B. Then display the value of A.

[(-)] 1 [÷] 2 [STO▶] [ALPHA] A [ENTER]
4 [÷] 5 [STO▶] [ALPHA] B [ENTER]
[ALPHA] A [ENTER]

```
-1/2→A
              -.5
4/5→B
               .8
A
              -.5
```

Expressions containing the variable can be easily evaluated.

Evaluate AB and $B - A^2$.

[ALPHA] A [ALPHA] B [ENTER]
[ALPHA] B [−] [ALPHA] A [x^2] [ENTER]

```
AB
              -.4
B-A²
              .55
```

ENTRY is a storage area containing the most recently entered expression or instruction. Selecting [ENTRY] ([2nd] [ENTER]) places the last instruction or expression at the current cursor location, and it can then be edited and evaluated. Repeatedly selecting ENTRY scrolls through recently entered expressions.

ANS is a storage area containing the most recent result. It can be used to extend a calculation or to store the result. Pressing [ANS] ([2nd] [(-)]) inserts the last result at the current cursor location.

Change the value of A to 2 and reevaluate the expression $B - A^2$.

2 [STO▶] [ALPHA] A [ENTER]
[ENTRY] [ENTRY] [ENTER]

Evaluate the expression $5(2 + B - A^2)$ and store the result as C.

5 [×] [(] 2 [+] [2nd] [(-)] [)] [STO▶] [ALPHA] C [ENTER]

```
2→A
                2
B-A²
             -3.2
5*(2+Ans)→C
               -6
```

2.3 Using the Distributive Property

When planning the purchase of new volleyball uniforms, the coach was able to find jerseys for \$27 and shorts for \$18. He calculated the total cost for the twelve uniforms in two ways.

$12(27 + 18)$	and	$12(27) + 12(18)$
$12(45)$		$324 + 216$
\$540		\$540

The fact that both methods produce the same answer illustrates the Distributive Property of Multiplication over Addition, an important and frequently used property of real numbers.

Distributive Property

For any real numbers a, b, and c, $a(b + c) = ab + ac$.

The Distributive Property is often used to simplify algebraic expressions. Observe how the parentheses are removed from each expression in the following examples.

Example 1

Distribute $4(a + 3b)$.

Answer $4(a + 3b) = 4(a) + 4(3b) = 4a + 12b$ Apply the Distributive Property and simplify.

Example 2

Distribute $-2(3x^2 - 4x + 5)$.

Answer
$-2(3x^2) - (-2)(4x) + (-2)(5)$ 1. Apply the Distributive Property.
$-6x^2 + 8x - 10$ 2. Simplify.

In Example 2, notice that the signs inside the parentheses change when the parentheses are removed because each term is multiplied by a negative number.

Mathematical expressions are written as simply as possible. If you are speaking of a group of apples and bananas, how would you most simply express 11 apples, 8 bananas, and 6 more apples? You would say 17 apples and 8 bananas. In other words, you would combine the two sets of apples into one set.

Likewise, to simplify an algebraic expression, you need to combine *like terms* by performing the indicated operations. In algebra, as in all math work, always express your final answer in simplest form.

Definitions

A **term** is a constant, a variable raised to a power, or the product or quotient of a constant and one or more such variables.

Like terms are terms that have the same variable (or variables) with corresponding variables having the same exponents.

$4 \qquad x \qquad x^3 \qquad 4x^3 \qquad \frac{x^3}{4}$

Each of the expressions above is an example of a term. Terms of an expression are separated by addition or subtraction. The expression $3x - 4y + 5x$ has three terms: $3x$, $-4y$, and $5x$. There are only two variables: x and y. The terms $3x$ and $5x$ are like terms because both have the same variable, x, with the same exponent, 1. Like terms can be combined into a simpler form, but unlike terms cannot. The expression in its simplest form is $8x - 4y$. When you combine like terms, you are actually applying the Distributive Property: $ab + ac = a(b + c)$.

Example 3

Simplify $3x - 4y + 5x$.

Answer		
	$3x + 5x - 4y$	1. Rearrange the terms (Commutative Property of Addition).
	$(3 + 5)x - 4y$	2. Apply the Distributive Property.
	$8x - 4y$	3. Add.

Since $8x$ and $-4y$ are unlike terms, they cannot be combined. Now the expression is in simplest form. Expressions with several variables are typically written with the terms in alphabetical order. When terms contain different powers, they are typically arranged in descending powers of the variable.

Example 4

Simplify $7x^2 + 8x - 4x^2 + 10x$.

Answer		
	$7x^2 - 4x^2 + 8x + 10x$	1. Rearrange the terms (Commutative Property of Addition).
	$3x^2 + 18x$	2. Apply the Distributive Property.

The terms containing x^2 and those containing x are unlike terms because the exponents of x are different.

Definition

The **numerical coefficient** (often referred to as the **coefficient**) is the numerical factor (constant) accompanying the variable(s) in a term.

In the expression $3x^2 - 7x^2 + x^2$, the coefficient of the first term is 3 and the coefficient of the second term is -7. If there is no expressed numerical coefficient of a variable, the coefficient is 1. Like terms can be combined by adding their coefficients: $3x^2 - 7x^2 + x^2 = [3 + (-7) + 1]x^2 = -3x^2$.

Combining Like Terms
1. Identify like terms.
2. Combine like terms by adding their coefficients.

Example 5

Simplify $8x^3 - x^2 + x^3 + 4x^3 - x^2 + 6x^3$.

Answer		
	$8x^3 - x^2 + x^3 + 4x^3 - x^2 + 6x^3$	1. Identify like terms. (Mark them if necessary.)
	$19x^3 - 2x^2$	2. Combine like terms.

Example 6

Simplify $4x^2y + 5xy - 10x^2y + 22xy - 8x + y$.

Answer		
	$4x^2y + 5xy - 10x^2y + 22xy - 8x + y$	1. Identify like terms. (Mark them if necessary.)
	$-6x^2y + 27xy - 8x + y$	2. Combine like terms.

Example 7

Simplify $4x^2yz + 2x^2y + xy^2 - 3xyz$.

Answer This expression cannot be further simplified since there are no like terms.

Some expressions require you to use the Distributive Property to eliminate parentheses before adding like terms. Carefully study the following examples, paying extra attention to the sign of each term.

Example 8

Simplify $(2x + 5) + (3x - 9) - (x - 6)$.

Answer

$2x + 5 + 3x - 9 - x + 6$	1. Remove the parentheses by distributing the coefficients of 1 and −1.
$4x + 2$	2. Combine like terms.

Example 9

Simplify $3k(k + 2) - k(k - 1)$.

Answer

$3k^2 + 6k - k^2 + k$	1. Apply the Distributive Property to remove the parentheses.
$2k^2 + 7k$	2. Combine like terms.

Example 10

Simplify $7x^2 - [3x + 4y(2x - 8y) - 4x] + 5xy$.

Answer

$7x^2 - [3x + 8xy - 32y^2 - 4x] + 5xy$	1. Apply the Distributive Property to remove the innermost parentheses.
$7x^2 - [-x + 8xy - 32y^2] + 5xy$	2. Combine like terms within the brackets.
$7x^2 + x - 8xy + 32y^2 + 5xy$	3. Remove the brackets by distributing the coefficient of −1.
$7x^2 - 3xy + x + 32y^2$	4. Combine like terms.

A. Exercises

For each algebraic expression, list all the numerical coefficients; then identify the like terms.

1. $3x + 4y - 5y$

2. $8x^2 + 3x + 6x^2$

3. $xy - 2xy - 9xy$

4. $x^2yz - xy^2z + 3x^2yz$

Use the Distributive Property to eliminate the parentheses.

5. $4(x + 2y)$

6. $-2(3z + 7)$

7. $5x(x + y)$

8. $\frac{2}{3}x\left(\frac{1}{2}y - \frac{9}{4}\right)$

Simplify.

9. $3a + 7a$

10. $19k - 13k$

11. $12xy + 4xy$

12. $12ab + 9ab - 7ab$

13. $d - 4d + 3f + 7f$

14. $2x^2 + 3x + 5x$

15. $y - 9y + 3y^2$

16. $6x^2 + 3x^2y - 4x + 8x^2$

B. Exercises

Simplify.

17. $4a + (2a + 9)$

18. $5 - (2x - 18)$

19. $2x - (8y + x)$

20. $(x + y) - (x - y)$

21. $(x + y) + (x - y)$

22. $(7b - 8c) - (3b + 2c)$

23. $3x + 2y + (5x - 7y)$

24. $-5a - (3a + b) + (2a - 7b)$

25. $(2a - b) - (3a + 4b) + (6a + b)$

26. $9m + 11n + (3n - m)$

27. $7b - 3(2b + 6)$

28. $-4(x + 3) + 5(x - 7)$

29. $2.46m + 1.6n - 8.4m$

30. $2.9x(x - 4) + 3.2x(x + 2)$

31. $\frac{1}{2}x + \frac{3}{2}x$

32. $4a - (2a^2 + 3z) + \frac{4}{5}(a^2 - 2a + 8)$

33. $x(x + y) - y(x + y)$

34. $x(x + y) + y(x + y)$

35. $4 - 3p - 6(p + 5) + 9$

36. $2r^2 + 5r + 3r - r(6r + 2)$

37. $2x^2 - 5x(4x - 3) - x(x^2 - 5x + 3)$

38. $4x^2(x - 9) - 3(x - 9) - 2x(5x^2 - 3x + 4)$

39. $-6[8x + 3(x - 4)]$

40. $7y + 3[y - 2(y + 1)] - 5$

C. Exercises

Find the expression that should be added to the equation to obtain the final expression.

41. $x^2 + x + 7 + ? = 5x^2 + 3x - 8$

42. $3(a^2 + 4a - 8) + ? = 4a^2 + 12a - 30$

Use the Distributive Property to rewrite the following formulas.

43. $A = \frac{1}{2}h(b_1 + b_2)$

44. $p = 2(l + w)$

45. $Ft = mv_1 - mv_2$

46. $C = \frac{5}{9}(F + 40) - 40$

Dominion Modeling

While many energy conservation strategies cost nothing to implement, others require an initial investment. The payback period is a simple method of evaluating the cost-effectiveness of conservation measures.

Example: By replacing a 65 W incandescent light bulb with a 15 W CFL bulb that produces the same amount of light, you can save $17.28 during one year of typical use. Determine the payback period of investing an extra $2 to replace a burned-out 65 W incandescent bulb costing $22.78/yr to operate with a 15 W CFL bulb producing the same amount of light that costs $5.26/yr to operate.

$$\text{savings} = \$22.78 - \$5.26 = \$17.28/\text{yr}$$

$$\text{payback period} = \frac{\$2}{\$17.28/\text{yr}} \approx 0.116 \text{ yr} \approx 1.4 \text{ mo}$$

$$\approx 42 \text{ days to recover the extra cost of the more efficient bulb}$$

47. Use the example above to write a generalized formula for the payback period.

48. A 4 W LED light bulb produces the same amount of light as a 25 W incandescent bulb but costs $35.

a. Determine the annual cost for running both bulbs continuously.

b. If this LED bulb costs $34 more than the incandescent bulb, calculate the payback period for this extra investment.

49. A typical household spends about $300 annually to heat water. Researchers have found that tankless water heaters can save 30% of this cost compared to using natural gas storage tank water heaters.

a. How much money can typically be saved annually by installing a tankless water heater?

b. If a tankless water heater costs an extra $550, what is the payback period for investing in a tankless water heater instead of a storage tank water heater?

CUMULATIVE REVIEW

Translate each word phrase into an algebraic expression. [2.1]

50. x increased by y

51. five degrees less than the outside temperature t

Simplify. [1.7]

52. $x \cdot x \cdot x \cdot x \cdot x$

53. $(5x)^0 + 5x^0 + 5^0x$

Simplify and then evaluate when $x = 2$ and $y = -3.7$. [2.2]

54. $4x + 3y - x + 6y - 8y$

55. $x^2 + 2x - 7x + 4$

State the property that justifies each numbered step in the following simplification. [1.3, 1.5]

$3x^2 + x + 2x^2 - 3x$

56. $= 3x^2 + 2x^2 + x - 3x$

57. $= 3x^2 + 2x^2 + 1x - 3x$

58. $= (3x^2 + 2x^2) + (1x - 3x)$

59. $= (3 + 2)x^2 + (1 - 3)x$

$= 5x^2 - 2x$ [addition]

MATH IN HISTORY

Isaac Newton (1642–1727)

Isaac Newton is the world's foremost mathematician and physicist. Math historians consider him one of the top three mathematicians of all time, along with Archimedes and Gauss. In an end-of-millennium poll of one hundred leading physicists, he was listed with Einstein and Maxwell as one of the top three physicists.

Isaac Newton was born on Christmas Day in 1642. Raised by his grandmother, he found more enjoyment from inventing things and from his personal studies than from his schoolwork. After graduating from Cambridge University, he served as its chair of mathematics for over twenty years. Newton was elected to the famed Royal Society, England's oldest scientific society, at age thirty and was annually reelected as its president from 1703 until his death over twenty years later. In 1701 he was elected to Parliament, and in 1705 he was knighted by Queen Anne.

Newton combined Galileo's advances in terrestrial motion and the results of Kepler's laws of celestial motion to formulate the universal law of gravitation and his three laws of motion. Despite his fear of criticism that made him reluctant to make his work public, his monumental scientific work, *Mathematical Principles of Natural Philosophy* (or *Principia*) was published in 1687 with the encouragement and financial assistance of Edmund Halley. This work confirmed the heliocentric theory of the universe and sparked the scientific revolution. The development of differential and integral calculus, the expansion of infinite series, and the binomial theorem top a long list of Newton's mathematical achievements.

Newton's religious beliefs were complex but important for Christians to consider. Many scientists today claim that theism (a belief in a personal God) is incompatible with good science. Newton would have strenuously disagreed. Commenting on the universe that he studied, he said, "This most beautiful system could only proceed from the dominion of an intelligent and powerful Being." Nevertheless, in Newton there were already the seeds of unbelief that have flowered among modern scientists. Newton thought it was unreasonable to believe in the deity of Christ and the Trinity; he therefore rejected these doctrines. Newton's life demonstrates that belief in God is not incompatible with great achievements in mathematics and science. But his life also presents a warning against allowing human reason a place above the Word of God.

2.4 Solving One-Step Equations

Cheetahs can cross 100 yd in 3 sec flat. Solve $d = rt$ for r to determine the cheetah's speed.

In algebra you will use mathematical equalities often. An equality contains an equal sign and two expressions, one on each side of the equal sign. When two expressions represent the same number, the expressions are *equal*. For example, 2^4 and $22 - 6$ have the same numerical value, 16. Therefore, $2^4 = 22 - 6$. Read the equal sign (=) as "equals," "is," or "is equal to."

Definition

An **equation** is a mathematical sentence stating that two expressions are equal.

If two quantities are equal, it does not matter which side of the equal sign they are placed on. This idea is formally stated in the following property.

Symmetric Property of Equality

If $a = b$, then $b = a$.

Numerical equations can be true or false. You can tell whether a numerical equation is true or false by simplifying both sides to see whether they are equal.

$3^2 - 5(7) = -2(13)$	$4(5) = 3 + 2(4)$
$9 - 35 = -26$	$20 = 3 + 8$
$-26 = -26$; true	$20 = 11$; false

Therefore, the first equation is true and the second is false. Equations containing a variable can also be true or false. The Distributive Property guarantees that $2(x + 1) = 2x + 2$ is always true. On the other hand, $x + 2 = x + 4$ is always false since there is no number whose sum with 2 is the same as its sum with 4.

An equation such as $x + 4 = 7$ whose truth depends on the value of the variable is a *conditional equation*.

If $x = 3$	If $x = 2$
$3 + 4 = 7$	$2 + 4 = 7$
$7 = 7$; true	$6 = 7$; false

You can see that $x + 4 = 7$ is true when x is 3 and false when x is 2 or any other number. Each number that makes an equation true is called a *solution* of the equation. To *solve* an equation, you must find the values that make the equation true. Can you find the number that should replace x to make each of the following equations true?

$x - 17 = -35$ $\quad$ $x + 10 = 142$ $\quad$ $\frac{x}{5} = 21$ $\quad$ $3x = 9$

To solve each of these equations, you have to undo an operation using its inverse operation.

Definition

An **inverse operation** is the operation that will reverse (undo) a given operation.

The following properties of equality require that you perform the same operation on both sides of the equation to produce an equivalent equation. *Equivalent equations* have the same solutions. Note how in each example a property of equality is applied to produce a simpler, equivalent equation.

PROPERTIES OF EQUALITY		
Operation	**Symbolization**	**Examples**
Addition	If $a = b$ and c is a real number, then $a + c = b + c$.	$x - 17 = -35$ $x - 17 + 17 = -35 + 17$ $x = -18$
		$x + 10 = 142$ $x + 10 - 10 = 142 - 10$ $x = 132$
Multiplication	If $a = b$ and c is a real number, then $ac = bc$.	$\frac{x}{5} = 21$ $\cancel{5}\left(\frac{x}{\cancel{5}}\right) = 5(21)$ $x = 105$
		$3x = 9$ $\frac{\cancel{3}x}{\cancel{3}} = \frac{\cancel{9}^{3}}{\cancel{3}}$ $x = 3$

To solve an equation like $x - 17 = -35$, first determine what is being done to the variable in the equation. Notice that 17 is being subtracted from x. The inverse of subtracting 17 is adding 17. If 17 is added to both sides of the equation, the Addition Property of Equality is being used to solve the equation.

Notice that there are no subtraction or division properties of equality, yet subtraction is used in the second example above and division is used in the fourth example. Because subtracting 10 in the second example can be thought of as adding -10, this is an application of the Addition Property of Equality. Likewise, division is defined as multiplying by the reciprocal, so dividing both sides by 3 is equivalent to multiplying by $\frac{1}{3}$ and is an application of the Multiplication Property of Equality.

Example 1

Solve $x - 12 = 15$.

Answer

$x - 12 = 15$ — 1. Observe that 12 is being subtracted from the variable.

$x - 12 + 12 = 15 + 12$ — 2. Add 12 to both sides.

$x = 27$ — 3. Simplify both sides.

Check $27 - 12 = 15$ — 4. Substitute your solution into the original equation. If the resulting equation is true, the solution is correct.

Example 2

Solve $x + \frac{2}{3} = -\frac{7}{12}$.

Answer

$$x + \frac{2}{3} = -\frac{7}{12}$$ 1. Observe that $\frac{2}{3}$ is being added to the variable.

$$x + \frac{2}{3} - \frac{2}{3} = -\frac{7}{12} - \frac{2}{3}$$ 2. Subtract $\frac{2}{3}$ from both sides.

$$x + \frac{2}{3} - \frac{2}{3} = -\frac{7}{12} - \frac{8}{12}$$ 3. Rename to a common denominator.

$$x = -\frac{15}{12}$$
$$x = -\frac{5}{4}$$ 4. Simplify and express the answer in lowest terms.

Check $-\frac{5}{4} + \frac{2}{3} = -\frac{15}{12} + \frac{8}{12} = -\frac{7}{12}$ 5. Substitute your solution into the original equation.

Example 3

Solve $\frac{x}{7} = -13$.

Answer

$$\frac{x}{7} = -13$$ 1. Observe that the variable is being divided by 7.

$$7\left(\frac{x}{7}\right) = 7(-13)$$ 2. Multiply both sides by 7.

$$x = -91$$ 3. Simplify.

Since you can check mentally, show the check only when directed.

Example 4

Solve $3x = 99$.

Answer

$$3x = 99$$ 1. Observe that the variable is being multiplied by 3.

$$\frac{3x}{3} = \frac{99}{3}$$ 2. Divide both sides by 3.

$$x = 33$$ 3. Simplify.

Solving Equations

1. Determine the operation performed on the variable in the equation.
2. Perform the inverse operation on both sides of the equation.
3. Simplify both sides of the equation.
4. Check your answer.

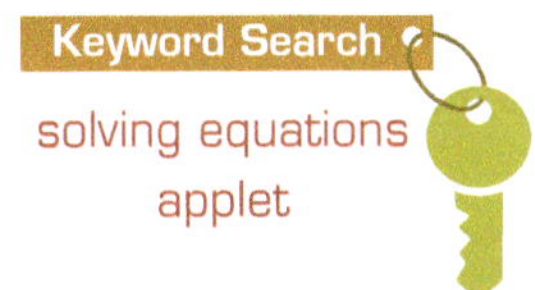

solving equations applet

In mathematics you are often asked to justify your work by giving reasons for each step of your argument or solution. This is called a *proof.* The following example states the property or definition that justifies each step in the solution.

Example 5

Justify each step in the solution of the equation $x - \frac{2}{9} = \frac{5}{3}$.

Answer

$x - \frac{2}{9} = \frac{5}{3}$	
$x - \frac{2}{9} + \frac{2}{9} = \frac{5}{3} + \frac{2}{9}$	1. Addition Property of Equality
$x + 0 = \frac{5}{3} + \frac{2}{9}$	2. Additive Inverse Property
$x = \frac{5}{3} + \frac{2}{9}$	3. Additive Identity Property
$x = \frac{15}{9} + \frac{2}{9}$	4. Multiplicative Identity Property
$x = \frac{17}{9}$	5. addition

Using an equation to model a problem situation provides a guide for which steps can be used to solve the problem. For instance, to find a number such that the sum of a number and 5 is 23, translate the problem into an equation. Let the variable n stand for the number. Next, translate the phrase "the sum of a number and 5" as $n + 5$. Since the word "is" means "equals," put $=$ in your algebraic sentence. The final number, 23, is the sum and is placed on the right side of the equal sign: $n + 5 = 23$.

Example 6

The difference of a number and 10 is 37. Find the number.

Answer

$n - 10 = 37$	1. Translate the sentence into an equation. Let the number be n. "Difference" indicates subtraction, and "is" means "equals."
$n - 10 + 10 = 37 + 10$	2. Solve.
$n = 47$	

Sometimes equations will contain more than one variable. These *literal equations* can be solved for the desired variable exactly as you would solve other equations.

Example 7

Newton's first law of motion relates the net force (F) on an object, its mass (m), and its resulting acceleration (a) in the formula $F = ma$. Solve this formula for a.

Answer

$F = ma$	1. Observe that a is being multiplied by m.
$\frac{F}{m} = \frac{\cancel{m}a}{\cancel{m}}$	2. Divide both sides by m.
$a = \frac{F}{m}$	3. Apply the Symmetric Property of Equality to place the desired variable on the left side.

A. Exercises

True or false

1. $5(6) = 2 + 3(6)$

2. $8 + 6 = 2(7)$

3. $3(5 + 2) = 11 + 5(2)$

4. $2 + 3(1 - 2) = 2^2 - \frac{1}{3}(-8 + 5) - 7$

Is the given number a solution to the equation?

5. $5x + 1 = 2$ if $x = 5$

6. $x - 7 = 9$ if $x = -2$

7. $-9x - 8 = -37.16$ if $x = 3.24$

8. $10 = 12x + 7$ if $x = \frac{1}{4}$

What should be done to both sides of each equation to solve it? Name the property of equality illustrated by this action.

9. $x + 7 = 15$

10. $x - 3 = 57$

11. $20 = 4x$

12. $\frac{x}{5} = \frac{2}{3}$

Solve. Show all steps and check your answers.

13. $x + 3 = 12$

14. $x - 4 = -23$

15. $b - 14 = 62$

16. $a + 7 = -3$

17. $2x = 5$

18. $\frac{y}{5} = -2$

19. $\frac{c}{-9} = -8$

20. $4x = 164$

B. Exercises

Justify each step in the solution of the equation $2x = 8$.

21. $\frac{1}{2}(2)x = \frac{1}{2}(8)$

22. $1x = \frac{1}{2}(8)$

23. $x = \frac{1}{2}(8)$

24. $x = 4$

Solve.

25. $x + \frac{1}{2} = \frac{3}{4}$

26. $x - \frac{18}{7} = \frac{4}{5}$

27. $\frac{4}{5}y = 24$

28. $\frac{z}{\frac{11}{2}} = 2$

29. $b - 3.89 = 0.49$

30. $a + 0.93 = 1.42$

31. $\frac{d}{1.3} = 1.24$

32. $0.8c = 0.24$

33. $2w = \frac{1}{9}$

34. $82 = x - 24$

35. $21.2 = y + 0.7$

36. $\frac{z}{12} = 4.69$

Write the equation you would use to solve each problem; then solve.

37. Four times a number is 20. Find the number.

38. The quotient of a number and 15 is 120. Find the number.

39. Emma made $37.80 for baby-sitting 6 hr. How much did she earn per hour?

40. Find the average high temperature in Madrid in October if Hong Kong's average high temperature is 84° and is 17° higher than Madrid's average high.

41. The sum of the angles of a triangle is 180°. Find the measure of the third angle if the sum of the other two angles is 115°.

42. If one of two complementary angles measures 37°, find the measure of the other angle.

C. Exercises

Solve each literal equation for x.

43. $x - b = d$

44. $a = bx$

45. $\frac{x}{n} = q$

46. $z = y + x$

Dominion Modeling

Return on investment (ROI) is another measure used to evaluate the effectiveness of an investment.

Example: Find the first-year ROI of spending an extra \$2 on a CFL light bulb that saves \$17.28 in electricity each year.

$$\text{ROI} = \frac{\$17.28}{\$2} = 8.64 = 864\%$$

The wisdom of using CFL bulbs to conserve energy is apparent when this ROI is compared to the typical ROI of other financial investments, which is generally well under 10%.

47. Use the example above to write a generalized formula for ROI.

48. Using your answers from Section 2.3, Dominion Modeling, exercise 48, determine the first-year ROI of installing a 4 W LED bulb instead of a 25 W incandescent bulb.

49. Using your answers from Section 2.3, Dominion Modeling, exercise 49, determine the first-year ROI of installing a tankless water heater instead of a storage tank water heater.

CUMULATIVE REVIEW

Draw a Venn diagram representing the real number system; then place each number within the innermost set to which it belongs. [1.1]

50. $2.\overline{3}$

51. $\sqrt{64}$

Simplify and then graph the number represented by each expression on the same number line. [1.2]

52. $|-3|$

53. $-[-(-2)]$

Simplify, leaving each answer in positive exponential form. [1.7]

54. $x^6y^4x^{-2}y^{-8}x$

55. $\frac{a^{-3}b^5}{b^7a^6}$

Evaluate. [1.8]

56. $-2 + 8(3 - 5)$

57. $\frac{1}{2} - \frac{2}{3}\left(3 - \frac{5}{6}\right)$

Simplify and then evaluate when $x = 2$ and $y = -3.7$. [2.2–2.3]

58. $3y - x^3 + 1 - y$

59. $(4x)^0 + xy^2 + x$

2.5 Solving Two-Step Equations

In many equations more than one operation is performed on the variable. Just as there is a standard order of operations that must be followed for simplifying expressions, the operations done to a variable must be "undone" using inverse operations in reverse order.

When you are solving the equation $2x + 6 = 21$, think of the order in which the operations are done to the variable to simplify the expression $2x + 6$. The multiplication by 2 would precede the addition of 6. Therefore, you should reverse the order and subtract 6 from both sides of the equation before dividing both sides by 2.

Example 1

Solve $2x + 6 = 21$.

Answer

$$2x + 6 = 21$$

1. Observe that x is multiplied by 2 and then 6 is added to the product.

$$2x + 6 - 6 = 21 - 6$$

2. Subtract 6 from both sides first.

$$2x = 15$$

$$\frac{2x}{2} = \frac{15}{2}$$

3. Divide both sides by 2.

$$x = \frac{15}{2}$$

Check

$$2\left(\frac{15}{2}\right) + 6 = 15 + 6 = 21$$

4. Check.

Example 2

Solve $\frac{a}{1.5} - 3.6 = -4.4$.

Answer

$$\frac{a}{1.5} - 3.6 = -4.4$$

1. Observe that a is divided by 1.5 and then 3.6 is subtracted from the quotient.

$$\frac{a}{1.5} - 3.6 + 3.6 = -4.4 + 3.6$$

2. Add 3.6 to both sides first.

$$\frac{a}{1.5} = -0.8$$

$$1.5\left(\frac{a}{1.5}\right) = 1.5(-0.8)$$

3. Multiply both sides by 1.5.

$$a = -1.2$$

Check

$$\frac{-1.2}{1.5} - 3.6 = -0.8 - 3.6 = -4.4$$

4. Check.

These two examples are in the form $ax + b = c$, in which the letters a, b, and c represent real numbers and x is an unknown variable. Solve this type of equation in two steps. Use inverse operations to undo any addition or subtraction and then to undo any multiplication or division.

Solving Equations of the Form $ax + b = c$

1. Undo the addition or subtraction of the constant term.
2. Undo the multiplication or division done to the variable.

Example 3

Solve $\frac{x}{3} - 7 = 2$.

Answer

$$\frac{x}{3} - 7 = 2$$

1. This equation is of the form $ax + b = c$, where $a = \frac{1}{3}$ and $b = -7$.

$$\frac{x}{3} - 7 + 7 = 2 + 7$$

2. Undo the subtraction of 7 by adding 7 to both sides.

$$\frac{x}{3} = 9$$

$$3\left(\frac{x}{3}\right) = 3(9)$$

3. Undo the division of 3 by multiplying both sides by 3.

$$x = 27$$

In mathematics you must also be prepared and have a plan of attack. Practice on many different problems and finish the job; if you quit partway through, you will never get the answer. The key to success is hard work and perseverance.

Using the following procedures will assist you in confronting a word problem. At first you will need to follow the procedures step by step, but with more experience you will know how to proceed more quickly and how to do some of the steps mentally.

Do you hide from problem solving like an ostrich?

Solving Word Problems
1. **Read** the problem, looking for operational words. a. *Draw a picture* (if possible) to be sure you understand the problem. b. *Assign a variable* to the main unknown.
2. **Plan** a method of attack. a. *Make a table* (if possible) to organize relevant information. b. *Translate word phrases into expressions* using the variable. Express each unknown quantity in terms of the variable.
3. **Solve** an equation. a. *Write an equation.* Find the mathematical verb and identify two quantities that are equal. b. *Solve the equation.*
4. **Check** your solution. Ask yourself these three questions. a. *Have I answered all of the questions in the problem?* b. *What do the numbers represent?* Interpret your answers (including units) in the context of the problem. c. *Are the answers reasonable?* Impossible and improbable answers are clues that you made a mistake (e.g., a negative distance is impossible, and a car's speed of 100 mi/hr is unlikely and usually illegal).

Example 4

Ninety-seven is twenty-three less than four times a number. Find the number.

Answer Let n = the unknown number.

1. **Read** the problem carefully, noting the operational words *is*, *times*, and *less than*. The unknown quantity is a number, so let the variable n represent the number.

Translate "twenty-three less than four times a number" as $4n - 23$.

2. **Plan** a strategy. You cannot make a table, but you can translate each phrase.

$$97 = 4n - 23$$
$$97 + 23 = 4n - 23 + 23$$
$$120 = 4n$$
$$\frac{120}{4} = \frac{4n}{4}$$
$$n = 30$$

3. **Solve**. The verb *is* identifies the equal expressions. Since $4n - 23$ is 97, write the equation and solve. Note that the multiplication by 4 could also be undone by multiplying both sides by the reciprocal, $\frac{1}{4}$. The last step applies the Symmetric Property of Equality.

Check $4(30) - 23 = 120 - 23 = 97$

4. **Check**. There is only one question. The answer 30 checks because 97 is 23 less than 4 times 30.

Example 5

How many quarters must be removed from \$2.64 to leave a total of \$1.89?

Answer q = the number of quarters

1. **Read** the problem carefully, noting that we are looking for the number of quarters.

$0.25q$ = the value of the quarters

2. **Plan** a strategy. The total value of the quarters is the number of quarters times the value of a quarter.

$$2.64 - 0.25q = 1.89$$
$$2.64 - 0.25q - 2.64 = 1.89 - 2.64$$
$$-0.25q = -0.75$$
$$\frac{-0.25q}{-0.25} = \frac{-0.75}{-0.25}$$
$$q = 3 \text{ quarters}$$

3. **Solve**. Write an equation modeling the situation. Undo addition first; then undo multiplication.

Check $2.64 - 0.25(3) = 2.64 - 0.75 = 1.89$

4. **Check**.

Example 6

The grain elevator weighed Mr. Young's truck at 17,760 lb before he dumped his wheat. The weight of the empty truck is 6180 lb. If the wheat weighed 60 lb/bu, how many bushels of wheat did he deliver to the elevator?

Answer b = bushels of wheat

1. **Read** the problem carefully, noting that the total weight of 17,760 lb consists of the weight of the wheat and the weight of Mr. Young's truck.

$60b$ = the weight of the wheat

2. **Plan** a strategy. The question asks for the number of bushels of wheat, b, and $60b$ represents the total weight of the wheat.

$$60b + 6180 = 17{,}760$$
$$60b + 6180 - 6180 = 17{,}760 - 6180$$
$$60b = 11{,}580$$
$$\frac{60b}{60} = \frac{11{,}580}{60}$$
$$b = 193 \text{ bu}$$

3. **Solve**. Write an equation modeling the situation. Undo addition first; then undo multiplication.

Check $60(200) + 6000 = 12{,}000 + 6000$
$= 18{,}000 \approx 17{,}760$

4. **Check**. Estimation shows that the answer is reasonable.

A. Exercises

Solve.

1. $3x - 10 = 14$
2. $5y + 24 = -36$
3. $17 = 3z + 2$
4. $12 = 2b - 28$
5. $-4c + 9 = 41$
6. $9d + 17 = 71$
7. $\frac{x}{5} - 13 = -84$
8. $4 + \frac{y}{6} = 9$
9. $\frac{p}{4} - 9 = 15$
10. $\frac{2}{5}n - 7 = 8$
11. $1.5r + 2.3 = 6.8$
12. $-1.8 = 4.6 - 0.4s$

Solve, using the procedures for solving word problems. Show all your work.

13. A number added to 27 gives -14. Find the number.
14. The difference of a number and 72 is 12. Find the number.
15. The difference of a number and -9 is -74. Find the number.
16. A number multiplied by 12 is 60. Find the number.
17. When Laura is six times as old as she is now, she will be 84. How old is she now?
18. Deedra canned 24 qt of beans, which was four times as much as Ava canned. How many quarts did Ava can?
19. Kent has three times as many stamps in his collection as Jen has in hers. If Kent has 243 stamps, how many stamps does Jen have?
20. Find the length of the third side of a triangle with a perimeter of 21 in. if the sum of the other two sides is 15 in.

B. Exercises

Solve.

21. $17 - w = -13$
22. $-24 - a = 78$
23. $6.8x - 4.7 = -17.62$
24. $1.2b - 1.8 = 47.4$
25. $\frac{y}{3.7} - 1.5 = -2.8$
26. $-\frac{c}{7.2} + 4.7 = 2.8$
27. $-\frac{2}{3}z - \frac{1}{2} = \frac{4}{5}$
28. $\frac{d}{3\frac{1}{2}} + \frac{1}{3} = -\frac{2}{5}$
29. $\frac{x - 5}{2} = 9$
30. $\frac{5 - y}{2} = 9$

Solve, using the procedures for solving word problems. Show all your work.

31. The sum of 47 and three times a number is 68. What is the number?
32. Five reduced by twice a number is two. Find the number.
33. How many nickels must be added to \$2.47 to make a total of \$3.12?
34. Of the \$395 that Carlos collected for the fundraiser, \$310 was in tens and twenties with the rest in fives. How many fives did he collect?

35. David is charged \$19/mo and \$0.05/min for his pay-as-you-go cell phone plan. If his bill was \$21.35, how many minutes did he use?

36. The class decided to purchase T-shirts from an online company that charged \$11.95 for each shirt and a shipping and handling fee of \$6.95. The bill was \$174.25. How many T-shirts did the class order?

37. A trucking company quoted the Fernandez family a rate of \$1782 to use 19 linear feet of trailer space for their cross-country move. An adjustment price of ±\$54/ft for extra or fewer feet used was also agreed upon. How many linear feet did the family use if the bill for their move was \$2052?

38. A camp offers a \$15 discount toward the weekly camp fee of \$255 for each new camper a returning teen brings with him. How many new campers does Lois need to bring if she wants to lower her fee to \$195?

39. The horizontal position of an object in uniform motion is given by the formula $d_f = vt + d_i$. How many seconds, t, does it take for an object moving at a velocity, v, of 88 ft/sec to move from an initial position, d_i, of -128 ft to a final position, d_f, of 532 ft?

40. The final velocity of an object experiencing uniform acceleration is given by the formula $v_f = at + v_i$. What is the acceleration, a (in m/s^2), of an object that takes 5 s to accelerate from an initial velocity, v_i, of 8 m/s to a final velocity, v_f, of 57 m/s?

C. Exercises

Solve, using the procedures for solving word problems. Show all your work.

41. Miss Miller teaches three times as many students during first hour as she does during third hour. Her second-hour class has twice as many as her third-hour class. If she has a total of 72 students in her three morning classes, how many does she have in each class?

42. A triangular lot has a perimeter of 590 ft. The road frontage is twice as long as the second side and 50 ft shorter than the third side. Find the lot's dimensions.

43. What should be done to both sides to solve the equation $x^2 = 81$?

44. Is -7 a solution to the equation $x^2 = 49$? Is 7 a solution to the equation $x^2 = 49$? Explain.

45. Is -5 a solution to the equation $x^2 = -25$? Is 5 a solution to the equation $x^2 = -25$? Explain.

46. What is the "inverse operation" for taking a term to a power?

Solve each literal equation for the stated variable.

47. $v_f = at + v_i$ for t

48. $S = \pi r^2 + \pi rl$ for l

Dominion Modeling

A homeowner needing to replace the water heater in his home researched his options at the local hardware store. He can purchase a \$338 electric water heater, estimated to use 5108 kWh of electricity annually, or a \$488 natural gas water heater, estimated to use 283 therms annually.

49. Determine the annual cost of electricity for the less expensive water heater if the power company charges \$0.12/kWh.

50. Determine the annual cost of natural gas for the more expensive water heater if the gas company charges \$1.30/therm.

51. Determine the payback period for investing in the more expensive natural gas water heater instead of the less expensive electric water heater.

52. Determine the first-year ROI of this investment.

CUMULATIVE REVIEW

Evaluate. [1.3–1.8]

53. $-12 + (-4) + (-3)$

54. $-12 - (-4) - (-3)$

55. $-12(-4)(-3)$

56. $\frac{-12(-3)}{-4}$

57. $\frac{1}{-2^{-2}} - (-3)^2$

58. $\frac{3 - 5(-2 + 6)}{|9 - 1|}$

Simplify. [2.3]

59. $6y(x - 2y)$

60. $7x^2 - 4x^2y - x^2 + 8x^2y$

61. $7 - (4x - 3)$

62. $-7(x + 5) + 3(x - 4)$

2.6 Simplifying Equations

The Distributive Property is a foundational concept of mathematics. It is often used to simplify the algebraic expression on one side of an equation prior to solving the equation. Equations are the foundation upon which algebra is built. To be successful in algebra and higher math, you need a good foundation in solving equations.

The largest monolith in the world, Ayers Rock in Australia, is 4 mi long by 1.5 mi wide.

Example 1

Solve $3x + 2 + 5x - 7 = 28$.

Answer

$8x - 5 = 28$ — 1. Combine like terms using the Distributive Property.

$8x - 5 + 5 = 28 + 5$ — 2. Solve.

$8x = 33$

$\frac{8x}{8} = \frac{33}{8}$

$x = \frac{33}{8}$

Check

$3\left(\frac{33}{8}\right) + 2 + 5\left(\frac{33}{8}\right) - 7$ — 3. Check.

$= \frac{99}{8} + \frac{16}{8} + \frac{165}{8} - \frac{56}{8} = \frac{224}{8} = 28$

If parentheses appear in the equation, simplify the equation by applying the Distributive Property before solving the equation. The properties of equality demand that the same operations be performed on both sides of the equation. These steps can be done mentally from now on, as shown in Example 2. Remember that you can check mentally also.

Example 2

Solve $3(x+1)-(x-2)=13$.

Answer	$3x+3-x+2=13$	1. Remove parentheses by applying the Distributive Property.
	$2x+5=13$	2. Combine like terms.
	$2x=8$	3. Subtract 5 from both sides.
	$x=4$	4. Divide both sides by 2.
Check	$3(4+1)-(4-2)=13$	5. Check mentally.

Example 3

John's mother is four times as old as John, and John's brother is twice as old as John. If you find the sum of their ages and add 3, you get 80. What are the ages of all three people?

Answer	Let x = John's age.	1. **Read** the problem carefully to find the main unknown. All ages are described based on John's age.
	$4x$ = John's mother's age $2x$ = John's brother's age	2. **Plan**. Define the other ages in terms of x.
	$4x+2x+x+3=80$ $7x+3=80$ $7x=77$ $x=11$	3. **Solve**. Write an equation and find its solution.
Check	mother's age: $4x=44$ brother's age: $2x=22$	4. **Check**. Find the other ages, interpret them, and notice that they are reasonable since $44+22+11+3=80$.

John is 11 years old, his mother is 44 years old, and his brother is 22 years old.

You may be surprised to learn how the properties of equality and equations can be used to convert a repeating decimal to a fraction.

Example 4

Express $0.\overline{36}$ as a quotient of integers.

Answer

$$x = 0.\overline{36}$$

1. Write an equation, letting x equal the repeating decimal.

$$\begin{array}{r} 100x = 36.\overline{36} \\ -\ x = 0.\overline{36} \\ \hline 99x = 36 \end{array}$$

2. Multiply both sides of the equation by $10^2 = 100$ (move the decimal to the right two places). Then subtract the original equation from the new equation.

$$\frac{99x}{99} = \frac{36}{99}$$

3. Divide both sides by 99.

$$x = \frac{4}{11}$$

4. Reduce the fraction.

Use the following steps to express a repeating decimal as a quotient of integers.

1. Write an equation in which x is equal to the repeating decimal.
2. Use the Multiplication Property of Equality to multiply both sides of this equation by 10^n, where n is the number of repeating digits.
3. Subtract the original equation from the new equation formed by the multiplication. This step is an application of the Addition Property of Equality since we are subtracting equal quantities from both sides of the equation.
4. Solve the resulting equation and reduce the fraction to lowest terms.

Example 5

Express $1.3\overline{6}$ as a quotient of integers.

Answer

$$x = 1.3\overline{6}$$

1. Write an equation, letting x equal the repeating decimal.

$$\begin{array}{r} 10x = 13.6\overline{6} \\ -\ x = 1.3\overline{6} \\ \hline 9x = 12.3 \end{array}$$

2. Multiply both sides of the equation by $10^1 = 10$ (move the decimal to the right one place). Then subtract the original equation from the new equation.

$$\frac{9x}{9} = \frac{12.3}{9} = \frac{123}{90}$$

$$x = \frac{41}{30}$$

3. Solve the resulting equation. Notice how the decimal is cleared from the fraction when both the numerator and the denominator are multiplied by 10.

Check $\frac{41}{30} \approx 1.366666667$

4. A calculator can be used to verify your solution.

This procedure can be used to convert any repeating decimal to an equivalent quotient of integers and demonstrates that all repeating decimals are rational numbers.

A. Exercises

Solve. Show all your work.

1. $4a + 8a = 12$

2. $-8b - 4b = 24$

3. $-2c + 2 - c = 20$

4. $5d + 9d - 4 - 6d = 12$

5. $4(w - 7) = 8$

6. $3(x + 2) = 66$

7. $-7(y + 3) = 77$

8. $3z + (2 + 4z) = 23$

9. $5m - (3 + 2m) = 9$

10. $2n - 4(n - 3) = 26$

11. $6 - 2(p + 4) = 108$

12. $3r + 9r + (12 - 2r) = 2$

13. $2(9s + 3) - 4s = 34$

14. $(t + 3) + 2(t + 5) = 34$

15. $3.2k + 4(k + 0.7) = 38.8$

16. $-2(j - 1) + 3(j + 4) = 8$

B. Exercises

Express each of the following repeating decimals as a quotient of integers.

17. 0.151515...

18. $0.\overline{81}$

19. $1.\overline{785}$

20. $2.4\overline{396}$

Write an equation and solve.

21. One number is 23 more than another number. The sum of the two numbers is 83. Find the two numbers.

22. Find two numbers whose difference is 12 and whose sum is 40.

23. The sum of three consecutive integers is 84. Find the integers.

24. Find three consecutive odd integers whose sum is 57.

25. If the freshman class of 63 students has 11 more boys than girls, determine how many of each are in the class.

26. The number of girls in Mrs. Conn's math class is two fewer than three times the number of boys. She has 26 students in her class. How many boys are in the class?

27. The sum of the angles of a triangle is 180°. $\angle A$ is three times as large as $\angle B$, and $\angle C$ is 120° more than $\angle B$. What is the measure of each angle?

28. Stan wants to make a rectangular garden. He has 34 ft of fencing to put around the perimeter. If the length is to be 5 ft more than the width, find the dimensions of the planned garden.

29. A calculator costs \$12 less than a camera. Together they cost \$87. What is the cost of each?

30. Jason and Bill share a bookshelf in their room. Jason has four more than twice as many books as Bill, and there are 22 books on the shelf. How many of the books belong to Jason and how many belong to Bill?

31. Find three numbers whose sum is 40 if the first number is four more than the product of six and the second number, and the third number is nine less than twice the second number.

32. Kenny is seven years older than his sister. The sum of their ages is 31. Find the age of each.

C. Exercises

There is often more than one way to solve a problem. Consider the task of finding two numbers whose sum is 85 and whose difference is 29.

33. Let x represent the larger number, and use the fact that the difference of the numbers is 29 to represent the smaller number. Then write an equation using the fact that their sum is 85.

34. Let x represent the first number, and use the fact that the sum of the numbers is 85 to represent the second number. Then write an equation using the fact that the numbers differ by 29.

35. Solve the equations from exercises 33 and 34 to find the numbers.

36. Compare the two methods used to solve the problem, state which method you prefer, and explain why you liked that method better.

Solve each literal equation for *x*.

37. $3x - (x + n) = m$

38. $a(x + c) + b(x + d) - ax = e$

Dominion Modeling

When faced with a \$3000 repair estimate for their central air conditioner, a family considers replacing the unit. They are quoted a replacement cost of \$6200 for a new energy-efficient unit that would qualify for a \$1500 tax credit (a government rebate). They estimate that the 40% increase in efficiency would translate into an average savings of \$500/yr.

39. Determine the amount invested by replacing the air conditioner instead of repairing it.

40. Determine the payback period for replacing instead of repairing the air conditioner.

41. List several other factors that might influence the family's decision to repair or replace the air conditioner.

CUMULATIVE REVIEW

Use the following sets to complete each set operation. [1.1]
$A = \{-13, -5, 1, 9, 17\}$; $B = \{10, 20, 30, \ldots\}$; $C = \{20, 15, 10, 5, 0, -5, \ldots\}$

42. $A \cap B$

43. $B \cap C$

44. $A \cup \{1, 4, 9\}$

45. $B \cup C$

46. $C \cap \mathbb{N}$

Evaluate when $x = 3$ and $y = 6$. [2.2]

47. $S = 3x^2 + 6xy$

48. $D = |x - y|$

49. $M = \frac{x + y}{2}$

50. $L = \sqrt{x + y}$

51. $D = \sqrt{(x + 1)^2 + (y - 3)^2}$

SEQUENCES

Arithmetic Sequences

Challenge **Can you supply the general-term formula describing each sequence of numbers?**

a. 6, 7, 8, 9, 10, …

b. −5, −3, −1, 1, 3, …

c. 1, 4, 7, 10, 13, …

The three sequences above are all examples of arithmetic sequences. Any two successive terms in an *arithmetic sequence* have a common difference. An arithmetic sequence can be determined if the first term, A_1, and the common difference, d, are given. Mathematicians often describe sequences *recursively*, based on previous terms. This is especially helpful for solving problems using a computer, where relationships are often modeled recursively. The *recursive formula* for any arithmetic sequence is $A_n = A_{n-1} + d$. The three sequences above are defined recursively as follows.

$A_n = A_{n-1} + 1; A_1 = 6$ $\qquad$ $A_n = A_{n-1} + 2; A_1 = -5$ $\qquad$ $A_n = A_{n-1} + 3; A_1 = 1$

Notice that the first term must be given in order for you to know how to begin the sequence. A disadvantage of the recursive formula is that a specific term, such as A_{75}, cannot be found unless you know the previous term.

An *explicit formula* or *general-term formula* allows you to directly calculate a general term from the characteristics of the sequence. Thus, the explicit formula is more powerful. However, it is often more difficult to identify the explicit formula than to identify the recursive formula. Did you get $A_n = n + 5$, $A_n = 2n - 7$, and $A_n = 3n - 2$ as the formulas for the three sequences in the Challenge? What steps of reasoning can you use to arrive at these answers? Consider the third sequence: 1, 4, 7, 10, 13, ….

$$1 = A_1 = A_1 + 0(3)$$
$$4 = A_2 = A_1 + 1(3)$$
$$7 = A_3 = A_1 + 2(3)$$
$$10 = A_4 = A_1 + 3(3)$$
$$13 = A_5 = A_1 + 4(3)$$

Keyword Search
arithmetic sequences series

This leads us inductively to the formula $A_n = A_1 + (n - 1)d$, which in this example is $A_n = 1 + (n - 1)3$. Distributing and simplifying this expression produces $A_n = 3n - 2$.

Exercises

List the first five terms of each of the following sequences.

1. $A_n = 3n - 1$ **2.** $A_1 = 5; d = 2$ **3.** $A_1 = -12; A_n = A_{n-1} - 7$

Find A_{30} for each of the following sequences.

4. $A_n = 5n - 2$ **5.** $A_n = -9n + 10$ **6.** $A_n = A_{n-1} + 4$

Find the specified formula for each of the following sequences.

7. 50, 45, 40, 35, 30, …; recursive

8. −5, 6, 17, 28, 39, …; recursive

9. 14, 24, 34, 44, 54, …; explicit

10. 3, 5, 7, 9, 11, …; explicit

2.7 Solving Multi-Step Equations

How are these equations different from the ones in prior sections?

$2x + 5 = x - 2$

$3(x - 4) = 4(x - 6)$

$5x + 3x - 4 = x + 2$

They have variables on both sides of the equal sign. The Addition Property of Equality is used to move all the variables to one side of the equation.

A careful "undoing" of residential construction

Example 1

Solve $2x + 5 = x - 2$.

Answer

$2x + 5 - x = x - 2 - x$ 1. To eliminate x from the right side, subtract x from both sides of the equation.

$x + 5 = -2$

$x = -7$ 2. Subtract 5 from both sides.

Check $2(-7) + 5 = -7 - 2$ 3. Check.

In more complicated equations, simplify both sides of the equation first.

Example 2

Solve $3(x - 4) = 4(x - 6)$.

Answer

$3x - 12 = 4x - 24$ 1. Remove the parentheses by applying the Distributive Property.

$3x - 12 - 3x = 4x - 24 - 3x$ 2. To move the variable terms to the right side, subtract $3x$ from both sides.

$-12 = x - 24$

$12 = x$ 3. Solve by adding 24 to both sides.

Will you get the same answer by moving the variable terms to the left side?

$3x - 12 = 4x - 24$ 1. Distribute to eliminate the parentheses.

$3x - 12 - 4x = 4x - 24 - 4x$ 2. Subtract $4x$ from both sides.

$-x - 12 = -24$

$-x = -12$ 3. Solve by adding 12 to both sides and then dividing both sides by -1.

$x = 12$

Yes, you will get the same answer regardless of which side you move the variable terms to.

Solving Multi-Step Equations

1. Simplify both sides by removing any parentheses and combining like terms.
2. Move the variable terms to the desired side of the equation.
3. Undo the addition or subtraction of the constant term.
4. Undo the multiplication or division done to the variable.

Example 3

Solve $5x + 3x - 4 = x + 2$.

Answer

$8x - 4 = x + 2$	1. Combine like terms.
$8x - 4 - x = x + 2 - x$	2. Subtract x from both sides.
$7x - 4 = 2$	
$7x = 6$	3. Undo the subtraction.
$x = \frac{6}{7}$	4. Undo the multiplication.

Simplifying the expression on either side of the equation or applying the Addition or Multiplication Property of Equality produces an equivalent equation, one having the same solution as the original equation.

Solving what appears to be a "typical" equation sometimes produces unexpected results.

Example 4

Solve $3(4x - 5) + 2 = 2(6x + 3)$.

Answer

$12x - 15 + 2 = 12x + 6$	1. Distribute to eliminate the parentheses.
$12x - 13 = 12x + 6$	2. Combine like terms.
$-13 = 6$	3. Subtract $12x$ from both sides.
no solution	4. No value of x will make the original equation true.

Since the last equation, $-13 = 6$, is false, the original equation to which it is equivalent must be false regardless of the value used for the variable.

On the other hand, an equation that is true regardless of the value of the variable is an *identity*, and its solution is listed as the set of real numbers. The next example illustrates the results of solving an identity.

Example 5

Solve $8 - (x - 4) = -2(x - 6) + x$.

Answer

$8 - x + 4 = -2x + 12 + x$	1. Distribute to eliminate the parentheses.
$-x + 12 = -x + 12$	2. Combine like terms.
The equation is an identity and the solution is $\mathbb{R}$, the set of real numbers.	3. Since the equation contains two equal expressions, it must be true for all values of x.

Example 6

Solve $2(x - 3) + 6 = 3x + 2 - (2 - x)$.

Answer

$2x - 6 + 6 = 3x + 2 - 2 + x$

$2x = 4x$

$2x - 4x = 4x - 4x$

$-2x = 0$

$x = 0$

1. Distribute to eliminate the parentheses.
2. Combine like terms.
3. Solve by subtracting $4x$ from both sides and dividing both sides by -2.

A. Exercises

Solve. Identify equations that are identities or that have no solution.

1. $3x + 5 = 8x - 5$

2. $7x - 20 = 2x + 25$

3. $3(x + 1) = 8x - 12$

4. $7(x - 9) = 3(x - 2)$

5. $3x = 5x - 2$

6. $3x = 9x$

7. $11x - 8 = 12x + 9 - x$

8. $4(x - 9) = 3x - 8x$

9. $6x - 1 = x + 4$

10. $82b - 9b + 7 = 73b + 19$

11. $5(x - 4) + 4x = 9x - 20$

12. $7n - 13 = 4n - 9$

13. $y + 9 = 2y - 4$

14. $3x - 7 = 2x$

15. $8a = 4(a - 9)$

16. $9a + 2 - 6(a + 2) = 3a - 10$

B. Exercises

Solve. Identify equations that are identities or that have no solution.

17. $4(m - 6) - m(3 + 4) = 8m - 2$

18. $8(2t + 1) - (t - 9) = -52 - 8t$

19. $0.25x - 3.09 = 0.75x + 8$

20. $5(y + 2) - 4y = 8y - (2 + y)$

21. $\frac{5}{4}x + 3 = \frac{3}{4}x - 8$

22. $0.3(6x - 9) + 2 = -8.2x$

23. $9(2.8a - 1.3) = 18.4a - 11.7 + 6.8a$

24. $7\left(\frac{6}{5} - b\right) + \frac{3}{5}b = \frac{2}{5}(b + 5)$

25. $-\frac{7}{8}x + 9 = \frac{4}{5}x - 17$

State the property that justifies each numbered step in the following solution.

$3(x - 5) + 4x = 6 - (5x - 3)$

26. $3x - 15 + 4x = 6 - 5x + 3$

27. $3x + 4x - 15 = -5x + 6 + 3$

28. $(3 + 4)x - 15 = -5x + 6 + 3$

$7x - 15 = -5x + 9$ [addition]

29. $12x - 15 = 9$

30. $12x = 24$

31. $x = 2$

Write an equation and solve.

32. Eight times a number equals the number increased by 504. Find the number.

33. Three subtracted from twice a number equals 31 more than three times the number. Find the number.

34. Find three consecutive integers such that the sum of the first and twice the third is equal to four times the second decreased by five.

35. Find two consecutive even integers such that four times the smaller is equal to four more than three times the larger.

36. The perimeters of two rectangles are equal. The width and length of one of the rectangles are x and $x + 14$. The other rectangle has dimensions of $x + 2$ by $7x$. Find the dimensions of each rectangle.

C. Exercises

Write an equation and solve.

37. The length of a rectangle is three times its width. The perimeter of the rectangle is equal to the perimeter of a triangle that has one side equal to the length of the rectangle. The sum of the other two sides of the triangle is 85. What are the dimensions of the rectangle?

38. Find two consecutive even integers such that three times the smaller is four more than eight times the larger.

39. Seema needs a certain length of ribbon to put around the outer edge of her craft project. She has several pieces of equal length. She discovers that if she uses one piece, it is 2 in. too short; and if she uses two pieces, they are 6 in. too long. What is the length that she needs for her project?

The break-even point in business is the point at which the cost of manufacturing a product equals the revenue brought in by the product when it sells. Find the break-even point for *x* items produced and sold for the given cost, *C*(*x*), and revenue, *R*(*x*), functions.

40. $C(x) = 14x - 22{,}200$; $R(x) = x + 1200$

41. $C(x) = 11{,}000x - 278{,}792$; $R(x) = 8x + 7000$

42. $C(x) = 6(x + 40) - 7240$; $R(x) = 24x - 34{,}000$

Dominion Modeling

The break-even point described in exercises 40–42 is similar to the payback period. Since there is no profit made or loss incurred at the break-even point, it is the lower limit of profitability for a business. Write separate expressions modeling revenue and expenses; then find the break-even point by solving the equation resulting from setting these expressions equal to each other.

43. Josh wants to begin a lawn-mowing business within his subdivision. Startup costs of purchasing a mower, edger, and trimmer total \$400, and he estimates that each lawn he mows will take \$4 worth of gas and other supplies. He plans on charging \$25 per lawn and estimates that he will receive a \$5 tip about half the times he mows a lawn.
    a. Write an expression for his expenses.
    b. Write an expression for his revenue.
    c. Write and solve an equation to find the number of lawns he needs to mow to break even.

44. As a fundraiser, the senior class wants to sell sweatshirts with the school mascot printed on the front. The sweatshirts cost \$19 each, and there is a printing setup fee of \$40 and a shipping fee of \$20.
    a. Write an expression for the expense of ordering s sweatshirts.
    b. Find the break-even point if they sell the sweatshirts at \$25 each.
    c. Find the break-even point if they sell the sweatshirts at \$29 each.
    d. Which price would you suggest for the sweatshirts? Explain your reasoning.

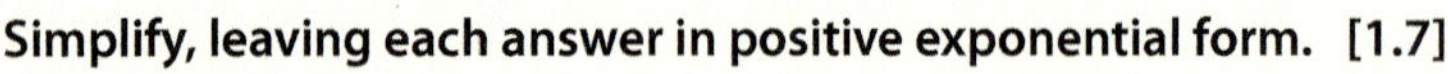

CUMULATIVE REVIEW

Simplify, leaving each answer in positive exponential form. [1.7]

45. $\frac{x^3(x^{-2})^2}{x^{-1}}$

46. $c^{-2}d^4c^5d^{-6}$

Evaluate. [1.8]

47. $-3^2 + (-3)^{-2}$

48. $-2 - 2^3(-2)^{-3}$

Use the formula $t = pC$ to calculate the sales tax (t) given the tax rate percentage (p) and cost (C). [2.2]

49. Find the tax on a \$23.95 shirt with a 5% tax rate.

50. How much sales tax is there on a \$2300 computer in a state with a 7% tax rate?

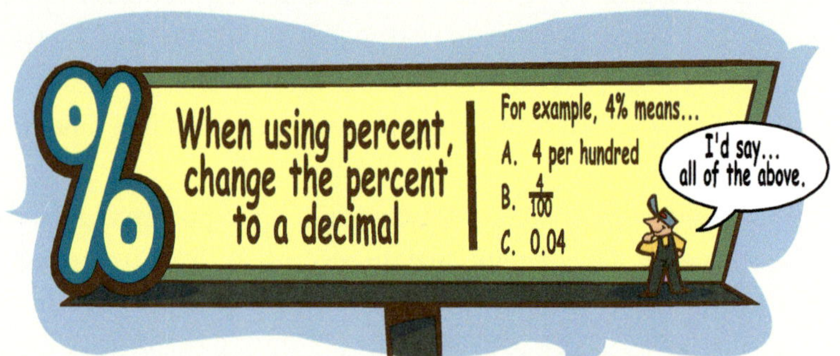

Write an equation and solve. [2.4, 2.6]

51. How many feet of fencing should be added to the 24 yd of fencing that the school already owns to enclose a field with a perimeter of 89 ft?

52. Two angles are supplementary, and one angle is one-fifth the other. Find the measure of the two angles.

Express each of the following repeating decimals as a quotient of integers. [2.6]

53. $1.\overline{5}$

54. $0.2\overline{09}$

2.8 Eliminating Fractions and Decimals

When an equation is written to solve a problem, it often contains fractions or decimals. Clearing an equation of fractions by multiplying both sides of the equation by the least common multiple of the denominators, the least common denominator (LCD), will simplify the process of solving the equation.

Example 1

Solve $\frac{x}{4} + x - \frac{1}{3} = 3$.

Answer

$$12\left(\frac{x}{4}\right) + 12x - 12\left(\frac{1}{3}\right) = 12(3)$$
$$3x + 12x - 4 = 36$$

1. Multiply both sides of the equation by 12, the LCD, and apply the Distributive Property.

$$15x - 4 = 36$$
$$15x = 40$$
$$x = \frac{40}{15} = \frac{8}{3}$$

2. Solve.

Example 2

Solve $\frac{x+2}{6} - \frac{x-1}{3} = \frac{1}{9}$.

Answer

$$18\left(\frac{x+2}{6} - \frac{x-1}{3}\right) = 18\left(\frac{1}{9}\right)$$
$$3(x+2) - 6(x-1) = 2$$
$$3x + 6 - 6x + 6 = 2$$

1. Multiply both sides of the equation by 18, the LCD, and apply the Distributive Property.

$$-3x + 12 = 2$$
$$-3x = -10$$
$$x = \frac{10}{3}$$

2. Solve.

Example 3

The coach has developed a drill that requires a rectangular field with the width being $\frac{1}{3}$ of the length. The perimeter of the field is 280 ft. What length and width should the field liner mark off?

Answer

Let x = the length of the field and $\frac{1}{3}x$ = the width.

$$x + x + \frac{1}{3}x + \frac{1}{3}x = 280$$

1. Write the equation.

$$3\left(x + x + \frac{1}{3}x + \frac{1}{3}x\right) = 3(280)$$
$$3x + 3x + x + x = 840$$

2. Clear the fractions by multiplying both sides by 3, and apply the Distributive Property.

$$8x = 840$$
$$x = 105$$
$$\frac{1}{3}x = 35$$

3. Solve.

The dimensions are 35 ft by 105 ft.

Solving Equations Containing Fractions

1. Find the LCD.
2. Eliminate fractions by multiplying both sides by the LCD.
3. Solve the resulting equation.
4. Check the solution in the original equation.

A decimal is a form of a fraction. The process of removing decimals from an equation is the same as eliminating fractions, but the multiplier will be a power of ten.

Example 4

Solve $2.6x + 12.24 = 0.27(x - 6)$.

Answer

$$100(2.6x + 12.24) = 100[0.27(x - 6)]$$

1. Since the hundredths place is the smallest decimal place value, multiply both sides by 100.

$$260x + 1224 = 27(x - 6)$$
$$260x + 1224 = 27x - 162$$

2. Apply the Associative Property of Multiplication before distributing on the right side.

$$260x + 1224 - 27x = 27x - 162 - 27x$$
$$233x + 1224 = -162$$
$$233x = -1386$$
$$x = -\frac{1386}{233} \approx -5.95$$

3. Solve and write the answer in decimal form rounded to the nearest hundredth.

How many coins do you have in your pocket or purse? Find their total value. If you have seven quarters, you must multiply the number of quarters (7) by the value of a quarter ($0.25) to find the total value of $1.75. Likewise, the value of q quarters can be represented by the expression $0.25q$.

Coin	Value per Coin	Number of Coins	Value of Coins
Pennies	$0.01	p	$0.01p$
Nickels	$0.05	n	$0.05n$
Dimes	$0.10	d	$0.10d$
Quarters	$0.25	q	$0.25q$

Suppose you have 20 coins in dimes and nickels. If 11 are dimes, how many are nickels? Yes, there are 9, but do you see that you subtracted a part (11) from the total (20) to obtain the other part? This principle is used with variables in problems in which a total amount is given and you are interested in the parts that make up that total. If there are d dimes, then there are $20 - d$ nickels. Using tables to organize information is especially helpful when you are solving this type of problem.

Example 5

Chris has 17 coins, all dimes and quarters. If he has a total of $2.45, how many of each coin does he have?

Answer Let q = the number of quarters and $17 - q$ = the number of dimes.

Coin	Number of Coins	Value of Coins
Quarters	q	$0.25q$
Dimes	$17 - q$	$0.10(17 - q)$

1. Use a table to organize expressions for the numbers and values of the coins.

$$0.25q + 0.10(17 - q) = 2.45$$

2. Write an equation expressing the total value.

$$100[0.25q + 0.10(17 - q)] = 100(2.45)$$

3. Multiply both sides by 100 to clear the decimals.

$$25q + 10(17 - q) = 245$$
$$25q + 170 - 10q = 245$$
$$15q + 170 = 245$$
$$15q = 75$$
$$q = \frac{75}{15} = 5$$

4. Solve.

There are 5 quarters and 12 dimes.

Solving Coin Problems

1. **Read**. Find the main unknown and assign a variable to it; then express any other unknowns in terms of it.
2. **Plan** and organize by making a table indicating the number and the total value of each type of coin.
3. **Solve** an equation obtained by using the information in the value column and any other information from the problem.
4. **Check** that you have answered all the questions and that your answers are reasonable.

Example 6

A coin bank contains four more quarters than nickels, twice as many dimes as nickels, and five more than three times as many pennies as nickels. If the bank contains \$22.25, how many of each coin are in it?

Answer

Read Let n = the number of nickels, $n + 4$ = the number of quarters, $2n$ = the number of dimes, and $3n + 5$ = the number of pennies.

Plan

Coin	Number of Coins	Value of Coins
Quarters	$n + 4$	$0.25(n + 4)$
Dimes	$2n$	$0.10(2n)$
Nickels	n	$0.05n$
Pennies	$3n + 5$	$0.01(3n + 5)$

Solve

$$0.25(n + 4) + 0.10(2n) + 0.05n + 0.01(3n + 5) = 22.25$$ 1. Write an equation.

$$25(n + 4) + 10(2n) + 5n + (3n + 5) = 2225$$ 2. Multiply by 100.

$$25n + 100 + 20n + 5n + 3n + 5 = 2225$$ 3. Distribute.

$$53n + 105 = 2225$$ 4. Combine like terms.

$$53n = 2120$$ 5. Solve.

$$n = \frac{2120}{53} = 40$$

$40 + 4 = 44$, $2(40) = 80$, and $3(40) + 5 = 125$

There are 44 quarters, 80 dimes, 40 nickels, and 125 pennies.

Check $44(0.25) + 80(0.10) + 40(0.05) + 125(0.01) = 22.25$

A. Exercises

Solve.

1. $\frac{x}{3} - \frac{x}{4} = 1$

2. $\frac{x}{3} + \frac{x}{8} = \frac{11}{24}$

3. $\frac{2x}{9} + \frac{1}{3} = 5$

4. $\frac{x}{6} - \frac{3x}{2} = \frac{16}{3}$

5. $\frac{y}{2} - \frac{2y}{7} = \frac{25}{7}$

6. $a + \frac{3a}{2} = \frac{15}{2}$

7. $\frac{b}{3} - \frac{8b}{7} = \frac{68}{21}$

8. $9m - \frac{3m}{8} = \frac{345}{8}$

9. $y = \frac{3y}{5} + 12$

10. $\frac{x + 2}{9} - \frac{x}{3} = 10$

11. $2b - 8 = 0.4b$

12. $0.5x = 0.3x + 9$

13. $1.4y + 0.09y = 1.49$

14. $0.67a + 3.06 = 0.84a$

15. $0.3x + 0.89 = 5.321$

B. Exercises

16. Which two properties are used to clear fractions from an equation?

17. Evaluate the following expressions when $a = 2$, $b = 3$, $c = 4$, and $d = 5$.

 a. $d[a(b + c)]$

 b. $da(b + c)$

 c. $da(db + dc)$

18. By what factor is $da(db + dc)$ larger than $d[a(b + c)]$ in exercise 17?

19. Which property can be used to prove that $d[a(b + c)] = da(b + c)$?

Solve.

20. $\frac{4y - 4}{8} + \frac{26}{2} = \frac{1}{4}$

21. $\frac{a - 3}{2} - \frac{a + 3}{4} + \frac{a}{8} = 6$

22. $3k + \frac{k - 4}{9} = \frac{1}{3}$

23. $\frac{107}{6} = \frac{7m}{2} - \frac{3m - 1}{3}$

24. $0.8(x + 4) - 0.3x = 1.7$

25. $1.4 - (2.56x - 3.5) = 8$

26. $0.35s + 0.6(30 - s) = 0.45(30)$

27. $0.15(70) + a = 0.2(70 + a)$

28. $\frac{t + 1}{2} = \frac{2t - 3}{6}$

29. $\frac{3x + 2}{4} = \frac{7x - 2}{6}$

Write an equation and solve.

30. Jill has \$9.96 in nickels, pennies, and dollar bills. She has twelve more pennies than nickels and five fewer dollar bills than nickels. How many of each does she have?

31. One Saturday Omar collected from his newspaper customers twice as many dollar bills as fives and one fewer ten than fives. If Omar collected \$58, how many tens, fives, and ones did he get?

32. Lynn has \$8.58 in quarters and pennies. If she has eight times as many pennies as quarters, how many of each does she have?

33. Josef has a jar of dimes, nickels, and pennies. He has one dime more than five times the number of nickels and four times as many pennies as nickels. How many of each coin does he have if their total value is \$7.77?

34. A bank teller knows she received 34 dimes and nickels combined from a customer. If the change totaled \$2.05, how many dimes and nickels did she receive?

35. In his pocket Chip has four more dimes than quarters. If the dimes and quarters total \$2.85, how many of each does he have?

36. Sam has five more dimes than quarters and four fewer nickels than quarters. If he has a total of \$4.70, how many of each does he have?

37. A teller has rolls of coins in his drawer, consisting of quarters (\$10 per roll), dimes (\$5 per roll), and nickels (\$2 per roll). He has three times as many rolls of quarters as nickels and twice as many rolls of dimes as nickels. The total is \$252. How many rolls of each does he have?

C. Exercises

Solve each literal equation for *x*.

38. $ax - \frac{2b}{c} = 3$

39. $pq + \frac{2x}{d} = 4b$

Dominion Modeling

A bicycle manufacturer can manufacture and ship unassembled bikes for \$149 per bike with fixed costs of \$225,000. To manufacture and ship assembled bikes requires \$250,000 in fixed costs and \$161 per bike. The manufacturer plans on charging \$179 for an unassembled bike and \$199 for an assembled bike.

40. Write expressions for the expense and for the revenue of producing and selling b unassembled bikes.

41. Find the break-even point for producing and selling unassembled bikes.

42. Write expressions for the expense and for the revenue of producing and selling k assembled bikes.

43. Find the break-even point for producing and selling assembled bikes.

44. Should the manufacturer ship its bikes assembled or unassembled? Explain your reasoning.

CUMULATIVE REVIEW

Simplify. [2.3]

45. $(3a - 2b) - (a + b) + (4a - 6b)$

46. $y(x - y) - x(y - x)$

Solve. [2.4–2.7]

47. $x - \frac{2}{3} = \frac{4}{7}$

48. $-\frac{4}{5}y = \frac{6}{7}$

49. $3.5x + 7.2 = -2.6$

50. $45.2 = 2g + 25.6$

51. $7 - 4(p + 3) = -3$

52. $(t + 5) - 2(t + 3) = 1$

53. $2b - 9b + 6 = 58b + 19$

54. $9x - 20 = 5(x - 4) - 4x$

CHAPTER 2 REVIEW

List the variables and constants in each expression.

1. $3x - 4$
2. $\frac{5ab}{c}$
3. Explain the difference between a constant and a numerical coefficient.

Translate each word phrase into an algebraic expression.

4. the product of three consecutive odd integers
5. the sum of an angle's complement and its supplement

Evaluate each expression for each member of the domain {−4, 2, 3, 8}.

6. $y^2 + 4$
7. $w^2 - 2w - 1$

Evaluate when $x = 3$, $y = -4$, and $z = 2$.

8. $4x^3y$
9. $2x + 3y - 6z$
10. $x^2 + yz$
11. $5x^2 + 2x + 7$

The formula for the distance that a dropped object falls when air resistance is negligible is $d = \frac{1}{2}gt^2$. Find the distance each object falls in 3 sec when dropped.

12. on the earth, where $g = 32$ ft/sec^2
13. on the moon, where $g = 5\frac{1}{3}$ ft/sec^2

Simplify.

14. $4x^2 + 3x - 8x^2$
15. $m + 2n - 4m - 6m + n$
16. $2r^2 + 3r - 8 + 4r^2 - 5r + 2$
17. $2a - (4a + 3b) - 6a$
18. $3x^2 + (2x + 7) - (5x + 6)$
19. $x + 3y - [4x + 2y - (8x - 9y)]$

Is the given number a solution to the equation?

20. $x^2 - x = 6$ if $x = -2$
21. $\frac{2}{3}y - \frac{3}{4} = -\frac{5}{12}$ if $y = \frac{1}{2}$

Solve. Show all steps and check your answers.

22. $4\frac{4}{7}x = -\frac{6}{5}$
23. $y + 0.38 = 8.7$
24. $\frac{z}{5} - 6 = 84$
25. $6a + 21 = 165$
26. $4b = 7 - 4b + 6$
27. $\frac{4k}{9} = 2$
28. $3(x + 2) = 4(x - 6) + 2(x + 5)$
29. $89d + 17 + 6d = 5d - 19$
30. $4(c + 10) - 146 = 6(2c + 9)$
31. $4v - 2(v + 9) + 10 = -14$
32. $\frac{x}{2} + 1 = \frac{4x}{3}$
33. $\frac{4x + 7}{4} - \frac{5x - 4}{20} = \frac{5 - x}{5}$
34. $0.8n + 0.3(12 - n) = 4.8$

State the property that justifies each step in the solution of the equation $\frac{2x - 8}{3} = \frac{x - 4}{6} + \frac{1}{2}$.

35. $6\left(\frac{2x - 8}{3}\right) = 6\left(\frac{x - 4}{6} + \frac{1}{2}\right)$
36. $4x - 16 = (x - 4) + 3$
37. $4x - 16 = x - 1$
38. $3x - 16 = -1$
39. $3x = 15$
40. $x = 5$

Write an equation and solve.

41. A paper boy delivers thirteen papers to an apartment complex. If these deliveries compose one-seventh of his route, how many papers does he deliver?

42. A candy bar costs $0.95, which is nineteen times the price of a piece of hard candy. What is the price of a piece of hard candy?

43. The music department has budgeted $350 for the purchase of CDs. How many CDs can be purchased if they cost $15 each and there is a $5 shipping fee for the order?

44. A local natural gas company charges small businesses a $22 monthly connection fee and $1.20/therm. How many therms were used by a company during a month in which its gas bill was $78.40?

45. One number is four times another, and the sum of the two numbers is 155. What are the two numbers?

46. Find three consecutive integers whose sum is 234.

47. Brandon wants to start a lawn-mowing service. He estimates that it will take $300 to purchase equipment and that he will have $5 of expenses for each lawn he mows. How many lawns must he mow at $25 per lawn before he breaks even (when his revenue equals his expenses)?

48. Find an angle whose supplement is three times its complement.

49. Jenica has a total of $3.40 in nickels and quarters. If she has twenty coins, how many are nickels and how many are quarters?

50. Dan opened up his piggy bank and found a total of thirty-five coins, made up of nickels, dimes, and quarters. If he has five more nickels than dimes and a total of $6, how many of each coin were in his bank?

3 Using Equations

Mr. Hill drives a work van that averages 10 mi/gal. The compact car his wife drives gets 30 mi/gal. They can afford either to replace Mr. Hill's work van with an SUV that gets 13 mi/gal or to replace Mrs. Hill's car with a hybrid that gets 60 mi/gal. Assuming they drive the same number of miles annually, which trade-in would save the most gasoline?

Intuitively, you would trade in the compact car for the hybrid and its much improved mileage. However, in order to wisely manage the resources God has given to them, the Hills need to carefully research which exchange will be more beneficial. By determining expressions that model the savings of each possible exchange, they can use equations to quickly compare a variety of options.

After this chapter you should be able to

1. solve literal equations for the indicated variable.
2. use proportions to solve problems.
3. find corresponding lengths, areas, and volumes of similar figures.
4. apply the percent formula to find the part, the percent, or the whole.
5. determine the percent change, percent error, or percent difference for two quantities.
6. solve problems related to cost, markup, retail price, discount, and sale price.
7. solve problems related to tips, commissions, and interest.
8. solve problems about related distances using $d = rt$.
9. solve problems involving mixtures.

3.1 Solving Literal Equations

Electricians solve Ohm's law, which states that $V = IR$, to determine the voltage (V), the current (I), or the resistance (R).

Sometimes an equation will contain several variables. This type of equation is called a *literal equation*. The process of solving a literal equation for the desired variable is the same as solving an equation with only one variable: Just treat the other variables as numbers (constants) in the solution. In the equation $5x = 10$, you would know to divide both sides by 5 to solve for x. Similarly, to solve for t in the literal equation $rt = d$, divide both sides by r, $\frac{\cancel{r}t}{\cancel{r}} = \frac{d}{r}$, to produce $t = \frac{d}{r}$.

Example 1

Solve $F = \frac{WH}{L}$ for H.

Answer

$F = \frac{WH}{L}$

$L \cdot F = \cancel{L}\left(\frac{WH}{\cancel{L}}\right)$ — 1. Multiply both sides by L to clear the fraction.

$LF = WH$

$\frac{LF}{W} = \frac{\cancel{W}H}{\cancel{W}}$ — 2. Divide both sides by W.

$H = \frac{LF}{W}$ — 3. Use the Symmetric Property of Equality to write the equation with the desired variable on the left side.

The formulas you saw in Section 2.2 are literal equations. It is often best to solve a formula for the desired variable before substituting values for the other variables.

Example 2

Solve $p = 2l + 2w$ for l and then find l when $p = 58$ and $w = 6$.

Answer

$p = 2l + 2w$

$p - 2w = 2l$ — 1. Subtract $2w$ from both sides.

$\frac{p - 2w}{2} = l$ — 2. Divide both sides by 2.

$l = \frac{58 - 2(6)}{2} = \frac{58 - 12}{2} = \frac{46}{2} = 23$ — 3. Evaluate to find the numerical value of l.

Recall that the Distributive Property is used to add like terms. For example, $3x + 5x = (3 + 5)x = 8x$. When solving literal equations, you may need to use the Distributive Property to change a sum of terms, each containing the desired variable, to a product. This is an important step in the solution of many literal equations.

Example 3

Solve $ax - 5b = 7x + 18b$ for x.

Answer

$ax - 5b = 7x + 18b$	
$ax - 5b - 7x = 18b$	1. Subtract $7x$ from both sides.
$ax - 7x = 23b$	2. Add $5b$ to both sides.
$(a - 7)x = 23b$	3. Apply the Distributive Property.
$x = \frac{23b}{a - 7}$	4. Divide both sides by $a - 7$.

Solving Literal Equations

1. Using the Distributive Property, clear the equation of fractions by multiplying both sides by the LCD.
2. Simplify both sides of the equation.
3. Collect all terms containing the desired variable on one side of the equation and all other terms on the other side.
4. Use the Distributive Property to isolate the desired variable.
5. Divide both sides by the coefficient of the variable.

Example 4

Solve $\frac{a + 2b}{c} = \frac{cd - a}{b}$ for a.

Answer

$\frac{a + 2b}{c} = \frac{cd - a}{b}$	
$b\cancel{c}\left(\frac{a + 2b}{\cancel{c}}\right) = \cancel{b}c\left(\frac{cd - a}{\cancel{b}}\right)$	1. Multiply both sides by bc to clear the equations of fractions.
$b(a + 2b) = c(cd - a)$	
$ab + 2b^2 = c^2d - ac$	2. Use the Distributive Property to remove the parentheses.
$ab + ac + 2b^2 = c^2d$	3. Add ac to both sides to collect all terms containing the desired variable, a, on the left side.
$ab + ac = c^2d - 2b^2$	4. Subtract $2b^2$ from both sides.
$(b + c)a = c^2d - 2b^2$	5. Use the Distributive Property to isolate the desired variable.
$a = \frac{c^2d - 2b^2}{b + c}$	6. Divide both sides by $b + c$.

Notice that the Addition Property of Equality was used twice in steps 3 and 4 of Example 4.

The manipulation of literal equations occurs in many disciplines, including finance, the physical sciences, and even the social sciences. By having a firm foundation of solving equations, you will be prepared to succeed in a variety of studies and occupations.

The next example is a typical application of literal equations. Subscripted variables, such as r_1 and r_2, are often used to clarify the meaning of formulas and are treated just like other variables. Chemists have noticed that the boiling point of a solution increases according to the number of particles dissolved in it. In order to raise the boiling point to a desired temperature, the chemist must determine the molality, or concentration, of dissolved particles.

Example 5

The boiling point of an ethanol solution, B_s, rises as sugar is added according to the formula $B_s = B_e + k_e m$, where B_e is the boiling point of pure ethanol. Solve the formula for m. Then find the molality, m, of a solution that has a boiling point (B_s) of 82.37°C if $k_e = 1.22\,\frac{°\text{C}}{\text{molal}}$ and $B_e = 78.26$°C.

Answer

$$B_s = B_e + k_e m$$

$$B_s - B_e = k_e m$$ 1. Subtract B_e from both sides.

$$\frac{B_s - B_e}{k_e} = m$$ 2. Divide both sides by k_e to isolate m.

$$m = \frac{82.37°\text{C} - 78.26°\text{C}}{1.22\,\frac{°\text{C}}{\text{molal}}} \approx 3.37 \text{ molal}$$ 3. Evaluate to find the numerical value of m.

A. Exercises

Solve each literal equation for the indicated variable.

1. $a = bx$ for x
2. $\frac{x}{n} = q$ for x
3. $\frac{x}{r} = r$ for x
4. $amx = 3a$ for x
5. $y = mx + b$ for b
6. $b - x = d$ for x
7. $I = Prt$ for t
8. $p = a + b + c$ for c
9. $d = rt$ for t
10. $c = 2\pi r$ for r
11. $p = 3s$ for s
12. $V = \pi r^2 h$ for h

B. Exercises

Solve each literal equation for the indicated variable.

13. $\frac{P}{6rn} = \frac{7}{9nx}$ for P
14. $f + \frac{y}{3} = \frac{y}{8}$ for f
15. $y = mx + b$ for x
16. $e = \frac{x - a}{a}$ for x
17. $m = \frac{y_2 - y_1}{x_2 - x_1}$ for x_1
18. $\frac{P_1V_1}{T_1} = \frac{P_2V_2}{T_2}$ for P_2
19. $S = 2\pi rH + 2\pi r^2$ for H
20. $ax + by = c$ for y
21. $ax + b + a(x + b) = b$ for x
22. $2y - 3x = ay - 2x$ for y
23. $\frac{nx}{b} - n = x$ for n
24. $Pnx = qx - r$ for x

Solve each literal equation for the indicated variable. Then substitute the values of the other variables into the derived expression and simplify.

25. Solve $d_f = vt + d_i$ for v. Then evaluate the resulting expression to find v when $d_f = 326$ ft, $t = 2$ sec, and $d_i = 150$ ft.

26. Solve $V = \frac{1}{3}\pi r^2 H$ for H. Then evaluate the resulting expression to find H when $V = 288\pi$ cm^3 and $r = 6$ cm.

27. Solve $S = c(h + r)$ for h. Then evaluate the resulting expression to find h when $S = 72\pi$ in.^{2}, $c = 6\pi$ in., and $r = 3$ in.

28. Solve $m = \frac{y_2 - y_1}{x_2 - x_1}$ for x_2. Then evaluate the resulting expression to find x_2 when $m = -\frac{2}{3}$, $x_1 = -1$, $y_1 = 4$, and $y_2 = 2$.

C. Exercises

Solve each literal equation for the indicated variable.

29. $\frac{nx}{b} - n = x$ for x

30. $\frac{2p}{m} - \frac{r}{s} = \frac{pr + q}{ms}$ for p

31. $\frac{x}{b} - \frac{x}{c} = x$ for b

32. $a(x + c) = 2ad + c$ for a

33. Solve $\frac{P_1V_1}{T_1} = \frac{P_2V_2}{T_2}$ for T_2. Then evaluate the resulting expression to find T_2 when $P_1 = 800$ torr, $V_1 = 2.5$ L, $T_1 = 300$ K, $P_2 = 760$ torr, and $V_2 = 1.5$ L.

34. Solve $E = mgh + \frac{1}{2}mv^2$ for m. Then evaluate the resulting expression to find m if $E = 580$ kg·m^2/s^2, $g = 9.8$ m/s^2, $h = 10$ m, and $v = 6$ m/s.

Dominion Modeling

35. Refer to the Chapter Opener to complete the first six calculations in the table below describing the gasoline consumption and savings for the trade-ins of Mr. and Mrs. Hill's vehicles. Assume that their vehicles will each be driven 15,000 mi annually.

36. Fill in the rest of the table by writing general expressions for the calculations. Use m to represent the number of miles driven each year, d for the mileage of the old vehicle, and n for the mileage of the new vehicle.

	Vehicle	Gallons Consumed	Gallons Saved
Mr. Hill	work van		
	SUV		
Mrs. Hill	compact car		
	hybrid		
General Expressions	old (d)		
	new (n)		

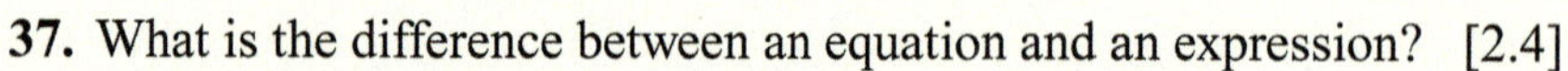

CUMULATIVE REVIEW

37. What is the difference between an equation and an expression? [2.4]

Translate each sentence into a literal equation. [2.1, 2.5]

38. The net force on an object (F) is equal to the product of its mass (m) and its resulting acceleration (a).

39. A business's net income (I) is calculated by subtracting its expenses (E) from its revenue (R).

Clear each equation of fractions or decimals and then solve. [2.8]

40. $\frac{1}{4} - x = \frac{x}{4}$

41. $\frac{y}{6} - \frac{y-3}{9} = \frac{16-y}{4}$

42. $0.15c + 0.35(20 - c) = 6$

43. $1.75(s - 5) - 1.25s = 12.25$

Simplify, leaving all variables in the numerator. [1.7]

44. $a^2b^4a^5b^{-8}$

45. $\frac{(c^2)^{-5}(d^{-3})^{-2}}{f^3}$

46. $\frac{m^8n}{n^5m^2}$

3.2 Ratios and Proportions

Definitions

A **ratio** is a comparison of two numbers by division and is often expressed as a fraction.

A **rate** is a ratio comparing numbers with different units.

A **unit rate** is a rate in which the denominator is one.

If a student gets 45 out of 60 questions right on a test, the ratio of correct answers to test questions can be expressed as $\frac{45}{60}$, which should be reduced to $\frac{3}{4}$. This ratio can also be written as 3 : 4 or 3 to 4. Two other ratios can be deduced from this statement: The ratio of incorrect answers to test questions is $\frac{15}{60}$, or $\frac{1}{4}$; and the ratio of correct answers to incorrect answers is $\frac{45}{15}$, or $\frac{3}{1}$. If the student was given 45 min to take the sixty-question test, the ratio $\frac{60 \text{ questions}}{45 \text{ min}}$, or $\frac{4 \text{ questions}}{3 \text{ min}}$, indicates the rate at which he must complete the test. Expressing this relationship using the unit rate $1\frac{1}{3}$ questions/min makes the rate easier to understand.

Example 1

A school of 312 students has 168 girls and 20 teachers. Find the following ratios.

a. girls to students **b.** boys to students

c. girls to boys **d.** unit rate of students per teacher

Answer

a. $\frac{\text{girls}}{\text{students}} = \frac{168}{312} = \frac{7}{13}$

b. $\frac{\text{boys}}{\text{students}} = \frac{144}{312} = \frac{6}{13}$ There must be 312 − 168 = 144 boys.

c. $\frac{\text{girls}}{\text{boys}} = \frac{168}{144} = \frac{7}{6}$

d. $\frac{\text{students}}{\text{teachers}} = \frac{312}{20} = 15.6$ students per teacher Complete the division to get a denominator of one.

A rate can be converted to an equivalent rate with different units in a process called *dimensional analysis* (or *unit analysis*). *Unit multipliers*, such as $\frac{1\text{ mi}}{5280\text{ ft}}$, are rates with different expressions representing the same quantity in the numerator and the denominator, so their value is one. Multiplying a rate by one or more unit multipliers will not change the ratio; instead, the ratio will be renamed as an equivalent ratio with different units.

Example 2

Use a unit multiplier to convert to the desired unit.

a. 28 ft to yards **b.** 62 in. to centimeters

Answer

a. $28\text{ ft}\left(\frac{1\text{ yd}}{3\text{ ft}}\right) = \frac{28}{3} \approx 9.3\text{ yd}$ Use the fact that 3 ft = 1 yd to make a fraction equal to one. Be sure that the quantity with the unit to be canceled is in the denominator.

b. $62\text{ in.}\left(\frac{2.54\text{ cm}}{1\text{ in.}}\right) = 62(2.54) = 157.48\text{ cm}$ Recall that 2.54 cm = 1 in. to make a fraction equal to one. Be sure that the units cancel.

Example 3

About 8 min into liftoff, a space shuttle is traveling 17,000 mi/hr. How fast is this in feet per second?

Answer

$17{,}000\text{ mi/hr} = \frac{17{,}000\text{ mi}}{1\text{ hr}}$ 1. Write the rate as a unit rate.

5280 ft = 1 mi 2. Find equivalent conversions for the units given compared to those needed.

60 sec = 1 min

60 min = 1 hr

$\frac{17{,}000\text{ mi}}{1\text{ hr}} \cdot \frac{5280\text{ ft}}{1\text{ mi}} \cdot \frac{1\text{ hr}}{60\text{ min}} \cdot \frac{1\text{ min}}{60\text{ sec}}$ 3. Multiply the given rate by unit multipliers made by considering the units that need to be canceled.

$\frac{17{,}000(5280)\text{ ft}}{60^2\text{ sec}} = 24{,}933.\overline{3}\text{ ft/sec}$ 4. Notice that the dimensions left in the problem are in feet per second. Use a calculator to convert to decimal form.

$\approx 25{,}000$ ft/sec

Definition

A **proportion** is a statement of equality between two ratios.

The proportion $\frac{a}{b} = \frac{c}{d}$ is read "a is to b as c is to d." The first and last numbers, a and d, are the extremes of the proportion; and the two middle numbers, b and c, are the means. When solving a proportion, multiply both sides of the equation by the common denominator to clear the equation of fractions.

$$\frac{a}{b} = \frac{c}{d}$$
$$bd\left(\frac{a}{b}\right) = bd\left(\frac{c}{d}\right)$$
$$ad = bc$$

Notice that the product of the extremes is equal to the product of the means. This property is frequently used to solve proportions.

Property of Proportions

In a proportion, the product of the extremes equals the product of the means.

Example 4

Solve the proportion $\frac{7}{8} = \frac{x}{4}$.

Answer

$\frac{7}{8} = \frac{x}{4}$

$8x = 7(4)$ — 1. Take the product of the extremes and the product of the means. Taking the product containing the variable first will place it on the left side of the equation.

$8x = 28$

$x = \frac{28}{8} = 3.5$ — 2. Divide both sides by 8.

The Property of Proportions can be combined with previously learned methods of solving equations to solve more complicated proportions.

Example 5

Solve the proportion $\frac{x + 9}{5} = \frac{x + 15}{7}$.

Answer

$\frac{x + 9}{5} = \frac{x + 15}{7}$

$7(x + 9) = 5(x + 15)$ — 1. Apply the Property of Proportions.

$7x + 63 = 5x + 75$ — 2. Distribute to remove the parentheses.

$2x + 63 = 75$ — 3. Collect all variable terms on the left side.

$2x = 12$ — 4. Undo addition by subtracting 63 from both sides.

$x = 6$ — 5. Undo multiplication by dividing both sides by 2.

Example 6

A car traveled 250 mi on 9 gal of gas. How many gallons are needed to travel 360 mi?

Answer

Let g = the number of gallons needed. — 1. Assign a variable to represent the unknown.

$\frac{250}{9} = \frac{360}{g}$ — 2. Write a proportion using miles per gallon.

$250g = 9(360)$ — 3. Apply the Property of Proportions.

$250g = 3240$

$g = \frac{3240}{250} = 12.96 \approx 13$ gal — 4. Divide both sides by 250.

You may need to determine the correct ratio to use when solving problems involving proportions.

Example 7

A car dealer typically sells new and used cars in a ratio of three new cars to two used cars. If he wants to sell a total of 325 cars next month, what goal should he set for the used car sales manager?

Answer

$\frac{3\text{ new}}{2\text{ used}}, \frac{3\text{ new}}{5\text{ total}},$ and $\frac{2\text{ used}}{5\text{ total}}$ — 1. Write the three stated and implied ratios.

$\frac{2}{5} = \frac{u}{325}$ — 2. Use the ratio relating the known quantity to the unknown quantity to write a proportion.

$5u = 2(325)$ — 3. Apply the Property of Proportions.

$5u = 650$

$u = \frac{650}{5} = 130$ used cars — 4. Divide both sides by 5.

A. Exercises

On a recent homework assignment, eight students received an A, ten received a B, six received a C, and there were no Ds or Fs. Write the following ratios in lowest terms using the word "to."

1. the ratio of As to Bs

2. the ratio of As to Cs

3. the ratio of As to students in the class

4. the ratio of Cs to students in the class

Find the unit rate.

5. 85 points scored in five games

6. \$27.50 for 11 gal of gas

7. \$370 for 40 hr of work

8. 20,000 bu of wheat from 800 acres

Convert the given rate to an equivalent rate.

9. 5¢/day to dollars per year

10. \$4/lb to cents per ounce

11. 55 mi/hr to feet per second

12. 100 km/hr to meters per second

Solve.

13. $\frac{5}{4} = \frac{70}{x}$

14. $\frac{y}{49} = \frac{10}{7}$

15. $\frac{9}{5} = \frac{1800}{z}$

16. $\frac{a}{8} = \frac{9}{50}$

17. $\frac{16}{7} = \frac{2}{b}$

18. $\frac{8}{7} = \frac{c}{16}$

B. Exercises

Solve.

19. $\frac{5}{2w} = \frac{45}{36}$

20. $\frac{3x}{25} = \frac{63}{175}$

21. $\frac{d-1}{18} = \frac{12}{27}$

22. $\frac{y+1}{6} = \frac{84}{63}$

23. $\frac{n+1}{15} = \frac{n-1}{12}$

24. $\frac{2p-3}{6} = \frac{3p+3}{4}$

25. $\frac{7}{4q+1} = \frac{14}{7-2q}$

26. $\frac{6r-5}{9r-3} = -\frac{1}{3}$

Find the unit rate that describes each situation.

27. buying a package of 12 rolls of paper towels for $16

28. buying a package of 20 rolls of paper towels for $38.15

29. buying a $36 shirt and getting a second shirt of equal or lesser value free

30. buying a $30 shirt and getting a second shirt of equal value for 50% off

Use proportions to solve.

31. Jeremy earned $76 for working 8 hr. How much will he earn if he works 20 hr?

32. If Ashima drove the first 348 mi of her trip in 6 hr, how long will it take her, if she drives at the same rate, to drive the remaining 145 mi?

33. A cookie recipe for three dozen caramel peanut butter bars uses 14 oz of caramel. How much caramel (to the nearest ounce) should be used to make 120 bars for a cookie exchange?

34. A random survey of 30 students found that 17 played a musical instrument. If the school has 450 students, predict how many do not play an instrument.

35. The ratio of soccer players to volleyball players on the bus was 5 : 3. If there were a total of 48 players on the bus, how many were volleyball players?

36. A transformer reduces or increases the voltage (V) of an alternating current based on the ratio of the coils in the input circuit (N_{in}) to the ratio of the coils in the output circuit (N_{out}) according to the following proportion: $\frac{V_{out}}{V_{in}} = \frac{N_{out}}{N_{in}}$. A transformer is used to reduce the 120 V household current to the 18 V used in model train sets. If the transformer has 180 coils on the input circuit, how many coils are on the output circuit?

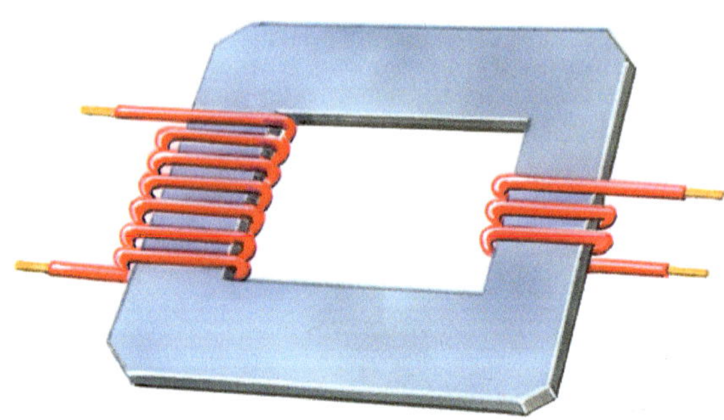

C. Exercises

37. If there are b boys in a classroom of s students, write the following ratios in terms of b and s.

    a. boys to girls

    b. girls to students

38. If the basketball team typically makes two out of every five three-point shots and makes one out of every two two-point shots, which type of shot should the team emphasize in its game plan? Explain your reasoning.

39. The density of a material is its mass-to-volume ratio. The density of copper (near room temperature) is 8.96 g/cm^3. Lincoln cents made from 1909 to 1982 were made almost entirely of copper. If one of these cents has a mass of 3.11 g, find its volume (to the nearest thousandth of a cubic centimeter).

40. Cents made beginning in 1982 have the same volume as those made before 1982, but they weigh only 2.50 g because they are made almost entirely of zinc (with a very thin copper plating). Using the results of exercise 39, determine the approximate density of zinc (to the nearest hundredth).

Dominion Modeling

The expression for the number of gallons saved annually can by rewritten as $m\left(\frac{n-d}{nd}\right)$, where m is the number of miles driven each year, d is the mileage of the old car, and n is the mileage of the new car.

41. Use this expression to confirm the number of gallons saved for each trade-in mentioned in the Chapter Opener.

    a. SUV for work van

    b. hybrid for compact car

42. Determine the number of gallons of gasoline saved if someone who drives an old sedan getting 18 mi/gal a distance of 15,000 mi/yr replaces the vehicle with a new sedan that gets 22 mi/gal.

CUMULATIVE REVIEW

Evaluate. [1.5–1.6]

43. $-\frac{14}{15}\left(-\frac{10}{21}\right)$

44. $-\frac{15}{14} \div \frac{10}{21}$

45. $\dfrac{\frac{15}{14}}{\frac{21}{10}}$

Solve. [2.5–2.6]

46. $0.4x + 3.1 = 2.5$

47. $a - (180 - a) = 90$

Solve each literal equation for the indicated variable. [3.1]

48. $mx + n = 2mx - n$ for x

49. $w = \dfrac{11(h-40)}{2}$ for h

Write an equation and solve.

50. After 11 mo of deputation, the Clarks had raised three-fourths of their monthly support. If their goal was $3800 per month, how much support had been raised? [2.4]

51. Find three consecutive odd integers such that twice the sum of the first and last is three more than the middle integer. [2.7]

52. Larry has twenty-five nickels and quarters with a total value of $3.05. How many of each coin does he have? [2.8]

SEQUENCES

Fibonacci and Golden Ratio Sequences

Challenge **Can you supply the next two numbers in each of the following sequences?**

a. 3, 4, 7, 11, 18, 29, ...

b. 1, 1, 2, 3, 5, 8, 13, ...

c. $1, 2, \frac{3}{2}, \frac{5}{3}, \frac{8}{5}, \frac{13}{8}, \ldots$

A *Fibonacci sequence* can be defined recursively as $A_n = A_{n-1} + A_{n-2}$, where A_1 and A_2 are any two integers. The first sequence above is a Fibonacci sequence. Notice that $3 + 4 = 7$, $4 + 7 = 11$, $7 + 11 = 18$, and so on. The second sequence is called the *standard Fibonacci sequence* and must have $A_1 = 1$ and $A_2 = 1$. This sequence was named after a thirteenth-century mathematician who studied the sequence of numbers as it related to the reproduction of rabbits. The Fibonacci sequence is so relevant mathematically that a magazine called the *Fibonacci Quarterly* presents new results related to this rich sequence and other recurrence sequences.

The third sequence above is called the *golden ratio sequence* and is defined recursively as $B_n = \frac{A_{n+1}}{A_n}$, where A_n is a Fibonacci sequence and the first golden ratio sequence term, B_1, is $\frac{A_2}{A_1}$. As n gets larger, the ratio $\frac{A_{n+1}}{A_n}$ gets closer to the *golden ratio*, which is approximately 1.618. (In the first sequence above, $\frac{A_8}{A_7} = \frac{76}{47} \approx 1.617$; in the second sequence, $\frac{A_9}{A_8} = \frac{34}{21} \approx 1.619$.)

Given the first two terms of a Fibonacci sequence, it is possible to list the terms of the related golden ratio sequence. For example, given $A_1 = 2$ and $A_2 = -3$, use the following two-step process. First, list the terms of the Fibonacci sequence: 2, −3, −1, −4, −5, −9, ... (Remember that, starting at the third term, each term is the sum of the previous two terms). Second, list the terms of the desired golden ratio sequence: $\frac{-3}{2}, \frac{-1}{-3}, \frac{-4}{-1}, \frac{-5}{-4}, \frac{-9}{-5}, \ldots,$ which simplifies to $-\frac{3}{2}, \frac{1}{3}, 4, \frac{5}{4}, \frac{9}{5}, \ldots.$

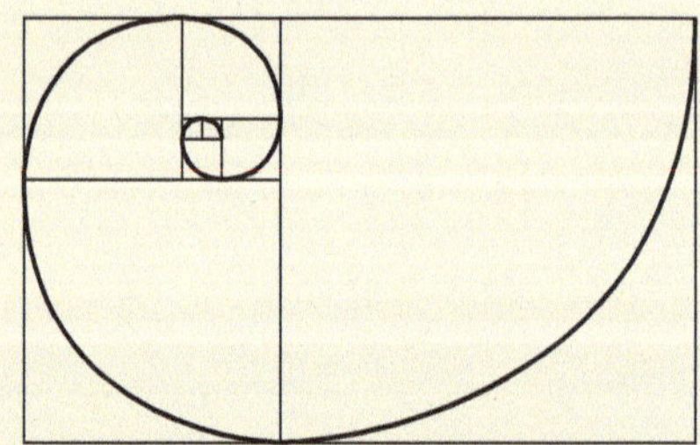

This Fibonacci spiral approximates the golden spiral.

From the spirals of pineapples to the proportions of the human body, the Fibonacci sequences and the golden ratio frequently appear in God's creation. The psalmist's response to the beauty and complexity of God's creation was to praise His awesome name (Ps. 8).

Exercises

List terms A_3 through A_7 for the given terms A_1 and A_2 of a Fibonacci sequence.

1. $A_1 = 2, A_2 = 5$ **2.** $A_1 = 8, A_2 = 1$ **3.** $A_1 = -4, A_2 = 3$

List the first five terms of a golden ratio sequence, B_n, for the given terms A_1 and A_2 of a Fibonacci sequence.

4. $A_1 = 2, A_2 = 5$ **5.** $A_1 = -1, A_2 = -1$ **6.** $A_1 = 8, A_2 = 13$

7. Given any Fibonacci sequence A_n, what number does the ratio $\frac{A_n}{A_{n+1}}$ approximate as n gets larger?

TECHNOLOGY CORNER (TI-84+ Family)

The sequence mode can be used to calculate and graph terms of up to three different sequences using the u, v, and w sequence variables. Select SEQ from the fourth line of the [MODE] screen and use [FORMAT] ([2nd] [ZOOM]) to make sure that Time and ExprOn are selected.

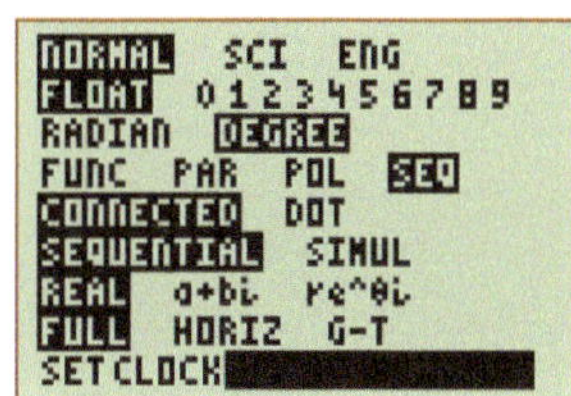

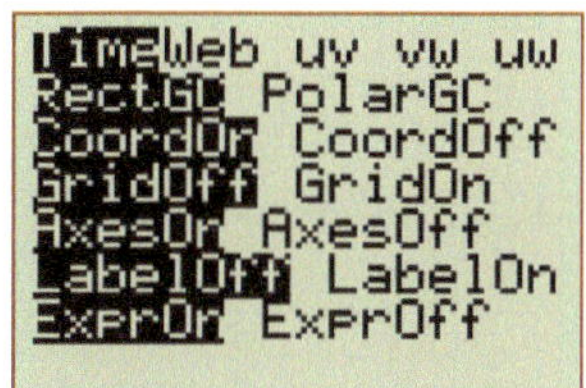

Select [Y=] to enter the sequence editor. The Fibonacci sequence 3, 4, 7, 11, … can be calculated by entering the recursive definition of the Fibonacci sequence, u(n) = u(n − 1) + u(n − 2). Use [2nd] 7 to enter u and [X,T,Θ,n] to enter n. Set u(nMin) = {4, 3} to define the first two terms of the sequence (in reverse order).

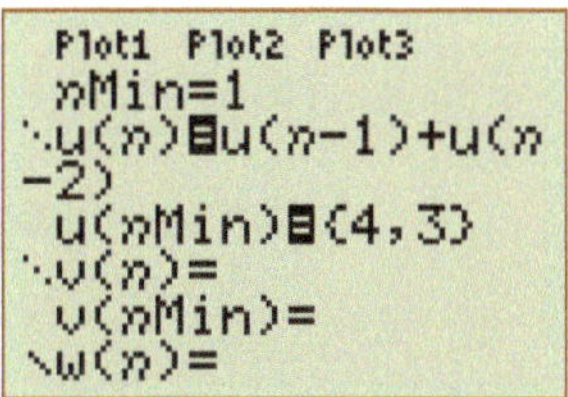

n	u(n)
1	3
2	4
3	7
4	11
5	18
6	29
7	47

n=1

The [TABLE] function ([2nd] [GRAPH]) will display the terms of the sequence. Use the [TBLSET] function ([2nd] [WINDOW]) to make sure TblStart = 1 and ΔTbl = 1.

A sequence may be defined based on the previous terms of another sequence. Entering v(n) as illustrated generates a golden ratio sequence. Scroll down the table to see how the terms of the sequence approximate the golden ratio.

Plot1 Plot2 Plot3
nMin=1
u(n)=u(n−1)+u(n−2)
u(nMin)={4,3}
v(n)=u(n−1)/u(n−2)
v(nMin)={0,0}

n	u(n)	v(n)
8	76	1.6207
9	123	1.617
10	199	1.6184
11	322	1.6179
12	521	1.6181
13	843	1.618
14	1364	1.618

n=14

3.3 Similar Figures and Scale Models

Geometric figures having the same shape, but not necessarily the same size, are said to be *similar* figures. Triangle ABC is similar to triangle XYZ, notated $\triangle ABC \sim \triangle XYZ$.

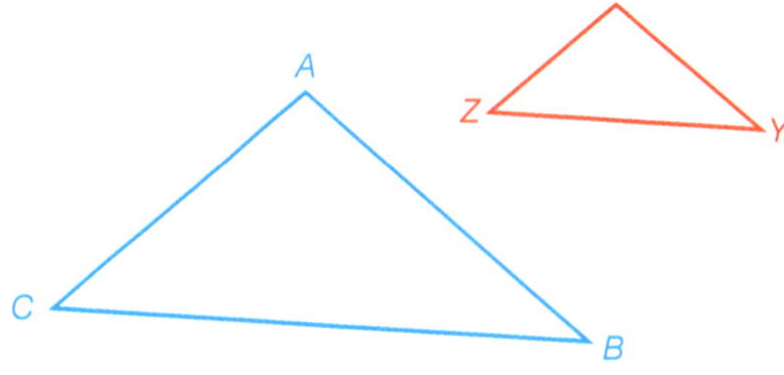

Definitions

Similar figures have the same shape, but not necessarily the same size.

Similar polygons are polygons in which corresponding angles are congruent and corresponding sides are proportional.

The order in which the notation is written is important. Corresponding vertices must be written in corresponding order. From the notation $\triangle ABC \sim \triangle XYZ$, we know the following.

$\angle A \cong \angle X$	Corresponding angles are congruent.
$\angle B \cong \angle Y$	
$\angle C \cong \angle Z$	
$\frac{AB}{XY} = \frac{BC}{YZ} = \frac{CA}{ZX}$	Corresponding sides are proportional.

Example 1

Find *XY*, given the measurements of the two similar triangles.

Answer

$\frac{CA}{ZX} = \frac{AB}{XY}$

$\frac{5}{2.5} = \frac{5.7}{m}$

$5m = 2.5(5.7)$

$5m = 14.25$

$m = \frac{14.25}{5} = 2.85$

1. Set up a proportion of corresponding sides of the triangles, as given in the diagram.
2. Apply the Property of Proportions and solve for *m*.

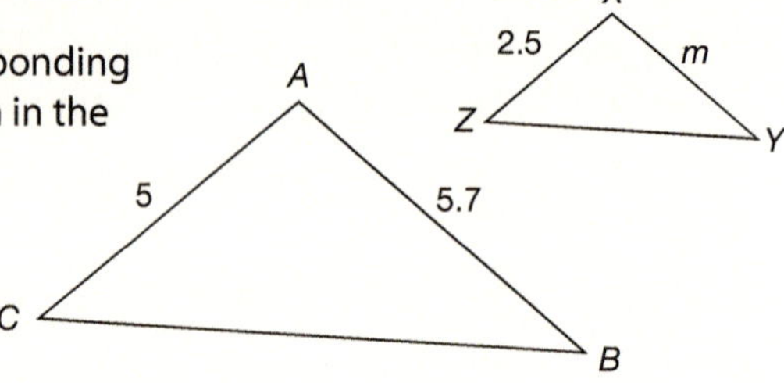

The side lengths of similar figures are proportional, but are the perimeters of similar figures in the same proportion? Example 2 extends our work with the same two similar triangles to determine whether the same ratio applies to the perimeters of these two figures.

Example 2

If segment *BC* measures 7.25, find *n*. Then compare the ratio of the triangles' perimeters to the ratio of their side lengths.

Answer

$\frac{BC}{YZ} = \frac{CA}{ZX}$, so $\frac{7.25}{n} = \frac{5}{2.5}$

$5n = 7.25(2.5)$

$5n = 18.125$

$n = \frac{18.125}{5} = 3.625$

$p_{ABC} = 5.7 + 7.25 + 5 = 17.95$

$p_{XYZ} = 2.85 + 3.625 + 2.5 = 8.975$

$\frac{p_{ABC}}{p_{XYZ}} = \frac{17.95}{8.975} = 2$

$\frac{CA}{ZX} = \frac{5}{2.5} = 2$

1. Set up a proportion to find *YZ*.
2. Apply the Property of Proportions and solve.
3. Determine the perimeters of the triangles.
4. Write the ratio of the perimeters and simplify.
5. Write the ratio of corresponding sides and simplify.

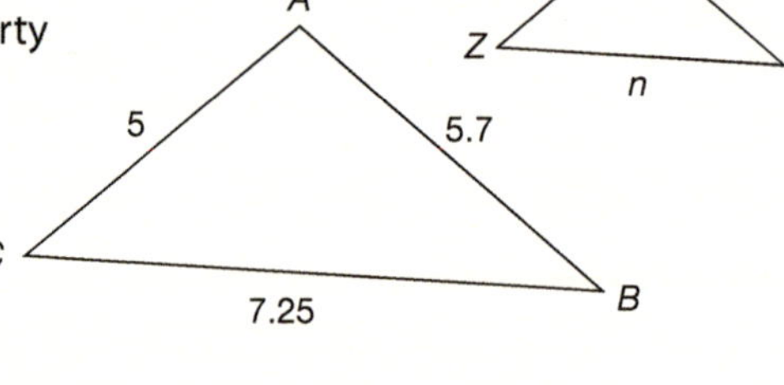

The ratio of the perimeters equals the ratio of corresponding sides.

It can be shown that any ratio of corresponding lengths of similar figures, including their perimeters, is equal to the ratio of their corresponding sides.

The fact that corresponding lengths of similar figures are proportional is often used to determine distances that are difficult to measure directly.

Example 3

Find the height of the flagpole at the Parliament House in Canberra, Australia, if it casts a shadow of 76 ft on level ground at the same time that Sydney, who is 5' 3" tall, casts a shadow of 1' 6" on level ground.

Answer

$5 + \frac{3}{12} = 5.25$ ft

$1 + \frac{6}{12} = 1.5$ ft

1. Draw a picture. Notice that the sun's rays form two similar right triangles. Convert Sydney's height and shadow length to feet.

$\frac{h}{5.25} = \frac{76}{1.5}$

2. Use the similar triangles to write a proportion.

$1.5h = 5.25(76)$

$1.5h = 399$

$h = \frac{399}{1.5} = 266$ ft

3. Apply the Property of Proportions and solve.

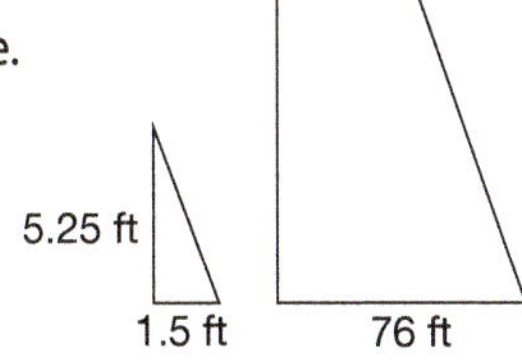

Since it can be proved that the perimeters of two similar figures are in the same ratio as corresponding sides, the question arises as to whether the same is true of areas and volumes. Consider two circles of radius a and radius b. Since the area of a circle is found by the formula $A = \pi r^2$, the areas of the respective circles are πa^2 and πb^2. The ratio of their areas is

$$\frac{\pi a^2}{\pi b^2} = \frac{a^2}{b^2} = \left(\frac{a}{b}\right)^2.$$

That is, the ratio of the areas of any two circles is equal to the square of the ratio of their radii. The same relationship can be shown to hold true for any two similar polygons: The ratio of their areas is equal to the square of the ratio of any two corresponding sides. Suppose two squares have sides in the ratio 5 : 4. Their actual side lengths can be represented by $5x$ and $4x$, and their areas are $(5x)^2$ and $(4x)^2$. So their areas are in the ratio

$$\frac{(5x)^2}{(4x)^2} = \frac{25x^2}{16x^2} = \frac{25}{16} = \frac{5^2}{4^2} = \left(\frac{5}{4}\right)^2.$$

Now consider the volumes of two similar cubes with sides in the ratio $a : b$. Their actual side lengths can be represented by ax and bx, and their volumes are in the ratio

$$\frac{(ax)^3}{(bx)^3} = \frac{a^3x^3}{b^3x^3} = \frac{a^3}{b^3} = \left(\frac{a}{b}\right)^3.$$

In summary, the ratio of the areas of similar figures is equal to the ratio of the corresponding sides squared, and the ratio of the volumes of similar figures is equal to the ratio of the corresponding sides cubed.

Keyword Search

similar triangles

Example 4

Two similar rectangular prisms have corresponding edges in the ratio 2 to 3. The area of the base of the smaller prism (A_S) is 36 units^2, and the volume of the larger prism (V_L) is 1215 units^3.

a. Find the ratio of their areas, and use a proportion to find the area of the larger prism's base, A_L.

b. Find the ratio of their volumes, and use a proportion to find the volume of the smaller prism, V_S.

Answer **a.** $\frac{A_S}{A_L} = \left(\frac{2}{3}\right)^2 = \frac{4}{9}$ — 1. The ratio of the areas is the ratio of corresponding edges squared.

$\frac{36}{A_L} = \frac{4}{9}$ — 2. Write a proportion for the areas of the bases.

$4A_L = 324$ — 3. Apply the Property of Proportions and solve.

$A_L = 81 \text{ units}^2$

b. $\frac{V_S}{V_L} = \left(\frac{2}{3}\right)^3 = \frac{8}{27}$ — 1. The ratio of the volumes is the ratio of corresponding edges cubed.

$\frac{V_S}{1215} = \frac{8}{27}$ — 2. Write a proportion for the volumes of the prisms.

$27V_S = 9720$ — 3. Apply the Property of Proportions and solve.

$V_S = 360 \text{ units}^3$

Similar figures are applied in two-dimensional *scale drawings* (such as maps, blueprints, and diagrams) and three-dimensional *models*. A scale drawing or model is one whose dimensions are all in the same ratio when compared to the actual object. This ratio is called the *scale*.

Definitions

The **scale** of a drawing or model is the ratio of the dimensions of the drawing or model to the corresponding dimensions of the actual object.

In a **scale drawing** or **model**, all the dimensions are in the same scale or ratio to the actual dimensions of the object.

Suppose you want to make a scale drawing of the floor of a rectangular building that is 50 ft long and 20 ft wide on an 11 in. × 17 in. piece of paper. If you choose a scale of 1 in. : 5 ft, then you can use the following proportions to determine the dimensions to use on the paper.

$$\frac{1 \text{ in.}}{5 \text{ ft}} = \frac{x}{50 \text{ ft}} \qquad \frac{1 \text{ in.}}{5 \text{ ft}} = \frac{y}{20 \text{ ft}}$$
$$5x = 50 \qquad 5y = 20$$
$$x = 10 \text{ in.} \qquad y = 4 \text{ in.}$$

The drawing would be 10 in. × 4 in.

Can a scale of 1 in. : 3 ft be used?

$$\frac{1 \text{ in.}}{3 \text{ ft}} = \frac{x}{50 \text{ ft}} \qquad \frac{1 \text{ in.}}{3 \text{ ft}} = \frac{y}{20 \text{ ft}}$$
$$3x = 50 \qquad 3y = 20$$
$$x \approx 16.7 \text{ in.} \qquad y \approx 6.7 \text{ in.}$$

The dimensions would be 16.7 in. × 6.7 in., leaving $\frac{17 - 16.7}{2} = 0.15$ in. margins the long way.

Calculating $\frac{1 \text{ in.}}{5 \text{ ft}} = \frac{1 \text{ in.}}{60 \text{ in.}} = \frac{1}{60}$ shows that the first drawing will be one-sixtieth of the size of the actual building. Notice that when the scale uses the same unit for the drawing and the actual object, the units cancel and the scale is expressed without any units.

Example 5

A toy manufacturer makes scale model tractors with a 1 : 8 scale. If a model has dimensions of 17.5 in. × 11.25 in. × 11.5 in., what are the corresponding actual dimensions of the tractor?

Answer

$\frac{17.5}{x} = \frac{1}{8}$ — 1. Set up a proportion of the model size to the actual size (length = 17.5 in.).

$x = 17.5(8) = 140$ — 2. Cross-multiply and solve.

$\frac{11.25}{x} = \frac{1}{8}$ — 3. Set up a proportion and find the actual width.

$x = 11.25(8) = 90$

$\frac{11.5}{x} = \frac{1}{8}$ — 4. Set up a proportion and find the actual height.

$x = 11.5(8) = 92$

The actual dimensions are 140 in. × 90 in. × 92 in., or approximately 11.7 ft × 7.5 ft × 7.7 ft.

A. Exercises

1. Given that $\triangle RSB \sim \triangle ZQM$, list the three pairs of congruent angles and state the three equal ratios of corresponding sides.
2. Given that $\triangle PQR \sim \triangle STU$, $m\angle P = 40°$, and $m\angle Q = 82°$, find $m\angle U$.
3. Given that $\triangle ABC \sim \triangle DEF$, find x and y.

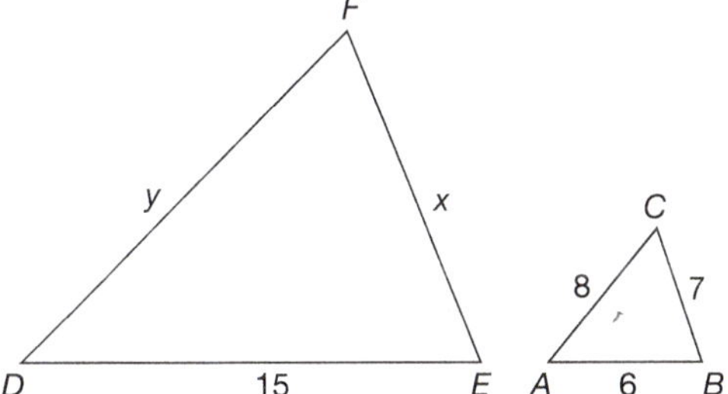

4. Given that $\triangle JKL \sim \triangle XYZ$, find a and b.

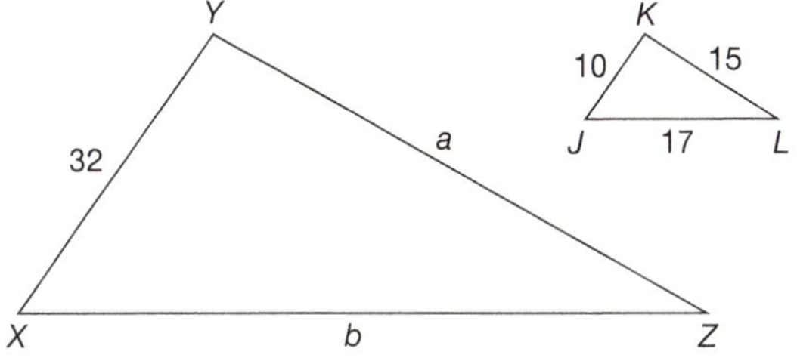

5. Using the ratio of the sides and the perimeter of $\triangle ABC$, write and solve a proportion to find the perimeter of $\triangle DEF$. Check your work using your answers from exercise 3.
6. Using the ratio of the sides and the perimeter of $\triangle JKL$, write and solve a proportion to find the perimeter of $\triangle XYZ$. Check your work using your answers from exercise 4.

Use proportions to solve.

7. A map of Kentucky has a scale of 1 cm : 10 mi. If Lexington and Louisville are 7 cm apart on the map, how far apart are the two cities?
8. The distance between Nashville and Knoxville is about 160 mi. How far apart are the cities on a map that has a scale of 1 in. : 10 mi?
9. The cities of Allentown and Philadelphia, which are about 50 mi apart, are drawn 2.5 in. apart on a map. Determine the scale of the map.
10. A map of the state of Washington shows the cities of Spokane and Kennewick separated by 4 in. when they are actually about 120 mi apart. Determine the scale of the map.
11. A scale drawing of a rectangle is 7 in. wide and 11.5 in. long. If it is drawn to a scale of 1 in. : 20 ft, what are the actual dimensions of the rectangle?
12. A scale drawing is to be made of a rectangular plot of land that is 120 yd × 53 yd. If it is drawn to a scale of 1 in. : 5 yd, what are the dimensions of the drawing?
13. A picture of a beetle 3 in. long and 2 in. wide is to be enlarged using a scale of 1 : 4. What should the dimensions of the enlargement be?

14. A picture of a cougar is $\frac{3}{4}$ in. long, and the legend below it says 1 : 80. How long is the actual cougar?

15. A toy soldier is to be manufactured with a scale of 1 : 30. How many inches tall should the toy be if it models a soldier 6 ft tall?

16. A model of a sports car is 7.5 in. long. If the scale of the model is 1 : 25, how many feet long is the sports car?

B. Exercises

17. Given that $\triangle MNO \sim \triangle RTS$, find MN and TS.

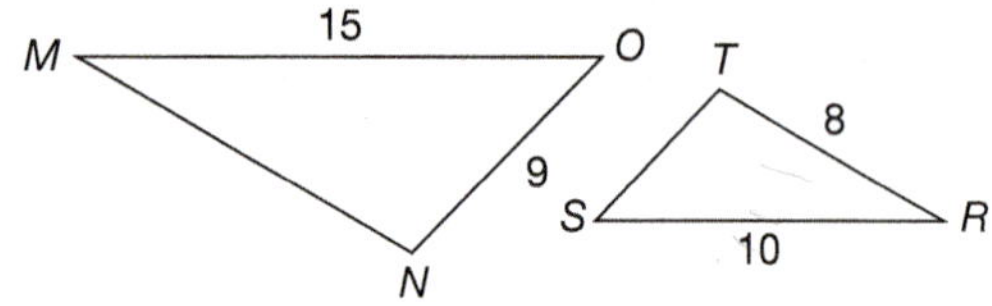

18. Given that $ABCDE \sim FGHIJ$, find DE and BC.

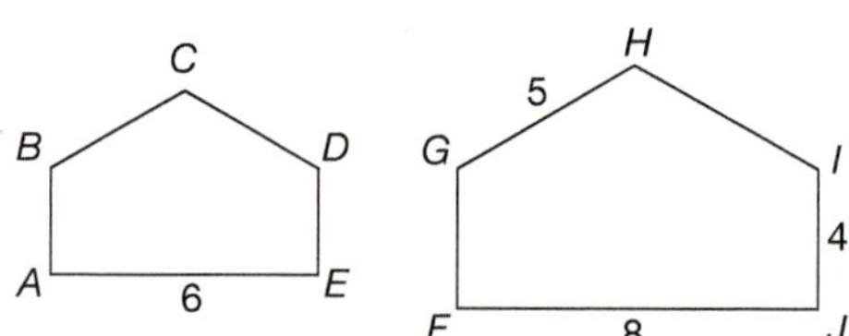

19. State the ratio of the area of $\triangle MNO$ to the area of $\triangle RTS$. Given that the area of $\triangle MNO$ is 54 units^2, write and solve a proportion to find the area of $\triangle RTS$.

20. State the ratio of the area of $ABCDE$ to the area of $FGHIJ$. Given that the area of $FGHIJ$ is 44 units^2, write and solve a proportion to find the area of $ABCDE$.

21. Given that the solids are similar, find h.

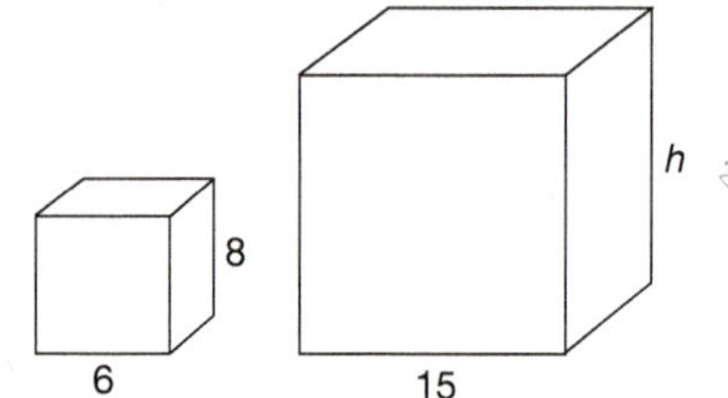

22. Given that the solids are similar, find s and l.

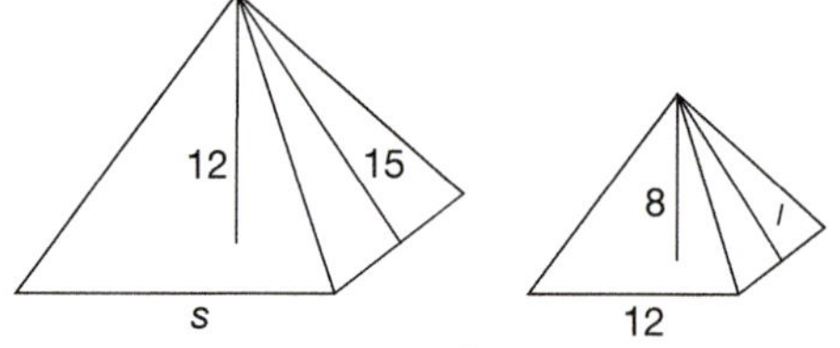

23. State the ratio of the volume of the smaller prism to the volume of the larger prism. Given that the volume of the smaller prism is 192 units^3, write and solve a proportion to find the volume of the larger prism.

24. State the ratio of the volume of the larger pyramid to the volume of the smaller pyramid. Given that the volume of the larger pyramid is 54 units^3, write and solve a proportion to find the volume of the smaller pyramid.

Use proportions to solve.

25. A triangle has sides of 5, 9, and 11 units. A similar triangle's shortest side is 12 units. To the nearest tenth, find the length of the second triangle's two other sides and perimeter.

26. The areas of two similar plane figures are in the ratio 36 : 25. What is the ratio of their sides? What is the ratio of their perimeters?

27. The ratio of the edges of two similar cubes is 2 : 5. What is the ratio of the areas of two corresponding faces? What is the ratio of their volumes?

28. The faces of two similar cubes have areas in the ratio 9 : 4. What is the ratio of the lengths of the cubes' edges? What is the ratio of the cubes' volumes?

29. The ratio of the sides of two similar pentagons is 5 : 7, and the area of the smaller one is 75 ft^2. Find the area of the larger one.

30. The ratio of the sides of two similar rectangles is 5 : 2, and the area of the larger one is 400 ft^2. What is the area of the smaller rectangle?

31. The ratio of the areas of two similar triangles is 25 : 16, and one side of the larger triangle is 200 ft. Find the corresponding side of the smaller triangle.

32. The ratio of the edges of two similar pyramids is 10 : 1, and the volume of the smaller one is 5 ft^3. Find the volume of the larger one.

33. Two similar cylinders have radii of 2 and 7 units respectively, and the smaller one has a volume of 24π units^3. What is the volume of the larger one?

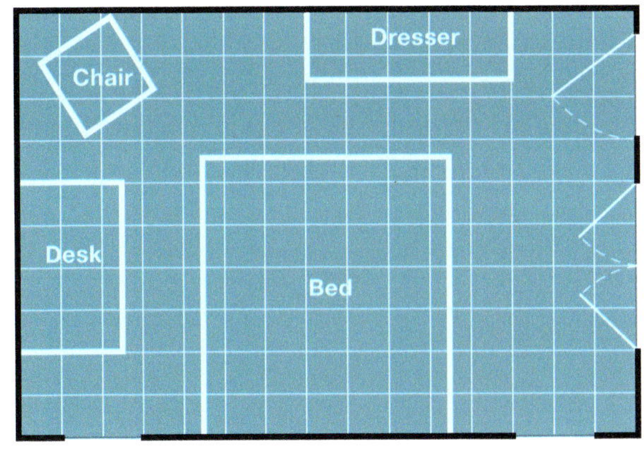

Use the scale drawing to the right, in which each gridline represents 1 ft, to determine the following dimensions in inches.

34. the width of each of the double closet doors

35. the length and width of the bed

36. the length and the width of the desk

37. the width of the single door

C. Exercises

Use proportions to solve.

38. Use a sheet of graph paper to make a scale drawing of your classroom or your bedroom. Include the location of doors, windows, and furniture. Be sure to state the scale of your drawing.

39. The sides of a golden rectangle are in the ratio $1 : \frac{1 + \sqrt{5}}{2}$. Find the length of a golden rectangle whose width is 2.5 units. Round to the nearest tenth.

40. Find the width of a golden rectangle whose length is 8 units. Round to the nearest tenth.

Dominion Modeling

The possible gasoline savings for two vehicle trade-ins can be compared by examining the ratio $\frac{n - d}{nd}$ (which gives the number of gallons per mile saved) for each trade-in.

41. What fraction of the original gasoline consumption is saved for each proposed trade-in?
   - **a.** SUV (13 mi/gal) for work van (10 mi/gal)
   - **b.** hybrid (60 mi/gal) for compact car (30 mi/gal)
   - **c.** new sedan (22 mi/gal) for old sedan (18 mi/gal)

42. Compare these fractions to determine which trade-in will result in the greatest fuel savings and which will result in the least fuel savings.

CUMULATIVE REVIEW

Classify each number as natural, whole, integer, rational, or irrational. Give the most specific set. [1.1]

43. 1.35

44. $\sqrt{8}$

True or false [1.1]

45. $\mathbb{Q} \cup \mathbb{Q}' = \varnothing$

46. $\mathbb{W} \subseteq \mathbb{Z}$

A segment has endpoints with the given coordinates. Find the length of the segment and the midpoint of the segment. [1.2]

47. 12 and -22

48. 2.8 and 3.5

Evaluate. [1.3–1.4]

49. $-\frac{3}{4} + \frac{1}{6}$

50. $3\frac{2}{7} + 2\frac{5}{6}$

51. $\frac{7}{12} - \left(-\frac{1}{5}\right)$

52. $2\frac{5}{6} - 3\frac{1}{7}$

3.4 The Percent Equation

Approximately what percent of the seats are filled?

Percentages are one of the most commonly used mathematical concepts. You have experienced them in the classroom, at stores, and in restaurants. They are extensively used in the scientific and business communities.

Definition

A **percent** is a ratio of a number to one hundred. The symbol for percent is %.

Percent can be expressed with the common percent symbol (40%), as a fraction $\left(\frac{40}{100} = \frac{2}{5}\right)$, or as a decimal (0.4).

Example 1

Write 82% as a fraction and a decimal.

Answer $82\% = \frac{82}{100} = \frac{41}{50}$

1. Express percent as a ratio per hundred and reduce to lowest terms.

$82\% = 0.82$

2. Move the decimal point two places to the left.

Example 2

Write $\frac{3}{5}$ as a decimal and a percent.

Answer $\frac{3}{5} = 0.6$

1. Convert $\frac{3}{5}$ to its decimal form by dividing.

$0.6(100\%) = 60\%$

2. Convert 0.6 to its percent form by multiplying by one in the form of 100%.

Example 3

Write 1.55 as a percent and a fraction.

Answer $1.55(100\%) = 155\%$

1. Convert 1.55 to its percent form by multiplying by one in the form of 100%.

$155\% = \frac{155}{100} = \frac{31}{20}$

2. Express percent as a ratio per hundred and reduce to lowest terms.

Problems involving percents can be translated into the percent equation. This single equation can be used to find a percent, to find the part of the whole represented by a given percent, or to find the original amount when you are given a percent and the part the percent represents.

The Percent Equation

The percent of the whole is equal to the part: *percent* × *whole* = *part.*

Example 4

What percent of 85 is 17?

Answer	$p \times 85 = 17$	1. Translate the question into the percent equation. Note that "of" is translated as "×" and "is" is translated as "=."
	$p = \frac{17}{85} = \frac{1}{5} = 0.2$	2. Solve.
	$p = 0.2(100\%) = 20\%$	3. Convert the decimal to a percent.
Check	20%, or $\frac{1}{5}$, of 85 is 17.	4. Check whether the answer is reasonable.

Example 5

What is 45% of 236?

Answer	$r = 0.45(236)$	1. Translate the question into the percent equation.
	$r = 106.2$	2. Evaluate.
Check	106 is a little less than half of 236.	3. Use estimation to check whether the answer is reasonable.

Example 6

78 is 130% of what number?

Answer	$1.3w = 78$	1. Translate the question into the percent equation.
	$w = \frac{78}{1.3} = 60$	2. Solve.
Check	$33\% \approx \frac{1}{3}$ and $\frac{1}{3}(60) = 20$, so 133% would be about 80.	3. Use estimation to check whether the answer is reasonable.

Many practical problems involve percents. Remember to convert a percent to an equivalent decimal or fraction when writing an equation to solve the problem. This problem-solving strategy will enable you to successfully solve a multitude of real-life problems.

Example 7

How many gallons of 2% milk does it take to yield 1 gal of butterfat?

Answer	2% of the milk is butterfat.	1. Carefully read the problem and identify the key facts.
	$0.02m = 1$	2. Write an equation.
	$m = \frac{1}{0.02} = 50$ gal	3. Solve.

50 gal of 2% milk contains 1 gal of butterfat.

Example 8

Miss Grandy received 17,490 of the 31,800 votes cast in the election. What percent of the votes did she receive? Did she win the election?

Answer

What percent of the vote did Miss Grandy receive?	1. Carefully read the problem to determine what is being asked.
$p \times 31{,}800 = 17{,}490$	2. Write an equation.
$p = \frac{17{,}490}{31{,}800} = 0.55$	3. Solve.
$p = 0.55(100\%) = 55\%$	4. Convert the decimal to a percent.
She received 55% of the vote, winning the election.	5. Interpret the answer in terms of the question.

A. Exercises

For each percent, fraction, or decimal, state the equivalent expressions in the other two forms.

1. 69% **2.** 187% **3.** 75% **4.** 14%

5. 0.25 **6.** 0.57 **7.** 3.5 **8.** $\frac{3}{5}$

9. $\frac{15}{2}$ **10.** $\frac{1}{8}$

Write the appropriate percent equation and solve. Round decimals to the nearest hundredth and percents to the nearest whole percent.

11. Find 84% of 150.

12. 18 is what percent of 76?

13. 58% of what number is 48?

14. What is 36% of 1980?

15. Find 50% of $\frac{3}{4}$.

16. 482 is what percent of 24?

17. 175% of 18,900 is what number?

18. 80% of what number is 52?

19. 10% of what number is 50?

20. 1600 is what percent of 3000?

B. Exercises

Write the appropriate percent equation and solve.

21. Mrs. Gardenghi noticed that there are 12 boys in her class of 30. What percent of her class are boys? What percent of her class are girls?

22. $\angle A$ in $\triangle ABC$ measures 80% of the total angle measures in the triangle. What is the measure of $\angle A$?

23. If 28% of the 842 respondents to a survey preferred candidate Conley over his opponent candidate Brewer, how many chose Conley?

24. Mrs. Chang knows that she spends $180 for groceries each week for her family. If this is 12% of the family's total income for the week, how much do they earn each week?

25. After four weeks of training at his new job, Brian will get a 7% increase in his hourly wage. If his increase amounts to $0.60/hr, how much per hour is he getting paid during training?

26. The school board requires that 65% of the faculty have a master's degree. If 18 of the 26 faculty members have a master's degree, what percent of the faculty hold a master's degree? Has the requirement been fulfilled?

27. Lydia got 64 out of 75 questions correct on her history exam. What percent did she earn?

28. Coach Wexler's teams have won 70% of the soccer games that he has coached. If he has a total of 49 victories in his career, how many games has he coached?

29. If the county allots 83% of its budget for capital projects and $17.1 million is going toward capital projects, what is the county's total budget?

30. Jared has $50 to spend on some new clothes for school. If the sales tax rate in his state is 6% and the pretax total comes to $48.92, determine both the sales tax and the total due. Will he have enough money to make the purchase?

The American Council on Exercise categorizes men's fitness levels according to their percent body fat in the following way.

Category Name	Percent Body Fat
Essential Fat	2–5%
Athletes	6–13%
Fitness	14–17%
Acceptable	18–25%
Obesity	26% or more

31. If Marcos falls in the Acceptable category and weighs 185 lb, how much of his weight is in body fat?

32. The exercise physiologist determines that Dave, who weighs 164 lb, has 20 lb of body fat. Determine his percentage of body fat and the category in which he falls.

33. A person's lean body mass (LBM) is his weight minus the weight of his body fat. What is Jesse's LBM if he weighs 195 lb and falls at the top of the Fitness category?

The USDA recommends that a healthy adult consume 10%–35% of his calories from protein, 20%–35% from fat, and 45%–65% from carbohydrates. Find the distribution for each of the following people and state whether each one is eating a healthy diet based on these recommendations.

34. Kara eats a diet of 160 calories from protein, 440 calories from fat, and 1400 calories from carbohydrates.

35. Eli eats a diet of 288 calories from protein, 504 calories from fat, and 1000 calories from carbohydrates.

36. Aaron eats a diet of 171 g of protein, 65 g of fat, and 145 g of carbohydrates. Carbohydrates and proteins provide 4 cal/g while fats contain 9 cal/g.

C. Exercises

Solve.

37. In February 2009 sales of passenger cars, buses, and trucks in China increased by 25% compared to the previous month. The sales in February were 827,600 vehicles. How many vehicles were sold in China in January of 2009?

38. The population of the United States in 2010 was approximately 310 million. If 74.5 million Americans had high blood pressure in 2010, what percent of Americans (to the nearest whole percent) had high blood pressure?

39. In the 2004 national election approximately 122,295,000 Americans voted, representing 55.3% of America's voting-age population. In 2008, the percentage increased to 57.4%, or about 132,619,000 Americans. What was the difference (to the nearest thousand) in the voting-age population of America between these four years?

40. The butcher sells 80% lean ground beef at $2.80/lb, 90% lean ground beef at $3.00/lb, and 95% lean ground beef at $3.25/lb. What is the cost per pound of the totally lean percentage of the beef in each case?

Dominion Modeling

A proportion can be used to find the mileage of a new vehicle that will produce the same gasoline savings as another trade-in.

Example: Determine the mileage of a replacement sedan for the old sedan that would produce the same gasoline savings as trading in the work van for an SUV.

Solve the proportion $\frac{n-18}{18n} = \frac{13-10}{13(10)}$.

$$\frac{n-18}{18n} = \frac{3}{130}$$
$$(n-18)(130) = 18n(3)$$
$$130n - 2340 = 54n$$
$$76n = 2340$$
$$n \approx 30.8 \text{ mi/gal}$$

41. Determine the mileage of a replacement vehicle for the compact car that would produce the same gasoline savings as trading in the work van for an SUV.

42. If we want to match the fuel savings obtained by trading in the compact car for a hybrid, what must the mileage be for a car replacing the (a) work van and (b) old sedan?

43. If we want to match the fuel savings obtained by trading in the old sedan for a new sedan, what must the mileage be for a car replacing the (a) work van and (b) compact car?

Summarize your results in a table similar to the one below.

Mileage of Vehicle Replacing the Following			Producing the Same Savings as the Following
Work Van	Old Sedan	Compact Car	
NA	30.8		SUV for work van
		NA	hybrid for compact car
	NA		new sedan for old sedan

CUMULATIVE REVIEW

Evaluate. Show each step in your work. [1.8]

44. $-2 + [5 + 2(-7)] - 4$

45. $3^0 - 2|5 - 3| + 2^3$

46. $\frac{1}{2}\left[\frac{2}{3} - \frac{5}{6}\left(\frac{3}{4}\right)\right]$

47. $-0.2(3.8) + 1.15$

Solve. [3.2]

48. $\frac{7}{12} = \frac{x}{9}$

49. $\frac{7}{9} = \frac{x}{12}$

Use the similar pentagons for the following exercises. [3.3]

50. Write and solve a proportion to find x.

51. Find the area of the larger pentagon if the area of the smaller pentagon is 44 in.^{2}.

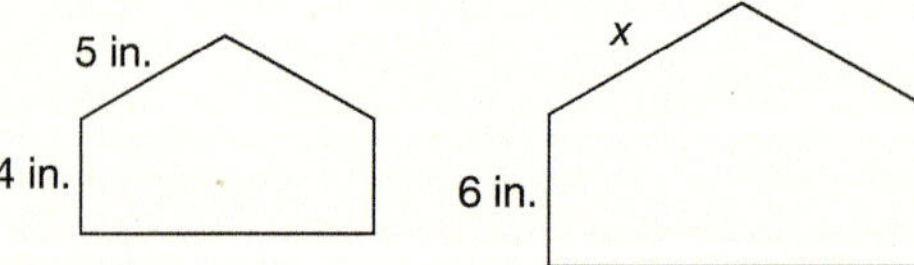

Solve each literal equation for x. [3.1]

52. $pq + \frac{2x}{d} = 4b$

53. $ax - \frac{2b}{c} = 3$

3.5 Percent Change and Error

A new car's depreciation in value can be expressed as a percent.

If the price of a product increased or decreased by \$10, would that be a significant change? Knowing the amount of change alone is not sufficient to answer this question. You must compare the change with the original amount. An increase of \$10 on a \$1000 computer is barely noticeable, while an increase of \$10 on a \$5 notebook would be considered outrageous. For this reason, the change in a quantity is often described using the *percent change*, the ratio of the change in value to the original value expressed as a percent.

Definition

Percent change $= \dfrac{\text{amount of change}}{\text{original amount}} \times 100\%$.

The percent change for the computer would be $\frac{10}{1000} \times 100\% = 1\%$, while the percent change for the notebook would be $\frac{10}{5} \times 100\% = 200\%$.

Example 1

A baker decided to raise the price of a loaf of his specialty bread from \$3 to \$4. This new price resulted in a large drop in sales, so he reduced the price back to \$3. Determine both the percent increase and the percent decrease in the price of bread during these changes.

Answer

a \$1 increase and a \$1 decrease — 1. Find the amount of change.

percent increase $= \frac{1}{3} \times 100\% \approx 33\%$ — 2. Divide the change by the original amount and convert to a percent.

percent decrease $= \frac{1}{4} \times 100\% = 25\%$ — 3. Divide the change by the original amount and convert to a percent.

Notice that the percents are not the same since they are based on different original numbers.

Example 2

Last year 1056 people voted in the city election. This year only 912 voted. Find the percent decrease, to the nearest tenth of a percent, in votes cast.

Answer

amount of change $= 1056 - 912 = 144$ — 1. Subtract to find the amount of change.

percent decrease $= \frac{144}{1056} \times 100\% \approx 13.6\%$ — 2. Divide the change by the original amount and convert to a percent.

13.6% fewer votes were cast this year than last year.

When a quantity increases by a given percent, the new amount can be calculated by adding the increase to the original amount. When a \$50 pair of shoes is purchased, a 5% sales tax adds \$2.50 to the cost, for a total bill of \$52.50. The total can be calculated directly using the fact that the total cost is 105% of the cost of the item. In general, an increase of x% implies that the new amount is $(100 + x)$% of the original. Similarly, a decrease of x% implies that the new amount is $(100 - x)$% of the original.

Example 3

If the city of Woodridge is projected to grow by 7% over the next decade and its current population is 35,000, how many people should be planned for at the end of the next ten years?

Answer

$n = 1.07(35{,}000) = 37{,}450$ — A 7% increase means that the new population will be 107% of the current population.

The city should plan for a population of 37,450.

Example 4

Through dedicated training, a runner was able to decrease his time in the marathon by 13% to a time of 2 hr 47 min. What was his time previously?

Answer

2 hr 47 min = 167 min — 1. Convert the time to a consistent unit.

$167 = 0.87r$ — 2. A 13% decrease means that his new time is 87% of his original time.

$r = \frac{167}{0.87} \approx 192$ — 3. Solve.

His previous time was 192 min = 3 hr 12 min.

The accuracy of an experiment is often judged by comparing an experimentally determined value to the correct known value. An experiment producing a result that is off by 0.5 g when the desired result is 50 g is much different from an experiment that is off by 0.5 g when the desired result is 5 g. Using the percent change formula with your *experimental error*, the difference between the experimental and the known values, produces a better model by which to judge the quality of an experiment.

Definition

$$\textbf{Percent error} = \frac{|\text{experimental value} - \text{known value}|}{\text{known value}} \times 100\%.$$

Example 5

In chemistry class, Jonah determines the density of copper to be 8.83 g/cm^3, while the accepted value for copper's density (near room temperature) is 8.96 g/cm^3. Determine both Jonah's experimental error and his percent error.

Answer

experimental error $= 8.83 - 8.96 = -0.13$ g/cm^3 — 1. Determine the amount of Jonah's experimental error.

percent error $= \frac{|-0.13|}{8.96} \times 100\%$ — 2. Apply the formula to determine the percent error.

$= \frac{0.13}{8.96} \times 100\% \approx 1.5\%$

His result is 0.13 g/cm^3 below the accepted value, which represents about 1.5% error.

How can the results of an experiment be verified when there is no known value to compare the results to? *Percent difference* is often used to indicate the precision of repeated measurements when the outcome is expected to be the same. It uses the average of the results in place of a known value.

Definition

Percent difference $= \dfrac{|\text{first value} - \text{second value}|}{\text{average of the values}} \times 100\%.$

Example 6

Suzanne is asked to determine the volume of a cylinder. Using water displacement, she arrives at an answer of 15.64 cm^3; then she measures the cylinder's dimensions and calculates a volume of 14.01 cm^3. Find the percent difference of her results.

Answer

$\dfrac{15.64 + 14.01}{2} = 14.825$

1. Find the average of the two values.

$\text{percent difference} = \dfrac{|15.64 - 14.01|}{14.825} \times 100\%$

$= \dfrac{|1.63|}{14.825} \times 100\% \approx 11.0\%$

2. Apply the formula to determine the percent difference.

A. Exercises

Find the percent change.

1. from 80 to 100
2. from 200 to 100
3. from 60 to 450
4. from 200 to 0.5

Find the new amount.

5. increase 700 by 40%
6. increase $20 by 70%
7. decrease 600 by 55%
8. decrease 1400 by 9%
9. increase 600 by 105%
10. increase 90 by 250%

Find both the experimental error and the percent error.

11. experimental value: 165; known value: 150
12. experimental value: 2.4; known value: 2.5

Find the percent difference.

13. 80 and 100
14. 200 and 100

B. Exercises

Complete the following table. Round percent change to the nearest tenth of a percent.

	Original Amount	Change	Percent Change	New Amount
15.	68	add 15		
16.	590	subtract 40		
17.		add 20		135
18.		subtract 33		87
19.	125			86
20.	163			210
21.	274		53% increase	
22.	70		22% decrease	
23.			75% decrease	350
24.			150% increase	48
25.		subtract 25	2% decrease	
26.		add 0.3	125% increase	

Write an equation and solve.

27. If a student uses a protractor to measure an angle of a triangle that is known to be 60° and gets a value of 58°, what is his percent error?
28. Each member of a math class was assigned the task of experimentally determining the value of π by measuring the circumference and diameter of a circular object and calculating the ratio $\frac{c}{d}$ to two decimal places. Find the percent error of a student answer of 3.10.
29. Calculate both the experimental error and the percent error if a student determines water's density to be 1.13 g/mL when its actual density is 1.00 g/mL.
30. Mary experimentally determines the boiling point of ethanol to be 72.2°C when it is actually 78.2°C. Determine both the experimental error and the percent error for Mary's experiment.
31. A student determines that 4.0 g of helium gas occupies a volume of 22.75 L. He repeats the experiment and determines the volume to be 22.90 L. Find the percent difference of his results.
32. Using litmus paper, Georgia recorded the pH of a solution as 7.6. The pH meter gave a reading of 7.3. Find the percent difference of these two measurements.

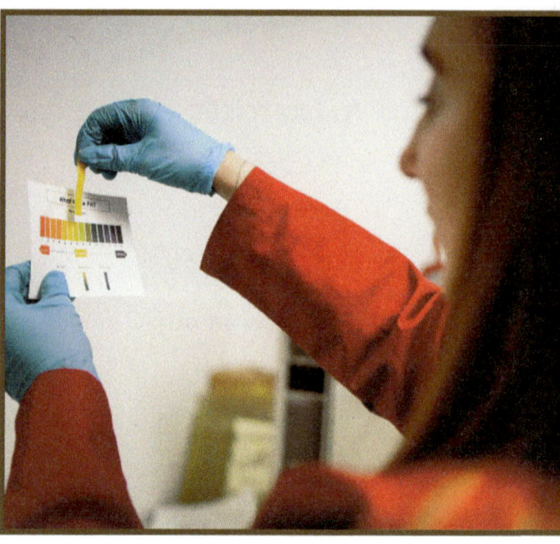

33. **a.** Increase 8 by 200%.

    b. Increase 8 to 200% of itself.
34. If a number x is increased by 400%, is the new number $4x$? Explain.
35. What is the result of decreasing 600 by 105%? Explain.

C. Exercises

36. If the attendance at the second game increased 50% from the first game and the third game's attendance increased 50% from the second game, what was the percent increase from the first game to the third game?
37. What percent change results from the following?

    a. increasing a number by 50% and then decreasing the result by 50%

    b. decreasing a number by 50% and then increasing the result by 50%
38. Will increasing a number by a given percent and then decreasing the result by the same percent yield a result that is greater than, less than, or equal to the original number? Explain.
39. If a number is increased by 25%, by what percent must the result be decreased in order to equal the original number?

Dominion Modeling

40. While a hybrid may not be the best vehicle for an owner's needs, the greatest fuel savings possible would obviously be obtained by trading in the work van getting 10 mi/gal for a hybrid getting 60 mi/gal. Determine the number of gallons saved over 15,000 mi by making this trade-in.
41. For each proposed trade-in, assuming a yearly total of 15,000 mi, use current gas prices to determine the average monthly savings.

    a. hybrid (60 mi/gal) for work van (10 mi/gal)

    b. SUV (13 mi/gal) for work van (10 mi/gal)

    c. hybrid (60 mi/gal) for compact car (30 mi/gal)

    d. newer sedan (22 mi/gal) for older sedan (18 mi/gal)

CUMULATIVE REVIEW

Solve. [2.5–2.6]

42. $1.2x + 1.58 = 2.3$ **43.** $6(y - 3) = 24$ **44.** $-2z + 1 - z - 8 = -24$

Write the appropriate percent equation and solve. [3.4]

45. What percent of 20 is 30? **46.** Find 20% of 30. **47.** 20% of what number is 30?

Answer the following exercises about similar figures. [3.3]

48. Two common sizes of index cards are 3" × 5" and 4" × 6". Are cards of these dimensions similar? Explain.

49. Find the maximum dimensions of a 4" × 6" picture enlarged to fit on an 8.5" × 11" sheet of paper.

50. What two properties do you use to clear an equation of fractions? [2.8]

51. Clear the following equation of fractions and decimals and then solve. [2.8]

$\frac{1}{3}(0.05)x + \frac{1}{4}(0.075)(10{,}000 - x) = 172.5$

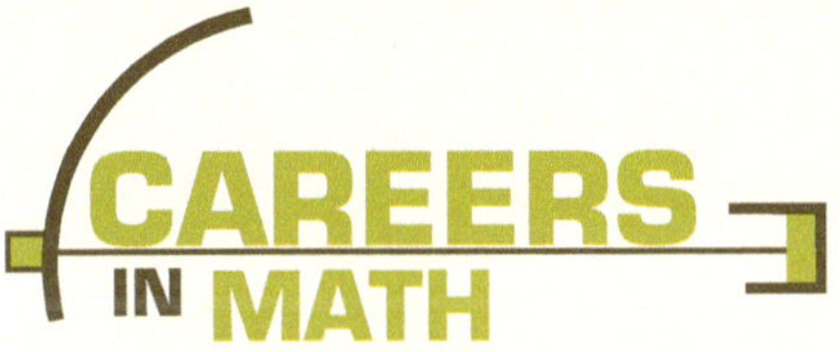

Actuary

The actuarial profession has been consistently rated as one of the top careers in the United States. An actuary evaluates the likelihood of future events and works to decrease the impact of undesirable events that do occur. The majority of actuaries are employed by insurance companies to determine risk and set premiums, but their job responsibilities are often much broader than that. An actuary deals with risk management, especially as it impacts the financial well-being of a business. Actuaries often help companies build strategies to improve their bottom line.

An actuary must have good math and analytical skills. Many colleges and universities offer an actuarial science major in their math department. An actuary determines the likelihood of uncertain events, such as the probability that a thirty-year-old man will live to be sixty-five.

James reminds Christians of the uncertainty of future events and instructs even businessmen to submit to whatever God wills (James 4:13–15). Nonetheless, the Bible teaches that the way of wisdom includes making prudent plans for the future (Prov. 22:3; 27:12).

Government and insurance industry opportunities for actuaries currently abound and are predicted to increase. In addition to working with insurance plans, pensions, and social welfare programs, actuaries work to design creative ways to reduce the likelihood of undesirable events. The Christian actuary can glorify God by using his God-given abilities to anticipate and plan for the difficult circumstances that may occur in the lives of others.

3.6 Money

The percent equation has frequent financial applications. God's Word contains many financial principles that we should implement in our lives. God states in 1 Timothy 6:10 that "the love of money is the root of all evil," but notice that the sin is not in possessing money but in loving it or coveting it. Money is a necessary commodity that should be used to serve God, not to hoard for ourselves (Matt. 6:19–21, 24; Luke 12:15). God's blessing is on those who give generously to Him and to those in need (Prov. 19:17; Mal. 3:8–11; Luke 6:38; 2 Cor. 9:6–7). The Bible also gives us guidance regarding earning interest and creating debt. In Jesus's parable in Luke 19, the nobleman asks the wicked servant why he did not put the money in the bank to earn interest (v. 23). But God also warns that when you borrow money, you become a servant to the lender (Prov. 22:7). An understanding of the proper use of money is important to living the godly life that is expected of a Christian.

Percentages are often used to determine retail and sale prices. After purchasing merchandise from a manufacturer at a given price (the *cost*), a store owner must determine the additional amount (the *markup*) he must add to arrive at the *retail price* he will charge the customer. (Markup is necessary so the merchant can cover his other costs of doing business and make a profit.) This can be expressed mathematically with the equation

$$\textit{cost} + \textit{markup} = \textit{retail price}.$$

The markup is a dollar amount and is often expressed as a percent of the cost, or *markup rate*, using the percent equation

$$\textit{markup} = \textit{markup rate} \times \textit{cost}.$$

Merchants frequently offer customers a *discount* off the retail price as an incentive for them to purchase goods. Retailers often advertise the *discount rate* (the "percent off"), leaving it to the customer to calculate the discount and the *sale price* using the following formulas:

$$\textit{discount} = \textit{discount rate} \times \textit{retail price}$$

and

$$\textit{retail price} - \textit{discount} = \textit{sale price}.$$

Definitions

The **cost** is the amount a merchant pays for the merchandise he will sell.

The **markup** is the amount the merchant adds to his cost to arrive at the retail price.

The **markup rate** is the amount of markup on an item as a percent of its cost.

The **retail price** is the merchant's cost plus markup; it is the regular amount a merchant asks his customer to pay for merchandise.

The **discount rate** is the percent by which the retail price is reduced.

The **discount** is the amount by which the retail price is reduced.

The **sale price** is the customer's cost after the retail price has been discounted.

Example 1

Tom & Harry's Sporting Goods marks up all clothing 75%. If their cost for a shirt is $15, what retail price will Richard have to pay?

Answer

markup = markup rate × cost
= 0.75(15) = 11.25

1. Use the appropriate formula to find the markup.

cost + markup = retail price
15 + 11.25 = $26.25

2. Use the appropriate formula to find the retail price.

By noting that the markup rate is the percent increase in the cost of the item, we can use a more direct calculation to solve the problem above.

$$\begin{aligned} \textit{retail price} &= \textit{cost} + 0.75(\textit{cost}) \\ &= (1 + 0.75)(\textit{cost}) \\ &= 1.75(\textit{cost}) \\ &= 1.75(15) = \$26.25 \end{aligned}$$

Example 2

Determine the retail price of a tennis racquet if Paulo saved $52.50 by buying it during the 30% off sale. How much did he pay for the racquet on sale?

Answer

discount rate × retail price = discount

1. Use the appropriate formula to find the retail price.

$0.3r = 52.50$

$r = \frac{52.5}{0.3} = \$175$

2. Write an equation and solve. The dollar amount he saved is the discount.

sale price = 175 − 52.50 = $122.50

3. Find the sale price by subtracting the discount from the retail price.

In Example 2, how would you find the retail price if you were given the discount rate (30%) and the sale price ($122.50)?

$$\begin{aligned} \textit{sale price} &= \textit{retail price} - \textit{discount} \\ &= \textit{retail price} - (\textit{discount rate})(\textit{retail price}) \\ &= \textit{retail price} - 0.3(\textit{retail price}) \\ &= 0.7(\textit{retail price}) \end{aligned}$$

This confirms what you probably already knew, that "saving" 30% means you are actually paying a sale price that is 70% of the retail price.

Example 3

Determine the retail price of a pair of running shoes and the discount if they are 40% off and have a sale price of $54.

Answer

sale price = 60% of the retail price

1. Recognize that saving 40% is the same as paying 60% of the retail price.

$54 = 0.6r$

$r = \frac{54}{0.6} = \$90$

2. Write an equation and solve for the retail price.

retail price − discount = sale price

$90 - d = 54$

$-d = -36$

$d = \$36$

3. Use the appropriate formula to write an equation and solve for the discount.

The retail price of the shoes is $90, and they were discounted $36.

The discount in Example 3, step 3, could also be calculated using the formula *discount* = *discount rate* × *retail price* = 0.4(90) = \$36.

Waiters and waitresses are usually paid a small hourly wage by their employer, but they rely on tips from their customers for the majority of their income. The tip typically ranges from fifteen to twenty percent of the total bill and can be calculated with another variation of the percent equation:

tip rate × *total bill* = *tip*.

A company often pays its salespeople a certain percentage of their total sales in addition to or instead of a regular salary. While the term "commission" can refer to either the percent or the amount earned, in this text *commission* will refer to the amount earned, and the percent will be referred to as the *commission rate* or simply the *rate*. The following variation of the percent equation is used to find the commission earned:

commission rate × *sales* = *commission*.

Definitions

A **tip** is the amount paid for a personal service.

Commission is the dollar amount an employee receives as a result of the sales he makes.

The **rate** is the percent used to determine a tip or commission based on the amount of the bill or an employee's sales.

Example 4

Find Jeff's commission rate if his total sales of electronic equipment for the month were \$8250 and his gross income for the month was \$3660, including a monthly salary of \$2175.

Answer

salary + commission rate × sales = commission

1. Choose the appropriate formula for this problem.

$2175 + 8250r = 3660$

$8250r = 1485$

$r = \frac{1485}{8250} = 0.18 = 0.18(100\%) = 18\%$

2. Write an equation and solve.

Another variation of the percent formula is used to calculate simple interest on a bank account or a loan:

$I = Prt$,

where P = the principal (the amount invested or borrowed), r = the annual interest rate, and t = the time (in years).

This formula can help you organize a table for simple interest finance problems. These problems are similar to the coin problems you studied in Section 2.8.

Example 5

Christopher invests the same amount of money in each of two accounts and receives a total of $36 interest at the end of one year. If the accounts earn 5% and 7% interest, how much was invested in each account?

Answer Let x = the amount in each account (principal).

1. **Read** and assign a variable to the unknown.

P	r	t	=	I
x	0.05	1	=	$0.05x$
x	0.07	1	=	$0.07x$

2. **Plan**. Make a table using $Prt = I$. Fill in the first three columns, and then multiply across to fill in the last column.

$$0.05x + 0.07x = 36$$
$$5x + 7x = 3600$$
$$12x = 3600$$
$$x = \$300$$

3. **Solve**. The total interest is $36. Multiply both sides of the equation by 100 to clear the decimals.

Check $0.05(300) = 15$ and $0.07(300) = 21$

$15 + 21 = 36$

Christopher invested $300 in each account.

4. **Check**.

Example 6

Robert borrowed $500 at a simple interest rate of 6%. Three months later he borrowed an additional $900 at 8%. If he paid off both loans at the same time and paid $33 in interest, for how many months did he borrow each amount of money?

Answer Let x = the time of the longer loan.

1. Assign a variable to the unknown.

P	r	t	=	I
500	0.06	x	=	$500(0.06)x$
900	0.08	$x - 0.25$	=	$900(0.08)(x - 0.25)$

2. Make a table. 3 mo can be written as $\frac{3}{12} = \frac{1}{4} = 0.25$ yr.

$$500(0.06)x + 900(0.08)(x - 0.25) = 33$$
$$500(6)x + 900(8)(x - 0.25) = 3300$$
$$3000x + 7200x - 1800 = 3300$$
$$10{,}200x = 5100$$
$$x = \frac{5100}{10{,}200} \approx 0.5 \text{ yr} = 6 \text{ mo}$$

3. The total interest is $33.
4. Solve, multiplying both sides of the equation by 100 to clear the decimals.

The $500 loan was for 6 mo, and the $900 loan was for 3 mo.

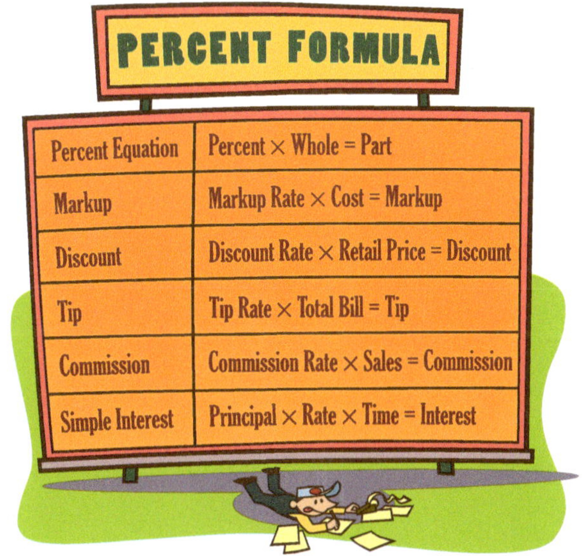

■ A. Exercises

Find the following tips.

	Bill	Tip Rate	Round Up to the Next	Tip Amount
1.	$26.41	15%	quarter	
2.	$17.25	15%	nickel	
3.	$64.32	20%	dollar	
4.	$78.19	18%	quarter	

Complete the following table. Round sales to the nearest dollar and rates to the nearest tenth of a percent.

	Sales	Rate	Commission
5.	$29,450	14%	
6.		17.5%	$23,108.75
7.	$6912		$483.84
8.	$40,009	35%	
9.		7.75%	$974.87
10.	$648,913		$16,222.83

Complete the following table. Round rates to the nearest whole percent.

	Cost	Markup Rate	Markup	Retail Price
11.	$75	20%		
12.	$24.50			$34.30
13.		15%	$18	
14.	$6.75		$2.36	
15.			$4.79	$24.74

Complete the following table. Round rates to the nearest whole percent.

	Retail Price	Discount Rate	Discount	Sale Price
16.	$77	30%		
17.			$1.01	$5.74
18.	$30,000			$28,200
19.		33%	$247.17	
20.	$24		$16.80	

Complete the following table.

	Principal	Annual Interest Rate	Time	Interest Earned
21.	$2500	4%	3 yr	
22.	$15,000	5%	5 yr	
23.		8%	10 yr	$4000
24.	$100,000		6 mo	$1250

B. Exercises

Write an equation and solve.

25. The Hamiltons celebrated Mother's Day by eating out. The pretax bill was $51.79. Determine the tip and the total amount spent if the tax rate is 8% and Mr. Hamilton tipped 18% of the total bill.

26. Jennifer, Leigh Anne, and Karyn went out to eat. Jennifer bought an entrée for $12.95 and split a $4.95 dessert with Karyn, who bought a sandwich for $7.95. Leigh Anne bought soup for $3.95, a salad for $6.95, and coffee for $1.70. Determine the total amount each should pay if tax is 6% and each one tips 15% of her individual bill rounded up to the next quarter.

27. A salesman earns $750 per week plus a 5% commission rate on his sales. Determine his total earnings for a week in which he sells $2020 worth of products.

28. A prospective employee is offered a commission rate of 8%. If he wants to make $7000 per month in commission, how much must he sell each month?

29. A prospective employee wants to earn at least $4000 per month in commission. He estimates he can sell $25,000 worth of merchandise each month. What is the minimum commission rate that he needs?

30. A real-estate agent gets a 6% commission on lots he lists and sells. He gets 3% on lots listed by someone else that he sells. One week he sold a $90,000 lot listed with another company and a $65,500 lot he had listed. Find his commission.

31. After purchasing basketball shoes at a cost of $30 and marking them up by 125%, the sports store owner sells them at a 50% off clearance sale. Find the retail price and the sale price.

32. A clothing retailer marks up his merchandise at a rate of 150% so that he can regularly offer a 25% off sale. If the retail price of a sweater is $60, determine the retailer's cost and the sale price.

33. A suit listed at $700 was reduced by 25% and then an additional 75%. Find the sale price and the total percent reduction based on the original retail price.

34. A dress priced at $299 was reduced 60% and then an additional 75%. Find the sale price and the total percent reduction based on the original retail price.

35. Mrs. Jones invested $10,000, divided between two separate accounts. One pays 5% interest and the other 6%. Her combined annual interest income is $575. How much does she have invested in each account?

36. An investor has $7500 invested at one rate and $6200 invested at a different rate. If the second investment is at a rate 2% lower than the first, find the rate for each investment if his combined annual interest income is $698.

37. The Halls borrowed $1800 at 12% and took out a second loan for $2500 at 9%. The time period for the second loan was twice as long as that for the first, and the total interest paid for both combined was $499.50. Find the length of each loan.

38. An investor placed her money in a venture paying 7% interest. If she had had another $1500, she would have been able to invest her money at 10% and would have gained an additional $255 in annual interest income. How much did she invest?

C. Exercises

39. Lending companies often offer better interest rates when you borrow a larger sum of money for a shorter period of time. A borrower is offered a loan of \$10,080 at 10% and a loan of \$7000 at 12%. If the larger loan is 6 mo shorter but the total interest paid is the same, find the length of each loan.
40. An automobile buyer is trying to decide whether to buy a new car and finance it for 5 yr at 7% interest or a used car costing \$11,500 less at 11% for 2 yr. The new car loan will cost \$5130 more in interest than the used car loan. Find the cost of each car.
41. What discount rate can be offered by a retailer who marks up his merchandise by 125% so that the sale price is the same as the cost of the merchandise?

CUMULATIVE REVIEW

Evaluate. [2.2]

42. $x^2 - x^3y$ when $x = -2$ and $y = 3$
43. $|c + d|d - c^0$ when $c = -4$ and $d = -1$
44. $a^0 - b(a - b^2)$ when $a = \frac{2}{3}$ and $b = \frac{1}{2}$
45. $-v(z - vz) + |z^2 - v|$ when $v = 2\frac{1}{3}$ and $z = \frac{3}{4}$

Simplify. [2.3]

46. $a^2b - ab^2 + 3ab^2 - 2a^2b$
47. $2x(x - y) - y(x - 2y)$

Exercises 48–49 refer to two similar triangles in which the ratio of the perimeters is 25 : 9. [3.3]

48. If a side of the smaller triangle is 45 in. long, what is the length of the corresponding side of the larger triangle?
49. If the area of the larger triangle is 3750 in.^{2}, find the area of the smaller triangle.
50. Find the percent difference if Kyle determines the volume of a solid to be 315 mm^3 by water displacement and 285 mm^3 by calculations based on measurements of its dimensions. [3.5]
51. Find the percent error if Lance determines the atomic mass of helium to be 3.75 amu when its actual atomic mass is 4.00 amu. [3.5]

3.7 Motion

A jet leaves O'Hare Airport in Chicago at 7:00 AM and flies south toward Atlanta, 606 air miles away, at an average speed of 420 mi/hr. At the same time, a jet leaves Atlanta and flies toward Chicago at an average rate of 500 mi/hr. Because of an error in calculations, the jets are flying at the same altitude and on the same route toward each other. How much time do the air traffic controllers have to change the direction or altitude of the jets before they collide?

This is a uniform motion problem. To solve it, you need to use the formula $d = rt$, in which d stands for distance, r for rate, and t for time.

An air traffic controller directs planes from the tower.

How far will a car travel if it goes 50 mi/hr for 3 hr?

$$d = rt$$
$$= 50(3)$$
$$= 150 \text{ mi}$$

How far will a car travel if it goes 50 mi/hr for t hr?

$$d = 50t \text{ mi}$$

At what rate is a car traveling if it goes 225 mi in 5 hr?

$$225 = r(5)$$
$$\frac{225}{5} = r$$
$$r = 45 \text{ mi/hr}$$

rate	×	time	=	distance
50	×	3	=	150
50	×	t	=	$50t$
r	×	5	=	$5r$
$\frac{d}{3}$	×	3	=	d
$\frac{150}{t}$	×	t	=	150

Often you can simplify motion problems if you make a diagram and a table and fill in the known quantities. When you have two quantities in any horizontal row, find the third quantity by using the formula $rt = d$, as written in the top row of the table.

Example 1

The junior-high and high-school Sunday school classes are going on a hike. The junior-high class leaves the church and travels east at 3 mi/hr. The high-school class leaves an hour later traveling at 5 mi/hr on the same route. How much time will pass before the high-school class catches up with the junior-high class?

Answer Let t = the time required for the high-school class to catch up.

1. **Read** the problem carefully.
   a. Identify the main unknown in the problem.
   b. Draw a diagram to represent the problem.

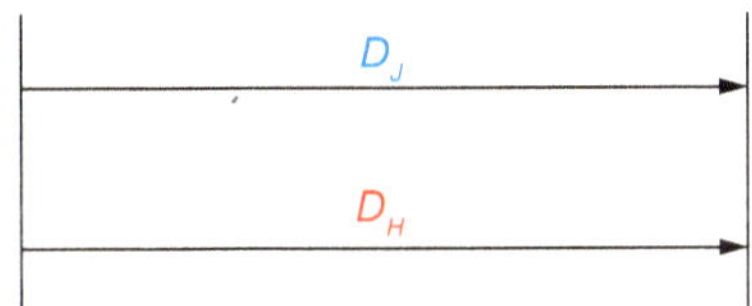

	r	·	t	=	d
junior high	3	·	$t + 1$	=	$3(t + 1)$
high school	5	·	t	=	$5t$

2. **Plan** your strategy. A table will help you organize the information.
   a. Since the junior-high class left one hour earlier, they traveled $t + 1$ hr.
   b. Fill in the rate column using the numbers from the problem.
   c. Using $rt = d$, fill in the distance column.

$$3(t + 1) = 5t$$
$$3t + 3 = 5t$$
$$3 = 2t$$
$$t = \frac{3}{2}$$

3. **Solve.** Since both groups start and end at the same place, the distances must be equal. Write an equation and solve.

Check The high-school class takes $1\frac{1}{2}$ hr to catch up with the junior-high class.

4. **Check.** Does this answer the question in the problem? In this problem, t represents the time the high-school class hikes. It takes them 1.5 hr to hike the 7.5 mi that the junior-high class hikes in 2.5 hr.

uniform motion problems

Use a diagram to help clarify the relationship between the distances in a uniform motion problem. These problems can be categorized according to the relationship between the distances.

1. The distances are the same when a person takes the same trip to and from a destination (roundtrip) or when a second person leaves later to catch up to someone who left the same place earlier.

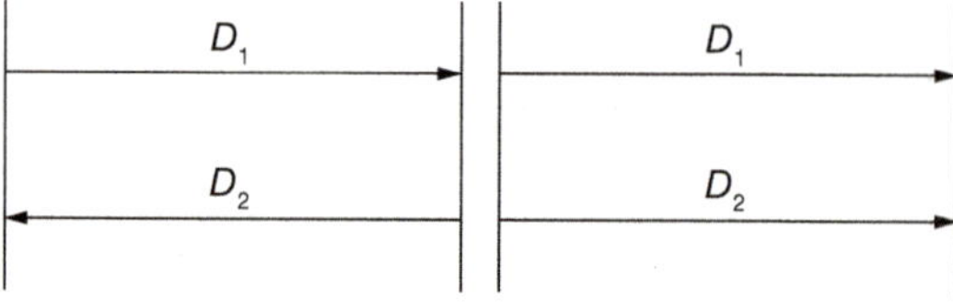

roundtrip or pursuit: $D_1 = D_2$

2. The sum of the distances is known when individuals travel from separate locations to meet or when they travel in opposite directions from a common location.

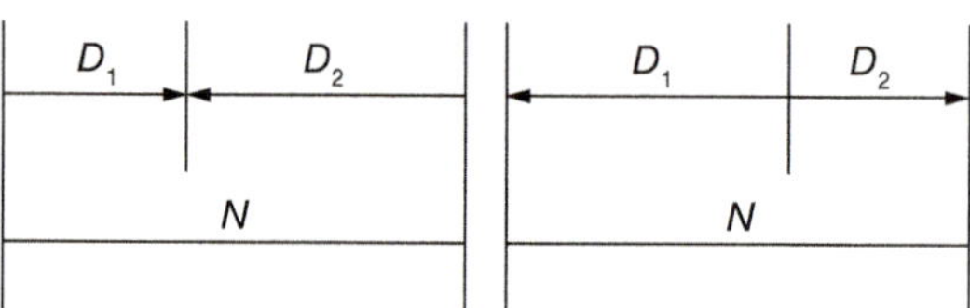

motion in opposite directions: $D_1 + D_2 = N$

3. A given distance must be added to a shorter distance to equal the longer distance.

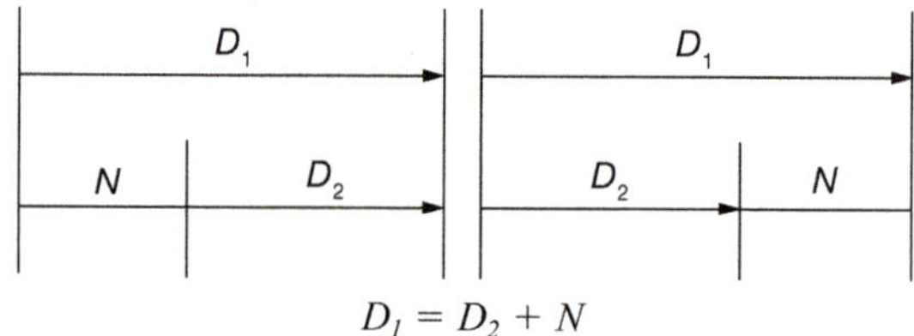

$D_1 = D_2 + N$

Solving Uniform Motion Problems

1. **Read** the problem carefully.
   a. Identify the main unknown in the problem and let the variable stand for this unknown.
   b. Draw a diagram to represent the problem.
2. **Plan** a strategy.
   a. Make a table and fill in the rate and time columns. One column will contain specific numbers from the problem, and the other will involve expressions using the variable.
   b. Using $rt = d$, fill in the distance column.
3. **Solve** an equation.
   a. Use your diagram to help you write an equation relating the distances.
   b. Solve the equation.
4. **Check**.
   a. Did you answer all the questions?
   b. Did you interpret the answer using proper units?
   c. Are your answers reasonable?

Example 2

Two trains leave Mattoon at the same time. The northbound train travels 5 mi/hr faster than the southbound train. What is the rate of each train if after 3 hr they are 333 mi apart?

Answer Let r = the rate of the southbound train and $r + 5$ = the rate of the northbound train.

1. **Read** the problem carefully.
   a. Identify the main unknown in the problem.
   b. Draw a diagram to represent the problem.

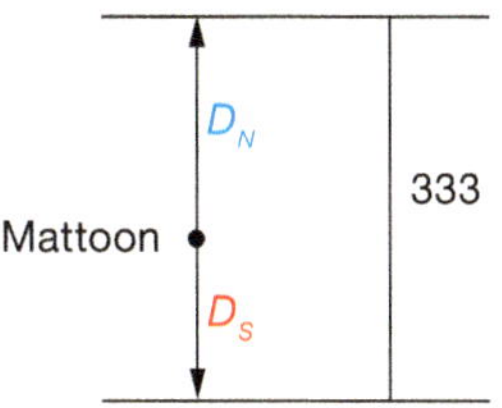

	r	·	t	=	d
northbound	$r + 5$	·	3	=	$3(r + 5)$
southbound	r	·	3	=	$3r$

2. **Plan** your strategy. A table will help you organize the information.
   a. Fill in the rate and time columns.
   b. Using $rt = d$, fill in the distance column.

$$3(r + 5) + 3r = 333$$
$$3r + 15 + 3r = 333$$
$$6r + 15 = 333$$
$$6r = 318$$
$$r = 53$$

3. **Solve**. The sum of the trains' distances is 333 mi. Write an equation and solve.

Check The southbound train travels at 53 mi/hr, and the northbound train travels at $r + 5 = 58$ mi/hr.

4. **Check**. Be sure to give the rate of each train and to check your solution: $58(3) + 53(3) = 333$.

Example 3

Jon, a senior on the track team, runs at a training rate of 375 m/min, while Paul, a freshman, trains at 325 m/min. How long does it take for Jon to lap Paul if they are training on a 400 m track? How many laps has each run during this time?

Answer Let t = the time of both Jon and Paul.

1. **Read** the problem carefully.
   a. Identify the main unknown in the problem.
   b. Draw a diagram to represent the problem. Lapping the slower runner requires the faster runner to run an extra 400 m.

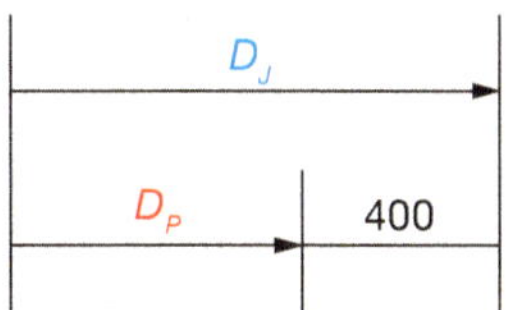

	r	·	t	=	d
Jon	375	·	t	=	$375t$
Paul	325	·	t	=	$325t$

2. **Plan** your strategy. A table will help you organize the information.
   a. Fill in the rate and time columns.
   b. Using $rt = d$, fill in the distance column.

$$375t = 325t + 400$$
$$50t = 400$$
$$t = 8 \text{ min}$$

3. **Solve**. The diagram illustrates that the longer distance equals the sum of the shorter distance and 400. Write an equation and solve.

Check $375(8) \div 400 = 7.5$ laps for Jon;
$325(8) \div 400 = 6.5$ laps for Paul

4. **Check**. Be sure to answer all of the questions posed in the problem. Find the distances run and the corresponding numbers of laps. Jon has lapped Paul in 8 min by running exactly one additional lap.

A. Exercises

Complete the following table.

	Rate	Time	Distance
1.	3	18	
2.		2	132
3.	19		d
4.	52	t	
5.		$\frac{3}{16}$	$\frac{15}{4}$
6.	35	$r + 2$	
7.		16	$16r$
8.	$r + 12$		$4(r + 12)$
9.	a	$7b$	
10.		16	d

Write an equation and solve.

11. Andy won a race by sailing 20 mi/hr for $6\frac{1}{4}$ hr. How far did he sail?

12. A train is traveling 62 mi/hr and has to go 372 mi. How long will it take the train to make the trip?

13. An airplane travels the 2800 mi from New York to Los Angeles in 7 hr. What is the average speed of the plane?

14. Sadie leaves Lewiston for Clarksville 388 mi away and drives 52 mi/hr. Monica leaves Clarksville for Lewiston at the same time and travels 45 mi/hr. How soon will the two meet?

15. Mr. Taylor rides his moped down the country road in front of his house. His wife and their two sons ride bicycles in the opposite direction on the same road. If they leave at the same time and Mr. Taylor travels 24 mi/hr while Mrs. Taylor and the boys travel 8 mi/hr, how much time will pass before they are 24 mi apart?

B. Exercises

Write an equation and solve.

16. Brantley and Shawna walk to Grandma's house at a rate of 4 mi/hr. They ride their bicycles back at 8 mi/hr over the same route that they walked. It takes 1 hr longer to walk than to ride. How long did it take them to walk to Grandma's?

17. Steve leaves the corner of First and Maple on his bicycle and travels an average speed of 9 mi/hr. Two hours later, his brother leaves from First and Maple, following him on a moped traveling 27 mi/hr. How long will it take him to catch up with Steve?

18. Ted runs two-thirds as fast as Frank. In 2 hr Frank runs 8 mi farther than Ted. How fast does each run?

19. A fishing boat leaves Tampa Bay at 4:00 AM and travels at 12 knots. At 5:00 AM a second boat leaves the same dock for the same destination and travels at 14 knots. How long will it take the second boat to catch up with the first one?

20. A passenger train leaves Charleston, WV, heading east at the same time a loaded coal train leaves Charleston heading west. The passenger train travels 23 mi/hr faster than the coal train, and in 4 hr the two trains are 412 mi apart. What is the speed of each train?

21. A roller coaster goes down the first grade eight times faster than it goes up the other side. The distance down is 100 ft more than the distance up, and it takes 70 sec to go up and 10 sec to come down. What is the speed in feet per second of the roller coaster when it goes up and when it comes down?

22. In 5 hr, how far can a car go somewhere and return if the average speed going is 48 mi/hr and the average speed returning is 52 mi/hr?

23. Mr. Thomas drove his old truck to the city at a rate of 40 mi/hr and drove back home on the interstate at a rate of 60 mi/hr. The total trip took 6 hr. How far is Mr. Thomas's home from the city?

24. A freight train leaves Centralia for Chicago at the same time a passenger train leaves Chicago for Centralia. The freight train travels at a speed of 45 mi/hr, and the passenger train travels at a speed of 64 mi/hr. If Chicago and Centralia are 218 mi apart, how long will it take for the trains to meet?

25. At the auto race one car travels 190 mi/hr while another travels 195 mi/hr. How long will it take the faster car to gain two laps on the slower car if the speedway track is $2\frac{1}{2}$ mi long?

C. Exercises

26. Zeke and Jay run the 220 yd dash, and Jay wins the race by 10 yd. If Jay runs the race in 30 sec, what is the rate of each boy in yards per second? How long did it take Zeke to run the 220 yd dash?

27. Joseph and Peter bicycled at 12 mi/hr to a lake and then canoed at 4 mi/hr across it. The entire trip took $4\frac{1}{2}$ hr. If they bicycled 6 mi more than three times the distance they canoed, how far did they travel by each method?

Dominion Modeling

In the previous chapter, the return on investment (ROI) and payback period were used to analyze the benefits of energy-efficient investments. Recall the following formulas:

$$\text{payback period} = \frac{\text{amount of investment}}{\text{savings per period}}$$

$$\text{first-year ROI} = \frac{\text{savings per year}}{\text{amount of investment}}$$

Using the monthly fuel cost savings calculated in Section 3.5, Dominion Modeling, and the vehicle values listed in the table below, determine the net cost, the payback period (in both months and years), and the first-year ROI (to the nearest tenth of a percent) for each of the following trade-ins.

Current Vehicle (age; miles)		Possible Trades			
		Used Vehicle (age; miles)		New Vehicle	
work van (3 yr; 45,000 mi)	\$12,000	small SUV (3 yr; 45,000 mi)	\$13,500	small SUV	\$25,000
compact car (3 yr; 45,000 mi)	\$11,500	hybrid (3 yr; 45,000 mi)	\$16,500	hybrid	\$25,000
sedan (5 yr; 75,000 mi)	\$7500			sedan	\$30,000

	Trade-in	Net Cost	Payback Period	ROI
28.	work van for a used small SUV			
29.	compact car for a used hybrid			
30.	work van for a new small SUV			
31.	compact car for a new hybrid			
32.	older sedan for a new sedan			
33.	work van for a new hybrid			

34. Which of the proposed trade-ins are justifiable on the basis of fuel cost savings? Explain your reasoning.

CUMULATIVE REVIEW

Solve. [2.4–2.7]

35. $\frac{w}{\frac{2}{3}} = 6$

36. $0.2x - 3 = -3.14$

37. $-2(y - 3) + 4y = 24$

38. $3(2z - 1) = 4z - 2$

39. The formula $v_f = at + v_i$ relates the initial and final velocities (v_i and v_f) of an object experiencing an acceleration (a) for a given amount of time (t). Solve the formula for t; then determine the amount of time it takes to accelerate at a rate of 32 ft/sec^2 from 40 ft/sec to 520 ft/sec. [3.1]

Write an equation and solve. [3.5–3.6]

40. If the population of Bridgetown increased 8% over the last decade to a population of 54,000, what was its population ten years ago?

41. If the annual profit of \$22,440 for Morgan's Painting, Inc, represents a decrease of 12% from the previous year, what was the company's annual profit during the previous year?

42. A \$200 coat was reduced 40% and then an additional 20% for a clearance sale. What is the clearance price of the coat?

43. What is the actual percent discount on the original retail price in exercise 42?

44. Explain why the actual discount is not 60% of the original retail price.

3.8 Mixtures

Mixture problems involve the combination of different substances. To solve such a problem, make a table similar to the ones you made for coin, interest, and motion problems. Read the following example and collect all the necessary facts.

Their feed must contain the correct mixture of nutrients.

Example 1

A feed store owner wants to mix corn and barley to make 150 bu of silage additive for cattle feed. If corn costs \$3.80/bu and barley costs \$2.60/bu, how many bushels of each grain would he mix to yield an additive worth \$3.00/bu?

Answer

Let c = the number of bushels of corn and $150 - c$ = the number of bushels of barley.

1. **Read** carefully to find the unknown quantities. We want to know how many bushels of corn and of barley should be combined to yield 150 bu. Name one part c and subtract from 150 to get the other part, just as you did in coin problems.

	Number of Bushels	Price per Bushel	Total Cost
Corn	c	3.80	$3.8c$
Barley	$150 - c$	2.60	$2.6(150 - c)$
Mix	150	3.00	450

2. **Plan**.
   a. Make a table and fill in the first two columns.
   b. Fill in the third column by multiplying the number of bushels by the price per bushel to get the total cost of each component.

$$3.8c + 2.6(150 - c) = 450$$
$$38c + 26(150 - c) = 4500$$
$$38c + 3900 - 26c = 4500$$
$$12c = 600$$
$$c = 50 \text{ bu of corn}$$

3. **Solve**.
   a. The total cost of the corn plus the total cost of the barley equals the total cost of the mixture. Write an equation.
   b. Multiply each term by 10 to clear the equation of decimals. Solve.

Check

The additive should contain 50 bu of corn and $150 - c = 100$ bu of barley.

4. **Check**: $50(3.80) + 100(2.60) = 190 + 260 = 450$.

Example 2

Nickel silver, often used for coating tableware, is an alloy containing 15% nickel along with copper and zinc. How much pure nickel must be melted with 70 lb of nickel silver to make an alloy that is 20% nickel?

Answer Let p = the number of pounds of pure nickel.

1. **Read** carefully to find the unknown quantity.
2. **Plan.**
   a. Make a table and fill in the first two columns.
   b. Use the percent equation to fill in the third column.

	Pounds of Metal	Percent of Nickel	Pounds of Nickel
Nickel Silver	70	15%	0.15(70)
Pure Nickel	p	100%	$1p$
Alloy	$70 + p$	20%	$0.2(70 + p)$

3. **Solve.**
   a. The amount of nickel in the nickel silver plus the amount of nickel in pure nickel equals the amount of nickel in the alloy. Write an equation.
   b. Multiply each term by 100 to clear the equation of decimals. Solve.

$$0.15(70) + 1p = 0.2(70 + p)$$
$$15(70) + 100p = 20(70 + p)$$
$$1050 + 100p = 1400 + 20p$$
$$80p = 350$$
$$p = 4.375 \text{ lb of pure nickel}$$

Check

4. **Check.**

$$0.15(70) + 4.375 = 0.2(70 + 4.375)$$
$$10.5 + 4.375 = 14.875$$

Solving Mixture Problems

1. **Read** the problem carefully to find the unknown quantities. Translate each into an algebraic expression.
2. **Plan.**
   a. Make a table to organize the information in the problem. Fill in all columns except the last one.
   b. Complete the last column without looking back at the problem.
3. **Solve.**
   a. Write an equation using the information in the last column.
   b. Solve the equation.
4. **Check.**
   a. Did you answer all the questions?
   b. Did you interpret the answer using proper units?
   c. Are your answers reasonable?

A. Exercises

Complete the following tables.

	Amount of Mixture	Percent of Salt	Amount of Salt
1.	20 gal	70%	
2.	x lb	65%	
3.	90 oz	8%	
4.	15 g		4.95 g
5.		18%	14.4 L

	Number of Items	Value per Item	Total Value
6.	20	$1.26	
7.	n	$5.31	

	Amount of Mixture	Percent of Ingredient	Amount of Ingredient
8.	80 L	20%	
9.	x	50%	
10.	$x + 80$	25%	

Make a table to help you answer each question.

11. If 10 gal of a 12% salt solution are mixed with 8 gal of an 80% salt solution, how much of the final solution do you have? How much salt is in the final solution? What percent of the final solution is salt?

12. If 4 gal of pure ethanol are added to 46 gal of 2% ethanol, how much of the final solution do you have? How much ethanol is contained in the mixture? What is the strength (percentage) of the ethanol in the mixture?

13. If you dilute 35 gal of a 40% acid solution with 7 gal of water, what is the amount and the strength of the mixture?

14. If you boil 3 gal of 10% salt water until there are only 2 gal left, what is the percent of salt when you finish?

15. Mr. Harper, the candy store manager, wants to mix butterscotch candies and cinnamon balls to make a deluxe mix to sell for $3/lb. Butterscotch candy sells for $3.40/lb and the cinnamon balls sell for $2.90/lb. How many pounds of each should he use to make 20 lb of the mix?

16. How much cream that is 30% butterfat must be mixed with milk that is 3% butterfat to make 45 gal of cream that is 12% butterfat?

17. Mr. Li owns a tea company and wants to make an exotic blend of tea. He has two brands to mix. One brand is an Indian tea that sells for $3.25/lb. The other is a Chinese tea that sells for $4/lb. If he wants to make 35 lb that will sell for $3.70/lb, how many pounds of each should he mix?

18. Chase has 3 gal of a solution that is 30% antifreeze that he wants to use to winterize his car. How much pure antifreeze should he add to this solution to produce a solution that is 65% antifreeze?

19. How many gallons of pure water must be added to 10 gal of 30% salt water to make a solution that is 25% salt water?

B. Exercises

Write an equation and solve.

20. Kim sells Pistachio Pleasure, containing 80% pistachio nuts, and Pistachio Surprise, containing 30% pistachio nuts. How much of each should she mix to make 12 lb of a new mix that is 40% pistachio nuts?

21. A chemist has 2 L of a solution that is 10% hydrochloric acid, and she wants a solution that is 25% hydrochloric acid. How much concentrated (37%) acid must she add?

22. How many gallons of pure alcohol must you add to 25 gal of a 28% alcohol solution to obtain a solution that is 40% alcohol?

23. A nurse wants 4 L of a 5% iodine solution. She already has 3 L of a 3% iodine solution. What percentage of iodine should the additional liter contain?

24. A butcher makes meat loaf by mixing hamburger, which sells for $1.89/lb, and sausage, which sells for $2.99/lb. How many pounds of each should he use if he makes 30 lb of the mix and sells it for $2.33/lb?

25. Jo makes fruit salad with grapefruit and pineapple. Grapefruit costs $1.25 per 12 oz can, and pineapple costs $0.83 per 12 oz can. How many cans should she mix to obtain 84 oz of a mixture that will sell for $1.01 for 12 oz?

26. How much water should you boil off of a 30 gal solution of 10% salt to increase it to 15% salt?

C. Exercises

Write an equation and solve.

27. Babbitt metal (named for its developer, Isaac Babbitt, who first made the metal in 1839) is used primarily for machine gearings. This metal alloy is composed of 90% tin, 7% antimony, and 3% copper. How many ounces of pure tin must be added to the alloy to make 6 oz of a metal that is 94% tin?

28. One 1700 lb load of feed mixture contains 25% corn and 40% bran; the rest is roughage. How many pounds of corn should be added to a load of feed mixture to make the mixture 40% corn?

Dominion Modeling

29. The table below lists typical prices of two different car models over the course of 6 yr. Use a calculator or a spreadsheet to complete the rows of the chart showing the yearly percent change in value for each car.

		New	1 yr old	2 yr old	3 yr old	4 yr old	5 yr old	6 yr old
Car A	Value	$27,370	$19,506	$16,453	$10,918	$8755	$7329	$6551
	Percent Change		29%					
Car B	Value	$29,045	$22,499	$18,829	$14,509	$14,031	$12,731	$11,112
	Percent Change							

30. Which car maintains its value better? Justify your answer mathematically.

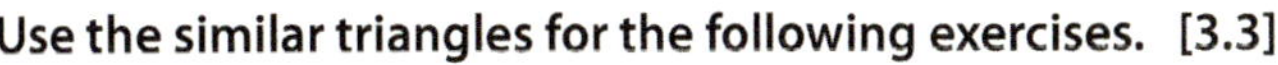

CUMULATIVE REVIEW

Use the similar triangles for the following exercises. [3.3]

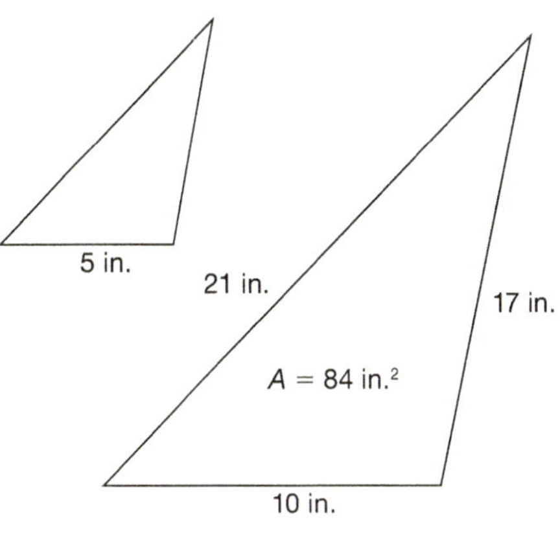

31. Find the perimeter of the smaller triangle.

32. Find the area of the smaller triangle.

Write an equation and solve.

33. A train going 80 mi/hr moves ten times as fast as George rides his bicycle. How fast can he ride his bicycle? [2.5]

34. The sum of twice a number and seven less than four times the number is fifteen. Find the number. [2.6]

35. The basketball team won $\frac{4}{7}$ of the games it played. If the team lost nine games in the season, how many did it win? [3.2]

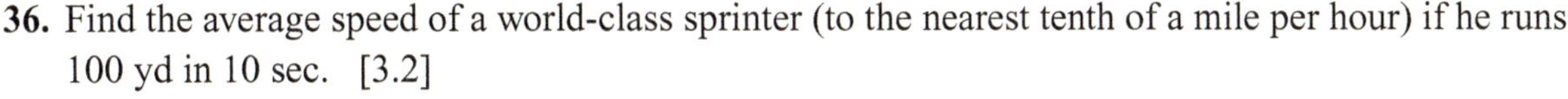

36. Find the average speed of a world-class sprinter (to the nearest tenth of a mile per hour) if he runs 100 yd in 10 sec. [3.2]

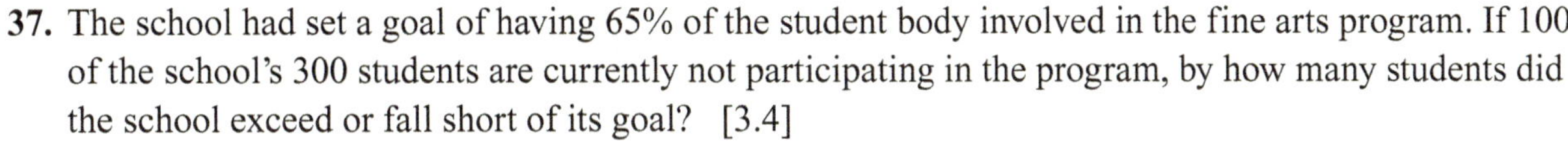

37. The school had set a goal of having 65% of the student body involved in the fine arts program. If 100 of the school's 300 students are currently not participating in the program, by how many students did the school exceed or fall short of its goal? [3.4]

38. Calculate both the experimental error and the percent error (to the nearest tenth) if Nancy determined the density of lead to be 12.0 g/cm^3 when its actual density is 11.3 g/cm^3. [3.5]

39. A merchant purchased a car for \$20,000 and marked it up 25%. When it did not sell, he reduced it 20%. How much did he make if the car sold at this discounted price? [3.6]

40. On vacation the Hart family traveled an average of 300 mi each day. How many days was their vacation if they traveled a total of 1800 mi? [3.7]

MIND OVER MATH

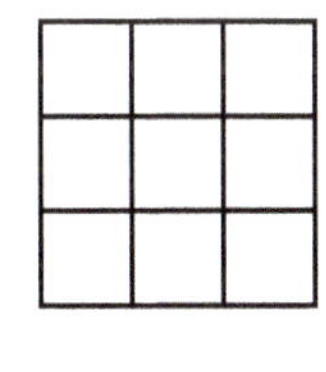

To the right is a 3 × 3 magic square. Use the numbers 1 through 9 to fill in the square so that the sum of each row, column, and diagonal is equal to 15.

You can make a magic square for any odd number by following the instructions given below. Try a 5 × 5 square.

1. Place a 1 in the center square of the top row.
2. Move up diagonally to the right. Since you have moved out of the top of the square, put 2 in the square at the bottom of that column.
3. Move up diagonally to the right from 2 and put 3 in the square.
4. Move up diagonally to the right again. Since you have moved out of the right side of the square, put 4 in the square at the left end of that row.
5. Move up diagonally to the right and put 5 in the square.
6. Since you cannot move up diagonally to the right from 5, put 6 in the square below 5. Then proceed with the steps as given above.

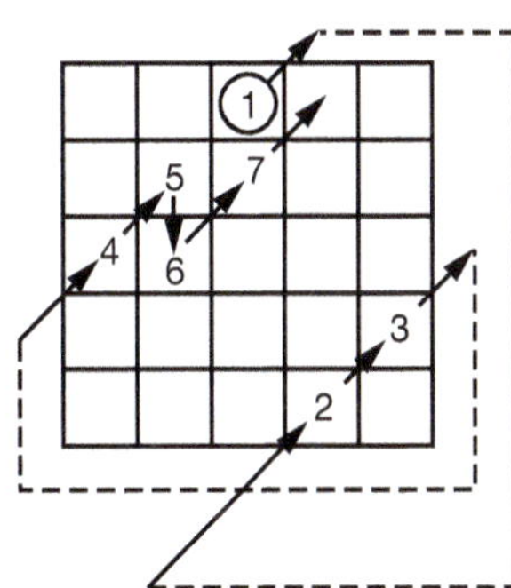

What is the sum of each row, column, and diagonal? Can you make a 7 × 7 magic square? What is the sum of each row, column, and diagonal of a 7 × 7 magic square?

CHAPTER 3 REVIEW

Solve each literal equation for the indicated variable.

1. $d = \frac{m}{v}$ for m

2. $y = mx + b$ for x

3. $S = c(h + r)$ for r

4. $\frac{a}{b} - \frac{1}{c} = d$ for c

5. $S = \frac{3A}{R - A}$ for A

6. Solve $a = \frac{v_2 - v_1}{t_2 - t_1}$ for v_2. Then evaluate the resulting expression to find v_2 when $a = 9.8$ m/s^2, $t_2 = 5$ s, $t_1 = 2$ s, and $v_1 = 19.6$ m/s.

Find the unit rate.

7. 1100 words in four pages

8. buying three tires at $80 each and getting a fourth tire free

Convert the given rate to an equivalent rate.

9. 88 ft/sec to miles per hour

10. 100 km/hr to meters per second

Solve.

11. $\frac{12}{25} = \frac{20}{y}$

12. $\frac{x - 1}{6} = \frac{x + 1}{9}$

Use proportions to solve.

13. If the ratio of the masses of hydrogen and oxygen in water is 1 : 8, how many grams of each are in 90 g of water?

14. What is the actual distance between two towns 5.25 in. apart on a map with a scale of 1 in. : 40 mi?

15. A popular manufacturer of die-cast metal toy cars uses a 1 : 64 scale. How many inches long is the model of a car that is 16 ft long?

16. The dimensions of a rectangular room are drawn at 7.5" × 6" on a plan that has a scale of 1 : 24. Determine the number of square feet of carpet needed for this room.

$\triangle ABC \sim \triangle DEF$; the area of $\triangle ABC$ is 210 cm^2.

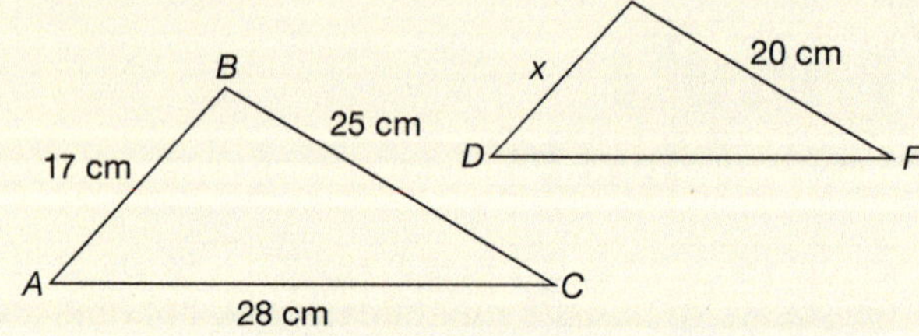

17. Write and solve a proportion to find x.

18. Find the perimeter of $\triangle DEF$.

19. Find the area of $\triangle DEF$.

For each percent, fraction, or decimal, state the equivalent expressions in the other two forms.

20. 0.5%

21. 0.5

22. $\frac{1}{5}$

Write the appropriate percent equation and solve.

23. 36% of what number is 54?

24. What percent of 16 is 76?

Complete the following table. Round answers to the nearest tenth.

	Original Amount	Change	Percent Change	New Amount
25.	900	add 36		
26.	150			100
27.		subtract 35	20% decrease	

Write the appropriate percent equation and solve.

28. If Orlando missed 5 out of 40 questions on the test, what percent of the questions did he get right?

29. Thirty percent of those in attendance had never seen a lacrosse match before. How many of the 400 in attendance were watching their first game of lacrosse? How many had previously seen a game of lacrosse?

The Zimmermans know that they usually average 55 mi/hr when driving the 715 mi to visit their grandparents. They leave at 6 AM.

30. What time should the Zimmermans tell their grandparents to expect them to arrive?
31. If they arrive at 9 PM, what was the Zimmermans' average speed for the trip?
32. On the way home, the Zimmermans left at 10:00 AM and drove at an average of 50 mi/hr before stopping at a hotel for the night at 6:00 PM. How many miles do they need to travel on the next day to get home?

Write an equation and solve.

33. The price of a stock rose from a yearly low of $9.68 to $18.61. What was the stock's percent increase from its low price that year?
34. Calculate both the experimental error and the percent error (to the nearest tenth of a percent) if Stephanie determined the density of aluminum to be 2.84 g/cm^3 when its actual density is 2.70 g/cm^3.
35. Terrance determined the volume of an object to be 5.6 cm^3 using water displacement, and Vic calculated it as 5.4 cm^3 after measuring its dimensions. Determine the percent difference (to the nearest tenth of a percent) between the two values.
36. A merchant bought a boat for $8000 and marked it up 15.5%. What is the retail price? What would the buyer have to pay if there is a 5% sales tax?
37. A traveler priced an airplane ticket at $650. He later was able to get it for $400. What percent did he save?
38. The retail price of a suit was $719. It was reduced 33% and later an additional 67%. What was the price after both reductions? What was the total percent reduction of the original retail price?
39. Ron earns a 15% commission on his sales in addition to a $1500 monthly salary. Determine his monthly sales goal if he budgets $3750 as his monthly income.
40. Determine the total amount Victor must repay if he borrows $8000 at 5% simple interest for 3 yr.
41. Timothy invested some of his savings in an account yielding 4% interest and twice as much in a second account yielding 6% interest. If Timothy earned $800 in interest last year, how much did he invest in each account?
42. Julie drives the 120 mi from Greenup to Newton at 46 mi/hr, while Beth drives from Newton to Greenup at 34 mi/hr. How long will it take for Julie and Beth to meet each other if they leave at the same time?
43. Jorge jogged the trail to the waterfall at 120 yd/min and walked back at 80 yd/min. If the entire trip took 2 hr, how long is the trail (in yards and to the nearest tenth of a mile)?
44. Brad is 15 mi down the road when Carl leaves the rest area driving at 65 mi/hr. How long will it take Carl to catch up if Brad is driving at 60 mi/hr?
45. A coffee merchant wants to make a mixture of 50 lb of Colombian and Brazilian coffees that will sell for $3.19/lb. If Colombian coffee sells for $3.35/lb and Brazilian coffee sells for $2.95/lb, how much of each type of coffee should be mixed together?
46. Tickets to the play cost $5 for children and $8 for adults. If the auditorium, which holds 350 people, was sold out for all three performances and the net income from ticket sales was $7455, how many children and how many adults attended the play?
47. How much water should be added to 20 L of a 20% salt solution to dilute it to a 15% salt solution?
48. How much salt should be added to 20 L of a 20% salt solution to increase it to a 25% salt solution?
49. How much water should be evaporated from 20 L of a 20% salt solution to increase it to a 25% salt solution?

4 Solving Inequalities

The First National Bank of Newton is holding its annual board meeting. Mr. Lee, the bank president, presents the annual report to the members of the board. Mr. Lee must make wise investments of the bank's funds so that the bank and its stockholders will make a profit. The bank creates income by lending its money to people or companies at a certain interest rate or by investing it in federal or municipal bonds. The profit the bank obtains from these investments pays the interest (calculated at a lower rate than the lending rate) on bank accounts for the use of the money deposited, and the rest is then divided among the stockholders as dividends.

The stockholders have a vested interest in the financial plans that Mr. Lee presents at the annual board meeting. Mr. Lee knows that he must gain their confidence in his plans for investing their money. He summarizes investments that will earn $3.4 million in the coming year. He needs to know how much money he must invest in municipal bonds that yield 8% annually so that the total interest income for the bank will be at least an additional $2.1 million.

As Mr. Lee solves these important business problems, he must work with inequalities. In the Dominion Modeling problems throughout this chapter, you will learn how to solve similar investment problems.

After this chapter you should be able to

1. use the Trichotomy Property to interpret negated inequalities.
2. graph inequalities on a number line.
3. use the Addition and Multiplication Properties of Inequality to solve inequalities.
4. solve conjunctions and disjunctions involving inequalities.
5. solve absolute value equations.
6. solve absolute value inequalities.
7. write and solve inequalities describing real-life situations.

4.1 Inequalities

Redwoods National Park, California, contains the tallest trees in the world. The redwood named Hyperion is over 378 ft tall.

An *inequality* is a statement comparing two quantities that are not equal. Scripture reminds us of some inequalities. God, the Creator, is greater than any of His created beings. Satan seeks equality with God: "I will be like the most High" (Isa. 14:14). Satan, in the form of a snake, told Eve that she could be like, or equal to, God (Gen. 3:5). Humanism is the continuation of that lie today. Humanism claims that man is his own god, but the Bible teaches that there is only one God (Deut. 4:39). No matter how hard Satan or humans try, they cannot become equal with God.

Definition

An **inequality** is a mathematical sentence stating that two quantities are not always equal.

Mathematical inequalities include many phrases that signify that the two quantities are not equal. These phrases include "is greater than," "is not equal to," "is less than," "is at least," "is at most," and so on. Symbols can be used to represent these and other important inequality phrases. The following table shows the five basic inequality symbols.

Symbol	Meaning	Example
$<$	is less than	$4 < 6$
$>$	is greater than	$5 > 3$
$\neq$	is not equal to	$-2 \neq 4$
$\leq$	is less than or equal to	$-1 \leq 8$
$\geq$	is greater than or equal to	$6 \geq 0$

Any comparison of two real numbers results in one of three possible outcomes. The Trichotomy Property formally states this idea.

Trichotomy Property
If a and b represent two real numbers, then $a = b$, $a > b$, or $a < b$.

Placing a slash through the symbols $>$, $<$, $\geq$, and $\leq$ results in their negation, just as the slash through $=$ means "is not equal to." The statement that $x \not> 5$ does not guarantee that $x < 5$. Instead, we know that $x \leq 5$. The inequality $x < 5$ can also be expressed as $x \not\geq 5$.

Example 1

State an inequality equivalent to $x \not\leq y$.

Answer $x > y$ Since the inequality symbol eliminates two of the three possibilities, less than or equal to, x must be greater than y.

Just like equations, inequalities can be true or false. The inequality $3^2 > 3(2)$ is true since $9 > 6$, while $x + 4 < x - 2$ is false, regardless of the value used to replace x. In conditional inequalities, the truth of the inequality depends on the value of the variable. Any value of the variable that makes the inequality true is a *solution* of the inequality. Consider the inequality $x - 3 < 8$. Does $x = -2$ make the inequality true? Is this the only value that will make the inequality true? How many solutions can you find to this inequality? As you consider the numbers that make this inequality true, you should understand that there are an infinite number of solutions.

Example 2

For each inequality, determine which of the following are solutions: {2, 3, 4}.

a. $a + 3 > 6$ **b.** $6 \geq 2b$ **c.** $1 - c \not< -2$

Answer

a. $a + 3 > 6$

$(2) + 3 \overset{?}{>} 6$

$5 \overset{?}{>} 6$; false

$(3) + 3 \overset{?}{>} 6$

$6 \overset{?}{>} 6$; false

$(4) + 3 \overset{?}{>} 6$

$7 \overset{?}{>} 6$; true

4 is a solution.

b. $6 \geq 2b$

$6 \overset{?}{\geq} 2(2)$

$6 \overset{?}{\geq} 4$; true

$6 \overset{?}{\geq} 2(3)$

$6 \overset{?}{\geq} 6$; true

$6 \overset{?}{\geq} 2(4)$

$6 \overset{?}{\geq} 8$; false

2 and 3 are solutions.

c. Rewrite as $1 - c \geq -2$.

$1 - (2) \overset{?}{\geq} -2$

$-1 \overset{?}{\geq} -2$; true

$1 - (3) \overset{?}{\geq} -2$

$-2 \overset{?}{\geq} -2$; true

$1 - (4) \overset{?}{\geq} -2$

$-3 \overset{?}{\geq} -2$; false

2 and 3 are solutions.

The numbers -3, 0, and 4.5 are all solutions to $5 > x$. Usually it is not possible to list all of the solutions to an inequality. Therefore, the solutions to inequalities are often graphed on a number line. To avoid confusion, inequalities such as $5 > x$ are usually rewritten as $x < 5$. The entire solution set is illustrated by shading all the numbers to the left of 5 on a number line and placing a circle at 5 to show that the number 5 itself is not part of the solution. Use an arrowhead to show that the graph continues indefinitely.

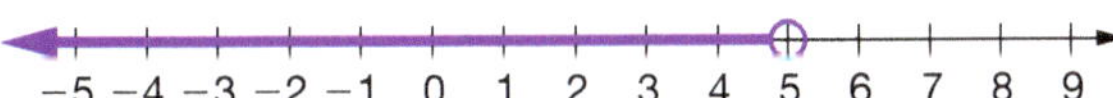

Example 3

Graph $x \geq -3$.

Answer

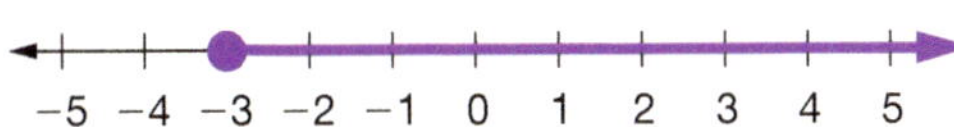

1. Since the inequality symbol includes equal to, place a solid dot at -3 to indicate that it is part of the solution.
2. Shade the portion of the number line to the right of -3.

Example 4

Graph $x \neq 2$.

Answer $x > 2$ or $x < 2$

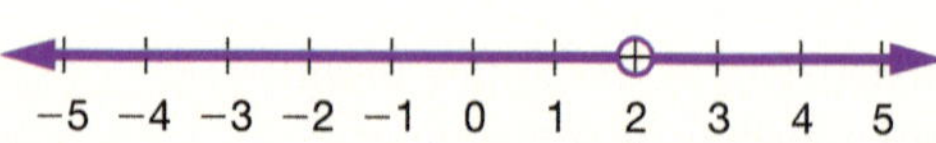

1. Since the inequality symbol excludes only equal to, any number greater than or less than is a solution.
2. Place a circle at 2 and shade the portions of the number line to the right and to the left of the circle.

Example 5

Write an inequality to describe the following graph; then write an equivalent negated inequality.

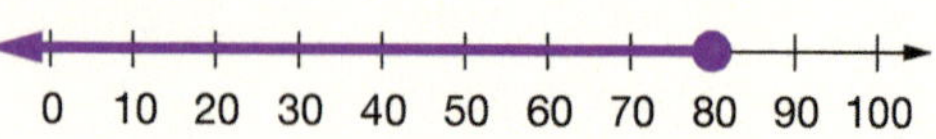

Answer $x \leq 80$

1. The solid dot means that 80 is part of the solution, and the shading to the left indicates numbers less than 80.

$x \not> 80$

2. Not greater than is equivalent to $\leq$.

Inequalities are often used to model real-life situations.

Example 6

Define a variable and write an inequality modeling each situation.

a.

b.

MAXIMUM
OCCUPANCY
NOT TO EXCEED
225 PERSONS

Answer **a.** Let $c =$ the price of an item; $c \geq 9.99$.

b. Let $p =$ the number of people; $p \leq 225$.

A. Exercises

Match each phrase with the equivalent inequality symbol.

1. is at most	**a.** $>$
2. is not greater than or equal to	**b.** $<$
3. is not less than	**c.** $\geq$
4. is at least	**d.** $\leq$
5. is not as much as	**e.** $\neq$
6. is greater than or is less than	
7. exceeds	
8. does not exceed	

Write an inequality equivalent to the given inequality.

9. $x \ngtr -4$ **10.** $y \nleq 1$ **11.** $z \ngeq 21$ **12.** $w \nless -7$

Determine whether each number is a solution of the inequality.

13. $2m < 4$
- **a.** -2
- **b.** 2
- **c.** 4

14. $n - 7 \geq -3$
- **a.** -1
- **b.** 4
- **c.** 8

15. $r^2 - 2 \ngtr -1$
- **a.** 0
- **b.** -1
- **c.** 2

16. $|7 - s| \nleq 13$
- **a.** 23
- **b.** 15
- **c.** -20

B. Exercises

Write an inequality to describe each of the following graphs; then write an equivalent negated inequality.

17. 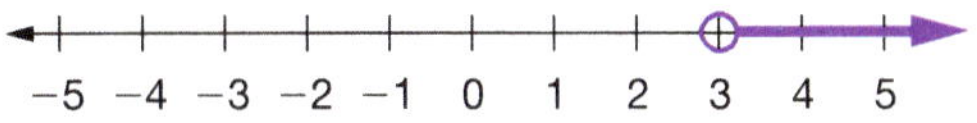

18. 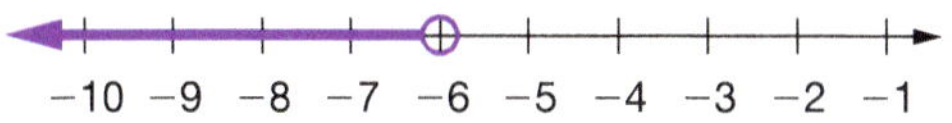

19. 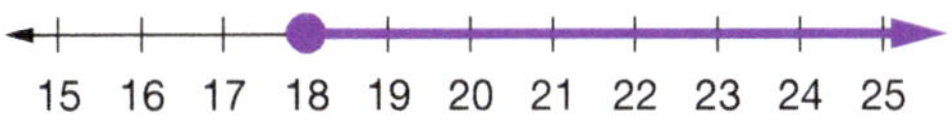

20. 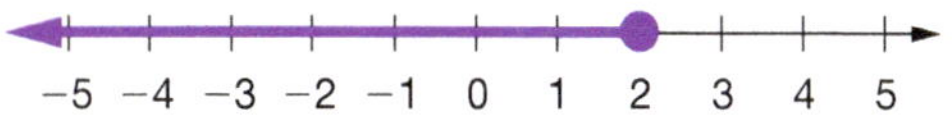

Graph each inequality on a separate number line.

21. $x < 4$ **22.** $x \geq -2$ **23.** $x \neq 5$ **24.** $-1 < x$

25. $x \leq 1.2$ **26.** $x \neq -55$ **27.** $100 \leq x$ **28.** $x \leq -0.8$

29. $x > 1.5$ **30.** $x \neq 4$ **31.** $x \ngtr 0$ **32.** $-3 \ngtr x$

Define a variable and write an inequality modeling each situation.

33. You must be at least 18 years old to vote.

34. the times of the other runners if the winner ran the race in 10.85 sec

35. maximum wattage: 60 W

36. The doctor recommended that Jordan restrict carbohydrates in his diet to 225 g per day.

37. The Institute of Medicine recommends a minimum daily intake of 75 mg of vitamin C for women.

38. distance to a star if the closest star, Proxima Centauri, is 4.2 light-years away from the sun

C. Exercises

Graph the union of each pair of inequalities.

39. $x \leq 1, x > 5$ **40.** $x \leq 2, x < -4$

Graph the intersection of each pair of inequalities.

41. $x < 6, x > -2$ **42.** $x > -3, x > 0$

43. Group the following symbols in pairs of opposites: $=, <, >, \leq, \geq, \neq$.

CUMULATIVE REVIEW

Evaluate. [1.3–1.4]

44. $-8 - 17 - (-14)$

45. $\frac{3}{4} - \frac{5}{7} + \frac{1}{2}$

46. $15.07 + (-4.6) - (-8.47)$

State the property that justifies each step in the solution of the equation $(7 + x) + 9 = 53$. [1.3, 2.4]

47. $(x + 7) + 9 = 53$

48. $x + (7 + 9) = 53$

49. $x + 16 - 16 = 53 - 16$

50. $x + 0 = 37$

51. $x = 37$

Solve. [2.4]

52. $-6x = 90$

53. $x - (-15) = 25$

4.2 Properties of Inequality

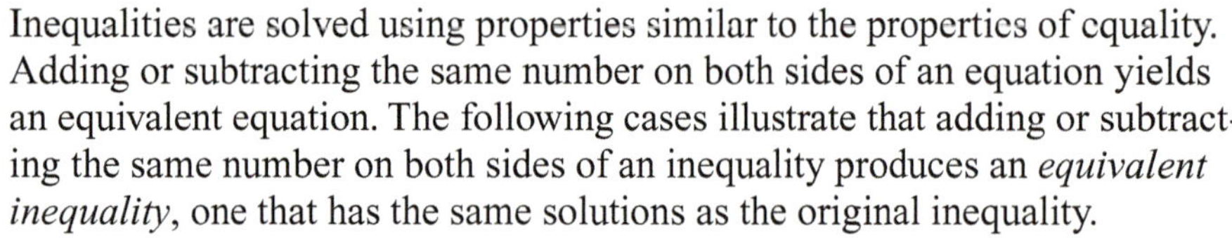

Inequalities are solved using properties similar to the properties of equality. Adding or subtracting the same number on both sides of an equation yields an equivalent equation. The following cases illustrate that adding or subtracting the same number on both sides of an inequality produces an *equivalent inequality*, one that has the same solutions as the original inequality.

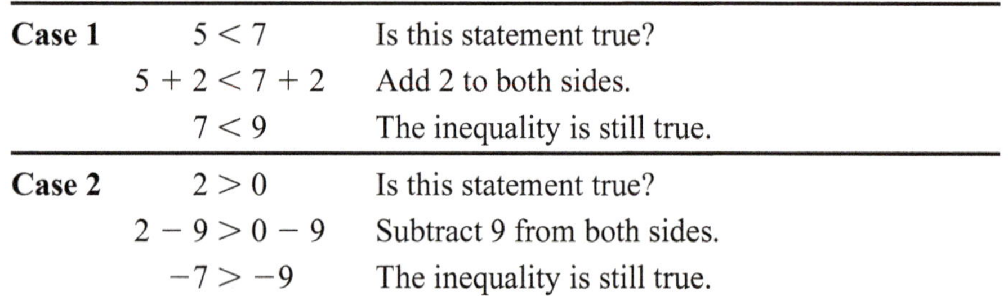

Case 1	$5 < 7$	Is this statement true?
	$5 + 2 < 7 + 2$	Add 2 to both sides.
	$7 < 9$	The inequality is still true.
Case 2	$2 > 0$	Is this statement true?
	$2 - 9 > 0 - 9$	Subtract 9 from both sides.
	$-7 > -9$	The inequality is still true.

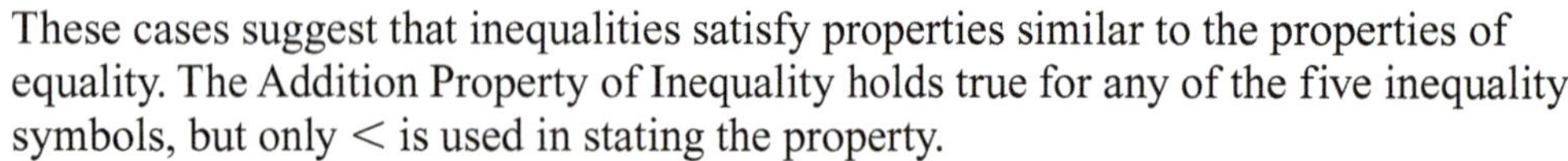

These cases suggest that inequalities satisfy properties similar to the properties of equality. The Addition Property of Inequality holds true for any of the five inequality symbols, but only $<$ is used in stating the property.

Addition Property of Inequality

If a, b, and c are real numbers or expressions such that $a < b$, then $a + c < b + c$.

A subtraction property of inequality is not stated because subtraction is the same as adding the opposite of a number. Therefore, the Addition Property of Inequality also encompasses subtraction. You can subtract the same number from both sides of the inequality or add a negative number to both sides of the inequality.

Example 1

Solve and graph the solution to each of the following inequalities.

a. $x + 7 < 8$ **b.** $x - 5 \neq 3$ **c.** $-0.6 \leq x - 0.4$

Answer

a. $x + 7 < 8$

$x + 7 - 7 < 8 - 7$ Add −7 (subtract 7) to both sides.

$x < 1$

−5 −4 −3 −2 −1 0 1 2 3 4 5

b. $x - 5 \neq 3$

$x - 5 + 5 \neq 3 + 5$ Add 5 to both sides.

$x \neq 8$

0 1 2 3 4 5 6 7 8 9 10

c. $-0.6 \leq x - 0.4$

$-0.6 + 0.4 \leq x - 0.4 + 0.4$ Add 0.4 to both sides.

$-0.2 \leq x$

$x \geq -0.2$ Rewrite the inequality with the variable on the left side.

−0.4 −0.3 −0.2 −0.1 0 0.1 0.2 0.3 0.4

Scientists use experiments to test a hypothesis. In mathematics, the truth of a hypothesis is examined in each possible case. Consider these test cases to see whether inequalities are always solved like equations.

Case 1	$3 < 5$	Is this statement true?
	$2 \cdot 3 < 2 \cdot 5$	Multiply both sides by 2.
	$6 < 10$	The inequality is still true.
Case 2	$4 < 6$	Is this statement true?
	$-2 \cdot 4 < -2 \cdot 6$	Multiply both sides by −2.
	$-8 < -12$	The inequality is no longer true.
	$-8 > -12$	The inequality must be reversed to make the statement true.

The first case illustrates that if you multiply both sides of an inequality by the same positive number, the resulting inequality is still true. Case 2 illustrates that if you multiply both sides of an inequality by the same negative number, you must reverse the inequality symbol to make an equivalent inequality.

Remember that multiplication and division are inverse operations. Dividing by 2 is the same as multiplying by $\frac{1}{2}$. Therefore, the principles that hold for multiplication will also hold for division. If both sides of an inequality are divided by the same positive number, a true statement is formed. If both sides are divided by the same negative number, the inequality symbol must be reversed to form a true statement.

Multiplication Property of Inequality

If a, b, and c are real numbers or expressions such that $a < b$ and
1. $c > 0$ (positive), then $ac < bc$.
2. $c < 0$ (negative), then $ac > bc$.

The Multiplication Property of Inequality above uses the $<$ symbol. The same results occur for $>$, $\leq$, or $\geq$. Remember to reverse the inequality symbol when multiplying or dividing both sides of an inequality by the same negative number.

Example 2

Solve and graph the solution to each of the following inequalities.

a. $-3x < -9$ **b.** $\frac{x}{12} > -4$ **c.** $-\frac{5}{2}x \geq 25$

Answer

a. $-3x < -9$

$\frac{-3x}{-3} > \frac{-9}{-3}$ Divide both sides by -3. Reverse the inequality symbol when you divide by a negative number.

$x > 3$

−5 −4 −3 −2 −1 0 1 2 3 4 5

b. $\frac{x}{12} > -4$

$12\left(\frac{x}{12}\right) > 12(-4)$ Multiply both sides by 12. The inequality symbol is not reversed when you multiply by a positive number.

$x > -48$

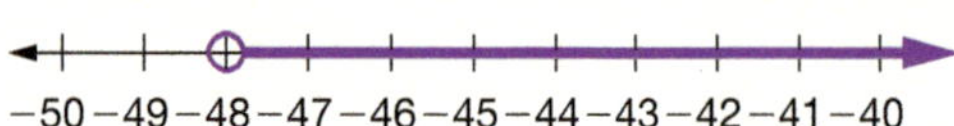

c. $-\frac{5}{2}x \geq 25$

$-\frac{2}{5}\left(-\frac{5}{2}\right)x \leq -\frac{2}{5}(25)$ Multiply both sides by $-\frac{2}{5}$. Reverse the inequality symbol when you multiply by a negative number.

$x \leq -10$

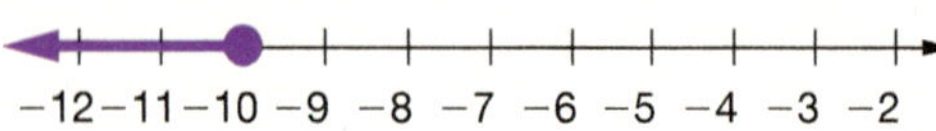

Inequalities can be used to model and solve questions from everyday life.

Example 3

How much money must be invested at 7% simple interest to produce a yearly income of at least \$875?

Answer Let P = the amount invested.

1. **Read** the problem carefully. Note that the phrase "at least" signifies $\geq$.

P	r	t	=	I
P	0.07	1	=	$0.07P$

2. **Plan**: Write the formula for simple interest: $I = Prt$. Make a table or substitute into the formula.

$0.07P \geq 875$

3. **Solve**: Write an inequality. The interest earned must be ≥ 875.

$7P \geq 87{,}500$ Multiply both sides by 100.

$P \geq 12{,}500$ Divide both sides by 7.

\$12,500 or more must be invested.

Check \$12,500(0.07)(1) = \$875

4. **Check**: Since this product is exactly \$875, any amount larger than \$12,500 will yield more than \$875 in interest. The answer is reasonable.

A. Exercises

Solve and graph the solution to each of the following inequalities.

1. $x + 3 > -4$
2. $x - 2 \le 3$
3. $\frac{x}{3} < -1$
4. $4x > -2$
5. $x + 1 \neq -2$
6. $-\frac{x}{2} < 4$
7. $-7x \ge -14$
8. $x - 8 \ge -8$

Solve.

9. $x + 18 \le 7$
10. $y - 26 > 82$
11. $z - 10 \neq 21$
12. $a + 5 < -6$
13. $5b > -25$
14. $-\frac{y}{4} > 3$
15. $\frac{x}{8} \ge 3$
16. $-3c \le -27$

B. Exercises

Solve.

17. $-\frac{3}{4}r > \frac{9}{2}$
18. $\frac{2}{5}s \le -\frac{4}{15}$
19. $\frac{8}{5} \le -2x$
20. $-\frac{3}{4}t < -\frac{9}{10}$
21. $3 + m + 4 \le 5$
22. $2 \cdot 4n < 6$
23. $10k - 4k \ge 3^2$
24. $7x \cdot 2 < -\frac{12}{5}$

Write an inequality and solve; then use a word phrase to state your answer.

25. Find all the numbers such that the sum of the number and 25 is less than 82.
26. Find all the numbers such that one-third of the number is at most five.
27. An elevator is rated with a maximum weight capacity of 750 lb. How many teenagers with an average weight of 150 lb can safely ride the elevator?
28. A season pass to the amusement park costs $117, and daily admission costs $39. Determine the number of trips to the park for which paying daily admission is more expensive than purchasing a season pass.
29. The current leader in the super combined skiing event had a downhill time of 84.96 sec and a slalom time of 45.12 sec. The final competitor had a downhill time of 84.49 sec in her first run. If the winner is determined by the total of the times in the two races, what slalom times will allow the final competitor to win the gold medal?

30. The bank offers a checking account without monthly fees if a minimum balance of $300 is maintained. Daniel had a balance of $538 and wrote a check for $325. How much should he transfer from his savings account to avoid paying a service fee on his checking account?
31. Pam wants to buy three times as many cans of green beans (62¢ per can) as cans of peas (69¢ per can). If she has no more than $10 to spend, what is the maximum number of cans of peas she can buy?
32. Each ounce of whole milk contains twice as many calories as an ounce of skim milk. A mixture of 4 oz of whole milk and 2 oz of skim milk contains at least 100 calories. What is the minimum number of calories in each ounce of whole and skim milk?

C. Exercises

Solve.

33. $4x - 3x + 12 < 4 \cdot 3$

34. $2x - (x + 4) \neq 7$

35. $5x - 4x > 3(2x - 4) - 6x$

36. $-2(x - 3) \leq 5 - x$

37. Water from the Dead Sea is eight times saltier than ocean water. If 50 gal of Dead Sea water and 100 gal of ocean water together contain at least 20 gal of salt, what is the percentage of salt in each water source?

Dominion Modeling

Some people think God is interested in only the "spiritual" parts of life and that He doesn't have much to say about the everyday details of life, including such things as borrowing and lending. However, the Bible says that "the borrower is servant to the lender" (Prov. 22:7). As long as a person is in debt, he is morally obligated to the person who is lending him money. For this reason, you should be very careful about borrowing money.

One way to prepare is to know about different types of loans, three of which are discussed below. Recall the interest formula $I = Prt$.

1. Simple interest loan: A single payment of interest and principal combined at the end of the specified time.
2. Add-on interest loan: Simple interest plus principal divided by the number of payments (usually monthly). Each payment applies the same set amount to the principal and the same amount to interest.
3. Discount loan: The interest on the desired amount to be borrowed is calculated and added to the amount that is borrowed at the beginning of the loan. This new amount owed is divided into equal payments throughout the life of the loan.

Compare these types of loans for $5000 borrowed at 6% interest for 2 yr in the three ways described above.

38. Simple interest loan: Find the amount of interest and the entire amount due at the end of the loan.

39. Add-on interest loan: Find the amount of interest, the total amount to repay, and the monthly payment.

40. Discount loan: Calculate the actual amount loaned for the borrower to receive $5000 using the formula $S = \frac{P}{1 - dt}$, where S = the amount that must be borrowed, P = the amount the borrower receives, d = the interest rate, and t = the time in years. Then find the amount of interest and the monthly payment.

CUMULATIVE REVIEW

41. The number $2.\overline{3}$ belongs to which subset(s) of the real numbers? [1.1]

42. How are $|a - b|$ and $|b - a|$ related? [1.2]

Solve. [2.4–2.7]

43. $\frac{2}{3}x = \frac{8}{9}$

44. $4y - 3 = 12$

45. $5z - (2z + 3) = 12$

46. $5(w + 3) = 2w - 5$

Clear each equation of fractions or decimals and then solve. [2.8]

47. $4.6a - 3.1 = 10.7$

48. $\frac{5}{3}x - \frac{1}{2} = 12$

Solve each literal equation for the indicated variable. [3.1]

49. $a + b + x = c$ for x

50. $3x - 2y = 6$ for y

4.3 Solving Inequalities

Van Gogh's Portrait of Dr. Gachet *sold in 1990 for $82.5 million, the highest price ever paid for a painting sold at auction at that time. Express the price, p, of any painting auctioned before 1990 as an inequality.*

Solving inequalities involving more than one operation requires you to "undo" the operations done to the variable using inverse operations in reverse order. This is similar to the process of solving equations. Simplify both sides of the inequality as much as possible. Then apply the properties of inequality to undo addition and subtraction before undoing multiplication and division. Remember to reverse the inequality symbol when you multiply or divide by a negative number. Study the examples below.

Example 1

Solve $3x + 5 > 20$.

Answer

$3x > 15$ — 1. Subtract 5 from both sides.

$\frac{3x}{3} > \frac{15}{3}$ — 2. Divide both sides by 3. The inequality symbol is not reversed when you divide by a positive number.

$x > 5$

Example 2

Solve $-\frac{x}{3} + 8 \geq 12$.

Answer

$-\frac{x}{3} \geq 4$ — 1. Subtract 8 from both sides.

$-3\left(-\frac{x}{3}\right) \leq -3(4)$ — 2. Multiply both sides by -3. Reverse the inequality symbol since the multiplier is negative.

$x \leq -12$

Example 3

Solve $4x - 3(2 + 2x) > 10$.

Answer

$4x - 6 - 6x > 10$ — 1. Use the Distributive Property to remove the parentheses.

$-2x - 6 > 10$ — 2. Combine like terms on the left side of the inequality.

$-2x > 16$ — 3. Add 6 to both sides.

$x < -8$ — 4. Divide both sides by -2. Reverse the inequality symbol.

Example 4

Solve $x + 4 \geq 5(x + 6) - 2x$.

Answer

$x + 4 \geq 5x + 30 - 2x$	1. Use the Distributive Property to remove the parentheses.
$x + 4 \geq 3x + 30$	2. Combine like terms on the right side.
$4 \geq 2x + 30$	3. Collect the variable terms on the right side by subtracting x from both sides.
$-26 \geq 2x$	4. Subtract 30 from both sides.
$-13 \geq x$	5. Divide both sides by 2. Do not reverse the inequality symbol.
$x \leq -13$	6. Rewrite the inequality with the variable on the left side.

Collecting the variable terms on the left side would require division by -2 and reversal of the inequality symbol, producing the equivalent statement $x \leq -13$.

Example 5

Geoffrey wants to fence in a rectangular garden and has up to 60 ft of fencing available. If the width must be 5 ft more than half the length, what are the possible dimensions of the garden?

Answer

length $= l$ width $= \frac{l}{2} + 5$	1. Choose a variable for the length and then express the width in terms of the length.
$2l + 2w \leq 60$	2. Write an equation. The perimeter of the garden (rectangle) is at most 60 ft.
$2l + 2\left(\frac{l}{2} + 5\right) \leq 60$	3. Substitute the variable expression used for the width of the rectangle.
$2l + l + 10 \leq 60$	4. Use the Distributive Property to remove the parentheses.
$3l + 10 \leq 60$	5. Combine like terms on the left side.
$3l \leq 50$	6. Subtract 10 from both sides.
$l \leq 16\frac{2}{3}$	7. Divide both sides by 3. Note that $\frac{2}{3}$ ft is the same as 8 in.

The length is at most 16 ft 8 in., and the width is at most half the length plus 5, or 13 ft 4 in.

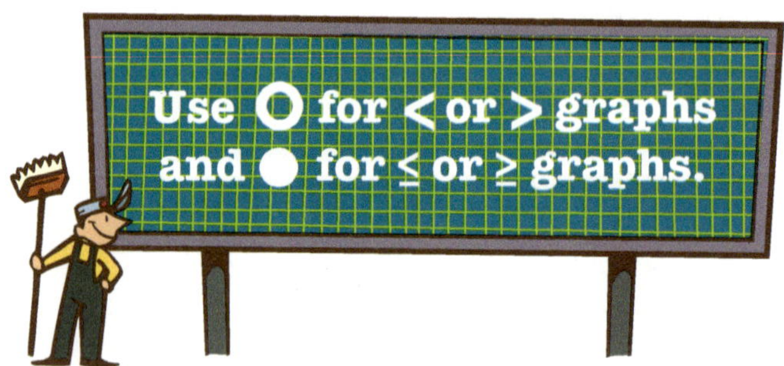

A. Exercises

Solve and graph the solution to each of the following inequalities.

1. $-5x + 3 \geq 13$
2. $8y - 7 > 17$
3. $-\frac{z}{2} + 4 \leq 2$
4. $4x - 6 < -6$
5. $-3y + 10 < 31$

Solve.

6. $\frac{w}{6} + 3 \leq -5$
7. $\frac{2}{3}z + 5 \neq 25$
8. $4a - 4 - 6a > 10$
9. $4c - 12 - 3c < -142$
10. $-4 < 5b - 10 - 3b$

B. Exercises

Solve.

11. $2m - 9 < 3m + 9$

12. $4n + 5 \geq 9n - 2$

13. $-6r + 3 \leq -5r - 2r + 6$

14. $-8s + 4s + 6 < 3s + 6$

15. $9y - 7 \leq 6y - 3 + 2y$

16. $x - 5 + 3x \neq 2x - 4$

17. $-\frac{8}{3}z + 2z - 6 \leq 3z + 5$

18. $-\frac{a}{4} - 6 + a \neq 9$

19. $\frac{11}{8}y + \frac{5}{6} \geq -\frac{1}{4}$

20. $\frac{1}{10} - \frac{3}{30}w < \frac{4}{15}$

21. $\frac{11}{16}c - \frac{1}{6} \leq \frac{7}{12}c + \frac{5}{8}$

22. $6(-2x + 4) \leq 3 - 5x + 7$

23. $-7(d - 1) + d > 9 - 2d$

24. $4 - (3y + 2) > 6 - 5y$

Write an inequality and solve.

25. Find the least positive integer such that 4 more than 7 times the number is greater than 95.

26. Find the greatest three consecutive integers such that the sum of the first and third is less than 20.

27. Mr. Sanders prefers to golf when it above 59°F. Use the formula $F = \frac{9}{5}C + 32$ to determine the temperatures in Celsius at which he will enjoy golfing as he vacations in Europe.

28. Heston earns twice as much per week as his sister Naomi. If she makes $30 per week more than her other brother Jake and the sum of their salaries is at most $330, what range can their salaries take each week?

29. Mr. Daniels is paid a $250 weekly salary and receives a 15% commission on any sales he makes. What is his weekly sales goal if he needs to make at least $700 a week?

30. The doctor has challenged Mr. Jonas, who weighs 240 lb, to achieve his ideal body weight of 186 lb by losing no more than 3 lb/week. How many weeks will it take for Mr. Jonas to achieve his ideal weight following his doctor's orders?

31. Jed bikes for 3 hr, and his brother bikes for 4 hr. Although Jed bikes 15 mi/hr faster, his total distance is at most twice as far as his brother's. How fast does each bike?

32. The president of First State Bank must invest the bank's funds so that the total interest income for the bank will be at least $1.5 million annually. If the bank's current investments will earn $1.3 million in interest this year, how much must he invest in municipal bonds that yield 8% interest annually?

C. Exercises

Solve.

33. $2(x + 8) - x \neq 5 - (x + 3)$

34. $3x + 3 - 5x \nleq 7(2x - 3) - (6x - 4)$

35. $6(-2x + 4) \ngtr 3 - 5x + 7$

36. Mr. Stevens invests $2500 between two accounts. One account earns 5% interest, and the other earns 8%. Since he cannot afford the early withdrawal penalty in the 8% account, he wants to invest as much as possible at 5% and still earn at least $150 this year in interest. What is the least amount that he can invest at 8% and still earn the desired amount of interest?

Dominion Modeling

Proverbs tells us that in order to please God in the practical areas of life, a person needs wisdom, the skill to know what to do in a given situation and the ability to do it. This wisdom is founded on the fear of the Lord (Prov. 9:10), but it also requires other skills. In the case of borrowing and lending, acting wisely requires math skills.

You have to pay interest when you borrow money, so it is wise to find the best rate. Interest rates are expressed in two ways: annual percentage rate (APR) and annual percentage yield (APY). The APR expresses annual interest without considering compound interest. The APY expresses annual interest, including compound interest.

When the lender receives monthly compounded interest on the remaining principal throughout the term of a loan, the actual interest rate is not equal to the simple interest rate (APR). Calculating the APY reveals that you are actually paying a higher rate than the APR.

$APY = \left(1 + \frac{APR}{n}\right)^n - 1$, where n is the number of compounding periods per year.

Example: Find the APY of a 5 yr \$8000 loan with a 7% APR compounded monthly.

Using $n = 12$, $APY = \left(1 + \frac{0.07}{12}\right)^{12} - 1 \approx (1.005833)^{12} - 1 \approx 0.0723 \approx 7.2\%$.

The actual interest rate is about 7.2%, a higher rate than the 7% APR.

To provide protection for borrowers, the Truth in Lending Act requires creditors to provide the borrower a document listing the total finance charges, the annual percentage rate (APR), and the annual percentage yield (APY).

Assuming monthly payments, find the APY for each of the following loans. Round to the nearest tenth of a percent.

37. Mr. Ellison was quoted an interest rate of 9% APR for a loan to expand his business.

38. A credit card company discloses an 18% APR for outstanding balances.

CUMULATIVE REVIEW

True or false [1.2]

39. $-11 > 3$

40. $|-11| < |3|$

Evaluate. [1.8]

41. $-2 - 3(17)$

42. $\frac{3}{16}\left(-\frac{2}{5}\right) + \frac{7}{4}$

State the property that justifies each step in the solution of the equation $\frac{1}{3}x + \frac{4}{9} = \frac{1}{6}(3 + x)$. [2.4–2.8]

43. $18\left(\frac{1}{3}x + \frac{4}{9}\right) = 3(3 + x)$

44. $6x + 8 = 9 + 3x$

45. $3x = 1$

46. $x = \frac{1}{3}$

Solve. [2.7]

47. $3x - 17 = 7x - 24$

48. $17z + 4 = 9z - 40$

SEQUENCES

Arithmetic Sums

Challenge **Can you supply the next two numbers in each of the following sequences?**

a. 1, 3, 6, 10, 15, …

b. 3, 8, 15, 24, 35, …

c. −1, −5, −12, −22, −35, …

Keyword Search
arithmetic sequence

Consider a student who works a part-time summer job at which he cannot exceed 35 hr in one week. If he works 8 hr on Monday, 6 hr on Tuesday, 8 hr on Wednesday, and 7 hr on Thursday, then the sequence 8, 6, 8, 7, representing hours worked each day, is not as helpful as the sequence representing total hours worked for the week. The sequence of sums, $8, 8 + 6, 8 + 6 + 8, 8 + 6 + 8 + 7$, which simplifies to 8, 14, 22, 29, is more meaningful in this 35 hr maximum work week situation. The student can work $35 - 29 = 6$ hr or less on Friday.

A sequence of *partial sums* is defined as

$$A_1, A_1 + A_2, A_1 + A_2 + A_3, A_1 + A_2 + A_3 + A_4, \ldots,$$

where A_n is a given sequence. The notation for the terms of a partial-sum sequence is $S_1, S_2, S_3, S_4, \ldots$.

List the first five partial sums for the following arithmetic sequences.

1, 2, 3, 4, 5, …

3, 5, 7, 9, 11, …

−1, −4, −7, −10, −13, …

Did you get the sequences given in the Challenge? If you were not successful before, you should now be able to supply the next two terms in each of those introductory partial-sum sequences.

The *nth partial sum* for a given sequence is denoted as S_n. Since an arithmetic sequence has a common difference between successive terms, the sum of the first n terms would be the number of terms n times the average of the first and last terms. For any arithmetic sequence A_n, we have $S_n = n \cdot \frac{A_1 + A_n}{2}$, or $S_n = \frac{n(A_1 + A_n)}{2}$. To find S_{20} for $A_n = 2n + 5$, we would first find $A_1 = 2(1) + 5 = 7$ and $A_{20} = 2(20) + 5 = 45$, then use the partial-sum formula to get $S_{20} = \frac{20(7 + 45)}{2} = 520$.

Exercises

List the first five partial sums for the following sequences.

1. 1, 3, 5, 7, 9, …

2. 9, 5, 1, −3, −7, …

3. −10, 0, 10, 20, 30

Find the specific term A_n and the partial sum S_n for the following sequences.

4. $2, 4, 6, \ldots, 2n, \ldots$; A_{20} and S_{20}

5. $A_n = 7n - 5$; A_{10} and S_{10}

Find the sum of the following.

6. the first 100 counting numbers

7. the first 100 odd counting numbers

4.4 Conjunctions

In mathematics, a *compound sentence* is made by connecting two statements using the word *and* or the word *or*. Compound sentences connected with the word *or* are called *disjunctions* and will be studied in the next section.

Double Arch is one of 2000 arches at Arches National Park, Utah.

Definition

A **conjunction** is a compound sentence consisting of mathematical statements connected by the word *and*.

In order for a conjunction to be a true statement, both parts of the conjunction must be true. When solving a conjunction, find all values of the variable that satisfy each of the simple statements. The intersection of these sets, those values that are in both solutions, forms the solution of the conjunction.

Example 1

Graph the solution set of $x > -1$ and $x \leq 4$.

Answer

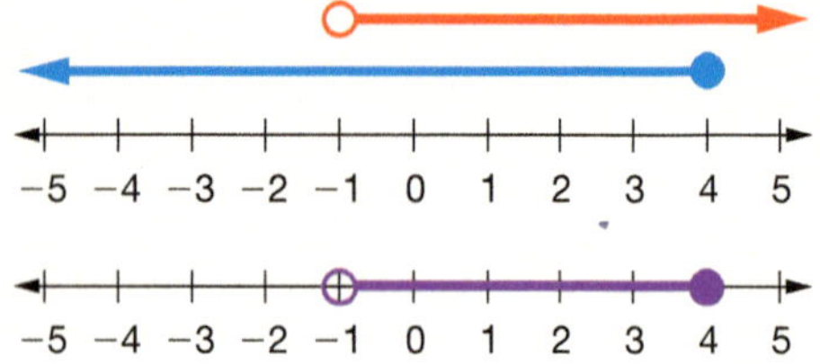

1. Graph each inequality above the number line.
2. Identify the intersection of the two graphs and graph it on the number line.

The intersection of two simple inequalities is often written as a single statement. The solution to Example 1 can be expressed as the combined conjunction $-1 < x \leq 4$, which can be read "-1 is less than x, which is less than or equal to 4." It identifies x as being any number to the right of -1 and at or to the left of 4 on the number line.

The symbol $\wedge$ is often used for the word *and* because of its similarity to the symbol for intersection, $\cap$.

Example 2

Graph and state the solution to $x \geq 5 \wedge x < 0$.

Answer

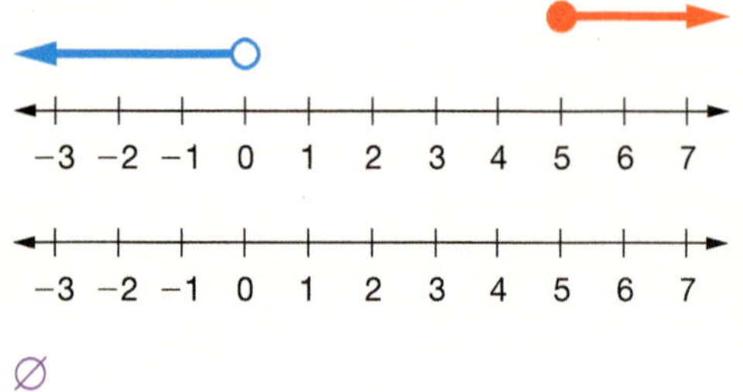

$\varnothing$

1. Graph each inequality above the number line.
2. The two graphs have no points in common, so there are no points in the solution set.
3. The solution is the empty set.

Example 3

Solve and graph the solution set of $2x + 5 \le 11$ and $-3x > 18$.

Answer

$2x + 5 \le 11$ and $-3x > 18$

$2x \le 6$ $\quad$ $x < -6$

$x \le 3$

1. Solve each inequality separately. Don't forget to reverse the inequality symbol when you divide by a negative.

$x \le 3 \wedge x < -6$

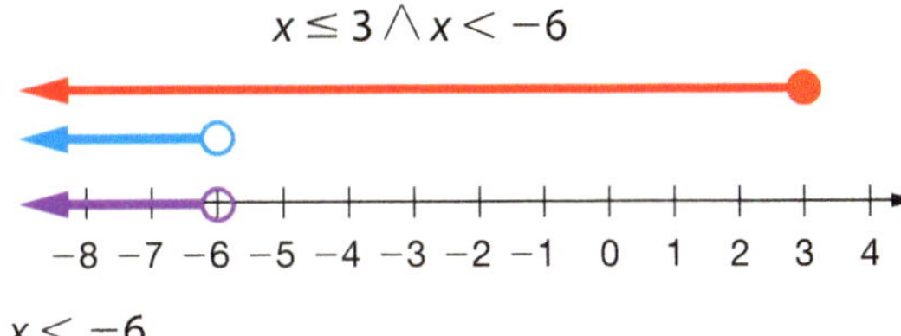

2. Graph each inequality above the number line. Then graph their intersection on the number line.

$x < -6$

3. Write the solution for the intersection.

The combined conjunction $-1 < 3x + 5 < 8$ can be separated into two inequalities joined by the word *and*.

$-1 < 3x + 5$ and $3x + 5 < 8$

Since the same steps would be used to solve each simple inequality, the properties of inequality can be used on each of the three parts of the combined conjunction.

Example 4

Solve and graph the solution set of $-1 < 3x + 5 < 8$.

Answer

$-1 < 3x + 5 < 8$

$-6 < 3x < 3$

1. Subtract 5 from each expression.

$-2 < x < 1$

2. Divide each expression by 3.

3. Graph the solution.

When the inequalities that form the combined conjunction require different steps in their solutions, you must rewrite them as two inequalities separated by the word *and* before solving.

Example 5

Solve and graph the solution set of $x - 3 \le 2x - 9 < 15$.

Answer

$x - 3 \le 2x - 9$ and $2x - 9 < 15$

1. Separate the combined conjunction into two inequalities separated by *and*.

$-3 \le x - 9$ $\quad$ $2x < 24$

$6 \le x$ $\quad$ $x < 12$

$x \ge 6$

2. Solve each inequality.

$x \ge 6 \wedge x < 12$

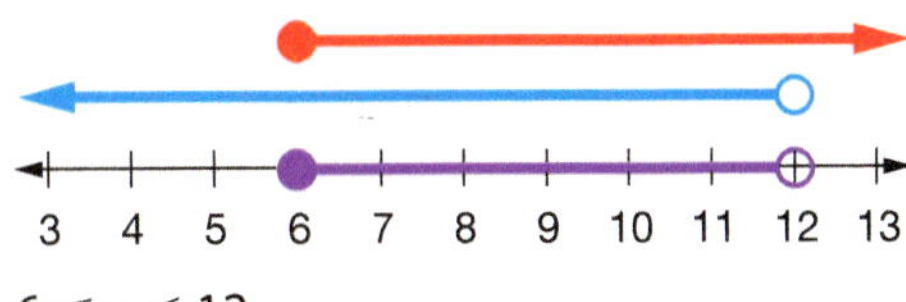

3. Graph each inequality above the number line. Then graph their intersection on the number line.

$6 \le x < 12$

4. Write the solution as a compound inequality.

Compound inequalities can be used to solve problems related to ranges of values. The grade assigned in a class, the percentage of income that is paid in taxes, and the strength of a hurricane are all based on a range of values.

Example 6

If Brittany received a 73 and an 85 on her first two tests, what grades can she get on her final test and get a B in the class?

Grade	Test Average
A	90
B	80
C	70
D	60
F	below 60

Answer Let t = the score on the third test.

1. **Read** the problem carefully. Define a variable for the unknown quantity.

$80 \leq \frac{73 + 85 + t}{3} < 90$

2. **Plan**: Write a combined inequality.

$240 \leq 158 + t < 270$

$82 \leq t < 112$

3. **Solve**: Multiply each expression by 3. Then subtract 158 from each expression.

Check She must get at least an 82.

A score of 112 needed to get an A is not possible.

4. **Check**: $82 + 73 + 85 = 240$ and $240 \div 3 = 80$. Any grade below an 82 results in an average less than 80.

A. Exercises

Graph each conjunction.

1. $x > 4$ and $x \geq 2$
2. $x \leq 3$ and $x < 6$
3. $x > -3 \wedge x \leq 1$
4. $x > 5 \wedge x \leq -2$
5. $x < -1 \wedge x \geq 4$
6. $x \leq 5 \wedge x \geq 3$

Solve. Express the answer as an inequality statement or the empty set.

7. $2x > -6$ and $x \leq 1$
8. $x \geq 5$ and $4x < 32$
9. $-5x + 13 > -7$ and $4x - 6 \leq 10$
10. $5x + 3 < 28 \wedge 6x < 12$
11. $2x - 4 \leq 4x$ and $x + 12 < 18$
12. $2x - 7 \geq 5 \wedge -7x < -49$

B. Exercises

Solve. Express the answer as an inequality statement, an equality, or the empty set.

13. $x - 3 \leq 4x$ and $2x + 6 < 18$
14. $9x + 2 < -7$ and $-2x - 8 \leq 12$
15. $2x - 1 < 5 \wedge 3x + 2 > -4$
16. $4x - 8 < 2x + 5$ and $4x + 4 > -2x - 4$
17. $4x - 2 - 5x > 10 \wedge 3x - 6 \leq 12$
18. $8x - 14 + x - 6 \leq 7 \wedge x - 16 \geq 0$
19. $x \neq -3 \wedge x = -3$
20. $4x \geq 8$ and $x + 6 \leq 8$
21. $4x + 6 \leq 26$ and $3x \neq 12$
22. $3x + 4 - 2x \leq 12$ and $2x + 6 \neq 8$

Solve each combined conjunction. Express the answer as an inequality statement or the empty set.

23. $-2 < x + 4 < 7$
24. $-4 \leq x + 3 \leq 8$
25. $|-11| < 3x - 1 < 20$
26. $3 < \frac{x}{2} + 5 \leq 7$
27. $-6 \leq -2x + 4 < -2$
28. $20 > 4 - 6x > -16$

The layers of the ocean can be classified as shown in the diagram to the right.

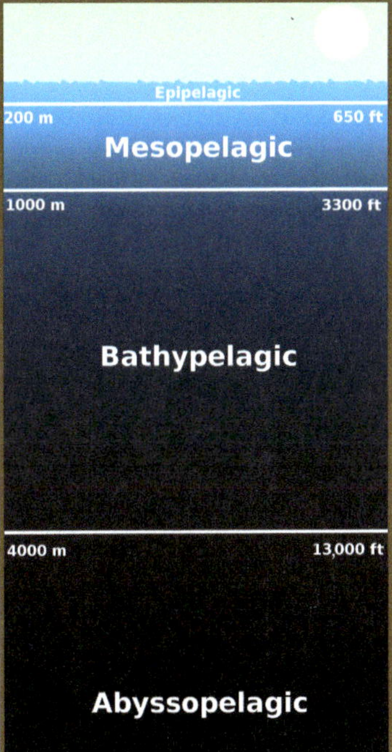

29. God's special design for several species of anglerfish that live in the ocean's bathypelagic zone includes a bioluminescent structure at the tip of a spine, which acts as bait for other fish. Write an inequality to describe the depth (in feet) at which these anglerfish live.

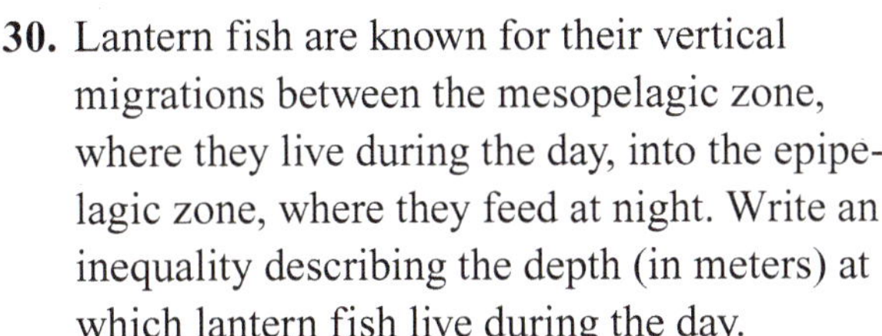

30. Lantern fish are known for their vertical migrations between the mesopelagic zone, where they live during the day, into the epipelagic zone, where they feed at night. Write an inequality describing the depth (in meters) at which lantern fish live during the day.

31. Arches National Park classifies an opening with a span of at least 3 ft as an arch. The longest arch in the park is Landscape Arch, 306 ft long. Write a compound inequality for arch spans in the park.

Write an inequality and solve.

32. Find the numbers such that four less than half the number is at least five and at most ten.

33. Find the angles whose supplements measure more than three times the angle and whose complements measure less than twice the angle.

34. The Saffir-Simpson Hurricane Scale rates a storm's intensity by its sustained wind speed. By the time a hurricane reached the coast, its sustained winds had decreased by 25 mi/hr, so it was downgraded to a Category 3 hurricane. Determine the possible sustained wind speeds of the hurricane at its peak.

Category	Wind Speed
1	74–95 mi/hr
2	96–110 mi/hr
3	111–130 mi/hr
4	131–155 mi/hr
5	156+ mi/hr

35. Hooke's law relates the distance a spring is stretched, d, to the force, F, applied to the spring by the formula $F = kd$, where k is an experimentally determined constant. An industrial scale has $k = 15$ lb/in. Determine the distances that the spring will be stretched as the scale weighs car parts between 240 lb and 360 lb.

C. Exercises

Solve.

36. $2x < x - 4 < 5$

37. $x + 7 < 3x - 5 < 2x + 10$

38. $\frac{2x}{3} - 1 < 10 \wedge \frac{4x}{5} - 2 > 6$

39. $3(x - 5) + 7 \leq 4 + x - 20 \wedge x - 3(x + 1) \leq 5$

40. A pen whose length is twice its width is to be fenced in with an area of at least 800 ft^2. No more than 150 ft of fencing is to be used. Find the range of widths and lengths that can be used.

41. A plumbing retailer sells three times as many of faucet A as he does of faucet B. Faucet A sells for \$75, and faucet B sells for \$60. How many of faucet B should he order if he wants to sell at least 100 faucets and if he wants to gross at least \$10,000?

Dominion Modeling

In the parable of the talents (Matt. 25:13–30), Jesus tells a story about servants who were to manage money for their master. The servant who did not do anything with the money entrusted to him was condemned as wicked. The servants who made wise investments were praised. Making wise financial investments is not the point of this parable, but Jesus certainly commends it.

In order to be a wise investor, you should know the meanings of APR and APY and the difference between them. In Section 4.3, Dominion Modeling, we calculated the APY on a loan from the commonly advertised APR. In this section we will focus on investors wanting to save money rather than borrow it. You will calculate the APR when given the APY, which is often advertised for investments because it includes the compounding effect.

Solving the formula from the previous section for *APR* produces the following formula.

$APR = n(1 + APY)^{\frac{1}{n}} - n$, where n is the number of compounding periods per year.

Example: Find the APR of a \$2000 30 mo CD with an advertised 3% APY if interest is compounded monthly.

Using $n = 12$, $APR = 12(1 + 0.03)^{\frac{1}{12}} - 12 \approx 12.0296 - 12 = 0.0296 = 2.96\%$.

Notice that there is little difference in the interest rates. But once again, the APY is greater because it includes the effect of compounding.

Assuming monthly compounding of interest, find the APR for each of the following investments. Round to the nearest tenth of a percent.

42. \$100,000 deposited into a 3 yr CD at 4.5% APY

43. a \$10,000 investment at 4% APY for 10 yr

44. the same \$10,000 investment at 6% APY for 10 yr

CUMULATIVE REVIEW

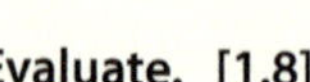

Evaluate. [1.8]

45. $\frac{175}{36} \cdot \frac{45}{140} - \frac{35}{64}$

46. $\frac{175}{36} \div \frac{45}{140} - \frac{64}{27}$

Simplify, leaving each answer in positive exponential form. [1.7]

47. $2x^3(3x^2y)$

48. $(5^{-1}z^2)^{-3}$

Solve. [3.2]

49. $\frac{x}{36} = \frac{7}{8}$

50. $\frac{x + 2}{5} = \frac{5}{12}$

Solve each literal equation for the indicated variable. [3.1]

51. $L = \pi rl$ for l

52. $c = 2ab$ for b

Use proportions to solve. [3.3]

53. Twelve of the forty-six trees on Mr. Schmidt's property are oak trees, and the rest are pine trees. What is the ratio of pine trees to oak trees?

54. A map has a scale of 1 in. : 80 mi. How far apart on the map are two towns that are 260 mi apart?

MIND OVER MATH

Use the clues to find the digits to solve each problem. No two letters represent the same number.

1. four
$+$ one
five

____

Clues
v = 2e
e = f − u
u = 5
$f = 2^3$

2. soft
$+$ ball
game

____

Clues
b = 2t
$f = a^2$
e = 8
l = 5

3. four
$-$ two
ten

____

Clues
w = 3
n = r + e
o = 8

4.5 Disjunctions

Visible light is a small range of the entire electromagnetic spectrum.

Definition

A **disjunction** is a compound sentence consisting of mathematical statements connected by the word *or*.

For a disjunction to be true, only one of the statements needs to be true. The solution to a disjunction will consist of any value that is part of the solution set to either inequality. To solve a disjunction, graph the two simple inequality statements above the number line. Then find the union of the two graphs by including any value that is found on one or both of the graphs. Graph the union on the number line.

Example 1

Graph the solution set of $x < -4$ or $x > -1$.

Answer

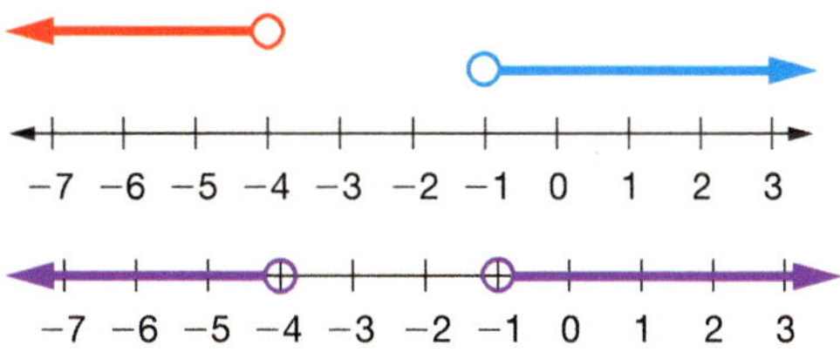

1. Graph each inequality above the number line.
2. Identify the union of the two graphs and graph it on the number line.

The solution to a disjunction cannot be written as a combined statement. Instead, the symbol $\vee$, which is similar to the $\cup$ used for union, can replace the word *or*.

$x < -4 \vee x > -1$

Only conjunctions, statements connected by the word *and*, can be written as combined statements.

Example 2

Solve and graph the solution set of $x < 4 \vee x \leq 2$.

Answer

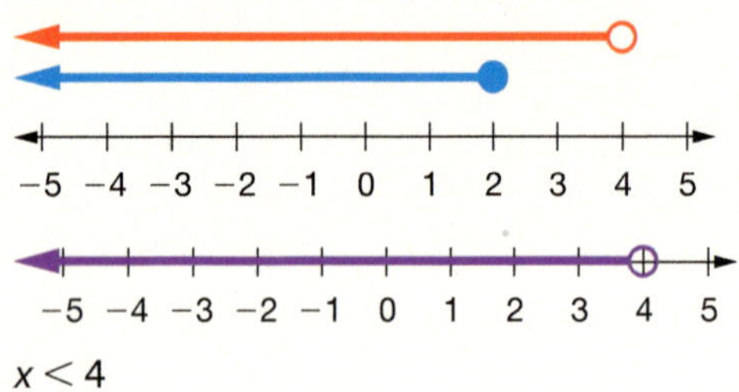

1. Graph each inequality above the number line.
2. Identify the union of the two graphs and graph it on the number line.

$x < 4$

3. Write the solution for the union.

Example 3

Solve and graph the solution set of $3x + 4 > 16$ or $-5x \leq 20$.

Answer

$3x + 4 > 16$ or $-5x \leq 20$

$3x > 12$ $\quad$ $x \geq -4$

$x > 4$

1. Solve each inequality separately. Don't forget to reverse the inequality symbol when you divide by a negative.

$x > 4 \vee x \geq -4$

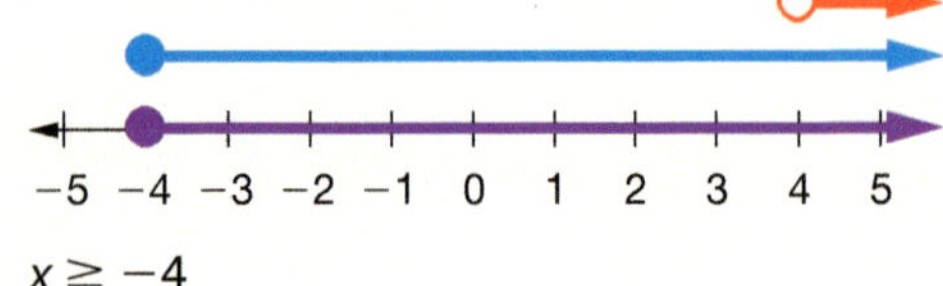

2. Graph each inequality above the number line. Then graph their union on the number line.

$x \geq -4$

3. Write the solution for the union.

Example 4

Solve and graph the solution set of $3x - 6 \leq 12 \vee 5x > -25$.

Answer

$3x - 6 \leq 12$ $\vee$ $5x > -25$

$3x \leq 18$ $\quad$ $x > -5$

$x \leq 6$

1. Solve each inequality separately.

$x \leq 6 \vee x > -5$

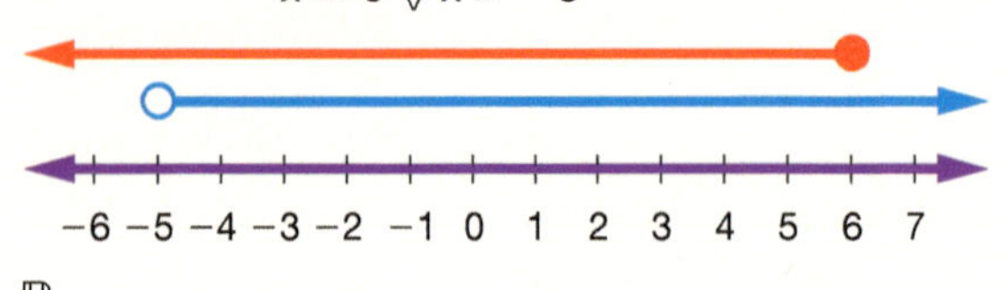

2. Graph each inequality above the number line. Then graph their union on the number line.

$\mathbb{R}$

3. Write the solution for the union. Since the entire number line is shaded, x can be any real number.

A. Exercises

1. In a conjunction, inequality statements are joined by the word ___.
2. In a disjunction, inequality statements are joined by the word ___.
3. The solution to a disjunction is the ___ of the two solution sets.
4. The solution to a conjunction is the ___ of the two solution sets.
5. The symbol $\vee$ indicates a(n) ___.
6. The symbol $\wedge$ indicates a(n) ___.

Graph each disjunction.

7. $x < 2 \vee x > 5$
8. $x < 2$ or $x < 5$
9. $x > 2 \vee x > 5$
10. $x > 2$ or $x < 5$
11. $x < -4 \vee x > -4$
12. $x < -3$ or $x \leq -2$

B. Exercises

Solve. Express the answer as an inequality statement or the set of real numbers.

13. $x \neq 6$ or $x \neq 3$
14. $2a + 5 > 17$ or $a + 2 \leq 8$
15. $y - 6 \leq -6 \vee 3y - 5 > 10$
16. $3z + 5 > 2z$ or $6z - 3z + 4 \geq 16$
17. $2b - 6 \geq 14 \vee 8b + 2b - 3 > 27$
18. $r - 7 > 0$ or $2r + 16 \leq 2$
19. $3x + 6 \leq -9$ or $x < 6$
20. $2s - 7 > -21 \vee 4s + 6 \neq 22$

Solve each compound inequality. Express the answer as an inequality statement, the set of real numbers, or the empty set.

21. $3 - 2x > -2 \vee 3x + 6 < 8$
22. $7y > 2y + 10$ and $2y > 5y - 15$
23. $4m - 4 \leq 6m - 8 \wedge -3m + 1 \geq 5m - 3$
24. $4x + 6 > 10 - x$ or $9 - 3x \geq -1$
25. $3n - 2 < 4 \vee 2n + 8 < 5n + 11$
26. $6 - 3x \geq 4 - 2x > 5 - 4x$
27. $2z + 3 \leq -2$ or $3z - 5 > 4$
28. $2c + 50 < 7c + 2 + 3c$ and $5 - 4c < 53$

Visible light is a very small part of the larger spectrum of electromagnetic radiation. Write a compound inequality describing each of the following.

29. the frequency of x-rays
30. the frequency of nonvisible electromagnetic radiation
31. the wavelength of nonultraviolet electromagnetic radiation
32. the wavelength of microwaves

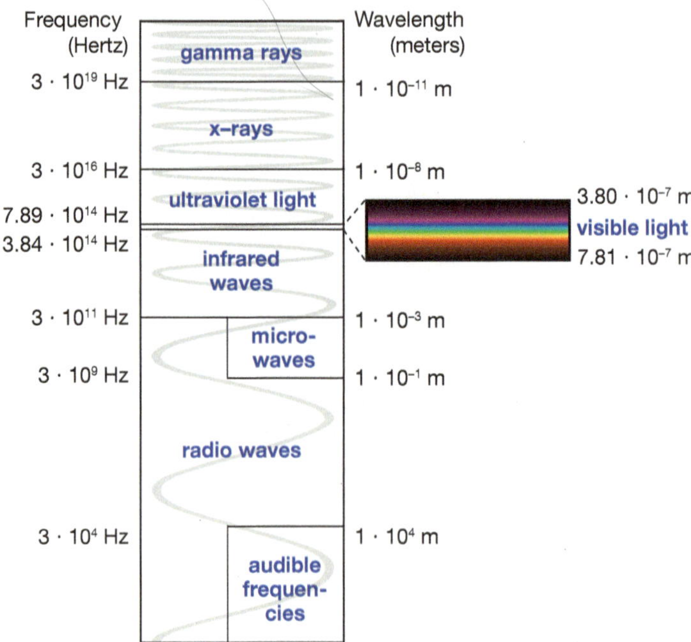

Write a compound inequality describing all the numbers that are *not* in the solution set of the following compound inequalities.

33. $-3 \leq x \leq 1$
34. $x > 4$ or $x < 2$
35. $x \leq -5 \vee x > 0$
36. $x > -8 \wedge x \leq 13$

C. Exercises

Solve.

37. $5x - 1 < 1$ or $3x - 2 \neq 5$
38. $\frac{5}{7}x + \frac{1}{2} < 2 \vee \frac{3}{4}x - \frac{1}{5} \leq \frac{1}{10}$
39. $3x + 1 \geq 6 \vee 9(4 - x) \leq 15$
40. A photo studio sells 5" × 7" portraits for $45 and 8" × 10"s for $75. They consistently sell about 150 more 5" × 7"s than 8" × 10"s each month. They want to meet at least one of the following goals next month: sell at least 400 portraits or gross at least $20,000. How many 8" × 10"s do they need to sell to meet one of their goals? Both goals?

Dominion Modeling

To be faithful with God's resources, we should be knowledgeable about amortized loans since they are so common; for instance, this is the type of loan used to buy a home. An amortized loan is one in which equal periodic payments are made throughout the life of the loan (usually monthly). Like the add-on loan, each payment is for the same amount. Unlike the add-on loan, the interest is figured on the unpaid balance. At first, most of the money goes to pay interest, with only a small amount going toward the principal. However, as the borrower nears the end of the contracted time, nearly all of the payment goes toward principal and only a small amount is interest.

The following two formulas are finance formulas. The first finds the monthly payment, R, you would need to make on an amortized loan.

$R = P\left[\frac{i(1+i)^n}{(1+i)^n - 1}\right]$, where P = the principal,

n = the total number of payments (payments per year × number of years), and

i = the periodic interest rate (APR ÷ payments per year).

The second formula determines the monthly deposit, D, you would need to make to accumulate S dollars in an account with compound interest (assuming no withdrawals).

$$D = S\left[\frac{i}{(1+i)^n - 1}\right]$$

41. Find the monthly payment on a 2 yr $5500 loan with 6% annual interest.
42. Find the monthly deposit you would need to make in order to save $5500 at 6% annual interest in 2 yr.
43. Why are the answers to exercises 41 and 42 different? The amounts are the same, the rates are the same, and the time is the same.

CUMULATIVE REVIEW

Simplify, leaving each answer in positive exponential form. [1.7]

44. $a^2b^6c^{-3}(a^3b^{-1}c^5)$
45. $(3xy^{-2}z^3)^{-3}$

Simplify. [1.8, 2.3]

46. $6 - 8(-3) \div 2$
47. $5x - 4(3x - 8) - (x - 1)$

Evaluate when $x = -2$, $y = 3$, and $z = -5$. [2.2]

48. $2xy - y^2z$
49. $(2yz - x)^2$

Perform the indicated operations. [1.2]

50. $|13 - 58|$
51. $|-45| - |61|$
52. Define the *absolute value* of a number. [1.2]
53. Find the length and the midpoint of a segment having endpoints at -4 and 17. [1.2]

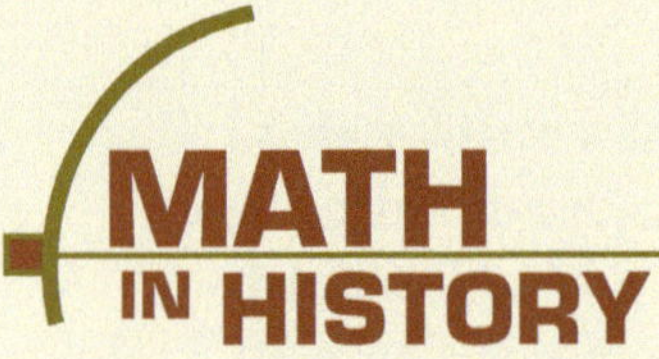

The Bernoulli Family

The Bernoulli family settled in Switzerland after fleeing religious persecution from Catholics in Belgium. This leading family of European mathematicians produced eight mathematicians within three generations.

While complying with his parents' desire that he obtain degrees in philosophy and theology, Jakob (1654–1705) independently studied mathematics and astronomy. He was appointed professor of mathematics at the University of Basel in 1687. Jakob's work in differential and integral calculus is applied today in electrical engineering, and his law of large numbers is widely used in statistics and insurance studies. It is applied in risk aversion, in particular by insurance companies, since they must pay off on some policies earlier than expected and on some much later than expected or not at all.

Unfortunately, the family is remembered for its internal conflicts as well as its many contributions to the field of mathematics. After failing as an apprentice in his father's business, Johann (1667–1748) entered the University of Basel to study medicine and was tutored by his older brother Jakob in mathematics. Johann quickly became Jakob's equal as they studied Leibniz's new calculus together. Johann's boasts led to unfair public criticism by Jakob, and their collaboration eventually turned to bitter rivalry. While Johann served as the chair of mathematics at Gronigen University, he tutored Guillaume de l'Hôpital in Leibniz's methods of calculus. In 1696 l'Hôpital published the first differential calculus book without acknowledging that the work was based on Bernoulli's lectures. Johann did not return to Basel until eight years later after Jakob's death, when he accepted the long-coveted position held by his brother. His accomplishments were in optical phenomena, analytical trigonometry, and exponential calculus.

Johann's nephew, all three of his sons, and two of his grandsons became famous mathematicians and scientists. Johann competed with his son Daniel (1700–1782), falsely dating his work on hydraulics in 1732, prior to Daniel's work entitled *Hydrodynamica* (completed in 1734).

Ecclesiastes 4:4 tells us that a person will never find satisfaction in his work if it is motivated by envy. While the members of this family contributed many advances in calculus, probability, differential equations, and science, how much more could have been done if they had cooperated with each other?

4.6 Absolute Value Equations

Turkey's capital city, Ankara, dates back to the time of the Hittites, circa 2000 BC. Its age is approximately $|2000 - (-2000)| = 4000$ years.

Recall that the absolute value of a number can be thought of as its graph's distance from the origin.

$|6| = 6$ and $|-6| = 6$

The absolute values of 6 and -6 are the same because their graphs are both 6 units away from 0 on the number line. A more precise definition of absolute value is required when more complicated expressions are included within the absolute value symbols.

Definition

The **absolute value** of a number, x, is defined as follows:

$$|x| = \begin{cases} x \text{ if } x \geq 0 \\ -x \text{ if } x < 0 \end{cases}.$$

This definition says that if x is a nonnegative number, such as 6, then $|x|$ is equal to the number itself. So $|6| = 6$. However, if x is a negative number, such as -6, then $|x|$ equals the opposite of the number. So $|-6| = -(-6) = 6$.

Consider the equation $|x| = 4$. Since we do not know whether x is positive or negative, the definition leads to two possible conclusions.

1. If $x \geq 0$, then $|x| = 4$ implies that $x = 4$.
2. If $x < 0$, then $|x| = 4$ implies that $-x = 4$ and that $x = -4$.

Thus, the equation $|x| = 4$ has two solutions. The notation $x = \pm 4$ is often used to indicate the positive and negative solutions.

What are the two solutions to the equation $|x| = 2.5$? Notice that the solution is both the positive and the negative values of the number that the absolute value equals.

$|x| = 2.5$ implies that $x = 2.5$ or $x = -2.5$.

The graph of the solution shows all the points that are 2.5 units from the origin on the number line.

−5 −4 −3 −2 −1 0 1 2 3 4 5

Will an absolute value equation always have two solutions? Consider the equation $|x| = 0$. This equation has only one value that will make the statement true since 0 is the only number whose absolute value is 0. Also, consider the equation $|x| = -4$. This equation has no solution since there is no number that has a negative absolute value.

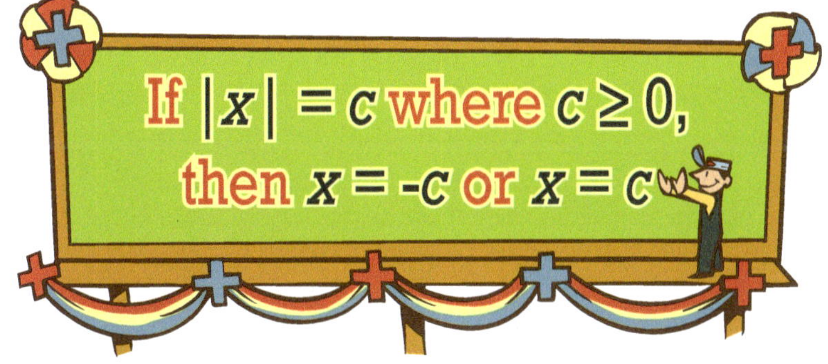

Example 1

Solve $|2x + 5| = 11$.

Answer $2x + 5 = 11$ or $2x + 5 = -11$ — 1. The expression within the absolute value symbols must equal 11 or −11.

$2x = 6 \qquad 2x = -16$ — 2. Solve each equation separately.

$x = 3 \qquad x = -8$

Check — 3. Check your solutions.

$|2(3) + 5| \qquad |2(-8) + 5|$

$= |6 + 5| \qquad = |-16 + 5|$

$= |11| = 11 \qquad = |-11| = 11$

The values for x are 3 or −8.

Solving Absolute Value Equations

1. Simplify within the absolute value symbols.
2. Isolate the absolute value on one side of the equation.
3. Use the definition of absolute value to write two equations.
4. Solve each equation separately.
5. Check your solutions.

Example 2

Solve $5 + |2y + 4 - 5y| = 21$.

Answer

$5 + |-3y + 4| = 21$ — 1. Simplify within the absolute value symbols.

$|-3y + 4| = 16$ — 2. Subtract 5 from both sides to isolate the absolute value.

$-3y + 4 = 16$ or $-3y + 4 = -16$ — 3. Use the definition of absolute value to write two equations.

$-3y = 12 \qquad -3y = -20$ — 4. Solve each equation separately.

$y = -4 \qquad y = \frac{20}{3}$

Check — 5. Check your solutions.

$5 + |-3(-4) + 4| \qquad 5 + \left|-3\left(\frac{20}{3}\right) + 4\right|$

$= 5 + |12 + 4| \qquad = 5 + |-20 + 4|$

$= 5 + |16| \qquad = 5 + |-16|$

$= 5 + 16 = 21 \qquad = 5 + 16 = 21$

$y = -4, \frac{20}{3}$

Be careful with equations such as $|x + 5| = -3$. Remember that the absolute value of any expression cannot equal a negative number, so state the solution as $\varnothing$. Incorrectly applying the definition of absolute value to state that

$$x + 5 = 3 \text{ or } x + 5 = -3$$

and solving to get $x = -8$ or -2 appears to produce two solutions. Substituting these values into the original equation shows that they are not solutions.

$$|(-8) + 5| = |-3| \neq -3 \text{ and } |(-2) + 5| = |3| \neq -3$$

Recall from Chapter 1 that the distance between two points on a number line can be determined using the absolute value of the difference of their coordinates, $|a - b|$. This distance can also be written as $|b - a|$ since these absolute values are equal.

The absolute value equation $|x - 5| = 3$ has two solutions, $x = 2$ or $x = 8$. By graphing these solutions, we can see that both solutions are a distance of 3 units from 5 on the number line.

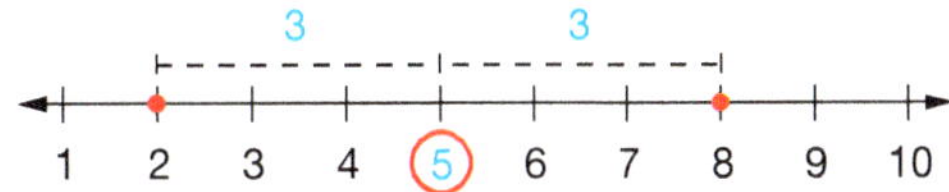

Notice that 5 is the coordinate of the midpoint of the segment having endpoints at 2 and 8, which are both 3 units from the midpoint.

Example 3

Write and solve absolute value equations to find the numbers that are as follows.

a. 4 units from 6 **b.** 2 units from -7

Answer

a. $|x - 6| = 4$

1. The distance is 4 units, so the absolute value equals 4. Since the distances are measured from 6, 6 is subtracted from x within the absolute value symbols.

$x - 6 = 4$ or $x - 6 = -4$

2. Use the definition of absolute value to write two equations.

$x = 10, 2$

3. 10 and 2 are both 4 units from 6.

b. $|x - (-7)| = 2$

1. The distance is 2 units, so the absolute value equals 2. Since the distances are measured from -7, -7 is subtracted from x within the absolute value symbols.

$x + 7 = 2$ or $x + 7 = -2$

2. Use the definition of absolute value to write two equations.

$x = -5, -9$

3. -5 and -9 are both 2 units from -7.

A. Exercises

Solve. Write ∅ if no solution exists.

1. $|x| = 3$
2. $|x| = 21$
3. $|x| - 6 = -1$
4. $4|x| = 44$
5. $-3|x| = -8$
6. $-5|x| = 35$
7. $|x| + 8 = 5$
8. $|x + 5| = 7$
9. $|3y| = 129$
10. $|7x| = 4$
11. $|x + 2| = -7$
12. $|x - 8| = 14$
13. $\left|x - \frac{1}{2}\right| = \frac{3}{2}$
14. $|y + 4| = 8.5$

B. Exercises

Solve. Write ∅ if no solution exists.

15. $|x + 1| + 2 = 8$
16. $|3z + 4 - z| = 16$
17. $|2y - 5| - 9 = 24$
18. $|2x - 3| + 7 = 2$
19. $|5y + 12| = 24$
20. $-1 + |7x + 9| = 50$
21. $19 = |12x - 5|$
22. $|3x - 4 + 4x| - 6 = 3$
23. $|2z - 10 + 5z| - 20 = -2$
24. $|x - 1.2| - 2 = 11$
25. $-36 = 2 - |7x + 3|$
26. $|3y + 4 - 6y| - 14 = 12$

Write an absolute value equation representing the following; then solve the equation.

27. numbers that are 4 units from 2
28. numbers that are 7 units from -3
29. numbers that are 8 units from -4
30. numbers that are 5 units from 15
31. numbers that are 15 units from 5
32. Explain why $|x - 4| = -2$ has no meaning in relation to the number line.

C. Exercises

Use the average of each pair of numbers to write an absolute value equation having the following solutions.

33. 7 and 13

34. -4 and -10

35. -15 and 3

36. -7 and 20

Dominion Modeling

Wise investors must distinguish among several investments to determine how they can best manage the resources entrusted to them. Determining the annual percentage yield (APY), as discussed in Section 4.3, Dominion Modeling, allows you to compare different investments.

Example: Which yields the better rate of return, 5% simple interest or 4.7% interest compounded daily?

5% simple interest: $APY = \left(1 + \frac{APR}{n}\right)^n - 1 = \left(1 + \frac{0.05}{1}\right)^1 - 1 = 1.05 - 1 = 0.05 = 5\%$

4.7% interest compounded daily: $APY = \left(1 + \frac{APR}{n}\right)^n - 1 = \left(1 + \frac{0.047}{365}\right)^{365} - 1 \approx 1.0481 - 1$
$= 0.048 = 4.8\%$

The 5% simple interest rate has a higher APY than 4.7% compounded monthly.

Keyword Search
APR APY calculator

Find the APY (to the nearest hundredth of a percent) for each of the following pairs of investments; then determine which has the greater yield.

37. 6% simple interest or 5% interest compounded quarterly

38. 4.4% interest compounded quarterly or 4.2% interest compounded monthly

39. 5.6% interest compounded monthly or 5.65% simple interest

40. 3.75% interest compounded daily or 3.8% interest compounded semiannually

CUMULATIVE REVIEW

Write an equation and solve.

41. 38 is what percent of 40? [3.4]

42. Increase $250 by 3.5%. [3.5]

43. What is the percent change (to the nearest percent) of a stock that dipped from $83 per share to $65 per share? [3.5]

44. What is the original retail price of a sweater that has been discounted by 40% for a sale price of $36? [3.6]

45. Determine the commission earned by Zach if he receives 15% on last week's sales of $2500. [3.6]

46. If the ratio of pine trees to other trees on Mr. Locke's property is 2 : 3 and he has a total of 60 trees, how many trees does he have that are not pine trees? [3.2]

47. If two congruent triangles have corresponding sides in the ratio 4 to 5 and the larger triangle has a perimeter of 20 in., what is the perimeter of the smaller triangle? [3.3]

48. Lawrence deposits $4000 in a CD (certificate of deposit) earning 4% simple interest. Determine the interest earned and the total amount in his CD at the end of 2 yr. [3.6]

49. A plane leaves Charlotte at 9:00 AM and flies north at 400 knots. At the same time, another plane flies south from Charlotte at 250 knots. How long is it before they are 1000 nautical miles apart? [3.7]

50. Coffee selling for $10/lb is mixed with coffee selling for $4/lb. To the nearest pound, how much of each is used to produce 50 lb of a mixture that sells for $8/lb? [3.8]

4.7 Absolute Value Inequalities

Recall that the absolute value of a number refers to its distance from the origin. This will help us interpret absolute value inequalities. To find the solutions to $|x| < 3$, we must find numbers whose distance from the origin is less than 3. For instance, the numbers -2, 1.5, 0, and -0.5 will make this sentence true. The graph illustrates all the solutions to this inequality.

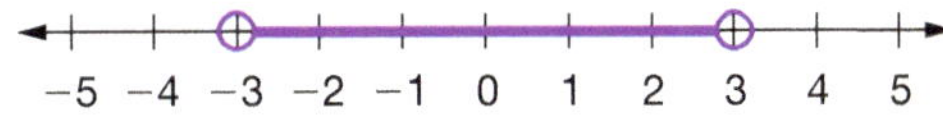

This inequality can also be written without the absolute value symbols as a conjunction using the word *and*,

$x > -3$ and $x < 3$,

or as the combined conjunction

$-3 < x < 3$.

The solution to the absolute value inequality $|x| \geq 3$ includes all numbers whose distance from the origin is greater than or equal to 3. The graph illustrates that this includes -3 and all numbers to its left as well as 3 and all numbers to its right.

Notice that this inequality is written as a disjunction using the word *or*,

$x \leq -3$ or $x \geq 3$.

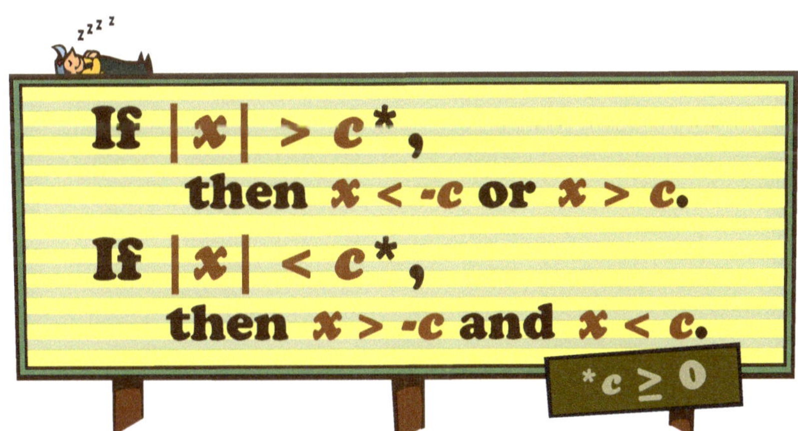

Example 1

Use a disjunction or a conjunction to state the solution to each inequality, and graph the solution set.

a. $|x| > 1$ **b.** $|x| \leq 4$

Answer

a. $x < -1$ or $x > 1$

A disjunction must be used to describe distances greater than 1 from the origin.

b. $-4 \leq x \leq 4$

A conjunction must be used to describe distances less than or equal to 4 from the origin.

Example 2

Solve and graph the solution set of $|x + 5| \leq 10$.

Answer

$|x + 5| \leq 10$

$x + 5 \geq -10$ and $x + 5 \leq 10$

$x \geq -15$ $\quad$ $x \leq 5$

$x \geq -15 \wedge x \leq 5$

−20 −15 −10 −5 0 5 10 15 20

$-15 \leq x \leq 5$

1. The symbol $\leq$ signifies a conjunction.
2. Write a conjunction and solve the two inequalities.
3. Graph each inequality above the number line. Then graph their intersection on the number line.
4. Write the solution as a combined conjunction.

Example 3

Solve and graph the solution set of $|-3x + 6| > 18$.

Answer

$|-3x + 6| > 18$

$-3x + 6 > 18$ or $-3x + 6 < -18$

$-3x > 12$ $\quad$ $-3x < -24$

$x < -4$ $\quad$ $x > 8$

$x < -4$ or $x > 8$

−5 −4 −3 −2 −1 0 1 2 3 4 5 6 7 8 9 10

1. The symbol $>$ signifies a disjunction.
2. Write a disjunction and solve the two inequalities.
3. Graph the union on the number line.

As with absolute value equations, be sure the absolute value is isolated before you write the two separate inequalities. The following table contrasts the key differences in absolute value inequalities.

Absolute Value Inequality	Type of Solution	Set Operation
$<$ or $\leq$	conjunction (and)	intersection
$>$ or $\geq$	disjunction (or)	union

Absolute value inequalities can be used to describe a range of values.

Example 4

Most human muscle cells function well within 2° of their optimal temperature, 98.6°F. Describe this optimal range of temperature with an absolute value inequality; then solve to find the range.

Answer

$|t - 98.6| \leq 2$

$t - 98.6 \leq 2$ and $t - 98.6 \geq -2$

$t \leq 100.6$ $\quad$ $t \geq 96.6$

$96.6 \leq t \leq 100.6$

1. The temperature must be within 2° of the optimal point.
2. Write a conjunction and solve the two inequalities.
3. Write the solution as a combined conjunction.

A. Exercises

Solve and graph the solution on a number line.

1. $|x| > 3$
2. $|x| < 10$
3. $|x| \leq 5$
4. $|x| \geq 7$
5. $|2x| > 16$
6. $|-3x| > 6$
7. $|x - 3| \leq 10$
8. $|x + 6| \geq 5$

B. Exercises

Solve.

9. $|2x + 6| \geq 14$
10. $|3x - 10| < 9$
11. $|4x + 17| > 7$
12. $\left|\frac{2x}{3} - 7\right| \leq 4$
13. $\left|\frac{5}{9}x + 6\right| > 5$
14. $|2x - 8| < 10$
15. $|4x - 7| \geq 29$
16. $|-4x - 8| > 4$
17. $|3x + 2| - 5 \leq 1$
18. $3|4x - 6| < 72$
19. $|7x + 9 - 6x| > 8$
20. $|2x - 6 + 5x| < 10$
21. Write an absolute value inequality representing all numbers that are at most 2 units from 5.
22. Write an absolute value inequality representing all numbers that are at least 2 units from 5.
23. Machined parts are manufactured with a given *tolerance*, the permissible limit of variation. Write an absolute value inequality representing the length of a bolt that is to be 3 cm long with a tolerance of ± 0.02 cm. Then solve the inequality to state the range of acceptable lengths.
24. The hole for the bolt in the previous exercise is designed to be 5 mm in diameter with a tolerance of ± 0.05 mm. Write an absolute value inequality representing the diameter of the hole. Then solve the inequality to state the range of acceptable diameters.
25. The USDA has defined the danger zone for meat, in which bacteria are most likely to grow, as between 40°F and 140°F. Meat should not be kept in this temperature zone for more than 2 hr. Write an absolute value inequality representing the danger zone.
26. Meat should be stored below 40°F or kept above 140°F after cooking, outside the danger zone described in the previous exercise. Write an absolute value inequality representing safe temperatures for meat.
27. To reduce storage costs, many manufacturers have implemented just-in-time delivery of the components that are assembled for their products. A car company needs parts delivered within 3 min of every half hour. Write an absolute value inequality representing the time of delivery. Then solve the inequality to state the range of acceptable times.

28. The results of a poll claiming a 4% margin of error indicated that a candidate was favored by 48% of the voters. Write an absolute value inequality representing the percent of voters favoring the candidate. Then solve the inequality to state the range of the candidate's support.

C. Exercises

Solve.

29. $|-4x| < -2$
30. $|4x + 2| \geq -1$
31. $|x + 7| \leq 0$
32. $|4x - 5| < 0$
33. $|3x - 1| > 0$

CUMULATIVE REVIEW

Solve each literal equation for *x*. [3.1]

34. $\frac{2}{3}x = a + 6$
35. $4x + b = 2c$
36. What is the percent increase from 80 to 100? [3.5]
37. Forty-five percent of what number is 90? [3.4]
38. Kendra went out for lunch and purchased a $6 entrée and a beverage for $2. She paid an 8% sales tax and tipped her server 20% of the pretax bill. What is the total amount she spent on lunch? [3.6]
39. Two towns are about 4.3 cm apart on a map. If the actual distance between them is 108 km, find the scale of the map. [3.3]
40. Two triangles have sides in the ratio 5 : 7, and the smaller one has an area of 50. What is the area of the larger one? [3.3]
41. One of the first calculations of the speed of light was 125,000 mi/sec. Determine the percent error (to the nearest percent) of this calculation using the currently accepted value of 186,300 mi/sec. [3.5]
42. In the first trial of an experiment, the temperature of the final solution was 90°C, and in the second trial it was 81°C. Determine the percent difference (to the nearest tenth of a percent) between the results of the two trials. [3.5]
43. Mr. Scurtu wants to invest some of his $10,000 in an aggressive stock fund that he expects to earn 12% annually and the rest in less risky bonds making 4% annually. If he hopes to earn $1000 in annual interest, how should he split his investment between the stock fund and the bonds? [3.6]

TECHNOLOGY CORNER (TI-84+ Family)

The [GRAPH] function can be used to graph the solutions to more complicated inequalities. Begin by selecting FUNC from the fourth line of the [MODE] screen and ZStandard from the [ZOOM] menu.

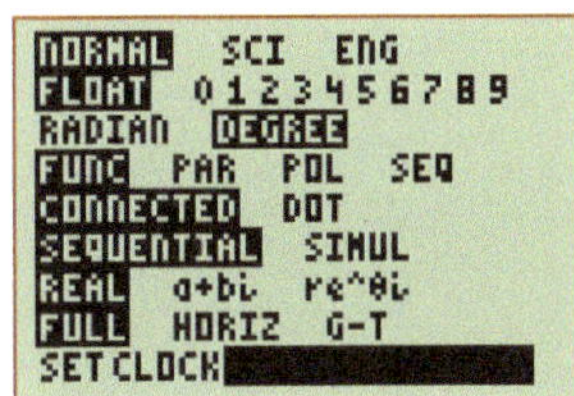

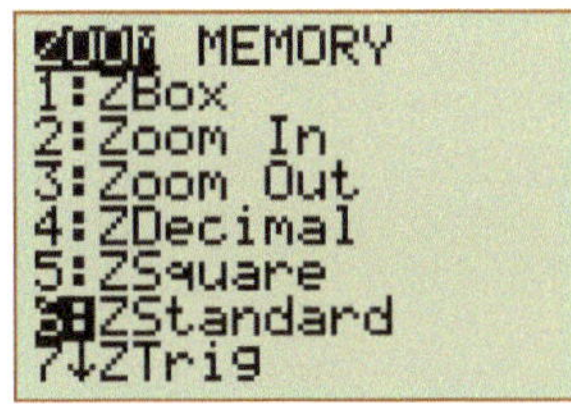

To graph the solution to $2x + 1 > 3$, select [Y=] to enter the inequality as Y_1, using the [TEST] menu ([2nd] [MATH]) to enter the inequality symbol.

2 [X,T,Θ,n] [+] 1 [TEST] [3] 3

Use [GRAPH] to plot the solution set; then select [TRACE] and use the [◄] and [►] keys to explore the values of x that are graphed. Notice that the function Y_1 equals 1 when the inequality is true and 0 when the inequality is false.

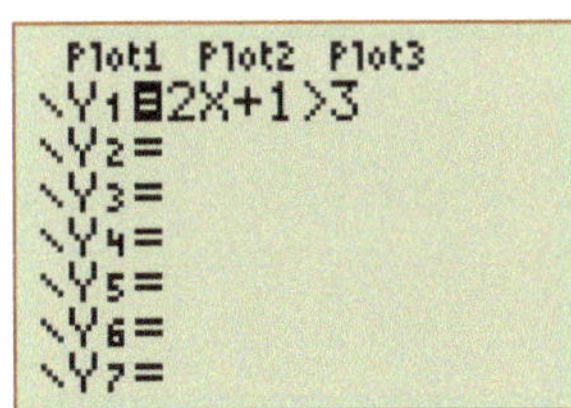

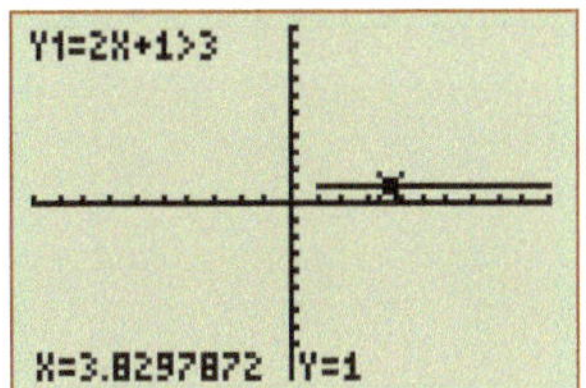

The LOGIC menu on the [TEST] screen can be used to enter disjunctions and conjunctions. Select [Y=] to return to the function editor and clear Y_1. Enter the combined conjunction $-3 \le x \le 3$ in Y_1 as two inequalities connected by the word *and*.

[X,T,Θ,n] [TEST] [4] [(-)] 3 [TEST] [►] [1] [X,T,Θ,n] [TEST] [6] 3

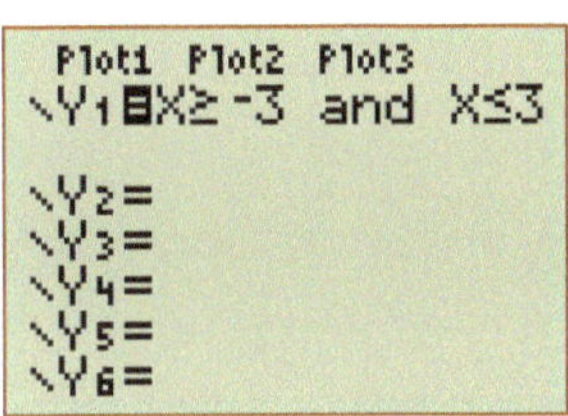

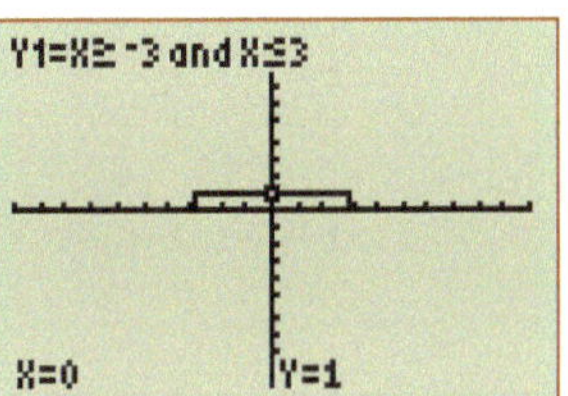

Use the NUM menu from the [MATH] screen to enter the absolute value function, abs(). To graph the solution set of $|x + 2| > 4$, return to the function editor using [Y=] and enter the following in Y_1.

[MATH] [►] [1] [X,T,Θ,n] [+] 2 [)] [TEST] [3] 4

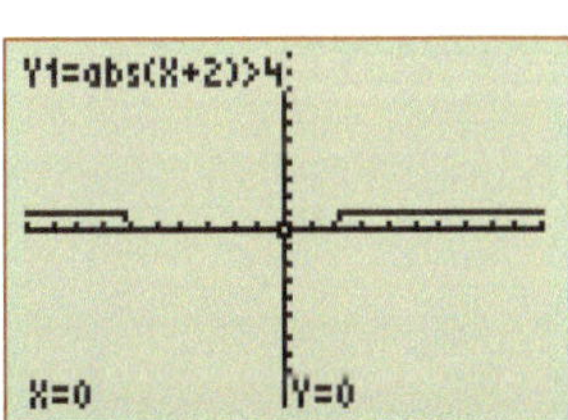

CHAPTER 4 REVIEW

Give the inequality symbol for each phrase.

1. is at least
2. does not exceed

Write an inequality equivalent to the given inequality.

3. $y \ngeq 3$
4. $z \nless -10$

Graph each inequality on a separate number line.

5. $x \geq -2$
6. $y < -3$

Solve.

7. $x - 17 < -6$
8. $-\frac{y}{6} < 8$
9. $-5z < 21$
10. $4a - 7 > 5$
11. $-5 \leq -3b + 7$
12. $2x - 12 > 5x - 4$
13. $-3(5x - 8) > 5 - 7x + 2$
14. $5(3x - 4) \leq 2(x - 10)$

Write an inequality and solve.

15. Find all the numbers such that adding 15 to 23 times the number is greater than 429.
16. Find the largest integer such that eight more than six times the integer is less than the opposite of 42.
17. Find all the numbers such that the product of seven and the number is more than the sum of seven and the number. What is the smallest integer that is a solution?
18. Samuel makes twice as much per hour as Kathy, but Kathy works three times as many hours per week as Samuel. Kathy works 36 hr, and the sum of their wages exceeds $480. How much does each of them make per hour?
19. Ali has test scores of 95, 78, and 93; and he needs to have a test average of at least 90 to get an A on his report card. What is the lowest score he can get on his final test and still have an A?
20. Two inequality statements joined by the word "and" are called a(n) ____.
21. The union of two solution sets is called a(n) ____.

Solve each compound inequality and graph its solution on a number line.

22. $x > 3$ or $x > -1$
23. $x > -2$ and $x \leq 4$
24. $x > 3 \wedge x < 7$
25. $x < 3 \vee x \geq -1$
26. $x \neq 5$ or $x < 2$
27. $4x - 12 \geq 28$ and $16x + 4x - 6 > 54$

Solve each compound inequality. Express the answer as an inequality, the set of real numbers, or the empty set.

28. $2x \leq -10$ or $2x > 8$
29. $6x + 5 < 29$ and $3x > 9$
30. $3x - 5 < 4 \wedge 3x + 2 > 12$
31. $5x + 6 - 2x \leq 12 \vee 8x < 40$
32. $7 < 4x - 9 < 19$
33. $5 - 3x < 10$ and $4x + 6 < 10$
34. $3x + 6 \leq 6$ or $2x + 8 < -3$
35. $5 - 3x < 10$ or $4x + 6 < 10$

Solve. Write $\varnothing$ if no solution exists.

36. $-5|x| = -125$
37. $-2|x| + 6 = 20$
38. $|x + 3| + 2 = 12$
39. $|4z - 20 + 10z| - 40 = -4$

Solve and graph the solution on a number line.

40. $|x| < 4$

41. $|x + 3| > 7$

42. $|3x - 8| \geq 11$

43. $|3x + 6| < 15$

Solve.

44. Write an absolute value equation representing the numbers that are 5 units from -2.

45. Write an absolute value equation representing the numbers that are 3 units from 4.

46. The proper pH of the water in swimming pools is 7.4 $\pm$0.2. Write an absolute value inequality representing this range. Then solve the inequality to state the range of the pH.

47. An investor has decided to sell his shares of stock if the stock's price increases or decreases more than \$14 from its present value of \$73 per share. Write an absolute value inequality representing the price at which he should sell his shares. Then solve the inequality to state the range of prices.

48. A person functions comfortably in temperatures from 16°C to 27°C. Write a compound inequality and an absolute value inequality representing a person's optimal temperature range.

5 Relations and Functions

On October 24, 1915, an Antarctic exploration team led by Ernest Shackleton found that their ship was being crushed by ice floes and was about to sink. The explorers abandoned ship and were left adrift with their supplies and three lifeboats on a large ice floe. The ship's captain was able to use his instruments to determine the latitude and longitude of the party as they floated along on the ice. When he could tell from his charts that they were near an island, the men took to the lifeboats and managed to get ashore. The ability to determine latitude and longitude was necessary to save the lives of the men on that expedition.

Latitude and longitude are still used today to notify rescue personnel of the location of ships or aircraft downed at sea. Today's aircraft, ships, land vehicles, and even missiles are guided by massive latitude and longitude calculations based on Global Positioning System (GPS) satellites. These satellites also track altitude, adding a third dimension. Effective dominion using modern technology demands a comprehensive understanding of latitude and longitude.

After this chapter you should be able to

1. graph points in the Cartesian coordinate plane.
2. represent relations and functions using sets of ordered pairs, tables, mapping diagrams, equations, or graphs.
3. use graphs of relations and functions to solve real-world problems.
4. write function rules from other representations of functions.
5. solve direct and inverse variation problems.
6. graph and write function rules for the absolute value function and its translations.

5.1 Points in the Coordinate Plane

The state capitol in Springfield, Illinois

While the Internet and GPS devices provide many mapping applications, on occasion you will need to find a certain location on a printed map. The map's index lists the general region in which certain features are located on a map. Look at the map shown and find the cities listed in the index.

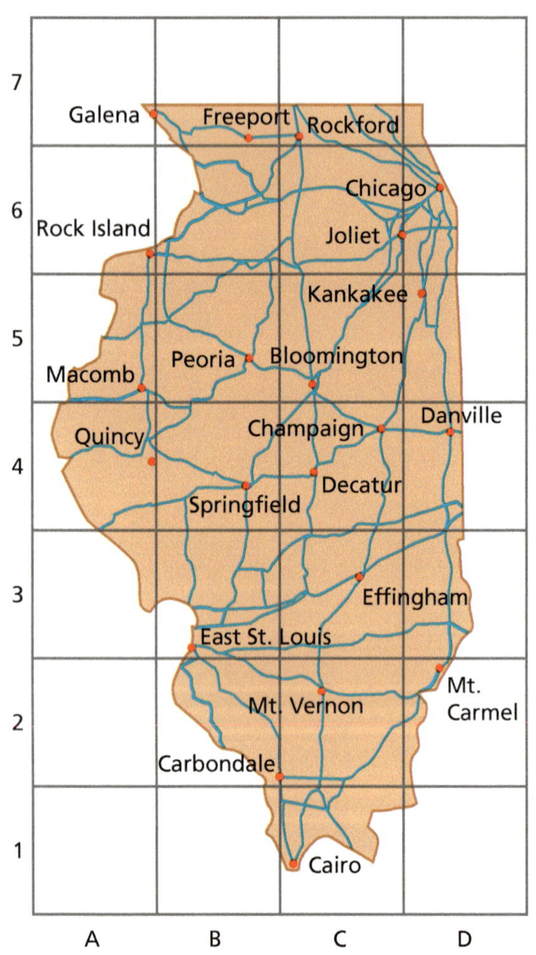

Index	
Bloomington	C-5
Cairo	C-1
Carbondale	B-2
Champaign	C-4
Chicago	D-6
Danville	D-4

In the index, two symbols are used to represent the location of the cities. The letter shows the horizontal position, and the number indicates the vertical position.

Remember that every point on a number line can be associated with a real number. René Descartes devised what is now called the *Cartesian coordinate system*, a method of associating every point in a plane with an ordered pair of numbers.

Definition

An **ordered pair** is a pair of numbers (x, y) describing a point in the Cartesian coordinate plane.

The two symbols in an ordered pair are called *coordinates*. For each city in the map of Illinois, the first coordinate is a letter from A to D. The second coordinate is a number from 1 to 7.

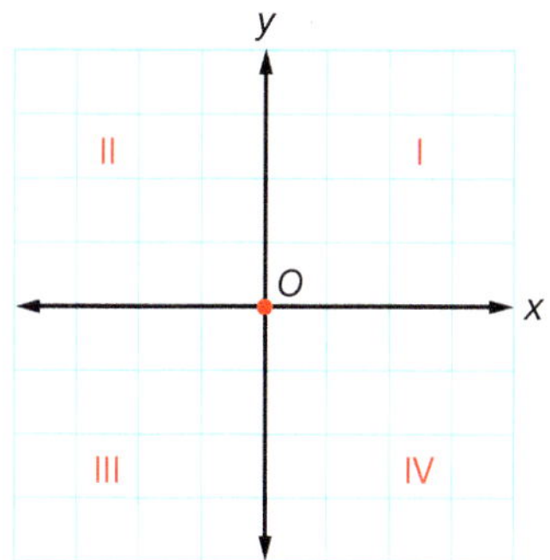

Two perpendicular number lines (one horizontal, the other vertical) divide the Cartesian coordinate plane into four sections called *quadrants*, which are numbered counterclockwise with Roman numerals as illustrated.

Definitions

The **x-axis** is the horizontal number line in the plane.

The **y-axis** is the vertical number line in the plane.

The **origin** (labeled O in the graph) is the point (0, 0), where the axes intersect.

The process of locating points in a plane is called plotting points or graphing ordered pairs. The Illinois map index gives the horizontal distance first and the vertical distance second. In the Cartesian plane, the horizontal distance (x) always comes first and the vertical distance (y) second. The order of the coordinates in an ordered pair is very important. Reversing them would name a different point.

Keyword Search
Cartesian coordinate plane

An ordered pair with unknown coordinates is usually expressed (x, y).

Definitions

The **x-coordinate** (abscissa) is the first coordinate of an ordered pair.

The **y-coordinate** (ordinate) is the second coordinate of an ordered pair.

Example 1

Graph and label the following ordered pairs.

a. (2, 1) **b.** (−3, 4) **c.** (0, −5)

Answer

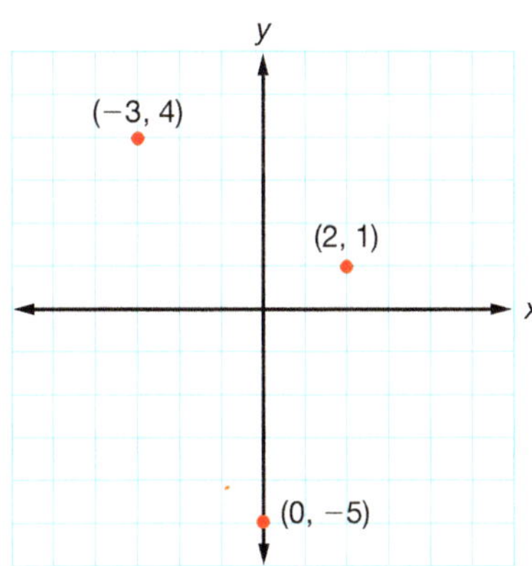

a. From the origin, move 2 units right and 1 unit up.

b. From the origin, move 3 units left and 4 units up.

c. From the origin, move 5 units down without moving horizontally.

Plotting Points (x, y)

1. Move x units right or left from the origin.
2. From that position, move y units up or down.
3. Graph the point and label it with its coordinates.

You will also need to state the coordinates of a point that is already plotted on a graph. This process is similar to plotting points.

Example 2

State the coordinates and name the quadrant or axis containing each of the following points.

a. *D* **b.** *E* **c.** *F*

Answer

a. *D* is located at $(-1, -2)$ and is in quadrant III. The point is 1 unit to the left of the origin and 2 units below the *x*-axis.

b. *E* is located at $(-3, 5)$ and is in quadrant II. The point is 3 units to the left of the origin and 5 units above the *x*-axis.

c. *F* is located at (4, 0) and is on the *x*-axis. The point is 4 units to the right of the origin between quadrants I and IV.

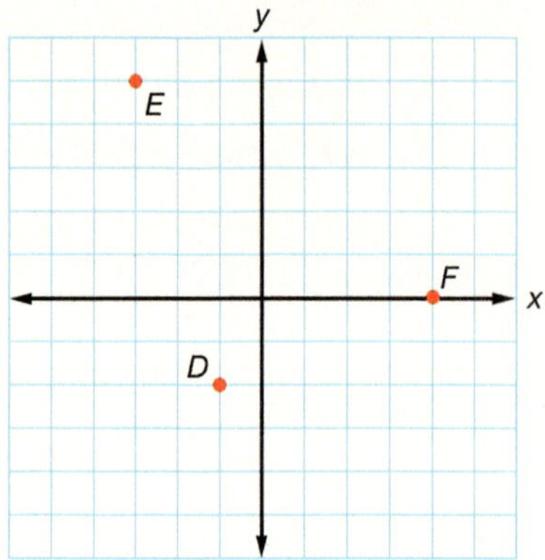

A *scatterplot* is a graph on a Cartesian plane that summarizes the relationship between two sets of numbers. A table will often summarize the data from a statistical study. Ordered pairs representing the data can be graphed so that a statistician can use the patterns formed by the points to determine the relationship between the variables in the study.

Example 3

Make a scatterplot of the heights and weights of students as recorded in the following table, and describe the relationship between the variables.

Weight/Height Comparison of Students										
Height (inches)	64	60	62	58	63	61	62	60	59	62
Weight (pounds)	130	124	138	98	159	136	142	126	101	145

Answer

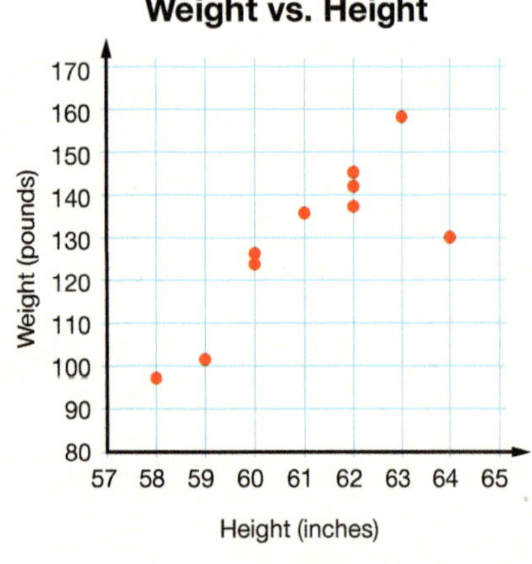

1. Adjust the scale of the axes and their intersection to fit the data.
2. Plot the ordered pairs.

In general, as height increases, so does weight.

3. The graph rises from left to right.

A. Exercises

For exercises 1–4, refer to the map of Illinois at the beginning of this section.

1. State the city with the location B-4.
2. State the city with the location B-3.
3. Give the location of Rock Island.
4. Give the location of Mt. Vernon.

State the coordinates of the following points.

5. A **6.** B **7.** C
8. D **9.** E **10.** F
11. G **12.** H **13.** I
14. J

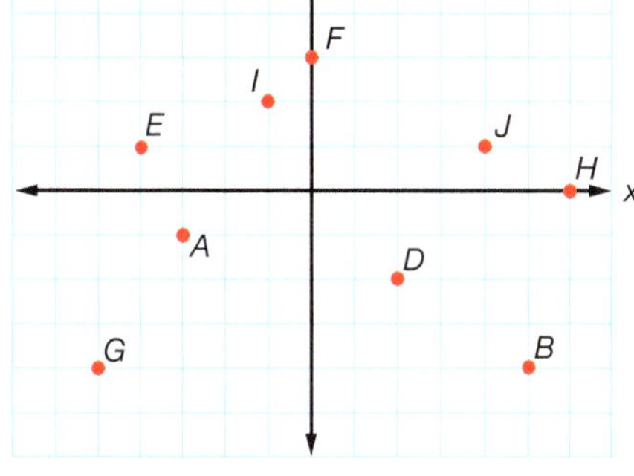

B. Exercises

Draw and label a Cartesian coordinate plane. Then graph and label the following points on it.

15. $(2, 4)$ **16.** $(5, -4)$ **17.** $(-1, 3)$ **18.** $(-2, -5)$
19. $(0, -4)$ **20.** $\left(-\frac{1}{2}, 0\right)$ **21.** $(-5, -2.5)$ **22.** $(-4, 1.75)$
23. $\left(\frac{3}{2}, -3\right)$ **24.** $(4.5, 2.25)$
25. What are the coordinates of the origin?

On one Cartesian plane, graph and label the following points. State the quadrant or axis containing each point.

26. $(-4, -6)$ **27.** $(5, -2)$ **28.** $(3, 0)$ **29.** $(-1, 6)$

In exercises 30–37, state the quadrant or axis containing each of the following points.

30. $(-3.22, 5)$ **31.** $(0, 1)$ **32.** $\left(2, -\frac{7}{4}\right)$
33. $(-5.9, 0)$ **34.** $(-1, -11)$
35. a point in which both coordinates are positive
36. a point in which both coordinates are negative
37. a positive abscissa and a negative ordinate
38. What is true about all the points that have 5 as their ordinate?
39. If rectangle $ABCD$ is placed on a coordinate plane with point A at the origin, state the coordinates of B, C, and D.
40. If rectangle $ABCD$ is placed on a coordinate plane with its center at the origin, state the coordinates of A, B, C, and D.

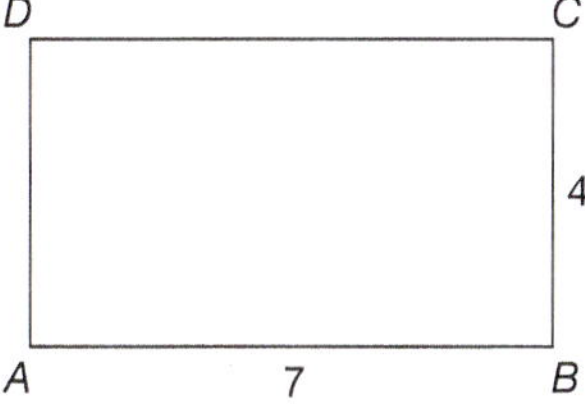

Create a scatterplot for each set of data and determine the general relationship between the variables.

41.

Daily Minutes of Exercise (x)	15	25	12	35	20	45	28	60
Resting Heart Rate (y)	72	68	75	66	71	58	70	56

42.

Calories per Day (x)	Weight (y)
2100	152
1750	130
1580	121
2640	165
1900	149
1975	138
2065	144
1490	119

C. Exercises

Exercising effective dominion over God's created earth often requires knowledge of an exact location on the planet. The earth has been divided into a grid of imaginary lines, allowing any point on the surface of the earth to be described by an ordered pair of measurements in this latitude-longitude coordinate system. Latitude measures the distance above or below the equator, while longitude measures the east-west position from a fixed line through Greenwich, England.

State the site in the Middle East represented by each of the following points.

43. 35.1° N, 33.3° E

44. 32.8° N, 35.1° E

45. 31.6° N, 35.5° E

Give the latitude-longitude ordered pair for each of the following locations.

46. Jerusalem

47. Sea of Galilee

48. Great Pyramids of Egypt

Dominion Modeling

Lines of latitude are also called *parallels* because they run east-west, parallel to the equator. The north-south lines of longitude all pass through both poles and are also called *meridians*. The *prime meridian* passes through Greenwich, England, and acts as the y-axis. Measures are given as 0° to 180° east and 0° to 180° west from this line. The *equator* acts as the x-axis, with distances measured from 0° to 90° north or south of the equator.

In the C. Exercises, distances were given using decimal degrees. Distances are also measured in units of degrees, minutes, and seconds, with 60 minutes in a degree and 60 seconds in a minute. You should also be able to convert measurements given in degrees and decimal minutes.

Example: Convert 42°34′18″ N (42 degrees, 34 minutes, 18 seconds) to degrees and decimal minutes; then convert it to decimal degrees.

$$18\ \cancel{\text{sec}}\left(\frac{1\text{ min}}{60\ \cancel{\text{sec}}}\right) = 0.3\text{ min; therefore, } 42°34'18''\text{ N} = 42°34.3'\text{ N}$$

$$34.3\ \cancel{\text{min}}\left(\frac{1\text{ degree}}{60\ \cancel{\text{min}}}\right) = 0.571\overline{6}\text{ degrees; therefore, } 42°34.3'\text{ N} \approx 42.5717°\text{ N}$$

49. Convert 26°13′24″ S, 157°10′15″ E to degrees and decimal minutes.

50. Convert 78°35′9″ N, 18°52′4″ W to decimal degrees.

51. Convert 41.7512° N, 88.0457° W to degrees, minutes, and seconds.

CUMULATIVE REVIEW

Write an equation and solve. [3.4–3.6]

52. What is 35% of 900?

53. 37 is what percent of 12?

54. Find the percent increase if admission to the amusement park is raised from $70 to $80.

55. In chemistry lab, Lori determined the melting point of iodine to be 120.0°C when the accepted melting point is 113.7°C. What is her percent error (to the nearest tenth of a percent)?

56. A banker loaned $5000 to one customer and $8000 to another. If the annual interest on the second loan was 2% higher than the first and the combined annual interest was $810, what was the interest rate for each loan?

57. $10,000 is invested, some at 6% interest and the rest at 7.5%. The combined annual interest income is $645. How much is invested at each rate?

Write an inequality equivalent to the given inequality. [4.1]

58. $x \not\geq -6$

59. $x < 4$

60. State the Trichotomy Property. [4.1]

61. The triangle inequality theorem states that the sum of the lengths of the two shorter sides of any triangle must be greater than the length of the third side. If the sides of a triangle from smallest to largest are a, b, and c, write the inequality implied by this theorem. [4.1]

5.2 Relations and Functions

Searching the Internet for the term *relation* produces a variety of meanings. The word can refer to a relative, someone related to you. It can also mean the connection or relationship between people.

The term *relation* can also refer more broadly to the connection or association between any two (or more) items. A common relation exists in a grocery store between items and their prices.

x	y
gum	$0.95
apple	$0.49
yogurt	$0.79

In mathematics, *relation* has a more specific meaning; it is an association between two sets. In algebra, its elements are ordered pairs of numbers.

Definition

Any set of ordered pairs is a **relation**.

A relation can be represented in a variety of ways. Relations are often listed as a set of ordered pairs and named with a capital letter.

$A = \{(-3, 5), (2, 2), (2, 4), (4, -1)\}$

A relation can also be illustrated using a table or a graph. The graph of a relation is the set of points in a plane that corresponds to the ordered pairs of the relation.

x	y
-3	5
2	2
2	4
4	-1

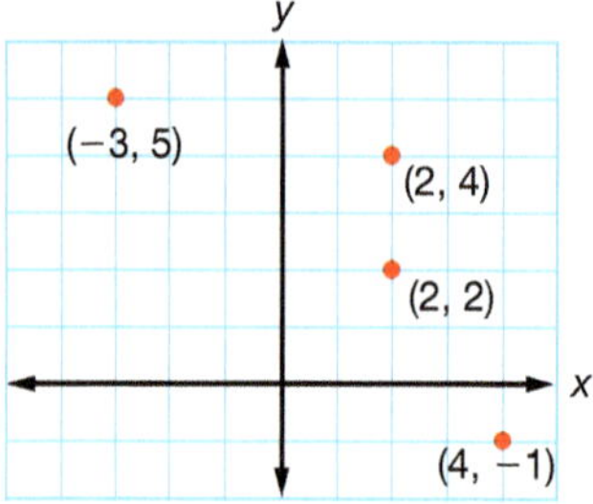

Definitions

The **domain** of a relation is the set of first elements (x-coordinates) of the ordered pairs.

The **range** of a relation is the set of second elements (y-coordinates) of the ordered pairs.

Example 1

State the domain, D, and range, R, of relation A shown above.

Answer $D = \{-3, 2, 4\}$ List the set of numbers that are x-coordinates.

$R = \{-1, 2, 4, 5\}$ List the set of numbers that are y-coordinates.

A fourth representation of a relation, a *mapping diagram*, emphasizes the domain and range. The arrows show the pairing of each number in the domain with its associated number in the range.

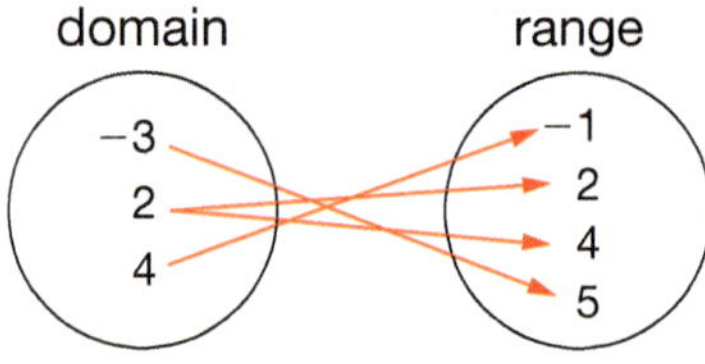

A *function* is a special type of relation in which every element in the domain is paired with one and only one element in the range. It cannot have two or more ordered pairs with the same first coordinate.

Definition

A **function** is a relation in which each x-coordinate is paired with one and only one y-coordinate.

Because a function is a relation, it has a domain and a range and can be illustrated using any of the representations of a relation: a set of ordered pairs, a table, a graph, or a mapping diagram. While any function must be a relation, not all relations are functions.

Example 2

Write a set of ordered pairs and draw a mapping diagram of the relation illustrated by the graph. Is this relation a function?

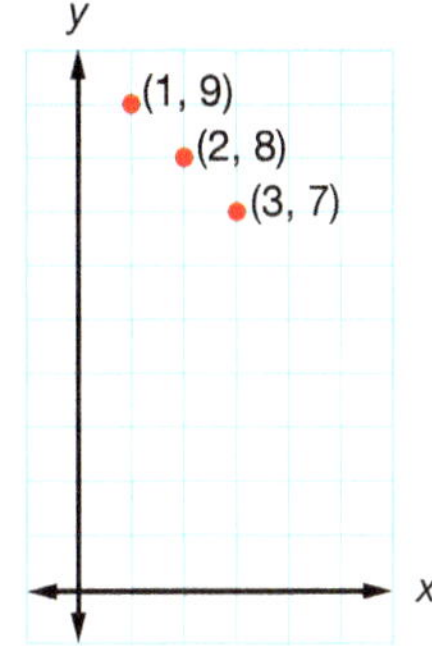

Answer $B = \{(1, 9), (2, 8), (3, 7)\}$

1. Write the set of ordered pairs.
2. Draw a mapping diagram.

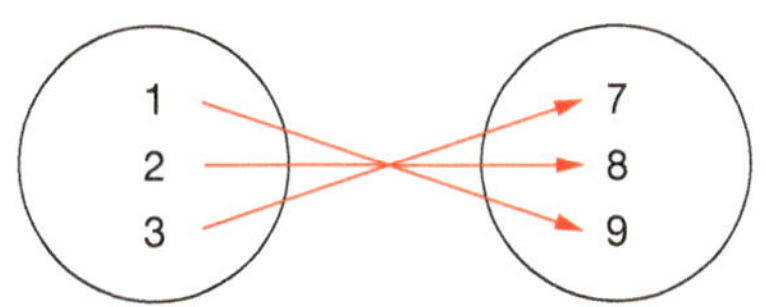

The relation is a function.

3. Each element of the domain is paired with a unique element of the range.

Example 3

Determine whether the following relations are functions.

a. $A = \{(-3, 5), (2, 2), (2, 4), (4, -1)\}$

b. $B = \{(0, 4), (-1, 2), (3, 4), (-4, -3)\}$

Answer **a.** A is not a function. The domain element 2 is paired with two different members of the range.

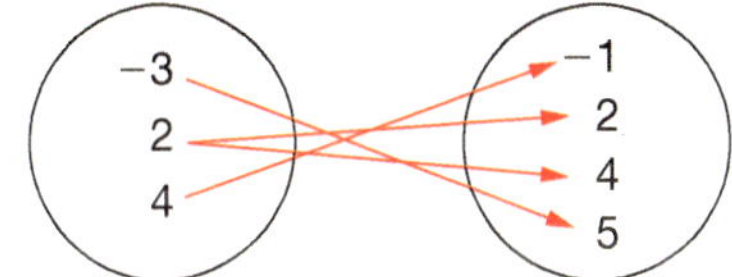

b. B is a function. Despite the fact that two members of the domain are paired with the same range element, 4, each member of the domain is paired with one and only one member of the range.

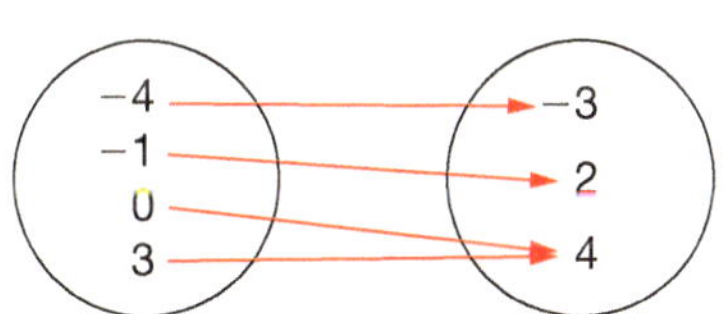

Lowercase letters, particularly f and g, are often used to name functions.

$f = \{(6, 9), (4, 12), (0, 4)\}$

The variable that represents the values of the domain, typically x, is called the *independent variable*. Because the value of the second coordinate, usually y, depends on the value chosen for the first coordinate, it is called the *dependent variable*. When a function is used to model the relationship between an item and its price, the independent variable is assigned to the item since it determines the price, the dependent variable.

x	y
6	9
4	12
0	4

or

x	$f(x)$
6	9
4	12
0	4

Functional notation emphasizes the fact that the value of the y-coordinate depends on the chosen value of the x-coordinate. It symbolizes the y-coordinate as $f(x)$, which is read "f of x." Since $y = f(x)$, the ordered pair (x, y) can be written $(x, f(x))$. Notice that $f(6) = 9$. Likewise, $f(4) = 12$ and $f(0) = 4$.

Functions and relations are often used to model relationships between two quantities in everyday life. Example 4 illustrates a simple application in a business setting.

Example 4

The local grocery store is having a sale on potato chips. One bag regularly sells for \$2, but if you buy two, you get a third one free. Represent the relationship between the number of bags purchased (from 0 to 12) and their cost using a table and a graph. Are these ordered pairs a relation? What is the domain? What is the range? Are these ordered pairs a function?

Answer

Bags, x	0	1	2	3	4	5	6	7	8	9	10	11	12
Cost in \$, $f(x)$	0	2	4	4	6	8	8	10	12	12	14	16	16

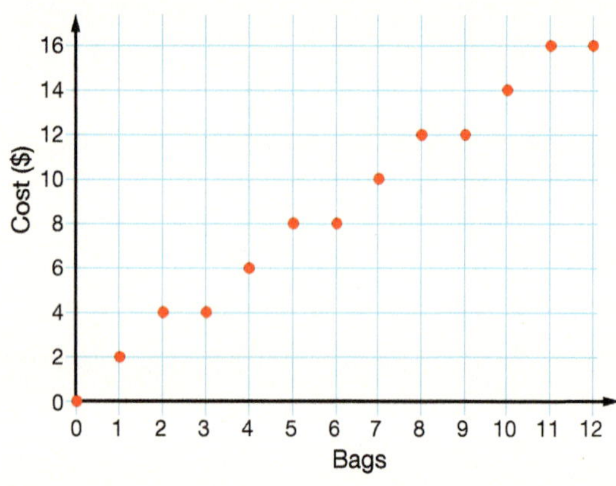

Yes, it is a relation. 1. Any set of ordered pairs is a relation.

$D = \{0, 1, 2, \ldots, 12\}$ 2. List the possible values of the independent variable.

$R = \{0, 2, 4, \ldots, 16\}$ 3. List the possible values of the dependent variable.

Yes, it is a function. 4. No element of the domain relates to more than one element of the range.

A. Exercises

State whether each set represents a relation.

1. $\{(2, 4), (5, 9), (8, 2)\}$

2. $\{3, 4, 7, 9\}$

3. $\{(-2, 5, 7), (5, -8, 2), (2.8, 6, 7)\}$

4. $\{(1, -2), (1, 3), (1, -4), (1, 5)\}$

Draw a mapping diagram of each relation and state whether it is a function.

5. $\{(-5, 1), (-5, 2), (2, 8), (5, 6)\}$

6. $\{(-6, -5), (-3, 4), (1, 7), (4, -3)\}$

7. $\{(1.4, 3.2), (4.2, 9.5), (6.8, 1.4), (8.4, 6.8)\}$

8. $\left\{\left(\frac{4}{3}, \frac{2}{5}\right), \left(\frac{4}{7}, \frac{8}{9}\right), \left(\frac{4}{3}, \frac{3}{8}\right)\right\}$

State the domain and range of each relation and whether it is a function.

9. $\{(6, 10), (-9, 14), (4, 2), (-9, 8)\}$

10. $\{(3, 2), (3, 3), (3, 4), (3, 5), (3, 6)\}$

11. $\{(1.3, 3.7), (1.4, 3.7), (1.5, 3.7), (1.6, 3.7)\}$

12. $\left\{\left(\frac{1}{2}, \frac{3}{5}\right), \left(\frac{2}{3}, \frac{7}{8}\right), \left(\frac{4}{5}, \frac{5}{9}\right), \left(\frac{5}{6}, \frac{3}{5}\right)\right\}$

For each set, choose the letter of the correct description.

13. $\{(2, 3), (4.7, -5)\}$

14. $\{(-2.6, 3), (4.9, 3)\}$

15. $\left\{2, 3, -\frac{4}{5}, 8.1\right\}$

16. $\{(2, 3), (3, 4), (3, 5)\}$

17. $\left\{(2, 3), \left(2, \frac{5}{4}\right)\right\}$

18. $\{(5, 9, 2), (-7, -3, 1), (6, 7, 0)\}$

a. both a relation and a function

b. a relation that is not a function

c. neither a relation nor a function

B. Exercises

A relation can be represented by a set of ordered pairs, a table, a graph, or a mapping diagram. For each relation, give the other three representations and state whether the relation is a function.

19. $\{(-6, -2), (-1, 0), (3, 4), (4, 3)\}$

20.

x	y
−2	0
−1	4
0	3
1	−1
1	4

21.

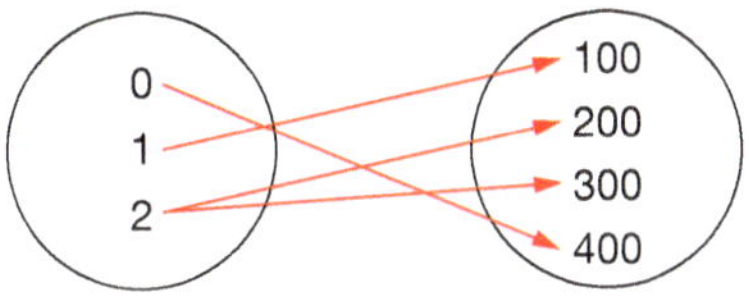

0
1
2

100
200
300
400

22.

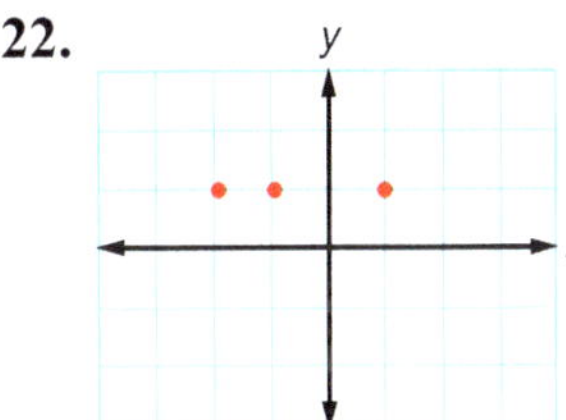

Given the function $f = \{(6, 28), (-7, -4), (3, -9), (0, -4), (-9, 2), (1, -10)\}$, find the following.

23. $f(-7)$

24. $f(0)$

25. $f(6)$

26. $f(1)$

For each of the situations described, state which quantity should be represented by the independent variable and which by the dependent variable.

27. modeling the time driven and the distance traveled if the driver maintains a constant speed

28. determining the total earnings of several student workers given the rate of $8/hr and the number of hours each student worked

29. graphing the ages and heights of students in a nutritional study

30. modeling the relationship between body mass index, which measures the level of physical fitness, and the number of minutes spent exercising each day

For exercises 31–34, $A = \{1, 6, 9\}$ and $B = \{3, 4\}$.

31. Make a relation in which the domain is A and the range is B. If possible, make this relation a function; if that is not possible, explain why not.

32. Make a relation in which the domain is B and the range is A. If possible, make this relation a function; if that is not possible, explain why not.

33. Make a function in which the domain is B and the range is a subset of A. Represent the function as a set of ordered pairs, a table, a graph, and a mapping diagram.

34. Make a function in which the domain is A and the range is a subset of B. Represent the function as a set of ordered pairs, a table, a graph, and a mapping diagram.

C. Exercises

Given the function $g = \{(a, m), (b, n), (c, o), (d, p), (e, q)\}$, find the following.

35. $g(c)$ **36.** $g(e)$ **37.** $g(a)$ **38.** $g(b)$

There is a function called the greatest integer function, symbolized by $[x]$. For any real number x, $[x]$ is the largest integer less than or equal to x.

39. Complete the following table of values for this function.

x	$4\frac{1}{2}$	5	$5\frac{1}{4}$	$6\frac{3}{4}$	$-\frac{1}{2}$	$-2\frac{1}{2}$	-5.75	$\frac{1}{3}$	$-\frac{1}{4}$
$[x]$									

40. Make a graph for $f(x) = [x]$ using all real numbers as the domain.

Dominion Modeling

The intersection of a sphere and a plane containing the sphere's center is called a *great circle*. The equator can be thought of as a great circle on the nearly spherical earth. Together, the meridians on the opposite sides of the earth form another great circle. The shortest distance between any two points on the surface of the earth follows the great circle determined by those points and the center of the earth. To save time and conserve resources, we need to know which of the two possible routes is shorter.

Determine the general direction that should be followed on the great circle to arrive at the given destination. Example: The shortest distance from 50° N, 97° W to 40° N, 87° E is achieved by heading north (over the Arctic circle).

41. from 86° N, 43° W to 24° S, 40° W

42. from 10° N, 150° W to 9° N, 90° E

43. from 12° S, 30° W to 33° N, 15° E

CUMULATIVE REVIEW

Evaluate. [1.8]

44. $6 - 3(-7) + (-5)^2$ **45.** $[(3 - 6)^3 - 8]^2$

State the property of equality used to solve each equation. [2.4]

46. $5x = 8$ **47.** $x - 15 = 39$

Solve. [2.6–2.7]

48. $2x + 8 + 4x - 1 = 25$ **49.** $2x + 5 + 7x + 1 = 42$

50. $3x - 2 = 5x + 8$ **51.** $3(4x - 1) = 2x + 4$

Express each of the following repeating decimals as a quotient of integers. [2.6]

52. $0.\overline{47}$ **53.** $0.4\overline{7}$

5.3 Graphs of Relations and Functions

This graph illustrates the relationship between the height of a ball thrown from the outfield and the time that it has been in the air. Since the height of the ball depends on the length of time it has been in the air, t is the independent variable and h is the dependent variable. Since there are an infinite number of points in this relation, the domain and range are best stated using inequalities.

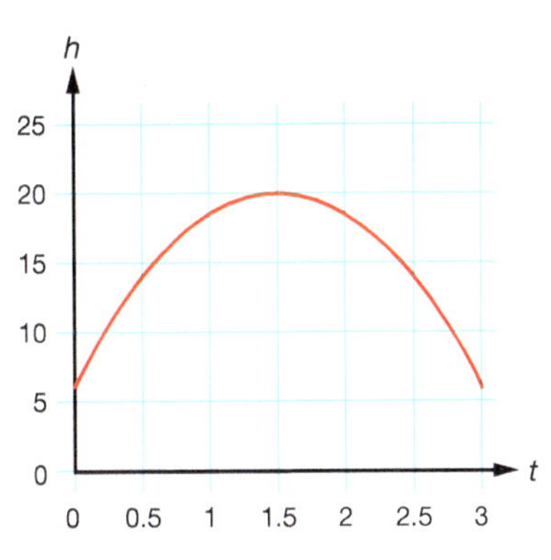

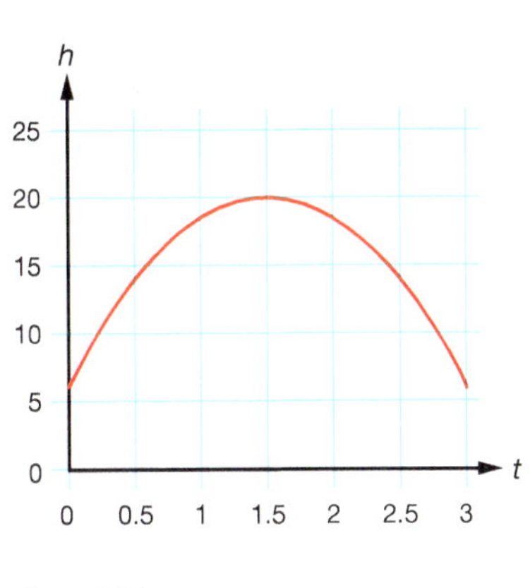
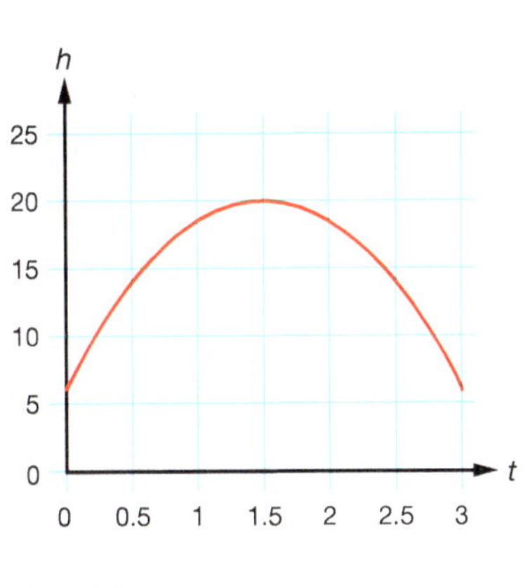

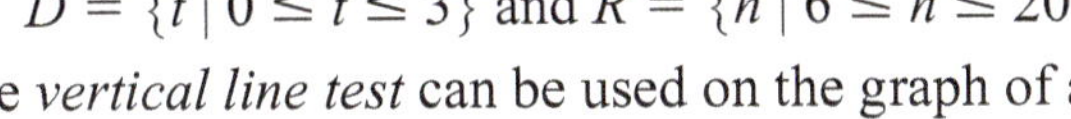

$$D = \{t \mid 0 \leq t \leq 3\} \text{ and } R = \{h \mid 6 \leq h \leq 20\}$$

The *vertical line test* can be used on the graph of a relation to determine whether it is a function. Think about moving an imaginary vertical line across the graph from left to right. If this vertical line intersects the relation's graph in more than one point, the relation is not a function because those ordered pairs have the same x-coordinate. The fact that no vertical line can be drawn containing more than one point of the relation shows that the relation is a function.

Example 1

State the domain and range of each relation. Then use the vertical line test to determine whether the relation is a function.

a.

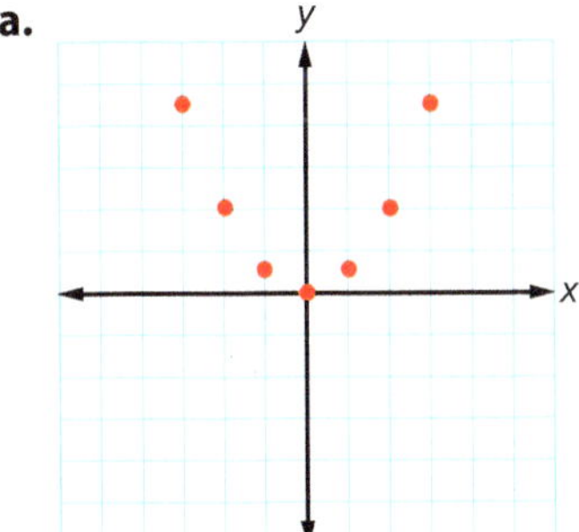

b.

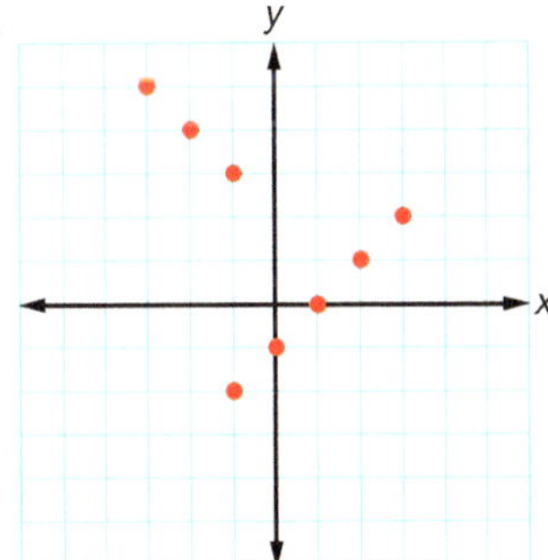

c.

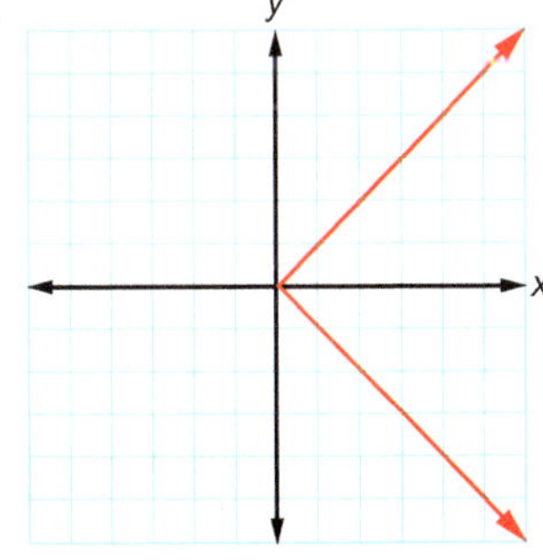

Answer

a. $D = \{0, \pm1, \pm2, \pm3\}$;
$R = \{0, 0.5, 2, 4.5\}$
The relation is a function. No vertical line intersects the graph at more than one point.

b. $D = \{0, \pm1, \pm2, \pm3\}$;
$R = \{-2, -1, 0, 1, 2, 3, 4, 5\}$
The relation is not a function. A vertical line crossing the x-axis at -1 contains both points $(-1, -2)$ and $(-1, 3)$.

c. $D = \{x \mid x \geq 0\}$;
$R = \mathbb{R}$
The domain is any nonnegative number. The range is the real numbers.
The relation is not a function. Any vertical line crossing the positive x-axis crosses the graph at more than one point.

An equation is often used to clearly define the rule that determines the values of the range associated with each member of the domain.

Example 2

Given the relation $A = \{(x, y) \mid y = x + 2; x = 1, 2, 3, 4\}$, make a table of ordered pairs to represent the relation. Then graph the relation and state whether it is a function.

Answer

x	$y = x + 2$	ordered pair
1	$y = 1 + 2 = 3$	(1, 3)
2	$y = 2 + 2 = 4$	(2, 4)
3	$y = 3 + 2 = 5$	(3, 5)
4	$y = 4 + 2 = 6$	(4, 6)

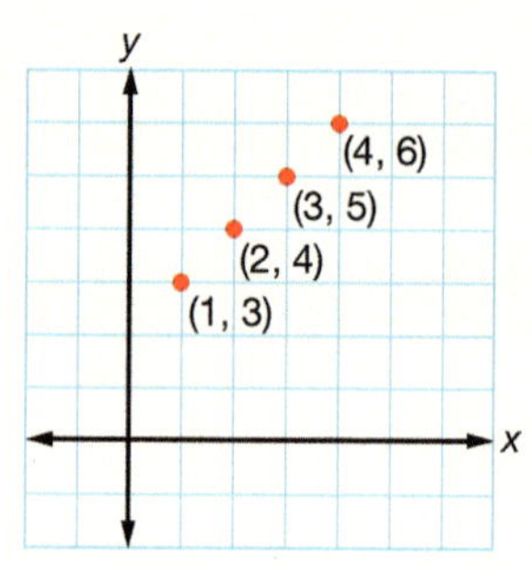

This relation is a function since no vertical line can be drawn that passes through more than one point.

What if the values for x were not stated in the previous example? When no domain is stated, it is assumed to be the real numbers. Therefore, the graph of the function $y = x + 2$ follows the pattern established by the ordered pairs already graphed. Since the coordinates of any point on the graphed line satisfy the equation, the line is the graph of the function.

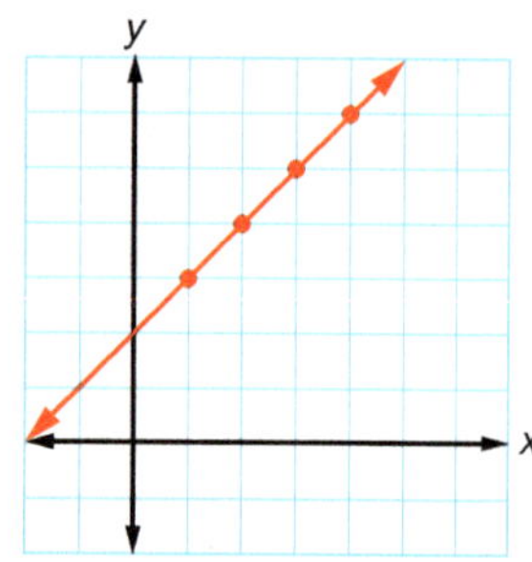

Example 3

Graph the relation $B = \{(x, y) \mid y = 3; 1 \leq x \leq 5, x \in \mathbb{Z}\}$. State whether the relation is a function.

Answer

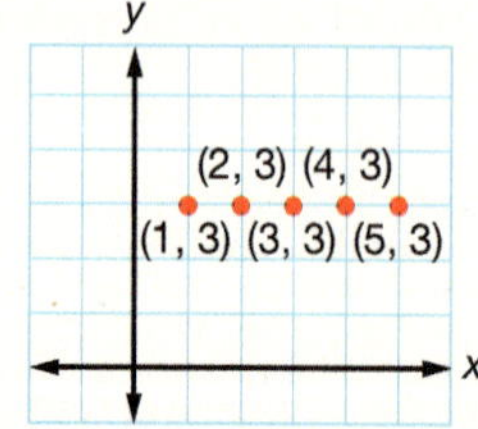

1. B is the set of ordered pairs (x, y) such that x is any integer from 1 to 5 (including 1 and 5) and y is equal to 3.
2. Write B in its set form: $\{(1, 3), (2, 3), (3, 3), (4, 3), (5, 3)\}$.
3. Graph the relation.

This relation is a function.

4. No vertical line can be drawn that passes through more than one point.

Without the restriction of the domain to integral values in Example 3, the graph would contain all possible real values of x from 1 to 5. The result is the segment from (1, 3) to (5, 3). Ordered pairs such as (2.1, 3), (π, 3), and $\left(3\frac{1}{2}, 3\right)$ are members of this relation. The vertical line test indicates that this is the graph of a function. Relations can often be described in terms of equations or inequalities.

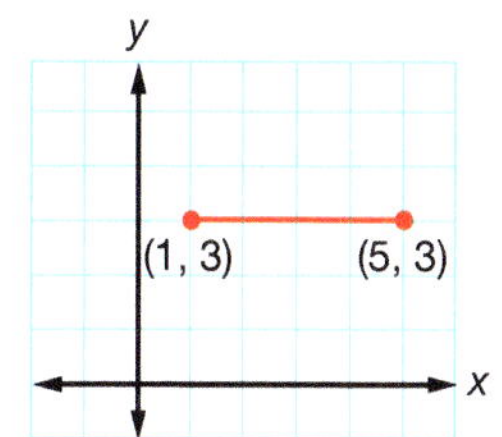

Example 4

Graph $D = \{(x, y) \mid y = \pm\sqrt{x};\, x = 1, 4, 9\}$. State its domain and range. Is this a function?

Answer

x	1	4	9
y	±1	±2	±3

1. Make a table to find the ordered pairs. Substitute the x values from the domain into the equation $y = \pm\sqrt{x}$. Each value of x produces two values for y.

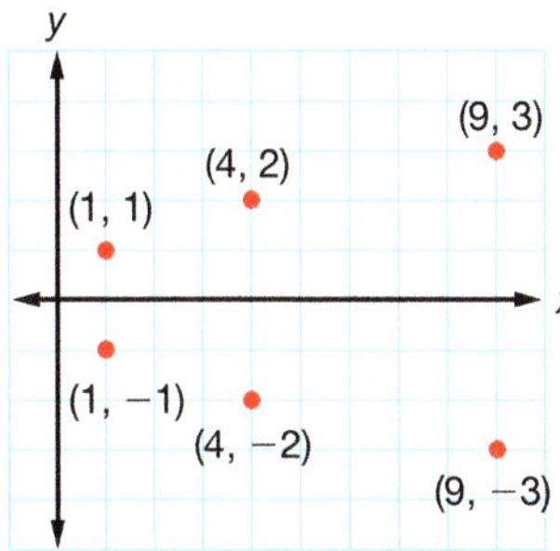

2. Graph the set of ordered pairs: $\{(1, 1), (1, -1), (4, 2), (4, -2), (9, 3), (9, -3)\}$.

$D = \{1, 4, 9\}$
$R = \{-3, -2, -1, 1, 2, 3\}$

3. The domain is the given set of x-coordinates. The range is the set of y-coordinates.

This relation is not a function.

4. The relation fails the vertical line test; each value of x generates two values for y.

Since functions are a special type of relation, they are graphed using the same methods. If the domain is not restricted, draw a smooth curve connecting the ordered pairs after you have graphed enough points to illustrate the general shape of the function.

Example 5

Graph the function $f(x) = -x^2$. Then state its domain and range.

Answer

x	$-x^2$	$f(x)$
-2	$-(-2)^2$	-4
-1	$-(-1)^2$	-1
0	-0^2	0
1	-1^2	-1
2	-2^2	-4

1. Make a table and plot enough ordered pairs so you can see the shape of the graph.

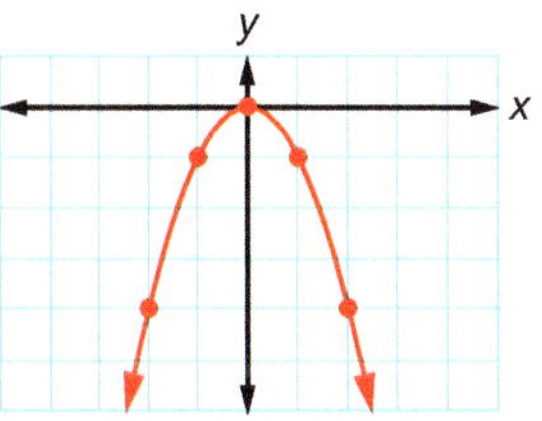

2. Connect the points with a smooth curve.

$D = \mathbb{R}$
$R = \{y \mid y \le 0\}$

3. Since there is no restriction on the domain, any real number can be used. The range consists of all nonpositive numbers.

A. Exercises

Is the graphed relation a function?

1.

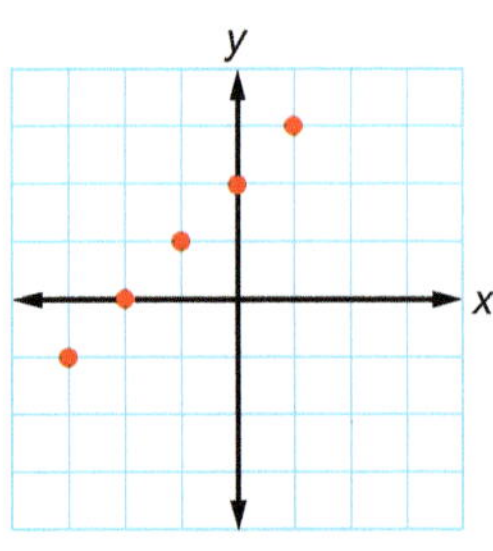

2.

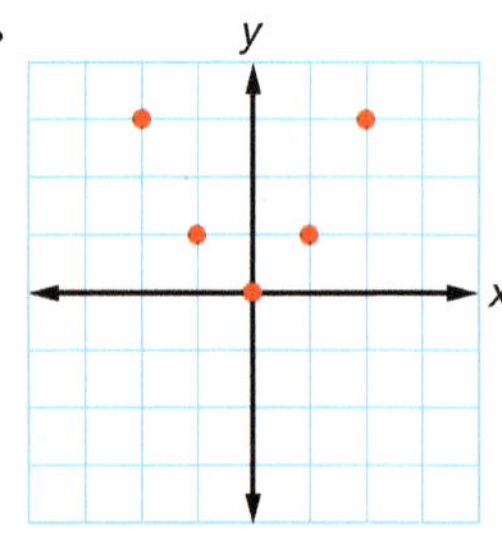

3.

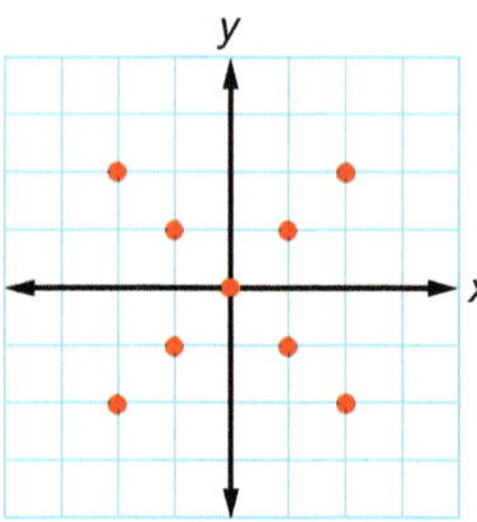

4.

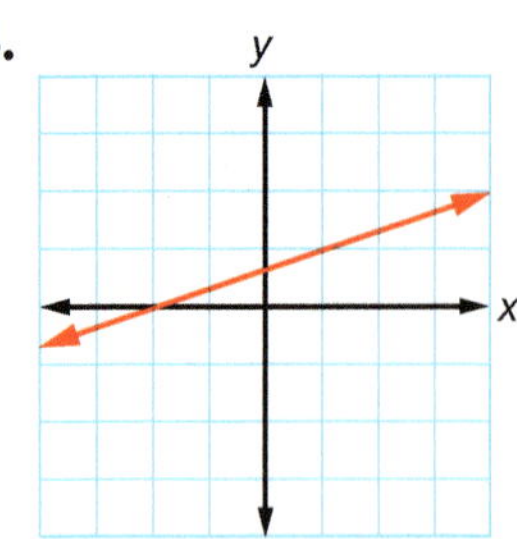

5.

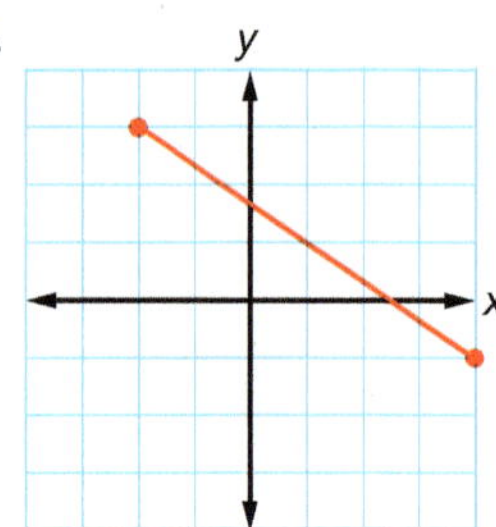

6.

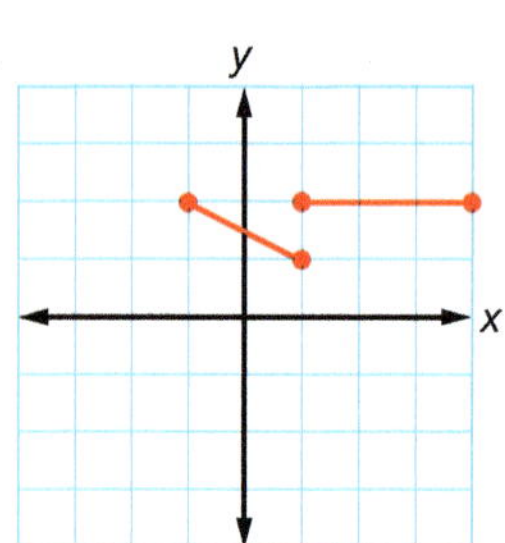

For each graphed relation, state its domain and range and whether it is a function.

7.

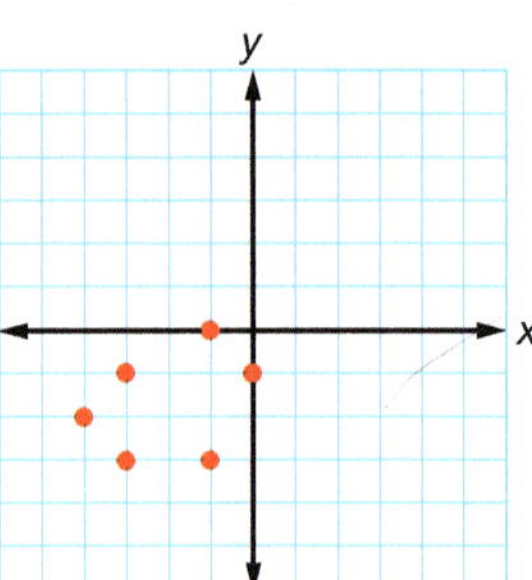

8.

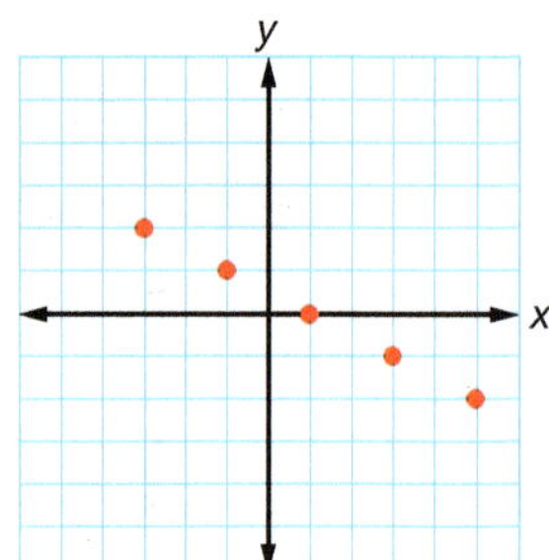

9.

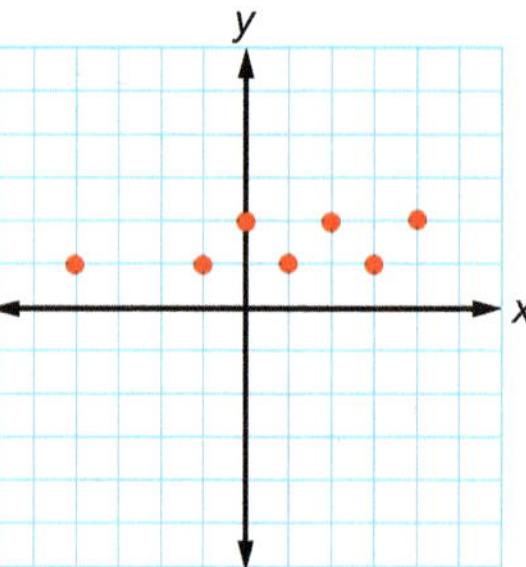

10.

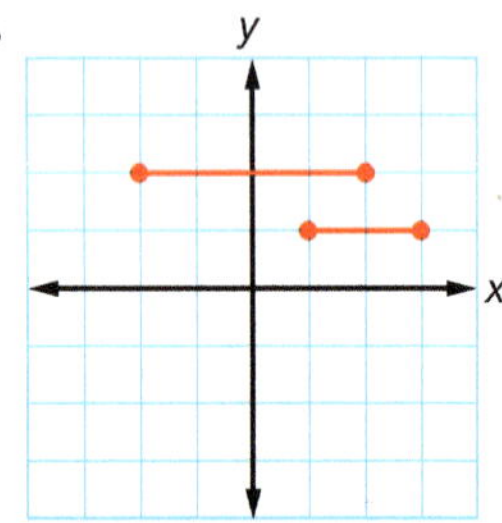

11.

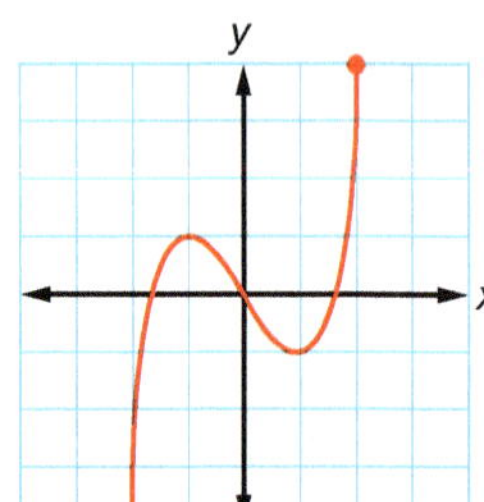

12.

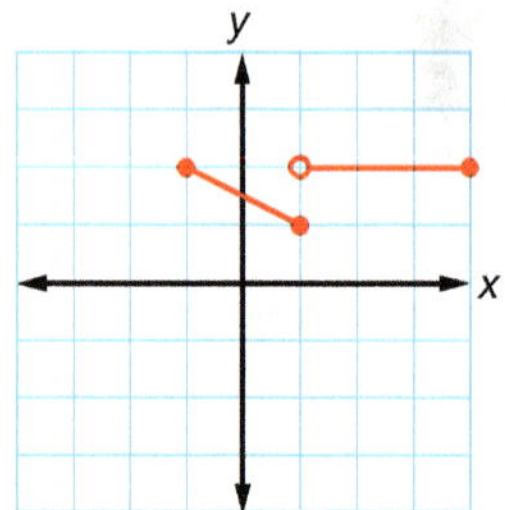

B. Exercises

Represent each function as a set of ordered pairs and graph the function.

13. $\{(x, y) \mid y = 3x; x = -1, 0, 1\}$

14. $\{(x, y) \mid y = x^2; x = -1, 0, 1, 3\}$

15. $\{(x, f(x)) \mid f(x) = 2x + 3; x = -2, 0, \frac{1}{2}, 1\}$

16. $\{(x, f(x)) \mid f(x) = -x - 4; x = -2.5, -1, 0, 1.5, 3\}$

Make a table of ordered pairs and graph each function. Then state the function's domain and range.

17. $\{(x, y) \mid y = 3x\}$

18. $\{(x, y) \mid y = x^2\}$

19. $\{(x, f(x)) \mid f(x) = 2x + 3\}$

20. $\{(x, f(x)) \mid f(x) = -x - 4\}$

Graph each relation and state whether it is a function.

21. $\{(x, y) \mid -1 \leq y \leq 3; x = 2\}$

22. $\{(x, y) \mid y = \pm x; x = -4, 2, 5\}$

23. $\{(x, y) \mid y = |x|; x = -5, -3, -1, 0, 3, 5\}$

24. $\{(x, y) \mid y = |x|\}$

The conversion of temperature from Celsius to Fahrenheit is given by the equation $F = \frac{9}{5}C + 32$.

25. Complete the table.

C	0	10	20	30
F				

26. Graph the relation.

27. Is the relation described by this formula a function?

28. Determine the Celsius equivalent of 500°F.

The perimeter of a regular hexagon can be expressed by the equation $p = 6s$, where s is the length of a side.

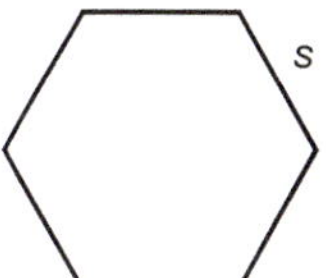

29. State a reasonable domain for this function.

30. Graph the function.

31. Determine the side length of a regular hexagon with a perimeter of 114 m.

The cost of replacing the carpet in a home depends on the number of square yards to be covered. If the set fee for labor is \$250 and the carpet (with padding) costs \$15 per square yard, then the total cost represented by C can be found by the following equation: $C = 15x + 250$, where x represents the number of square yards needed.

32. State a reasonable domain for this function.

33. Graph the function.

34. Determine the cost of carpeting 1800 ft^2 of a home.

A construction company is able to buy 10 ft long 2 in. by 4 in. framing studs at the listed bulk rates.

More than	Up to	Price
	100	\$3.00
100	500	\$2.75
500	1000	\$2.50
1000		\$2.25

35. Which graph best represents the relationship between the number of studs purchased and the price for each stud?

a.

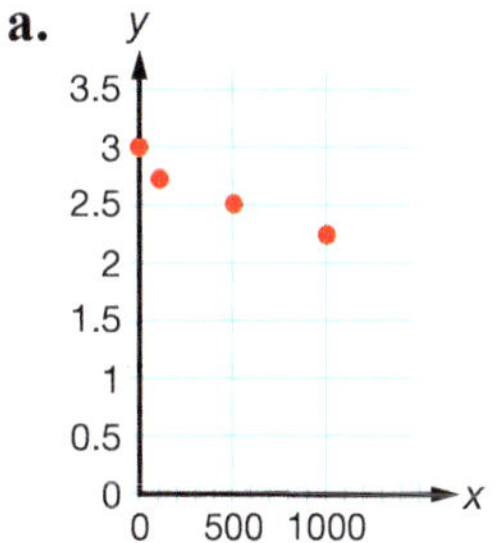

b.

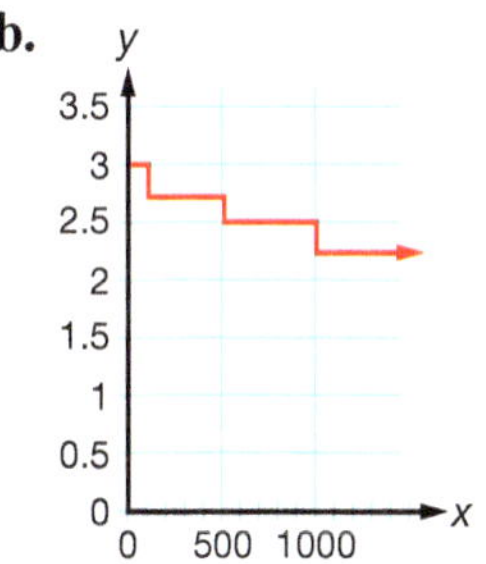

c.

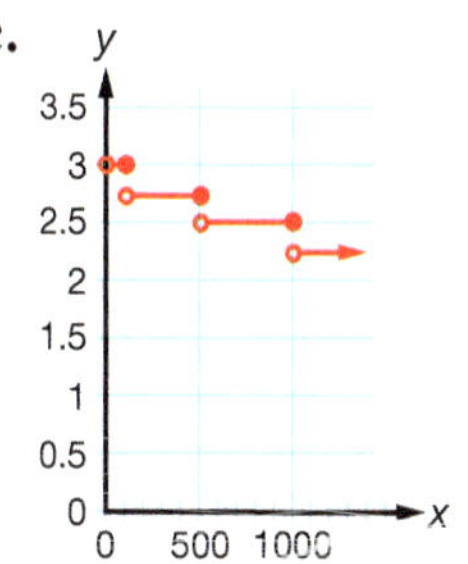

36. Is this relation a function?

C. Exercises

Graph each relation and state whether it is a function.

37. $\{(x, y) \mid -3 \le y \le 2; 0 \le x \le 4, x \in \mathbb{Z}\}$

38. $\{(x, y) \mid -3 \le y \le 2; 0 \le x \le 4\}$

Graph each set of functions on one Cartesian plane. Assume that the domain is the set of real numbers.

39. $y = x$
$y = 2x$
$y = 3x$

40. $y = |x|$
$y = |2x|$
$y = |3x|$

41. Make a conjecture about the effect of the coefficient of x in exercises 39–40.

Dominion Modeling

Distance on land is measured in miles or kilometers. Speed of land vehicles is measured in miles per hour (mi/hr) or kilometers per hour (km/h). Distance at sea and in the air is measured in nautical miles. A nautical mile is one-sixtieth of one degree (one minute) of the earth's circumference, or about 6076 ft. Ships and planes often measure speed in nautical miles per hour, or *knots*. A nautical mile is longer than a land mile (1 nautical mile $\approx$ 1.151 land miles).

42. Convert 30 nautical miles into land miles (to the nearest tenth).

43. Convert 150 land miles into nautical miles (to the nearest tenth).

44. A ship left harbor traveling at 22 knots. If a light plane left the same harbor 4.75 hr later, how fast would it have to travel to catch up to the ship in 2 hr?

CUMULATIVE REVIEW

Clear each equation of fractions or decimals and then solve. [2.8]

45. $\frac{x}{6} - \frac{5}{7} = 2$

46. $\frac{3}{16}x + \frac{2}{3} = \frac{5}{8}$

47. $2.5x - 0.65 = 3.1$

48. $0.2x + 0.78 = 12$

Solve each literal equation for *x*. [3.1]

49. $3axb + c = d$

50. $\frac{dx}{c} = \frac{b}{2a}$

Graph. [4.4–4.5]

51. $x \ge -3$ and $x < 4$

52. $x < 2$ or $x < -1$

Solve. [4.4]

53. $10 < 3x + 1 < 28$

54. $x > -5$ and $x > 2$

TECHNOLOGY CORNER (TI-84+ Family)

Your graphing calculator can be used to display several representations of a function. First, be sure that the calculator is in function mode by selecting FUNC on the [MODE] menu. Then select [Y=] to enter the function editor, and enter the function rule $f(x) = 2x - 1$ as Y_1.

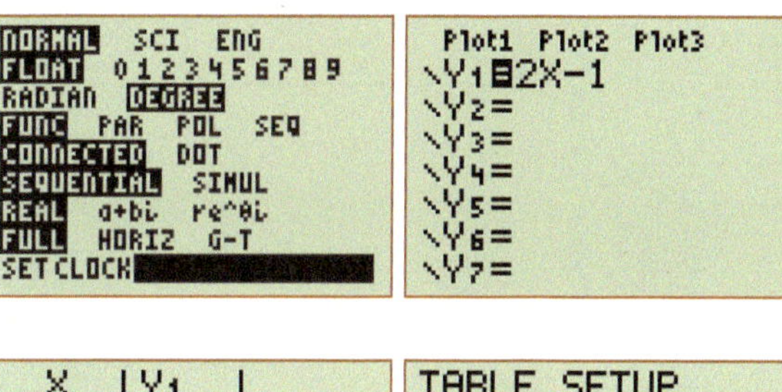

Use the [TABLE] command ([2nd] [GRAPH]) to display a table of ordered pairs for the defined function. The [TBLSET] command ([2nd] [WINDOW]) produces the TABLE SETUP menu. TblStart and ΔTbl are used to change the initial value and increment for the independent variable. (Setting Indpnt: to Ask allows you to enter individual values for x into the table.)

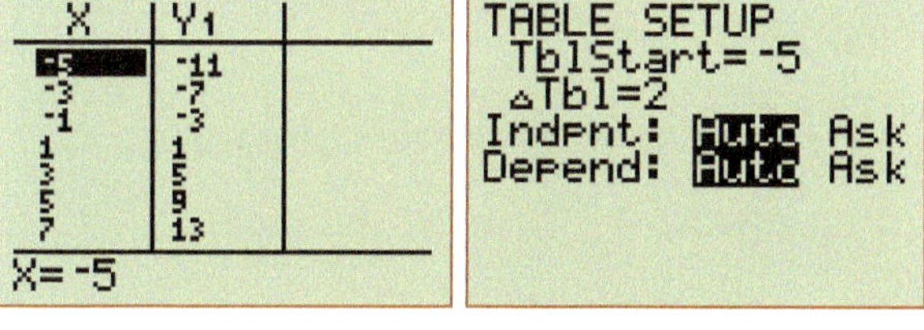

Use the [GRAPH] command to display a graph of the function and use the [ZOOM] and [WINDOW] commands to adjust the portion of the graph that is viewed. Selecting ZStandard from the [ZOOM] menu causes the function to be graphed in a window with $-10 \leq x \leq 10$ and $-10 \leq y \leq 10$. Notice that this causes each axis to have a different scale.

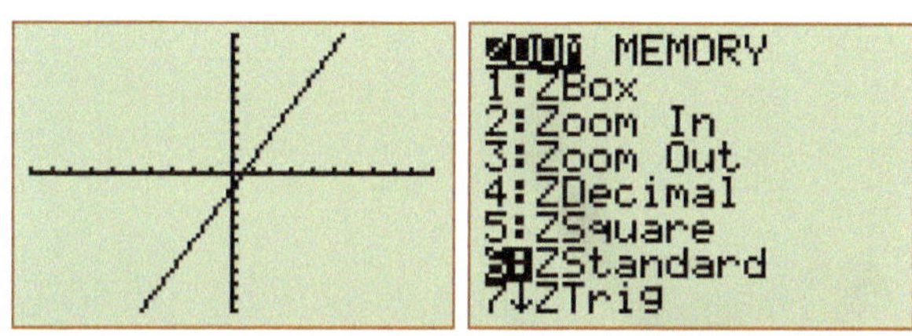

Selecting ZSquare from the [ZOOM] menu generates an axis with the same scale but different minimum and maximum values.

You can enter desired limits and scale for each axis using the [WINDOW] menu. With experience, you will be able to quickly determine the best window to display the key characteristics of a function's graph.

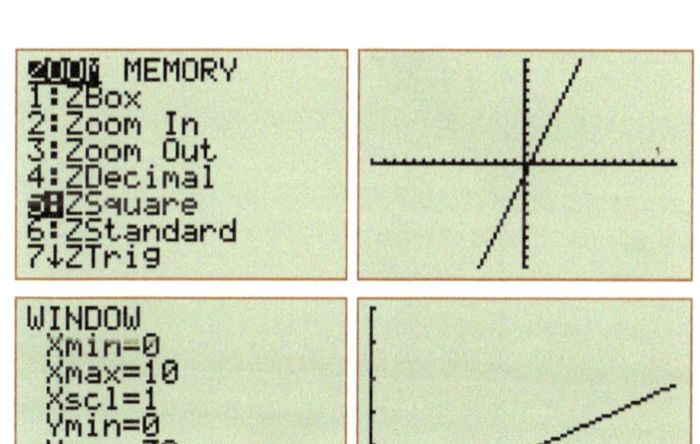

5.4 Using Graphs

Graphs are often used to visualize the relationship between variables. They can illustrate a great variety of phenomena, such as the growth of a fish or the motion of a Ferris wheel. You will often need to draw or interpret graphs that model real-life situations.

Growth of a Largemouth Bass	
Age (years)	**Length (inches)**
1	6.5
2	11
3	14
4	16
5	17
6	17.5

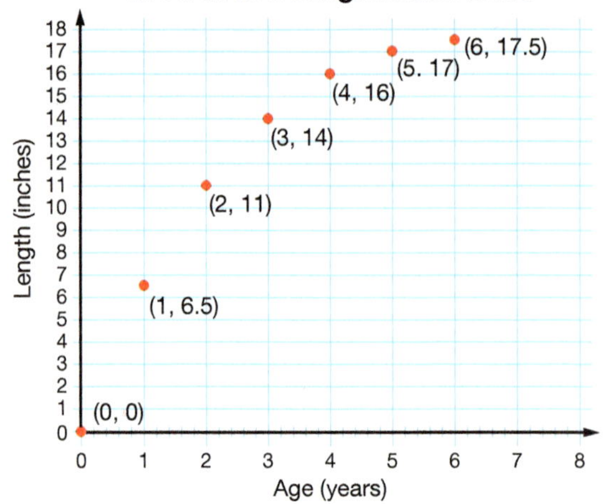

The table gives the approximate length of a largemouth bass depending on its age. A graph can be used to show the relation between length and age. Since the age of a fish is not affected by the fish's length, it would be graphed on the horizontal axis as the independent variable. The fish's length is the dependent variable since it is affected by the fish's age. It is graphed on the vertical axis.

By drawing a smooth curve that approximates the shape of the plotted points, you can discover general characteristics and make predictions. Rapid growth occurs when the fish is young, but growth slows down as the fish gets older. We could predict that a $1\frac{1}{2}$ yr old bass would be about 9 in. long and that a 15 in. long bass is about $3\frac{1}{2}$ yr old. We could also estimate the length of an 8 yr old bass to be almost 18 in. This process of graphing data to study the shape of its distribution is called *mathematical modeling*.

Growth of a Largemouth Bass

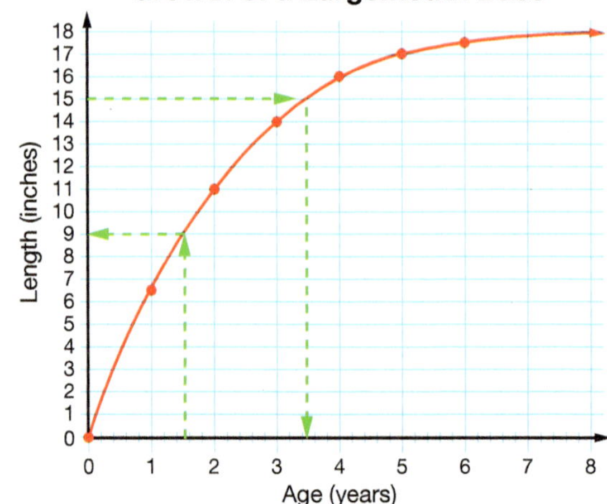

The purpose of a mathematical model, such as a graph, is to illustrate a mathematical idea and make it easier to comprehend. In a parallel fashion, Christians are to model their lives after Jesus Christ for the purpose of showing the world what He is like. Peter exhorts Christians to follow the steps of Christ's example because they have been chosen by God to "shew forth the praises of him who hath called you out of darkness into his marvellous light" (1 Pet. 2:9, 21). How well do you imitate Christ? Do others see Christ in your life?

Example 1

Model the first-class postage rate data given in the table and state the general characteristics of the rate of increase of postage over the years. Predict what the postage rate will be in 2050.

Year	Stamp Cost (cents)
1865	6
1885	2
1900	3
1930	3
1960	4

Year	Stamp Cost (cents)
1965	5
1970	6
1975	13
1985	22
1990	25

Year	Stamp Cost (cents)
1995	32
2000	33
2005	37
2010	44

Answer

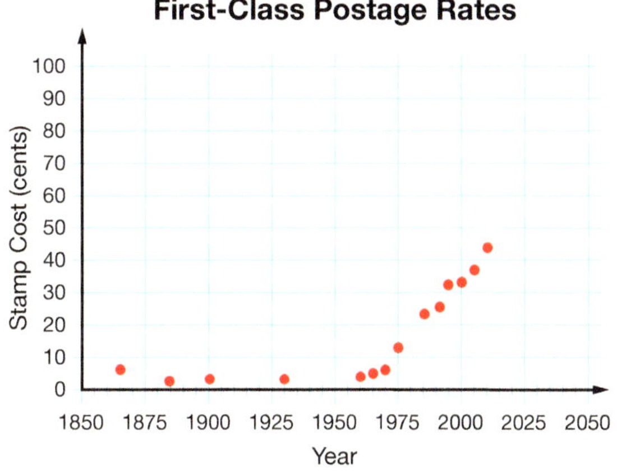

1. The year is the independent variable and is graphed on the horizontal axis. The price is the dependent variable and is graphed on the vertical axis.
2. Early in the history of postage stamps, the rate decreased slightly and then stayed relatively steady for almost one hundred years. Rates have increased dramatically in recent years.

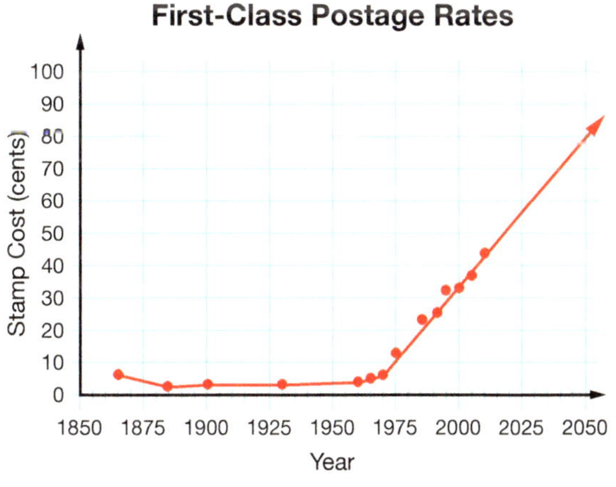

3. Draw a smooth curve that shows the general shape of the postal rate distribution over the time period pictured on the graph. If the price continues to increase at a similar rate, by 2050 it will cost about 80 cents to mail a letter.

You must be able to interpret the data represented on a graph and to analyze the general relationship between the variables. In this section you will state the general relationship between the variables. In future sections you will learn how to determine the equations that describe the behavior pictured by a graph.

The following graphs model the distance fallen and the speed of a skydiver during his fall.

Example 2

Use the distance-time graph to answer the following questions.

a. How far has the skydiver fallen after 10 sec?

b. How long does it take the skydiver to fall 2000 ft?

c. Describe the fall of the skydiver in the first 5 sec, the next 15 sec, and the final 10 sec illustrated.

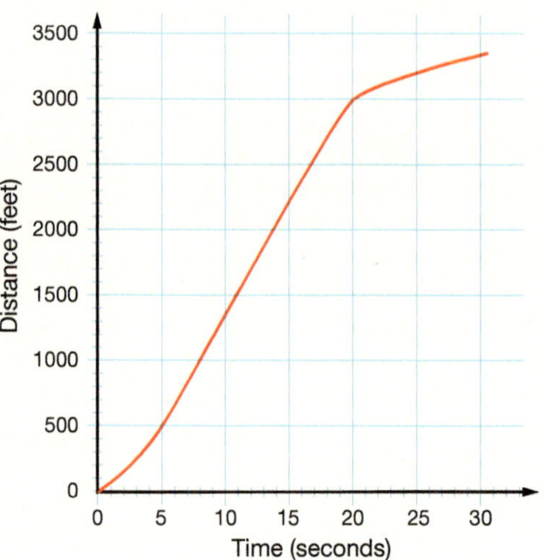

Answer

a. At 10 sec he has fallen about 1300 ft.

b. It takes about 14 sec to fall 2000 ft.

c. The skydiver falls greater distances each second for the first 5 sec.

He then falls about the same number of feet each second for the next 15 sec.

He then falls at a much slower, constant rate during the last 10 sec.

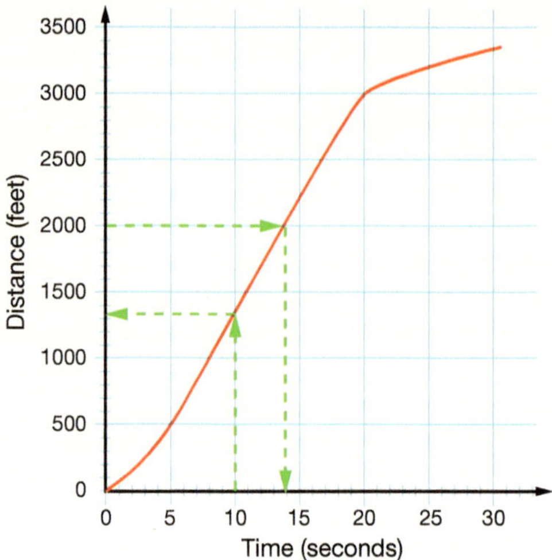

Graphs without numbered axes can be drawn to illustrate the relationship between variables. The graph to the right illustrates the relative speeds at which the skydiver falls. A more thorough investigation of the distance-time graph above would indicate that he is falling approximately 167 ft/sec once he reaches a constant speed and about 33 ft/sec after he deploys his parachute.

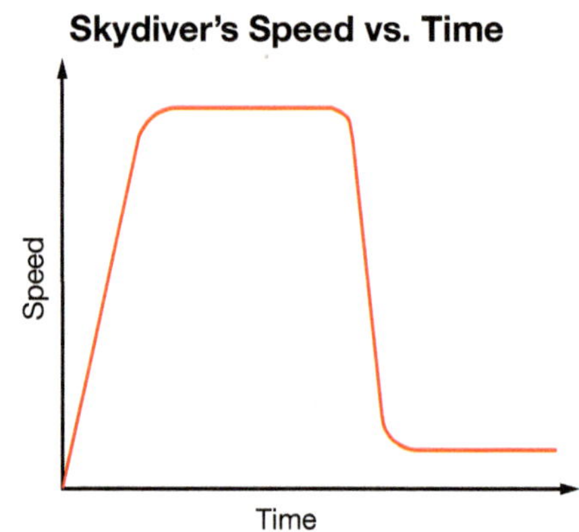

Example 3

The Wadsworths spent a day hiking in the mountains. Mrs. Wadsworth's GPS-enabled sports watch records her heart rate and position, including altitude. Analyze the following graphs produced from their trip to describe their hike during each period of time.

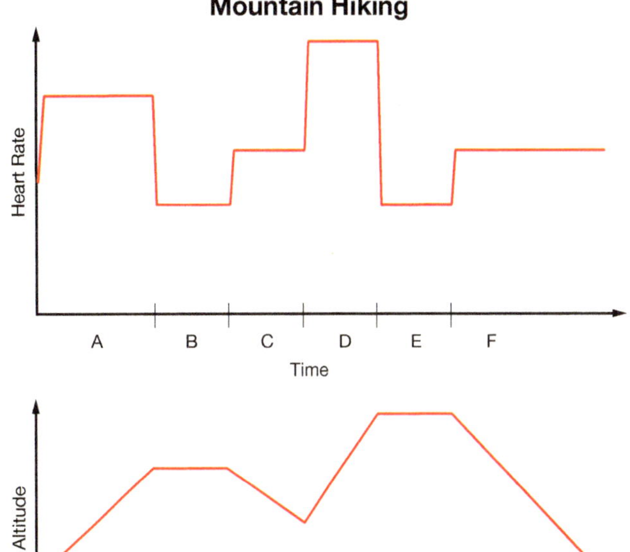

Answer

A. Her heart rate is elevated as they ascend the first slope.
B. Her heart rate returns to normal while they are at the top of the mountain.
C. Her heart rate is elevated slightly as they walk partway down the mountain.
D. Her heart rate is elevated greatly as they climb a steeper slope.
E. At the top of the second mountain, her heart rate returns to normal.
F. As she descends to the trailhead, her heart rate remains slightly elevated.

A. Exercises

Interpret each graph, explaining each labeled section.

1. **Daily Savings Account Balance**

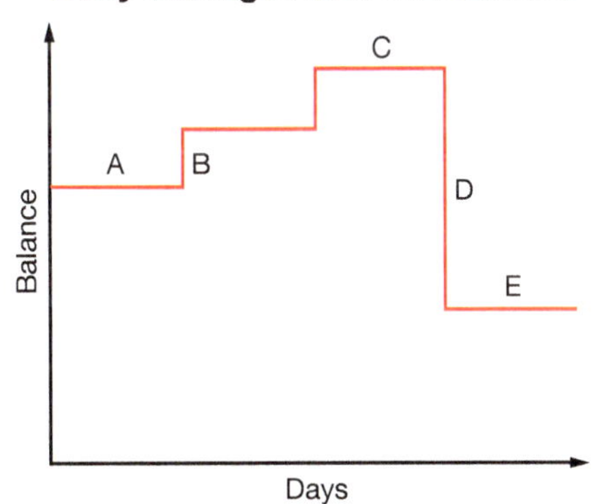

2. **Cars Passing Through an Intersection**

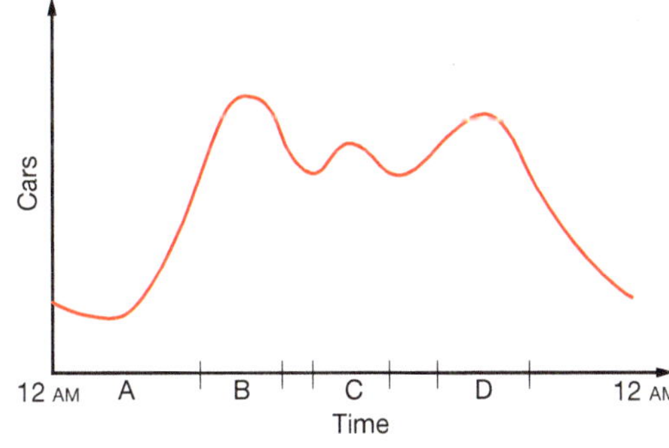

3. **Altitude During a Hot Air Balloon Ride**

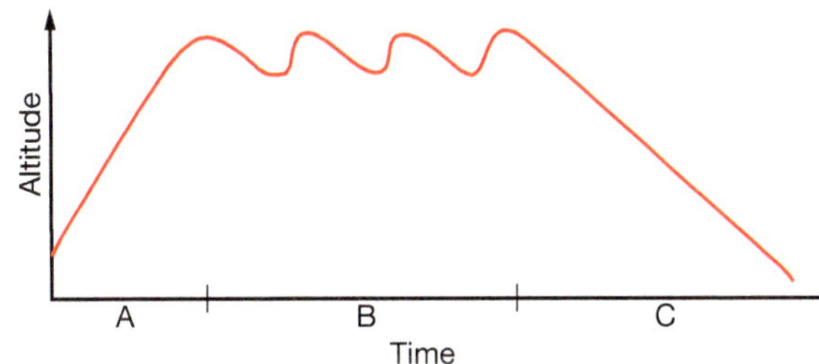

4. **Speed of a Bus**

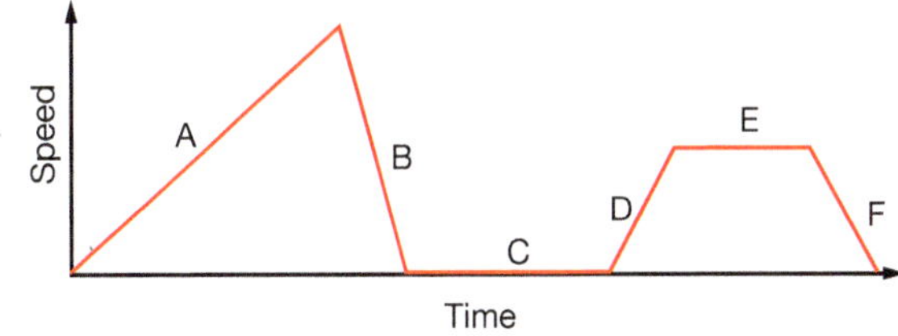

5. Matching: Which graph would represent the height of water in each container as water is poured into it at a continuous rate?

a. 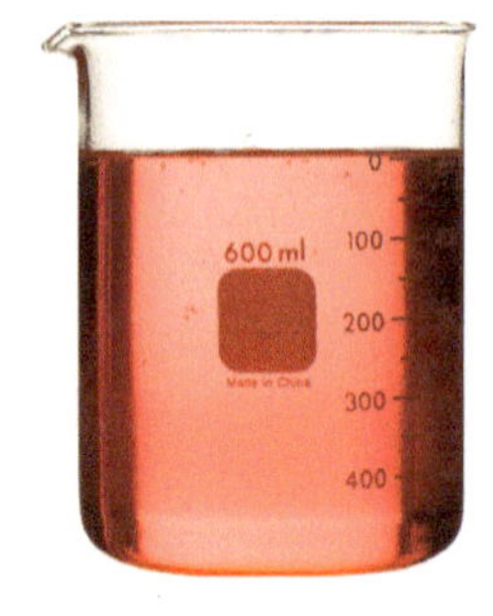

b. 

c. 

I. 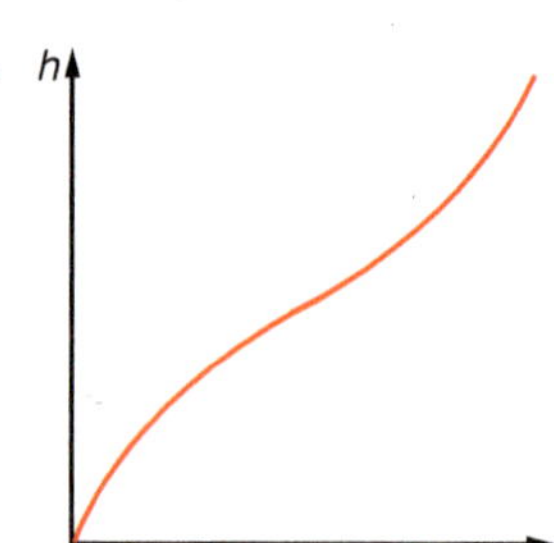

II. 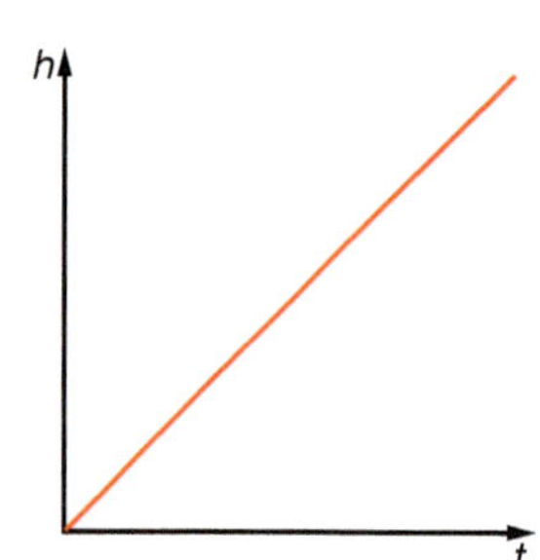

III. 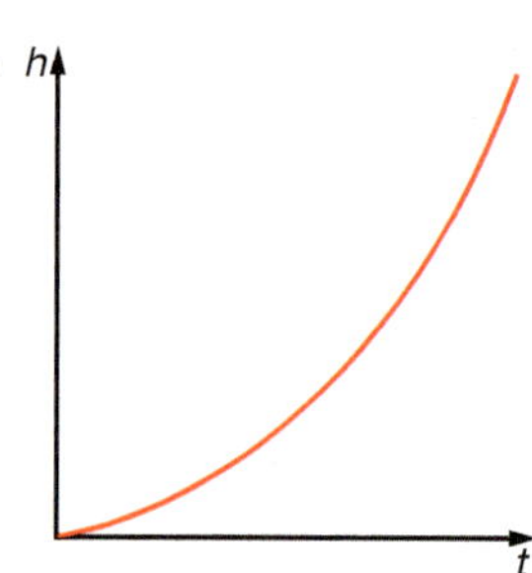

6. Which of these graphs would represent the speed of a ball thrown straight up in the air?

a. 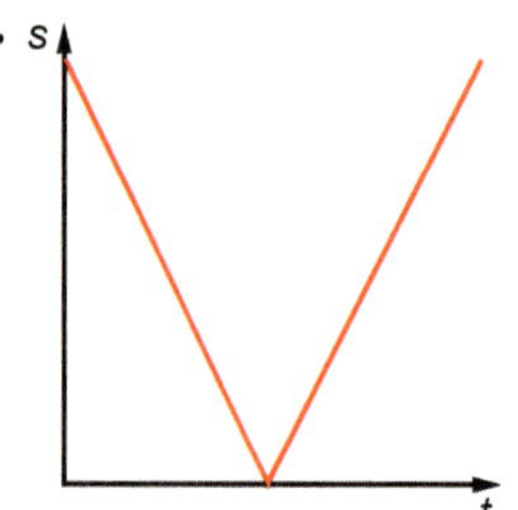

b. 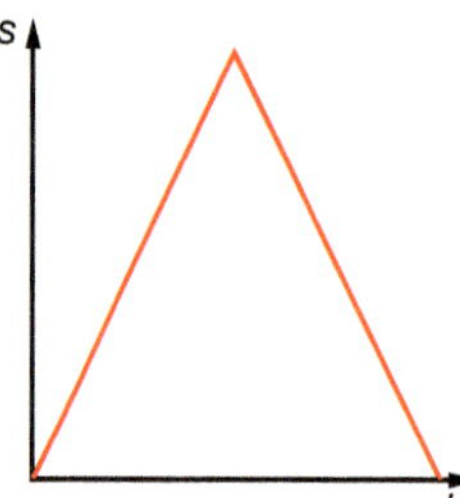

c. 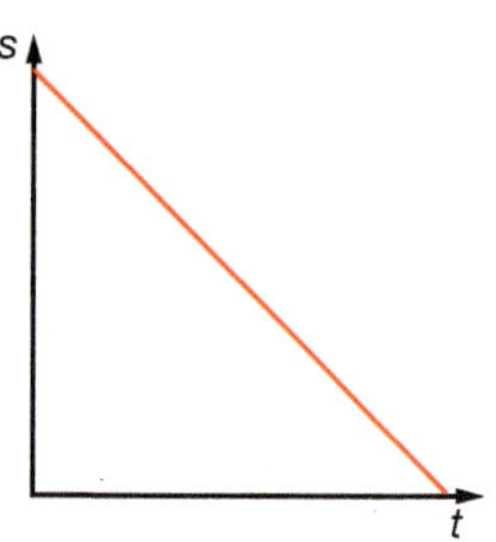

d. 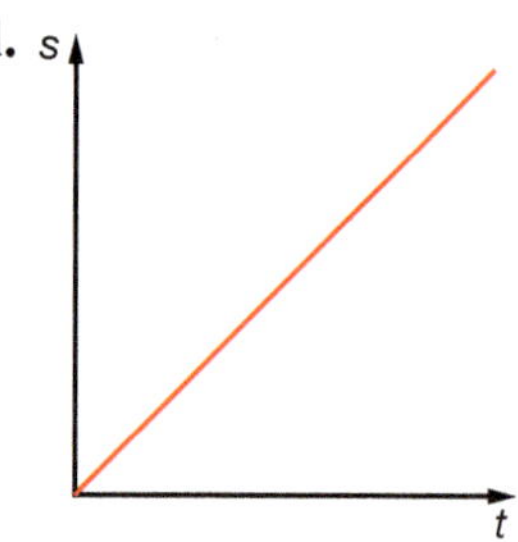

Sketch a graph (without numbered axes) illustrating each of the following situations.

7. Susan's heart rate as she drives to the track, warms up, runs three miles, cools down, and then drives home
8. Ken's distance from home as he drives to the grocery store, shops for half an hour, and then returns home
9. Gary's total distance driven as he drives to the bank, deposits a check and withdraws some cash there, and then drives home
10. yesterday's outside temperatures
11. the height of a punted football during its flight through the air
12. the speed of a dragster during a quarter-mile race

B. Exercises

Use the following table for exercises 13–16.

13. Graph the data and draw a smooth curve that models the relationship between Grant's income and his offering.
14. Describe the general trend of the data.
15. Use your graph to determine his offering if he made $300.
16. Use your graph to determine his income if he gave an offering of $60.

Income ($)	Offering ($)
50	5
100	10
250	25
500	50
750	75
1000	100

Use the following table for exercises 17–20.

Minutes Playing Basketball	Calories Burned
10	94
20	187
30	280
40	375
50	470
60	565

17. Graph the data and draw a smooth curve that models the relationship between the number of calories Jonathan burned and the time he spent playing basketball.

18. Describe the general trend of the data.

19. How many minutes does Jonathan need to play basketball in order to burn 150 calories?

20. How many calories does Jonathan burn if he plays basketball for 45 min?

Use the following table for exercises 21–24.

Boiling Point of Water (°C)	101.2	101	100	99.8	97.9	96.2	96.1	95.7
Elevation (feet)	0	70	100	350	700	1500	2300	3500

21. Which quantity should be represented by the independent variable? Explain why.

22. Graph the data and draw a smooth curve that models the relationship between elevation and water's boiling point.

23. Describe the general trend of the data.

24. Use your graph to determine water's boiling point at an elevation of 1000 ft.

Use the following table for exercises 25–28.

Diameter (cm)	Area (cm^2)
2	3
4	13
6	28
10	79
16	201
20	314

25. Graph the data and draw a smooth curve that models the relationship between the diameter of a circle and its area.

26. Describe the general trend of the data.

27. Use your graph to determine the area of a circle whose diameter is 5 cm.

28. Use your graph to determine the diameter of a circle with an area of 140 cm^2.

Use the following table for exercises 29–32.

Elapsed Time (hours)	0	2	4	6	8	10	12	14	16
Concentration (mg/L)	0	30	50	42	36	28	20	18	12

29. Graph the data and draw a smooth curve that models the relationship between the concentration of medication in the blood and the amount of time after the medication is taken.

30. Describe the general trend of the data.

31. Use your graph to determine the concentration of the medication after 9 hr.

32. When is the concentration 45 mg/L?

Give a detailed description of the behavior modeled by each graph. Include significant quantities in your description.

33. Describe the motion of a student.

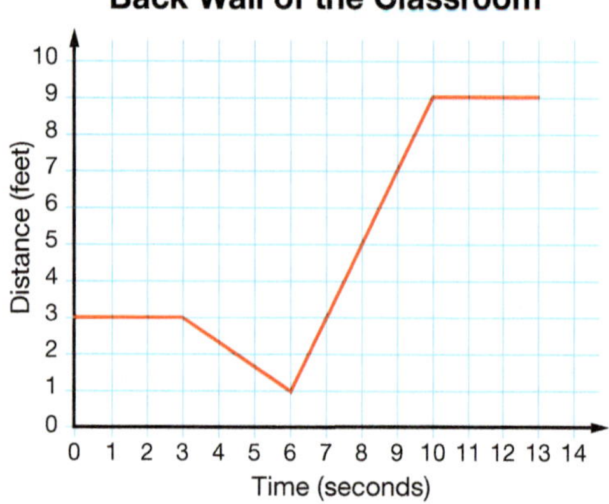

34. Describe the growth of bacteria in a Petri dish.

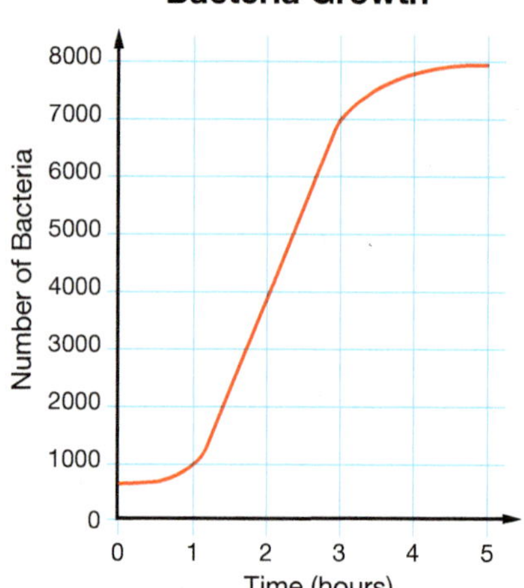

35. Describe the motion of a dropped ball.

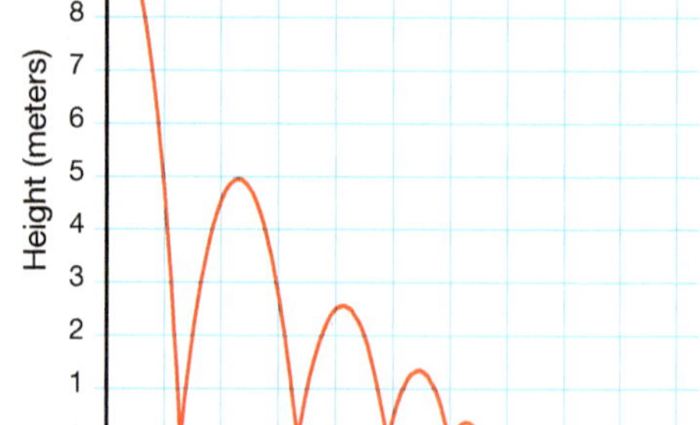

36. Describe the occupancy of the gymnasium on a typical Friday.

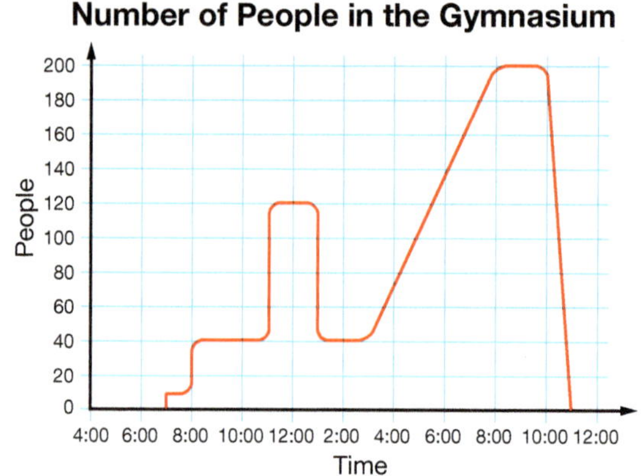

C. Exercises

Use the following table describing the height of a Ferris wheel rider for exercises 37–40.

37. Graph the data and draw a smooth curve that models the height of the rider and the time of the ride.

38. Describe the general trend of the data.

39. Use your graph to determine the height of the rider 15 sec into the ride.

40. At what times is the rider 100 ft high?

Time (seconds)	Height (feet)
0	5
10	27
20	80
30	133
40	155
50	133
60	80

Time (seconds)	Height (feet)
70	27
80	5
90	27
100	80
110	133
120	155
130	133

Dominion Modeling

41. What other name is given to one-sixtieth of one degree of the earth's circumference?

42. Find the number of miles (to the nearest hundredth) in one degree along the equator, using 24,901 mi as the circumference of the earth.

43. Is this distance the same along other lines of latitude besides the equator? Explain.

44. If the earth were a perfect sphere, would this distance be the same along the lines of longitude?

45. What is the distance between 0° N, 63°20′ W and 0° N, 17°30′ E (to the nearest mile)?

CUMULATIVE REVIEW

True or false [1.1]

46. The set of rational numbers is a subset of the irrationals.

47. The set of irrational numbers is a subset of the real numbers.

Translate each word phrase into an algebraic expression. [2.1]

48. five more than twice a number

49. twenty decreased by three times a number

Solve. [4.6]

50. $|x - 3| = 8$

51. $|2x + 5| = 17$

Solve and graph the solution on a number line. [4.7]

52. $|x| > 3$

53. $|x - 4| < 6$

Solve. [4.7]

54. $|2x - 3| < 8$

55. $|4x + 3| > 77$

Plot the following ordered pairs on a graph and connect them with line segments in the order they are listed (working down each column). Start a new portion of the graph after the word *end*.

(−4, 5)
(−4, 1)
(−1, 1)
(−1, 5)
(−4, 5)
end

(−1, −1)
(−1, −2)
(1, −2)
(1, −1)
(−1, −1)
end

(1, 5)
(4, 5)
(4, 1)
(1, 1)
(1, 5)
end

(−5, 5)
(−3, 7)
(−2, 7)
(−1, 6)
end

(6, 10)
(6, −7)
(5, −8)
(2, −9)
(−2, −9)
(−5, −8)
(−6, −7)
(−6, 10)
end

(4, 2)
(3, 2)
(3, 1)
end

(−4, −3)
(−3, −4)
(−2, −5)
(2, −5)
(3, −4)
(4, −3)
(2, −4)
(−2, −4)
(−4, −3)
end

(−6, 10)
(−5, 11)
(−3, 12)
(3, 12)
(5, 11)
(6, 10)
(5, 9)
(2, 8)
(−2, 8)
(−5, 9)
(−6, 10)
end

(1, 6)
(2, 7)
(3, 7)
(5, 5)
end

(−1, 2)
(−2, 2)
(−2, 1)
end

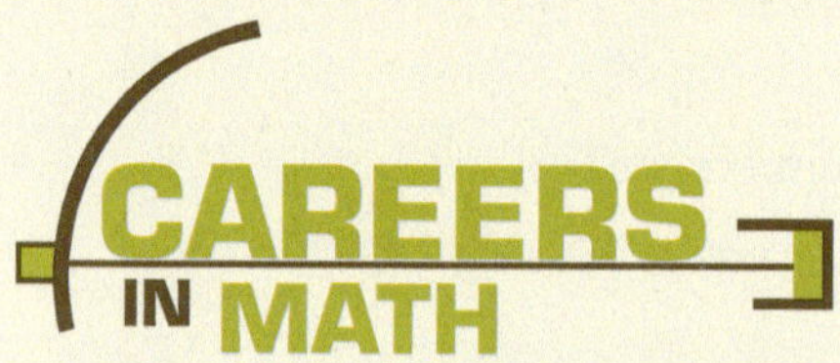

Market Research Analyst

A 2006 *Money Magazine* article based on an online survey of 26,000 workers ranked the job of a market research analyst sixth in its ranking of top jobs. Like other statisticians, market research analysts gather and analyze data; but they focus on helping businesses market their products to consumers or to other businesses for maximum profit. A Christian might wonder whether it would be worthwhile to spend his life helping businesses make greater profits. However, the successful marketing of beneficial products enables the business to employ more people and to research other means of satisfying consumer needs.

Large corporations and government agencies hire their own market research analysts. Many other analysts are employed by consulting firms and marketing research firms that serve small companies, which often prefer to contract out work rather than support their own marketing department. People with good communicative and quantitative skills will have the best opportunities in the field, and a strong background in math and computer science is especially helpful.

Market research analysts supervise the collection of data through a variety of methods, including surveys, interviews, and focus groups. A market research analyst must know how to calculate an appropriate sample size and how to take a sample from a larger population. After compiling and interpreting the data collected, an analyst provides information and recommendations that will help the business make decisions concerning product design, pricing, and advertising. This information may also lead to other decisions, such as promotional rebates or new store locations. An analyst must also be able to calculate and to state the degree of certainty for his conclusions.

A market research analyst makes life easier for the consumer by making the things the consumer wants more easily available. He also increases the business's profit through his advice on which items to promote. A Christian market research analyst should also take into account the ethical questions that arise in such work. Just because something can be marketed does not mean it should be, and just because the market may want a product does not mean a business should supply it. In addressing these ethical issues, the Christian has a great advantage over his secular counterpart.

5.5 Function Rules

The function is a fundamental concept of all higher mathematics. It can be thought of as a machine, as illustrated to the right. Values are chosen for x, put into the machine, manipulated by a function rule, and put out as corresponding y values.

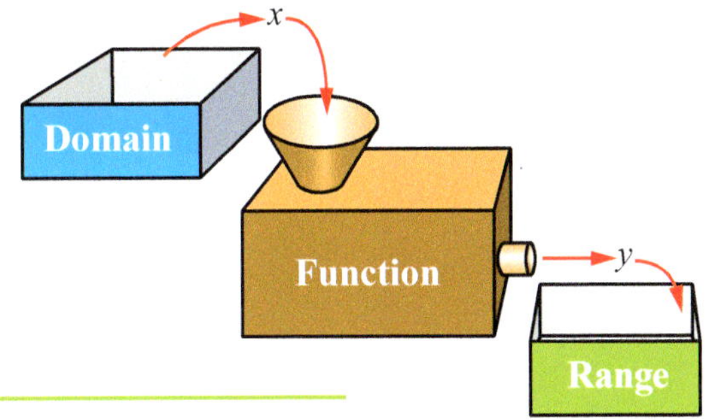

Definition

A **function rule** is an equation that pairs each member of the domain with an element of the range.

If a function rule and the domain are given, then the entire function is defined and can be listed as a set, written as a table, graphed, or illustrated with a mapping diagram.

Example 1

Given the domain $\{-3, -1, 0, 2, 4\}$ and the function rule $f(x) = 3x + 2$, represent the function as a table, a set of ordered pairs, a graph, and a mapping diagram.

Answer

x	$3x + 2$	$f(x)$
-3	$f(-3) = 3(-3) + 2 = -7$	-7
-1	$f(-1) = 3(-1) + 2 = -1$	-1
0	$f(0) = 3(0) + 2 = 2$	2
2	$f(2) = 3(2) + 2 = 8$	8
4	$f(4) = 3(4) + 2 = 14$	14

1. Use the function rule to find the element of the range associated with each given element of the domain.

$f = \{(-3, -7), (-1, -1), (0, 2), (2, 8), (4, 14)\}$

2. Write the function as a set of ordered pairs.
3. Create a graph from the ordered pairs.
4. Draw a mapping diagram.

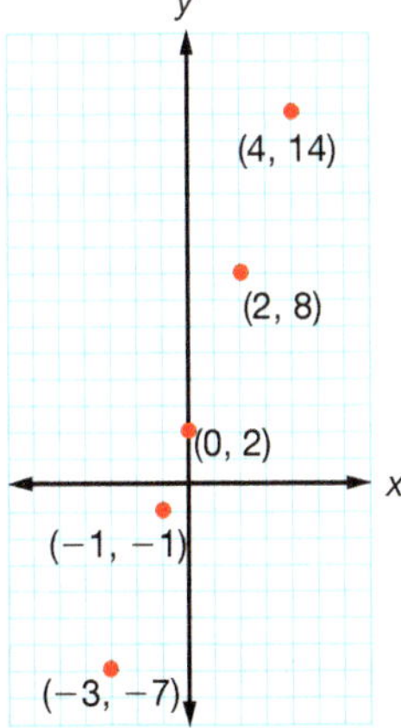

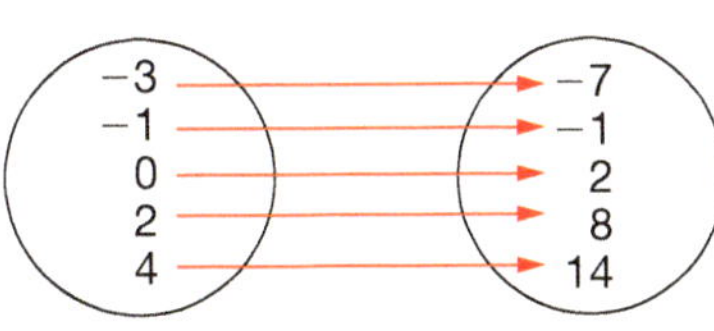

Example 2

Graph the function $y = |x|$.

Answer

x	$\|x\|$	y
-3	$f(-3) = \|-3\| = 3$	3
-1	$f(-1) = \|-1\| = 1$	1
0	$f(0) = \|0\| = 0$	0
2	$f(2) = \|2\| = 2$	2
5	$f(5) = \|5\| = 5$	5

1. Select several values of the domain to make a table of values for the function.

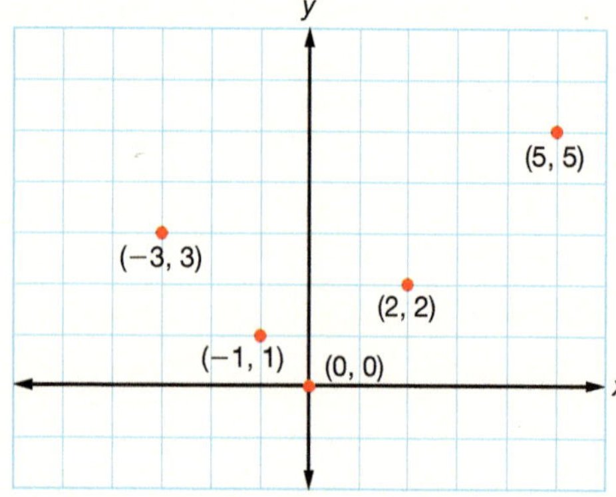

2. Plot enough ordered pairs to determine the general shape of the graph.

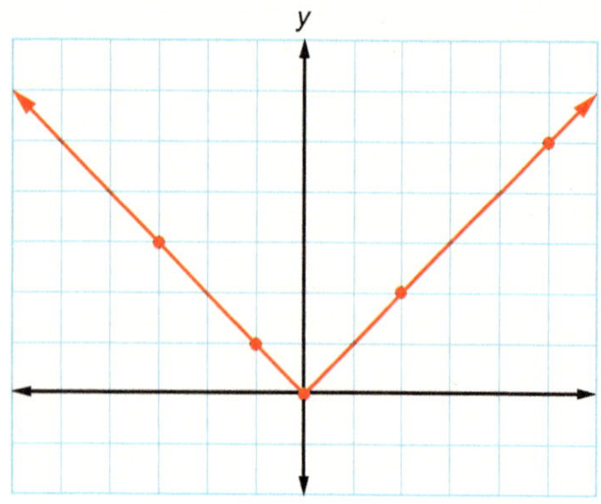

3. Since no domain is given, assume that all real numbers are members of the domain. Connect the graphed points to illustrate the general shape of the function.

You can find the function rule from the other representations of a function. Carefully examine the domain elements and their related range elements in order to determine a consistent pattern. Each element of the range must be the result of the same function rule.

x	$f(x)$
-1	3
0	4
2	6
3	7
4	8

The table to the right represents a function. By comparing the coordinates, you can observe that each y-coordinate is 4 more than its x-coordinate. So the function rule that describes this pattern is $f(x) = x + 4$. Check to make sure that this rule gives the correct range value for each domain value.

Example 3

Find the function rule for the table.

x	-2	-1	0	1	2
$f(x)$	-6	-3	0	3	6

Answer

(differences in x: 1, 1, 1, 1)

x	-2	-1	0	1	2
$f(x)$	-6	-3	0	3	6

(differences in $f(x)$: 3, 3, 3, 3)

1. As the values for x increase by 1, the values for y increase by 3.

$f(x) = 3x$

2. The ratio of y to x remains constant. Check that the function rule is true for all the ordered pairs.

Example 4

Find the function rule for the graph.

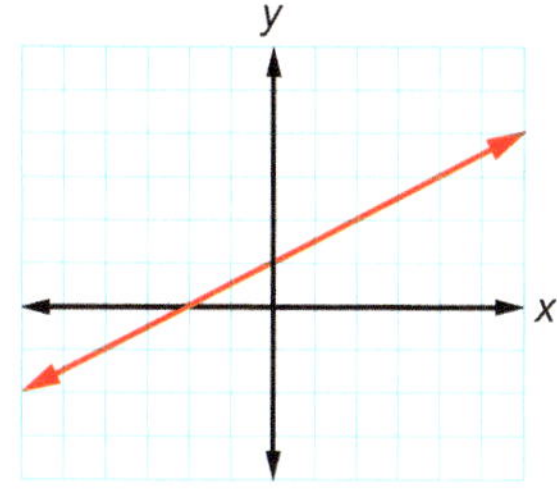

Answer

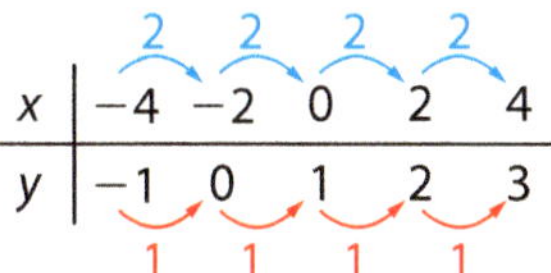

(differences in x: 2, 2, 2, 2)

x	-4	-2	0	2	4
y	-1	0	1	2	3

(differences in y: 1, 1, 1, 1)

1. Make a table using several ordered pairs from the graph. Notice that the x values differ by 2, while the y values differ by 1. This suggests that $y = \frac{1}{2}x$ may be the function rule.

x	-4	-2	0	2	4
$\frac{1}{2}x$	-2	-1	0	1	2
y	-1	0	1	2	3

2. A check of this function rule produces values that are always 1 less than y.

$y = \frac{1}{2}x + 1$

3. Revise your function rule by adding 1.

Example 5

Find the function rule for the ordered pairs given in the graph.

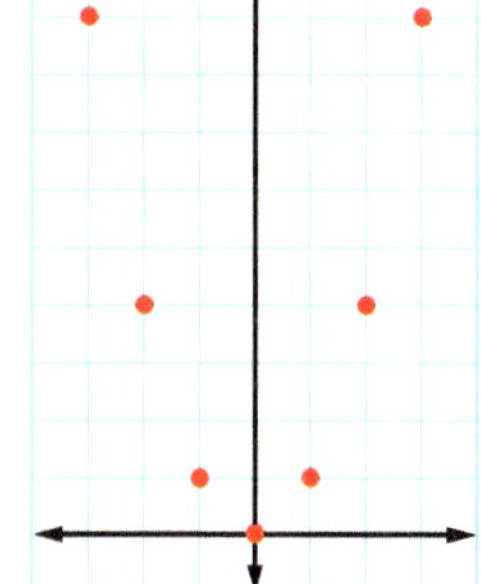

Answer

x	-3	-2	-1	0	1	2	3
y	9	4	1	0	1	4	9

1. List the ordered pairs in table form.

$f(x) = x^2$

2. Each value of the domain has been squared to obtain its associated range value.

Functions abound in real-life applications. Your experience in translating word problems into equations will help you write function rules for these situations.

Example 6

Caitlyn is comparing two cell phone plans. Plan A costs \$47/mo with unlimited minutes. Plan B costs \$28/mo with a fee of \$0.04/min. State a function rule for both plan A and plan B, graph the two functions, and then determine which plan is better if Caitlyn typically uses 480 minutes in a month.

Answer

$A(m) = 47$

$B(m) = 28 + 0.04m$

1. Express the cost of each plan as a function of the minutes used.

2. Graph both functions on the same Cartesian plane.

$A(480) = 47$

$B(480) = 28 + 0.04(480) = 47.20$

Plan A is slightly less expensive for 480 min.

3. Find the cost of 480 min for each plan and compare.

From the graph, you can see that the two plans intersect at 475 min. This indicates that plan A is less expensive if Caitlyn talks more than 475 min/mo and that plan B is less expensive if she talks less than 475 min/mo.

A. Exercises

Given the function rule and domain, make a table and write a set of ordered pairs to represent the function.

1. $f(x) = x - 4; D = \{-2, 0, 3\}$

2. $g(x) = -2x - 3; D = \{-3, 0, 2, 5\}$

Graph each function.

3. $y = -3x; D = \{-1, 0, 2\}$

4. $y = \frac{1}{2}x + 1; D = \{-4, 0, 3, 6\}$

5. $h(x) = 2x + 1$

6. $f(x) = \frac{1}{3}x$

Use *f*(*x*) notation to write the rule for each function.

7. $\{(2, 8), (1, 4), (0, 0), (1, 4), (5, 20)\}$

8. $\{(-4, 2), (-2, 1), (0, 0), (2, -1), (4, -2)\}$

9. $\{(-3, -5), (-1, -3), (2, 0), \left(\frac{1}{3}, \frac{-5}{3}\right), (5, 3)\}$

10. $\{(-8, -5), (-4, -1), (-3, 0), (2, 5), (4, 7)\}$

11.

x	-5	-2	-1	3	5
y	-15	-6	-3	9	15

12.

x	-3	-1	0	1	3
y	-4	-2	-1	0	2

13.

x	100	50	0	20	30
y	-98	-48	2	-18	-28

14.

x	-30	-20	-10	0.4	1
y	-15	-10	-5	0.2	0.5

15.

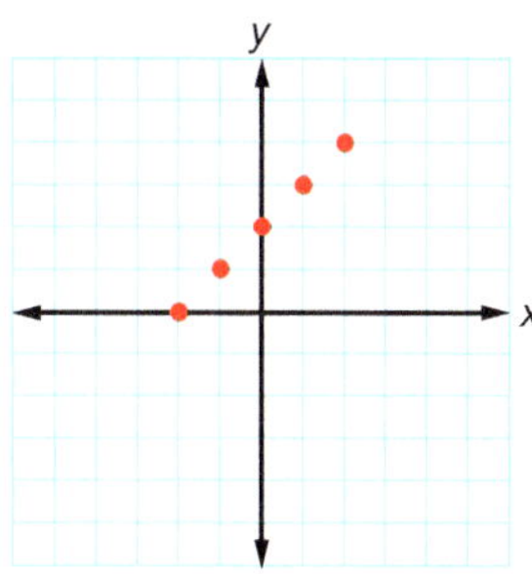

16.

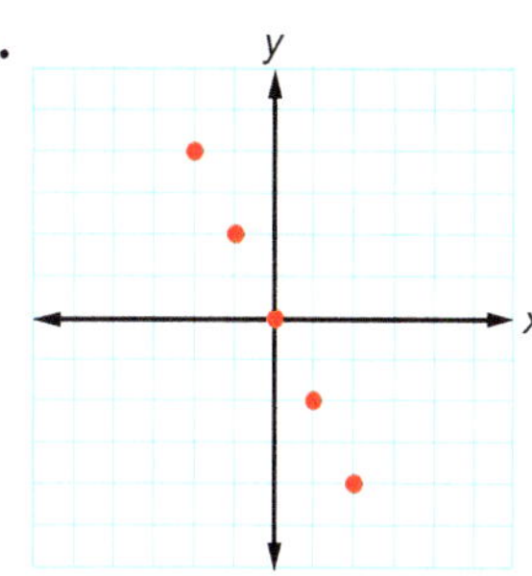

17.

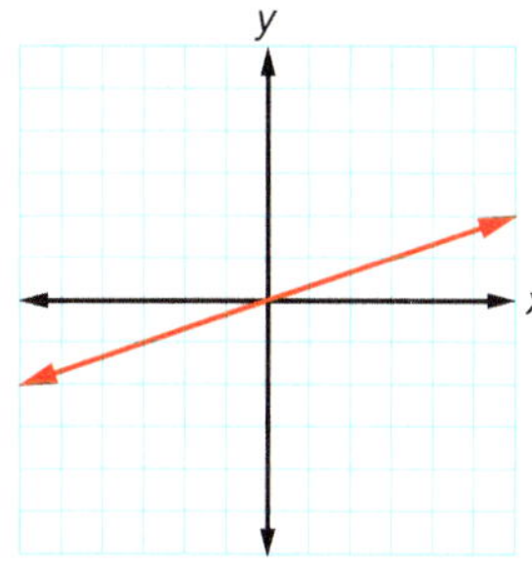

18.

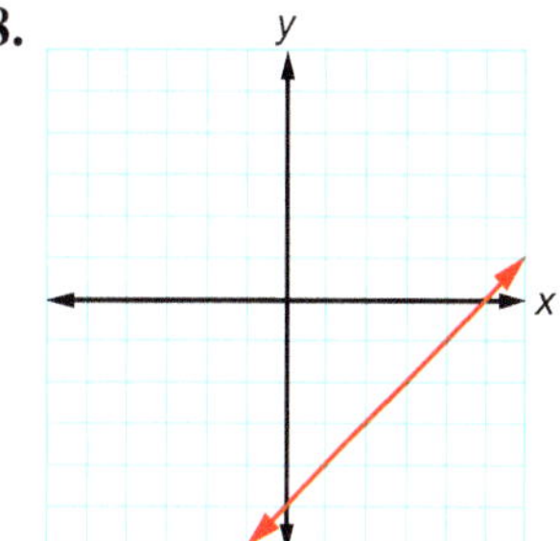

B. Exercises

Use function notation to write the rule for each function.

19. $f = \{(0, 1), (1, 3), (2, 5), (3, 7), (4, 9)\}$

20. $g = \{(-4, -10), (-2, -4), (0, 2), (2, 8), (4, 14)\}$

21.

x	-3	-1	1	3	5
$r(x)$	-13	-3	7	17	27

22.

x	-2	-1	0	1	2
$s(x)$	8	1	0	-1	-8

23.

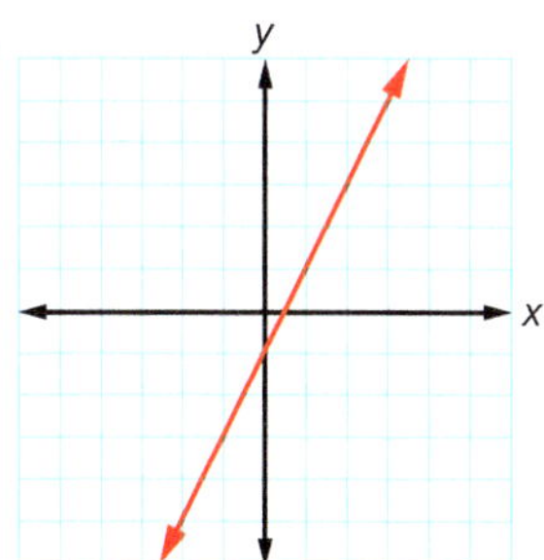

24.

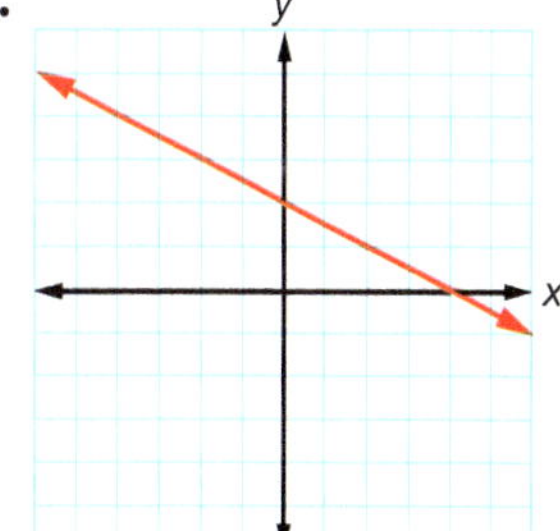

Write a function rule to describe each of the following scenarios.

25. Carissa is using a copier to reduce the size of some of her drawings. She wants them to each be one-fourth of their original size. If d is the length of a side of one of her original drawings, express the length of the reduced side as $f(d)$.
26. If 1 in. of rain is equivalent to 10 in. of snow, express the number of inches of snow that is equivalent to r inches of rain.
27. Noel is two years older than Angelica. Express Noel's age as a function of Angelica's age, a.
28. the perimeter of a regular pentagon with side length s
29. the profit that you make selling c candy bars if each candy bar sold produces $0.25 profit
30. the number of cups of flour required for a triple batch of a recipe that calls for c cups per batch
31. the time it takes for Dmitri to travel distance d at 50 mi/hr
32. the height of a tree in yards if you know that it is f ft tall
33. the area of a semicircle with radius r
34. the cost of playing r games of racquetball each month at a fitness club that charges a $35 monthly membership fee plus $2 per game of racquetball

C. Exercises

Use function rules to answer each of the following exercises.

35. Mr. Gemson is the manager of a business that manufactures and sells tennis balls for $1.05 each. The production costs are $0.48 per ball plus fixed costs of $1200 per month for the lease and utilities. Write function rules for the revenue, $R(x)$, and the cost, $C(x)$. Graph each function on the same set of axes to determine how many tennis balls must be made and sold each month for the company to break even [i.e., $R(x) = C(x)$].
36. JD Electronics offers Jamal a base salary of $150 per week plus $15 for each computer he sells. CompuStore offers him $200 per week plus $12 for each computer he sells. Write a function rule for each and graph the functions to determine which store pays better under what circumstances.
37. A Canadian study indicated that the more money a man earns, the less sleep he gets on average. For every $1000 increase in his annual salary, he sleeps one minute less. On average, a man who is unemployed and has no income sleeps 485 min (8 hr 5 min) each night. If x represents thousands of dollars of annual income, write a function rule that determines the amount of sleep a man gets based on his income. Graph the function and estimate from it how much sleep he will get if he makes $60,000 per year and how much he would earn when he is getting no sleep. Will this pattern produce reasonable results indefinitely?

Dominion Modeling

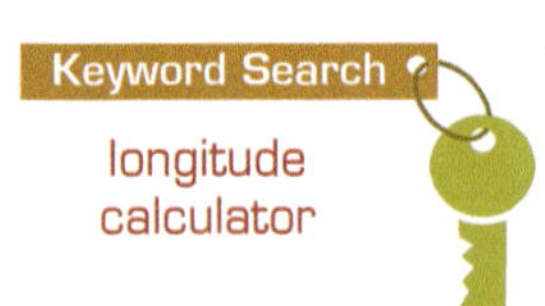

38. Determine the distance from 45° N, 90° W (near Wausau, WI) to 30° N, 90° W (New Orleans, LA) to the nearest mile (1 nautical mile ≈ 1.151 statutory miles).
39. Use the same method to determine the distance from 40° N, 75° W (near Philadelphia, PA) to 40° N, 105° W (near Denver, CO) to the nearest mile.
40. The actual distance from 40° N, 75° W to 40° N, 105° W is approximately 1575 mi. Explain why the method used to determine the north-south distance in exercise 38 will not produce an accurate result when applied to the east-west distance in exercise 39.

CUMULATIVE REVIEW

Evaluate. [1.5–1.6]

41. $1\frac{1}{2} \times \frac{3}{4}$ **42.** 2.751(4.3) **43.** $210 \div 0.005$ **44.** $\frac{6}{19} \div \frac{8}{3}$

Simplify. [1.7, 2.3]

45. $\frac{xy^4}{x^3y^{-2}}$ **46.** $\frac{2^{-3}}{5 \cdot 3^{-3}}$ **47.** $3x - (x - 5)$

48. $1.5(3x + 6) + 0.5(3 - x)$

Evaluate when $x = -3$, $y = 2$, and $z = \frac{1}{2}$. [2.2]

49. $\frac{y^3}{xz}$ **50.** $3x^2yz$

5.6 Direct and Inverse Variations

Items Sold	Profit ($)
1	3
2	6
3	___
___	12
5	15

Can you fill in the missing values in the table? Each ordered pair listed in the table indicates that there is a $3 profit per item sold. Selling three items would yield $9 in profit, and four items must be sold to make $12. This relationship can be expressed using the equation $y = 3x$, where y is the profit in dollars and x is the number of items sold. Notice that the ratio $\frac{y}{x} = 3$ for every ordered pair.

Definitions

A **direct variation** is a function in which the ratio of variables is a nonzero constant, k. That is, $\frac{y}{x} = k$, where $k \neq 0$. The quantities are said to *vary directly* or to be **directly proportional**.

The constant k is called the **constant of variation** or the *constant of proportionality*.

A direct variation can also be stated using the general equations $y = kx$ or $f(x) = kx$. When variables vary directly, one variable is a constant multiple of the other. Most applications of direct variation involve a positive constant of proportionality ($k > 0$), in which an increase in one of the variables causes an increase in the other variable.

The perimeter of a square is directly proportional to the length of one side of the square. What is the constant of variation? Since the perimeter equals four times the length of one side, the constant of variation is 4.

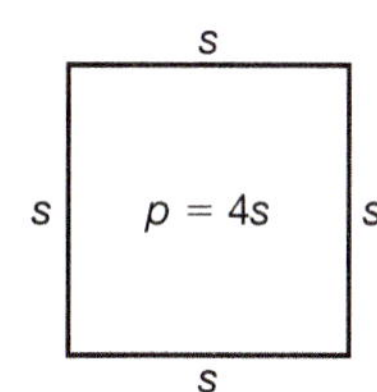

Example 1

Does the table represent a direct variation? If so, find the constant of variation and write the function rule.

x	y
1	7
2	14
3	21
4	28

Answer The table represents a direct variation since the ratio $\frac{y}{x} = 7$ for each ordered pair. The constant of variation is 7, and the function rule is $f(x) = 7x$.

Example 2

Find the constant of variation if y varies directly with x and $y = 6$ when $x = 3$. Write the function rule for the variation.

Answer $k = \frac{y}{x} = \frac{6}{3} = 2$ — 1. In a direct variation, the ratio $\frac{y}{x} = k$.

$y = 2x$ — 2. Write the function rule for the direct variation.

Whenever two quantities vary directly, their ratio is constant. In other words, for two different ordered pairs, (x_1, y_1) and (x_2, y_2),

$$\frac{y_1}{x_1} = k \text{ and } \frac{y_2}{x_2} = k.$$

The Transitive Property of Equality states that two expressions equal to a third expression must be equal to each other. Therefore, in any direct variation,

$$\frac{y_1}{x_1} = \frac{y_2}{x_2}.$$

This explains why quantities that vary directly are often simply said to be proportional.

Example 3

If the distance traveled varies directly with time and Jose rode 12 mi in 45 min, how far can he ride in 2 hr?

Answer

$\frac{d_1}{t_1} = \frac{d_2}{t_2}$ — 1. Write a proportion for the quantities that are directly related.

$\frac{12}{45} = \frac{d_2}{120}$ — 2. Use consistent units in the proportion.

$45d_2 = 1440$ — 3. Solve the proportion.

$d_2 = 32$ mi

You could also solve Example 3 by finding the constant of variation. Since Jose rode 12 mi in $\frac{3}{4}$ hr, $\frac{d}{r} = k = 16$ mi/hr. Then use the equation $d = 16t = 16(2) = 32$ mi.

The graph of this function illustrates a key characteristic of a direct variation when $k > 0$. An increase in one variable causes a proportional increase in the other variable.

Does the following table of base and height measurements of a series of rectangles represent a direct variation?

Base (b)	1	2	3	4	6	12
Height (h)	12	6	4	3	2	1

Notice that the ratio of b to h is not constant. Instead, the product of the base and the height is constant: $bh = 12$. The data in this table models an *inverse variation.*

Definitions

An **inverse variation** is a relation in which the product of the two variables is a nonzero constant, k. That is, $xy = k$, where $k \neq 0$. The quantities are said to *vary inversely* or to be **inversely proportional**.

The constant k is once again called the **constant of variation**.

When the equation $xy = k$ is solved for y, general equations for inverse variations can be stated: $y = \frac{k}{x}$ or $f(x) = \frac{k}{x}$. As with direct variations, most applications of inverse variation involve a positive constant, k. In such inverse variations, an increase in one variable causes a decrease in the other.

Example 4

Does the table represent a direct variation, an inverse variation, or neither? If it is a variation, find the constant of variation and write the function rule for the variation.

x	y
2	72
4	36
8	18
12	12
16	9
24	6
48	3

Answer

This is an inverse variation. — 1. The ratio $\frac{y}{x}$ is not constant, but the product xy is.

$k = 144$ — 2. For each ordered pair, the product $xy = 144$.

$y = \frac{144}{x}$ or $f(x) = \frac{144}{x}$ — 3. Divide both sides of $xy = 144$ by x.

Whenever two quantities vary inversely, their product is constant. For any two different ordered pairs, (x_1, y_1) and (x_2, y_2),

$x_1y_1 = k$ and $x_2y_2 = k$.

Therefore,

$x_1y_1 = x_2y_2$.

This equation can help you solve problems involving inverse variations.

Example 5

The number of census workers varies inversely with the number of days it takes to complete the census in Pleasantburg. If it takes 24 workers 15 days to complete the census, how many workers are needed to complete the census in 10 days?

Answer

$w_1d_1 = w_2d_2$ — 1. Write an equation for the quantities that are inversely proportional.

$24(15) = w_2(10)$ — 2. Replace the variables with known values and solve.

$360 = 10w_2$

$w_2 = 36$ workers

This solution is equivalent to determining the constant of variation ($k = wd = 360$ workers · days) and then solving the equation $10w = 360$.

	Direct Variation	Inverse Variation
Definition	Ratio is constant.	Product is constant.
Constant	$\frac{y}{x} = k$	$xy = k$
Implies	$\frac{y_1}{x_1} = \frac{y_2}{x_2}$	$x_1y_1 = x_2y_2$
Function Rule	$y = kx$	$y = \frac{k}{x}$
Graph ($k > 0$)		

You will come across many real-life applications of direct and inverse variations. God has created consistently and proportionally. We have no need to worry about whether gravity will disappear or whether the sun will come up tomorrow. After the Flood, God promised Noah that the world would operate consistently until its end (Gen. 8:22).

Example 6

Hooke's law states that the length of the stretch in a spring is directly proportional to the amount of force applied to the spring. If a spring stretches 8 in. when a weight of 20 lb is put on it, how far will it stretch when a 50 lb weight is put on it?

Answer

$$\frac{s_1}{w_1} = \frac{s_2}{w_2}$$

1. Write a proportion for the quantities that are directly proportional.

$$\frac{8}{20} = \frac{s_2}{50}$$

2. Replace the variables with known values and solve.

$$20s_2 = 400$$

$$s_2 = 20 \text{ in.}$$

A. Exercises

State whether each of the following represents a direct variation, an inverse variation, or neither.

1. $a = 2b$

2. $y = \frac{7}{x}$

3. $xy = -3$

4. $y = \frac{x}{3}$

5.

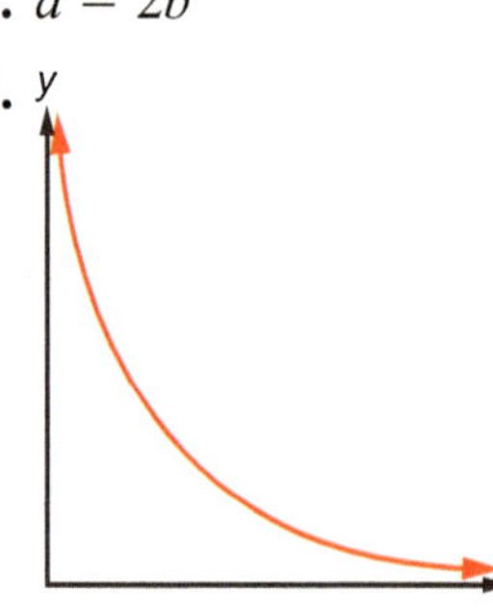

6.

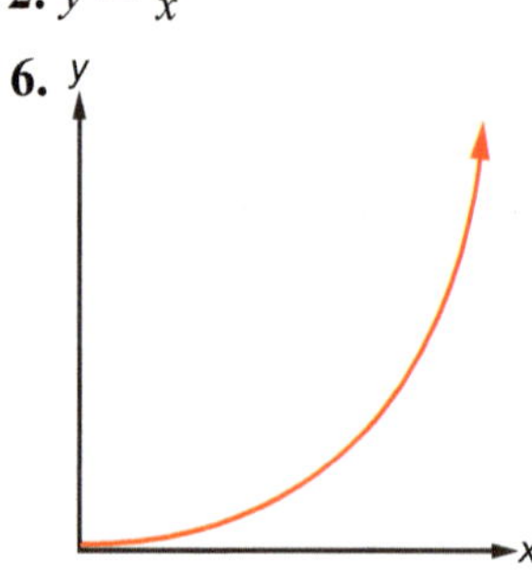

7.

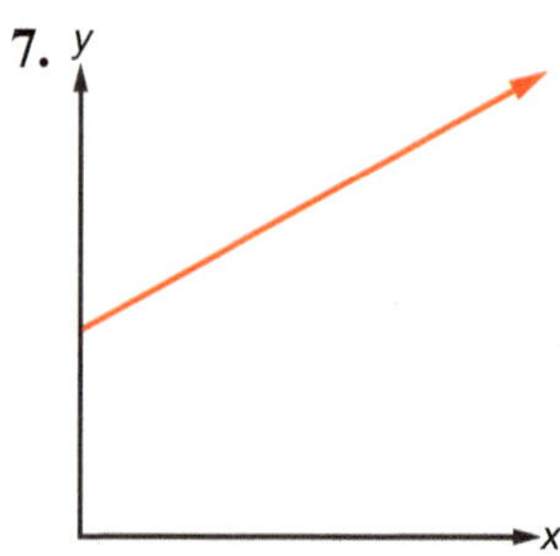

8.

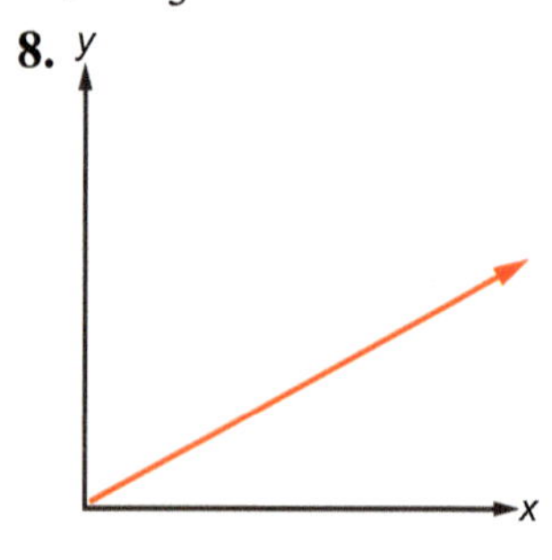

For each variation, find the constant of variation and write the function rule.

9. y varies directly with x and $y = 18$ when $x = 9$

10. y varies inversely with x and $y = 12$ when $x = 6$

11. m is inversely proportional to a and $m = 9$ when $a = 4$

12. c is directly proportional to v and $v = 9$ when $c = 6$

13. r is inversely proportional to s and $r = 2$ when $s = 2.5$

14. g varies directly with h and $h = 5$ when $g = 0.5$

B. Exercises

State whether each table represents a direct variation, an inverse variation, or neither. For each variation, write the function rule and graph the first quadrant of the function.

15.

x	y
1	6
2	12
3	18
4	24

16.

x	y
2	18
4	36
8	72
10	90

17.

x	y
2	9
3	6
6	3
9	2

18.

x	y
3	12
6	6
12	3
18	2

19.

x	y
2	5
3	$\frac{10}{3}$
4	2.5
5	2

20.

x	y
16	64
8	32
4	16
2	8

21.

x	y
20	30
15	$22\frac{1}{2}$
10	15
5	$7\frac{1}{2}$

22.

x	y
1	25
2	16
3	9
4	4
5	1

23.

x	y
1	2
2	3
3	4
4	5

24.

x	y
0.125	12
0.5	3
2	0.75
6	0.25

Find the missing value for each variation.

25. If y is directly proportional to x and $y = 16$ when $x = 4$, find y when $x = 10$.

26. If y varies inversely with x and $y = 4$ when $x = 2$, find y when $x = 8$.

27. If y is inversely proportional to x and $y = 0.25$ when $x = 4$, find y when $x = 10$.

28. If y varies directly with x and $y = 3$ when $x = 2$, find y when $x = 7$.

29. If y varies inversely with x and $y = 4$ when $x = 3$, find y when $x = 2$.

30. If y is directly proportional to x and $y = 5$ when $x = 1$, find y when $x = 20$.

State whether each of the following represents a direct or inverse variation, and find the constant of variation.

31. The formula for the circumference of a circle is $c = \pi d$, where d is the diameter of the circle.

32. The formula for the additional pressure resulting from being underwater is $P = 0.433d$, where d is the depth below the surface.

33. Boyle's law states that the product of the pressure, P, exerted by a gas and its volume, V, is constant when there is no change in temperature. Gene was able to determine that 5 m^3 of a gas exerted a pressure of 0.8 kilopascals.

34. Charles's law states that the volume, V, of a gas increases proportionally as the temperature in Kelvin, T, increases if the pressure remains constant. At a temperature of 500 K, a given amount of gas occupies 6 m^3.

Write an equation expressing the variation described; then use it to solve the problem.

35. Find the distance, d, a car travels at the rate of 52 mi/hr if the distance varies directly with the time, t, in hours. How far will the car travel in 2.5 hr?

36. Find the sales tax, t, paid on purchases at the rate of 7% if tax varies directly with the total purchase price, p. How much sales tax is paid on a $495 computer purchase?

37. The width, w, of a rectangle with an area of 20 square units varies inversely with its length, l. Find the width when the length is 8.

38. It takes one person 18 hr to paint a fence. If the time, t, that it takes to paint a fence varies inversely with the number of painters, p, how long will it take four people to paint the fence?

39. The average monthly attendance, a, at an amusement park varies inversely with the number of days it rains during the month, r. If the attendance in July was 104,000 and it rained three days, what would you predict the attendance to be for August if it rains six days?

40. The height, h, of a triangle varies inversely with its base, b. If the base of a triangle is 6 and its height is 6 also, find the height of a triangle whose base is 4.

C. Exercises

Use the following table for exercises 41–43.

Time, t (seconds)	Distance Fallen, d (meters)
1	5
2	20
3	45
4	80
5	125

41. Graph the data and determine whether the function is a direct variation, an inverse variation, or neither.

42. Make a table of time squared and distance. Graph your data and determine whether the relation between t^2 and d is a direct variation, an inverse variation, or neither.

43. Write a function rule for the relationship between t^2 and d and use it to find the following.

a. the distance fallen after 7.5 sec **b.** the time it takes to fall 500 m

Write an equation expressing the variation described; then use it to solve the problem.

44. The weight of an object varies inversely with the square of its distance from the center of the earth, which has a radius of 4000 mi. If a boy weighs 140 lb on the surface of the earth, find his weight 10,000 mi from the center of the earth.

45. The volume of a ball varies directly with the cube of its diameter. If the volume of a ball 4 in. in diameter is 33.5 in.^{3}, find the volume of a ball with a 6 in. diameter (to the nearest cubic inch).

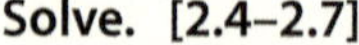

CUMULATIVE REVIEW

Solve. [2.4–2.7]

46. $2.5x = 10$ **47.** $x - 25.4 = 57.29$ **48.** $\frac{2}{x} = 10$

49. $\frac{300}{x} = 12$ **50.** $3(2x - 5) = -(x - 8)$ **51.** $a - 2(a - 3) = 20$

Write an equation and solve. [3.5–3.7]

52. Find the discount rate (to the nearest whole percent) of a $42 item that is on sale for $28.

53. Find the percent increase (to the nearest tenth of a percent) in attendance from 15,000 for the first game to 17,500 for the second.

54. Two cars left the same point at the same time traveling in opposite directions at 60 mi/hr and 46 mi/hr respectively. How long will it be before they are 371 mi apart?

55. Suppose the cars in the previous exercise were traveling in the same direction and the slower one left 2 hr before the faster one. How long will it take for the faster car to overtake the slower car?

SEQUENCES

Geometric Sequences

Challenge **Can you supply the next two terms in each of the following sequences?**

a. 3, 6, 12, 24, 48, ...

b. $2, 1, \frac{1}{2}, \frac{1}{4}, \frac{1}{8}, \ldots$

c. $\frac{1}{25}, -\frac{1}{5}, 1, -5, 25, \ldots$

Recall that the next term in an arithmetic sequence is obtained by adding the common difference, d. Their recursive formula is $A_n = A_{n-1} + d$, and their explicit (general-term) formula is $A_n = A_1 + (n - 1)d$. The common difference can be found using $d = A_n - A_{n-1}$, where $n > 1$.

Sequences like those in the Challenge above, in which the next term is obtained by multiplying by a constant, are called *geometric*. Geometric sequences have the recursive formula $A_n = rA_{n-1}$, where r is the common ratio between successive terms and can be calculated using $r = \frac{A_n}{A_{n-1}}$, where $n > 1$. The recursive definitions for the sequences above are as follows.

a. $A_n = 2A_{n-1}; A_1 = 3$ **b.** $A_n = \frac{1}{2}A_{n-1}; A_1 = 2$ **c.** $A_n = -5A_{n-1}; A_1 = \frac{1}{25}$

Remember that a recursive definition must include the first term in addition to the recursive formula.

To find the general-term formula for 3, 6, 12, 24, 48, ..., we can express the terms as $A_1 = 3, A_2 = 3 \cdot 2, A_3 = 3 \cdot 2^2, A_4 = 3 \cdot 2^3, A_5 = 3 \cdot 2^4, \ldots$, leading us inductively to $A_n = 3r^{n-1}$. Thus, for a geometric sequence, the general-term formula is $A_n = A_1r^{n-1}$.

To find A_{20} for 3, 6, 12, 24, 48, ..., we use the formula $A_{20} = 3 \cdot 2^{20-1}$, which simplifies to 1,572,864. Since geometric sequences involve exponentiation (repeated multiplications by r), the terms increase rapidly when the common ratio $r > 1$, as in this example. This example also involves *deductive reasoning* rather than inductive reasoning since we are going from a general formula to a specific result of that formula.

Exercises

List the first five terms of each of the following geometric sequences.

1. $A_n = 6 \cdot 3^{n-1}$ **2.** $A_1 = 10; r = 5$ **3.** $A_n = -2A_{n-1}; A_1 = \frac{1}{4}$

Find A_{10} for each of the following geometric sequences.

4. $A_n = 6 \cdot 3^{n-1}$ **5.** $A_1 = 10; r = 5$ **6.** $A_n = -2A_{n-1}; A_1 = \frac{1}{4}$

Find the specified formula A_n for each of the following sequences.

7. 2, −6, 18, −54, 162, ...; recursive

8. $2, 1, \frac{1}{2}, \frac{1}{4}, \frac{1}{8}, \ldots$; general term

9. $\frac{1}{25}, -\frac{1}{5}, 1, -5, 25, \ldots$; general term

10. 11, 7, 3, −1, −5, ...; recursive

5.7 Graphing Absolute Value Functions

Recall the following definition of absolute value.

$$|x| = \begin{cases} x \text{ if } x \geq 0 \\ -x \text{ if } x < 0 \end{cases}$$

This definition can be used to construct a variety of representations of the absolute value function, $f(x) = |x|$.

Example 1

Represent the absolute value function $f(x) = |x|$; $D = \{-3, -2, -1, 0, 1, 2, 3\}$ as a table, a set of ordered pairs, and a graph.

Answer

x	$\lvert x\rvert$	$f(x)$
-3	$f(-3) = \lvert -3\rvert = 3$	3
-2	$f(-2) = \lvert -2\rvert = 2$	2
-1	$f(-1) = \lvert -1\rvert = 1$	1
0	$f(0) = \lvert 0\rvert = 0$	0
1	$f(1) = \lvert 1\rvert = 1$	1
2	$f(2) = \lvert 2\rvert = 2$	2
3	$f(3) = \lvert 3\rvert = 3$	3

1. Evaluate the function rule for each member of the domain to make a table of values for the function.

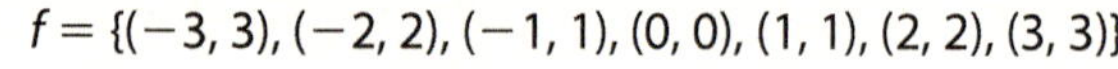
$f = \{(-3, 3), (-2, 2), (-1, 1), (0, 0), (1, 1), (2, 2), (3, 3)\}$

2. Write the function as a set of ordered pairs.

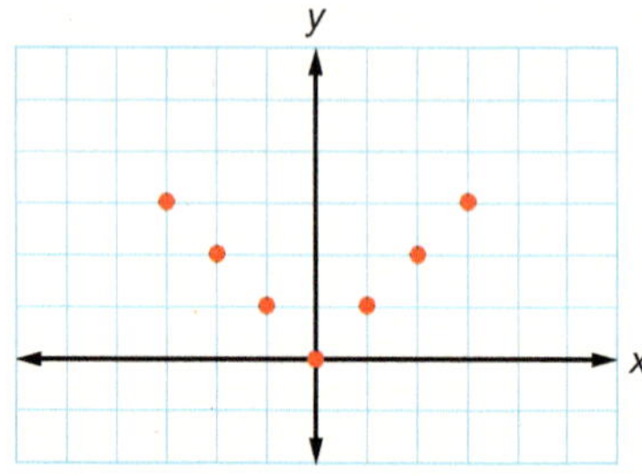

3. Plot the ordered pairs to create a graph of the function.

If no specific domain is given, the domain is understood to be the real numbers. The graph of the absolute value function $f(x) = |x|$ forms a V with the turning point at the origin. This point is called the *vertex*.

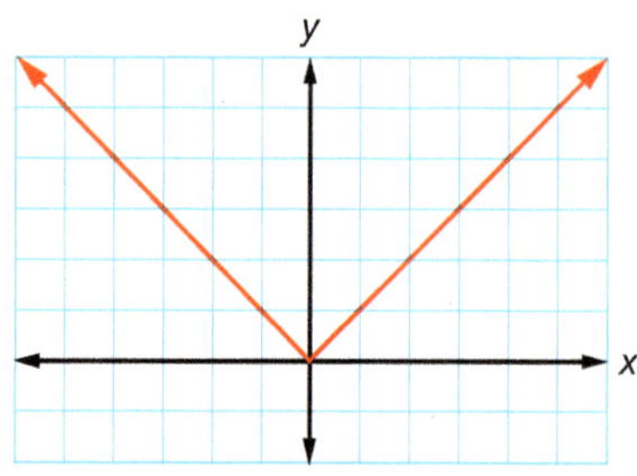

This graph is the basic absolute value function graph. This basic graph can move up or down and left or right in the plane when numbers are added to or subtracted from the absolute value function. This movement of the graph is called a *translation*.

Example 2

Graph $f(x) = |x| + 4$. Then compare this graph with the basic absolute value graph to determine the effect of adding 4 to the absolute value function rule.

Answer

x	$\lvert x\rvert + 4$	$f(x)$
-3	$f(-3) = \lvert -3\rvert + 4 = 3 + 4 = 7$	7
-2	$f(-2) = \lvert -2\rvert + 4 = 2 + 4 = 6$	6
-1	$f(-1) = \lvert -1\rvert + 4 = 1 + 4 = 5$	5
0	$f(0) = \lvert 0\rvert + 4 = 0 + 4 = 4$	4
1	$f(1) = \lvert 1\rvert + 4 = 1 + 4 = 5$	5
2	$f(2) = \lvert 2\rvert + 4 = 2 + 4 = 6$	6
3	$f(3) = \lvert 3\rvert + 4 = 3 + 4 = 7$	7

1. Evaluate the function rule for several members of the domain to make a table of values for the function.

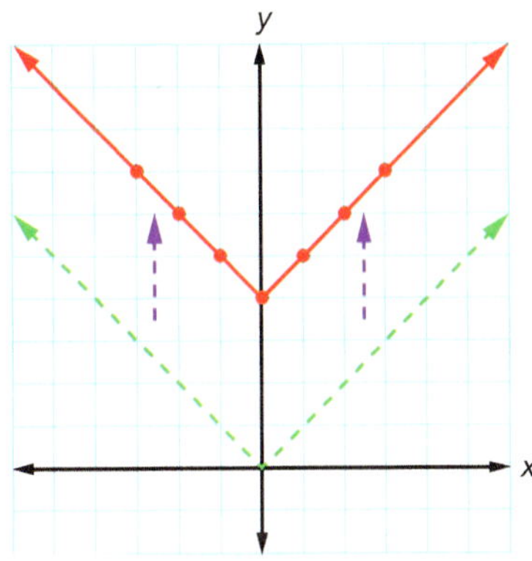

2. Plot enough ordered pairs to determine the shape of the graph, and then connect the points to create a graph of the function.
3. Sketch a graph of the basic absolute value function.

The entire graph has moved 4 units up.

4. Every value of the range has been increased by 4. Its vertex is at (0, 4).

In general, the graph of $y = |x| + k$ (where k is a positive constant) is the result of translating the graph of the basic absolute value function up by k units. Similarly, the graph of $y = |x| - k$ is a translation of k units down.

Example 3

Write the function rule that describes the following graph.

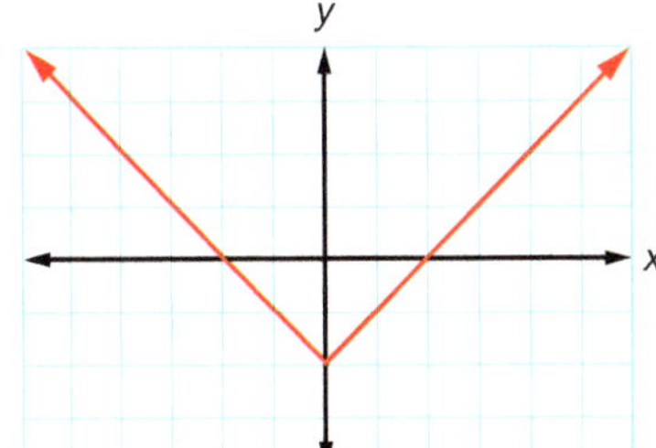

Answer

$f(x) = |x|$

1. Notice that the graph has the shape of the basic absolute value function. Begin by writing its basic function rule.

$f(x) = |x| - 2$

2. The graph has been translated 2 units down. To shift the graph 2 units down, 2 must be subtracted from the function.

When a constant is added or subtracted within the absolute value symbols, a horizontal translation occurs. The vertex is moved to the x value that causes the expression within the absolute value symbols to be equal to zero.

Example 4

Graph $f(x) = |x + 3|$. Then compare this graph with the basic absolute value graph to determine the effect of adding 3 within the absolute value function.

Answer

$x + 3 = 0$
$x = -3$

1. Set the expression within the absolute value function equal to zero. Solve to find the value of x that is used to find the vertex.

x	$\|x + 3\|$	$f(x)$
−6	$f(-6) = \|-6 + 3\| = \|-3\| = 3$	3
−5	$f(-5) = \|-5 + 3\| = \|-2\| = 2$	2
−4	$f(-4) = \|-4 + 3\| = \|-1\| = 1$	1
−3	$f(-3) = \|-3 + 3\| = \|0\| = 0$	0
−2	$f(-2) = \|-2 + 3\| = \|1\| = 1$	1
−1	$f(-1) = \|-1 + 3\| = \|2\| = 2$	2
0	$f(0) = \|0 + 3\| = \|3\| = 3$	3

2. Evaluate the function rule to find the vertex and several ordered pairs on either side of the vertex.

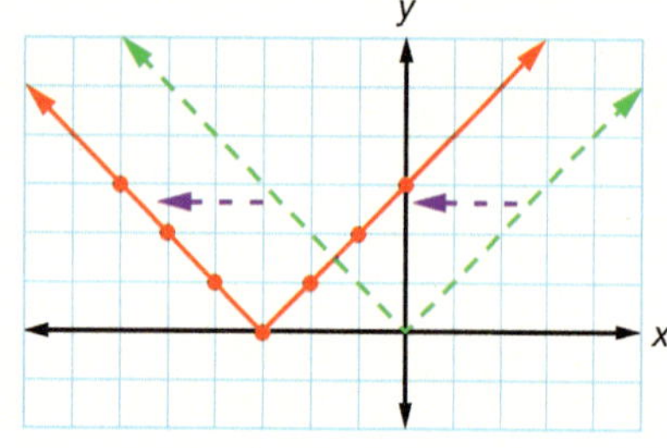

3. Plot enough ordered pairs to determine the shape of the graph, and then connect the points to create a graph of the function.
4. Sketch a graph of the basic absolute value function.

The entire graph has moved 3 units left.

5. Its vertex is at $(-3, 0)$.

In general, the graph of $y = |x + h|$ (where h is a positive constant) is the result of translating the graph of the basic absolute value function h units to the left. Unlike vertical translations, the direction is opposite that of the sign on the x-axis. The graph of $y = |x - h|$ is a translation of h units to the right.

Example 5

Write the function rule that describes the following graph.

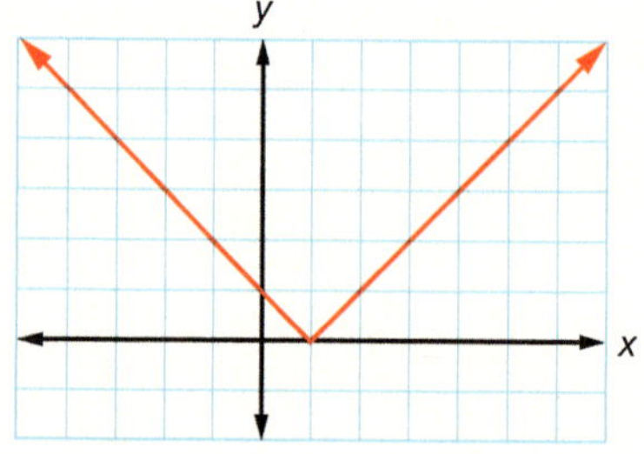

Answer $f(x) = |x|$

1. Notice that the graph has the shape of the basic absolute value function. Begin by writing its basic function rule.

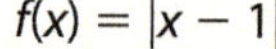

$f(x) = |x - 1|$

2. The graph has been translated 1 unit right.

Check

$f(-1) = |-1 - 1| = 2$
$f(4) = |4 - 1| = 3$
$(-1, 2)$ and $(4, 3)$ are on the graph.

3. Evaluate several ordered pairs and make sure that the points are on the graph.

Both vertical and horizontal translations may occur within the same function. The location of the vertex can be determined by examining the function rule. You should be able to look at an absolute value function rule and determine the location of the V-shaped graph.

Example 6

Graph $f(x) = |x - 2| + 1$.

Answer $f(x) = |x|$

$f(x) = |x - 2| + 1$

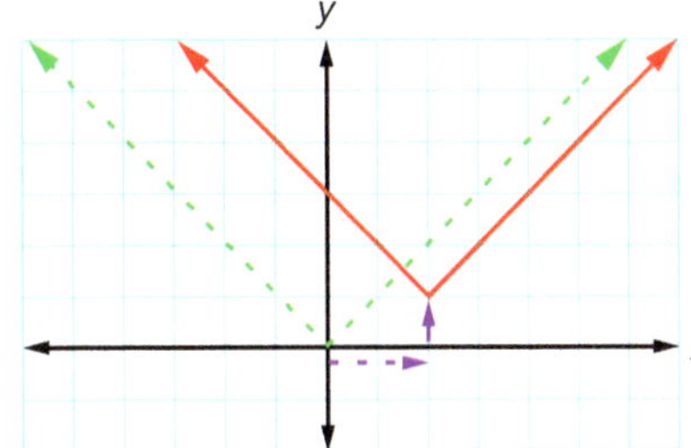

1. Compare the given function rule with the basic absolute value function rule.
2. Notice that 2 has been subtracted within the absolute value function. This causes the graph to be translated 2 units right. Also, 1 has been added to the absolute value function rule, causing the graph to be translated 1 unit up. Its vertex is at (2, 1).

Check $f(0) = |0 - 2| + 1 = 2 + 1 = 3$

$f(5) = |5 - 2| + 1 = 3 + 1 = 4$

(0, 3) and (5, 4) are on the graph.

3. Evaluate several ordered pairs and make sure that the points are on the graph.

A. Exercises

Make a table of at least seven ordered pairs and graph each function.

1. $f(x) = |x| - 4$ **2.** $f(x) = |x| + 1$ **3.** $y = |x + 3|$

4. $f(x) = |x - 5|$ **5.** $f(x) = |x - 2| - 3$ **6.** $y = |x + 2| + 4$

Describe the translation of the graph of $y = |x|$ that produces the graph of each function.

7. $f(x) = |x| + 4$ **8.** $y = |x + 3|$ **9.** $f(x) = |x - 10|$

10. $y = |x| - 6$ **11.** $f(x) = |x| + \frac{1}{2}$ **12.** $f(x) = \left|x - \frac{5}{2}\right|$

Write the function rule for each translation of $y = |x|$.

13. 2.5 units down **14.** 3 units up

15. 4 units left **16.** 7 units right

B. Exercises

Describe the translations of the graph of $y = |x|$ that produce the graph of each function.

17. $f(x) = |x - 3| - 4$

18. $y = |x + 2| - 6$

19. $y = |x + 7| + 1$

20. $f(x) = |x - 1| + 3$

State the coordinates of the vertex of each function.

21. $y = |x + 5| + 3$

22. $f(x) = |x - 5| - 7$

23. $f(x) = |x - a| + b$

24. $y = |x + a| - b$

Write the function rule for each translation of $y = |x|$.

25. 2 units left, 6 units up

26. 6 units down, 3 units right

27. 5.9 units up, 6.23 units right

28. $\frac{8}{9}$ units left, $\frac{2}{5}$ units down

Use translations and the location of the vertex to graph each function without evaluating ordered pairs.

29. $f(x) = |x + 3.5|$

30. $f(x) = |x| - 1.5$

31. $y = |x + 3| - 2$

32. $y = |x - 4| - 3$

33. $f(x) = \left|x - \frac{1}{2}\right| + 2$

34. $y = |x + 2| - \frac{2}{3}$

C. Exercises

Make a table of at least seven ordered pairs and graph each function.

35. $f(x) = -|x|$

36. $y = -|x| + 3$

37. $y = -|x| - 2$

38. $y = -|x - 2|$

39. $f(x) = -|x + 1|$

40. $f(x) = -|x + 4| - 1$

41. From your observations in the preceding exercises, what is the effect of the negative sign on the graph of the absolute value function?

CUMULATIVE REVIEW

Complete each of the following set operations. [1.1]

42. $\mathbb{Q} \cap \mathbb{Z}$

43. $\mathbb{Q} \cup \mathbb{Q}'$

44. State the Trichotomy Property using x and y. [4.1]

45. Write an inequality symbol equivalent to $\not>$. [4.1]

46. If $4x - 3 < 32$, which of the following are solutions: $\frac{26}{3}$, 9.1, 20, or -6? [4.1]

Solve. [4.2–4.5]

47. $-3x < 51$

48. $3x - 6 \geq -10$

49. $7x + 12 < 8$

50. $3x + 6 < 81$ and $2x > 10$

51. $4x < -8$ or $x + 20 < 30$

CHAPTER 5 REVIEW

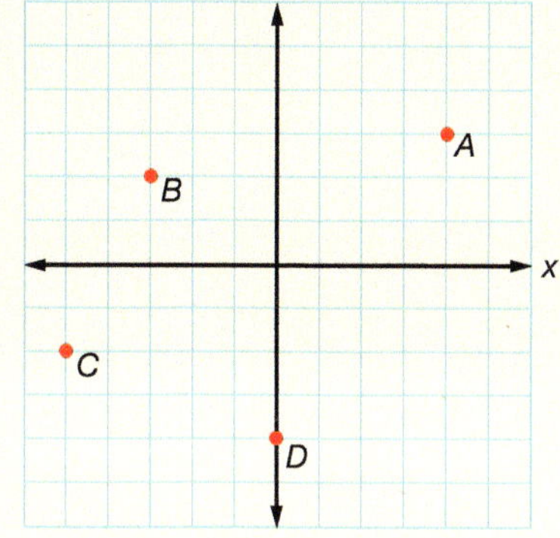

1. State the coordinates of each point.
2. Graph the following points in a single coordinate plane: $E(3, 0)$, $F(2, -4)$, $G(-5, 2)$, and $H(2, 3)$.
3. State the quadrant or axis containing each of the following points: $W(-3, -3)$, $X(-10, 4)$, $Y(3, 9)$, $Z(0, 2)$.

The following table relates the number of pages in a book to the cost of the book.

Pages	600	330	430	380	340	670
Cost	$42.50	$18.50	$21.50	$20.00	$19.00	$49.00

4. Which quantity should be represented by the independent variable? Explain why.
5. Create a scatterplot of the data in the table.
6. Describe the relationship between the number of pages and the cost of the book.
7. State the domain and range of the following relation: $G = \{(-7, 5), (-3, 4), (-2, 3), (1, 7), (3, 8), (5, 6), (8, 5)\}$.
8. Draw a mapping diagram of relation G.
9. Is relation G a function?
10. Is the relation $\{(x, y) \mid y = 3\}$ a function?
11. Make a table of at least five ordered pairs for relation $H = \{(x, y) \mid y = 3x - 5\}$. Is it a function?
12. If $f(x) = 2x - 3$, find $f(4)$.
13. When no domain is stated for a function, what is the assumed domain?

State the domain and range of each relation and whether it is a function.

14.

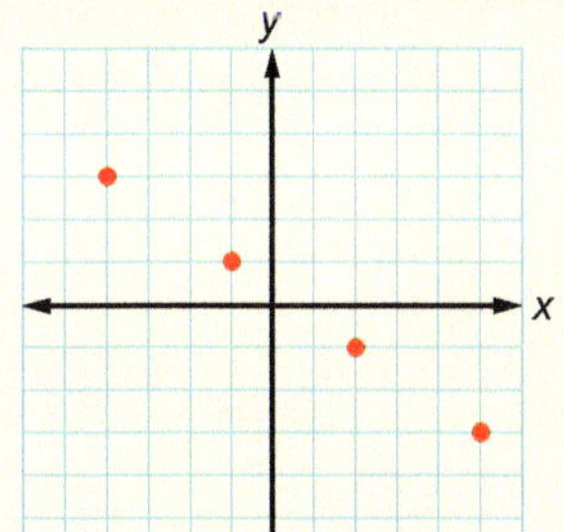

15.

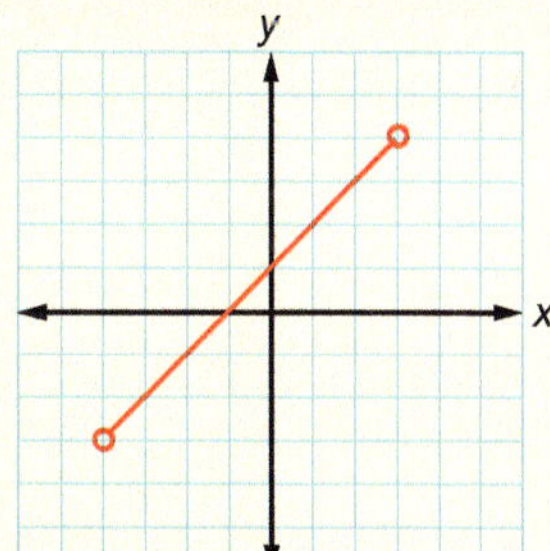

16.

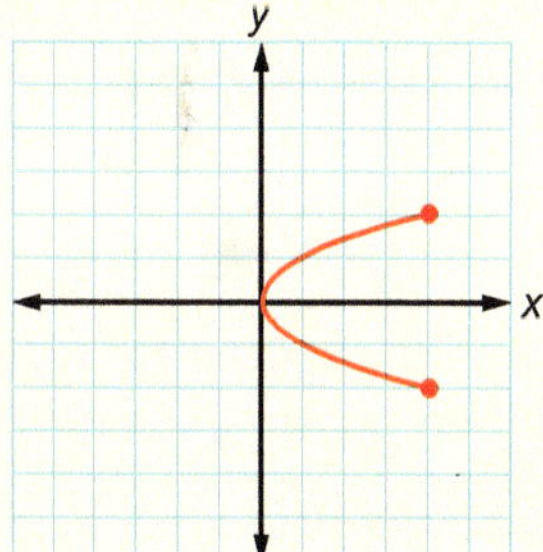

Graph each relation and state whether it is a function.

17. $A = \{(x, y) \mid y = 2x - 4; x = 0, 2, 5\}$

18. $B = \{(x, y) \mid x = 2; 2 \leq y \leq 5\}$

Interpret each graph, explaining each labeled section.

19.

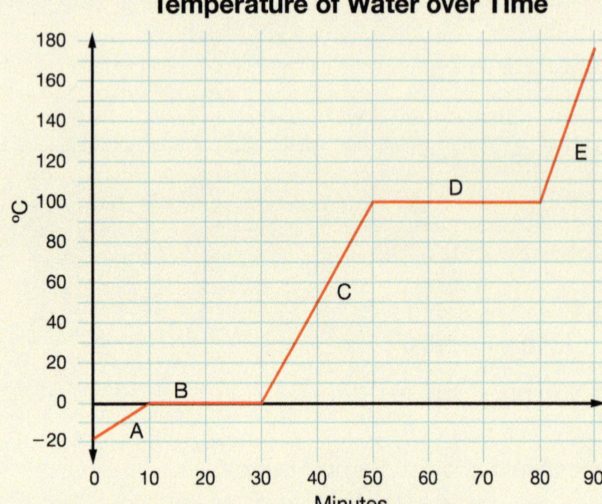

20.

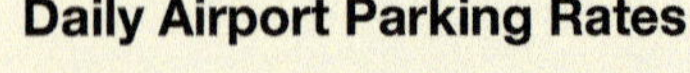

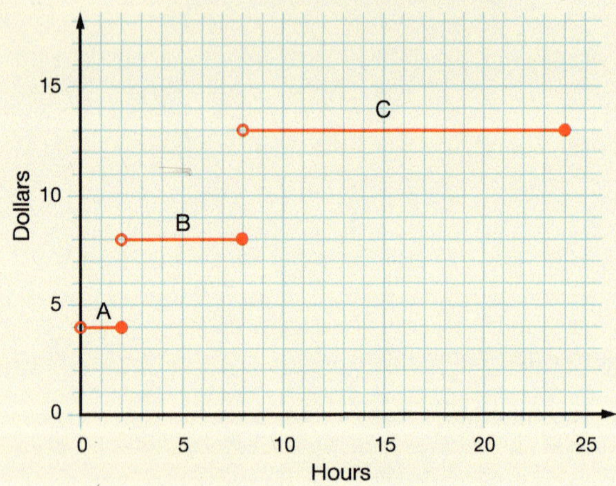

21. Sketch a graph of a man's heart rate over the course of an hour in which he mows the lawn with a push mower for 20 min, rests for 15 min, lifts weights for 10 min, and then reads for the rest of the hour.

22. Illustrate the relationship between the amount of gas used and the number of miles traveled for Mr. Blake's car by graphing the data in the table and drawing a smooth curve that models the relation.

Miles Traveled	Gallons of Gas Used
22	1
44	2
88	4
154	7
330	15

23. Use your graph to estimate the amount of gas used if he travels 200 mi.

24. How far can Mr. Blake travel using 40 gal of gas?

For exercises 25–27, $g(x) = x^2 - 3$; $D = \{-2, -1, 0, 1, 2\}$.

25. Make a table to represent the function.

26. Write the function as a set of ordered pairs.

27. Graph $g(x)$.

Write the rule for each function.

28.

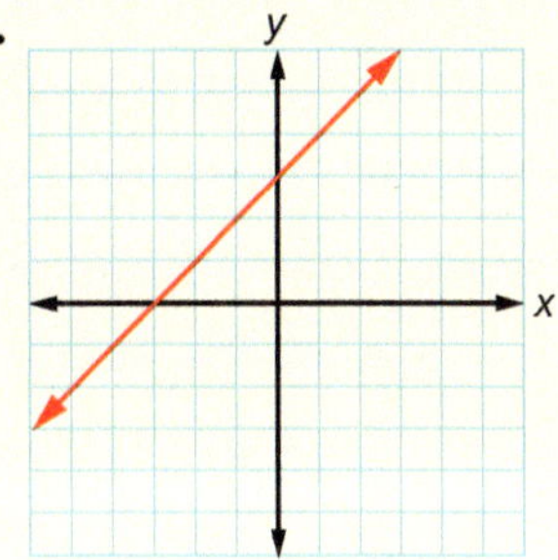

29. $\{(-2, 1), (0, 0), (2, -1), (4, -2), (6, -3)\}$

30.

x	-2	-1	0	1	2
$h(x)$	-8	-5	-2	1	4

State whether each table represents a direct variation, an inverse variation, or neither. For each variation, write the function rule.

31.

x	y
2	9
4	18
6	27
8	36
12	54

32.

x	$f(x)$
3	12
4	9
6	6
8	4.5
10	3.6

33. If y varies directly with x and $y = 8$ when $x = 5$, find the constant of variation and write the function rule.

34. If the amount of paint varies directly with the area to be painted and 2 gal covers 690 ft^2, how much paint (to the nearest tenth of a gallon) will it take to cover the remaining 970 ft^2?

35. If the volume of a gas, V, varies inversely with pressure, P, and $V = 400$ mL when $P = 1.8$ atm, what pressure is needed to reduce the volume to 375 mL?

36. Write the function $f(x) = |x|$; $D = \{-4, -2, 0, 2, 4\}$ as a set of ordered pairs.

37. Draw a mapping diagram of the function $h(x) = |x| - 2$; $D = \{-4, -2, 0, 2, 4\}$.

38. Make a table to represent the function $g(x) = |x - 3|$; $D = \{-2, -1, 0, 1, 2\}$.

Graph each function.

39. $f(x) = |x| + 1$

40. $f(x) = |x + 2|$; $D = \{-2, -1, 0, 1, 2\}$

41. $f(x) = |x - 1| - 3$; $D = \{-2, -1, 0, 1, 2\}$

42. $f(x) = |x + 3| + 2.5$

6 Linear Functions

Cheryl's father had always wanted her to marry a man who was going to be in full-time service for the Lord. Eventually she met Brad, a construction worker who recognized his need for more Bible training. She could have never envisioned at that time the many ways in which Brad was going to be used to serve the Lord as a construction worker. Brad's desires and gifts in construction, along with his understanding of math and Scripture, led him into a life of service to Christian camps, missionaries, churches, high-school and university students, and individual families by constructing buildings and teaching Scripture and skills that would last for generations. Cheryl and her father soon recognized that full-time service for Brad would be found in the secular workplace as well as in Christian ministries.

Brad's service for the Lord required a good understanding of safe construction practices. Interestingly, many safety factors in construction involve the careful mathematical calculation of slopes. Incorrect slopes caused by mathematical error can result in serious injuries or even fatalities. In this chapter you will learn critical concepts necessary for calculating slopes, as well as how to help yourself and your neighbor by creating safe slopes.

After this chapter you should be able to

1. graph linear equations and determine their x- and y-intercepts.
2. find the slope of a given line.
3. write linear equations in slope-intercept form.
4. graph linear equations from slope-intercept form.
5. write the equation of a line when given key characteristics of the line.
6. determine whether two lines are parallel or perpendicular by analyzing their slopes.
7. determine a trend line from a scatterplot.
8. apply linear equations to real-world problems.
9. graph linear inequalities.

6.1 Graphing Lines

Interceptor fighters like this F-14A rely on speed to meet and destroy enemy aircraft.

The graph of a relation or a function can be sketched by graphing enough ordered pairs to indicate the shape of the graph and then connecting the points. This is the method used by graphing calculators. By recognizing key characteristics of the function, you can quickly determine the shape and location of the graph.

Functions such as $f(x) = 2x + 7$ can also be written as $y = 2x + 7$. They are called *linear functions* or *linear equations* because all the points that satisfy them form a straight line.

Example 1

Graph $\{(x, y) \mid y = x + 5\}$.

Answer

x	$x + 5$	(x, y)
-2	$(-2) + 5 = 3$	$(-2, 3)$
-1	$(-1) + 5 = 4$	$(-1, 4)$
0	$(0) + 5 = 5$	$(0, 5)$
1	$(1) + 5 = 6$	$(1, 6)$
2	$(2) + 5 = 7$	$(2, 7)$

1. Make a table of ordered pairs using convenient values for x.

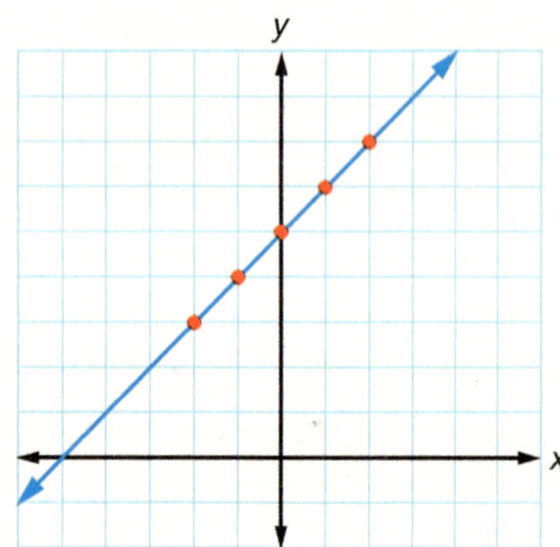

2. Plot the points. Notice that the points are on a straight line.
3. Connect the points with a straight line.

The *standard form* of a linear equation is $Ax + By = C$, where A, B, and C are real numbers. The equation in Example 1 can be converted to standard form by subtracting x from both sides: $-x + y = 5$. Several different standard-form equations are possible for the same linear function:

$$-2x + 2y = 10;\ x - y = -5;\ \text{and } 2x - 2y = -10.$$

If a linear equation in standard form contains fractional or decimal coefficients, multiply each term by a common denominator or power of ten to change A, B, and C to integers. It is also common for A to be nonnegative.

Example 2

Write the linear function $f(x) = -\frac{5}{7}x + 2$ as a linear equation in standard form.

Answer

$y = -\frac{5}{7}x + 2$	1. Write the function as an equation.
$\frac{5}{7}x + y = 2$	2. Add $\frac{5}{7}x$ to both sides.
$5x + 7y = 14$	3. Multiply each term by 7.

Example 3

Graph $x + 3y = 4$.

Answer

$x + 3y = 4$

$3y = -x + 4$

$y = \frac{-x+4}{3} = -\frac{x}{3} + \frac{4}{3}$

1. Solve the equation for y.

x	$\frac{-x+4}{3} = -\frac{x}{3} + \frac{4}{3}$	(x, y)
0	$-\frac{0}{3} + \frac{4}{3} = \frac{4}{3}$	$\left(0, \frac{4}{3}\right)$
1	$\frac{-1+4}{3} = \frac{3}{3} = 1$	$(1, 1)$
4	$\frac{-4+4}{3} = \frac{0}{3} = 0$	$(4, 0)$

2. Make a table of ordered pairs using convenient values for x.

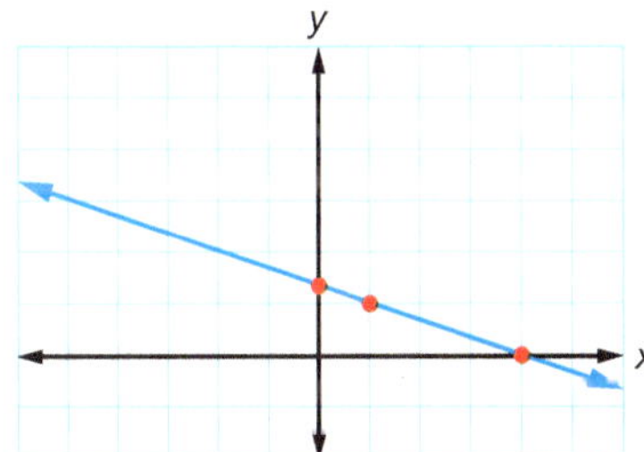

3. Graph the ordered pairs.
4. Connect the points with a straight line.

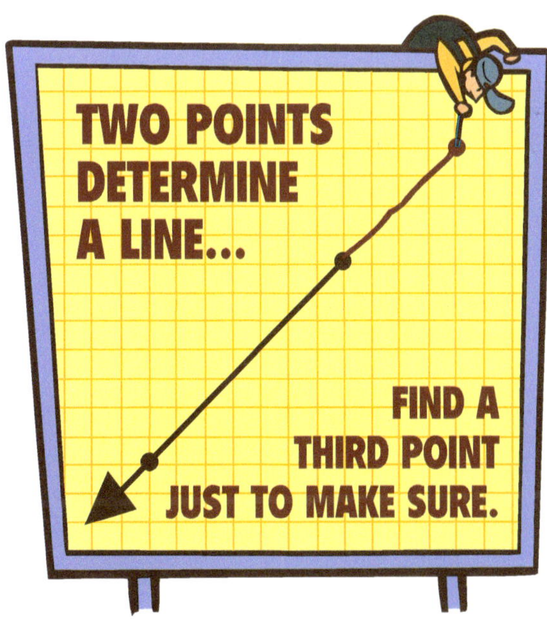

Graphing a Linear Equation

1. Solve the linear equation for y.
2. Make a table of at least three ordered pairs using convenient values for x.
3. Graph the ordered pairs on a Cartesian plane.
4. Connect the points with a straight line.

From the graph in Example 3, you can identify where the graph crosses the axes. These points are called *intercepts*. The y-intercept is $\left(0, \frac{4}{3}\right)$, and the x-intercept is $(4, 0)$.

Definitions

The ***y*-intercept** is the point where the graph intersects the y-axis.

The ***x*-intercept** is the point where the graph intersects the x-axis.

You can also find each intercept from the equation without graphing. The y-intercept always has zero as its x-coordinate, and the x-intercept always has zero as its y-coordinate. It is often convenient to find and graph the intercepts as two of your three points.

Example 4

Use the x- and y-intercepts to graph $2x + y = 6$.

Answer

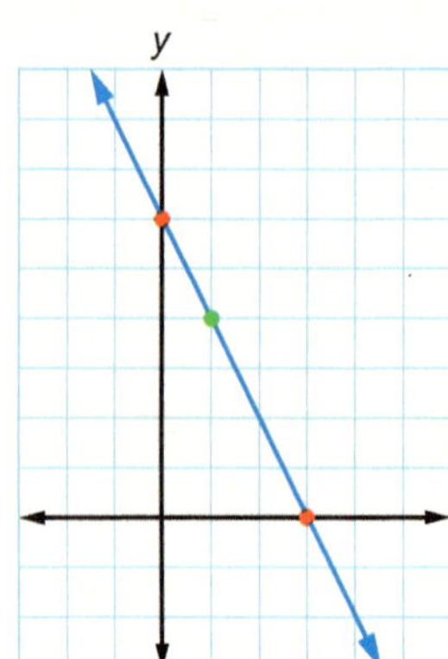

$2x + y = 6$	1. Substitute 0 for y and solve for x.
$2x + 0 = 6$	
$2x = 6$	
$x = 3$	2. Plot the x-intercept: (3, 0).
$2x + y = 6$	3. Substitute 0 for x and solve for y.
$2(0) + y = 6$	
$y = 6$	4. Plot the y-intercept: (0, 6).
	5. Graph the line.

Check

$2(1) + 4 = 6$	6. Choose a convenient point on the line, (1, 4), and check that it is a solution.

The equations $y = c$ and $x = c$ have graphs that are horizontal and vertical lines respectively. They are special cases of the standard form $Ax + By = C$ in which either $A = 0$ or $B = 0$.

Example 5

Graph each equation; then determine whether it is a function.

a. $y = -2$ **b.** $x = 3$

Answer

a.

x	y
-3	-2
0	-2
4	-2

1. $y = -2$ regardless of the value chosen for x.

2. Graph the points. Then draw and label the line.

$y = -2$ is a function.

3. The graph passes the vertical line test.

b.

x	y
3	-2
3	0
3	4

1. $x = 3$ regardless of the value chosen for y.

2. Graph the points. Then draw and label the line.

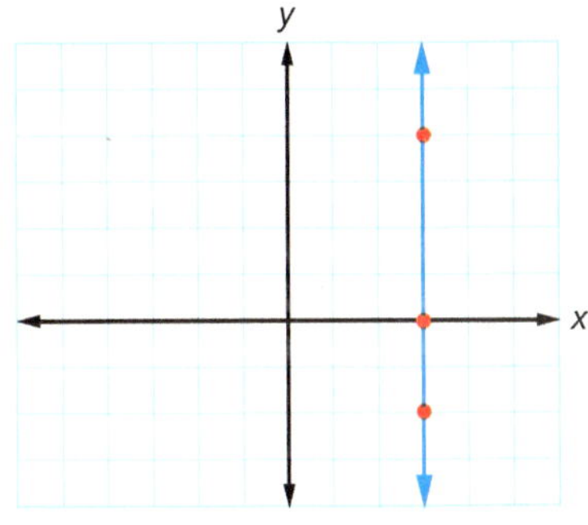

$x = 3$ is not a function.

3. The graph fails the vertical line test.

A. Exercises

Make a table of at least three ordered pairs; then graph each linear equation on its own set of axes.

1. $y = x - 6$

2. $y = -x + 8$

3. $y = -2x + 5$

4. $y = -4x + 1$

5. $y = 4$

6. $x = -1$

7. $y = \frac{2}{3}x - 2$

8. $y = \frac{x - 3}{2}$

Solve each linear equation for *y*.

9. $5x + y = 20$

10. $3x - y = 8$

11. $8x - 2y = 14$

12. $x + 4y = -12$

Write each linear equation in standard form.

13. $y = -x + 9$

14. $3x = 18 + y$

15. $y = \frac{4x - 2}{7}$

16. $y = \frac{5}{8}x - 3$

B. Exercises

Find the x- and y-intercepts of each linear equation; then use them to graph the equation.

17. $x - 2y = 4$ **18.** $x - 3y = -3$ **19.** $y = -x + 4$

20. $x = 3y + 6$ **21.** $x = 2y - 3$ **22.** $3x + y = 1$

23. $x = 2$ **24.** $y = -3$

Graph each linear equation.

25. $y = 2x + 3$ **26.** $4x + 2y = 8$ **27.** $y = -\frac{1}{2}x$ **28.** $x = 3y$

29. $x - y = 8$ **30.** $y = x - 9$ **31.** $y = 8x + 12$ **32.** $7x + 2y = 5$

33. What is the x-coordinate of every y-intercept?

34. What is the y-coordinate of every x-intercept?

35. Is the graph of $y = c$ a horizontal or a vertical line?

36. Which axis does the graph of $x = c$ intersect?

C. Exercises

37. Find general expressions for the x- and y-intercepts of the linear equation $Ax + By = C$.

38. Use the general expressions developed in the previous exercise to find the x- and y-intercepts of $3x + 2y = -12$. Then check your results.

39. Can a function ever have two x-intercepts? two y-intercepts? Explain.

40. Graph the following equations on the same set of axes: $y = x + 1$; $y = 2x + 1$; $y = -2x + 1$; and $y = -\frac{1}{2}x + 1$. What do the equations have in common? What do the graphs have in common?

41. Graph the following equations on the same set of axes: $y = 2x$; $y = 2x - 4$; $y = 2x + 3$; and $y = 2x + 8$. What do the equations have in common? What do the graphs have in common?

42. Faith buys a phone plan that costs $32/mo plus $0.02/min for each call made or received. Write an equation to describe the total cost, c, based on the number of minutes used, m. Graph this equation and find the y-intercept. What is the meaning of the y-intercept? Does the x-intercept have any meaning?

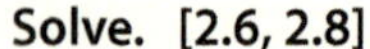

CUMULATIVE REVIEW

Solve. [2.6, 2.8]

43. $3(x - 1) - (x - 7) = 1$ **44.** $\frac{2x - 3}{5} = x + 2$

45. $1.24x + 8 = 8.6$ **46.** $\frac{1}{3}(x - 0.05) = 13$

Solve each literal equation for the indicated variable. [3.1]

47. $S = b^2 + 2bl$ for l **48.** $C = \frac{5}{9}(F - 32)$ for F

Use proportions to solve. [3.3]

49. A triangle has sides of 3 cm, 5 cm, and 6 cm. A similar triangle's shortest side is 4.5 in. Find the lengths of the other two sides.

50. Two similar figures have corresponding sides of 26 mm and 65 mm. What is the ratio of their sides?

A drawing is made with the scale 1 cm : 12 km. [3.3]

51. A length of 4.5 cm represents an actual distance of ___ km.

52. A distance of 99 km is represented by a line ___ cm long.

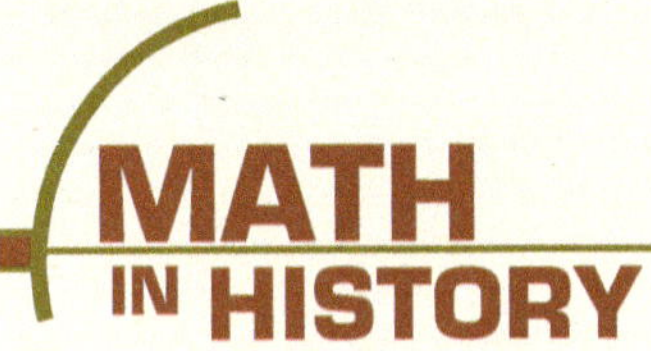

René Descartes (1596–1650)

René Descartes was a prominent French mathematician and philosopher. He was formally educated from age 8 to 16 at a Jesuit school, where he came to view mathematics as the only source of certain knowledge. In mathematics Descartes is most famous for his development of the Cartesian plane and analytic geometry, the study of equations and their graphical representations. Descartes wrote mathematical papers and books but was initially afraid to publish them, fearing the Catholic church would declare them heretical as it had done with Galileo's discoveries. In 1637 Descartes' friends convinced him to publish his masterpiece on mathematics, *Discourse on the Method of Reasoning Well and Seeking Truth in the Sciences*, which revealed analytic geometry to the world.

In *Discourse on the Method* Descartes also presented a new method for arriving at truth. He concluded that since even famous teachers disagreed among themselves, relying on what others taught was not a good way to come to truth. He proposed that a person doubt everything that he had ever been taught until he could demonstrate its truth with reason. He believed that, even if he doubted everything, his doubting demonstrated that he was thinking. This is the origin of Descartes' famous saying, "I doubt, therefore I think; I think, therefore I am."

Descartes did not attack Christianity. In fact, he attempted to use his new method to prove the existence of God. But Descartes' method itself is dangerous to Christianity. Reason, instead of being a servant in the quest to understand the revealed Word of God, is made into a judge of what may or may not be true. This makes the individual's mind, rather than revelation from God, the final authority on what is truth.

René Descartes serves as a warning that even highly intelligent teachers who are right on many points may make grave errors. Christians must test the ideas they encounter not by reason alone, but by the Word of God (John 17:17; Acts 17:11). In the end, Descartes' method has not provided a way to certain knowledge. Philosophers still wrangle over the questions that puzzled the Greeks. But the Word of God remains sure and reliable.

6.2 Slope

The skier negotiates a descent of about 25°. The slope is $m = -\frac{1}{2}$.

The word *slope* probably reminds you of a hill. If you are a downhill skier, you know that the slope of the hill affects your speed going down. The steeper the slope, the faster you descend. In mathematics, the *slope* of a line describes the incline or steepness of the line.

Definition

The **slope** of a line, m, is the ratio of the *rise* (vertical change) to the corresponding *run* (horizontal change).

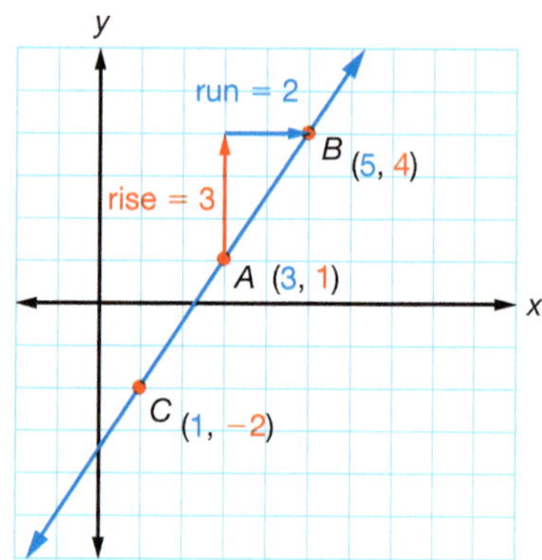

Going from point A to point B on the graph, the line rises up 3 as it runs to the right 2. Therefore, the line's slope is $\frac{3}{2}$ or 1.5. This ratio is the same for any two points on the line. To move from B to C, you must go down 6 and left 4. This ratio is $\frac{-6}{-4} = \frac{3}{2}$ or 1.5. Notice that the slope is the vertical change of the line for each unit of horizontal change from left to right.

The coordinates of points can also be used to find the rise and the run. Using A (3, 1) and B (5, 4),

$$m = \frac{\text{rise}}{\text{run}} = \frac{4 - 1}{5 - 3} = \frac{3}{2}.$$

Example 1

Find the slope of the line.

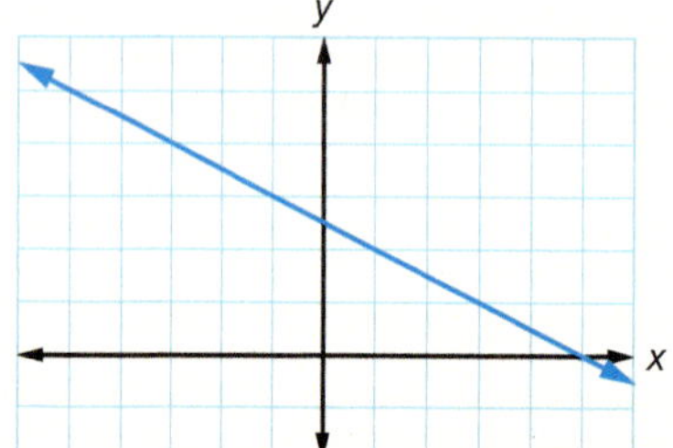

Answer

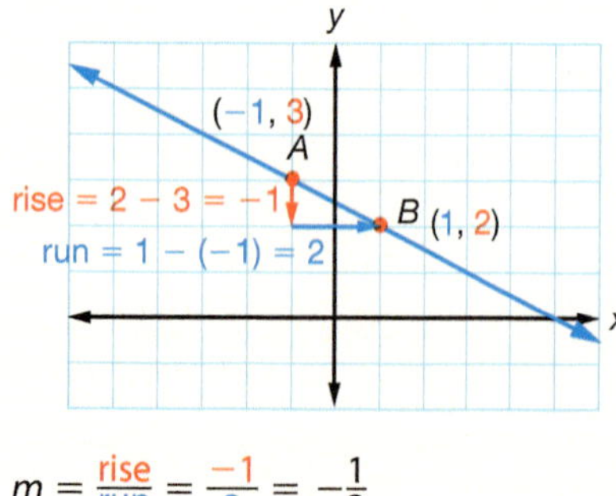

1. Choose two convenient points: A (−1, 3) and B (1, 2).
2. Find the rise (change in y). Moving from A to B, you must go down 1.
3. Then move right 2 to point B to find the run (change in x).
4. Find the slope.

$$m = \frac{\text{rise}}{\text{run}} = \frac{-1}{2} = -\frac{1}{2}$$

If you had started with point B and moved to A, you would have gone up 1 and left 2, but the slope is the same.

$$m = \frac{\text{rise}}{\text{run}} = \frac{1}{-2} = -\frac{1}{2}$$

A more generalized example using two arbitrary points, $P_1\ (x_1, y_1)$ and $P_2\ (x_2, y_2)$, gives the general formula for finding the slope of a line through two points.

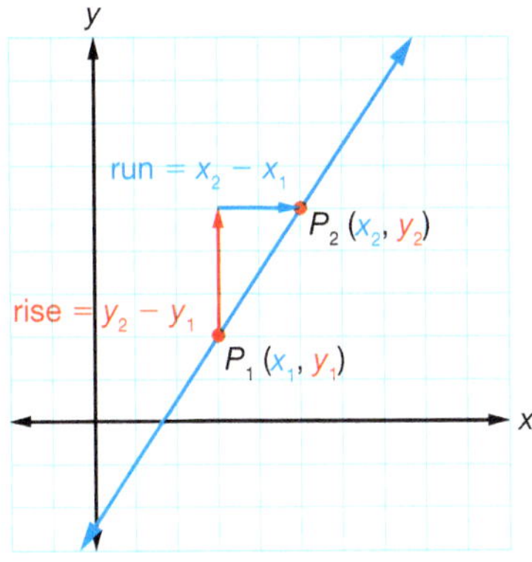

$$m = \frac{\text{rise}}{\text{run}} = \frac{y_2 - y_1}{x_2 - x_1}$$

This formula can be used to find the slope of a line without its graph if you know the coordinates of any two points on the line.

Example 2

Find the slope of the line passing through (2, −3) and (4, 1).

Answer

$P_1\ (x_1, y_1) = (2, -3)$ — 1. Label the points.

$P_2\ (x_2, y_2) = (4, 1)$

$m = \frac{y_2 - y_1}{x_2 - x_1} = \frac{1 - (-3)}{4 - 2} = \frac{4}{2} = 2$ — 2. Substitute values into the slope formula and evaluate.

When using the slope formula, be sure that the order of subtraction is the same in the numerator and the denominator. In Example 2, choosing P_1 to be (4, 1) and P_2 to be (2, −3) produces the same slope:

$$m = \frac{y_2 - y_1}{x_2 - x_1} = \frac{-3 - 1}{2 - 4} = \frac{-4}{-2} = 2.$$

The slope of this line is 2, but writing it as $\frac{2}{1}$ illustrates that the rise is 2 and the run is 1.

Finding the Slope of a Line
When given the graph, determine the rise and run from one point to another point on the line. Then write the ratio of rise over run and reduce if possible.
When given two points, label the points $P_1\ (x_1, y_1)$ and $P_2\ (x_2, y_2)$. Then use the slope formula, $m = \frac{y_2 - y_1}{x_2 - x_1}$.

In the previous examples the lines were slanted. What is the slope of a horizontal line?

Example 3

Find the slope of the line $y = 4$.

Answer

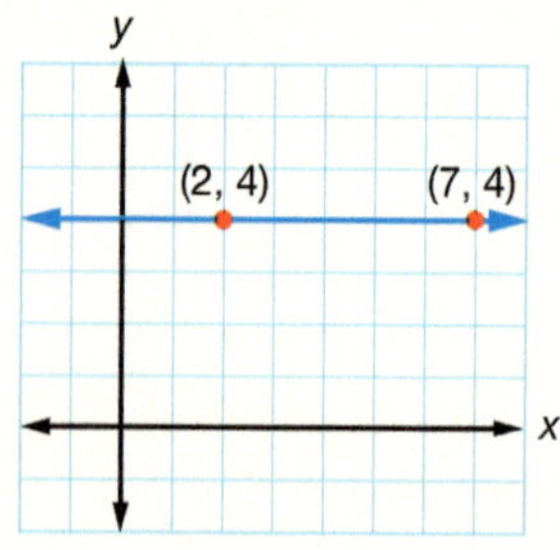

1. Graph the horizontal line $y = 4$.
2. Choose two convenient points: (2, 4) and (7, 4).
3. The rise (change in y) is 0.
4. The run (change in x) is 5.

$m = \frac{\text{rise}}{\text{run}} = \frac{0}{5} = 0$

or $m = \frac{y_2 - y_1}{x_2 - x_1} = \frac{4 - 4}{7 - 2} = \frac{0}{5} = 0$

5. Find the slope.

The slope of a horizontal line is always zero because there is no vertical change. Now consider a vertical line.

Example 4

Find the slope of the line $x = 3$.

Answer

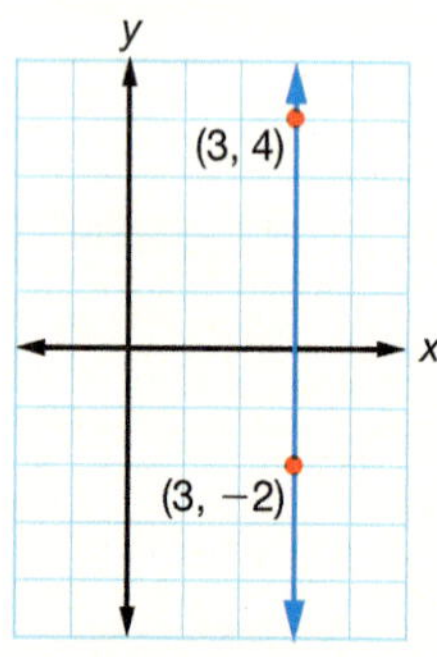

1. Graph the vertical line $x = 3$.
2. Choose two convenient points: (3, −2) and (3, 4).
3. The rise (change in y) is 6.
4. The run (change in x) is 0.

$m = \frac{\text{rise}}{\text{run}} = \frac{6}{0}$

or $m = \frac{y_2 - y_1}{x_2 - x_1} = \frac{4 - (-2)}{3 - 3} = \frac{6}{0}$

The slope is undefined.

5. Division by zero is undefined.

Example 5

Graph the line passing through the point (−1, 2) with a slope of $-\frac{1}{3}$.

Answer

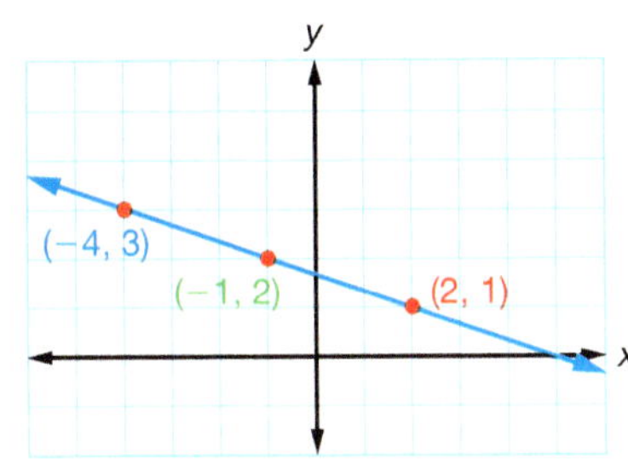

1. Plot the point (−1, 2).
2. Using the slope, $m = -\frac{1}{3} = \frac{-1}{3}$, plot another point on the line. From (−1, 2), move down one unit and then right three units to plot the point (2, 1).
3. Use an equivalent form of the slope, $m = \frac{1}{-3}$, to plot a third point. From (−1, 2), move up one unit and then left three units to plot the point (−4, 3).
4. Draw a line through the points.

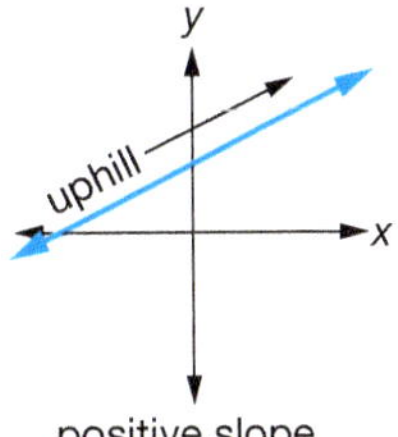

positive slope

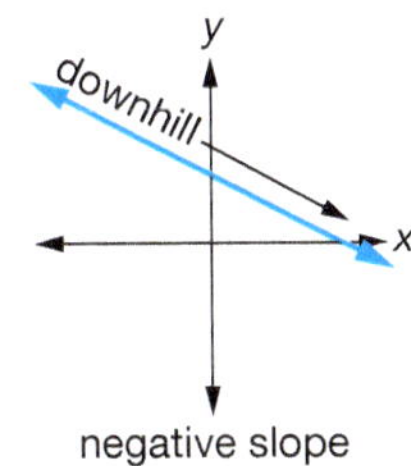

negative slope

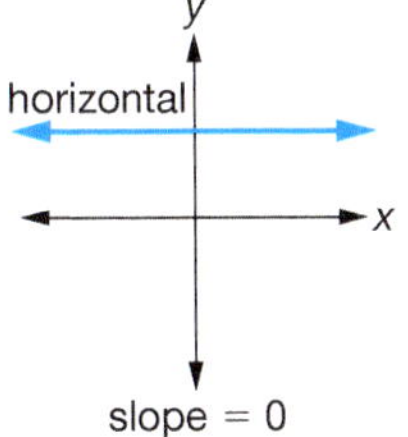

slope = 0

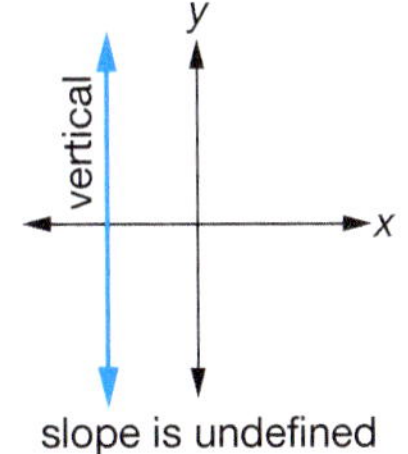

slope is undefined

The sign of the slope indicates whether the line slants up or down moving from left to right. The absolute value of the slope indicates the "steepness" of the line. Slope has been defined as $\frac{\text{rise}}{\text{run}}$ in order that steeper lines would have slopes with larger absolute values.

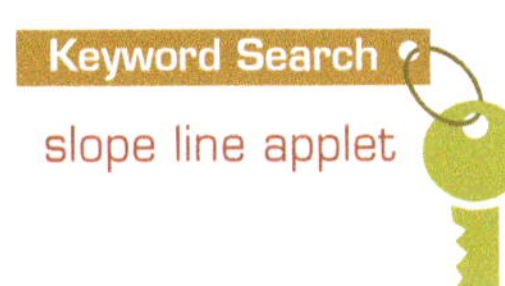

There are many different notations for slope.

$$m = \frac{\text{rise}}{\text{run}} = \frac{y_2 - y_1}{x_2 - x_1} = \frac{\Delta y}{\Delta x} = \frac{\text{change in } y}{\text{change in } x}$$

The Greek letter delta, Δ, is used to symbolize the change in a quantity. Using this notation emphasizes that slope measures a *rate of change*. Rates of change compare two changing quantities. For instance, the slope of a distance-time graph indicates the change in distance per unit of time, or the speed of the object.

Example 6

Determine the speed of the train for each section of the graph.

a. 0–2 hr

b. 2–4 hr

c. 4–7 hr

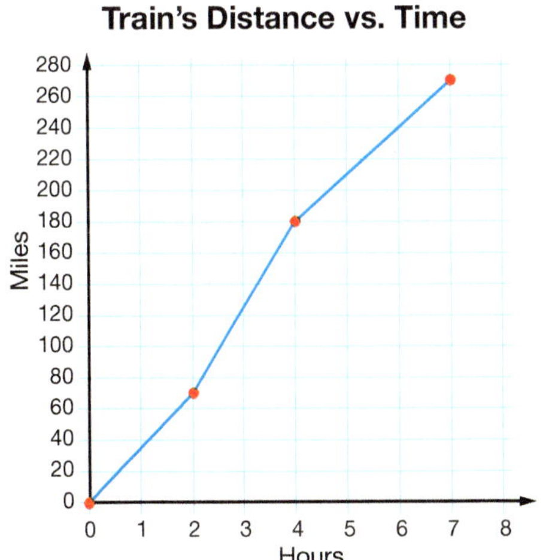

Answer

a. $m = \frac{70 - 0}{2 - 0} = \frac{70}{2} = 35$ mi/hr

b. $m = \frac{180 - 70}{4 - 2} = \frac{110}{2} = 55$ mi/hr

c. $m = \frac{270 - 180}{7 - 4} = \frac{90}{3} = 30$ mi/hr

A. Exercises

Use the following graphs for exercises 1–6.

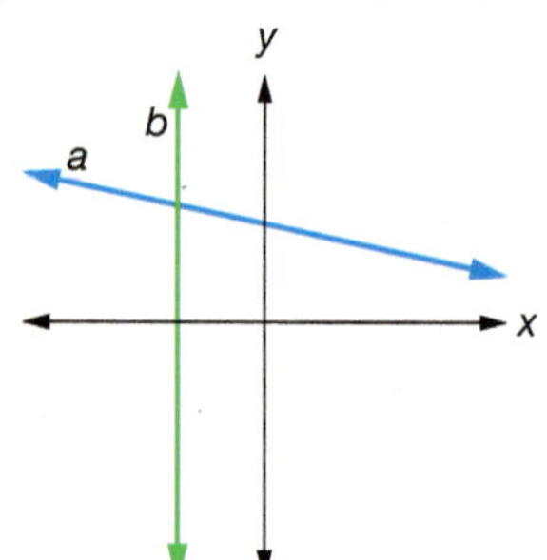

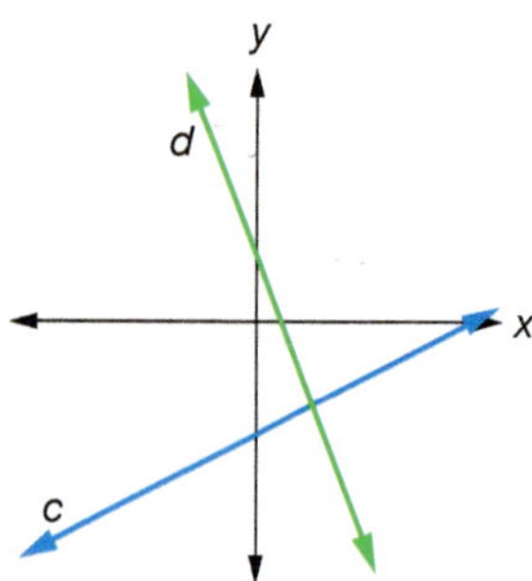

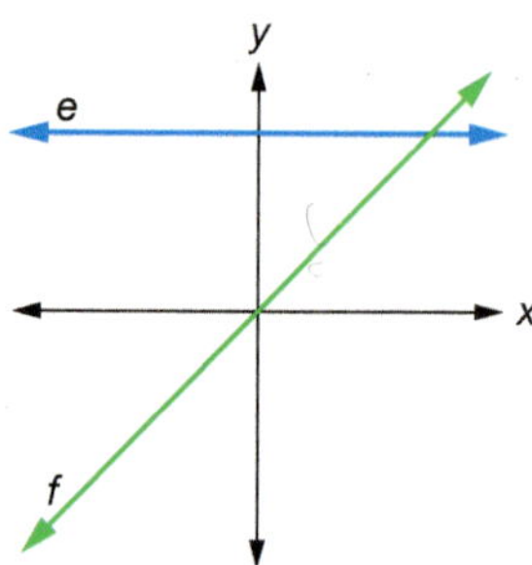

1. Which line(s) have a negative slope?
2. Which line(s) have an undefined slope?
3. Which line(s) have a slope of zero?
4. Which line(s) have a positive slope?
5. Which line has the slope with the greatest absolute value?
6. Which line has the slope with the least nonzero absolute value?

Find the slope of each line.

7. 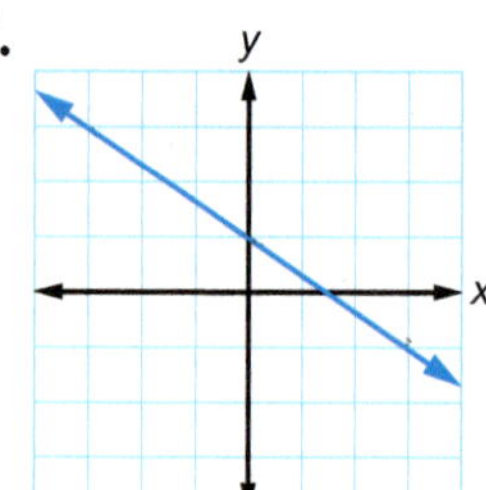

8. 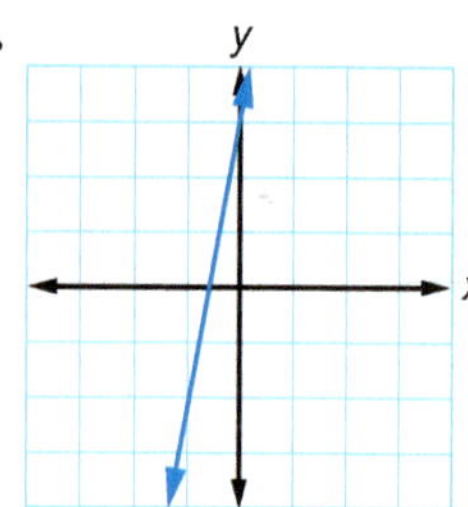

9. 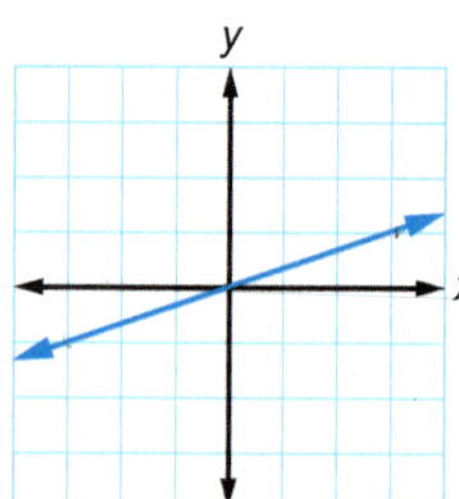

10. 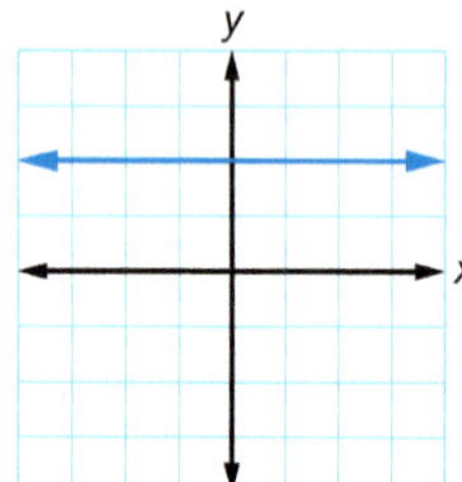

11. 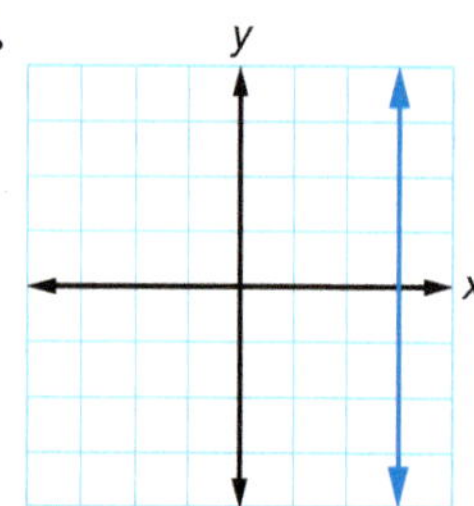

12. 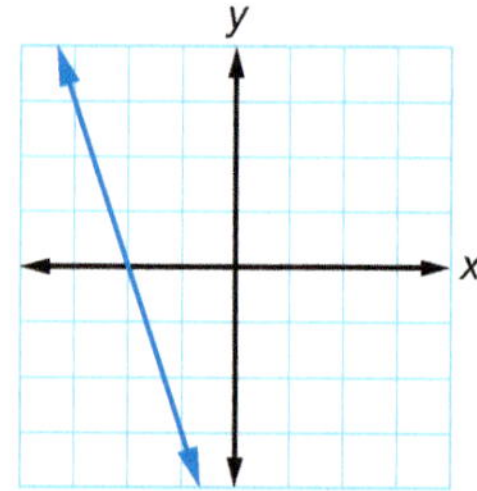

Find the slope of the line passing through the given points.

13. (2, 5), (3, 9)
14. (3, 6), (1, 9)
15. (2, 7), (3, 7)
16. (2, 5), (2, 3)

B. Exercises

Find the slope of the line passing through the given points.

17. (3, 6), (−4, −1)
18. (−4, −6), (−3, −2)
19. (5, 7), (1, −3)
20. (−2, −5), (1, 0)
21. (−1, 7), (−1, 2)
22. (−4, −1), (−1, 7)
23. (4.75, −2.5), (7.25, 4.5)
24. $\left(\frac{2}{3}, -\frac{3}{2}\right), \left(\frac{7}{3}, -4\right)$

Graph the line passing through the given point with the given slope.

25. (3, 1); $m = -\frac{1}{4}$

26. (−4, 0); $m = 6$

27. (2, −5); $m = \frac{3}{2}$

28. (0, 1); $m = -\frac{5}{3}$

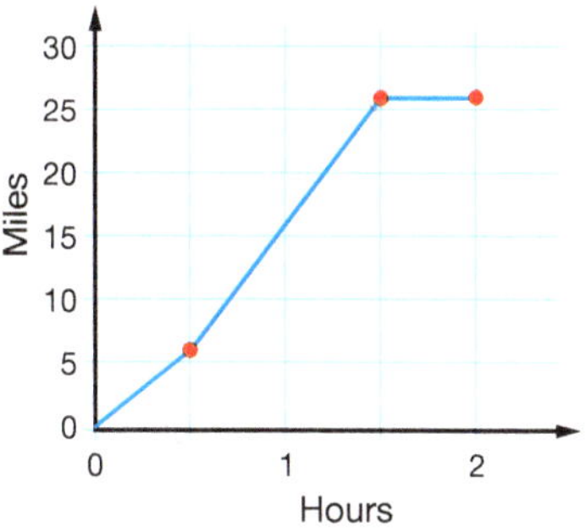

29. Use the distance-time graph to determine the speed of the bicyclist for each section of the graph.

a. from 0 to 0.5 hr

b. from 0.5 to 1.5 hr

c. from 1.5 to 2 hr

d. Describe the motion of the bicyclist from 1.5 to 2 hr.

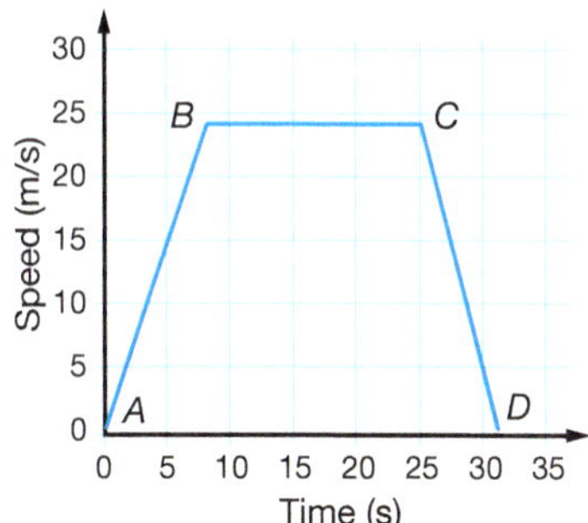

30. The slope of a speed-time graph indicates the change in speed per unit of time, or the *acceleration* of the object. Determine the acceleration (in m/s^2) of the car for each section of the graph.

a. from *A* to *B*

b. from *B* to *C*

c. from *C* to *D*

d. Describe the motion of the car from *B* to *C*.

31. Pediatricians often track a patient's height. Determine the growth rate (in cm/yr) of a typical girl for each time period.

a. 0–2 yr

b. 2–8 yr

c. 8–13 yr

d. 13–18 yr

Age (yr)	Height (cm)
0	50
2	86
8	134
13	159
18	164

32. The price of gasoline often fluctuates dramatically. Determine the rate of change (in $/mo) of the cost of a gallon of gasoline for each time period.

a. from $2.96/gal in January 2008 to $4.04/gal in July 2008

b. from $4.04/gal in July 2008 to $1.64/gal in January 2009

c. from $1.64/gal in January 2009 to $2.60/gal in January 2010

C. Exercises

The slope of a steep road is often given as a percent to warn motorists. In exercises 33–35 assume that distances measured in miles represent horizontal distances.

33. If a sign states "6% downgrade next 3 miles," how much elevation will be lost (to the nearest tenth of a foot)?

34. Express the slope of a road that descends 800 ft over 2 mi as a fraction. Then express this downgrade as a percent (to the nearest tenth).

35. An engineer is planning a road from Barren Pass, elevation 6030 ft, to Fern Creek, elevation 4502 ft. How long (to the nearest tenth of mile) must the road be in order to avoid grades over 5%?

A 6% downgrade implies a descent of 6 ft for every 100 ft horizontally, or $-\frac{6}{100}$.

Dominion Modeling

While on vacation, Brad's construction crew enjoyed skiing and snowboarding. Ski slope ratings are based on the angle of the incline and other factors of difficulty. The table lists some common ratings of slope difficulty.

●	Easy	$< 25°$
■	Intermediate	$25°–40°$
◆	Difficult	$> 40°$

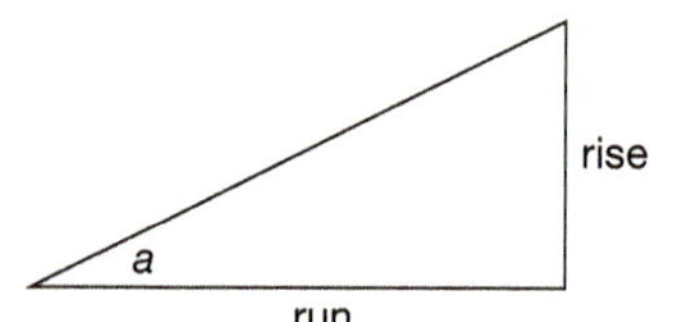

Slope, the ratio of rise over run, may also be expressed as a ratio (e.g., 1 : 2) or in degrees (e.g., 27°). The trigonometric function *tangent* relates the angle of inclination, a, to the slope. The Tan (tangent) function on your calculator gives the decimal equivalent of the slope for a given angle. The Tan^{-1} (inverse tangent) function can be used to find the angle of inclination that corresponds to a given slope.

Example: Given a slope ratio of 1 : 2, find the measure of the angle of inclination.

After checking that your TI-84+ calculator is in degree mode, use the following keystrokes to find $\text{Tan}^{-1}\frac{1}{2}$: [2nd] [TAN] 1 [÷] 2 [ENTER]. The angle of inclination is approximately 26.6°.

Find the angle of inclination (to the nearest degree) and use the chart to classify each slope.

36. Brad's son Michael went snowboarding on a slope that dropped 20 ft for every 60 ft of run.

37. Michael's friend Peter was injured on a slope that dropped 2 ft for every foot of run.

CUMULATIVE REVIEW

Solve.

38. $3x + |-6| = |12|$ [1.2, 2.5]

39. $|2x - 5| = 10$ [4.6]

40. 108% of what number is 64.8? [3.4]

41. What is 0.4% of 190? [3.4]

Write an equation and solve. [3.6, 3.8]

42. Find the sale price of a $70 item reduced by 40%.

43. Find the sale price of a $250 item reduced by 30% and then by another 60%.

44. How many pounds of nuts selling for $4/lb should be mixed with 20 lb of nuts selling for $8/lb to make a mix selling for $7/lb?

45. How much pure antifreeze should be mixed with 5 qt of a 30% antifreeze solution to obtain a solution that is 40% antifreeze?

46. How much water must be added to 5 qt of a 45% antifreeze solution to dilute it to a solution that is 35% antifreeze?

47. Express $0.\overline{9}$ as a simplified quotient of integers. [2.6]

6.3 Slope-Intercept Form of a Line

The tallest building in San Francisco, the Transamerica Pyramid, was among the five tallest buildings in the world upon its completion in 1972. It has a square base 145 ft on each side and is 853 ft tall. Determine the slope of each side.

Equations of lines can be written in several different forms by applying the properties of equality. You have already learned the standard form of a linear equation, $Ax + By = C$. Another form for a line is the *slope-intercept form*.

The slope-intercept form of a linear equation is $y = mx + b$. In this form m, the coefficient of x, is the slope of the line. By substituting $x = 0$ and simplifying to $y = b$, you can see that b is the y-coordinate of the y-intercept, the point $(0, b)$.

Example 1

Write $3x + 2y = 6$ in slope-intercept form; then state the slope and y-intercept.

Answer

$3x + 2y = 6$ — 1. Solve for y.

$2y = 6 - 3x$

$\frac{2y}{2} = \frac{6}{2} - \frac{3x}{2}$

$y = 3 - \frac{3}{2}x$

$y = -\frac{3}{2}x + 3$ — 2. Apply the Commutative Property of Addition to rearrange the right side so that the x term is first.

$y = mx + b$ — 3. Compare this equation to the slope-intercept form to find m and b.

$m = -\frac{3}{2}$; $b = 3$

slope: $-\frac{3}{2}$ — 4. Be sure to use the ordered pair $(0, b)$ to name the y-intercept.

y-intercept: (0, 3)

Three equations can be used to describe this line.

$3x + 2y = 6$	standard form
$y = -\frac{3}{2}x + 3$	slope-intercept form
$f(x) = -\frac{3}{2}x + 3$	function form

All linear equations, except for vertical lines, can be written in function notation.

Changing a Linear Equation to Slope-Intercept Form

1. Solve the linear equation for y.
2. Arrange terms on the right side with the variable term first.

The slope-intercept form of a line can be used to quickly graph the line. Plot the y-intercept $(0, b)$ and then use the slope, m, to plot several other points. While it takes only two points to determine a line, graphing at least three points serves as a check of your work.

Example 2

Graph $y = \frac{1}{2}x + 4$.

Answer

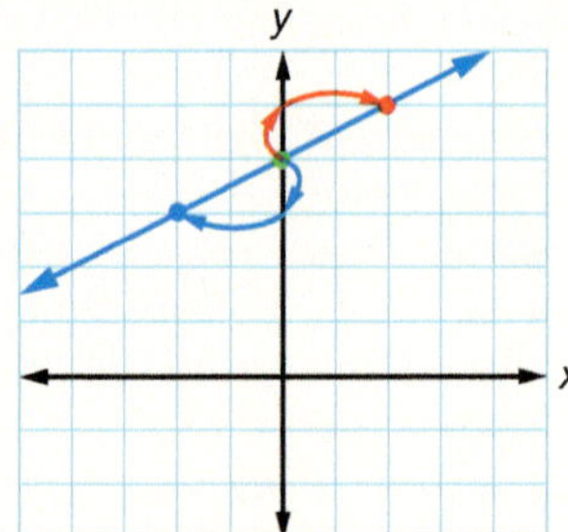

1. Plot the y-intercept. Since $b = 4$, the y-intercept is $(0, 4)$.
2. Using the slope, $m = \frac{1}{2}$, plot another point on the line. From $(0, 4)$, the y-intercept, move up one unit and then right two units to plot the point $(2, 5)$.
3. Use an equivalent form of the slope, $m = \frac{-1}{-2}$, to plot a third point. From $(0, 4)$, move down one unit and then left two units to plot the point $(-2, 3)$.
4. Draw a line through the points.

Using Slope-Intercept Form to Graph a Linear Equation

1. Write the equation in slope-intercept form $(y = mx + b)$.
2. Plot the y-intercept, $(0, b)$, on the graph.
3. Starting at the y-intercept, use the slope to plot several other points.
4. Draw a line through the points.

Example 3

Write $2x + 3y = -3$ as a linear function; then graph the function.

Answer

$$2x + 3y = -3$$
$$3y = -2x - 3$$
$$\frac{3y}{3} = \frac{-2x}{3} - \frac{3}{3}$$
$$y = -\frac{2}{3}x - 1, \text{ so } f(x) = -\frac{2}{3}x - 1$$

1. Write the equation in slope-intercept form and as a linear function.

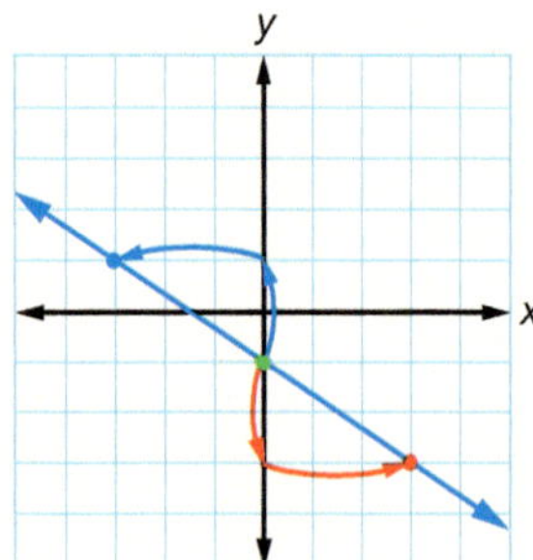

2. Plot the y-intercept, $(0, -1)$.
3. Using the slope, $m = -\frac{2}{3} = \frac{-2}{3}$, plot another point on the line. From $(0, -1)$, move down two units and then right three units to plot the point $(3, -3)$.
4. Use an equivalent form of the slope, $m = \frac{2}{-3}$, to plot a third point. From $(0, -1)$, move up two units and then left three units to plot the point $(-3, 1)$.
5. Draw a line through the points.

Example 4

Write $y = 3$ in slope-intercept form, state the slope and y-intercept, and graph the equation.

Answer

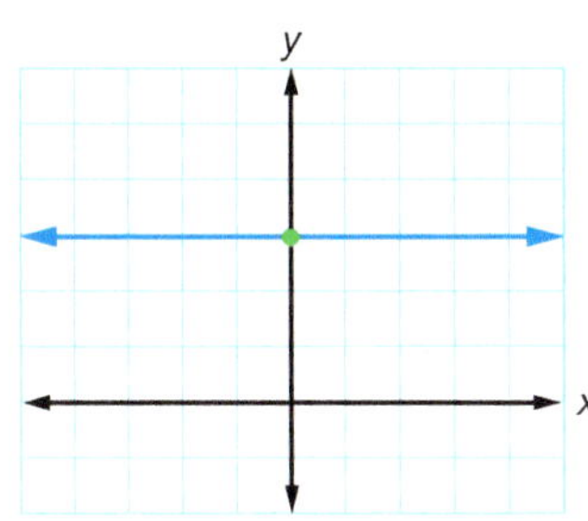

1. Compare $y = 3$ with $y = mx + b$. Since there is no x term in the equation, rewrite $y = 3$ as $y = 0x + 3$.
2. Since $b = 3$, the y-intercept is (0, 3). Since $m = 0$, the line is horizontal.
3. Draw and label the horizontal line through (0, 3).

Since any vertical line intersects the graph of $y = 3$ at only one point, the equation is a function and can be written as $f(x) = 3$. It is called a *constant function*.

Many of the functions you have studied are linear functions of the form $f(x) = mx + b$. The initial value of the function is b, and the rate of change is m.

Example 5

An online company produces photo albums from uploaded photo files. If Sarah chooses to pay a membership fee of $25, a premium photo album will cost her $1.25 per page with free shipping. Write a linear function rule for the cost and graph the function. Then determine how much it would cost Sarah for a membership and a 23 page photo album.

Answer

Let n = the number of pages.

$C(n) = 1.25n + 25$

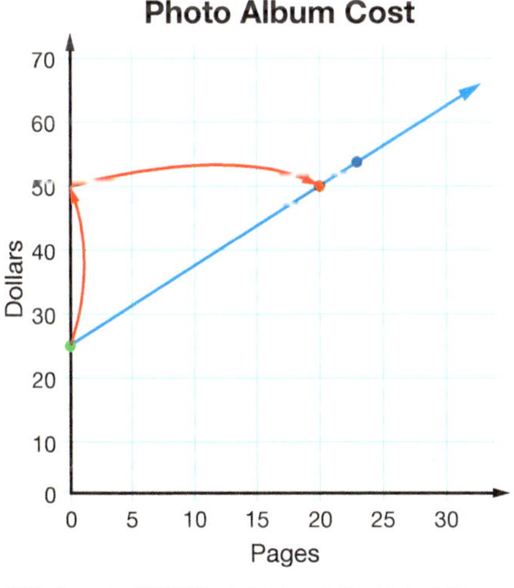

$C(23) = 1.25(23) + 25 = 28.75 + 25$
$= \$53.75$

1. Identify the independent variable.
2. Write the linear function. $b = 25$ (the initial fee), and $m = 1.25$, the rate of change (the cost per page).
3. Graph the function using the y-intercept, (0, 25), and the slope, $m = \frac{125}{100} = \frac{5}{4}$ or $\frac{25}{20}$.
4. From the graph, you can see that a 23 page album would cost slightly less than $55.
5. Find an exact solution by substituting 23 for the number of pages in the function rule and simplifying to find the cost.

A. Exercises

State the slope and y-intercept of each linear equation.

1. $y = 3x + 4$ **2.** $y = -\frac{1}{3}x - 2$ **3.** $y = -x$ **4.** $y = 6$

Match each equation with its graph.

a.
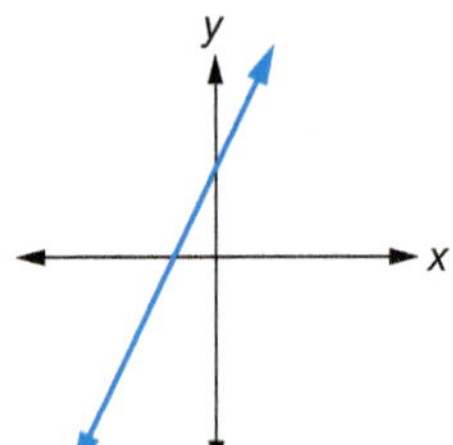

b.
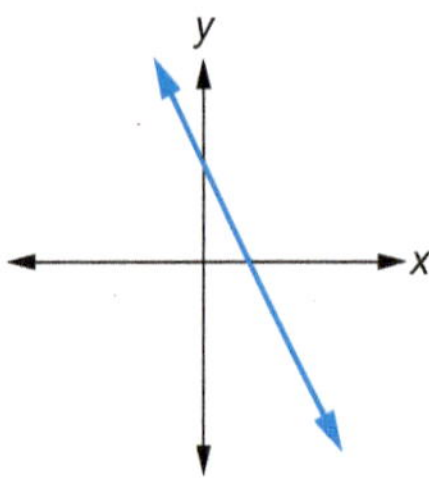

c.
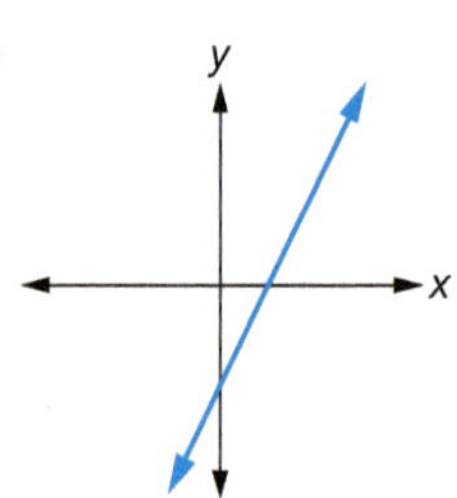

d.
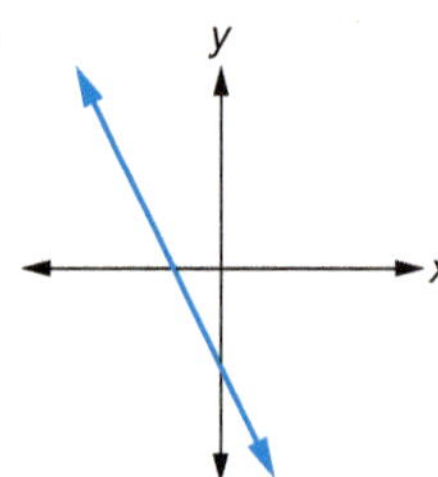

5. $y = 2x - 1$ **6.** $y = -2x - 1$ **7.** $y = -2x + 1$ **8.** $y = 2x + 1$

Graph each equation using its slope and y-intercept.

9. $y = -2x + 4$ **10.** $y = 4x - 3$ **11.** $y = -\frac{1}{4}x + 5$

12. $y = -2$ **13.** $y = \frac{2}{3}x + 2$ **14.** $y = -\frac{3}{5}x + 1$

B. Exercises

Write each equation in slope-intercept form; then state the slope and y-intercept.

15. $3x + y = 5$ **16.** $x - y = -6$

17. $x + 7y = 14$ **18.** $3x + 7y = 21$

19. $8x - 3y = 9$ **20.** $x = 5y$

21. $dx + y = k$ **22.** $ry = sx + t$

Write each equation in slope-intercept form; then graph the equation.

23. $2x + y = 5$ **24.** $x + y = 2$ **25.** $x + 2y = 8$

26. $\frac{1}{2}y = x - 3$ **27.** $3x + 4y - 8 = 0$ **28.** $4x - 5y - 10 = 0$

29. $x = 3y - 9$ **30.** $3x = -2y + 8$ **31.** $3x + 2(y - 5) = 7x$

32. $3(2x - 3) + 1 = 2(y - 1)$

Solve each problem by writing a linear function.

33. The enrollment at Green Mountain Christian School is increasing at a rate of 18 students per year. If the enrollment was 226 in the fall of 2011, write a linear function that models the enrollment. Use your function to predict the number of students enrolled in the fall of 2031.

34. When a new car is purchased, it immediately begins to lose value, or depreciate. If its depreciation is a linear function and a \$28,550 new car depreciates to a value of \$13,500 over a seven-year period, what is the rate of change? Write a linear function that models the value of the car over time. Use your function to predict the value of the car when it is ten years old.

35. Luke's motorcycle has a 4.5 gal gas tank and uses 1.4 gal to travel 63 mi. At what rate is the gasoline consumed (in gal/mi)? Write a linear function, g, that models the number of gallons left in the tank after he travels m miles, assuming the tank was full when he started. How many gallons are left in the tank after he travels 180 mi?

C. Exercises

36. Write $Ax + By = C$ in slope-intercept form. Then find general expressions for the slope and y-intercept of a line written in standard form.

37. Use the general expressions developed in the previous exercise to find the slope and y-intercept of $6x - y = 5$. Then check your results by solving the equation for y.

Write each equation in slope-intercept form; then graph the equation.

38. $4(x - 2) + 7y = 5 - (2x - 3y)$

39. $-2x + 8(3x - 1) = 7x + y - 2(3y + 4)$

Dominion Modeling

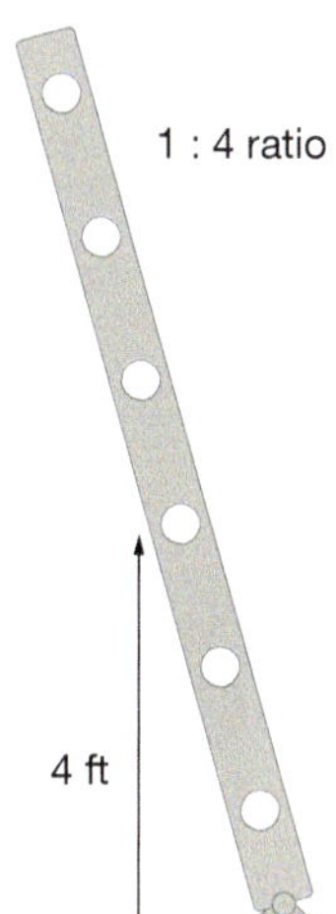

Setting up a ladder at the correct angle is crucial to working safely. Emergency rooms in the US attend to more than 150,000 people annually who have been injured while using ladders. While most slopes are described as rise over run, the slope (or pitch) of a ladder is often described using a run-to-rise ratio. The run is the horizontal distance from the legs of a ladder to a building, and the rise is the vertical height at which the ladder contacts the building. Recommended ladder pitch may vary, but a 1 : 4 ratio, or 75°, is common.

Use the recommended 1 : 4 ratio in exercises 40–41.

40. If Brad's ladder contacts a house at a point 24 ft above the ground, how far should the bottom of the ladder be from the side of the house?

41. If the bottom of a ladder is 4 ft from a wall, at what height on the wall should the ladder make contact?

42. If a ladder contacts a building 15 ft above the ground and the bottom of the ladder is 30 in. from the building, what is the ladder's pitch?

43. What could happen if the distance from the wall to the bottom of the ladder is too short?

44. Name two things that could happen if the distance from the wall to the bottom of the ladder is too far.

CUMULATIVE REVIEW

Perform the indicated operations. [1.3–1.7]

45. $2.346 + \frac{3}{16}$

46. $7.2 + 1.\overline{9}$

47. $-16 + \frac{2}{3} - \frac{1}{6}$

48. $4.3 - 2.95827$

49. $4\frac{4}{5}\left(\frac{5}{24}\right)$

50. $0.\overline{6}(0.\overline{3})$

51. $4\frac{7}{8} \div 2\frac{1}{9}$

52. $1533.1825 \div 365$

53. $2^3 + 5^{-2}$

54. $[x^3y^{-2}(x^{-5}y^{-8})]^2$

6.4 Writing Linear Equations

You have seen how to graph a linear equation using a table of ordered pairs, using the x- and y-intercepts, and using the slope-intercept form of a line. We will now focus on how to write a linear equation when given information about the line. You should be able to write the equation of a line if you are given the following information:

1. the slope and y-intercept of the line,
2. the slope and a point on the line,
3. two points on the line, or
4. the graph of the line.

On a steep slope, engineers extend the horizontal distance with a pigtail curve, as on Iron Mountain Road in South Dakota.

To find the equation of a line when the slope and y-intercept are known, substitute the given values for the appropriate variables into the slope-intercept form of the equation.

Example 1

Write the equation of the line with a slope of $\frac{2}{3}$ and a y-intercept of $(0, -5)$.

Answer $y = \frac{2}{3}x - 5$ Since $m = \frac{2}{3}$ and $b = -5$, substitute these values into the equation $y = mx + b$.

Whenever the slope and any other point on the line are known, the slope-intercept form can be used to find the y-intercept. Substitute the slope, m, and the point's coordinates, (x, y), into slope-intercept form and solve for b.

Example 2

Write the equation of the line passing through the point $(1, 5)$ with a slope of 3.

Answer

$y = mx + b$	1. Substitute the slope, $m = 3$, and the point $(1, 5)$ into the slope-intercept form of the equation.
$5 = 3(1) + b$	
$5 = 3 + b$	2. Solve for b.
$b = 2$	
$y = 3x + 2$	3. Since $m = 3$ and $b = 2$, insert these values into the equation $y = mx + b$.

Example 3

Write the equation of the line passing through the point $(-5, 6)$ with a slope of $-\frac{1}{2}$.

Answer

$y = mx + b$

$6 = -\frac{1}{2}(-5) + b$

1. Substitute the values of m, x, and y into slope-intercept form.

$6 = \frac{5}{2} + b$

$b = \frac{12}{2} - \frac{5}{2} = \frac{7}{2}$

2. Solve for b. Be sure to use common denominators when adding or subtracting fractions.

$y = -\frac{1}{2}x + \frac{7}{2}$

3. Insert $m = -\frac{1}{2}$ and $b = \frac{7}{2}$ into the equation $y = mx + b$.

Example 4

Write the equation of the line passing through the point $(1, -6)$ with a slope of 0.

Answer

$y = mx + b$

$-6 = 0(1) + b$

$b = -6$

1. Substitute the values of m, x, and y into slope-intercept form; then solve for b.

$y = 0x - 6$

$y = -6$

2. Insert $m = 0$ and $b = -6$ into the equation $y = mx + b$ to confirm that this line is a horizontal line.

When given two points on a line, use the slope formula to find m. Then use the slope and the coordinates of either point to find b. Finally, insert these values into slope-intercept form to write the equation of the line.

Example 5

Write the equation of the line passing through $(1, 3)$ and $(4, -2)$.

Answer

$m = \frac{y_2 - y_1}{x_2 - x_1} = \frac{-2 - 3}{4 - 1} = \frac{-5}{3} = -\frac{5}{3}$

1. Find the slope of the line.

$y = mx + b$

$3 = -\frac{5}{3}(1) + b$

$3 = b - \frac{5}{3}$

$b = \frac{9}{3} + \frac{5}{3} = \frac{14}{3}$

2. Use one of the points, $(1, 3)$, and the slope to find b. Note that using the point $(4, -2)$ produces the same value for b.

$y = -\frac{5}{3}x + \frac{14}{3}$

3. Use the slope and b to write the equation in slope-intercept form.

The second point, $(4, -2)$, can be used to check your work since it must be a solution to the equation.

$$-\frac{5}{3}(4) + \frac{14}{3} = -\frac{20}{3} + \frac{14}{3} = -\frac{6}{3} = -2$$

Example 6

Write the equation of the line passing through $(7, 4)$ and $(7, 2)$.

Answer

$m = \frac{y_2 - y_1}{x_2 - x_1} = \frac{2 - 4}{7 - 7} = \frac{-2}{0}$

1. Since the slope of the line is undefined, the line must be vertical.

$x = 7$

2. To write the equation of a vertical line through a point, set x equal to the x-coordinate.

If the slope and y-intercept can be easily determined from the graph of a line, substitute the values of m and b into slope-intercept form to write the equation. If not, use the slope and another convenient point or use the coordinates of two convenient points to determine the line's equation, as was done in the previous examples.

Example 7

The planning committee for the city of Brownsville used census data to project the growth of the city. Determine the estimated rate of growth and write the function rule describing this population estimate.

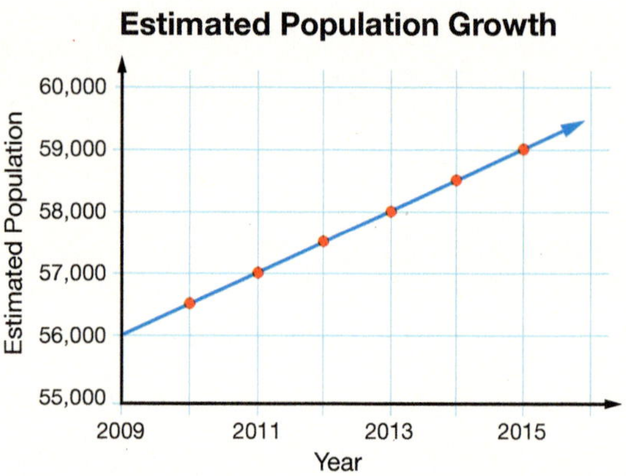

Answer

P_1 (2010, 56,500)

P_2 (2015, 59,000)

1. Choose two convenient points from the graph.

$$m = \frac{59{,}000 - 56{,}500}{2015 - 2010} = \frac{2500}{5} = 500\ \frac{\text{people}}{\text{year}}$$

2. Use the slope formula to determine the estimated rate of growth.

$56{,}500 = 500(2010) + b$

$56{,}500 = b + 1{,}005{,}000$

$b = -948{,}500$

3. Use one of the points, (2010, 56,500), and the slope to find b.

$y = 500x - 948{,}500$

4. Use the slope and b to write the equation in slope-intercept form.

$P(t) = 500t - 948{,}500$

5. Rewrite the equation showing population as a function of time.

This function acts as a model of the population growth. While the model may be useful for years close to the data, making predictions well beyond the data points often causes errors. Estimating the founding of the city to be 1897 by setting the population to zero and solving for t assumes a constant rate of growth over all those years. Predictions of negative population in years prior to 1897 are obviously meaningless.

A. Exercises

Write the equation of the line with the given slope and y-intercept.

1. $m = -5$; $(0, 10)$

2. $m = 4$; $(0, -9)$

3. $m = -\frac{1}{3}$; $(0, 3)$

4. $m = 6$; $\left(0, \frac{1}{2}\right)$

5. $m = 0$; $(0, 7)$

6. $m = -\frac{3}{2}$; $(0, 0)$

Write the equation of each line in slope-intercept form.

7. 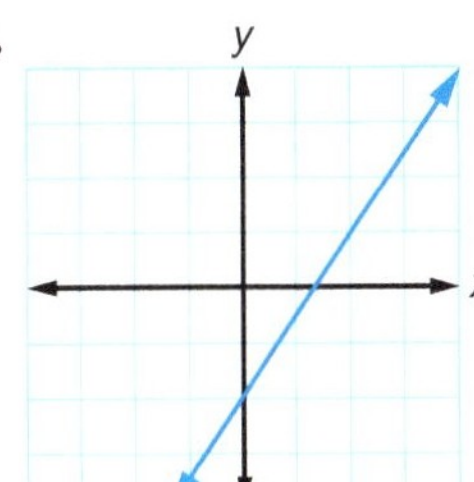

8. 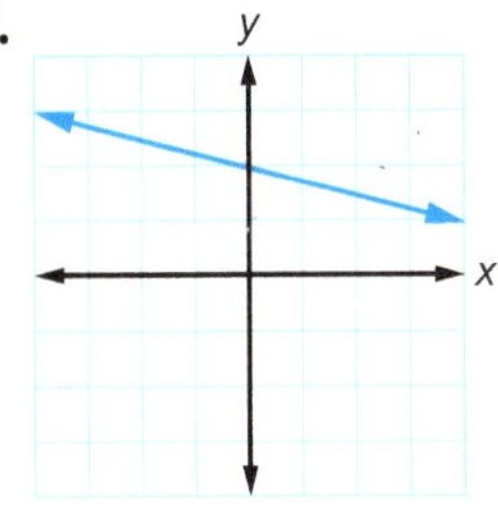

9. 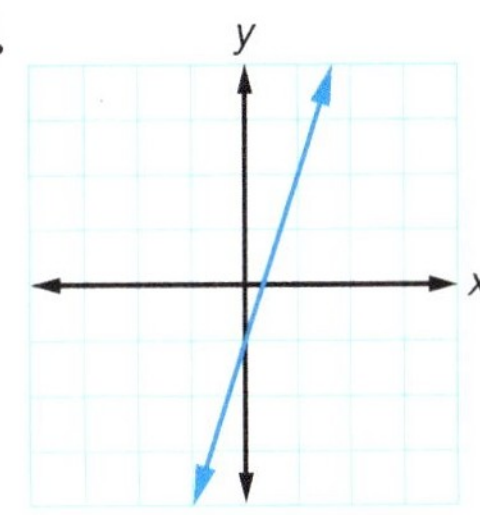

10. 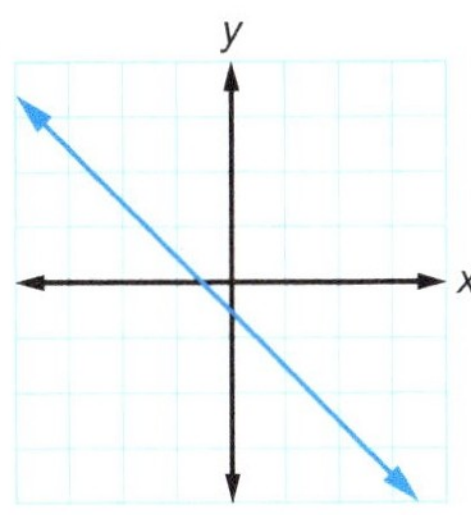

B. Exercises

Write the equation of the line passing through the given point with the given slope.

11. $(2, -5)$; $m = -3$

12. $(-3, 2)$; $m = 8$

13. $(-3, -1)$; $m = \frac{5}{3}$

14. $(-5, 2)$; $m = -\frac{4}{5}$

15. $(1, 6)$; $m = -\frac{1}{4}$

16. $(-9, 0)$; $m = -\frac{1}{8}$

17. $(0, 4)$; $m = \frac{3}{5}$

18. $(3, 7)$; $m = \frac{1}{4}$

Write the equation of the line passing through the given points.

19. $(4, 2), (-3, 9)$

20. $(0, 4), (2, 7)$

21. $(1, 4), (0, 0)$

22. $(1, 3), (2, 5)$

23. $(-4, 2), (7, 3)$

24. $(-3, 2), (10, -6)$

25. $(-5, -4), (3, -4)$

26. $(2, -3), (2, 4)$

Write the equation of each line.

27. a line passing through the point $(-1, -3)$ with a slope of 0.4

28. a line passing through $(1, 3)$ and $(-4, -9)$

29. a line passing through $(2, 0)$ and $(6, -5)$

30. a line passing through the point $(13, 9)$ with a slope of $-\frac{1}{2}$

31. Use the graph of Dr. Jokowski's salary for each year of his practice to write a linear function modeling his salary. Use the function to predict how much he will make during his twentieth year of practice.

32. A hot air balloon takes off from Buckingham, PA, and ascends at a linear rate. Use the graph to find its rate of ascent; then write an equation representing the altitude of the balloon as a function of time.

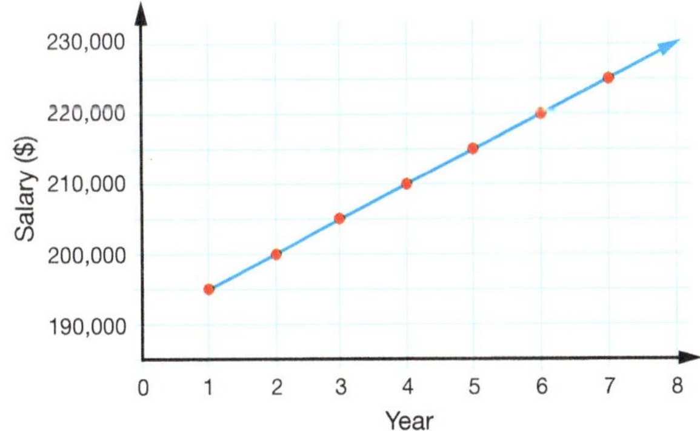

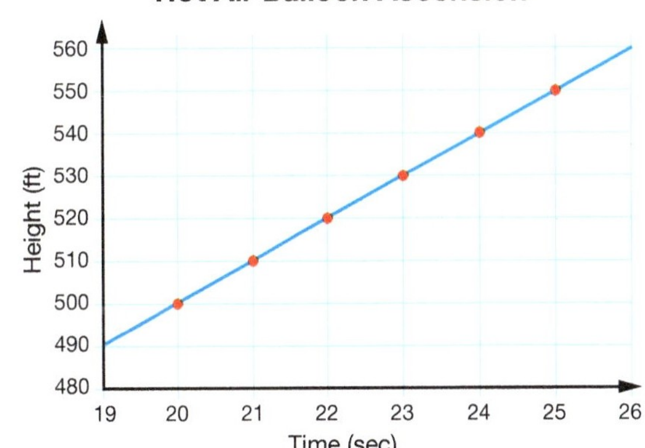

33. Two years ago the enrollment of Hope Christian Academy was 520 students; and the students ate, on average, 419 hot lunches per week. Last year the 540 students consumed an average of 444 hot lunches per week. Write a linear function modeling the weekly consumption of hot lunches based on the school's enrollment. Then use this function to predict the number of hot lunches the school should provide each week if the enrollment is 584 students this year.

34. Justin is filling a 50 gal aquarium with water. After 5 min the aquarium has 3 gal of water in it, and Justin increases the inflow rate to a constant rate of 4 gal/min. How much water will be in the tank 8 min after he began filling it?

35. Mr. Rodriguez works in a clothing store. He notices that when shirts are priced at \$60 each, he sells 20 shirts per week. For each \$10 increase in the price of the shirts, he sells three fewer shirts that week. Find the rate of change in the number of shirts sold per dollar increase in price. Then write a linear function, s, modeling the number of shirts sold based on the cost of the shirts, c. Use the x-intercept to determine the price at which no shirts will sell.

C. Exercises

Write the equation of the line (in standard form) passing through the given point with the given slope.

36. $(0, 8);\ m = \frac{1}{5}$

37. $(3, 0);\ m = \frac{9}{4}$

38. $(6, -1);\ m = -4.5$

39. $\left(\frac{15}{7}, -\frac{3}{7}\right);\ m = -\frac{3}{5}$

Dominion Modeling

Brad's construction work involved a lot of time on roofs. Once he experienced a severe fall from a roof, breaking his back. Although he fully recovered from this fall, many do not. One of the most significant factors to consider when working on a roof is its slope (or pitch). A 4/12 pitch is generally considered to be walkable. This means that the roof rises 4 in. for every foot of horizontal distance. Note that roof pitch ratios, instead of being reduced to lowest terms, are always expressed with a denominator of 12 (inches per foot).

40. On a roof with a 4/12 pitch, how high does the roof rise for every 36 in. of horizontal run?

41. On a 7/12 pitched roof, how many inches does the roof rise for every 4 ft of run?

42. Using Tan^{-1}, find the angle of inclination of a 7/12 pitched roof (to the nearest degree).

43. To determine the slope of his roof, Tyler went into the attic and held a two-foot level against the roof sheathing. He measured from the other side of the level to the roof. The measurement was exactly 14 in. What is the pitch of Tyler's roof?

roof
14 in.
level

44. Brad's crew built a small church in Alaska with an 8/12 pitched roof. If the width of the building, including the eaves, is 120 ft, find the height to the ridge, h. Then use the Pythagorean theorem to find the width of the roof from the eaves to the ridge, w, to the nearest inch.

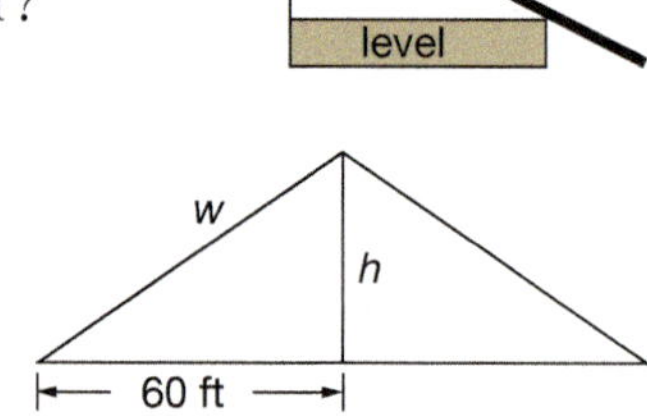

CUMULATIVE REVIEW

Simplify. [2.3]

45. $3 - (2x + 5) + (-x - 3)$

46. $-5(x - 3) + 2(4x - 8)$

Solve. [4.2–4.3, 4.6]

47. $4x < -261$

48. $3x - 77 > -14$

49. $|2x - 12| = 8$

50. $|14 - 3x| = 2$

Write an equation and solve. [2.8]

51. If there are three more quarters than dimes and the total is $5.65, how many of each coin are there?

52. Linda has fifty coins, made up of nickels and dimes, worth a total of $4.65. How many of each coin does she have?

Last year's dog show had twenty entrants, including five poodles. This year's show has thirty-five entrants, but only three poodles. [3.5]

53. What is the percent increase in contestants?

54. What is the percent decrease in poodles?

Sudoku is a game of logic in which the digits 1–9 are placed in each row, each column, and each smaller 3 × 3 grid without repetition. Copy the partially filled grid and complete this puzzle. General hints and more difficult puzzles are readily available on the Internet.

7				4		9		
	6	5			8			
				5	6	2		
8	9					1		4
		6	4		3			9
1		4				5	2	
	4					8		1
		7				6		
				7	5			

6.5 Parallel and Perpendicular Lines

The parallel tracks do not appear to be parallel in this photo.

Linear equations are usually written in standard form, $Ax + By = C$, or in slope-intercept form, $y = mx + b$. The slope of a line is very important. It is the first thing you identify whenever you want to find the equation of a line. In the previous section you used the slope-intercept form of a line and another point on the line to determine the y-intercept.

The formula for slope is the basis for a third form of a linear equation, the *point-slope* form. This form provides an alternative method of determining the equation of a line from the slope, m, and any given point, (x_1, y_1), on the line.

Let (x, y) represent any general point on the line.

$$\frac{y - y_1}{x - x_1} = m$$

Multiplying both sides by $x - x_1$ produces the point-slope form.

$$y - y_1 = m(x - x_1)$$

Example 1

Use the point-slope form of a linear equation to write the equation of the line passing through $(-5, 2)$ and $(3, -4)$ in both slope-intercept form and standard form.

Answer $m = \frac{y_2 - y_1}{x_2 - x_1} = \frac{-4 - 2}{3 - (-5)} = \frac{-6}{8} = -\frac{3}{4}$

1. Use the slope formula to find the slope between the two points.

$$y - y_1 = m(x - x_1)$$
$$y - (-4) = -\frac{3}{4}(x - 3)$$

2. Substitute the slope and one of the points on the line, $(3, -4)$, into the point-slope form.

$$y + 4 = -\frac{3}{4}x + \frac{9}{4}$$
$$y + 4 - 4 = -\frac{3}{4}x + \frac{9}{4} - \frac{16}{4}$$
$$y = -\frac{3}{4}x - \frac{7}{4} \text{ (slope-intercept form)}$$

3. Solve for y to write the equation in slope-intercept form. Notice that 4 has been renamed as $\frac{16}{4}$ on the right side of the equation.

$$4y = -3x - 7$$
$$3x + 4y = -7 \text{ (standard form)}$$

4. Convert to standard form by multiplying each term by 4 and then adding $3x$ to both sides.

Check $3(-5) + 4(2) = -15 + 8 = -7$

5. Substitute the coordinates of the other point, $(-5, 2)$, into the equation.

Definitions

Parallel lines are lines in the same plane that have no points in common.

Perpendicular lines meet at right angles (90°).

Determine the slope of each line in the graph below. Notice that the lines appear to be parallel.

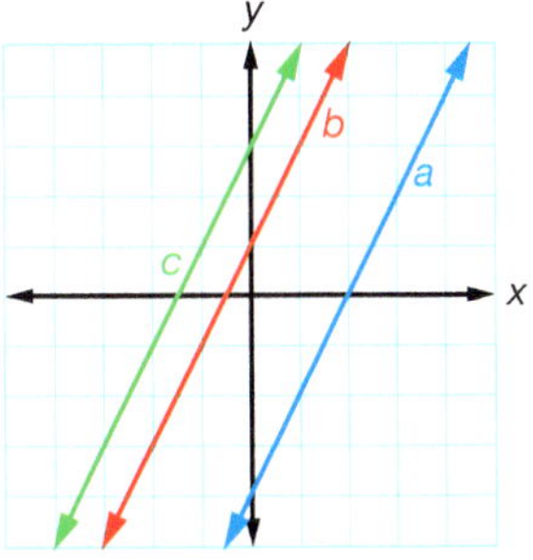

line	slope	y-intercept	equation
a	2	(0, −4)	$y = 2x - 4$
b	2	(0, 1)	$y = 2x + 1$
c	2	(0, 3)	$y = 2x + 3$

The slant is the same, but the lines' vertical locations are different. The equation of each line can be determined by looking at the graph. The graphs of linear equations are parallel if they have the same slope but different y-intercepts.

Example 2

Determine whether the lines described by $3x - y = 6$ and $-9x + 3y = 12$ are parallel.

Answer

$3x - y = 6$
$-y = -3x + 6$
$y = 3x - 6$

$-9x + 3y = 12$
$3y = 9x + 12$
$y = 3x + 4$

1. Write each equation in slope-intercept form.

$m_1 = 3$
y-intercept: (0, −6)

$m_2 = 3$
y-intercept: (0, 4)

2. Determine the slope and y-intercept of each line.

The lines are parallel.

3. The lines have the same slope but different y-intercepts.

Notice that the graphs of the equations in Example 2 are parallel.

The fact that parallel lines have the same slope can be used to find the equation of a line parallel to a given line and passing through a given point.

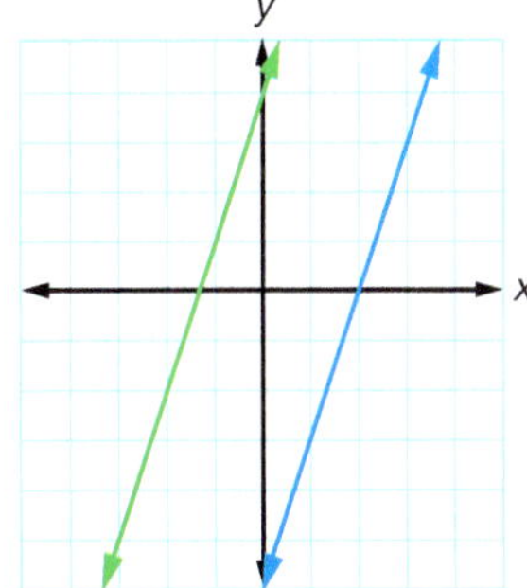

Example 3

Write the slope-intercept form of the equation of the line parallel to $4x + 3y = 17$ and passing through $(-5, 9)$.

Answer

$$4x + 3y = 17$$
$$3y = -4x + 17$$
$$y = -\frac{4}{3}x + \frac{17}{3}$$

1. Write the equation in slope-intercept form.

$$m = -\frac{4}{3}$$

2. Determine the slope of the given line.

$$y - y_1 = m(x - x_1)$$
$$y - 9 = -\frac{4}{3}[x - (-5)]$$

3. Since the lines are parallel, substitute the slope of the given line and the point $(-5, 9)$ into the point-slope form of a line.

$$y - 9 = -\frac{4}{3}x - \frac{20}{3}$$
$$y - 9 + 9 = -\frac{4}{3}x - \frac{20}{3} + \frac{27}{3}$$
$$y = -\frac{4}{3}x + \frac{7}{3}$$

4. Simplify and write the equation in slope-intercept form.

Slopes of Parallel and Perpendicular Lines
Parallel lines have the same slope. Perpendicular lines have slopes that are negative reciprocals of each other.

The term *negative reciprocal* may be new to you. The reciprocal of a number is found by interchanging its numerator and denominator. The negative reciprocal is the opposite of the reciprocal. For instance, the negative reciprocal of $\frac{5}{9}$ is $-\frac{9}{5}$. The negative reciprocal of -8 is $\frac{1}{8}$.

Example 4

State the negative reciprocal of each number.

a. $\frac{8}{5}$ **b.** 9 **c.** $-\frac{13}{7}$

Answer

a. The reciprocal of $\frac{8}{5}$ is $\frac{5}{8}$. The negative reciprocal of $\frac{8}{5}$ is $-\frac{5}{8}$.

b. Write 9 as $\frac{9}{1}$. The reciprocal of $\frac{9}{1}$ is $\frac{1}{9}$. The negative reciprocal of 9 is $-\frac{1}{9}$.

c. The reciprocal of $-\frac{13}{7}$ is $-\frac{7}{13}$. The negative reciprocal of $-\frac{13}{7}$ is $\frac{7}{13}$.

Example 5

Write the slope-intercept form of the equation of the line perpendicular to $y = -4x + 9$ and passing through (2, 7).

Answer

$m_1 = -4$ — 1. Determine the slope of the given line.

$m_2 = \frac{1}{4}$ — 2. The slope of any line perpendicular to the given line is the negative reciprocal of -4. Find this new slope, m_2.

3. Since the lines are perpendicular, substitute the new slope and the point (2, 7) into the point-slope form of a line.

$$y - y_1 = m(x - x_1)$$
$$y - 7 = \frac{1}{4}(x - 2)$$
$$y - 7 = \frac{1}{4}x - \frac{1}{2}$$
$$y - 7 + 7 = \frac{1}{4}x - \frac{1}{2} + \frac{14}{2}$$
$$y = \frac{1}{4}x + \frac{13}{2}$$

Example 6

Are the graphs of the following lines parallel, perpendicular, or neither?

$2x + 5y = 9$ $\quad$ $4x + y = 1$

Answer

1. Write each equation in slope-intercept form.

$$2x + 5y = 9 \qquad 4x + y = 1$$
$$5y = -2x + 9 \qquad y = -4x + 1$$
$$y = -\frac{2}{5}x + \frac{9}{5}$$

2. Determine the slope of each line.

$m_1 = -\frac{2}{5}$ $\quad$ $m_2 = -4$

3. The slopes are neither equal nor negative reciprocals.

The lines are neither parallel nor perpendicular.

A. Exercises

State the negative reciprocal of each slope.

1. $m = 7$

2. $m = -8$

3. $m = -\frac{3}{5}$

4. $m = \frac{1}{9}$

5. $m = \frac{4}{7}$

State the slope of a line parallel to the given line.

6. $5x - y = 3$

7. $-4x + 6y = 7$

8. $2x - y = -5$

9. $-8x + 5y = 7$

10. $x + 9y = -4$

State the slope of a line perpendicular to the given line.

11. $2x - 10y = 3$

12. $x + 9y = -3$

13. $4x - y = 13$

14. $7x + 3y = 8$

15. $5x - 9y = 1$

B. Exercises

Use the point-slope form to write the slope-intercept form of the equation of the line parallel to the given line and passing through the given point.

16. $4x + y = 9$; $(-2, 5)$

17. $x + y = -4$; $(7, 0)$

18. $2x + 7y = 4$; $(6, 2)$

19. $x - 9y = -1$; $(-1, -8)$

20. $6x + 5y = -8$; $(-4, 10)$

Use the point-slope form to write the slope-intercept form of the equation of the line perpendicular to the given line and passing through the given point.

21. $x + y = 8$; $(0, 4)$

22. $3x - y = 7$; $(-6, 1)$

23. $x = 8$; $(2, 5)$

24. $y = 9x + 2$; $(8, -3)$

25. $5x + 7y = -12$; $(3, 0)$

Determine whether the graphs of each pair of equations are parallel, perpendicular, or neither.

26. $-4x + 6y = -24$
$2x - 3y = -15$

27. $x - 5y = -5$
$5x - y = -6$

28. $3x - y = 1$
$x + 3y = -15$

29. $5x - 9y = -9$
$18x + 10y = -40$

30. $18x - 3y = 3$
$6x - y = -5$

Write the equations of each pair of lines and determine whether they are parallel, perpendicular, or neither.

31.

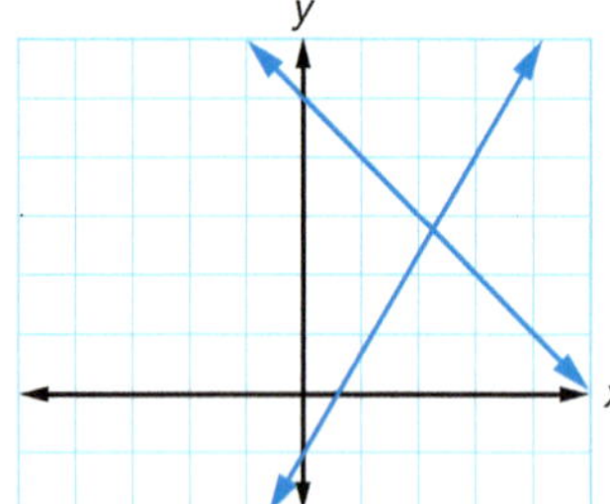

32.

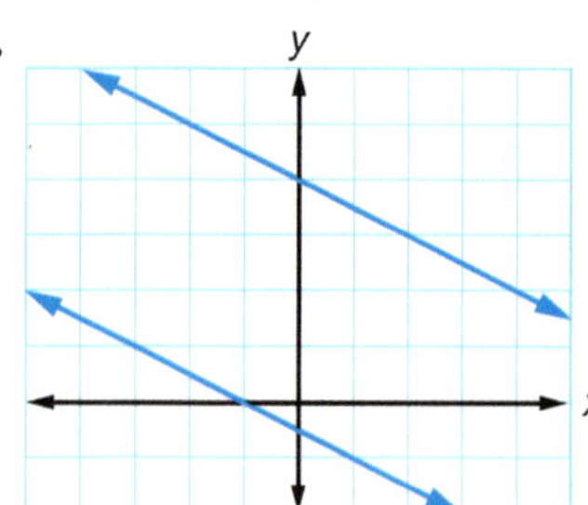

33.

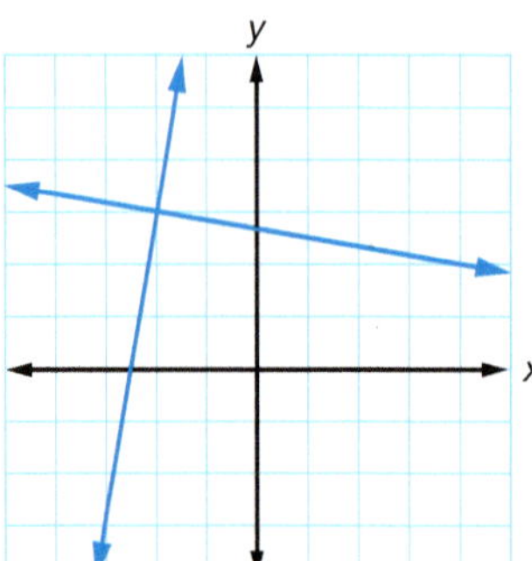

34.

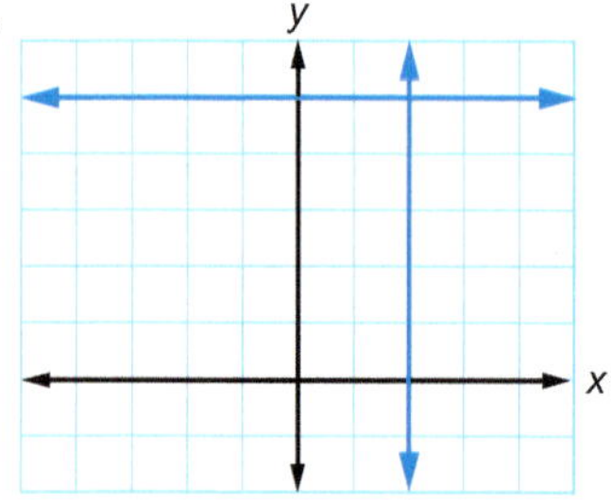

C. Exercises

35. Write the standard-form equation of the line perpendicular to $4x + 3y = 1$ and passing through its y-intercept.

36. Write the standard-form equation of the line perpendicular to $x + 7y = 4$ and passing through its x-intercept.

37. If $3x + 7y = 2$ and $Ax + 2y = -9$ are parallel lines, what is the value of A?

38. If $6x + By = 3$ and $8x + y = 5$ are perpendicular lines, what is the value of B?

The vertices of a square are $A\ (2, 2)$, $B\ (-1, -2)$, $C\ (-5, 1)$, and $D\ (-2, 5)$.

39. Find the slope of each side of the square. Use the slopes of the sides to show that the opposite sides are parallel and the consecutive sides are perpendicular.

40. Find the slope of each diagonal of the square (the segment connecting opposite vertices). What relationship do the diagonals of the square have?

41. A triangle's vertices are $R\ (-6, 2)$, $S\ (3, -5)$, and $T\ (4, -1)$. Use the slopes of the triangle's sides to determine whether it is a right triangle. Justify your conclusion.

Dominion Modeling

Solomon's temple contained a spiral staircase. First Kings 6:8 states, "They went up with winding stairs into the middle chamber, and out of the middle into the third." Winding staircases are constructed to save space. Building any set of stairs can be mathematically challenging and extremely frustrating if errors are made.

The *tread* is the horizontal portion we step on, and the *riser* is the vertical portion of a step. Building codes regulate these dimensions to encourage safe stairways.

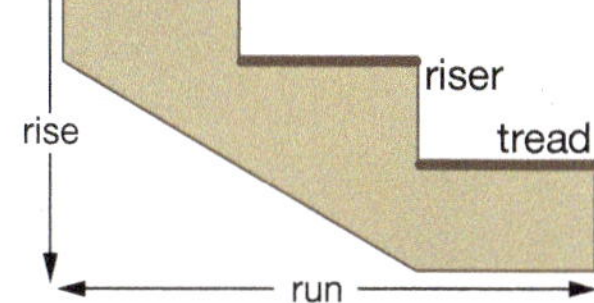

Use the following fixed-stairs building code.

Riser: 6.5–9.5 in. Tread: 8–11 in. Slope: maximum 1.2

Brad's crew flew to St. Thomas to help rebuild after severe hurricane damage. They constructed concrete stairs to one of the new buildings. The plan proposed a rise of 118 in. and a run of 150 in.

42. Determine the slope of the planned stairs and state whether the plan meets code.

43. How many risers are needed if Brad chooses a riser height as close to 7.5 in. as possible?

44. The building code requires all risers to be the same height. Using the results of exercise 43, determine the height of each riser.

45. Using the results of exercise 43, determine the width of each tread. Does this width meet code?

CUMULATIVE REVIEW

Evaluate when $x = -4$, $y = 7$, and $z = -\frac{1}{3}$. [2.2]

46. $3x^2 - 2xy + 5z$

47. $-\frac{3xy^2}{7z}$

Simplify. [2.3]

48. $3(2x^2 - 4x + 8) - (2x^2 - x + 3)$

49. $3(12x - 8y) - 2(x + 6y)$

Translate each sentence into a compound inequality. [4.4–4.5]

50. A number is at least -2 and not more than 5.

51. A number is not less than 7, or it is less than 3.

Solve and graph the solution on a number line. [4.7]

52. $|0.5x + 3| < 15$

53. $|2x + 3| \geq 12$

Write an equation and solve. [3.7]

54. A plane flies west from Wichita at 350 knots at the same time another plane flies east from the same airport at 430 knots. How long (to the nearest minute) will it take for the planes to be 1500 nautical miles apart?

55. Joshua leaves home for Memphis at 8:00 AM, but his sister Savannah cannot leave until 9:30 AM. If she travels, on average, 10 mi/hr faster than Joshua and catches up to him at 5:30 PM, how fast (to the nearest mile per hour) was each car traveling?

SEQUENCES

Converging and Diverging Sequences

Challenge

State the integer that each sequence is getting closer and closer to as *n* gets larger and larger.

a. $\frac{1}{2}, \frac{1}{4}, \frac{1}{8}, \frac{1}{16}, \ldots$

b. $\frac{1}{2}, \frac{2}{3}, \frac{3}{4}, \frac{4}{5}, \ldots$

c. $\frac{1}{3}, \frac{4}{7}, \frac{3}{4}, \frac{8}{9}, \ldots$

A sequence A_n *converges* if the terms get closer and closer to some finite number as n gets larger and larger. All three examples above are converging sequences. In the first sequence the terms approach zero as n gets larger. In the second sequence the terms are getting closer to one.

The third sequence is more challenging. First rewrite the sequence in nonsimplified form as $\frac{2}{6}, \frac{4}{7}, \frac{6}{8}, \frac{8}{9}, \ldots$ to reveal the patterns in the numerators and denominators of the terms. The patterns make it clear that the fifth term is $\frac{10}{10}$ and the sixth term is $\frac{12}{11}$, so the answer is not one. Notice that the general-term formula is $\frac{2n}{n+5}$, so $A_{10} = \frac{20}{15}$, $A_{100} = \frac{200}{105}$, and $A_{1000} = \frac{2000}{1005}$. It is now obvious that the terms of the third sequence are getting closer to two as n gets larger. (For many sequences, finding the general-term formula and then checking the hundredth or thousandth term is much better than simply considering the first several terms.)

If a sequence does not converge, then it *diverges*. The sequences $-2, -7, -12, -17, -22, \ldots$ and $1, 3, 9, 27, 81, \ldots$ both diverge. All arithmetic sequences diverge since the terms approach ∞ (infinity) or $-\infty$. Geometric sequences sometimes converge and sometimes diverge.

Although converging sequences may seem irrelevant and impractical to you right now, the concept is crucial in the development and application of calculus, the most important branch of higher mathematics.

Exercises

State the finite number that each sequence is getting closer and closer to as *n* gets larger and larger.

1. $1, \frac{1}{4}, \frac{1}{9}, \frac{1}{16}, \ldots$

2. $-\frac{3}{2}, -\frac{4}{3}, -\frac{5}{4}, -\frac{6}{5}, \ldots$

3. $\frac{3}{4}, \frac{2}{3}, \frac{5}{8}, \frac{3}{5}, \ldots$

4. $\frac{3}{2}, 2, \frac{9}{4}, \frac{12}{5}, \ldots$

Classify the following sequences as converging or diverging.

5. $3, 10, 17, 24, 31, 38, \ldots$

6. $8, -4, 2, -1, \frac{1}{2}, -\frac{1}{4}, \ldots$

7. $1, 1, 2, 3, 5, 8, \ldots$

8. $1, 2, \frac{3}{2}, \frac{5}{3}, \frac{8}{5}, \frac{13}{8}, \ldots$

9. Geometric sequences diverge if the common ratio is greater than what number (i.e., $r > x$)?

10. Geometric sequences diverge if the common ratio is less than what number (i.e., $r < x$)?

6.6 Trend Lines and Correlation

Scatterplots are used to illustrate the relationship between two groups of data. While these points rarely lie on a straight line, a *trend line* can often be drawn on a scatterplot to show the relationship more clearly. This trend line is drawn as close to as many of the points as possible. Though the exact method of drawing a trend line is quite involved, a carefully drawn estimate can produce useful results. Graphing calculators and spreadsheets can find the exact equation of the *line of best fit*.

You have already seen how a scatterplot illustrates data, such as that in the table below listing the sales revenues of an electronics company. It can also be used to write a linear function modeling the data.

Year	Sales Revenues
1966	$5,000,000
1971	$6,000,000
1976	$5,000,000
1988	$11,000,000
2000	$15,000,000
2001	$16,000,000
2004	$17,000,000

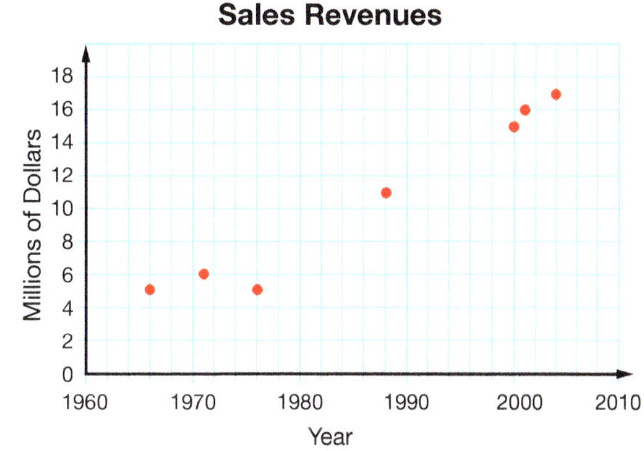

Example 1

Draw and determine the equation of a trend line that models the sales revenues.

Answer

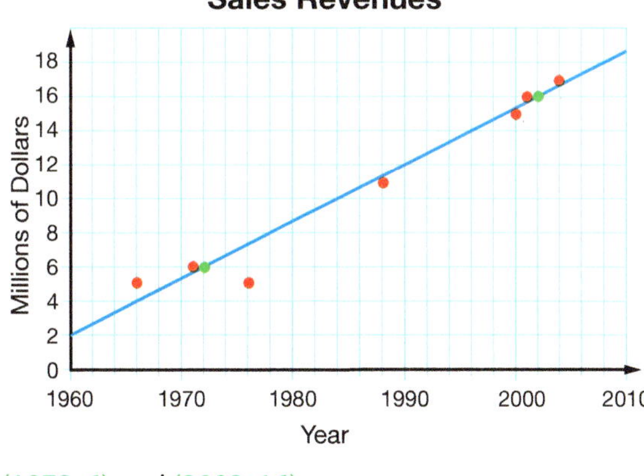

1. Carefully estimate and draw a trend line that by visual inspection minimizes the distances to the data points.

(1972, 6) and (2002, 16)

2. Choose two points on the trend line.

3. Find the equation of the line passing through those two points.

$m = \frac{y_2 - y_1}{x_2 - x_1} = \frac{16 - 6}{2002 - 1972} = \frac{10}{30} = \frac{1}{3}$

$y = mx + b$

$6 = \frac{1}{3}(1972) + b$

$b = \frac{18}{3} - \frac{1972}{3} = -\frac{1954}{3} = -651.\overline{3}$

$y = \frac{1}{3}x - \frac{1954}{3}$ or $y = 0.\overline{3}x - 651.\overline{3}$

You can use the equation of a trend line to make predictions related to the data. These predictions are just estimates. *Interpolation* is the process of making predictions within the range of the existing data points, while *extrapolation* makes predictions outside this range. Interpolation tends to result in more accurate predictions than extrapolation.

Example 2

Use the results of Example 1 to estimate the sales revenues for 1995 and predict the year in which the company will have sales of $24 million. Which prediction is more likely to be accurate?

Answer

$y = \frac{1}{3}x - \frac{1954}{3}$

$y = \frac{1}{3}(1995) - \frac{1954}{3} = \frac{1995 - 1954}{3}$

$y = \frac{41}{3} = 13.\overline{6}$

1. Substitute 1995 for x in the trend-line equation and evaluate.

The model estimates sales of $14 million in 1995.

$y = \frac{1}{3}x - \frac{1954}{3}$

$24 = \frac{1}{3}x - \frac{1954}{3}$

$72 = x - 1954$

$x = 2026$

2. Substitute 24 for y in the trend-line equation. Solve for x after clearing the equation of fractions.

The model predicts sales of $24 million in 2026.

Interpolating $14 million in sales in 1995 is likely more accurate than extrapolating $24 million in sales in 2026.

The *linear correlation* of the data describes how closely the data points are clustered along the trend line. When the majority of the points are close to the line, there is a strong correlation. Data with a trend line from lower left to upper right has a positive correlation, whereas data with a trend line from upper left to lower right has a negative correlation. Trend lines from data with stronger correlations are more likely to produce accurate predictions.

strong positive correlation

little or no correlation

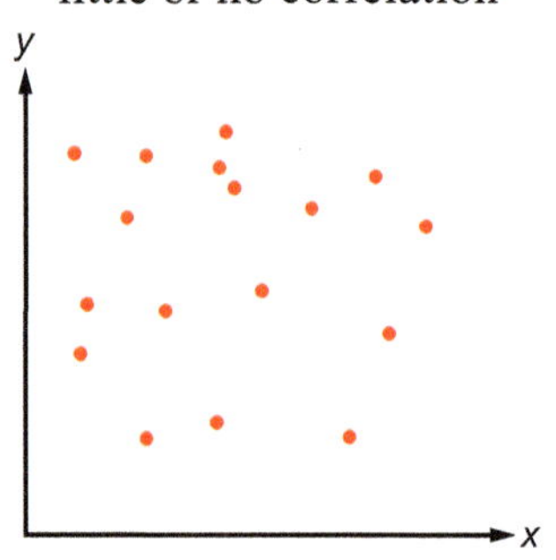

strong negative correlation

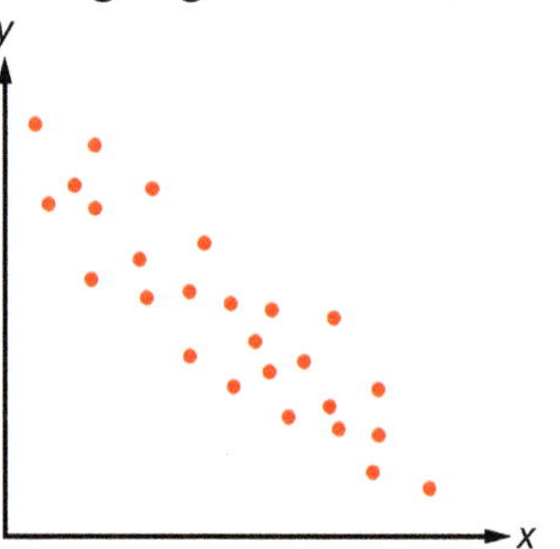

weak positive correlation

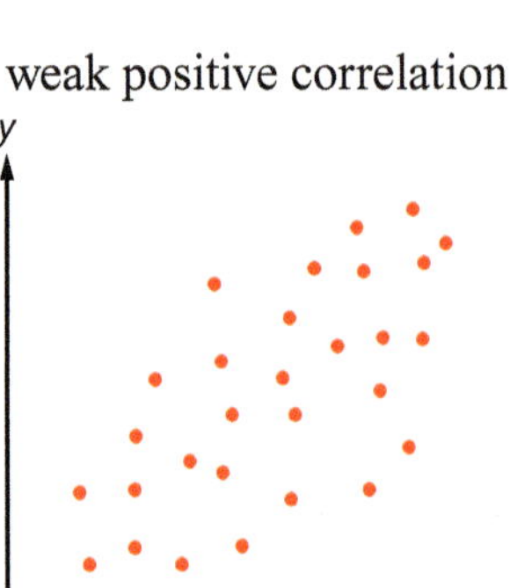

weak negative correlation

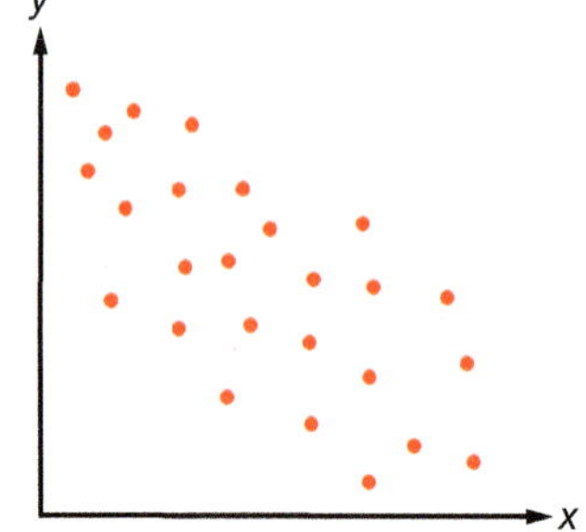

TECHNOLOGY CORNER (TI-84+ Family)

You can use the STAT menu of your graphing calculator to produce a scatterplot of a data set and determine the line of best fit.

Pressing STAT ENTER selects the first option, Edit, which allows you to enter x values under L1 and y values under L2. Use ENTER and the arrow keys, ▶, ▼, ▲, and ◀, to enter the data from Example 1 in this section.

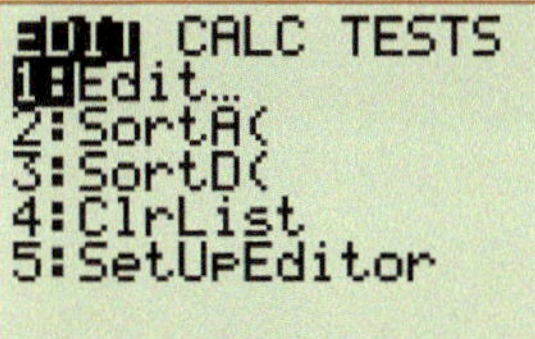

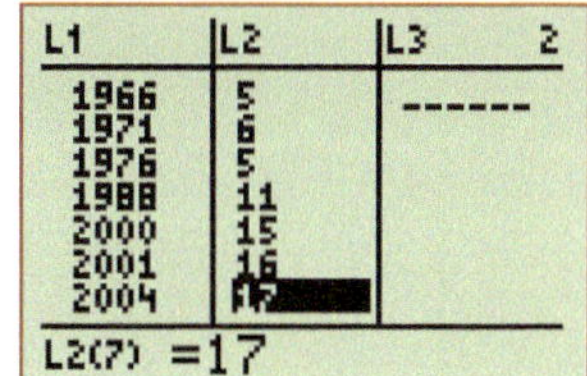

The STAT PLOTS menu is used to define the data's scatterplot. Press Y= and use CLEAR to erase any functions that have been entered previously. Then make sure all plots are off by selecting [STAT PLOTS] (2nd Y=) 4 ENTER, which returns the user to the home screen.

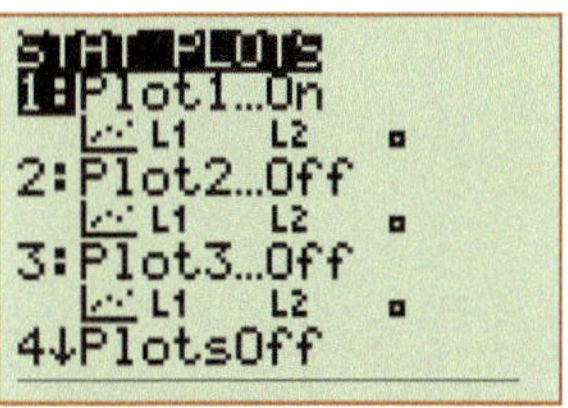

Select [STAT PLOTS] again and then hit ENTER to select Plot1. Hit ENTER to turn Plot1 on. The default option for Type: is the scatterplot, so leave it as is. Be sure the Xlist: and Ylist: options are set to their default values of L_1 (2nd 1) and L_2 (2nd 2) respectively. The Mark: option allows the user to choose how to display the plotted data points.

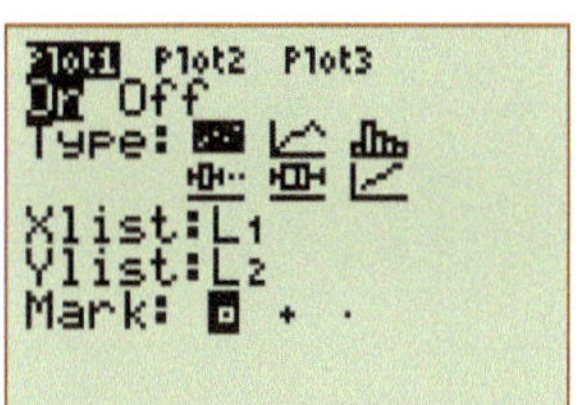

Press ZOOM 9 to select ZoomStat. This function draws the scatterplot and automatically sizes the window to include all the data points. WINDOW can be used to view and adjust the scale and range of the axes.

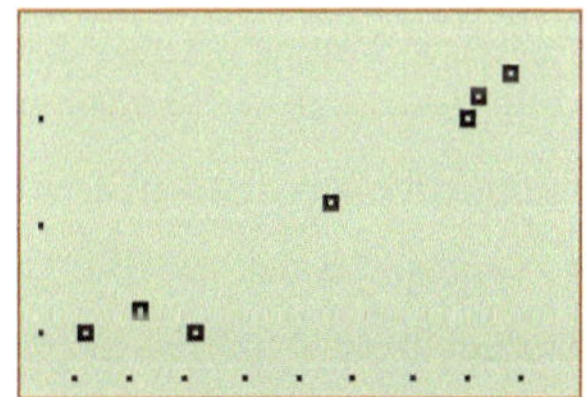

The CALC option of the STAT menu is used to determine the equation of the line of best fit. Select 4:LinReg(ax+b) to place the linear regression command on the home screen. Selecting ENTER produces values that can be used to write the equation in slope-intercept form: $y = 0.34x - 657.65$. Your calculator's guidebook contains many other options that can be explored.

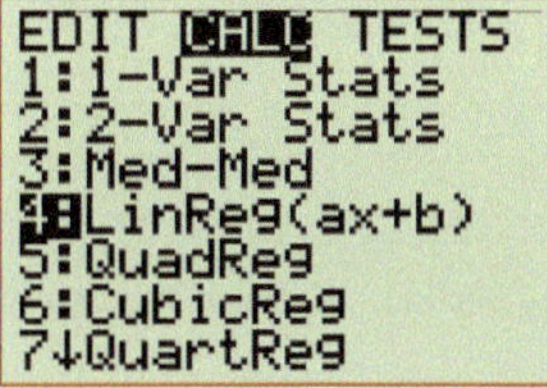

LinReg
y=ax+b
a=.3364395263
b=-657.6468647

A. Exercises

Matching: Describe the linear correlation of each data set.

a. strong positive correlation **b.** weak positive correlation **c.** little or no correlation

d. weak negative correlation **e.** strong negative correlation

1.

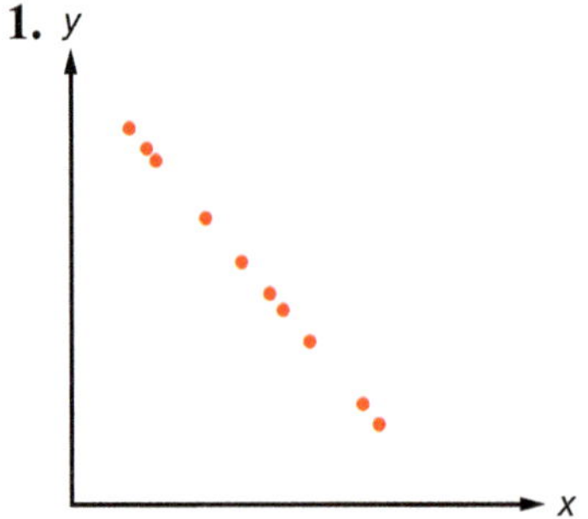

2.

3.

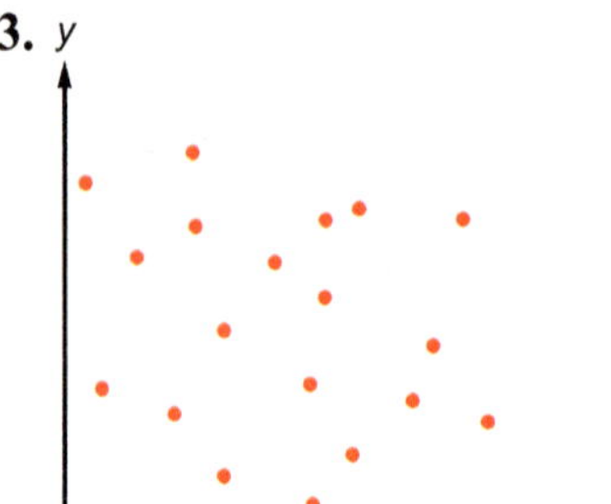

4.

5.

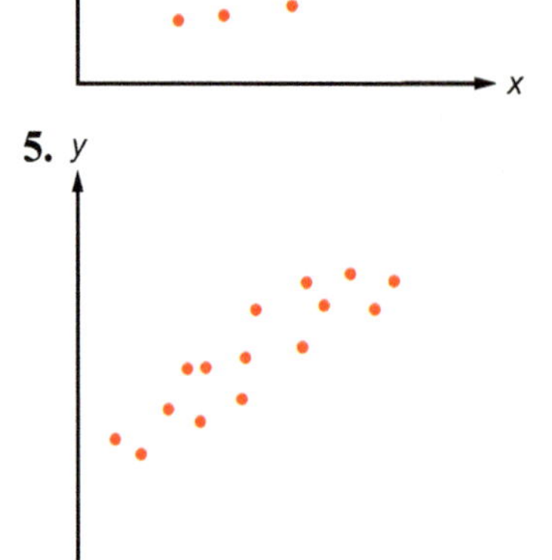

Use the following set of data for exercises 6–10.

x	y
1	27.5
4	35
7	42.5
10	50
13	57.5

6. Make a scatterplot of the data and describe the linear correlation of the data.

7. Draw and determine the equation of a trend line that models the data set.

8. Use your equation to predict the value of y when $x = 3.5$.

9. Use your equation to predict the value of y when $x = 20$.

10. Does the correlation of the data suggest predictable results?

Use the following set of data for exercises 11–15.

x	y
1	7
2	8
3	6
4	3
5	3
6	1

11. Make a scatterplot of the data and describe the linear correlation of the data.

12. Draw and determine the equation of a trend line that models the data set.

13. Use your equation to predict the value of y when $x = 3.5$.

14. Use your equation to predict the value of y when $x = 17$.

15. Does the correlation of the data suggest predictable results?

B. Exercises

Use the table listing the final history and English grades of eight students for exercises 16–20.

History Grade	English Grade
87	93
75	84
81	78
91	86
69	71
86	88
74	78
95	97

16. Make a scatterplot of the data using the x-axis for history grades and the y-axis for English grades.

17. Draw and determine the equation of a trend line that models the data set.

18. Use your equation to predict the English grade of a student who earned an 83 in history.

19. Use your equation to predict the history grade of a student who earned an 83 in English.

20. Describe the correlation of the data. Does a high grade in one class cause a high grade in the other class? Explain.

Use the table giving the population of Rhode Island for exercises 21–25.

Year	Population of Rhode Island (1000s)
1920	604
1930	687
1940	713
1950	792
1960	859
1970	945
1980	947
1990	1003

21. Make a scatterplot of the data and describe the linear correlation of the data.

22. Draw and determine the equation of a trend line that models the data set.

23. Use your equation to predict the population of Rhode Island in 2020.

24. Use your equation to predict when the population of Rhode Island was zero.

25. Which of these predictions is more likely to be accurate? Explain.

Use the table listing the size (by the radius of the wheels) and the weight of mountain bikes for exercises 26–30.

Wheel Size (in.)	Weight (lb)
19	32
20	30
18	31
21	31
15	29
19	27
17	30

26. Make a scatterplot of the data using the x-axis for wheel size and the y-axis for weight.

27. Draw and determine the equation of a trend line that models the data set.

28. Use your equation to predict the weight of a 16" bike.

29. Use your equation to predict the wheel size of a bike weighing 31 lb.

30. Describe the correlation of the data. What does this correlation imply about the accuracy of the predictions made by the equation modeling the data?

C. Exercises

Use the table listing the speeds and stopping distances of a car for exercises 31–35.

Stopping Distance (ft)	Speed (mi/hr)
16	10
59	25
80	30
154	45
183	50
216	55
251	60

31. Which quantity should be defined as the independent variable? Which is the dependent variable?

32. Make a scatterplot of the data. Would a line or a curve best model the data? By visual inspection, sketch a trend line or curve to model the data.

33. Make a table listing the speed squared for each stopping distance. Then make a scatterplot using the x-axis for speed squared and the y-axis for stopping distance.

34. Draw a trend line on the scatterplot made in the previous exercise and determine its equation. Since the x-axis represents speed squared, v^2, and the y-axis represents stopping distance, d, write the final form of this equation as $d = mv^2 + b$ instead of $y = mx + b$.

35. Use your equation to predict the stopping distances at 40 mi/hr and at 70 mi/hr.

CUMULATIVE REVIEW

Complete each of the following set operations. [1.1]

36. $\mathbb{Q} \cup \mathbb{Q}'$

37. $\mathbb{Z} \cap \mathbb{W}$

Solve. [2.4–2.6]

38. $\frac{2x}{3} = 4.5$

39. $7x = -31$

40. $12x + 11 = 56$

41. $\frac{5}{2}x - 27 = 48$

42. $3(2x - 9) = 4x + 5$

43. $9 - (4x - 29) = 2(x + 7)$

State whether each of the following represents a direct variation, an inverse variation, or neither. [5.6]

44. $y = \frac{2}{x}$

45. $y = 3.5x$

6.7 Graphing Linear Inequalities

Linear inequalities are similar to linear equations but contain an inequality sign instead of an equal sign. Remember that inequalities use the signs $\neq$, $<$, $>$, $\geq$, or $\leq$ and have an infinite number of solutions.

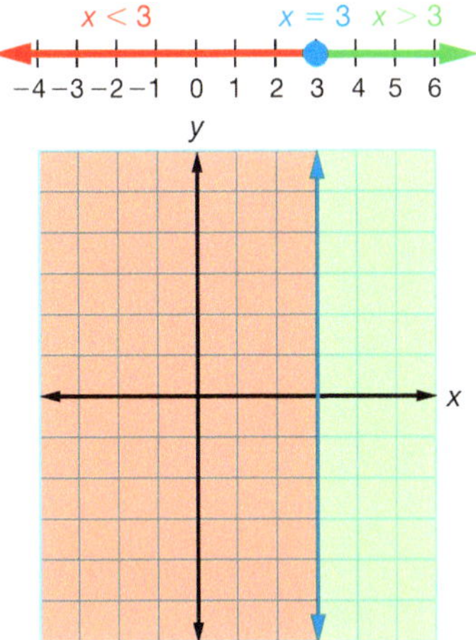

The graph of $x = 3$ on a number line separates the line into three distinct sets of points.

The graph of the linear equation $x = 3$ similarly separates the coordinate plane into three distinct sets of points: the points on the line $x = 3$; the region to the left of the line, $x < 3$; and the region to the right of the line, $x > 3$.

Any linear equation, such as $2x + y = 3$, divides the coordinate plane into three sets of points: the line itself, which acts as a boundary; the set of points on one side of the line; and the set of points on the other side of the line. The solutions to a linear inequality such as $2x + y > 3$ will lie on one side of the line. Depending upon the type of inequality, the points on the line may or may not be solutions.

Example 1

Graph $2x + y > 3$.

Answer

$2x + y > 3$

$y > -2x + 3$

$m = -2$

y-intercept: $(0, 3)$

1. Solve the inequality for y in order to identify the slope and y-intercept of the corresponding linear equation $y = -2x + 3$.

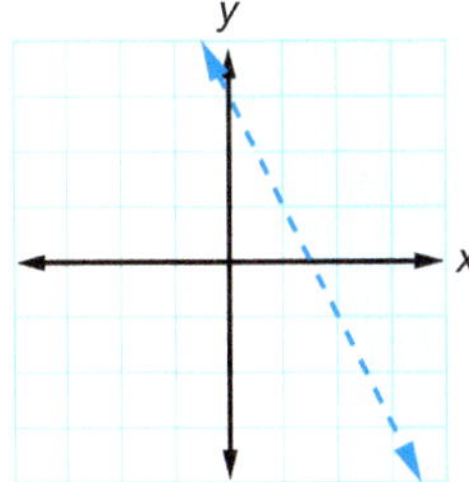

2. Since $y > -2x + 3$ is an inequality, not an equation, the solution set does not include the points on the line. Graph the equation $y = -2x + 3$ with a dashed line to represent this boundary.
3. Use a convenient point to determine which side of the line contains solutions to the inequality.
   a. The origin, $(0, 0)$, is an easy choice.
   b. Substitute the x and y values of the point into the inequality. Since $2(0) + 0 \not> 3$, the point is not a solution.
   c. Since the test point is not a solution, neither is any other point on that side of the line. The solutions must be on the other side of the line.
4. Shade the side of the line that contains the solution set.

If the point you choose makes the inequality true, all points on that side of the line will also make the inequality true.

Graphing an Inequality
1. Solve for y and identify the slope and y-intercept of the corresponding equality.
2. Graph the corresponding linear equation. a. If the inequality is $<$ or $>$, use a dashed line to indicate that points on the line are not solutions. b. If the inequality is $\leq$ or $\geq$, use a solid line to indicate that points on the line are solutions.
3. Choose a test point that is not on the line and substitute it into the inequality. a. If the resulting inequality is true, shade the side containing the test point. b. If the inequality is false, shade the other side.

Example 2

Graph $6x - 2y \geq -10$.

Answer

$$6x - 2y \geq -10$$
$$-2y \geq -6x - 10$$
$$\frac{-2y}{-2} \leq \frac{-6x}{-2} + \frac{-10}{-2}$$
$$y \leq 3x + 5$$

$m = 3$

y-intercept: (0, 5)

1. Solve the inequality for y and identify the slope and y-intercept. Remember to reverse the inequality sign when multiplying or dividing by a negative number.

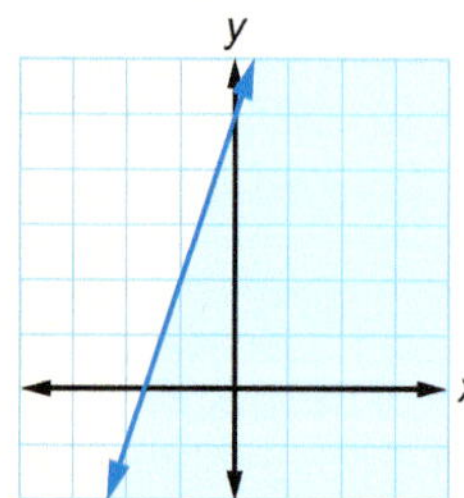

2. Since the inequality sign is $\leq$, graph the equation $y = 3x + 5$ with a solid line.
3. Use a test point, (0, 0), to determine which side of the line contains solutions to the inequality. Since $6(0) - 2(0) \geq -10$, the point (0, 0) and all points on that side of the line are part of the solution set.
4. Shade below the line to indicate the solution set.

As a check, substitute some points in the portion you shaded into the original inequality. Do they make the inequality true?

If the inequality is solved for y, it is easy to determine which side of the line should be shaded. On the y-axis, larger values of y are higher and smaller values of y are lower. Therefore, if $y > mx + b$, shade above the line; and if $y < mx + b$, shade below the line. In Example 2, $y \leq 3x + 5$ indicates that the solution is all the points on and below the line.

Linear inequalities are often used to model real-life situations in which there are acceptable ranges of values. The domain and range of these applications are often limited to nonnegative numbers.

Example 3

Jessica is on a strict weight-loss diet and wants to limit her total caloric intake from carbohydrates and fats to 720 calories each day (she gets the rest of her calories from protein). Each gram of carbohydrates contains 4 calories, and each gram of fats contains 9 calories. Write and graph a linear inequality that models her plan. List several acceptable amounts for daily consumption of carbohydrates and fats.

Answer

Let x = grams of carbohydrates and y = grams of fats.

1. Assign variables to the unknowns.

total calories ≤ 720

calories from carbohydrates $= 4x$

calories from fat $= 9y$

$4x + 9y \leq 720$

2. Write an inequality describing the limited quantity.

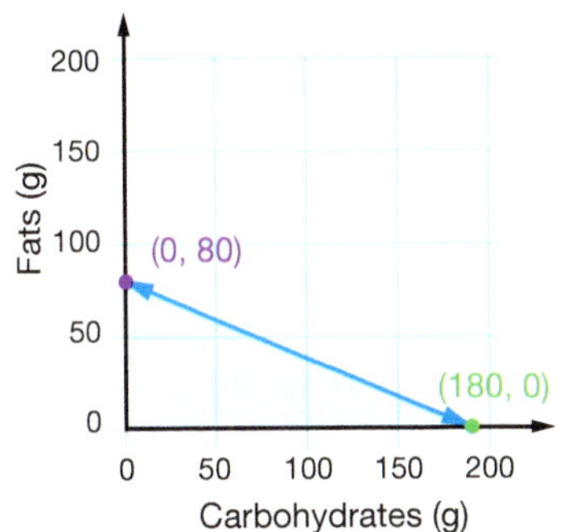

3. Graph the corresponding equation $4x + 9y = 720$ with a solid line using the x-intercept (180, 0) and the y-intercept (0, 80).
4. The point (0, 0) is clearly a solution, so shade below the line to include this point.
5. Theoretically, Jessica could stay within her plan by eating no carbohydrates or fats; or she could eat 180 g of carbohydrates and no fats or no carbohydrates and 80 g of fats. A point within the shaded region, such as (100, 30), provides a more likely (and healthful) diet of 100 g of carbohydrates and 30 g of fats. Other acceptable amounts include 115 g of carbohydrates and 25 g of fats or 130 g of carbohydrates and 20 g of fats.

A. Exercises

State whether a solid or a dashed line is used in the graph of each inequality.

1. $y \geq -5x + 3$ **2.** $y < -7x - 9$ **3.** $2x - 13y > 8$ **4.** $5x - y \leq 1$

Determine which ordered pairs are solutions to the given inequality.

5. $y < 2x - 9$; $A\,(0, 0)$, $B\,(4, 7)$, $C\,(-2, -13)$, $D\,(3, -5)$

6. $y \geq 3x + 4$; $F\,(0, 0)$, $G\,(-5, -7)$, $H\,(2, -5)$, $J\,(0, 4)$

7. $2x - 5y > 1$; $K\,(0, 0)$, $L\,(3, 1)$, $M\,(-2, -5)$, $N\,(-7, 3)$

8. $7x - y \leq 1$; $P\,(0, 0)$, $Q\,(5, 8)$, $R\,(-1, 0)$, $S\,(-8, -3)$

Multiple choice: Determine which side of the line you should shade.

9. $-5x + 8y \geq 7$; (0, 0)

10. $y < 2x + 4$; (0, 0)

11. $y \leq 5x - 6$; (3, 1)

12. $8x - y \geq 4$; (−4, 3)

a. the side that includes the given test point

b. the side that is opposite the given test point

Match each inequality with its graph.

a.

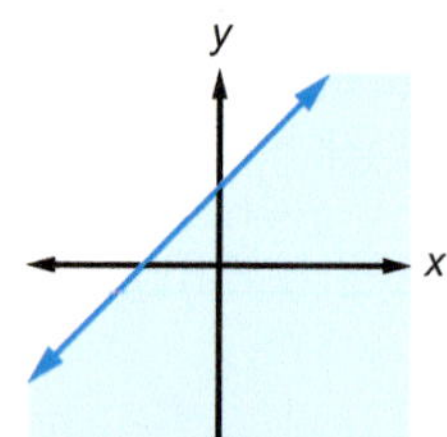

b.

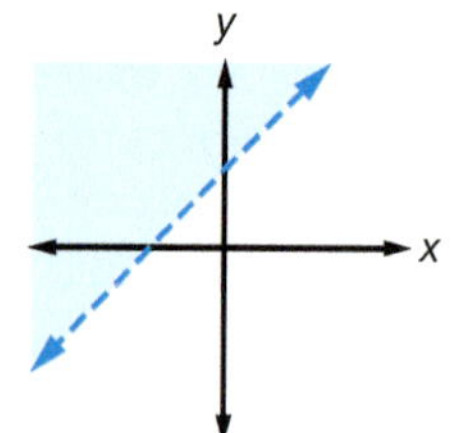

c.

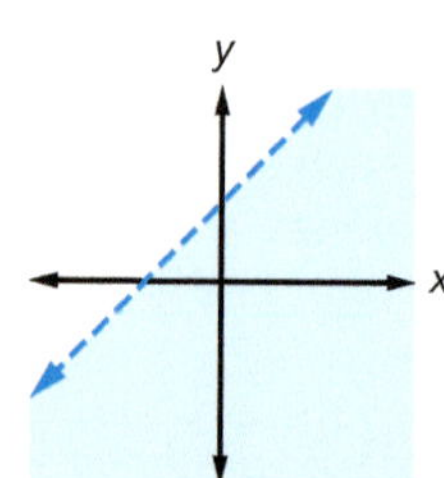

d.

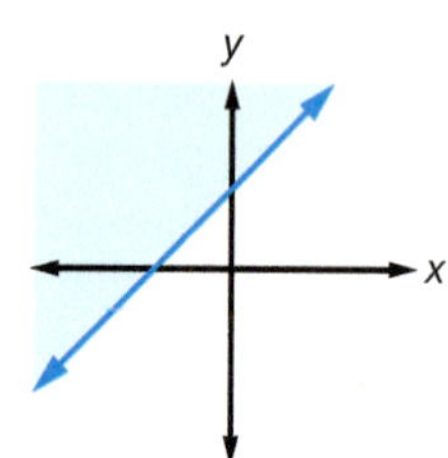

13. $y > x + 2$
14. $y < x + 2$
15. $y \leq x + 2$
16. $y \geq x + 2$

B. Exercises

Graph each inequality.

17. $x < -4$
18. $y \geq 5$
19. $y > 3x + 2$
20. $y \geq 2x - 6$
21. $2x - y \leq 4$
22. $3x - 2y \geq 6$
23. $x + y > -2$
24. $2x - y > -9$
25. $x + 3y \leq 12$
26. $x - 4y \geq 16$
27. $3y < 7$
28. $-x \leq -2$

Write a linear inequality describing each graph.

29.

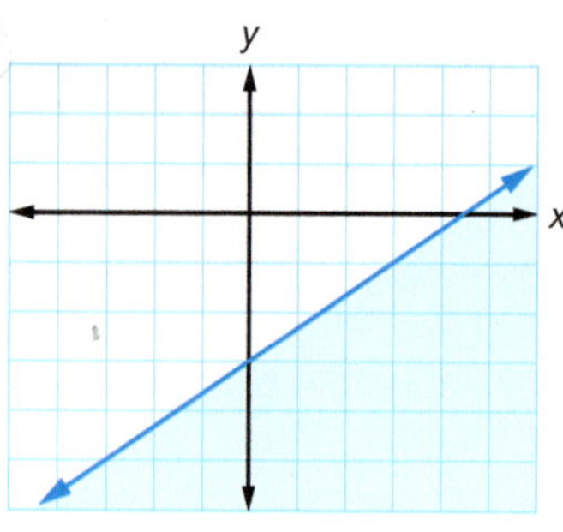

30.

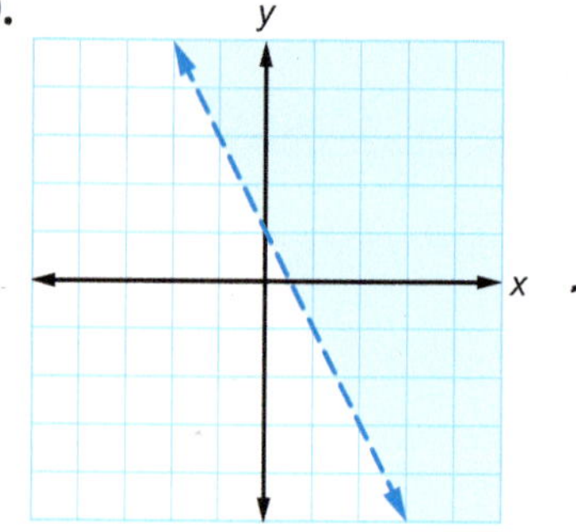

31. How many solutions are there to a linear inequality?
32. Are linear inequalities relations? Are they functions?

Translate each statement into a linear inequality; then graph the inequality.

33. A number, y, is at most five times another number, x.
34. The sum of two numbers, x and y, is at least five.

Write and graph a linear inequality to model each situation.

35. Amber is making a party mix of mixed nuts and chocolate candies for her sister's wedding reception. She has a budget of \$60 and can buy bags of mixed nuts for \$4/lb and bags of chocolate candies for \$3/lb. Use x to represent the number of pounds of nuts and y for the number of pounds of candy.
36. Trey earns a commission of \$25 on each computer and \$5 on each calculator that he sells. He needs to make at least \$325 in commission this week. Use x to represent the number of computers sold and y for the number of calculators sold.

C. Exercises

Graph each conjunction or disjunction, shading the region that satisfies it.

37. $x > 1$ and $x \leq 4$
38. $y \geq 2x + 3$ or $y \leq 2x - 1$

39. Amber's sister suggested that the party mix described in exercise 35 would be too salty if it exceeded a 3 : 1 ratio of nuts to candy and too sweet if it had less than a 1 : 1 ratio. Write two direct variation function rules modeling these ratios and add their graphs to the graph of the inequality from exercise 35. Then state three reasonable combinations of nuts and candies for the wedding reception.

40. Last year's sales records indicate that Trey's calculator-to-computer sales ratio ranged from 2 : 1 to 4 : 1. Write two direct variation function rules modeling these ratios and add their graphs to the graph of the inequality from exercise 36. Then state three reasonable combinations of calculator and computer sales that would allow Trey to meet his goal.

Dominion Modeling

The parable of the Good Samaritan in Luke 10 teaches us to show compassion when others experience times of need. Brad depended on his math skills, his knowledge of construction, his crew, and his relationship with God to minister to the needs of many people. Brad and his crew demonstrated the love of God by using the abilities they had received from God to help others (1 John 3:17).

The installation of wheelchair ramps helps people in need by allowing greater accessibility for those with limited mobility. Codes commonly limit wheelchair ramps to a maximum slope of 1 : 12.

Brad's crew wants to construct a wheelchair access ramp at the home of an elderly couple in their church. Michael takes measurements and finds that the distance from their front porch to their driveway is 40 ft and the floor of the porch is 32 in. high.

41. Express the code's maximum ramp slope in decimal form.

42. Using Tan^{-1}, express the code's maximum ramp slope to the nearest degree.

43. Find the slope ratio of Michael's proposed 40 ft ramp. Would this ramp meet code?

44. What is the shortest ramp they can construct and remain in code?

CUMULATIVE REVIEW

Evaluate. [1.8]

45. $3 - 4(1 - 8)^2 - \frac{18}{6}$

46. $1 - 3(4 - 6)^3$

Solve. [2.8]

47. $\frac{3}{8}x - \frac{5}{9} = \frac{4}{3}$

48. $3.5x - 12 = 5.01$

Solve each literal equation for *p*. [3.1]

49. $S = pH + 2B$

50. $S = \frac{1}{2}pl + B$

Use proportions to solve. [3.2–3.3]

51. If the number of bushels of grain harvested is directly proportional to the number of acres planted, and 18 acres produced 414 bu of grain, how many bushels could be produced on 640 acres?

52. If the number of cars sold is directly proportional to the number of salesmen, and three salesmen sold eleven cars, how many could five salesmen have sold? Round to the nearest whole number.

53. If a blueprint's scale is 1 in. : 25 ft, how many feet are represented by $8\frac{7}{8}$ in. (to the nearest foot)?

54. Two rectangles are similar. If one is 6 cm by 8 cm and the short side of the other is 13.9 cm, what is the length of its other side (to the nearest tenth of a centimeter)?

CHAPTER 6 REVIEW

Solve each linear equation for *y*.

1. $5x - y = 6$
2. $3x - 5y = 5$

Make a table of at least three ordered pairs; then graph each linear equation.

3. $\{(x, y) \mid y = x - 3\}$
4. $x - 5y = 6$

Find the *x*- and *y*-intercepts of each linear equation; then use them to graph the equation.

5. $x - 4y = 4$
6. $3x - 2y = 8$

Find the slope of each line.

7.

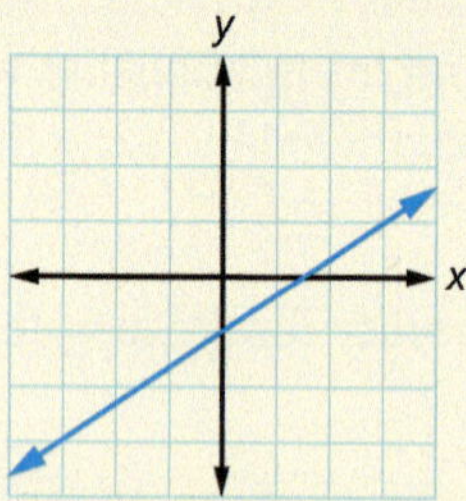

8.

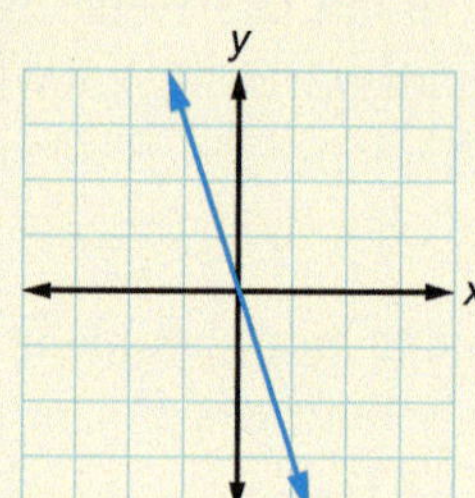

Find the slope of the line passing through the given points.

9. (5, 9), (3, 2)
10. (−5, 6), (4, 2)
11. (5, 9), (3, 9)
12. The slope of every vertical line is ____.
13. Graph the line passing through the point (−4, 2) with a slope of $-\frac{3}{4}$.

State the slope and *y*-intercept of each linear equation.

14. $y = \frac{5}{6}x - 3$
15. $x = 4$

Graph each equation using its slope and *y*-intercept.

16. $y = \frac{2}{3}x + 2$
17. $3x + y = -6$

Write each equation as a linear function; then graph the function.

18. $5x + 4y = 8$
19. $2x - 3y = 9$

Write the equation of each line in slope-intercept form.

20. slope of $\frac{5}{2}$ and *y*-intercept of (0, −1)
21. slope of 2 and *y*-intercept of (0, 3)
22. passing through (1, 4) with a slope of $-\frac{1}{2}$
23. passing through (−3, −2) with a slope of $\frac{3}{2}$
24. passing through (−2, 3) with a slope of 0
25. passing through (−4, 6) and (2, 3)
26. passing through (3, −7) and (5, −1)
27. passing through (4, −3) and (7, −3)
28. passing through (0, 0) and (4, 3)
29. parallel to $2x - 7y = 12$ and passing through (3, 4)
30. perpendicular to $4x + y = 7$ and passing through (5, 4)
31. Write the standard-form equation of the graphed line.

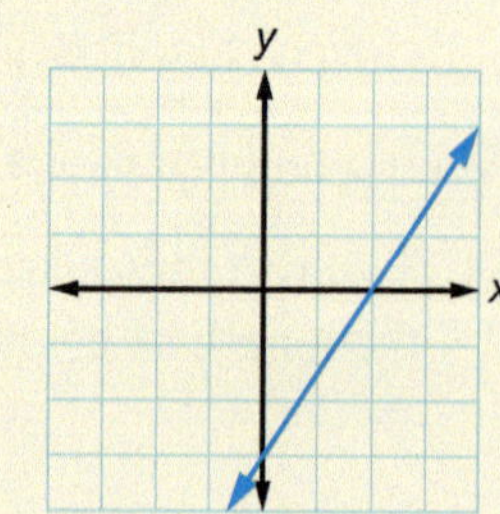

32. Write the standard-form equation of the line parallel to $2x + y = 1$ and passing through (−2, 6).

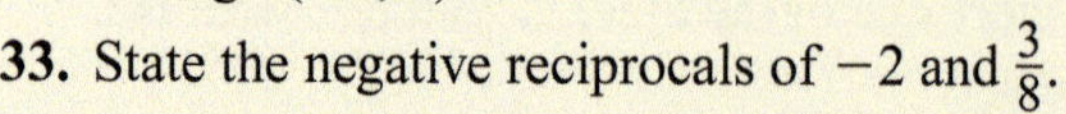

33. State the negative reciprocals of −2 and $\frac{3}{8}$.

Determine whether the graphs of each pair of equations are parallel, perpendicular, or neither.

34. $3x + y = 8$
$-2x + 6y = 18$

35. $4x + 7y = 12$
$5x + 9y = 18$

Find the slopes of each pair of lines and determine whether they are parallel, perpendicular, or neither.

36.

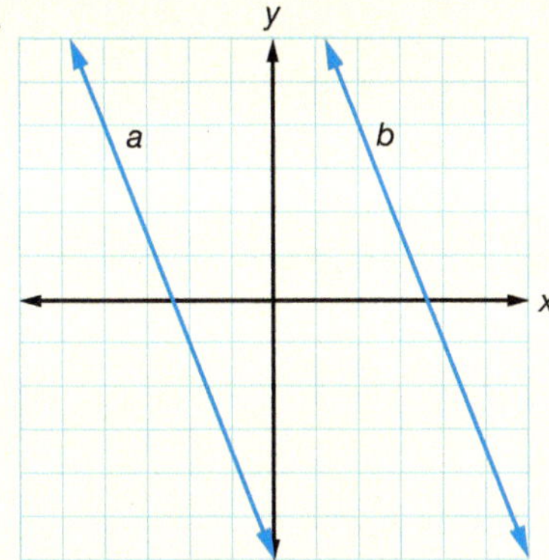

37.

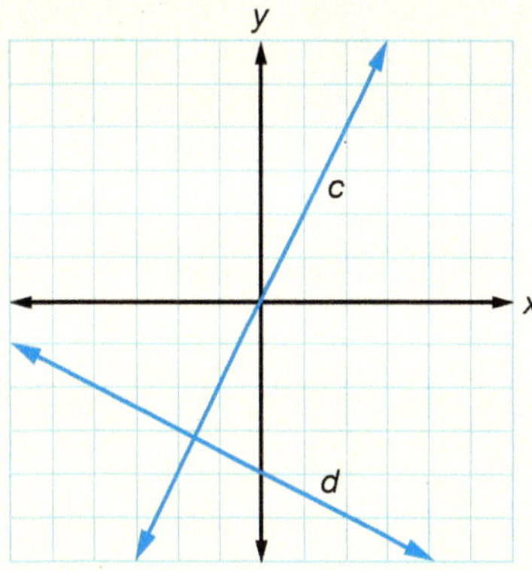

38. Which of the following ordered pairs are solutions to the inequality $y < 3x - 11$?
(0, 0), (−3, −19), (6, 8), (5, 3)

39. The graph of $y > mx + b$ is drawn by graphing $y = mx + b$ with a (*solid* or *dashed*) line and shading the region (*above* or *below*) the line.

Graph each inequality.

40. $y > -2x + 6$

41. $5x - 3y \geq 10$

42. Explain why a linear inequality cannot be a function.

43. A spring that is normally 8 in. long stretches $\frac{1}{4}$ in. per pound of applied force. Write a linear function, f, modeling the length of the spring based on the applied force, x. How many pounds of force are required to stretch the spring to a total length of $13\frac{1}{2}$ in.?

44. The US national debt at the beginning of 2008 was about $9.5 trillion, and at the beginning of 2010 it was about $12.8 trillion. Write a linear function, d, to predict the national debt for any year, x. Then predict the debt at the beginning of 2020. Explain why you think your linear function is or is not a good model for the national debt.

The fork length of a shark is the distance from the tip of the nose to the fork in the tail. Use the following set of data for exercises 45–48.

Fork Length (cm)	Weight (kg)
188	92
197	116
169	59
180	77
158	45
117	22
136	39

45. Make a scatterplot of the data using the x-axis for fork length and the y-axis for weight. Describe the linear correlation of the data.

46. Draw and determine the equation of a trend line that models the data set.

47. Use your equation to predict the weight of a 160 cm shark.

48. Use your equation to predict the fork length of a shark weighing 130 kg.

49. The predictions made from trend lines tend to be most accurate when the data exhibits which type of correlation?

a. positive **b.** negative **c.** strong **d.** weak **e.** little or none

50. True or false: Predictions made by interpolation tend to be more accurate than those made by extrapolation.

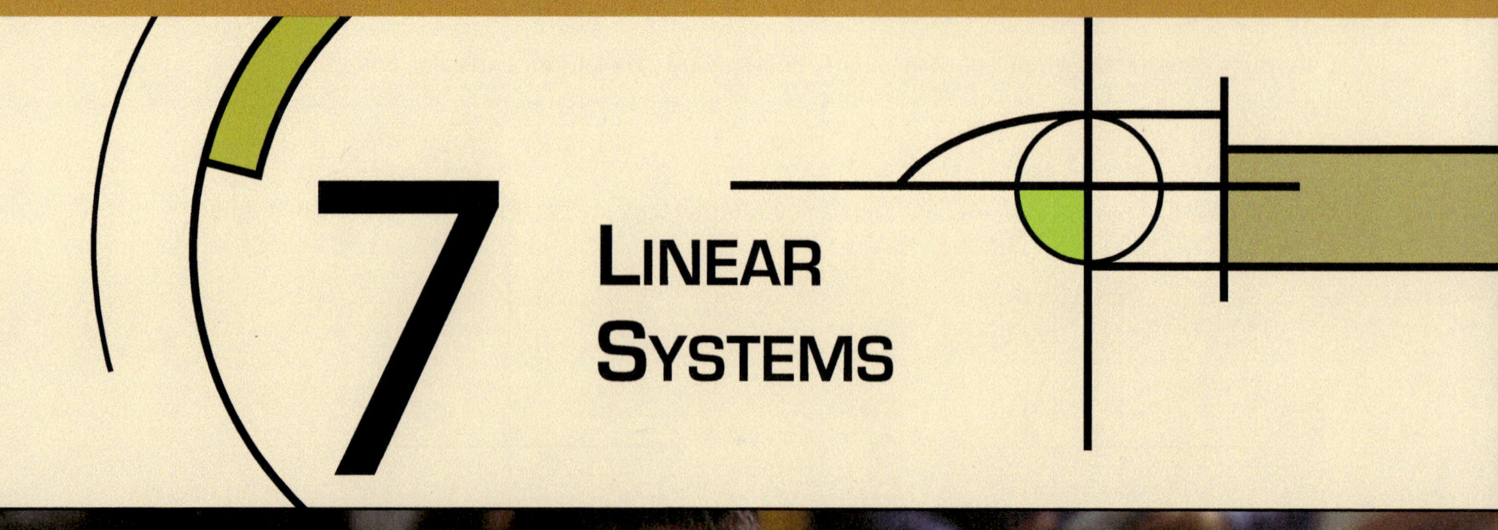

7 Linear Systems

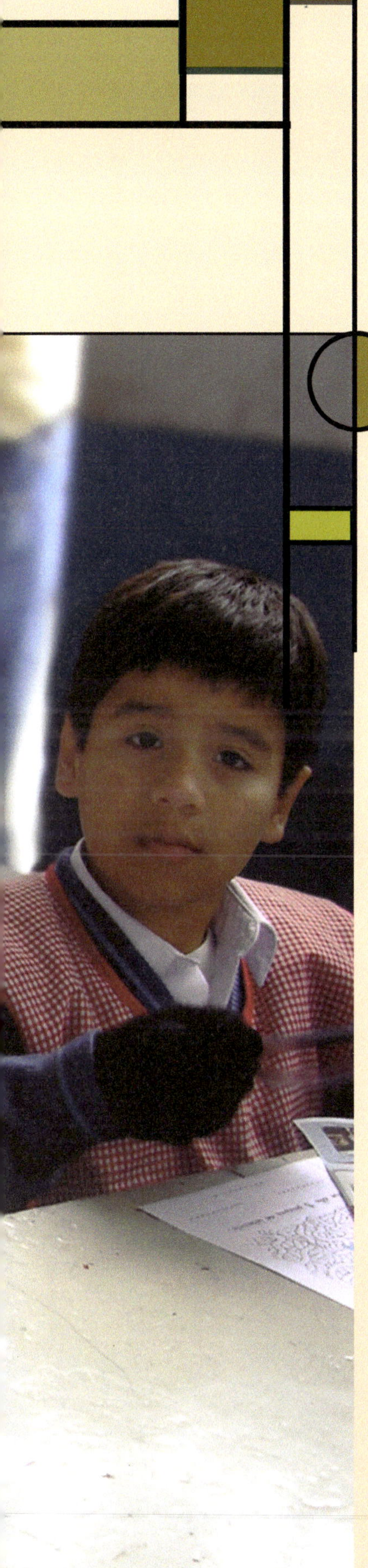

As the populations of Spanish-speaking countries continue to grow, missionaries look for more ways to reach the people in their communities. Although many of these countries are poor, it is amazing to observe people's ability to obtain technology. Their homes may not have running water, but many have electricity and access to radio/CD players.

Mark Ring had a vision. Could a series of children's Bible stories—which clearly presents the gospel story from the Fall of man to Christ's provision for man's redemption—be produced as a CD in Spanish? As he shared his idea with missionaries in Peru, they thought that the CDs could be distributed to children in schools. Since then the CDs have also been given to the general public at large community events and even door to door. With church information provided on the CD jackets, these CDs have proved to be a valuable tool for missions.

While this project had great potential to advance the cause of Christ among the Spanish-speaking population, much planning was necessary to successfully complete it. The ability to solve the linear systems studied in this chapter enabled Mr. Ring to make wise decisions during this process. As you will see, mathematics played a crucial role in fulfilling his vision.

After this chapter you should be able to

1. solve a system of linear equations by using three different methods: graphing, substitution, and elimination.
2. classify a system of equations as consistent or inconsistent, dependent or independent.
3. solve word problems using systems of equations.
4. solve a system of inequalities.

7.1 Graphing Systems of Equations

Up to this point, your work in algebra has consisted of working with one equation at a time. You have solved one-variable equations, such as $3x - 9 = 18$. You have solved literal equations, such as $2l + 2w = 24$ for either l or w. You have also graphed equations in two variables, such as $x - 2y = 8$.

In this chapter you will be working with more than one equation at a time. A *system of linear equations* is made up of two or more linear equations. Any ordered pair that makes all of the equations true is a solution to the system of equations.

The slope of a side of this pyramid, El Castillo at Chichén Itzá, Mexico, is greater than the slope of the staircase on that side.

Example 1

Determine whether each ordered pair is a solution to the system below.

$x - y = 9$
$2x + y = 9$

a. (11, 2)
b. (6, −3)

Answer

a. Substitute 11 for x and 2 for y in both equations.

$$x - y = 9 \qquad 2x + y = 9$$
$$11 - 2 \stackrel{?}{=} 9 \qquad 2(11) + 2 \stackrel{?}{=} 9$$
$$9 \stackrel{?}{=} 9;\text{ true} \qquad 24 \stackrel{?}{=} 9;\text{ false}$$

(11, 2) is not a solution since it does not satisfy both equations.

b. Substitute 6 for x and −3 for y in both equations.

$$x - y = 9 \qquad 2x + y = 9$$
$$6 - (-3) \stackrel{?}{=} 9 \qquad 2(6) + (-3) \stackrel{?}{=} 9$$
$$9 \stackrel{?}{=} 9;\text{ true} \qquad 9 \stackrel{?}{=} 9;\text{ true}$$

(6, −3) is a solution since it satisfies both equations.

The ordered pair (6, −3) is the only point that satisfies both of these equations.

The Old Testament prophets wrote about Jesus the Messiah hundreds of years before He was born. In Micah 5:2 we read that Messiah would be born in Bethlehem. Isaiah 53:7 indicates that He would not answer His accusers. Isaiah 35:4–6 discusses the power of Messiah to heal the blind, deaf, and lame. These and many other prophecies that needed to be satisfied are similar to the equations of a system. Many men would satisfy any one condition; but when you consider all of the prophecies mentioned in over three hundred passages, they all point to just one man, Jesus the Messiah.

With the simplest systems, some people can intuitively determine the ordered pair that satisfies both equations. As systems get more challenging, you need an organized approach to efficiently solve systems.

In this section you will graph both equations on the same coordinate plane. Since the solution to the system is normally an ordered pair that satisfies both equations, the answer will be the intersection of the two line graphs.

Solving a System by Graphing

1. Graph both equations on the same coordinate plane.
2. Identify the solution, the coordinates of the point of intersection.
3. Check the solution by substituting it into each equation.

Example 2

Solve the system by graphing.

$y = -x - 2$

$y = x - 4$

Answer $y = -x - 2$ and $y = x - 4$

1. Both equations are already in slope-intercept form.

$m = -1$ | $m = 1$

y-intercept: $(0, -2)$ | y-intercept: $(0, -4)$

2. Determine the slope and y-intercept of each line.
3. Graph both lines on the same coordinate plane.

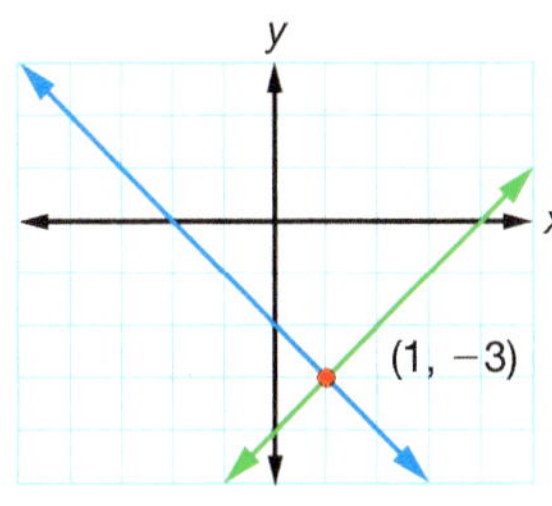

solution: $(1, -3)$

4. Identify the point of intersection.

Check $y = -x - 2$ and $y = x - 4$

$-3 = -1 - 2$ | $-3 = 1 - 4$

5. The solution checks in both equations.

Example 3

Solve the system by graphing.

$x - y = -5$

$3x + y = -7$

Answer $x - y = -5$ and $3x + y = -7$

$-y = -x - 5$ | $y = -3x - 7$

$y = x + 5$

1. Write each equation in slope-intercept form.
2. Graph both lines on the same coordinate plane.

$m = 1$ | $m = -3$

y-intercept: $(0, 5)$ | y-intercept: $(0, -7)$

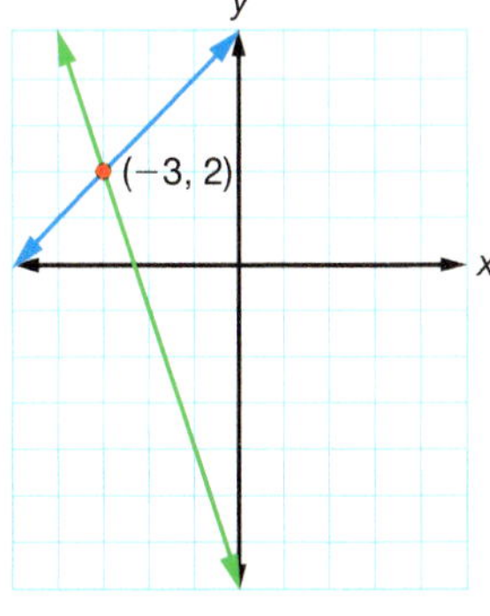

solution: $(-3, 2)$

3. Identify the point of intersection.

Check $x - y = -5$ and $3x + y = -7$

$-3 - 2 = -5$ | $3(-3) + 2 = -7$

4. The solution checks in both equations.

The solution to each system in Examples 2–3 is a single ordered pair where the lines intersect. Some systems, however, contain equations that name parallel lines or the same line. When the graphs are parallel lines, the lines share no common point. Therefore, the system has no solutions. When the graphs are the same line, they are said to *coincide*. The solution is the set of all the points on the line.

Example 4

Solve the systems by graphing.

a. $x + y = 3$
$2y = -2x + 6$

b. $x + y = 3$
$x + y = 7$

Answer

a. $x + y = 3$ and $2y = -2x + 6$

1. Write each equation in slope-intercept form.

$y = -x + 3$ $\quad$ $y = -x + 3$

2. Graph both lines on the same coordinate plane.

$m = -1$, y-intercept: (0, 3) $\quad$ $m = -1$, y-intercept: (0, 3)

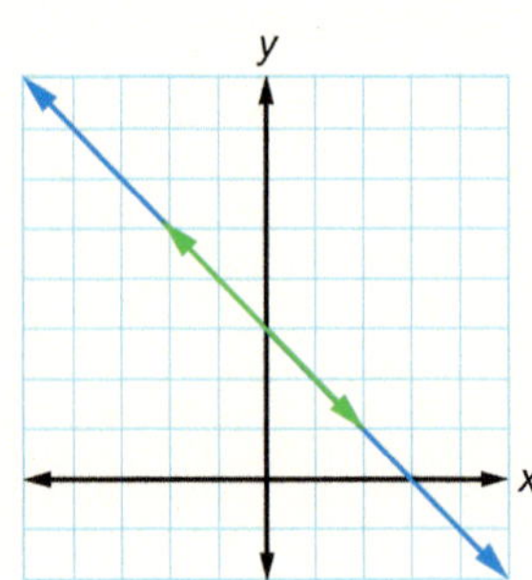

3. Identify the solution.

solution: $y = -x + 3$

b. $x + y = 3$ and $x + y = 7$

1. Write each equation in slope-intercept form.

$y = -x + 3$ $\quad$ $y = -x + 7$

2. Graph both lines on the same coordinate plane.

$m = -1$, y-intercept: (0, 3) $\quad$ $m = -1$, y-intercept: (0, 7)

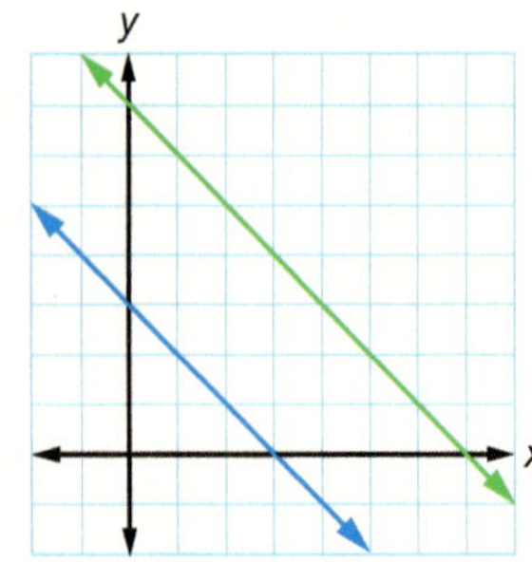

3. Identify the solution.

solution: $\varnothing$

Possible Solutions to a System

	intersecting lines	parallel lines	coinciding lines
Graph			
Number of Solutions	one solution	no solution	infinite solutions
Answer	ordered pair	$\varnothing$	all the points on a line

Example 5

Drew is considering two options for financing a used car. The loan from bank A requires \$3000 down and \$250/mo. The loan from bank B has the same interest rate but requires \$2000 down and \$350/mo. Write a function rule to describe each financing option. Then graph the system to find when Drew will have spent the same amount of money.

Answer

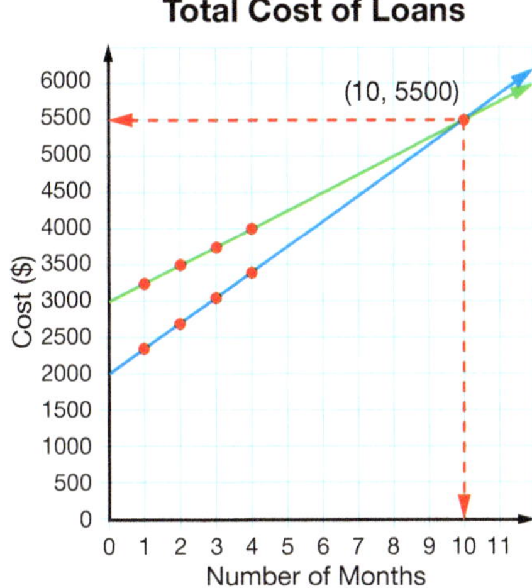

solution: $x = 10$ mo

1. Graph both functions on the same coordinate plane. Since the numbers of months and costs are always positive in both equations, you need to consider the graph only in the first quadrant.
2. Loan A: $a(x) = 250x + 3000$
   Plot the y-intercept, (0, 3000), and use the slope to find several more points: (1, 3250), (2, 3500), and so on.
3. Loan B: $b(x) = 350x + 2000$
   Plot the y-intercept, (0, 2000), and use the slope to find several more points: (1, 2350), (2, 2700), and so on.
4. For both options, Drew will have spent \$5500 at 10 mo. Note that loan B will pay the car off at a faster rate as long as the price of the car is more than \$5500.

A. Exercises

Identify the solution to each sytem from its graph. If the solution is a line, state the line's equation.

1.

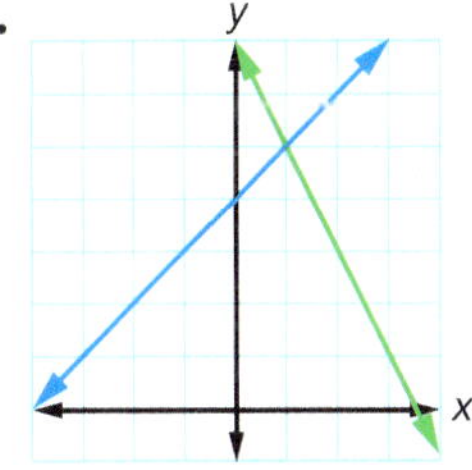

2.

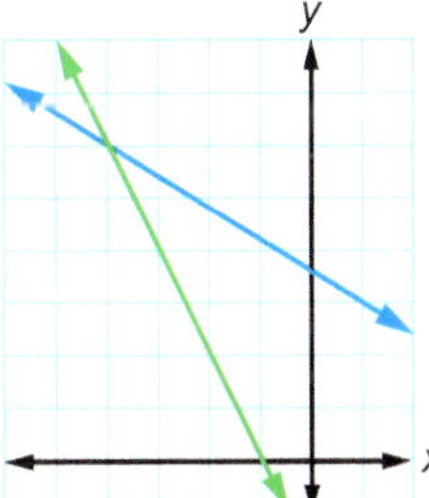

3.

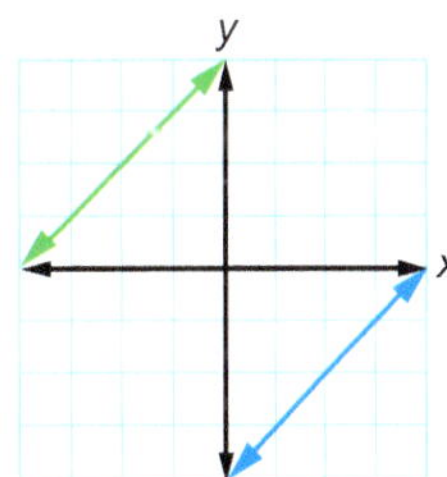

4.

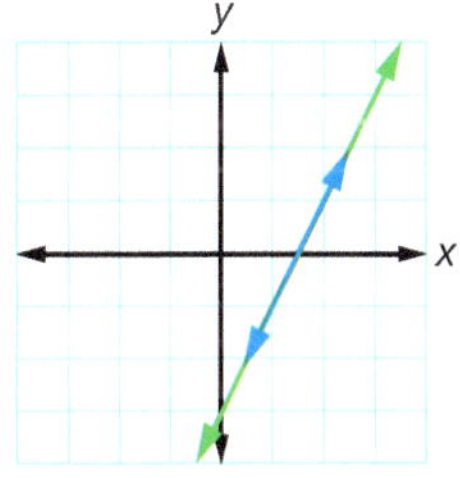

Determine whether (−2, 5) is a solution to each system.

5. $x + y = 3$
$2x + y = 1$

6. $x + 2y = 8$
$3x + y = 1$

7. $x - y = 7$
$2x + 3y = 4$

8. $3x + 2y = 4$
$5x + 3y = 5$

For each system, determine whether the given ordered pair is a solution.

9. $x + 3y = 13$
$6x + y = 7$
$(2, -5)$

10. $2x - 5y = 9$
$x + y = 8$
$(7, 1)$

11. $x + 2y = 1$
$3x - 2y = 5$
$(-3, -2)$

12. $3x - y = -7$
$x - 5y = -21$
$(-1, 4)$

13. $5x + 6y = 0$
$2x + y = 7$
$(6, -5)$

14. $y = -2$
$y = 4x$
$(-2, -8)$

Solve each system by graphing.

15. $y = -x + 5$
$y = x - 1$

16. $y = x + 2$
$y = 4x - 4$

17. $y = -6x + 7$
$y = 1$

18. $y = 3x + 4$
$x = -2$

B. Exercises

Solve each system by graphing.

19. $4x + y = 7$
$x - 3y = 18$

20. $x + y = 9$
$-2x + y = -3$

21. $x + 2y = 10$
$9x + 2y = -6$

22. $8x - 4y = -16$
$4x - y = -8$

23. $x - 5y = -30$
$2x - y = 3$

24. $y = -3x + 5$
$3y = -4x + 15$

25. $2y = -x + 10$
$2y = -3x + 14$

26. $-2y = -x + 4$
$2y = -3x - 12$

27. $5x - 3y = -12$
$2x + 3y = -9$

28. $2x - 7y = 21$
$3x + 7y = 14$

29. $8x - 2y = 6$
$y = 4x + 3$

30. $x + 2y = 8$
$y = -\frac{x}{2} + 4$

31. $x = -3$
$x = 4$

32. $x - y = 6$
$-5y = -5x + 30$

33. The sum of two numbers is -7. The difference of the same two numbers is 3.

a. Write two equations to represent these conditions.

b. Graph the equations and name the point of intersection.

34. Marie's Deli charges a \$45 fee plus \$5 per plate. Just Right Catering charges \$14 per plate.

a. Write a function rule describing the cost of catering for x people using each caterer.

b. Graph the equations to determine the number of people that would result in the same cost. What is that cost?

c. Analyze the graph and determine when Marie's Deli would be cheaper.

35. Miguel sells jackets for \$30 each. He pays \$20 for each jacket and estimates overhead expenses at \$120.

a. Write a function rule, $r(x)$, describing his revenue if he sells x jackets.

b. Write a function rule, $e(x)$, describing his expenses if he buys x jackets and has an overhead of \$120.

c. Graph the equations to determine the number of jackets Miguel must sell to break even (the intersection of the two graphs).

C. Exercises

36. Solve the system by graphing: $9x - 2y = 7$ and $3x + 4y = 14$.

37. Mrs. Seibert is ready to purchase a cell phone for her daughter. Plan A offers 200 min for \$35/mo plus \$0.35/min for additional minutes. Plan B offers 300 min for \$45/mo plus \$0.25/min for additional minutes. Graph the cost of each plan on the same set of axes. Then use the graph to approximate the number of minutes at which the two plans cost the same.

Dominion Modeling

Mr. Ring focused on areas in which the need was the greatest and found missionaries eager to support the Bible story CD project. With half of its population under the age of 21, Mexico is the largest Spanish-speaking country. Consider the populations of the two largest Spanish-speaking countries.

Year	Mexico	Colombia
1950	26,280,000	11,590,000
1960	34,999,000	15,950,000
1970	51,910,000	21,330,000
1980	68,870,000	26,890,000
1990	83,400,000	33,200,000
2000	99,530,000	39,770,000
2010	111,000,000	46,000,000

38. Make a new table with each population rounded to the nearest million.

39. Using two different colors, make a scatterplot of each data set on the same set of axes.

CUMULATIVE REVIEW

Simplify.

40. $\frac{3}{5} + \frac{12}{35} \cdot \frac{49}{9}$ [1.5]

41. $\left|\frac{7}{26} - \frac{14}{39}\right|$ [1.4]

42. $\frac{6^{-2}}{3} \cdot \frac{2^{-1}}{9^{-3}}$ [1.7]

Solve. [2.7]

43. $5x - 11 = 2x + 4$

44. $3(x - 5) = 2(7 - x)$

Solve each literal equation for the indicated variable. [3.1]

45. $3x + 4y = 5$ for x

46. $x + 3y = 21$ for y

Solve.

47. $4(x - 3) \leq 8x$ [4.3]

48. $x \leq 4$ and $3x \geq 2x + 4$ [4.4]

49. $x < 5$ or $3(x + 2) \geq 2x$ [4.5]

TECHNOLOGY CORNER (TI-84+ Family)

You can use your graphing calculator to solve systems of equations. First, you will have to solve the equations for y to enter them into the function editor. After using the MODE menu to make sure that the calculator is in function mode, press Y= and enter the equations as shown. Make sure that all stat plots are turned off. If Plot1, Plot2, or Plot3 is highlighted, arrow up and press ENTER to turn off the stat plot.

Selecting 6:ZStandard from the ZOOM menu graphs the lines in a window with $-10 \leq x \leq 10$ and $-10 \leq y \leq 10$. The intersection point cannot be seen within this window. Choose 3:Zoom Out from the ZOOM menu. A cursor will be flashing at the point from which the calculator will zoom out. This point can be adjusted if needed. Press ENTER a second time.

To find the coordinates of the point of intersection, press 2nd TRACE to access the CALCULATE menu. Choose 5:intersect. It will ask for verification of the first curve and the second curve and will give an option to guess the point of intersection. Press ENTER three times. The solution will be given along the bottom of the screen.

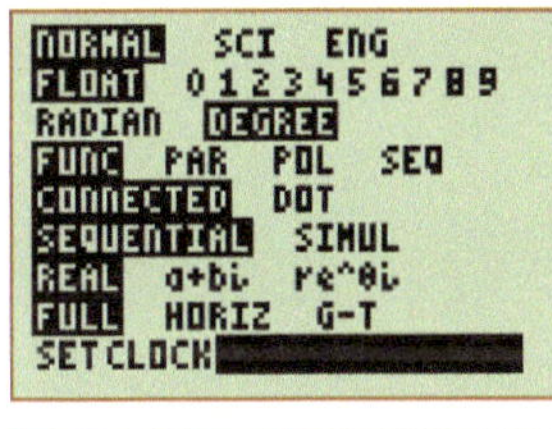

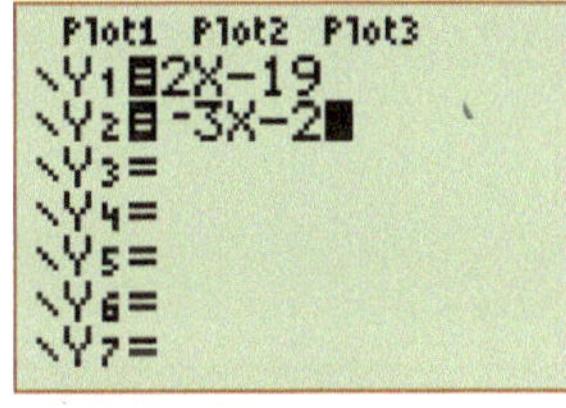

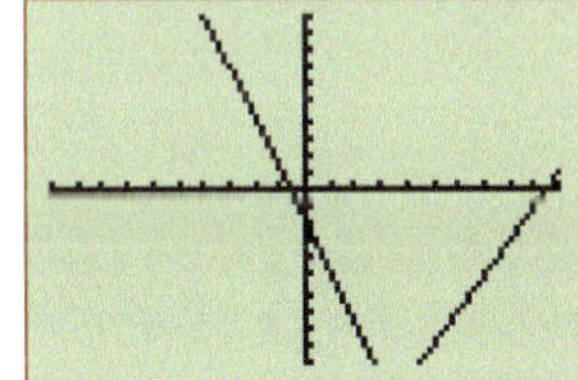

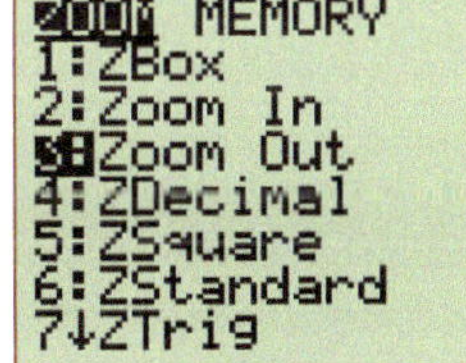

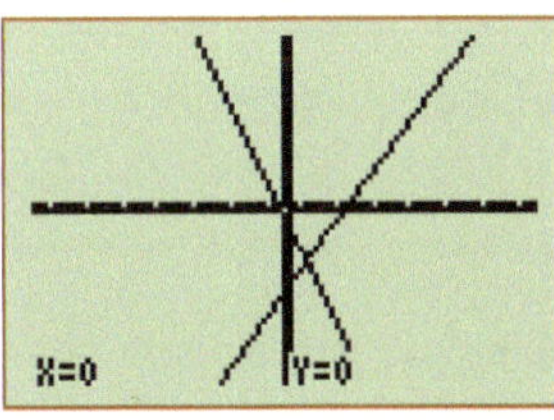

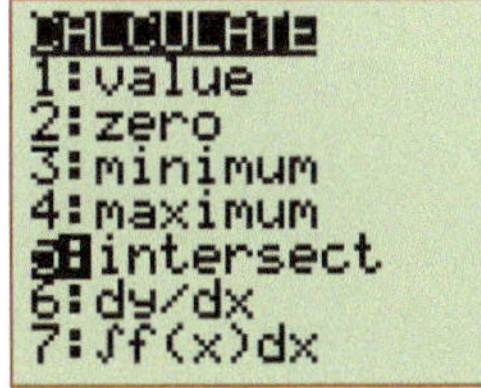

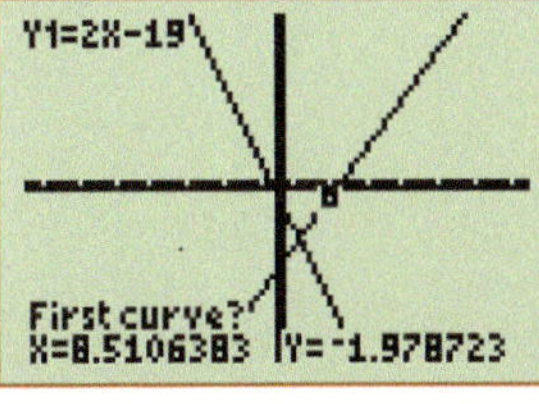

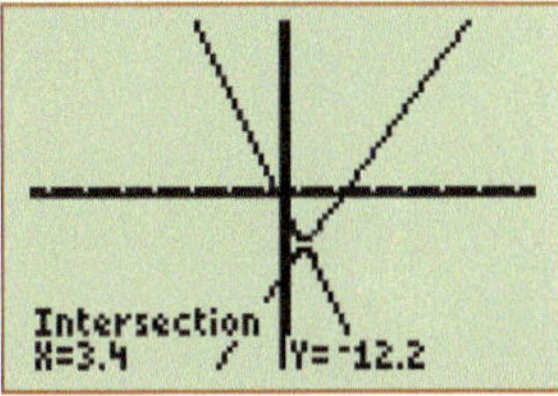

7.2 Solving Simple Systems by Substitution

The graphing method of solving systems of linear equations works well if the coordinates of the solution are small integers, but it becomes awkward when the solutions contain large integers or fractions.

The *substitution method* can be used to find a single variable equation by replacing one of the variables with an equivalent expression. Once this equation is solved, the result is substituted into one of the equations to find the value of the other variable.

Example 1

Solve the system by substitution.

$x + y = 5$

$y = -3x + 7$

Answer $x + y = 5$ and $y = -3x + 7$ — 1. The second equation is already solved for y.

$x + (-3x + 7) = 5$ — 2. Substitute the expression for y in the first equation.

$-2x + 7 = 5$ — 3. Solve for x.

$-2x = -2$

$x = 1$

$y = -3(1) + 7$ — 4. Substitute 1 for x in either equation and solve for y. Using the second equation is simpler.

$y = 4$

solution: (1, 4) — 5. Write the solution as an ordered pair.

Check $x + y = 5$ and $y = -3x + 7$ — 6. The solution checks in both equations.

$1 + 4 = 5$ $\quad 4 = -3(1) + 7$

The graphs of the equations in Example 1 intersect at the point (1, 4).

Example 2

Solve the system by substitution.

$x + y = -4$

$x - y = 10$

Answer $x + y = -4$ and $x - y = 10$ — 1. Solve the first equation for x. (You could choose to solve either equation for either variable.)

$x = -y - 4$

$(-y - 4) - y = 10$ — 2. Substitute the expression for x in the second equation.

$-2y - 4 = 10$ — 3. Solve for y.

$-2y = 14$

$y = -7$

$x = -(-7) - 4$ — 4. Substitute -7 for y in either equation and solve for x.

$x = 3$

solution: (3, −7) — 5. Write the solution as an ordered pair.

Check $x + y = -4$ and $x - y = 10$ — 6. The solution checks in both equations.

$3 + (-7) = -4$ $\quad 3 - (-7) = 10$

Solving a System by Substitution

1. Solve either equation for either variable. Choose the easiest variable to solve for.
2. Substitute the solution into the other equation to obtain an equation in one variable; then solve.
3. Use the value of the known variable to solve for the other variable.
4. Write the solution as an ordered pair.
5. Check the solution by substituting it into each equation.

Example 3

Solve the system by substitution.

$x - 4y = 0$

$3x + 2y = 8$

Answer

$x - 4y = 0$ and $3x + 2y = 8$ — 1. Solve the first equation for x.

$x = 4y$

$3(4y) + 2y = 8$ — 2. Substitute the expression for x in the second equation.

$12y + 2y = 8$ — 3. Solve for y.

$14y = 8$

$y = \frac{8}{14} = \frac{4}{7}$

$x = 4\left(\frac{4}{7}\right)$ — 4. Substitute $\frac{4}{7}$ for y in either equation and solve for x.

$x = \frac{16}{7}$

solution: $\left(\frac{16}{7}, \frac{4}{7}\right)$ — 5. Write the solution as an ordered pair.

Check

$x = 4y$ and $3x + 2y = 8$ — 6. The solution checks in both equations.

$\frac{16}{7} = 4\left(\frac{4}{7}\right)$

$3\left(\frac{16}{7}\right) + 2\left(\frac{4}{7}\right) = 8$

$\frac{48}{7} + \frac{8}{7} = 8$

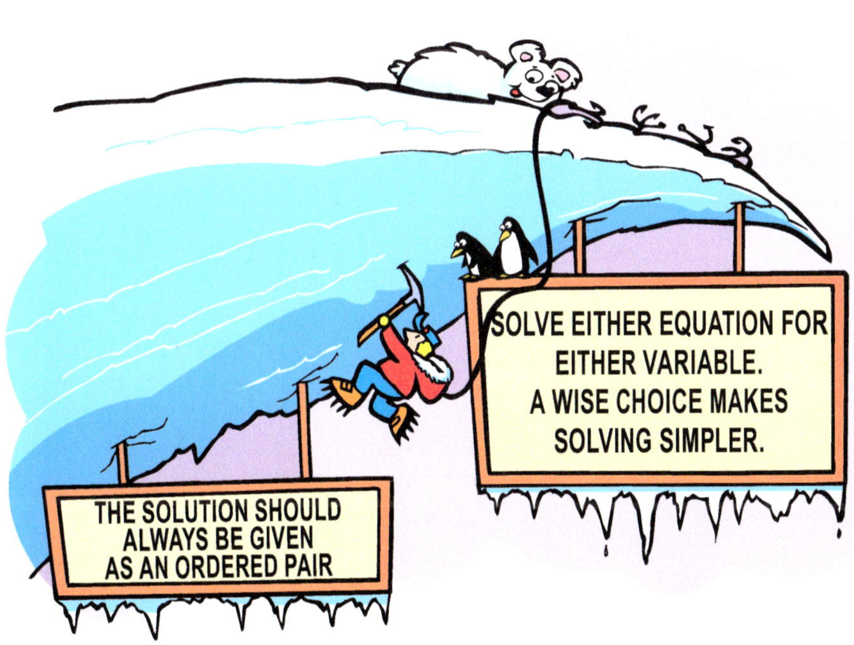

Example 4

Bryan has 84 ft of fencing and wishes to fence a rectangular area so that the length is 6 ft longer than the width. Find the dimensions.

Answer

Let l = the length and w = the width. — 1. Assign variables to the unknowns.

$2l + 2w = 84$ and $l = w + 6$ — 2. Write a system of equations. The perimeter formula is used for the first equation.

$$2(w + 6) + 2w = 84$$
$$2w + 12 + 2w = 84$$
$$4w + 12 = 84$$
$$4w = 72$$

— 3. Substitute the expression for l in the other equation and solve for w.

$$w = 18$$

— 4. The width is 18 ft.

$$l = 18 + 6$$

— 5. Substitute for w to find l.

$$l = 24$$

— 6. The length is 24 ft.

Check

$p = 2l + 2w = 2(24) + 2(18)$
$= 48 + 36 = 84$ — 7. The perimeter is 84 ft.

A. Exercises

Matching: Choose the best initial step in solving each system.

a. Solve the first equation for x.
b. Solve the first equation for y.
c. Solve the second equation for x.
d. Solve the second equation for y.

1. $x + 2y = 11$
$3x + 5y = 17$

2. $3x - 2y = 14$
$2x + y = 11$

3. $2x - y = 5$
$x + 3y = 7$

4. $y - x = 19$
$3x + 2y = 7$

Solve each system by substitution.

5. $y = 3x$
$4x + y = 49$

6. $y = 4x$
$x - 2y = 56$

7. $y = 3x + 2$
$y = 5x - 6$

8. $y = -2x + 8$
$y = 3x + 13$

9. $a - b = 5$
$a + b = 11$

10. $x + y = 5$
$x - y = 3$

11. $x + y = -8$
$x - y = 4$

12. $3c + d = 9$
$c - d = -13$

13. $2x + y = -5$
$x - 2y = -10$

14. $r - s = 12$
$2r + s = 3$

15. $v + w = 11$
$2v - w = 16$

16. $2x - y = -8$
$3x + y = 13$

B. Exercises

Solve each system by substitution.

17. $x + 4y = 30$
$2x + 5y = 36$

18. $x + 2y = 9$
$x + y = 28$

19. $x - 5y = 17$
$2x - 7y = 22$

20. $2x + 3y = 25$
$x - y = -5$

21. $g - 3h = -5$
$2g - 5h = -9$

22. $x + 8y = 23$
$x + 3y = 8$

23. $3x + 5y = 13$
$x + 8y = 36$

24. $13x + 7y = 88$
$x + 8y = -38$

25. $y = \frac{1}{2}x$
$x + 3y = 25$

26. $a = \frac{3}{5}b$
$10a + 3b = 6$

27. $x + 3y = 14$
$x = 2y$

28. $2x - y = -4$
$x + y = -6$

Write a system of equations and solve the system by substitution.

29. The sum of two numbers is 63, and their difference is 13. Find the two numbers.

30. Two numbers total 996. The difference of the larger and twice the smaller is 33. Find the two numbers.

31. The perimeter of a garden is 90 ft. Its width is 7 ft less than its length. Find the dimensions.

32. Picture Perfect Studios charges \$8 per sheet of photos plus a \$15 sitting fee. The Portrait Place charges \$13 per sheet with no sitting fee. For what number of photo sheets will the costs be the same?

33. Thomas needs to make twelve pies on a budget of \$62. If an apple pie costs \$4 to make and a pecan pie costs \$6, how many of each should he plan on making?

C. Exercises

Solve each system by substitution.

34. $2x - 3y = 16$
$5x + 9y = -59$

35. $3x + 5y = 8$
$7x - 2y = 10$

Dominion Modeling

36. Using the scatterplots and data from Section 7.1, Dominion Modeling, draw and write the equations for the trend lines that model the populations for Mexico and Colombia.

37. If the trends continue, will the population of Colombia catch up with that of Mexico? Why?

CUMULATIVE REVIEW

Multiple choice [1.1]

38. Which number is not an integer?
- **a.** -7
- **b.** $\sqrt{4}$
- **c.** $4\frac{1}{2}$
- **d.** 0
- **e.** $\frac{6}{3}$

39. Which number is not rational?
- **a.** -7
- **b.** $\sqrt{5}$
- **c.** $\frac{11}{5}$
- **d.** 0
- **e.** $2.\overline{34}$

Simplify. [2.3]

40. $(2x + 3y + 5) + (-2x + 4y + 9)$

41. $(3x - 4y + 9) - (2x - 4y + 2)$

42. $\frac{1}{2}(2x^2 + 4x + 2)$

43. $\frac{2x^2 + 4x + 2}{2}$

44. $\frac{24x + 51}{3}$

45. $6\left(\frac{x + 7}{3}\right)$

Evaluate when $x = \frac{3}{4}$ and $y = \frac{2}{3}$. [2.2]

46. $8x + 6y$

47. $8x - 12y$

7.3 Solving Advanced Systems by Substitution

In the previous section most systems contained at least one term with a coefficient of one. Several techniques can be used to simplify the process of solving more complex systems with larger coefficients.

Careful examination of an equation may reveal a common factor. Dividing both sides of an equation by the greatest common factor will produce a simplified equation, making the system easier to solve.

Example 1

Solve the system by substitution.
$3x + 7y = 34$
$2x + 4y = 20$

Answer

$3x + 7y = 34$ and $2x + 4y = 20$

$x + 2y = 10$

$x = -2y + 10$

1. Dividing both sides of the second equation by 2 makes it easier to solve for x.

$3(-2y + 10) + 7y = 34$

$-6y + 30 + 7y = 34$

$y + 30 = 34$

$y = 4$

2. Substitute the expression for x in the other equation and solve for y.

$x = -2(4) + 10$

$x = 2$

3. Substitute 4 for y in either equation and solve for x.

solution: (2, 4)

4. Write the solution as an ordered pair.

Check

$3(2) + 7(4) = 34$ and $2(2) + 4(4) = 20$

5. The solution checks in both equations.

You cannot always avoid fractions when solving by substitution. If the coefficient of a term is a factor of the like term in the other equation, solve for that variable. Though the result may contain fractions, the denominators will cancel when the substitution is made into the second equation.

Example 2

Solve the system by substitution.
$2x + 3y = 16$
$4x + 5y = 28$

Answer

$2x + 3y = 16$ and $4x + 5y = 28$

$2x = -3y + 16$

$x = \frac{-3y + 16}{2}$

1. Solve the first equation for x. The $2x$ term was chosen since 2 is a factor of 4 (the coefficient of its like term).

$\overset{2}{\cancel{4}}\left(\frac{-3y + 16}{\cancel{2}}\right) + 5y = 28$

$-6y + 32 + 5y = 28$

$-y + 32 = 28$

$-y = -4$

$y = 4$

2. Substitute the expression for x in the other equation and solve for y.

Example 2, continued

$$x = \frac{-3(4) + 16}{2}$$

$$x = \frac{-12 + 16}{2}$$

$$x = 2$$

3. Substitute 4 for y in either equation and solve for x.

solution: $(2, 4)$

4. Write the solution as an ordered pair.

Check $2(2) + 3(4) = 16$ and $4(2) + 5(4) = 28$

5. The solution checks in both equations.

There is often a clear way to begin solving a system that makes the process simpler. In systems where there is no obvious starting point, you may want to solve for the term with the smallest coefficient.

Example 3

Solve the system by substitution.

$4x + 3y = 3$
$5x + 2y = 1$

Answer $4x + 3y = 3$ and $5x + 2y = 1$

$$2y = -5x + 1$$

$$y = \frac{-5x + 1}{2}$$

1. Solve the second equation for y since 2 is the smallest coefficient.

$$4x + 3\left(\frac{-5x + 1}{2}\right) = 3$$

$$4x + \frac{-15x + 3}{2} = 3$$

$$2\left(4x + \frac{-15x + 3}{2}\right) = 2(3)$$

$$8x - 15x + 3 = 6$$

$$-7x + 3 = 6$$

$$-7x = 3$$

$$x = -\frac{3}{7}$$

2. Substitute the expression for y in the other equation and solve for x.
   a. Distribute the factor of 3 to remove the parentheses.
   b. Clear the fraction by multiplying both sides of the equation by 2.

$$4\left(-\frac{3}{7}\right) + 3y = 3$$

$$-\frac{12}{7} + 3y = 3$$

$$7\left(-\frac{12}{7} + 3y\right) = 7(3)$$

$$-12 + 21y = 21$$

$$21y = 33$$

$$y = \frac{33}{21} = \frac{11}{7}$$

3. Substitute $-\frac{3}{7}$ for x in either equation and solve for y. (When a fractional answer is obtained, it is often simpler to substitute into one of the original equations.)

solution: $\left(-\frac{3}{7}, \frac{11}{7}\right)$

4. Write the solution as an ordered pair.

Check $4\left(-\frac{3}{7}\right) + 3\left(\frac{11}{7}\right) = 3$ and $5\left(-\frac{3}{7}\right) + 2\left(\frac{11}{7}\right) = 1$

5. The solution checks in both equations.

Example 4

The aquarium offers a family discount for five or more people. If the Moore family paid \$82 for two adults and three children and the Smythe family paid \$130 for three adults and five children, what was the discounted price for children and adults?

Answer

Let a = the discounted adult ticket price and c = the discounted child ticket price.

1. Assign variables to the unknowns.

the Moores and the Smythes

$2a + 3c = 82$ and $3a + 5c = 130$

2. Write and solve a system of two equations.

$$2a = -3c + 82$$
$$a = -\frac{3}{2}c + 41$$

a. Solve the first equation for a.

$$3\left(-\frac{3}{2}c + 41\right) + 5c = 130$$
$$-\frac{9}{2}c + 123 + 5c = 130$$
$$2\left(-\frac{9}{2}c + 123 + 5c\right) = 2(130)$$
$$-9c + 246 + 10c = 260$$
$$c + 246 = 260$$
$$c = 14$$

b. Substitute the expression for a in the other equation and solve for c.

$$a = -\frac{3}{2}(14) + 41$$
$$a = -21 + 41$$
$$a = 20$$

c. Substitute 14 for c in the first equation and solve for a.

The discounted adult tickets were \$20 and the discounted child tickets were \$14.

3. State the solution to the problem.

Check

$2(20) + 3(14) = 82$ and $3(20) + 5(14) = 130$

4. The solution checks in both equations.

Many of the word problems solved using tables in Chapter 3 can be solved using two variables. When two variables are used, write a system of two equations to solve the problem.

Example 5

Mr. Peterson invested a total of \$20,000 between two accounts. If the interest rates were 2% and 4.5% and the total interest earned in 2 yr was \$1175, how much was invested in each account?

Answer Let x = the principal for the 2% account and y = the principal for the 4.5% account.

1. Assign variables to the unknowns.

P	r	t	I
x	0.02	2	$0.04x$
y	0.045	2	$0.09y$

2. Make a table using the formula principal × rate × time = interest.

Total principal is \$20,000. Total interest is \$1175.

$x + y = 20{,}000$ $\quad$ $0.04x + 0.09y = 1175$

3. Write and solve a system of two equations.

$y = 20{,}000 - x$

a. Solve the first equation for y.

$$4x + 9y = 117{,}500$$
$$4x + 9(20{,}000 - x) = 117{,}500$$
$$4x + 180{,}000 - 9x = 117{,}500$$
$$-5x + 180{,}000 = 117{,}500$$
$$-5x = -62{,}500$$
$$x = 12{,}500$$

b. Clear the second equation of decimals. Substitute the expression for y and solve for x.

$y = 20{,}000 - 12{,}500$

$y = 7500$

c. Substitute 12,500 for x in the first equation and solve for y.

Mr. Peterson invested \$12,500 at 2% and \$7500 at 4.5%.

4. State the solution to the problem.

Check $0.04(12{,}500) + 0.09(7500) = 500 + 675 = 1175$

5. The total interest earned is \$1175.

A. Exercises

Matching: Choose the best initial step in solving each system. Then explain your reasoning. Do not solve the system.

a. Solve the first equation for x.

b. Solve the first equation for y.

c. Solve the second equation for x.

d. Solve the second equation for y.

1. $2x + 5y = 11$
$3x + 6y = 12$

2. $4x - 2y = 9$
$5x + 6y = 7$

3. $3x + 5y = 4$
$8x - 4y = 20$

4. $x + 4y = 5$
$5x - 2y = 7$

5. $3x + 2y = 7$
$6x - 7y = 15$

6. $3x - 5y = 11$
$2x + 6y = 5$

7. $2x - y = 9$
$5x + 3y = 7$

8. $2x + 6y = 8$
$3x - 5y = 13$

9. $4x - 7y = 11$
$7x + 3y = 9$

10. $2x + 3y = 8$
$5x + 3y = 7$

Solve each system by substitution.

11. $2x - 3y = -13$
$x - 11y = 3$

12. $x + 3y = 18$
$3x - y = -6$

13. $9x + y = 80$
$5x + 7y = 38$

14. $2x - 4y = 10$
$3x + 5y = 4$

15. $3c - 4d = 31$
$2c + 6d = 12$

16. $2s + 6t = -4$
$9s - 3t = 12$

B. Exercises

Solve each system by substitution.

17. $2x + 5y = 7$
$3x - 5y = 8$

18. $x + y = 14$
$3x + 2(y - 5) = 21$

19. $2x - 8y = 7$
$3x + y = -2$

20. $4x + 7y = 2$
$5x - 3y = -21$

21. $2a + 5b = 7$
$4a - 3b = -25$

22. $4f + 5g = 18$
$8f + 3g = -6$

23. $2x + 3y = 9$
$2x - 3y = 1$

24. $3x - 5y = -3$
$3x + 15y = 11$

25. $2x + 10y = 9$
$3x - 6y = -4$

26. $5j - 4k = 2$
$5j + 12k = -4$

27. $4n + 3p = 9$
$n - 5p = 6$

28. $3x + 5y = 8$
$2x - 4y = 7$

Write a system of equations using two variables and solve the system by substitution.

29. Katie has two photo package options. The first one contains sixteen wallet-sized and two 5" × 7" photos and costs $18. The second costs $19 and contains eight wallets and three 5" × 7"s. What is the cost for each size of photo?

30. Andrew borrowed $4800 to purchase a car. He borrowed the down payment from his father at 5% interest for 6 mo and the rest from the bank at 8% for 3 yr. If the total interest he paid was $980, how much was the down payment and how much did he borrow from the bank?

31. Christy is investing $6000 in an account that earns 4% interest and $10,000 in an account that earns 5%. If she plans to keep the 5% investment one year longer than the 4% investment, how long should she leave the money in each account to earn $5000 in interest?

32. Thomas bakes pies for a local caterer. On Tuesday he made three peach pies and six cherry pies, costing him $48 in expenses. Friday he spent $54.50 to make seven peach and four cherry pies. What is the cost of each peach and each cherry pie?

33. Thomas is making cheesecakes and chocolate pies. Each cheesecake serves twelve people and costs $10 to make. Each chocolate pie serves eight and costs $6 to make. He needs to serve 88 people on a budget of $70. Assuming each guest will eat only one serving of dessert, how many of each dessert should he make?

C. Exercises

Solve each system by substitution.

34. $3x + 2y = 3$
$4x - 7y = -\frac{13}{4}$

35. $2x + 5y = 3$
$4x + 3y = -2\frac{2}{5}$

36. $x^2 + y = 5$
$8x^2 - 2y = 0$

37. Shlomo invested $4200 in a CD that guarantees a 3% return as long as he leaves the money in for at least a full year. One year later he invested $10,800 in a mutual fund that historically has had an 8% return. How long (to the nearest year) would he have to leave each investment in place for the money to double?

Dominion Modeling

CD Creators will produce the Bible story CDs for \$0.18 each with a one-time stamping fee of \$50. Digital Dzign will produce the CDs for \$0.20 each and waive the stamping fee.

38. Write a function rule describing the cost of producing x CDs using each company.

39. Use substitution to determine the number of CDs at which the cost is the same for both companies.

40. If Mr. Ring needs 10,000 CDs, which company should he choose? How much will they cost?

CUMULATIVE REVIEW

Solve. [4.6–4.7]

41. $|x - 4| = 1$

42. $|x + 8| = 0$

43. $|-5x + 4 + 2x| + 3 = 17$

44. $|x - 4| > 1$

45. $|x + 8| < 0$

46. $|x + 8| > 0$

Write the rule for each function. [5.5]

47. $f = \{(-4, -7), (0, 1), (1, 3), (2, 5)\}$

48. $g = \{(-5, -13), (-2, -4), (0, 2), (3, 11)\}$

Find the slope of the line passing through the given points. [6.2]

49. $(-1, 8), (3, 4)$

50. $(-2, 4), (-5, -2)$

MIND OVER MATH

In 2004 Tetsuya Miyamoto invented KenKen, a logic game similar to Sudoku but with an arithmetic twist. In the puzzle to the right, the object is to place the numbers 1–4 exactly once in each row and column of the 4 × 4 grid. The twist is that the numbers in each outlined region of cells must produce the indicated result using the indicated mathematical operation. A number may be repeated within an outlined region as long as it is not repeated in a row or column.

9 +	7 +		2 ×
2 ÷		2 −	12 ×
2 −			

SEQUENCES

Limit of a Sequence

Challenge **State the number that each sequence is getting closer and closer to as *n* gets larger and larger. (Hint: Substitute a large number for *n*.)**

a. $A_n = \frac{1}{n}$

b. $A_n = \frac{3n}{n + 1}$

c. $A_n = \frac{2n + 5}{3n - 2}$

The finite number that a converging sequence gets closer and closer to as n gets larger and larger is called the *limit* of the sequence. (Limits are used in defining the two main concepts of calculus.) In the third sequence above the limit of A_n is $\frac{2}{3} = 0.\overline{6}$ since $A_{1{,}000{,}000} = \frac{2{,}000{,}005}{2{,}999{,}998}$ (which is approximately 0.666668778).

Converging sequences have a limit; diverging sequences do not have a limit. Therefore, arithmetic sequences, which always diverge, do not have a limit. (Exception: A constant sequence such as $A_n = 5$ can be considered arithmetic with $d = 0$, in which case it has a limit.) Geometric sequences diverge (and thus do not have a limit) if $|r| > 1$.

Listing the first several terms of a sequence is often helpful in determining whether the sequence has a limit. Consider the sequence $A_n = (-1)^n$. The terms are $(-1)^1, (-1)^2, (-1)^3, (-1)^4, (-1)^5, (-1)^6, \ldots$, which simplify to the sequence $-1, 1, -1, 1, -1, 1, \ldots$. Since the terms are not getting closer and closer to a single finite number, the sequence diverges and has no limit.

Exercises

Do the following recursively defined sequences have a limit? (Hint: Listing the first several terms of each sequence may be helpful.)

1. $A_n = A_{n-1} + 3; A_1 = 2$

2. $A_n = -3A_{n-1}; A_1 = 2$

3. $A_n = \frac{1}{2}A_{n-1}; A_1 = 2$

4. $A_n = A_{n-1} + A_{n-2}; A_1 = -1, A_2 = 2$

State the limit of each sequence.

5. $A_n = \frac{1}{n^2}$

6. $A_n = \frac{3n - 4}{4n + 1}$

7. $A_n = \frac{1 - 2n}{n - 4}$

8. $A_n = \frac{5n^2}{2n - 7}$

State the limit of each golden ratio sequence (to the nearest thousandth).

9. $1, 2, \frac{3}{2}, \frac{5}{3}, \frac{8}{5}, \frac{13}{8}, \frac{21}{13}, \frac{34}{21}, \ldots$

10. $1, \frac{1}{2}, \frac{2}{3}, \frac{3}{5}, \frac{5}{8}, \frac{8}{13}, \frac{13}{21}, \frac{21}{34}, \ldots$

7.4 Solving Systems by Elimination

The *elimination method*, also known as the addition method, is a third method used to solve systems of equations. By adding or subtracting the equations or multiples of the equations, you can eliminate one of the variables. Solving the resulting equation produces the value of the remaining variable. This value can then be substituted into either original equation to find the value of the eliminated variable.

Sand dunes are classified using systems. If the prevailing wind direction parallels the dune length, seif dunes result. Otherwise, barchan dunes form, such as this dune in the Sahara.

Example 1

Solve the system by elimination.

$4x - 2y = 13$

$8x + 2y = 23$

Answer

$$\begin{aligned} 4x - 2y &= 13 \\ 8x + 2y &= 23 \\ \hline 12x \quad\; &= 36 \\ x &= 3 \end{aligned}$$

1. Since the coefficients of y are opposites, adding the two equations will eliminate the y term.

2. Solve for x.

$$\begin{aligned} 8x + 2y &= 23 \\ 8(3) + 2y &= 23 \\ 24 + 2y &= 23 \\ 2y &= -1 \\ y &= -\frac{1}{2} \end{aligned}$$

3. Substitute 3 for x in either original equation and solve for y.

solution: $\left(3, -\frac{1}{2}\right)$

4. Write the solution as an ordered pair.

Check $4(3) - 2\left(-\frac{1}{2}\right) = 13$ and $8(3) + 2\left(-\frac{1}{2}\right) = 23$

5. The solution checks in both equations.

The system in Example 1 was easy to solve by elimination since the coefficients of y were opposites. When like terms have the same coefficients, the variable can be eliminated by subtracting one equation from the other.

Example 2

Solve the system by elimination.

$3x + 2y = 8$
$3x + 5y = 14$

Answer

$$\begin{array}{r} 3x + 2y = 8 \\ -(3x + 5y = 14) \\ \hline -3y = -6 \end{array} \quad \text{or} \quad \begin{array}{r} 3x + 2y = 8 \\ -3x - 5y = -14 \\ \hline -3y = -6 \end{array}$$

1. Because the coefficients of the x terms are the same, subtract the equations. Note: This is the same as multiplying both sides of the second equation by -1 and then adding the equations.

$$y = 2$$

2. Solve for y.

$$\begin{aligned} 3x + 5y &= 14 \\ 3x + 5(2) &= 14 \\ 3x + 10 &= 14 \\ 3x &= 4 \\ x &= \tfrac{4}{3} \end{aligned}$$

3. Substitute 2 for y in either original equation and solve for x.

solution: $\left(\frac{4}{3}, 2\right)$

4. Write the solution as an ordered pair.

Check $2(2) + 3\left(\frac{4}{3}\right) = 8$ and $3\left(\frac{4}{3}\right) + 5(2) = 14$

5. The solution checks in both equations.

In many systems, the coefficients of the like terms do not have the same absolute value. The Multiplication Property of Equality can be applied to one or both equations in such a way that causes the coefficients of one of the variables to have the same absolute value.

Example 3

Solve the system by elimination.

$2x + 3y = 5$
$4x - 9y = -10$

Answer

$$\begin{array}{rcr} -2(2x + 3y = 5) & \Rightarrow & -4x - 6y = -10 \\ 4x - 9y = -10 & \Rightarrow & 4x - 9y = -10 \\ \hline & & -15y = -20 \end{array}$$

1. To create opposite coefficients, multiply both sides of the first equation by -2. Then add the equations. Note: The first equation could have been multiplied by 3 to eliminate the y term.

$$y = \frac{-20}{-15} = \frac{4}{3}$$

2. Solve for y.

$$\begin{aligned} 2x + 3y &= 5 \\ 2x + 3\left(\tfrac{4}{3}\right) &= 5 \\ 2x + 4 &= 5 \\ 2x &= 1 \\ x &= \tfrac{1}{2} \end{aligned}$$

3. Substitute $\frac{4}{3}$ for y in either original equation and solve for x.

solution: $\left(\frac{1}{2}, \frac{4}{3}\right)$

4. Write the solution as an ordered pair.

Check $2\left(\frac{1}{2}\right) + 3\left(\frac{4}{3}\right) = 5$ and $4\left(\frac{1}{2}\right) - 9\left(\frac{4}{3}\right) = -10$

5. The solution checks in both equations.

Sometimes both equations must be manipulated. To obtain coefficients with the same absolute value, find the least common multiple of the coefficients.

Example 4

Solve the system by elimination.

$2x + 3y = 6$
$3x - 2y = 8$

Answer

$$2(2x + 3y = 6) \Rightarrow 4x + 6y = 12$$
$$3(3x - 2y = 8) \Rightarrow \underline{9x - 6y = 24}$$
$$13x = 36$$

1. The coefficients of *y* have a least common multiple of 6. To create opposite *y* coefficients, multiply the first equation by 2 and the second by 3.

$$x = \frac{36}{13}$$

2. Solve for *x*.

$$3(2x + 3y = 6) \Rightarrow 6x + 9y = 18$$
$$2(3x - 2y = 8) \Rightarrow \underline{-(6x - 4y = 16)}$$
$$13y = 2$$

3. Since the *x* value is a fraction, use the elimination method a second time to find *y*. Multiply the first equation by 3 and the second by 2. Then subtract the equations.

$$y = \frac{2}{13}$$

4. Solve for *y*.

solution: $\left(\frac{36}{13}, \frac{2}{13}\right)$

5. Write solution as an ordered pair.

Check $2\left(\frac{36}{13}\right) + 3\left(\frac{2}{13}\right) = 6$ and $3\left(\frac{36}{13}\right) - 2\left(\frac{2}{13}\right) = 8$

6. The solution checks in both equations.

Solving a System by Elimination
1. Multiply both sides of one or both equations so that the coefficients of one of the variables have the same absolute value.
2. Add or subtract the equations to eliminate a variable.
3. Solve the resulting equation.
4. Find the value of the eliminated variable by a. substituting the result into an original equation and solving or b. eliminating the other variable.
5. Write the solution as an ordered pair.
6. Check the solution by substituting it into each equation.

A. Exercises

Determine the multipliers needed to a) eliminate the *x* terms and b) eliminate the *y* terms. Sometimes a multiplier is not needed; if so, state "just add." Do not solve.

Example: $2x + 3y = 3$; $6x - 3y = 1$
Answer: a) 1st equation by -3; b) just add

1. $7x + y = 35$
$x + y = 5$

2. $x + 4y = 6$
$9x - 4y = -66$

3. $2x - 7y = 26$
$x - 7y = 18$

4. $2x + 6y = 5$
$2x - 3y = -4$

Solve each system by elimination.

5. $x + y = 5$
$x - y = -3$

6. $2x + y = 9$
$-2x - 3y = 13$

7. $x - y = -2$
$6x + y = -19$

8. $x + 2y = 13$
$x + 6y = 45$

9. $2f + 9g = 30$
$2f + g = -2$

10. $2j + 3k = 19$
$2j + 7k = 47$

B. Exercises

Determine the multipliers needed to a) eliminate the *x* terms and b) eliminate the *y* terms. Do not solve.

11. $2x + 3y = 10$
$4x - 5y = -24$

12. $x + y = 14$
$2x - 3y = -17$

13. $6x - 11y = -26$
$x + 3y = -14$

14. $9x + 4y = 10$
$3x - 10y = 26$

Solve each system by elimination.

15. $2y + x = 9$
$3x - 2y = 7$

16. $x - 3y = 10$
$3y + x = 8$

17. $2x + y = 15$
$x + 2y = 18$

18. $2b + 3a = 18$
$5a - 3b = 11$

19. $5x - 4y = 7$
$7x - 3y = 2$

20. $6x - 5y = -38$
$4x - 17y = 2$

21. $9x + 7y = 12$
$7x + 3y = 24$

22. $2x + 5y = -45$
$19x - 8y = -39$

23. $5r + s = 17$
$3r + s = 12$

24. $x + 2y = 9$
$3x + y = 16$

25. $2x + 4y = 5$
$3x - 2y = 7$

26. $5x - 4y = 3$
$2x + 3y = 17$

27. $3x - 2y = 5$
$4x + 3y = 8$

28. $\frac{x}{2} + y = 9$
$x + y = 14$

Write a system of equations and solve the system by elimination.

29. The sum of two numbers is 621, and their difference is 109. Find the two numbers.

30. The difference of two numbers is 62, and the sum of twice the larger and three times the smaller is 304. Find the two numbers.

31. The senior class sold 245 tickets to their class play. The adult tickets sold for \$4 each and the student tickets for \$2.50 each. The total income from the ticket sales was \$725. How many of each type of ticket were sold?

32. The exam will have true/false and short answer questions. The short answer questions are worth four points each, and the true/false questions are worth two points each. There will be 50 questions worth a total of 150 points. How many of each type of question are on the exam?

33. Westside Christian School's athletic director ordered practice balls for the baseball and softball teams. He cannot remember how many of each he ordered. He knows that he ordered a total of 43 balls, costing the athletic department \$165. If baseballs cost \$3 each and softballs cost \$5 each, how many of each type of ball did he order?

C. Exercises

Write a system of equations and solve the system by elimination.

34. Four less than five times a number is twelve more than another number. Three times the first number less twice the second number is eight. Find the two numbers.

35. Jackie was mailing 2 oz and 3 oz letters. The total weight of the letters was 88 oz. A 2 oz letter cost $0.61 to mail, and a 3 oz letter cost $0.78 to mail. She spent $24.95 to mail the letters. How many of each type did she mail?

Solve each system by elimination.

36. $3x - y = \frac{23}{5}$
$15x + 25y = 90$

37. $\frac{1}{2}x + \frac{1}{5}y = 1$
$\frac{3}{2}x + \frac{1}{5}y = -7$

Dominion Modeling

Mr. Ring distributes the CDs in a heavy paper jacket. An artist friend has donated his time to do the artwork for the jacket, so now he needs to choose a printer.

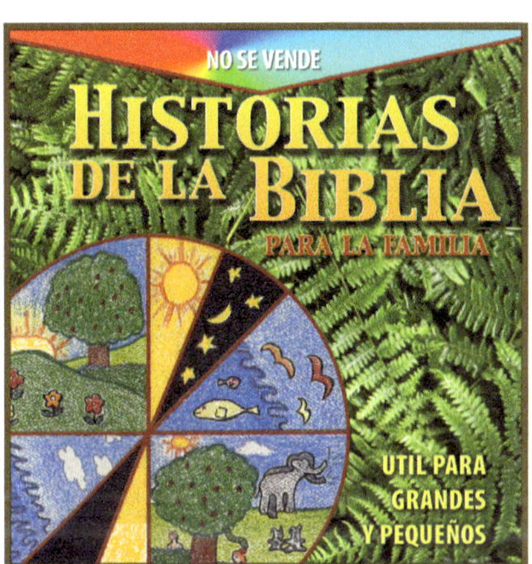

38. Express Press, an online company, charges $0.049 per jacket for printing, while the Copy Center, a local company, charges $0.055 per jacket.

a. Determine each company's printing fees for 10,000 CD jackets.

b Express Press also charges a shipping and handling fee of $1.15/lb. Determine the total cost of having them provide the CD jackets if each jacket weights 0.4 oz.

39. The youth group from a local church stuffs the CDs into their jackets and boxes them for transport to the mission field. In
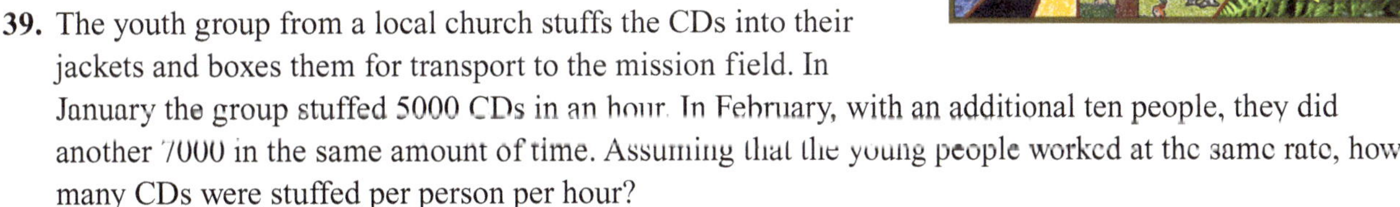
January the group stuffed 5000 CDs in an hour. In February, with an additional ten people, they did another 7000 in the same amount of time. Assuming that the young people worked at the same rate, how many CDs were stuffed per person per hour?

40. How many young people would be needed to stuff the 10,000 newly produced CDs during the one-hour youth meeting?

CUMULATIVE REVIEW

Evaluate. [1.8]

41. $6 - 2^3 \div 4 + 3(2)$

42. $-3 + (-13) - 25 \div 5 + 4^2$

Simplify. [2.3]

43. $5x - x^2 + 3x + 4 - 7x + 2 - x^2$

44. $3 + 4x + 5x^2 - (2x - 2)$

Solve each literal equation for *x*. [3.1]

45. $ax + 3b = 17b + c$

46. $\frac{a}{2x} = 5c$

Write the appropriate percent equation and solve. [3.4]

47. 3.5 is what percent of 8?

48. 45% of what number is 18?

49. 45 L of a 30% acid solution contains how much water?

50. What percent salt is a salt solution if 30 gal of solution contains 1.2 gal of salt?

7.5 Special Systems

All three methods—graphing, substitution, and elimination—yield the same solution for a system of linear equations. You must decide which method is best for each problem.

Method	Disadvantage	Advantage
Substitution	The arithmetic can be challenging.	Is easiest when a single variable is isolated.
Elimination	Equations may need to be rearranged and multiplied.	Works well when a coefficient is a factor of the other coefficients.
Graphing	May not produce precise answers when the solution does not consist of integers.	Visual representation may be helpful for analyzing real-world problems.

Example 1

Solve the system.

$x + y = -3$
$x - 3 = -2$

Answer $x + y = -3$ and $x - 3 = -2$, $x = 1$

1. Since the second equation is easily solved for x, use substitution.

$1 + y = -3$
$y = -4$

2. Substitute 1 for x in the other equation and solve for y.

solution: $(1, -4)$

3. Mentally check the solution in both equations.

Example 2

Solve the system.

$3x + 4y = 3$
$2x - 7y = 31$

Answer

$$2(3x + 4y = 3) \Rightarrow \quad 6x + 8y = 6$$
$$-3(2x - 7y = 31) \Rightarrow \quad \underline{-6x + 21y = -93}$$
$$29y = -87$$
$$y = -3$$

1. Since solving for any of the variables would result in a fraction, use elimination.

$$3x + 4y = 3$$
$$3x + 4(-3) = 3$$
$$3x - 12 = 3$$
$$3x = 15$$
$$x = 5$$

2. Substitute -3 for y in either equation and solve for x.

solution: $(5, -3)$

3. Mentally check the solution in both equations.

Example 3

Solve the system.

$y = 3x + 5$
$y = x - 3$

Answer

$y = 3x + 5$
$y = x - 3$

1. Since the equations are in slope-intercept form, graphing may be acceptable.

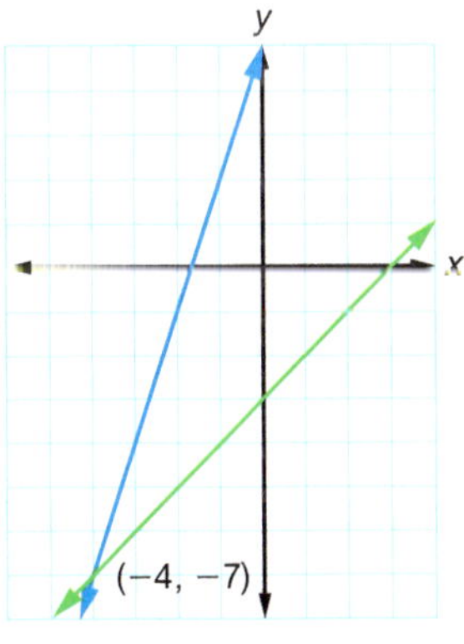

2. Graph both lines on the same coordinate plane.

   Equation 1: $m = 3$ and $b = 5$
   Equation 2: $m = 1$ and $b = -3$

   Note: Since you cannot predict whether the intersection will have coordinates that are integers, you should not normally choose the graphing method.

solution: $(-4, -7)$

3. Be sure to check your answer in both equations since your graphed solution may not be precise.

Systems of linear equations can be classified according to the number of their solutions. Systems with no solution are called *inconsistent* because there is no ordered pair that makes both equations true. Linear systems are inconsistent when the graphs are parallel lines. *Consistent* systems have one or more ordered-pair solutions that satisfy all equations in the system. If a consistent system has only one solution, it is *independent*. Systems whose graphs are intersecting lines are independent consistent systems. A consistent system with an infinite number of ordered-pair solutions is called *dependent*. When the graphs of the equations coincide, the system is a consistent dependent system.

Sections 7.2–7.4 contained only consistent independent systems whose solutions were a single ordered pair. When substitution or elimination is used to solve the other types of systems, all the variable terms can be eliminated from the equation. If the remaining equation is a false statement, such as $0 = -5$, the system is inconsistent. When these equations are graphed, they form parallel lines, indicating that the solution is the empty set.

Example 4

Solve the system.

$x + y = 7$
$x + y = 5$

Answer

$$\begin{array}{r} x + y = 7 \\ -(x + y = 5) \\ \hline 0 = 2 \end{array}$$

1. Subtracting the second equation from the first produces a false statement.

system type: inconsistent
solution: $\varnothing$

2. When a false statement results, the system is inconsistent and there are no solutions.

Solving the system by graphing illustrates why there is no solution.

$x + y = 7$ and $x + y = 5$
$y = -x + 7$ $\quad$ $y = -x + 5$

1. Write each equation in slope-intercept form.

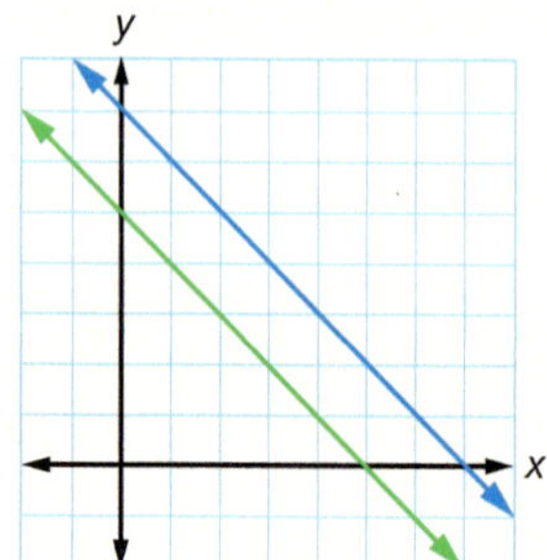

2. Graph both lines on the same coordinate plane.
   Equation 1: $m = -1$ and $b = 7$
   Equation 2: $m = -1$ and $b = 5$

By analyzing the slope-intercept forms of the equations, you can see that the lines have the same slope and different y-intercepts. Therefore, the lines are parallel, and there is no point of intersection that satisfies both equations.

If the resulting statement is a true statement, such as $0 = 0$, then the equations represent coinciding lines and the solution consists of all the points on the line. This type of solution occurs in a system that is consistent dependent.

These F-16 jets are flying in parallel courses.

Example 5

Solve the system.

$x = -y + 2$

$3x + 3y = 6$

Answer

$x = -y + 2$ and $3x + 3y = 6$

1. Since x is already isolated in the first equation, substitution will work well.

$3(-y + 2) + 3y = 6$

2. Substitute the expression for x in the second equation.

$-3y + 6 + 3y = 6$

$6 = 6$

3. Solve. The result is a true statement.

system type: consistent dependent

4. When a true statement results, the system is consistent dependent and the graphs of the equations coincide.

solution: $y = -x + 2$

5. Use the slope-intercept form of the line to state the solution.

When both equations are written in slope-intercept form, the equations are identical: $y = -x + 2$.

Types of Systems

System	consistent independent	consistent dependent	inconsistent
Number of Solutions	one ordered pair	all the points on a line (infinite number)	no solution
Graph	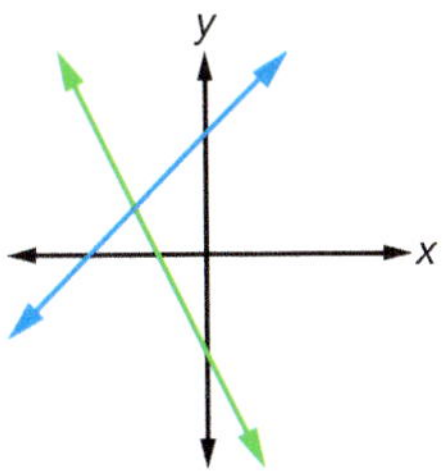	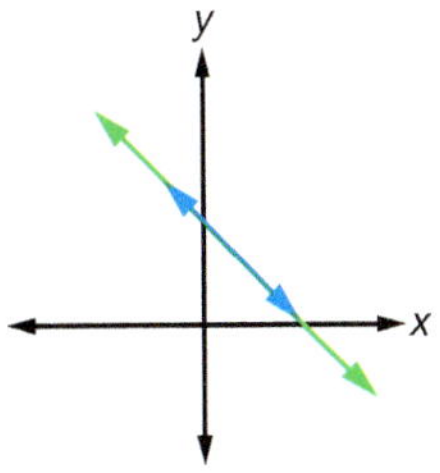	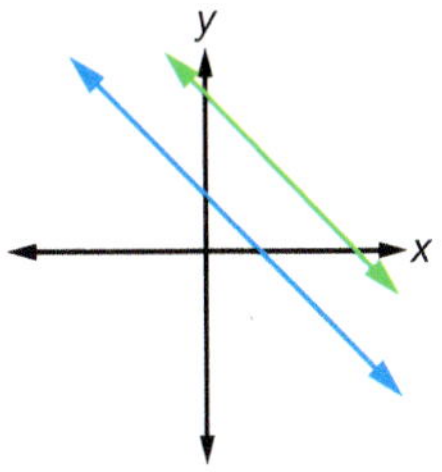
Slope-Intercept Comparison	$m_1 \neq m_2$	$m_1 = m_2$ $b_1 = b_2$	$m_1 = m_2$ $b_1 \neq b_2$
Results from Substitution or Elimination	an ordered pair (x, y)	a true statement $(0 = 0)$	a false statement $(0 = -2)$

Example 6

Which equations form a consistent dependent system? Which form an inconsistent system?
Equation 1: $3y = -3x + 9$
Equation 2: $x + y = 4$
Equation 3: $-6x = 6y - 24$

Answer

$3y = -3x + 9$ | $x + y = 4$ | $-6x = 6y - 24$
$y = -x + 3$ | $y = -x + 4$ | $-6y = 6x - 24$
 | | $y = -x + 4$

1. Write each equation in slope-intercept form.

$m_1 = -1$ | $m_2 = -1$ | $m_3 = -1$
$b_1 = 3$ | $b_2 = 4$ | $b_3 = 4$

2. Determine m and b for each line.

consistent dependent system:
$x + y = 4$ and $-6x = 6y - 24$

3. Equations 2 and 3 describe the same line since $m_2 = m_3$ and $b_2 = b_3$.

inconsistent system:
$3y = -3x + 9$ and $x + y = 4$

4. Equations 1 and 2 describe parallel lines since $m_1 = m_2$ and $b_1 \neq b_2$.

$3y = -3x + 9$ and $-6x = 6y - 24$ also form an inconsistent system since their graphs are parallel.

A. Exercises

Multiple choice

1. Which system is easiest to solve by graphing?
   a. $3x - 5y = 9$, $2x + 8y = 11$ **b.** $x + y = -2$, $x - y = 4$ **c.** $3x - 4y = 2$, $2x - 5y = 8$
2. Which system is easiest to solve by substitution?
   a. $3x - 9y = -21$, $2x + 7y = 13$ **b.** $4x - 6y = 18$, $7x + 3y = 12$ **c.** $4x + y = 12$, $x - 5y = 2$
3. Which system is best solved by elimination?
   a. $y = -2x + 4$, $y = x + 2$ **b.** $4x - 3y = 8$, $2x + 3y = 4$ **c.** $x = 8$, $y = 2x + 1$

Matching: More than one answer may be needed.

a. consistent **b.** dependent **c.** inconsistent **d.** independent

4. a system with two parallel lines
5. a system with one ordered-pair solution
6. a system with two coinciding lines
7. a system with no solution
8. a system with two intersecting lines
9. a system with an infinite number of solutions (all the points on a line)

Matching: In exercises 10–14, choose the best method for solving each system.

a. graphing **b.** substitution **c.** elimination

10. $x + 3y = 7$
$5x + 11y = 3$

11. $2x - 5y = 3$
$4x + 7y = 9$

12. $5x - 3y = 3$
$2x + 7y = 6$

13. $y = -4$
$2x + 5y = 13$

14. Sally is comparing the costs of two dressmakers for her bridesmaids' dresses. Given the differing costs per dress and consultation fee for two different seamstresses, compare the two choices and decide which seamstress would be the better choice according to how she increases or decreases her wedding party.

B. Exercises

Solve each system using the method of your choice. Then classify it as consistent independent, consistent dependent, or inconsistent. If the solution is an entire line, write its equation in slope-intercept form.

15. $8x - 3y = -12$
$y - 4 = -8$

16. $x + 2y = 8$
$-x + 5y = 6$

17. $x + y = -4$
$x + y = 8$

18. $2x + 2y = 14$
$x + y = 7$

19. $a - b = -2$
$4a - b = 4$

20. $c = -2d + 10$
$9c + 2d = -6$

21. $4x + 3y = 7$
$4x = 7 - 3y$

22. $x + y = 18$
$2x + 2y = 10$

23. $2x - y = 11$
$6x - 3y = -24$

24. $h = -4g + 7$
$g - 3h = 18$

25. $2x + 4y = 10$
$3x - y = 18$

26. $x + y = 10$
$2y = 20 - 2x$

27. $4r + 3s = 9$
$r - 5s = 6$

28. $5x - y = -42$
$-44 + y = 5x$

29. $15x - 24y = 7$
$32y + 9 = 20x$

30. $10x - 12y = 14$
$35 + 30y = 25x$

31. $2(x + y) - 5(x + 1) = 3$
$-3x + 2y = 8$

32. What is true about the slopes and y-intercepts of two linear equations that form an inconsistent system?

33. What is true about the slopes and y-intercepts of two linear equations that form a consistent dependent system?

34. What is true about the slopes and y-intercepts of two linear equations that form a consistent independent system?

Write and solve a system of equations for each word problem.

35. An isosceles triangle (two congruent sides) has a perimeter of 48 cm. The noncongruent side is 3 cm more than half of one of the other sides. Find the length of each side of the triangle.

36. Jessie and Kayla decided to purchase books and DVDs for the youth library at their church. Jessie purchased eight books and four DVDs, and Kayla purchased five books and six DVDs. Jessie's total cost was $80 and Kayla's was $92. If all the books sold for the same amount and all the DVDs had the same price, how much was the cost per item?

37. Charles wants to rent a car for the weekend. After researching the costs, he narrows his choice to the following two options: 1) Easy Rental: $50 per day + $0.15 per mile driven, and 2) Clean Rental: $40 per day + $0.20 per mile driven. Write a system of equations for the costs of renting a car from each company. Graph the system and determine when the companies will cost the same amount. When would it be better to use Easy Rental?

C. Exercises

38. Explain why graphing is not a good choice for solving this system.

$9x - 2y = 7$

$3x + 4y = 14$

39. Explain why substitution is not a good choice for solving this system.

$9x - 2y = 7$

$3x + 4y = 14$

40. Jackson is trying to make a decision between two colleges. College A's tuition for a semester is $325 per credit hour plus a $450 fee. College B's tuition is $350 per credit hour with no additional fees. At how many credit hours would both college tuition costs be the same? When would College A be less expensive?

CUMULATIVE REVIEW

If *y* varies directly with *x*, use the given values of *x* and *y* to find the constant of variation and write the function rule. [5.6]

41. $x = 9, y = 18$

42. $x = 7, y = 4$

43. What do the graphs of the two equations in exercises 41 and 42 have in common?

The weight of the atmosphere at sea level is 14.7 pounds per square inch (psi). This is also referred to as one atmosphere (atm). Underwater pressure increases by 14.7 psi for every 33 ft of depth. Absolute pressure at ocean depths includes the atmospheric pressure at sea level.

44. Determine the rate of change in pressure per foot of ocean depth in psi/ft (to the nearest thousandth). [6.2]

45. Write a function rule modeling the absolute pressure, $p(d)$, in pounds per square inch at a given ocean depth, d feet. [6.3]

46. The Mariana Trench of the western Pacific contains the deepest spot in Earth's oceans, Challenger Deep, which is 35,840 ft below sea level. Find the absolute pressure at this depth (to the nearest whole number). [6.3]

47. Convert the pressure found in the previous exercise to atmospheres. [1.5]

48. The crush level of a submarine (the depth at which the outside pressure will crush the submarine's steel hull) has been estimated to be 2400 ft. Determine the pressure that is estimated to crush the submarine (to the nearest whole number). [6.3]

49. Convert the pressure found in the previous exercise to atmospheres. [1.5]

Seawolf-class submarines are described as the fastest, quietest, and best-armed subs in the world, costing about $2.8 billion each.

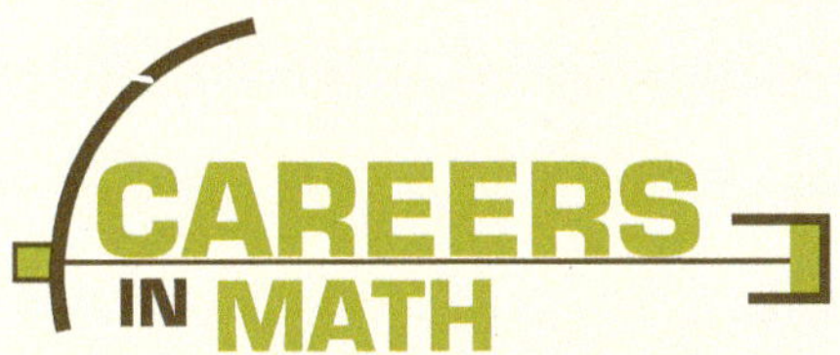

Computer Programmer/Software Engineer

While we are familiar with the increasing power of computers and cell phones, we may not realize the importance of computers to our daily lives. Computers help our cars run more efficiently and handle slippery road conditions. Computers enable retailers to efficiently deliver low-cost goods across the nation according to demand. Police use computers to track the location of crimes so that officers can recognize patterns and adjust their patrols. But none of these computers can function without programmers.

A computer programmer writes code, the logical sequence of instructions for a computer, called a program. A college degree in computer science is recommended since programmers need to be familiar with multiple computer languages and various programming techniques. Programmers write, debug, and maintain code for a wide range of applications, including education, gaming, finances, robotic control, and data management. Mathematical and analytical skills learned in high-school math provide a good foundation for the pursuit of a programming career.

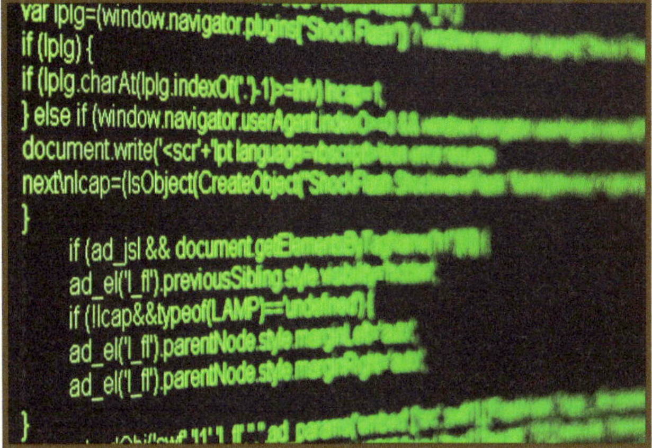

A software engineer's main responsibility is the development of computer programs. A software engineer (SE) may or may not write the actual code for the software. The SE is involved in defining and assessing needs and then ensuring the writing and implementation of a program that best meets those needs. He works closely with programmers to ensure the success of their efforts. Most software engineering jobs require a degree in computer science or computer information systems.

Advanced technologies continue to create growing demands for programmers and software engineers in every business and ministry. These occupations help provide effective management of our God-given resources.

7.6 Motion Problems

Motion problems were solved using one variable in Chapter 3. Often these problems are easier to solve using two variables, and sometimes they can be solved only by using a system of equations with more than one variable. Remember the basic formula $rt = d$.

Example 1

Two cars start toward each other at the same time from towns 612 mi apart and pass in 6 hr. One travels 2 mi/hr faster than the other. How fast is each car traveling?

Answer

r_f = the rate of the faster car
r_s = the rate of the slower car

1. Meaningful subscripted variables can help to clarify the problem.

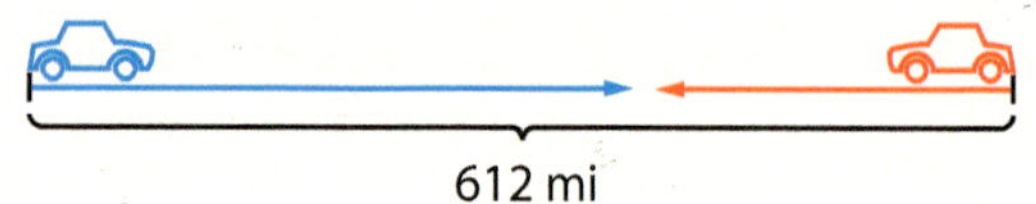

	r	t	d
Faster Car	r_f	6	$6r_f$
Slower Car	r_s	6	$6r_s$

2. Make a drawing and a table. Fill in the first two columns and then use the formula $rt = d$ to fill in the third column.

$r_f = r_s + 2$ and $6r_f + 6r_s = 612$

3. Write the system.
   Equation 1: The faster car is 2 mi/hr faster than the slower car.
   Equation 2: The total distance is 612 mi.

$r_f = r_s + 2$

$$6(r_s + 2) + 6r_s = 612$$
$$6r_s + 12 + 6r_s = 612$$
$$12r_s + 12 = 612$$
$$12r_s = 600$$
$$r_s = 50$$

4. Solve the system using substitution since the first equation is already solved for r_f.

$r_f = 50 + 2$
$r_f = 52$

5. Substitute 50 for r_s in the first equation and solve for r_f.

The slower car travels at 50 mi/hr and the faster at 52 mi/hr.

6. Write the solution.

Problems in which the rates are affected by wind or water currents can be solved using two variables. Traveling into a headwind or upstream against a current decreases your typical rate of travel. Moving downstream or with a tailwind increases your normal speed.

No Wind or Current	Headwind 25 mi/hr	Tailwind 25 mi/hr	Downstream 3 mi/hr current	Upstream 3 mi/hr current
r	$r - 25$	$r + 25$	$r + 3$	$r - 3$

Example 2

When his Sunday school class went canoeing, Brent paddled 12 mi down the river in 3 hr. After paddling back upstream for 2 hr, he had traveled 4 mi. How fast does Brent paddle in still water? What is the rate of the current?

Answer

b = Brent's paddling rate
c = the rate of the current

1. Read carefully to define the variables.

	r	t	d
Downstream	$b + c$	3	12
Upstream	$b - c$	2	4

2. Make a table and fill it in with the known information.

$3(b + c) = 12$ and $2(b - c) = 4$

3. Use the distance equation $rt = d$ to produce the two equations for the system.

$$\frac{\cancel{3}(b + c)}{\cancel{3}} = \frac{12}{3} \Rightarrow b + c = 4$$
$$\frac{\cancel{2}(b - c)}{\cancel{2}} = \frac{4}{2} \Rightarrow b - c = 2$$
$$2b = 6$$
$$b = 3$$

4. Both equations can be simplified before solving the system. Eliminate the c terms by adding the equations.

$b + c = 4$
$3 + c = 4$
$c = 1$

5. Substitute 3 for b in the first simplified equation and solve for c.

Brent's paddling rate is 3 mi/hr.
The rate of the current is 1 mi/hr.

6. Write the solution.

The term *knot* is the unit of speed referring to nautical miles per hour (1 knot $\approx$ 1.151 mi/hr). The unit is often used to measure speeds in aviation and sea navigation.

Example 3

When flying the 1900 nautical miles to California, James flew nonstop for 4 hr. Flying home by the same route and under the same weather conditions, the flight took 3 hr 30 min. What rate would the airplane have flown with no wind? What was the wind speed?

Answer

p = the rate of the airplane with no wind
w = the wind speed

1. Read carefully to define the variables.

	r	t	d
Flight to CA	$p - w$	4	1900
Flight Home	$p + w$	3.5	1900

2. Make a table and fill it in with the known information.

$4(p - w) = 1900$ and $3.5(p + w) = 1900$

3. Use the distance equation $rt = d$ to produce the two equations for the system.

$$\frac{4(p-w)}{4} = \frac{1900}{4}$$
$$p - w = 475$$

$$3.5p + 3.5w = 1900$$
$$35p + 35w = 19{,}000$$

4. Solve. Since 1900 is divisible by 4, simplify the first equation. In the second equation, however, 1900 is not divisible by 3.5. Remove the parentheses and clear the decimals.

$$35(p - w = 475) \Rightarrow 35p - 35w = 16{,}625$$
$$35p + 35w = 19{,}000 \Rightarrow \underline{35p + 35w = 19{,}000}$$
$$70p = 35{,}625$$
$$p \approx 509$$

5. Multiply the first equation by 35 and eliminate the w terms. The answer is rounded to the nearest knot.

$$p - w = 475$$
$$509 - w \approx 475$$
$$-w \approx -34$$
$$w \approx 34$$

6. Substitute 509 for p in the first simplified equation and solve for w. Since the airplane's rate was approximated, the wind speed will be as well.

The airplane's rate of speed is about 509 knots.
The wind speed is about 34 knots.

7. Write the solution.

A. Exercises

Write an algebraic expression that models each rate.

1. a bicyclist riding against a wind of w mi/hr if his speed on a calm day is 12 mi/hr
2. a blimp that usually goes r mi/hr with no wind when a 10 mi/hr tailwind is blowing
3. a canoe going downstream in a current of c mi/hr if the canoe averages 4 mi/hr on a calm lake
4. A boat travels upstream on Pine Creek, which has a current of 2 mi/hr. The boat usually goes z mi/hr in still water.
5. A plane that flies 500 knots with no wind flies into a headwind of w knots.

Use each table to write a system of equations. Do not solve.

6. Pedro traveled a total of 230 mi in 7 hr.
   a. Define the variables.
   b. Write an equation describing the times.
   c. Write an equation describing the distances.

	r	t	d
Before Lunch	30	t_b	$30t_b$
After Lunch	40	t_a	$40t_a$

7. Marissa went biking on a day when the wind speed was 8 mi/hr. Each leg of the trip was 36 mi.

   a. Define the variable.

   b. Write an equation describing the distances.

   c. Why would we need only one equation to find her normal biking speed?

	r	t	d
With Wind	$b + 8$	1.125	36
Against Wind	$b - 8$	2.25	36

Make a table for each problem. Do not solve.

8. A blimp travels 2 hr with the wind and 5 hr against it. Let b represent the rate of the blimp with no wind and let w represent the wind speed.

9. Blake rows 6 hr upstream against a current of 3 mi/hr and returns with the same current in 4 hr. Let b represent the rate of the boat in still water.

10. A pilot flies 150 knots in still air. He flies with the wind 4 hr to Albuquerque and then against the wind 9 hr to Miami. Let w represent the wind speed.

Use each table to write and solve a system of equations.

Terry traveled a total of 150 mi on his bicycle. The difference in the rates was 10 mi/hr.

	r	t	d
Going Against the Wind	$x - w$	5	$5(x - w)$
Returning with the Wind	$x + w$	2	$2(x + w)$

11. Define the variables.
12. Write an equation describing the rates.
13. Write an equation describing the distances.
14. Solve to find the rate of the bicycle with no wind and the wind speed. (Round to the nearest tenth.)

A plane flew to Omaha and then returned. The speed of the plane with no wind was 100 mi/hr more than the wind speed.

	r	t	d
With Wind	$x + w$	2	$2(x + w)$
Against Wind	$x - w$	3	$3(x - w)$

15. Define the variables.
16. Write an equation describing the rates.
17. Write an equation describing the distances.
18. Solve to find the rate of the plane with no wind and the wind speed.

B. Exercises

Write and solve a system of equations for each word problem.

19. The hiking club went on a hike one Saturday. In the morning they walked at an average speed of 3 mi/hr; in the afternoon they averaged 2 mi/hr for 1 hr more than the number of hours they had walked in the morning. How many hours did they walk in the morning if they walked a total of 22 mi?

20. Bob leaves Centerville at 11:00 AM and travels west at 45 mi/hr. Lane leaves Centerville at 1:00 PM the same day and travels east at 50 mi/hr. At what time will the two be 375 mi apart?

21. Flying with the wind, a plane travels 1200 nautical miles in 4 hr. Flying against the wind, it travels 500 nautical miles in 2 hr. What is the speed of the plane with no wind? What is the wind speed?

22. Mrs. Vaughn and her son Martin live 184 mi apart. If they leave their homes at the same time and head toward each other, they will meet in 2 hr. Martin travels 12 mi/hr faster than his mother. How fast does each travel?

23. A southbound train travels for 2 hr and meets a northbound train that has been traveling on a parallel track for 4 hr. The trains started from cities 350 mi apart. What is the speed of each train if the northbound train travels 10 mi/hr slower than the southbound?

24. Find the rate of two cars traveling in the same direction if they leave the same place at the same time. One car is one and a half times as fast as the other, and after 4 hr the faster car is 80 mi ahead of the slower car.

25. On a calm day Dylan can bike to Janesville in 3 hr. Yesterday he made it in 2 hr with a 12 mi/hr tailwind. How fast does he bike on a completely calm day?

26. A plane leaves the Ligonier Airport and travels west at 230 knots. A second plane leaves 3 hr later and travels east at 215 knots. How long has each plane been in the air when the two planes are 1135 nautical miles apart?

27. A car leaves Middletown and heads toward Rochester 248 mi away at the same time a car leaves Rochester and heads toward Middletown. If the rate of the first car is 6 mi/hr faster than the rate of the second and if the cars meet in 2 hr, how fast is each traveling?

28. The scouts can raft 15 mi downstream from one camp to another in 3 hr, but it takes 5 hr to paddle back to their camp upstream. What is the scouts' paddling rate in still water? What is the rate of the current in the stream?

29. Two trains leave the station at the same time and travel in opposite directions. The express train travels 14 mi/hr faster than the local train. After 3 hr they are 315 mi apart. What is the rate of each train?

C. Exercises

Write and solve a system of equations for each word problem.

30. A train leaves City View at 11:00 AM and heads east at 38 mi/hr. At 2:00 PM another train leaves, traveling east at 53 mi/hr. How long will it take the second train to come within 54 mi of the first train?

31. Mr. Brannon took his Sunday school boys on a camping trip. They bicycled to the trailhead and then hiked to their camp. They rode at an average speed of 9 mi/hr and walked at a speed of 2 mi/hr in the 6 hr they traveled. How long did they hike if they traveled a total distance of 40 mi?

32. Salmon hatch in freshwater streams and then migrate 70 mi downstream to the ocean in 8 hr. After reaching spawning age, they return upstream to fresh water to spawn and usually die soon after. If the salmon travel 58 mi in 8 hr during the upstream trip, find the rate of the salmons' swimming in still water and the rate of the current.

Dominion Modeling

Operation Salvation was a project of distributing 500,000 CDs in Mexico during the country's two-hundredth anniversary celebration. Mr. Ring raised the money for the project through church and individual donations. Christians in the United States were encouraged to sponsor a box of CDs, and Mexican churches donated toward the cost of the jackets.

33. \$70,000 was needed for the first 400,000 CDs, including the shipping cost of \$4300. Write and solve an equation to find the cost of producing each CD (to the nearest tenth of a cent).

34. Find the cost per CD after shipping is added.

CUMULATIVE REVIEW

Evaluate. [1.7]

35. $(-2)^5$

36. 2^{-3}

37. $\frac{1}{3^{-4}}$

38. $\frac{1}{(-2)^{-4}}$

Simplify, leaving each answer in positive exponential form. [1.7]

39. $2^4 \cdot 2^{-2}$

40. $3^6 \div 3^{-1}$

41. $(3^2)^{-3}$

42. $2^3 \cdot 2^{-2} \div 2^4$

43. $\frac{2^4 \cdot 3^0}{2 \cdot 3^5}$

44. $\frac{2^2 \cdot 3^3 \cdot 5^2}{2^6 \cdot 3^7 \cdot 5^9}$

7.7 Mixture Problems

Mixture problems can also be solved with two variables. Some mixture problems combine items with different values. Using a variation of the following basic equation, you can create a table to organize the information in the problem.

quantity × cost per item = total cost

Almonds and other nuts for sale at a bazaar in Tunis, Tunisia

Example 1

The Nut Shop sells cashews for \$8.15/lb and almonds for \$6.95/lb. The manager wants to make 80 lb of a mix of cashews and almonds that sells for \$7.25/lb. How many pounds of each kind of nut should he mix together?

Answer

c = the number of pounds of cashews
a = the number of pounds of almonds

1. **Read** the problem carefully and assign variables to represent the unknowns.

	Number of Pounds	Price per Pound	Total Cost
Cashews	c	8.15	$8.15c$
Almonds	a	6.95	$6.95a$
Mix	80	7.25	580

2. **Plan**: Use a table to organize the information about the cashews, the almonds, and the mixture.

$8.15c + 6.95a = 580$ and $c + a = 80$

3. **Solve** a system to answer the question.
   a. The first equation relates the total costs, and the second equation relates the number of pounds.

$$100(8.15c + 6.95a = 580) \Rightarrow 815c + 695a = 58{,}000$$
$$-815(c + a = 80) \Rightarrow -815c - 815a = -65{,}200$$
$$-120a = -7200$$
$$a = 60$$

   b. Multiply the first equation by 100 to clear the decimals. Multiply the second equation by −815 to obtain opposite coefficients.

$$c + a = 80$$
$$c + 60 = 80$$
$$c = 20$$

   c. Substitute 60 for a in the second equation and solve for c.

Mix 60 lb of almonds and 20 lb of cashews.

   d. Write the solution.

Check $8.15(20) + 6.95(60) = 580 = 80(7.25)$

4. **Check** by substituting the solution in the other original equation.

Solving Word Problems with Two Variables

1. **Read** the problem carefully and look for the unknown quantities. Use variables to represent the unknowns.
2. **Plan** by making a table. Fill in as much information as possible.
3. **Solve** a system of equations obtained from information in the table and in the problem.
4. **Check** your answer in the context of the original statement of the problem.

Example 2

Emmanuel Christian School sent out 150 pieces of mail. Some required $0.60 postage while others required $0.84. If the total bill was $108.48, how many pieces of each type were mailed?

Answer

x = the number of $0.60 pieces of mail
y = the number of $0.84 pieces of mail

1. **Read** the problem carefully and assign variables to represent the unknowns.

	Pieces of Mail	Price per Item	Total Cost
$0.60	x	60	$60x$
$0.84	y	84	$84y$
Total	150		10,848

2. **Plan**: Use a table to organize the information. Using cents instead of dollars simplifies the system to be solved.

3. **Solve** a system to answer the question.

$x + y = 150$ and $60x + 84y = 10{,}848$

a. The first equation relates the pieces of mail, and the second equation relates the total costs.

$x = -y + 150$

$$60(-y + 150) + 84y = 10{,}848$$
$$-60y + 9000 + 84y = 10{,}848$$
$$24y + 9000 = 10{,}848$$
$$24y = 1848$$
$$y = 77$$

b. Solve the first equation for x and substitute its value in the second equation.

$x + 77 = 150$
$x = 73$

c. Substitute 77 for y in the first equation and solve for x.

The school mailed 73 pieces of mail that cost $0.60 and 77 that cost $0.84.

d. Write the solution.

Check

$73(0.60) + 77(0.84) = \$108.48$

4. **Check** the solution.

Other problems involve combining two mixtures of different strengths in order to get a final mixture of a desired strength. When 10 gal of a 5% salt solution is combined with 20 gal of an 8% salt solution, the total amount of salt in the 30 gal can be found by adding the amount of salt from each component.

$$0.05(10) + 0.08(20) = 0.5 + 1.6 = 2.1 \text{ gal of salt}$$

The concentration of the resulting mixture can then be determined.

$$\frac{2.1}{30} = 0.07 = 7\% \text{ salt solution}$$

Consider the difference between adding pure water and adding pure salt to a solution. When adding pure water, you are adding a 0% salt solution. Adding pure salt can be thought of as adding a 100% salt solution.

Example 3

If milk that is 3% fat is mixed with fat-free milk to produce 5 gal of 2% milk, how much of each should be used?

Answer

x = the number of gallons of 3% milk
y = the number of gallons of fat-free milk

1. **Read** the problem carefully and assign variables to represent the unknowns.

	Number of Gallons	Percent of Fat	Gallons of Fat
3%	x	0.03	$0.03x$
Fat-free	y	0	0
2%	5	0.02	0.1

2. **Plan**: Use a table to organize the information. The percent formula is applied to determine the expressions in the third column.

$x + y = 5$ and $0.03x + 0 = 0.1$

$0.03x = 0.1$

$3x = 10$

$x = \frac{10}{3}$ or $3\frac{1}{3}$

3. **Solve** a system to answer the question.
   a. The first equation relates the number of gallons, and the second equation relates the gallons of fat.
   b. Eliminate the decimal in the second equation and solve for x.

$\frac{10}{3} + y = 5$

$y = \frac{15}{3} - \frac{10}{3}$

$y = \frac{5}{3}$ or $1\frac{2}{3}$

   c. Substitute $\frac{10}{3}$ for x in the first equation and solve for y.

$3\frac{1}{3}$ gal of 3% milk should be mixed with $1\frac{2}{3}$ gal of fat-free milk to produce 5 gal of 2% milk.

   d. Write the solution.

Check

$0.03\left(\frac{10}{3}\right) + 0 = 0.1$

4. **Check** by substituting the solution in the other original equation.

The interest problems that we have already discussed in this chapter are a third variation of mixture problems. You will continue to practice these problems in the following exercises.

A. Exercises

Complete the following tables.

	Quantity	Cost per Item	Total Cost
1.	80 lb of flour	\$0.50/lb	
2.	27 apples		\$17.55
3.		\$7 per bag	\$574
4.	16 oz	d dollars per oz	
5.	x cases	\$30/case	

	Amount of Solution	Concentration	Amount of Ingredient
6.	20 gal	15% acid	
7.	25 L		7 L of salt
8.		7% acid	10.5 L of acid
9.	x gal	12% acid	
10.	200 mL	p% salt	

11. What is the percent of salt in pure water?

12. What is the strength of pure acid?

Write and solve a system of equations for each word problem.

13. Kimberly wants to make a mix of nuts and raisins for a party. Nuts cost $0.20/oz and raisins cost $0.12/oz. How many ounces of each should she buy if she wants 48 oz of a mix that costs $0.15/oz?

14. A grocer mixes oranges and bananas to make a 10 lb fruit basket. The oranges cost $0.75/lb and the bananas cost $0.60/lb. How many pounds of each should he use if the basket is to cost $6.90?

15. Caramels cost $3.50/lb and butterscotch candies cost $4.50/lb. A 5 lb bag of mixed candy is to sell for $4.20. How many pounds of each kind of candy should be mixed together?

16. Maria plans to invest a total of $7500 between two types of stocks, one yielding 6% interest and the other 15%. How much does she invest in each type of stock if the total interest gained annually is $585?

B. Exercises

Write and solve a system of equations for each word problem.

17. Shawn has a solution that is 30% salt. How many gallons of pure water and solution should he mix to make 40 gal of a diluted solution that is 9% salt?

18. The florist advertises roses for $2 a bud and carnations for $0.75 a flower. If Sebastien pays $20.50, excluding tax, for a bouquet of fourteen flowers, how many of each flower are in the bouquet?

19. A pharmacist is making an antiseptic of iodine and alcohol. The solution that he now has is 5% iodine. He wants a mixture that is 6% iodine. How much of a 10% iodine solution and how much of his original antiseptic should he mix to make 50 mL of the new antiseptic?

20. Raisin Rich breakfast cereal is 10% protein. A dietician wants to raise the percentage of protein per serving to 20%. The high-protein grain that will be mixed with the cereal is 40% protein. How many grams of each should be mixed together to make a 12 g serving of cereal with the proper percentage of protein?

21. The gardener plans to seed the front lawn with a new mix of grass seed. He has a mix that is 80% fescue and 15% winter rye. How much fescue and how much of his original mix should he blend together to create 50 lb of an 86% fescue mix?

22. Concrete is made of cement, water, and aggregate (sand and gravel). Aggregate should make up 75% of the total mass of concrete. If you have a concrete mix that is 60% aggregate, how many kilograms of this mix and how many kilograms of aggregate should be combined to form 48 kg of the concrete mix with the proper percentage of aggregate?

23. A chemist has two HNO_3 acid and water solutions. He knows that the percentage of acid in solution I is twenty percentage points less than the percentage of acid in solution II. If he mixes 40 mL of solution I with 60 mL of solution II and obtains a 27% acid solution, what percentage of acid is in each solution?

24. Becky makes an antiseptic of alcohol and water. If she buys a solution that is 40% alcohol to make a 12.5% alcohol solution, how much water and alcohol solution should she mix to obtain 128 mL of her homemade antiseptic?

25. Hunter invested a total of $10,000 between two accounts. One account yields 10% interest and the other 9.5%. How much did he invest in each account if his total annual interest was $971?

26. Miss Parker invested a certain amount of money at 9.2% interest and another amount, $700 more than the first, at 10.4%. If the total annual interest is $425.60, how much money is in each investment?

Write and solve a system of equations for each word problem. Round to the nearest tenth if necessary.

27. A factory sells two types of antifreeze. Type A is 40% and type B is 52% ethylene glycol. An auto mechanic wants 200 gal of antifreeze that protects radiators to $-25°F$. To give that protection, the solution must be 45% ethylene glycol. How many gallons of each type should be mixed to get the correct strength?
28. A butcher wants to grind round steak and mix it with regular ground beef to make a deluxe ground beef. The round steak sells for \$3.89/lb and the ground beef for \$2.79/lb. How many pounds of each should he mix together to make 50 lb of the deluxe ground beef that will sell for \$3.29/lb?
29. An ice cream company is going to mix cream that is 40% fat with milk that is $3\frac{1}{2}\%$ fat to produce 60 gal of a mixture that is 10% fat. How many gallons of cream and how many gallons of milk should they use?
30. Lee mixes corn syrup and molasses together to make a special pancake syrup. She can afford to pay \$0.35/oz for the mix. If corn syrup costs \$0.215/oz and molasses costs \$0.52/oz, how many ounces of each should she mix together to make 24 oz of syrup?
31. A pet store owner wants to mix two different kinds of seeds together to form a high-grade birdseed. Seed A sells for \$0.28/lb, seed B sells for \$0.60/lb, and the mixture will sell for \$0.50/lb. If he wants to make 85 lb of the seed mixture, how much of each seed should he mix together?
32. Will is mixing one coffee brand that has an excellent taste with a second coffee that has a nice aroma. The more expensive coffee costs \$16.77/lb, and the second coffee costs \$4.75/lb. If he needs 50 lb that cost \$7.50/lb, how much of each type should he use in the mixture?

C. Exercises

Write and solve a system of equations for each word problem. Round to the nearest tenth if necessary.

33. A businessman invests \$20,000 in a dry cleaning establishment and \$30,000 in a restaurant. What is his expected rate of return (or rate of interest) on each investment if he expects a total of \$4600 annually and the average of the two rates is 9%?
34. Mr. Johnson ran for mayor in last week's election. In district 12 he received 85% of the vote, and in district 13 he received 43% of the vote. From the two districts he received 10,253 votes. How many voted for Mr. Johnson in each district if a total of 16,460 people voted in the two districts?
35. In ordinary ocean water, sodium chloride (the chemical name for ordinary table salt) is 2.8% by mass. The Dead Sea is about 12% sodium chloride. If a chemist takes a sample of water from the Dead Sea and wants to dilute it with ordinary ocean water, how much ocean water and how much Dead Sea water should he combine to form 12 L of a 10% sodium chloride solution?
36. A farmer can sell 50 bu of beans and 100 bu of corn for \$810, or he can sell 20 bu of beans and 205 bu of corn for \$868.50. What is the selling price of beans and corn per bushel?

Dominion Modeling

The producer needs to ship 200,000 CDs from California to Texas. The maximum weight for each skid is 2000 lb. Each box contains 500 CDs and weighs about 16.5 lb with packaging.

37. Write and solve an inequality that describes the number of boxes that can be placed on each skid. What is the maximum number of boxes per skid?
38. How many skids will be needed and how many boxes should be on each skid if the same number of boxes are to be packed on each skid?
39. Each skid itself weighs 50 lb and must be considered in the shipping cost in addition to the weight of the potential freight. Determine the total weight of a fully loaded skid.
40. If each CD weighs 0.5 oz, what percent of the total shipping weight is due to the packaging of the CDs and shipping materials (to the nearest tenth of a percent)?

CUMULATIVE REVIEW

Solve and graph the solution on a number line. [4.2]

41. $\frac{12x}{3} < -4$

42. $6y > -10 - 8$

43. $3x - 4 \not> x + 2$

Solve and graph the solution on a coordinate plane. [6.7]

44. $y - 2 > x + 2$

45. $y - 2 + 3y \leq -6$

46. $8 > 5 - (x - 4)$

Graph each inequality on a coordinate plane. [6.7]

47. $y \leq -5x + 7$

48. $3x + y > 4$

Solve each system by graphing. [7.1]

49. Where will the graphs of $y = 6$ and $x + 3y = 12$ intersect?

50. Where will the graphs of $x = -3$ and $x + 2y = 7$ intersect?

7.8 Solving Systems of Inequalities

In Chapter 6 you learned how to graph an inequality with two variables. When you graph an inequality, the shaded region represents all possible solutions. The solution to a *system of inequalities* consists of the ordered pairs that satisfy all the inequalities of the system. This set of points is the intersection of the graphs of the inequalities in the system.

Example 1

Solve the system of inequalities.

$3x + y > 4$
$x - y \leq 8$

Answer

$3x + y > 4$ and $x - y \leq 8$

$y > -3x + 4$ $\quad -y \leq -x + 8$

$y \geq x - 8$

1. Write each inequality in slope-intercept form.

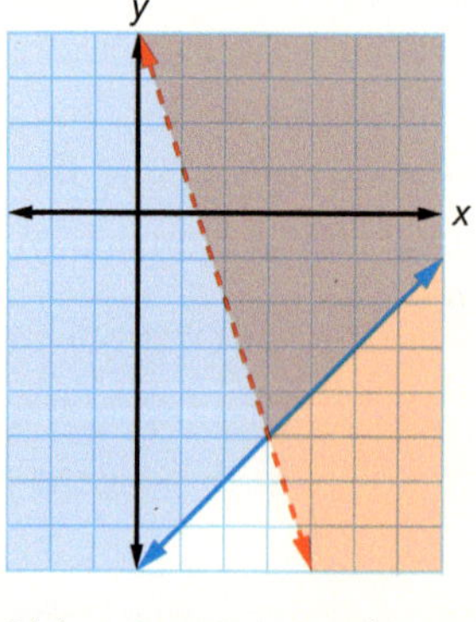

2. Graph each inequality on the same set of axes.

y-intercept: (0, 4)
$m = -3$
Shade above the dashed line.

y-intercept: (0, −8)
$m = 1$
Shade above the solid line.

3. Solution: The purple region indicates the points in the intersection of the two graphs.

Check $3(4) + 0 > 4$ and $4 - 0 \leq 8$

4. Pick a point in the region, (4, 0), to test the solution to the system. The ordered pair satisfies both inequalities.

Solving a System of Inequalities

1. Graph all inequalities on the same set of axes.
2. Find the solution set where the shaded regions intersect.
3. Check the result.

Example 2

Solve the system of inequalities.

$x > 0$

$y > 0$

$2x - y > 1$

Answer

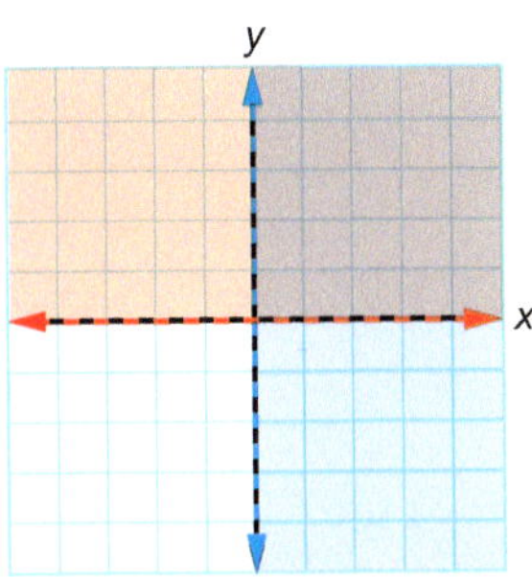

1. Graph the first two inequalities.

$x > 0$	$y > 0$
Shade to the right of the dashed line $x = 0$ (the y-axis).	Shade above the dashed line $y = 0$ (the x-axis).

2. Note that the first two inequalities limit our solutions to the first quadrant.

$2x - y > 1$

$-y > -2x + 1$

$y < 2x - 1$

3. Write the third inequality in slope-intercept form and graph.

y-intercept: $(0, -1)$

$m = 2$

Shade below the dotted line.

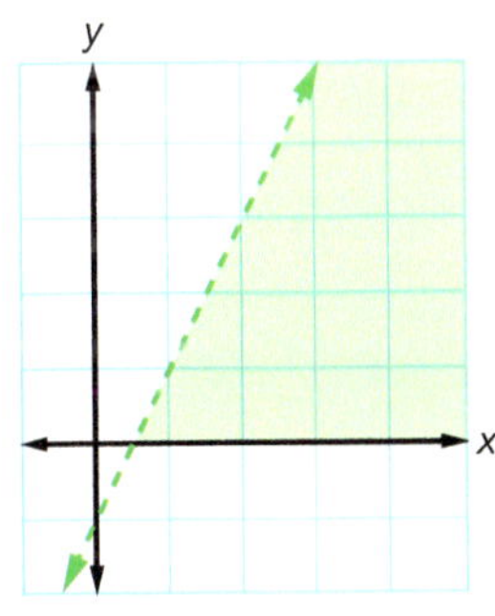

4. The solution is all first-quadrant points in the green shaded area.

Check $4 > 0$ $3 > 0$ $2(4) - 3 > 4$

$5 > 4$

5. Pick a point in the region, $(4, 3)$, to test the solution to the system. The ordered pair satisfies all three inequalities.

The avalanche pictured is on a nearly vertical slope. The graph below shows that such avalanches are rare and indicates the relative risks for different slopes. We can write a system modeling this high-risk area.

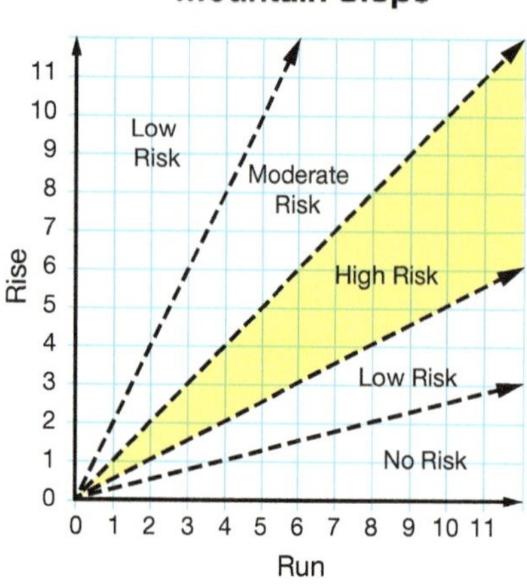

The lower boundary's y-intercept is the origin, and its slope is $\frac{1}{2}$. Since the shaded area is above this boundary line, the inequality is $y > \frac{1}{2}x$. The y-intercept of the upper boundary is also (0, 0), but its slope is 1. The shaded area is under this boundary, so the inequality is $y < x$. The high-risk area can be modeled by the following system.

$$y > \frac{1}{2}x$$
$$y < x$$

Example 3

Write the system of inequalities represented by the graph.

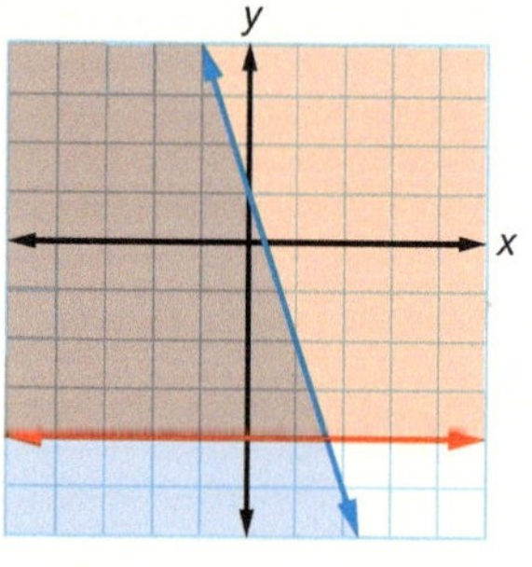

Answer

y-intercept: $(0, -4)$		y-intercept: $(0, 1)$	1. Find the equations of the boundary lines.
$m = 0$		$m = -3$	
$y = -4$		$y = -3x + 1$	
$y \geq -4$	and	$y \leq -3x + 1$	2. Write the inequalities. The solution includes points above the solid red line and points below the solid blue line.

Since the lines are solid in Example 3, we know that the points on the boundary lines are included in the set of solutions. Remember that if the points on a boundary line are not to be included, then a dashed line is used.

A. Exercises

Without graphing, identify which quadrant each system describes.

1. $x > 0$
$y > 0$

2. $x < 0$
$y < 0$

3. $x < 0$
$y > 0$

4. $x > 0$
$y < 0$

For each system, determine whether the given ordered pair is a solution.

5. $2x + 3y \geq 5$
$x + y \leq 7$
$(2, 3)$

6. $x + 3y < 15$
$x + 3y > 6$
$(3, 0)$

7. $y \leq -x + 6$
$x < 4$
$(4, 2)$

8. $2x + 3y < -9$
$x - 2y \geq -4$
$(-4, -6)$

Graph each system of inequalities. List one point in your solution set.

9. $y \leq x - 2$
$y \geq 3x - 5$

10. $y > \frac{1}{2}x - 1$
$y > -3x - 3$

11. $y \geq -2x + 6$
$y < \frac{1}{3}x - 5$

12. $x \geq 0$
$y < x + 5$

13. $y > 4$
$y \leq -2x + 3$

14. $x \leq 5$
$y > -\frac{1}{2}x + 6$

B. Exercises

Write the system represented by each graph.

15.

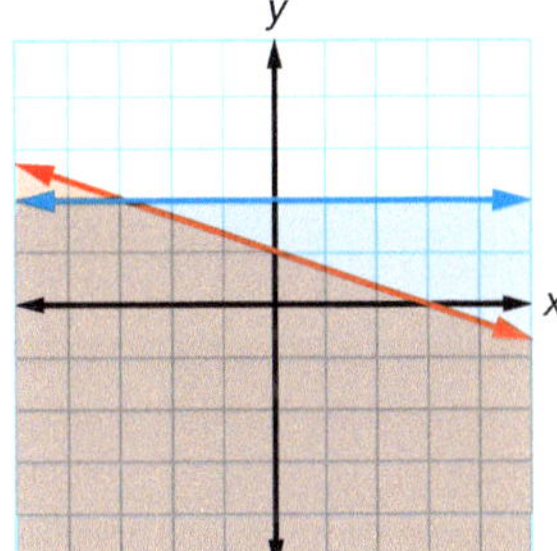

16.

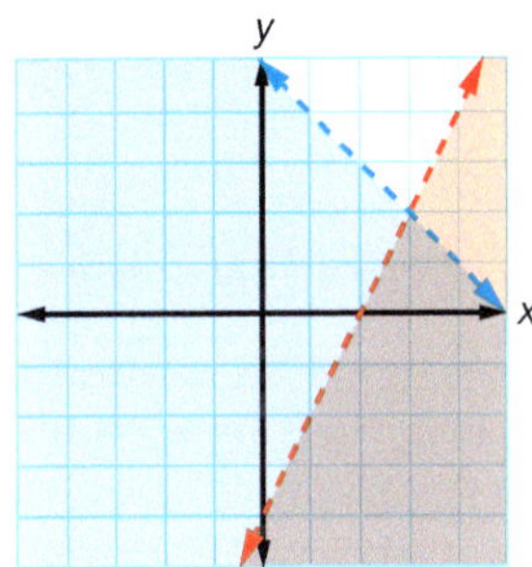

17.

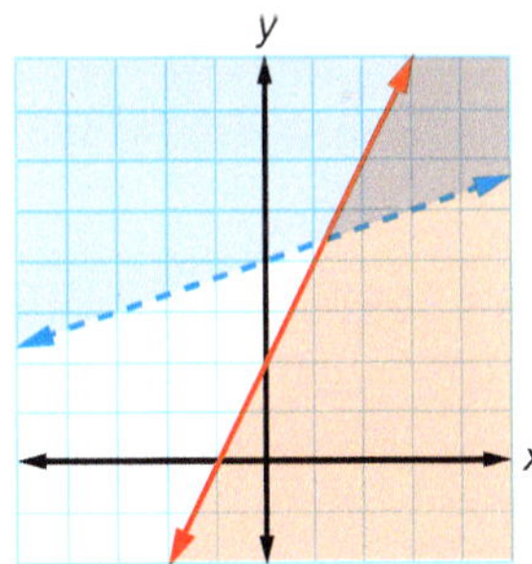

18.

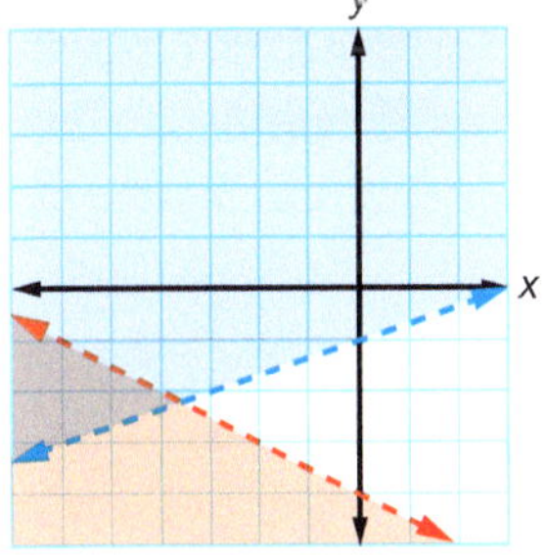

Graph each system of inequalities.

19. $x + y \geq 2$
$x - y < 4$

20. $5x + y > 2$
$2x + y \leq 1$

21. $3x - 2y \geq 4$
$x + y \leq 3$

22. $2x - y < 5$
$3x + y \geq 4$

23. $x + 5y < 15$
$3x + 2y \leq 8$

24. $x + 2y \geq -4$
$x + 2y < 6$

25. $x \geq 0$
$y \geq 0$
$5x + 6y \leq 30$

26. $x \geq 0$
$y \leq 2$
$x + y > 1$

27. $x \geq 0$
$y \geq 0$
$3x + y \leq 8$
$x + 2y \leq 6$

28. $x \geq 0$
$y \geq 0$
$2x + y \leq 4$
$x + 4y \leq 8$

Write the system represented by each graph.

29.

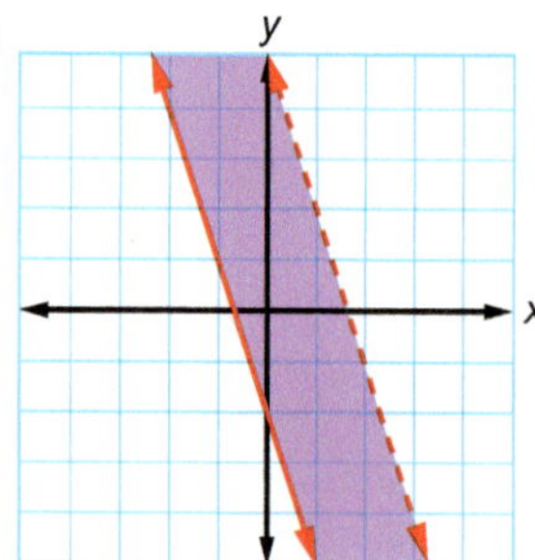

30.

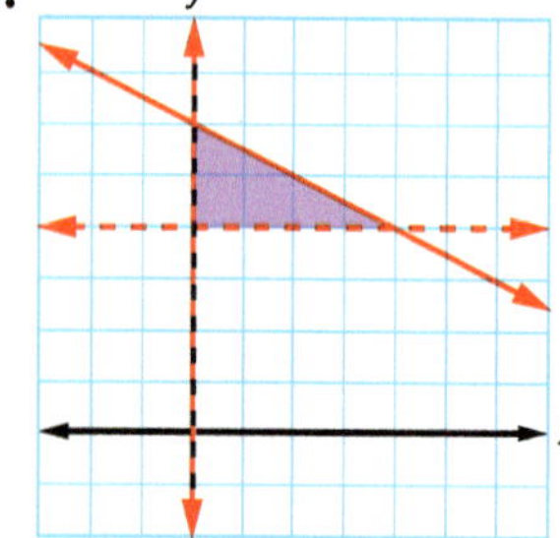

31.

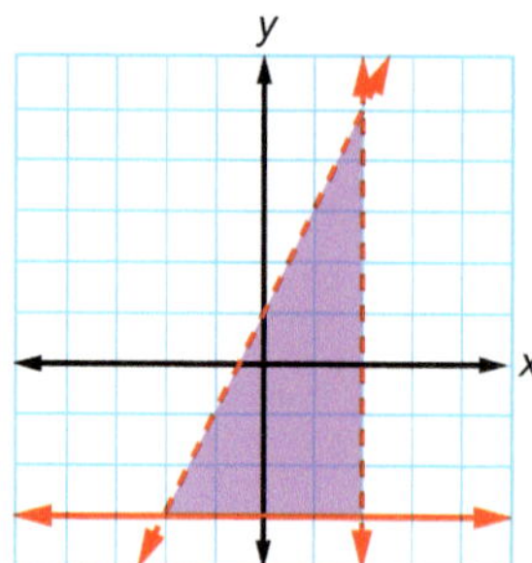

32.

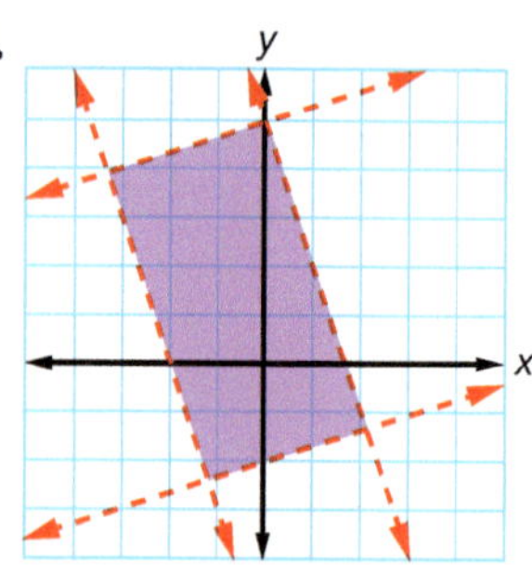

Write a system of inequalities and solve to find possible solutions.

33. Thomas is making desserts for a banquet. A chocolate silk pie costs $12 to make and serves 8. A small lemon chiffon cake costs $7 and serves 12. He must serve at least 72 people and can spend no more than $70.

a. Let x represent the number of chocolate silk pies, and let y represent the number of lemon chiffon cakes. Write one inequality describing the cost, another describing the number of servings, and then two more assuming that the number of pies and cakes cannot be negative.

b. Graph the system of inequalities.

c. Assuming that Thomas makes at least two of each dessert, complete the first two columns of the table by listing his options.

d. Complete the table by determining the number of servings and the cost for each option.

Chocolate Silk	Lemon Chiffon	Number of Servings	Cost

34. Phillip is investing at most $16,000 between two different accounts. The first investment is averaging a 5% return, and the second yields 8%. He wishes to make at least $5500 in a five-year time span.

a. Fill in the table using the $I = Prt$ formula.

b. Write a system of four inequalities, three desribing the principal and one describing the interest.

c. Graph the system of inequalities. (Hint: Use a scale of 1000.)

d. List three of Phillip's options. Stay with answers that are multiples of $2000.

P	*r*	*t*	*I*

35. Tristan is building a rectangular dog pen. The width, x, must be more than 10 ft, and the length, y, must be more than 12 ft; but the perimeter must not be more than 60 ft. Write and graph a system of inequalities that models these constraints and give one possible option.

C. Exercises

36. Explain why the system $4x + 5 < y$ and $4x - y > 3$ has no solutions.

37. Elizabeth makes doll clothes to sell online. She can make a dress for a 12 in. doll in 2.5 hr and a dress for a 19 in. doll in 2 hr. The materials for a smaller dress cost \$2, and the materials for a larger dress cost \$3. If she wishes to spend no more than 80 hr sewing and must spend no more than \$100, write and graph a system of inequalities that models her options. If she sells the dresses in sets of five, which option is closest to both limits of cost and time?

Dominion Modeling

Mr. Ring often has volunteers who travel with him and help distribute CDs. If the CDs are carried onto a plane, the size of each suitcase and its weight are important. A typical carry-on suitcase weighs 6 lb when empty and has interior dimensions of 12" by 12" by 8.5". Each CD in its jacket weighs 0.9 oz and takes up 1.5 in.^{3}.

38. Write an inequality that models the number of CDs that can be placed in a suitcase without exceeding a 25 lb weight limit for a carry-on.

39. Write a second inequality that models the number of CDs that will fit in the suitcase.

40. Write a simplified compound inequality that models the number of CDs that will fit in the suitcase. State the largest number of CDs that should be packed in it.

CUMULATIVE REVIEW

Solve each system by substitution. [7.2–7.3]

41. $y = x + 8$
$y = 2x - 3$

42. $x = y - 8$
$y = 2x + 9$

43. $x + 4y = 10$
$3x - y = -9$

44. $3x + y = 7$
$2x + 3y = -14$

45. $4x + 5y = 6$
$2x - 3y = 14$

Solve each system by elimination. [7.4]

46. $x + y = 3$
$x - y = 9$

47. $3x + 7y = 3$
$3x + 4y = 12$

48. $3x + 4y = -8$
$6x + 5y = -1$

49. $2x - 7y = 6$
$3x - 2y = -8$

50. $3x - 2y = 11$
$4x + 3y = -8$

CHAPTER 7 REVIEW

Determine whether (−2, 3) is a solution to each system.

1. $3x + 4y = 6$
   $y = 1 - x$

2. $5x - 2y = 7$
   $x + y = 1$

3. $x + y < 2$
   $x + 2y \geq 4$

Solve each system by graphing.

4. $x + y = 2$
   $2x - y = -5$

5. $2x + y = 1$
   $x + 3y = -12$

6. $x + 3y = 5$
   $y = 2$

Solve each system by substitution.

7. $x + 2y = 4$
   $2x + 3y = 8$

8. $5x + y = -14$
   $4x + 3y = -20$

9. $5x + y = 8$
   $-10x - 2y = -3$

Solve each system by elimination.

10. $x - 2y = -11$
    $3x + y = 16$

11. $5x - 3y = -25$
    $6x + 2y = -2$

12. $2x + 3y = 9$
    $4x - y = 2$

Solve each system using the method of your choice.

13. $6x + y = 7$
    $y = 1$

14. $8x - 3y = -12$
    $y - 4 = -8$

15. $x - y = -2$
    $4x - y = 4$

16. $y = 2x - 3$
    $y = \frac{1}{3}x + 7$

17. $x - 2y = -42$
    $3x - y = -31$

18. $2x - 7y = 3$
    $-4x + 14y = -6$

19. $2x + 5y = 8$
    $x - 5y = 4$

20. $0.5x - 2.1y = 7.2$
    $1.4x + 0.3y = 7.8$

21. $x + 3y = 11$
    $x + 3y = 7$

22. $3x + y = -2$
    $5x - 3y = 12$

23. Describe the relationship of two lines that have the same slope but different y-intercepts.
24. Describe the relationship of two lines that have different slopes.
25. Describe the relationship of two lines that have different slopes but the same y-intercept.
26. Describe the relationship of two lines that have the same slope and the same y-intercept.

Determine whether each system has one solution, no solutions, or an infinite number of solutions. Then classify the system as consistent independent, consistent dependent, or inconsistent.

27. $x + y = 5$
    $x + y = 2$

28. $8x - 3y = -12$
    $y - 4 = -8$

29. $x + y = 10$
    $2y = 20 - 2x$

30. $3x + 5y = 1$
    $4x - 3y = 11$

31. $3x + y = 6$
    $\frac{2}{3}y = 5 - 2x$

32. $3x + y = 4$
    $2y = 8 - 6x$

Write an algebraic expression that models each statement.

33. the distance traveled by a jet in 12 hr at x mi/hr
34. the upstream speed of a kayaker who goes 3 mi/hr in still water if he is traveling in a current of c mi/hr
35. the speed of a jet flying with a tailwind of 40 knots if its speed with no wind is r mi/hr
36. the annual interest received from an investment of d dollars at 7%

Using two variables, write and solve a system of equations for each word problem. Use a table when needed.

37. The sum of two numbers is -23, and their difference is 5. Find the two numbers.

38. Eleven is the sum of twice a number and three times a second number. The second number is equal to the difference of two and the first number. Find the two numbers.

39. The cost of renting the ice skating rink is \$200 plus \$10 per skater. The cost of renting the roller skating rink is \$300 plus \$7 per skater. For what number of people would the cost be almost the same?

40. Two cars leave from the same location but travel in opposite directions. The first car travels an average of 50 mi/hr and leaves 15 min before the second car, which travels an average of 55 mi/hr. How long will each car travel before they are 200 mi apart?

41. Two fishing boats leave Sandy Cove at the same time traveling in the same direction. One boat is traveling three times as fast as the other boat. After 5 hr the faster boat is 80 mi ahead of the slower boat. What is the speed of each boat?

42. A plane travels 800 mi to Chicago in 4 hr against a headwind. The return trip with a tailwind of the same speed takes 2.5 hr. Find the speed of the plane with no wind and the average wind speed.

43. Tessa wants to mix chocolates that cost \$4/lb with nuts that cost \$6/lb to make 6 lb of a mixture that costs \$31. How many pounds of each should she include in the mix?

44. Adrien needs 5 gal of a 23% salt solution. He has a 35% salt solution. How much pure water and how much 35% solution should he mix to get the desired solution? Round to the nearest tenth.

45. Sugar Tooth Candy Company needs 300 gal of a 32% sucrose solution for a certain kind of candy. The company has a solution that is 60% sucrose and a solution that is 25% sucrose. How many gallons of each should the company mix together to obtain the desired solution?

46. Mr. Arnold invested \$1550 between two accounts. One paid 7.5% interest and the other 8.25%. How much did he invest in each account if his first year's return on his investment was \$121.20?

Graph each system of inequalities.

47. $2x + y < 6$
$3x - y \geq 4$

48. $5x + 2y \leq 8$
$x + y > -5$

49. $3x + 5y \geq 12$
$2x - 3y < -6$

50. $y \leq \frac{1}{3}x + 4$
$y > 2$

8 EXPONENTS

The Internet is an extremely powerful tool, providing valuable information in countless areas. Whether you are looking for help on a math concept or trying to find a good church, college, or job, Internet searches can produce the valuable information you need. The Internet also enables us to reach more people with the gospel than was ever possible before in the history of the world.

The power of the Internet is in its vastness. This vastness provides both true and false information, both godly and ungodly content. God has given us dominion, resources, and liberties, not so that we would use them as opportunities to sin, but so that we would serve one another in love and serve God with fear (Gal. 5:13; Heb. 12:28).

How vast is the Internet? The amount of information can be expressed in zettabytes, or 10^{21} bytes, where each byte may represent a single character. Exponential and scientific notation are used to simplify the calculations involved with very large and very small numbers. In the Domain Modeling exercises in this chapter, you will discover some fascinating numbers related to the Internet.

After this chapter you should be able to

1. multiply, divide, and find powers of expressions containing exponents.
2. simplify expressions with negative exponents.
3. use scientific notation to express and compute very small and very large numbers.
4. graph and translate the power functions $y = x^2$ and $y = x^3$.
5. graph exponential functions.
6. write and evaluate functions modeling exponential growth and decay.

8.1 Products and Powers

These people are learning biblical principles at a Sunday morning service in Togo, Africa.

The application of learned concepts is an important part of a successful life. We must take scriptural principles from the Bible, learn them well, hide them in our hearts (Ps. 119:9–11), and then properly apply them to a variety of situations. Likewise, in math we must take learned principles and properly transfer them to a variety of applications. In this chapter you will expand previously learned concepts to some new situations.

The principles used to simplify numerical expressions also apply to algebraic expressions. Review the following properties of exponents covered in Chapter 1.

Properties of Exponents		
Product Property	$x^a \cdot x^b = x^{a+b}$	To multiply powers with like bases, add the exponents.
Power Property	$(x^a)^b = x^{ab}$	To raise a power to a power, multiply the exponents.

Example 1

Simplify.

a. $x^5 \cdot x^2$ **b.** $(x^5)^2$ **c.** $y^4(y^3)^2$

Answer

a. $x^5 \cdot x^2 = x^{5+2} = x^7$ — Apply the Product Property of Exponents.

b. $(x^5)^2 = x^{5(2)} = x^{10}$ — Apply the Power Property of Exponents.

c. $y^4(y^3)^2 = y^4y^6$
$= y^{4+6} = y^{10}$

1. Simplify the power first by applying the Power Property of Exponents.
2. Then simplify the product by applying the Product Property of Exponents.

The Commutative and Associative Properties of Multiplication are used for multiplying algebraic expressions.

Example 2

Simplify $3x^2 \cdot 5x$.

Answer

$(3 \cdot 5)(x^2 \cdot x)$
$= (3 \cdot 5)(x^{2+1})$
$= 15x^3$

1. Apply the Commutative and Associative Properties of Multiplication.
2. Apply the Product Property of Exponents.
3. Simplify.

Example 3

Simplify $-8x^2yz^3(xyz^2)$.

Answer

$-8(x^2 \cdot x)(y \cdot y)(z^3 \cdot z^2)$ — 1. Apply the Commutative and Associative Properties of Multiplication.

$= -8x^{2+1}y^{1+1}z^{3+2}$ — 2. Apply the Product Property of Exponents.

$= -8x^3y^2z^5$ — 3. Simplify.

Example 4

Simplify $(-ab^2)^3$.

Answer

$(-ab^2)^3 = (-ab^2)(-ab^2)(-ab^2)$ — 1. Rewrite the power as repeated multiplication.

$= -1(-1)(-1)(aaa)(b^2b^2b^2)$ — 2. Apply the Commutative and Associative Properties of Multiplication.

$= -1a^{1+1+1}b^{2+2+2}$ — 3. Apply the Product Property of Exponents.

$= -a^3b^6$ — 4. Simplify.

The Commutative and Associative Properties can be used to show that the power of a product is equal to the product of the powers. In general, this can be stated

$$(abc)^n = a^nb^nc^n.$$

Example 5

Simplify $(-2m^4n)^3\left(\frac{1}{6}mn^2\right)^2$.

Answer

$(-2)^3(m^4)^3n^3\left(\frac{1}{6}\right)^2m^2(n^2)^2$ — 1. The power of a product is the product of the powers.

$= -8m^{12}n^3\left(\frac{1}{36}\right)m^2n^4$ — 2. Apply the Power Property of Exponents.

$= -8\left(\frac{1}{36}\right)m^{12+2}n^{3+4}$ — 3. Apply the Product Property of Exponents.

$= -\frac{2}{9}m^{14}n^7$ — 4. Simplify.

Remember that only like terms can be added when you are simplifying.

Example 6

Simplify $x^2xy^2 + 3xyx^2 - 5x^3y^2$.

Answer $x^3y^2 + 3x^3y - 5x^3y^2$ — 1. Apply the Power Property of Exponents.

$= -4x^3y^2 + 3x^3y$ — 2. Add like terms.

A. Exercises

Simplify.

1. x^3x^5
2. y^3y^2
3. $x^2y^3x^5y^2$
4. $m^3n^6m^2np^4$
5. $3w^3 \cdot 4w^4$
6. $-5z^7 \cdot 7z^{10}$
7. $(a^4)^3$
8. $(b^7)^2$
9. $c^5(c^2)^4$
10. $d(d^4)^5d^6$
11. $(-2b^2cd^5)^2$
12. $(-a^4)^3$

B. Exercises

Simplify.

13. $2a^2(-5b^3)$
14. $-3x^4(-12x^2)$
15. $4abc(-3a^3bc)$
16. $x^4y^3z(-2xy^5z^2)$
17. $(5x^3)^3(2x)^2$
18. $2x^6y^2(5x^6y^3)^3$
19. $(-3x^3yz^8)^2(-5x^3yz^6)$
20. $(6x^3y^7)^2\left(-\frac{1}{2}x^2y^3\right)^3$
21. $\left(\frac{1}{3}x^3y^2z\right)^2(2x^3y^2z)^3(18x)$
22. $(4xy^2)^2(0.25x^3y^5)(-2x^7)^4$
23. $3b^2cc - c^2b + 2bbc^2$
24. $2h^2g^4h^5g - 7hg^2h^6g^3 - 4g^2h^3h^4g^3$
25. $5x^5y^4 - 5x^3y^2xy^3 + 2y^2x^2y^2x^3 - x^4y^5$
26. $2wzw^2z + w^2z^3 - 2wz^3w + 3wz^2w^2$

Write a simplified algebraic expression for each quantity.

27. area

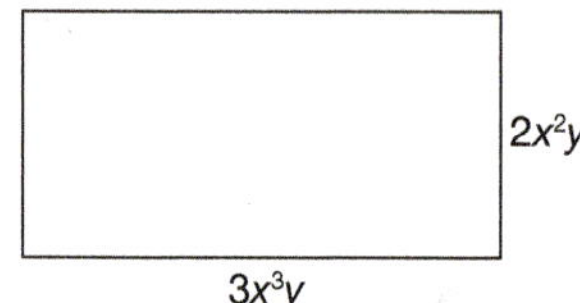

28. area

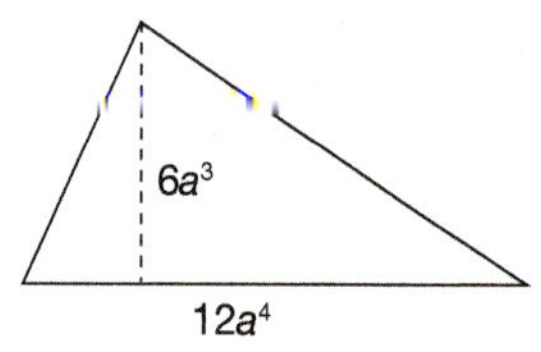

29. volume

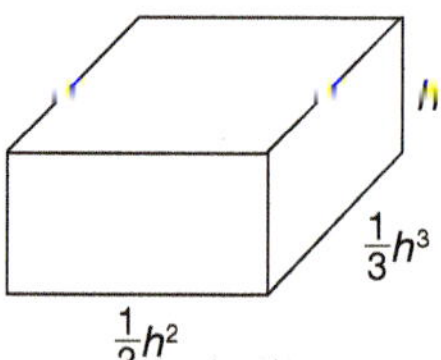

30. the volume of a cube with an edge of length $3a^3b$

Simplify.

31. $x^n \cdot x^3$
32. $2^x \cdot 2^{5-x}$
33. $a^x \cdot a^{x+2}$
34. $(y^z)^5 \cdot y^3$

C. Exercises

Use the Distributive Property to simplify each expression.

35. $xy^2(3x - 2y)$
36. $2a^3b(5a^2 + 7b^4)$
37. $\frac{1}{2}rs^2(-8r^3s + 2rs^3)$
38. $-\frac{5}{6}m^6n^2(3m^2n^4 - 4n^7)$
39. Use the Commutative and Associative Properties to show that $(abc)^3 = a^3b^3c^3$.

Dominion Modeling

Computers process, store, and transfer information in binary form, sets of 0s and 1s. This binary engineering is the reason data storage is expressed in powers of two. A byte of information is a set of 8 bits (0s and 1s) and can represent a single character. A kilobyte is 1024 or 2^{10} bytes. A megabyte is 1024 kilobytes; it is therefore 1024(1024) bytes or $2^{10}(2^{10}) = 2^{20}$ bytes. These names are derived from the base 10 approximations of the exact number of bytes. For example, the actual 2^{10} or 1024 bytes in 1 KB can be estimated as 1000 or 10^3 bytes.

40. Annual global Internet Protocol (IP) traffic approached 1 zettabyte in 2010. To gain more appreciation for the size of a zettabyte, follow the patterns to complete the table below.

	Exact Number of Bytes		Approximate Number of Bytes	
	Power of 1024	Power of Two	In Words	Power of Ten
Kilobyte (KB)	1024^1	2^{10}	1 thousand	10^3
Megabyte (MB)	1024^2	2^{20}	1 million	10^6
Gigabyte (GB)	1024^3			
Terabyte (TB)				
Petabyte (PB)				
Exabyte (EB)				
Zettabyte (ZB)				

CUMULATIVE REVIEW

41. Define b^x, where x is a natural number. [1.7]

Write using repeated factors; then evaluate. [1.7]

42. $\left(\frac{2}{3}\right)^3$

43. $(-2)^{-4}$

44. $\frac{1}{-3^{-4}}$

45. $-\left(\frac{4}{5}\right)^{-2}$

Explain the concept of "none" in each context below.

46. There are no apples in the baskets. [1.1]

47. There is no written coefficient of a term. [2.3]

48. A variable has no written exponent. [1.7]

49. A vertical line has no slope. [6.2]

50. A compound inequality has no solution. [4.4]

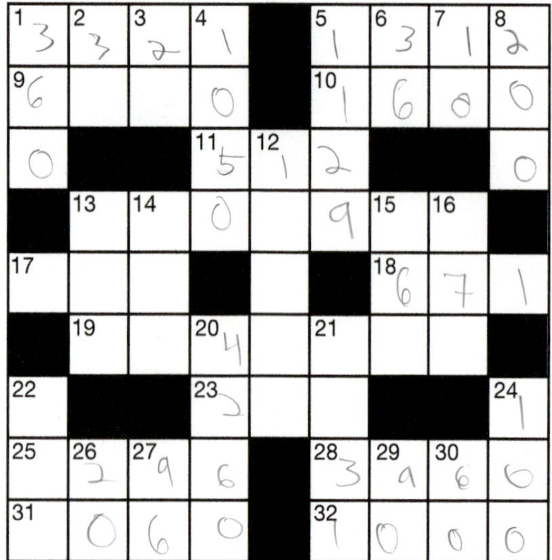

ACROSS

1. 36 plus the number of days in nine regular years
5. cups in 82 gal
9. digits > 5
10. White House street number
11. 2^9
13. date the *Eagle* landed on the moon
17. chapters in the New Testament
18. $\sqrt{450{,}000}$ to the nearest integer
19. date Lincoln was shot
23. Fahrenheit temperature at which water boils
25. square inches in a square yard
28. feet in $\frac{3}{4}$ mi
31. sum of all primes < 100
32. pounds in a ton

DOWN

1. degrees in a circle
2. normal body temperature (Celsius)
3. days in a lunar month
4. middle year of the twentieth century
5. 37^2 minus minutes in 4 hr
6. inches in a yard
7. base of the decimal system
8. America's age in 1976
12. 31 in the binary system
13. a multiple of 191
14. CCI
15. the number of the Beast
16. weeks in $18\frac{3}{4}$ yr
20. 1020 subtracted from feet in a mile
21. UCDB on the phone
22. product of two primes < 40 whose difference is 34
24. centimeters in a meter
26. years in two decades
27. eight dozen
29. degrees in a right angle
30. seconds in a minute

8.2 Quotients

In literature and Scripture we find various forms of the word *divide*. Although similar, these references often have a slightly different meaning from the mathematical concept of division. In literature, the term *divide* can mean to separate or organize into distinct parts, groups, classes, or categories; or to distinguish. God distinguished the light from the darkness, the day from the night (Gen. 1:4–5). Through the Word, we learn to distinguish good from evil, right from wrong, godly from ungodly (Heb. 5:14; Isa. 5:20). Distinguishing these elements in our lives and dividing mathematical expressions are important to our spiritual and intellectual growth.

In this section you will increase your skills and understanding of positive and negative exponents as you work with quotients containing exponents.

Example 1

Simplify $32a^3 \div 4a$.

Answer $\frac{32a^3}{4a} = \frac{32}{4} \cdot \frac{\cancel{a}aa}{\cancel{a}} = 8a^2$

The Quotient Property of Exponents was applied to numeric expressions in Chapter 1. It can also be applied to algebraic expressions.

Quotient Property of Exponents	
$\frac{x^a}{x^b} = x^{a-b}$ for $x \neq 0$	To divide with like bases, subtract the exponents.

Example 2

Simplify $\frac{24x^2y^6z^7}{-8xy^2z^5}$.

Answer $\frac{24x^2y^6z^7}{-8xy^2z^5} = -3x^{2-1}y^{6-2}z^{7-5} = -3xy^4z^2$

Recall that negative exponents were defined in Chapter 1 as indicating a quotient.

$$x^{-n} = \frac{1}{x^n}$$

This definition is consistent with the Quotient and Product Properties of Exponents.

$$a^2a^{-6} = a^{2+(-6)} = a^{-4} \text{ and } a^2a^{-6} = \frac{a^2}{a^6} = \frac{\not{a}\not{a}}{\not{a}\not{a}aaaa} = \frac{1}{a^4}, \text{ so } a^{-4} = \frac{1}{a^4}.$$

$$\frac{b^3}{b^5} = b^{3-5} = b^{-2} \text{ and } \frac{b^3}{b^5} = \frac{\not{b}\not{b}\not{b}}{\not{b}\not{b}\not{b}bb} = \frac{1}{b^2}, \text{ so } b^{-2} = \frac{1}{b^2}.$$

Notice the result of applying this definition to a negative exponent in the denominator.

$$\frac{1}{x^{-n}} = \frac{1}{\frac{1}{x^n}} = 1 \cdot \frac{x^n}{1} = x^n, \text{ so } \frac{1}{x^{-n}} = x^n.$$

A quantity with a negative exponent is equivalent to the reciprocal of that power with a positive exponent. This definition implies that $a^{-2} = \frac{1}{a^2}$ and that $\frac{1}{5^{-2}} = 5^2 = 25$.

Example 3

Write an equivalent expression in positive exponential form.

a. $x^{-4}y^2$ **b.** $2x^{-2}$ **c.** $(-3x^{-2})^3$ **d.** $\left(\frac{1}{3}x^{-3}\right)^{-2}$

Answer

a. $x^{-4}y^2 = \frac{1}{x^4}(y^2) = \frac{y^2}{x^4}$

b. $2x^{-2} = 2\left(\frac{1}{x^2}\right) = \frac{2}{x^2}$

Use the definition of a negative exponent to write the expression using only positive exponents.

c. $(-3x^{-2})^3 = (-3)^3x^{-2(3)} = -27x^{-6} = -27\left(\frac{1}{x^6}\right) = -\frac{27}{x^6}$

Apply the Power Property of Exponents.

d. $\left(\frac{1}{3}x^{-3}\right)^{-2} = \left(\frac{1}{3}\right)^{-2}x^{-3(-2)} = 3^2x^6 = 9x^6$

The meaning of a power with an exponent of zero is also defined to be consistent with the Quotient Property.

$$x^0 = 1 \text{ since } \frac{x^3}{x^3} = x^{3-3} = x^0 \text{ and } \frac{x^3}{x^3} = 1.$$

Example 4

Divide $34x^{-3}y^2z^{-5}$ by $-18x^{-6}y^2z^{-2}$ and express the answer in positive exponential form.

Answer

$$\frac{34x^{-3}y^2z^{-5}}{-18x^{-6}y^2z^{-2}} = \frac{34}{-18}x^{-3-(-6)}y^{2-2}z^{-5-(-2)}$$

1. Apply the Quotient Property of Exponents.

$$= -\frac{17}{9}x^3y^0z^{-3}$$

$$= -\frac{17x^3(1)}{9z^3} = -\frac{17x^3}{9z^3}$$

2. Simplify and write the expression using only positive exponents.

By viewing the power as repeated multiplication, we can show that the power of a quotient is equal to the quotient of the powers.

$$\left(\frac{a}{b}\right)^n = \frac{a^n}{b^n}$$

Example 5

Simplify $\left(\frac{y}{x^2}\right)^3(x^3y^{-4})^2$, leaving all variables in the numerator.

Answer

$\left(\frac{y}{x^2}\right)^3(x^3y^{-4})^2 = \frac{y^3}{x^6}(x^6y^{-8})$ — 1. Simplify each power.

$= x^{6-6}y^{3+(-8)}$ — 2. Apply the Product and Quotient Properties of Exponents.

$= x^0y^{-5} = y^{-5}$ — 3. Simplify, leaving the variable in the numerator.

An expression containing negative exponents in the denominator is not considered to be simplified. An expression may be considered simplified if all variables are listed in the numerator. However, it is generally preferable to use only positive exponents in a simplified expression.

Example 6

Simplify.

a. $\frac{-2x^{-2}(-3xy^{-2})}{12y^3}$ **b.** $\frac{8x^2z^{-3}}{y}\left(\frac{yz^{-4}}{2x^5}\right)^{-2}$

Answer

a. $\frac{-2x^{-2}(-3xy^{-2})}{12y^3} = \frac{-2(-3)}{12}x^{-2+1}y^{-2-3}$ — 1. Apply the Product and Quotient Properties of Exponents.

$= \frac{1}{2}x^{-1}y^{-5} = \frac{1}{2xy^5}$ — 2. Rewrite the expression using only positive exponents.

b. $\frac{8x^2z^{-3}}{y}\left(\frac{yz^{-4}}{2x^5}\right)^{-2} = \frac{8x^2z^{-3}}{y} \cdot \frac{y^{-2}z^8}{2^{-2}x^{-10}}$ — 1. Multiply exponents to find the power of a power.

$= 8(4)x^{2-(-10)}y^{-2-1}z^{-3+8}$ — 2. Apply the Product and Quotient Properties of Exponents.

$= 32x^{12}y^{-3}z^5 = \frac{32x^{12}z^5}{y^3}$ — 3. Rewrite the expression using only positive exponents.

or

$\frac{8x^2z^{-3}}{y}\left(\frac{yz^{-4}}{2x^5}\right)^{-2} = \frac{8x^2z^{-3}}{y} \cdot \frac{y^{-2}z^8}{2^{-2}x^{-10}}$ — 1. Multiply exponents to find the power of a power.

$= \frac{8x^2}{yz^3} \cdot \frac{2^2x^{10}z^8}{y^2}$ — 2. Use the definition of a negative exponent to write the expression using only positive exponents.

$= \frac{32x^{12}z^5}{y^3}$ — 3. Apply the Product and Quotient Properties of Exponents.

A. Exercises

Simplify, leaving each answer in positive exponential form.

1. $\frac{18x^7}{2x^2}$
2. $\frac{36y^5}{24y}$
3. $\frac{-12a^3b^6}{16a^2b}$
4. $\frac{-3r^5s^9t}{18rs^8t}$
5. $\frac{1}{z^{-2}}$
6. x^{-5}
7. $2x^{-3}$
8. $\frac{x^{-1}}{y^{-1}}$
9. $\frac{x^{-2}}{x^{-3}}$
10. $\frac{-z^2}{-z^{-3}}$
11. $\frac{a^4a^2}{a^5a}$
12. $\frac{-j(-k)^{-2}}{j^2k}$
13. $(x^2y^{-5})^3$
14. $(x^2y^{-1})^{-3}$
15. $\left(\frac{x}{y}\right)^3$
16. $\left(\frac{x}{y}\right)^{-3}$

B. Exercises

Simplify, leaving all variables in the numerator.

17. $16x^6y^{-4} \div 2xy^5$
18. $-27a^3b^6 \div 9a^4b^{-2}$
19. $4a^6b^{-7}c^2 \div 20a^6b^2c^7$
20. $\frac{x^3y^{-5}}{(2x)^{-2}}$
21. $\frac{10b^{-2}c^4}{-2a^{-2}b^3}$
22. $\frac{a^{-1}b^{-2}c^{-3}}{a^3b^2c}$

Simplify, leaving each answer in positive exponential form.

23. $\frac{-6x^5y^2z^3}{7x^2y^2z^2}$
24. $\frac{38x^2y^3z^{-1}}{2x^{-2}y^{-3}}$
25. $\frac{-3^{-1}ab^{-4}c^2}{3^{-1}a^{-5}b^7c^{-3}}$
26. $\frac{6^{-1}x^3y^6}{2^{-2}x^{-1}y^2}$
27. $\frac{ab^{-2}}{2b}\left(\frac{a^{-4}}{b^{-3}}\right)$
28. $\frac{ax^2}{b^2y} \div \frac{cx^2}{by^3}$
29. $\left(\frac{x^2y^3}{y}\right)^{-2}$
30. $x^2(x^2y^3)^{-3}$
31. $\frac{(abc^2)^2}{-3bc}$
32. $\left(\frac{3x}{y^4}\right)^{-2}\left(\frac{x^2}{y}\right)^{-1}$

Write a simplified algebraic expression for each quantity.

33. height

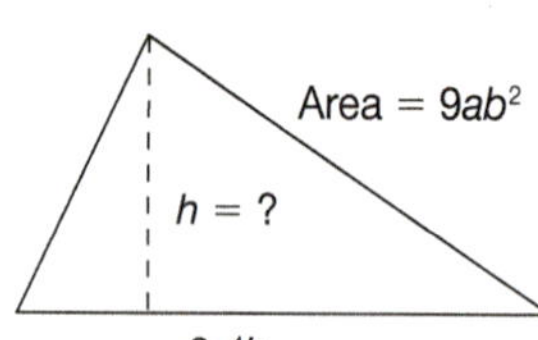

34. width

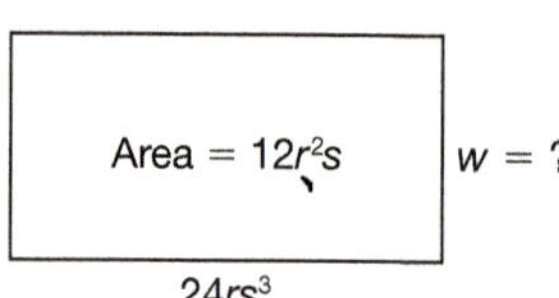

35. height

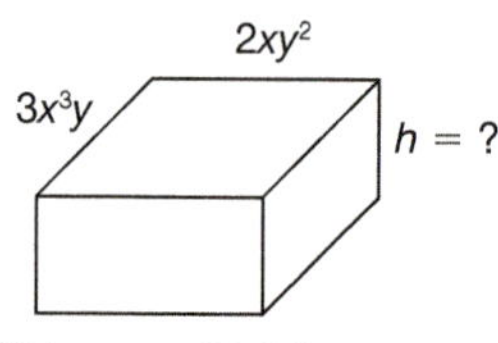

36. the ratio of the volume of a sphere ($V = \frac{4}{3}\pi r^3$) to the volume of a circular cylinder containing that sphere ($V = \pi r^2H$ with $H = 2r$)

Simplify.

37. $\frac{a^{z+2}}{a^z}$
38. $\frac{y^{3a}}{y^a}$
39. $\frac{x^{2a}}{x^b}$
40. $\frac{(b^3)^x}{b^{-5}}$

C. Exercises

Use the Distributive Property to simplify each expression.

41. $\frac{x}{y^2}\left(\frac{x}{y^2} - \frac{y^2}{x}\right)$
42. $\frac{a^2}{bc^3}\left(\frac{cb}{a^2} + \frac{ab^4}{c^2}\right)$
43. $\frac{1}{2}rs^{-3}(2r^{-2}s - 6r^{-1}s^3)$
44. $m^{-6}n^2(3m^2n^{-4} + 4m^9n^{-2})$
45. Show that $\left(\frac{a}{b}\right)^3 = \frac{a^3}{b^3}$ (i.e., the power of a quotient is the quotient of the powers).

Dominion Modeling

Bandwidth is the amount of data that can be transferred from one point to another in a given amount of time. These rates of transfer are usually expressed in bits per second. Older dial-up modems typically transferred data at a rate of 56 Kbps. Broadband technology transmits data simultaneously over multiple frequencies (a "broad band"). A T3 ultra high-speed network connection (also known as DS3) can transfer data at a rate up to 45 Mbps, equivalent to using more than 600 ordinary telephone lines. Fiber optic connections such as OC-3 have a transfer rate of 155 Mbps, and newer technologies allow even faster data transfer rates.

A text file containing the entire King James Version of the Bible (with a brief introduction but no images or extra commentary) is about 2000 pages long and contains 4,720,600 characters, including chapter and verse numbers. This file is 4.8 MB (about 38.4 megabits) in size.

46. Assuming 1 megabit = 1000 kilobits, how long would it take (to the nearest tenth of a second) to download the entire 38.4 megabit KJV Bible file with an older 56 Kbps (kilobits per second) modem connection?

47. How long would it take (to the nearest hundredth of a second) to transfer the entire 38.4 megabit KJV Bible file across a 45 Mbps T3 connection?

48. How long would it take (to the nearest hundredth of a second) to transfer the entire 38.4 megabit KJV Bible file across a 155 Mbps OC-3 connection?

Actual download times are almost always longer because the stated maximum theoretical rates are seldom achieved, even under ideal conditions.

CUMULATIVE REVIEW

State the domain and range of each relation and whether it is a function. [5.2]

49. $\{(-2, 8), (4, 9), (8, 9), (9, 4)\}$

50. $\{(-19.2, -3.5), (1.7, -18), (6.4, 18.2)\}$

Rebekah and Justine notice that the time it takes to travel the 300 mi to their destination is determined by the rate at which they travel. [5.2, 5.6]

51. Define the independent variable and the dependent variable in this relation.

52. Is the relation a direct variation, an inverse variation, or neither?

53. Write a function rule modeling the relation between the two variables.

Write the equation of each line in slope-intercept form. [6.4]

54. a line with a y-intercept of $(0, -3)$ and a slope of $\frac{1}{2}$

55. a line passing through the point $(1, 4)$ with a slope of -2

56. a line passing through $(-5, -2)$ and $(-2, 4)$

57. a vertical line passing through $(-3, 7)$

58.

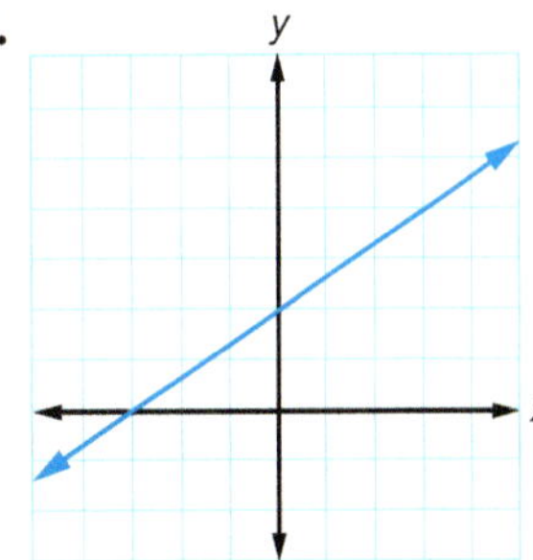

MATH IN HISTORY

Karl Friedrich Gauss (1777–1855)

German mathematician and physicist Karl Gauss is viewed by math historians as one of the three greatest mathematicians of all time. While still a teenager, he solved three previously unsolved math problems, including becoming the first to construct a regular heptadecagon (seventeen-sided polygon) using only a compass and a straightedge.

He also developed the method of least squares, a crucial advancement in statistics. On a scatterplot, the vertical distances from each point to a curve are measured. All the measurements are then squared to eliminate the negative values from points below the curve. The sum of these squares measures how well the curve fits the data. Gauss used this method to relocate a newly discovered asteroid, Ceres, after it had disappeared behind the sun. This method is used today to determine the equation that best fits a set of data. From such an equation, industrial and financial decisions are made to help us manage the earth's resources and improve people's lives.

His third discovery was a proof of the law of quadratic reciprocity, which was later included in his 1801 number theory masterpiece, *Arithmetical Investigations*. That treatise developed modular arithmetic, which today is used to prevent credit card fraud and to track books in large libraries, among many other beneficial uses.

Gauss also gave the first full and correct proof of the Fundamental Theorem of Arithmetic, which states that every natural number has a unique prime factorization. His doctoral thesis discussed the Fundamental Theorem of Algebra, which states that every algebraic equation has at least one solution. His broad range of interests enabled Gauss to make important contributions to all existing branches of mathematics.

8.3 Scientific Notation

Are you familiar with the abbreviation "slap"? Imagine communicating without abbreviations such as lol, ugtbk, jk, or slap (sounds like a plan). Texting abbreviations and other symbolic writing methods increase speed and save space.

Scientists also use an abbreviated notation when writing numbers. The estimated number of atoms in a 154 lb human body, 7,000,000,000,000,000,000,000,000,000, can be written in shorter scientific notation as 7×10^{27}. The mass of an electron, 0.0000000000000000000000000000009109 kg, is written in scientific notation as 9.109×10^{-31} kg. These numeric abbreviations save time and space.

Definition

A number is written in **scientific notation** when it is expressed as a product in the form $a \times 10^n$, where $1 \leq a < 10$.

To write 4,200,000 in scientific notation, move the decimal point six places to the left to form a number between one and ten. This is equivalent to dividing by one million, so the result must be multiplied by one million (10^6) to express the same number in a different form: 4.2×10^6. Notice that 4.2×10^6 can be read or written as 4.2 million.

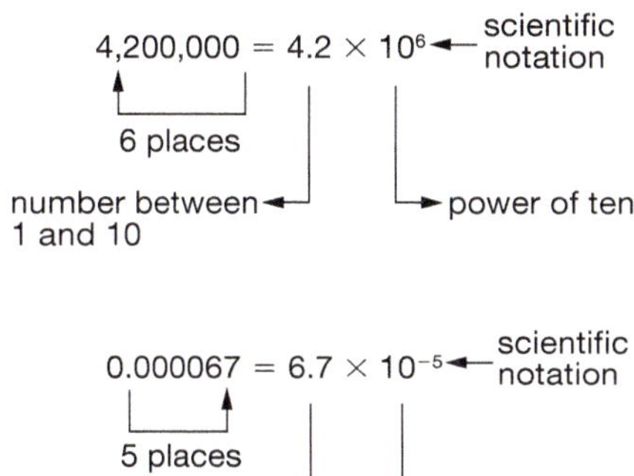

To express a number such as 0.000067 in scientific notation, the decimal point must be moved five places to the right. This is equivalent to multiplying by 10^5. The result, 6.7, must then be divided by 10^5, the equivalent of multiplying by 10^{-5}, to express the original number in scientific notation as 6.7×10^{-5}.

Example 1

Express each number in scientific notation.

a. 29,500 **b.** 0.000000043

Answer

a. $29{,}500 = 2.95 \times 10^4$ — Move the decimal point four places to the left (dividing by 10^4) and multiply the result by 10^4.

b. $0.000000043 = 4.3 \times 10^{-8}$ — Move the decimal point eight places to the right (multiplying by 10^8) and multiply the result by 10^{-8}.

To convert a number written in scientific notation into standard notation, complete the indicated multiplication or division by moving the decimal the indicated number of decimal places to the right or left.

Example 2

Express each number in standard notation.

a. 2.1×10^5 **b.** 6.8×10^{-6}

Answer

a. $2.1 \times 10^5 = 210{,}000$ — Multiply by 10^5 by moving the decimal point five places to the right.

b. $6.8 \times 10^{-6} = 0.0000068$ — Divide by 10^6 (the equivalent of multiplying by 10^{-6}) by moving the decimal point six places to the left.

Use the Commutative and Associative Properties of Multiplication and the properties of exponents to calculate the products, powers, and quotients of numbers in scientific notation. Be sure to write the result of such calculations in correct scientific notation.

Example 3

Simplify.

a. $(7.5 \times 10^{-5})(2 \times 10^3)$ **b.** $(1.2 \times 10^{-4})^2$ **c.** $(2.4 \times 10^7) \div (6 \times 10^2)$

Answer

a. $(7.5 \times 10^{-5})(2 \times 10^3)$

$= (7.5 \times 2)(10^{-5} \times 10^3)$ 1. Reorder and regroup.

$= 15 \times 10^{-2}$ 2. Add exponents to multiply the powers of ten.

$= (1.5 \times 10^1) \times 10^{-2}$ 3. Express the result in scientific notation.

$= 1.5 \times 10^{-1}$

b. $(1.2 \times 10^{-4})^2$

$= 1.2^2 \times (10^{-4})^2$ 1. The power of a product is the product of the powers.

$= 1.44 \times 10^{-8}$ 2. Multiply exponents to find the power of a power. Notice that the result is already in scientific notation.

c. $\frac{2.4 \times 10^7}{6 \times 10^2} = \frac{2.4}{6} \times \frac{10^7}{10^2}$ 1. Write the division problem as a product of fractions.

$= 0.4 \times 10^5$ 2. Subtract exponents to divide the powers of ten.

$= (4 \times 10^{-1}) \times 10^5$ 3. Express the result in scientific notation.

$= 4 \times 10^4$

Before adding or subtracting numbers in scientific notation, make sure that both numbers are expressed using the same power of ten. It is usually easier to rename the number with the lower power of ten to match the higher power of ten.

Example 4

Simplify.

a. $(2.3 \times 10^6) + (5 \times 10^6)$ **b.** $(5.4 \times 10^5) - (4.7 \times 10^4)$

Answer

a. $(2.3 \times 10^6) + (5 \times 10^6)$

$= (2.3 + 5) \times 10^6$

$= 7.3 \times 10^6$

The Distributive Property implies that $ac + bc = (a + b)c$.

b. $(5.4 \times 10^5) - (4.7 \times 10^4)$

$= (5.4 \times 10^5) - [(4.7 \times 10^{-1}) \times (10^4 \times 10^1)]$

$= (5.4 \times 10^5) - (0.47 \times 10^5)$

$= (5.4 - 0.47) \times 10^5$

$= 4.93 \times 10^5$

1. Multiplying 10^4 by 10^1 to get 10^5 requires the decimal factor to be multiplied by 10^{-1}.
2. The Distributive Property can be used when the powers of ten are the same.

The prefixes in the metric system are another way to shorten the notation used to represent a number. Knowing the prefixes and values in this chart will simplify calculations involving very large or very small numbers.

Prefix	10^n	Value
tera (T)	10^{12}	trillion
giga (G)	10^9	billion
mega (M)	10^6	million
kilo (K or k)	10^3	thousand
milli (m)	10^{-3}	thousandth
micro (μ)	10^{-6}	millionth
nano (n)	10^{-9}	billionth

Example 5

The mass of Earth is about 6×10^{24} kg, the mass of Mars is about 6×10^{23} kg, and the mass of Jupiter is about 2×10^{27} kg.

a. What is the difference in the masses of Earth and Mars?

b. Approximately how many times more massive is Jupiter than Earth?

Answer

a. $6 \times 10^{23} = (6 \times 10^{-1}) \times (10^{23} \times 10^{1}) = 0.6 \times 10^{24}$

$(6 \times 10^{24}) - (0.6 \times 10^{24}) = 5.4 \times 10^{24}$ kg

Rename the expression with the lower power of ten before subtracting.

b. $\frac{2 \times 10^{27} \text{ kg}}{6 \times 10^{24} \text{ kg}}$

1. Since the units are the same, divide the numbers.

$= \frac{1}{3} \times \frac{10^{27}}{10^{24}} = 0.\overline{3} \times 10^{3}$

2. Simplify.

$\approx (3 \times 10^{-1}) \times 10^{3}$

$= 3 \times 10^{2}$

Jupiter is about 300 times more massive than Earth.

A. Exercises

Express each number in scientific notation.

1. 390,000,000

2. 273,000

3. 0.000052

4. 0.0032

5. 13.6×10^{7}

6. 1067×10^{-8}

7. 235.6×10^{-4}

8. 135.8×10^{8}

Express each number in standard notation.

9. 2.54×10^{8}

10. 3.4×10^{7}

11. 1.53×10^{-4}

12. 2.89×10^{-6}

B. Exercises

Simplify, expressing each answer in scientific notation.

13. $(2 \times 10^{2})(3 \times 10^{5})$

14. $(6 \times 10^{7})(4 \times 10^{7})$

15. $(3 \times 10^{5})^{4}$

16. $(9 \times 10^{8}) \div (3 \times 10^{3})$

17. $(4 \times 10^{11}) \div (8 \times 10^{4})$

18. $(2 \times 10^{2})^{-3}$

19. $(4.3 \times 10^{3}) + (5.2 \times 10^{3})$

20. $(9.5 \times 10^{6}) - (7 \times 10^{5})$

21. $(2.8 \times 10^{9}) - (3 \times 10^{7})$

22. $(1.2 \times 10^{8}) + (4.7 \times 10^{6})$

23. $(3.5 \times 10^{9})(8.6 \times 10^{-7})$

24. $(1.25 \times 10^{-9}) \div (5 \times 10^{-4})$

25. $(7.5 \times 10^{-15})^{2}$

26. $(4 \times 10^{13}) + (5.5 \times 10^{15})$

27. $(3.4 \times 10^{7}) - (6.7 \times 10^{6})$

28. $(8.5 \times 10^{6}) - (5 \times 10^{5})$

Express each answer in scientific notation. Round to two digits after the decimal point and refer to the table of prefixes as needed.

29. The diameter of the smallest capillary in the mature human body is about 0.008 mm. The diameter of the largest artery, the aorta, is about 25,000 μm. Using scientific notation, express each of these measurements in meters.

30. Saturn's average distance from the sun is 887 million miles, and Earth's average distance from the sun is 93 million miles. How many miles closer to the sun is Earth?

31. The speed of light is approximately 300 million meters per second. Earth is 1.5×10^8 km from the sun. How many seconds does it take light from the sun to reach Earth?

32. A pinhead has a diameter of about 1.5 mm, and a red blood cell has a diameter of about 7.5 μm. Approximately how many red blood cells will fit across the diameter of a pinhead?

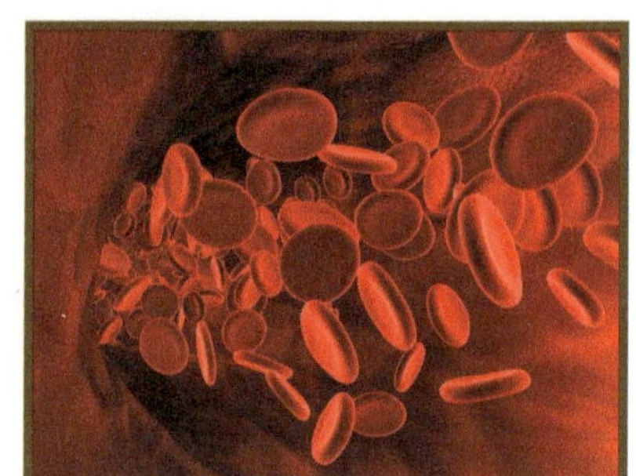

33. The human body contains about 30 trillion red blood cells with a diameter of about 7.5 μm each. If these cells were lined up end-to-end, determine the length of the line in meters and in kilometers.

34. An Internet search engine returned 150,000,000 hits from its database for the keyword "math" in 0.18 sec.

    a. What was the average time per hit?

    b. Express this time in nanoseconds.

C. Exercises

Simplify.

35. $\dfrac{(3 \times 10^7)(7 \times 10^3)}{3 \times 10^5}$

36. $\dfrac{(5 \times 10^6)(9 \times 10^2)}{3 \times 10^{-2}}$

37. $\dfrac{(4 \times 10^5)^2}{(8 \times 10^{-3})(20 \times 10^{12})}$

38. $\dfrac{(2.4 \times 10^{-2})(25 \times 10^{12})}{(2.5 \times 10^2)(1.2 \times 10^3)}$

39. $\dfrac{(4 \times 10^8)(4.5 \times 10^{11})}{(1.5 \times 10^8)(2 \times 10^6)}$

40. $\dfrac{(6 \times 10^{10})(3 \times 10^7)}{(2 \times 10^9) - (2 \times 10^8)}$

Engineering notation is a variation of scientific notation in which a number is written as a product in the form $a \times 10^n$, where $1 \le a < 1000$ and n is a multiple of 3. Engineering notation allows the power of ten to be replaced with a label.

Example: The number 12,500,000 would be written as 12.5×10^6 and read as 12.5 million.

Express each number in engineering notation; then state the number with the appropriate label in place of the power of ten.

41. 3.14×10^5

42. 189,000,000,000

43. 5.385×10^{-5}

44. 0.0001

Dominion Modeling

Data storage on the Internet is growing by zettabytes per year. Massive amounts of text, image, and video files are added to cyberspace each day. The amount of digital data on the Internet has grown to well over 800 billion zettabytes, or 800 sextillion gigabytes.

This information can be extremely valuable if used properly. For example, you can go to a college's website to research available majors and to find descriptions of the courses offered. You can also apply for enrollment and for financial aid. You can locate owner's manuals for appliances or find out how to make a repair on your car. The Internet is a valuable tool for communicating with many relatives and friends simultaneously. It allows you to witness or to mentor someone anywhere in the world. The profitable possibilities are endless. Wise use of the Internet can provide greater knowledge and understanding for a prudent person (cf. Prov. 14:18).

Use the conventional approximations for data measurements from Section 8.1, Dominion Modeling, to complete the following exercises.

45. Express the number of bytes in 1 zettabyte in standard notation.

46. Express the number of bytes in 1 zettabyte in scientific notation.

47. Express the number of bytes in 800 billion zettabytes in scientific notation.

48. The population of the world in 2010 was approximately 6,864,000,000. Using scientific notation, calculate the approximate amount of digital data on the Internet per person in the world. Express the number of bytes in scientific notation and then express the answer in zettabytes.

49. Do an Internet search on the college of your choice and list three courses required for the major of your choice.

CUMULATIVE REVIEW

State the negative reciprocal of each number. [6.5]

50. -7

51. $\frac{2}{3}$

Write the equation of each line in standard form. [6.5]

52. a line parallel to $5x + 2y = -3$ and passing through $(-4, 5)$

53. a line perpendicular to $x + 4y = 7$ and passing through $(-1, 6)$

Solve each system by graphing. [7.1]

54. $y = -x + 7$
$y = \frac{2}{3}x + 2$

55. $y = -\frac{1}{2}x - 1$
$y = x - 4$

Solve each system by substitution or elimination. [7.2–7.4]

56. $y = -x + 2$
$2x - 3y = -6$

57. $2x - 3y = -5$
$x = -3y + 11$

58. $4x + 3y = 6$
$3x - 3y = -6$

59. $3x - 2y = 4$
$4x + y = 9$

8.4 Translating Power Functions

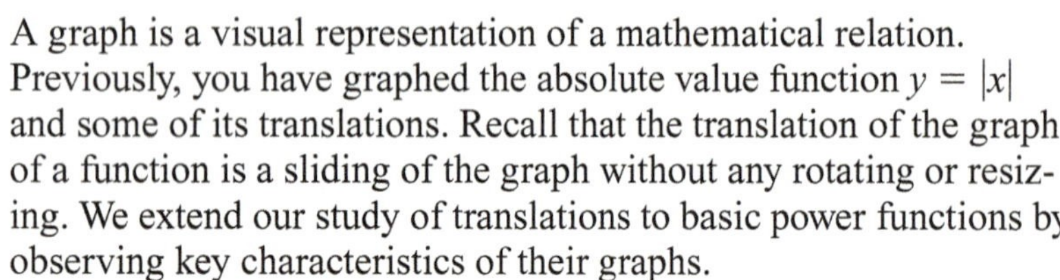

A graph is a visual representation of a mathematical relation. Previously, you have graphed the absolute value function $y = |x|$ and some of its translations. Recall that the translation of the graph of a function is a sliding of the graph without any rotating or resizing. We extend our study of translations to basic power functions by observing key characteristics of their graphs.

Consider the functions $y = x^2$ and $y = x^3$. These are called *power functions*.

Definition

A **power function** is a function of the form $y = ax^n$, where the base x is a variable while the exponent n and the coefficient a are constants.

Example 1

Graph the function $y = x^2$.

Answer

x	−3	−2	−1	0	1	2	3
x^2	9	4	1	0	1	4	9

1. Make a table of ordered pairs.

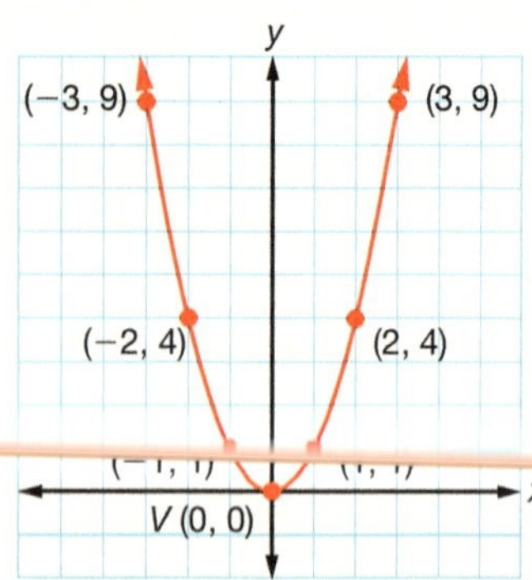

2. Plot the points and connect them with a smooth curve.

Notice that the graph of a power function, unlike that of a linear function, is curved. The curved graph of $y = x^2$ is called a *parabola*. A parabola has a turning point called the *vertex*. The vertex, V, on the graph above is at the point (0, 0). This turning point represents the minimum value of the function $y = x^2$. The vertex of a parabola can be used to determine values such as minimum cost or maximum height in real-life applications.

Example 2

Graph the functions $y = x^2$, $y = x^2 + 2$, and $y = x^2 - 3$ on the same coordinate plane. Compare their graphs.

Answer

x	−3	−2	−1	0	1	2	3
x^2	9	4	1	0	1	4	9
$x^2 + 2$	11	6	3	2	3	6	11
$x^2 - 3$	6	1	−2	−3	−2	1	6

1. Make a table of ordered pairs for each function.

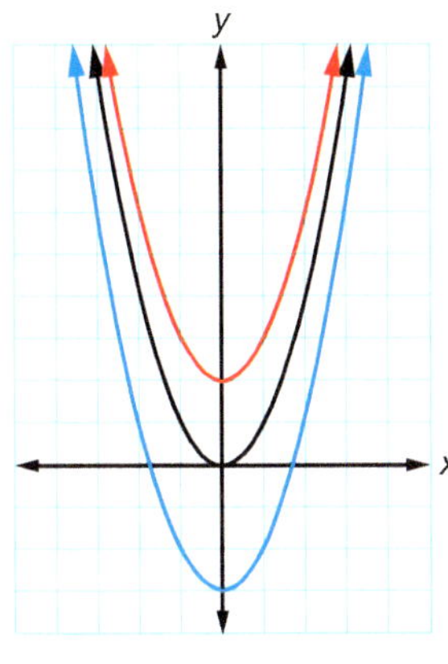

$y = x^2 + 2$

$y = x^2$

$y = x^2 - 3$

2. Use the ordered pairs to determine the location of each graph.

$y = x^2 + 2$ The graph shifts 2 units up when 2 is added to the base function.

$y = x^2 - 3$ The graph shifts 3 units down when 3 is subtracted from the base function.

3. Compare the graphs.

Recall that a function can be thought of as a machine that, when given an input value, processes the input according to a rule and produces an output value. For example, the function $y = x^2$ squares the input. Putting 3 into the machine will result in an output of 9.

A *translation* of a function results when a number is added to or subtracted from either the input or the output of the function.

In the function $y = x^2 + 2$ from the previous example, 2 has been added to the output of $y = x^2$, causing the graph to shift up. Subtracting 3 from the output values of $y = x^2$ causes the graph of $y = x^2 - 3$ to be 3 units lower than the graph of $y = x^2$. Adding to or subtracting from the output of the base function results in a vertical translation. In general, given $y = x^2 + k$, the graph will shift k units vertically.

Example 3

Graph the functions $y = x^2$, $y = (x - 2)^2$, and $y = (x + 3)^2$ on the same coordinate plane. Compare their graphs.

Answer

x	−4	−3	−2	−1	0	1	2	3	4
x^2	16	9	4	1	0	1	4	9	16
$(x - 2)^2$	36	25	16	9	4	1	0	1	4
$(x + 3)^2$	1	0	1	4	9	16	25	36	49

1. Make a table of ordered pairs for each function.

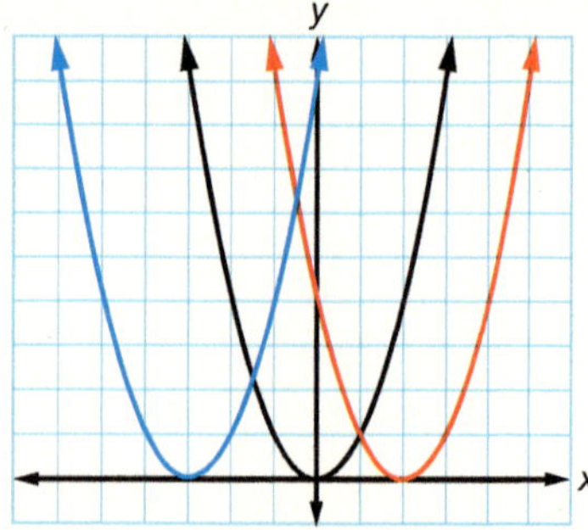

2. Use the ordered pairs to determine the location of each graph.

$y = (x - 2)^2$ The graph shifts 2 units right when 2 is subtracted within the squared expression.

$y = (x + 3)^2$ The graph shifts 3 units left when 3 is added within the squared expression.

3. Compare the graphs.

You can see that the y values of $y = (x - 2)^2$ are shifted 2 units right of the base function $y = x^2$. Additionally, the output values of $y = (x + 3)^2$ are shifted 3 units left of the base function.

In general, adding to or subtracting from the input before performing the base function results in a horizontal translation. Adding to the input of the function shifts the graph left, and subtracting from the input shifts the graph right. That is, given $y = (x - h)^2$, the graph will shift h units horizontally.

Type of Translation	Change to Base Function	Translation of Graph
Vertical	$+ k$ after performing the base function	k units up
	$- k$ after performing the base function	k units down
Horizontal	$+ h$ before performing the base function	h units left
	$- h$ before performing the base function	h units right

Given $y = (x - h)^2 + k$, the graph will shift h units horizontally and k units vertically, and the vertex (turning point) of the function will be located at (h, k). Knowing the location of the vertex makes it much easier to graph the function.

The graphed functions illustrate both horizontal and vertical translations.

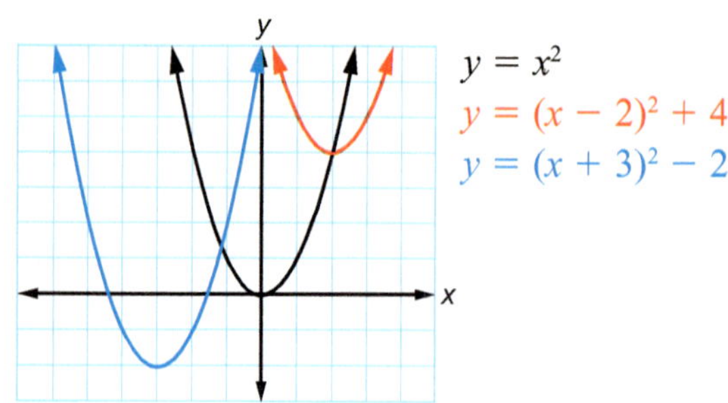

Translated Function	Change to Base Function	Type of Translation	Translation of Graph	Vertex (h, k)
$y = (x - 2)^2 + 4$	$-$ 2 from input	horizontal	2 units right	(2, 4)
	$+$ 4 to output	vertical	4 units up	
$y = (x + 3)^2 - 2$	$+$ 3 to input	horizontal	3 units left	$(-3, -2)$
	$-$ 2 from output	vertical	2 units down	

Now consider the power function $y = x^3$. The graph of $y = x^3$ can be drawn using a table of ordered pairs as illustrated.

x	-2	-1	0	1	2
x^3	-8	-1	0	1	8

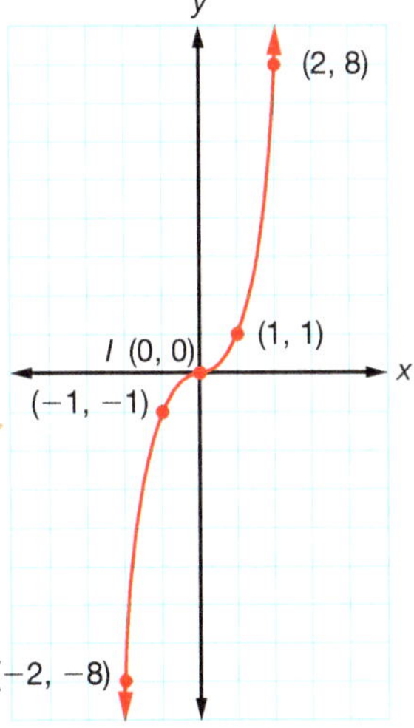

The *point of inflection* of a function is similar to the vertex of a parabola, except that the graph continues in the same basic direction instead of curving back. In the graph of $y = x^3$, the point of inflection, I, is at (0, 0). In general, the point of inflection on the graph of $y = (x - h)^3 + k$ is located at (h, k).

Example 4

Study the graphs of the following functions and complete the table describing the translations in reference to the base function $y = x^3$.

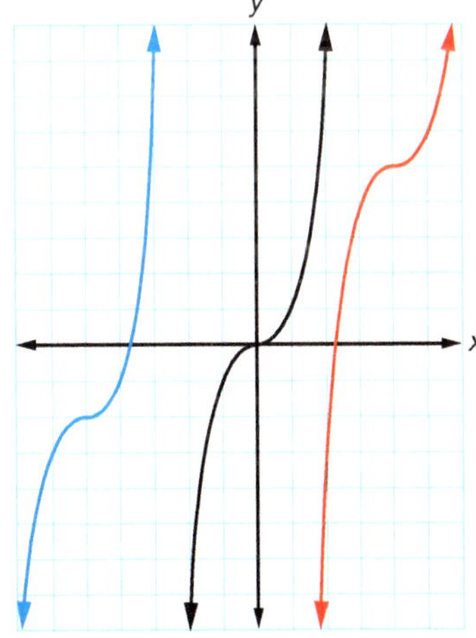

$y = (x - 4)^3 + 5$

$y = x^3$

$y = (x + 5)^3 - 2$

Answer

Translated Function	Change to Base Function	Type of Translation	Translation of Graph	Point of Inflection (h, k)
$y = (x - 4)^3 + 5$	$-$ 4 from input	horizontal	4 units right	(4, 5)
	$+$ 5 to output	vertical	5 units up	
$y = (x + 5)^3 - 2$	$+$ 5 to input	horizontal	5 units left	$(-5, -2)$
	$-$ 2 from output	vertical	2 units down	

A. Exercises

Describe the translation(s) of the graph of $y = x^2$ or $y = x^3$ that produce the graph of each function.

1. $y = x^2 + 5$
2. $y = (x - 4)^3$
3. $y = (x + 7)^2$
4. $y = x^2 - 12$
5. $y = (x - 5)^2 + 6$
6. $y = (x + 1)^3 - 9$
7. $y = (x - h)^2 + k$
8. $y = (x + h)^2 - k$

State the coordinates of the vertex or point of inflection of each function.

9. $f(x) = (x - 3)^2 + 1$
10. $f(x) = (x - 15)^2 - 9$
11. $f(x) = (x + 8)^3 + 6$
12. $f(x) = (x + 3)^3 - 1$
13. $f(x) = x^3 + 19$
14. $f(x) = (x + 12)^2$
15. $f(x) = (x - 7)^3$
16. $f(x) = x^2 - 5$
17. $f(x) = (x - h)^2 - k$
18. $f(x) = (x + h)^2 + k$

B. Exercises

State the coordinates of the vertex or point of inflection. Then create a table of at least four more ordered pairs and graph each function.

19. $f(x) = x^2 + 3$
20. $f(x) = (x - 1)^2$
21. $f(x) = (x + 2)^3$
22. $f(x) = x^3 - 2$
23. $f(x) = (x - 4)^2 + 1$
24. $f(x) = (x + 1)^2 - 5$

Write the function rule for each translation of the base function $y = x^2$.

25. 10 units right
26. 5 units down
27. 3 units left and 12 units down
28. 7 units right and 2 units up
29. 2 units right and 1 unit up
30. 6 units left and 3 units down

Write the equation of each graph.

31. 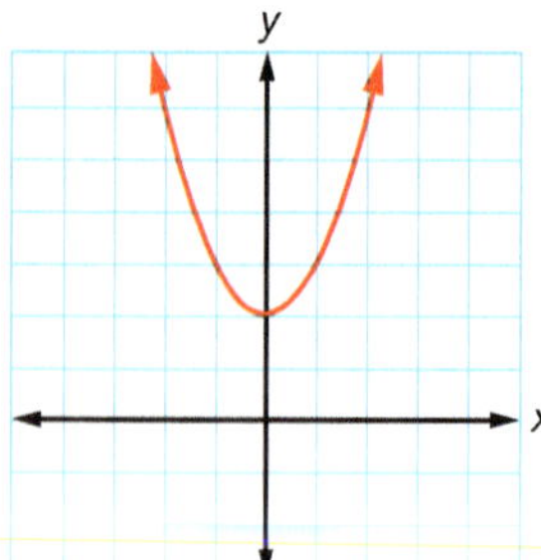

32. 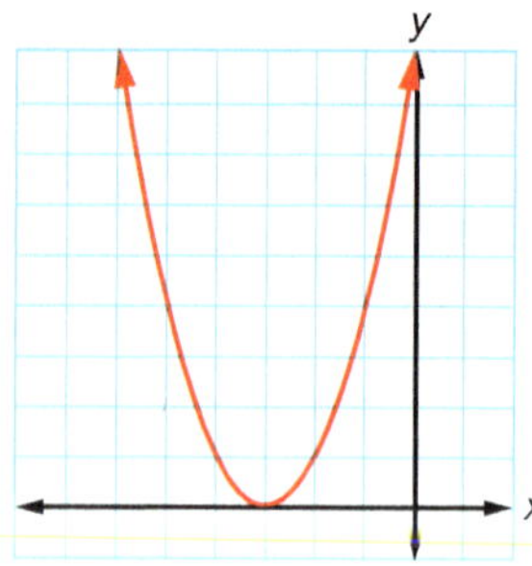

33. 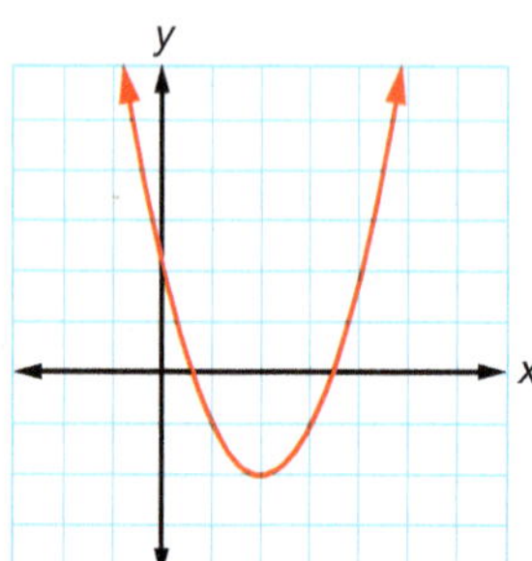

34. 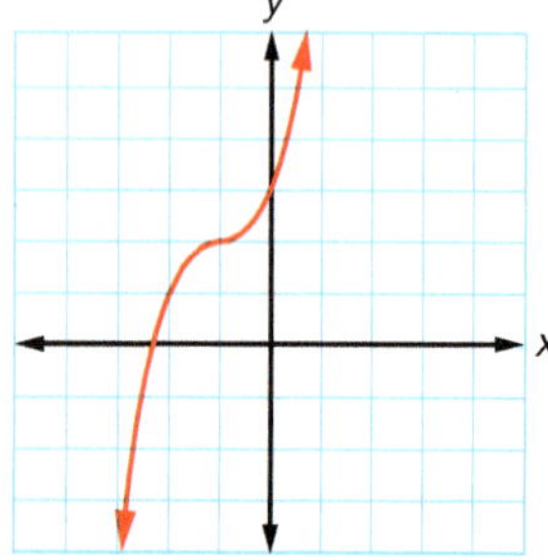

C. Exercises

Create a table of at least five ordered pairs and graph each function.

35. $y = 3x^2$

36. $y = \frac{1}{2}x^2$

37. $y = -x^2$

Describe the effect on the graph of $y = ax^n$ when each of the following occurs.

38. as $|a|$ increases

39. as $|a|$ decreases

40. when a is negative

CUMULATIVE REVIEW

Describe the translation(s) of the graph of $y = |x|$ that produce the graph of each function. [5.7]

41. $f(x) = |x| - 2$

42. $y = |x - 5|$

43. $f(x) = |x + 12| + 3$

44. $f(x) = |x - 15| - 1$

Describe the linear correlation of each data set as (a) strong positive, (b) weak positive, (c) little or none, (d) weak negative, or (e) strong negative. [6.6]

45.

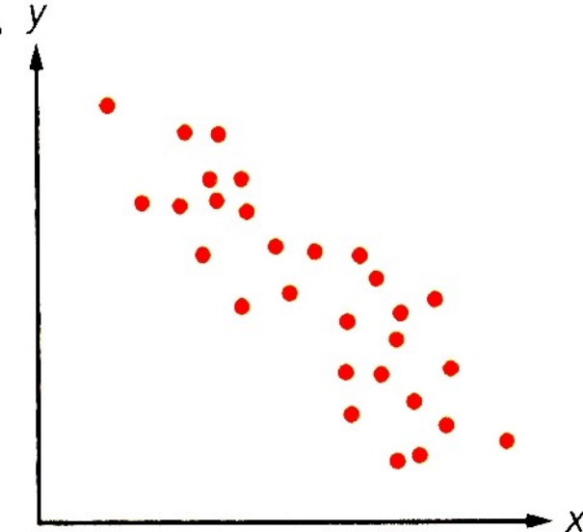

46.

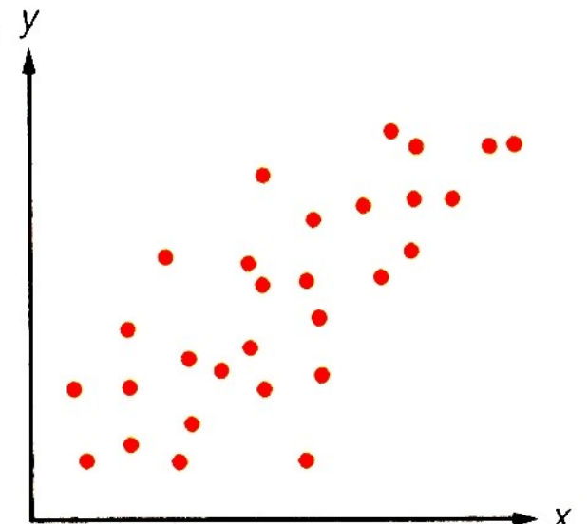

47.

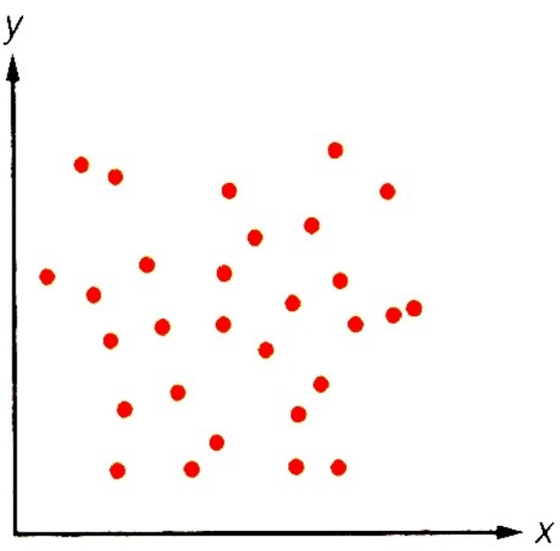

Solve each system. [7.5]

48. $x + y = 9$
$x - 3y = 5$

49. $4a - 4b = -4$
$-3a + 5b = -1$

50. $\frac{1}{2}x + \frac{1}{3}y = 8$
$2x + 3y = 47$

8.5 Exponential Functions

Gordon Moore, cofounder of Intel, predicted in 1965 that the number of transistors per square inch on integrated circuits would double every 18 mo.

While linear functions can be used to model many situations, exponential functions are powerful tools capable of representing many scientific and financial phenomena. In a linear function, the change in the dependent variable is constant during each unit of time. In an exponential function, the dependent variable increases by the same factor during each unit of time. These relationships can be modeled using equations similar to $y = 2^x$, in which the base is a constant and the exponent is the independent variable.

Definition

An **exponential function** is a function of the form $f(x) = ab^x$, where $b > 0$ and $b \neq 1$.

Example 1

Create a table of ordered pairs and graph each function.

a. $y = 2^x$ **b.** $y = \left(\frac{1}{2}\right)^x$

Answer

a.

x	2^x	Ordered Pairs
-3	$2^{-3} = \frac{1}{2^3} = \frac{1}{8}$	$\left(-3, \frac{1}{8}\right)$
-2	$2^{-2} = \frac{1}{2^2} = \frac{1}{4}$	$\left(-2, \frac{1}{4}\right)$
-1	$2^{-1} = \frac{1}{2}$	$\left(-1, \frac{1}{2}\right)$
0	$2^0 = 1$	(0, 1)
1	$2^1 = 2$	(1, 2)
2	$2^2 = 4$	(2, 4)
3	$2^3 = 8$	(3, 8)

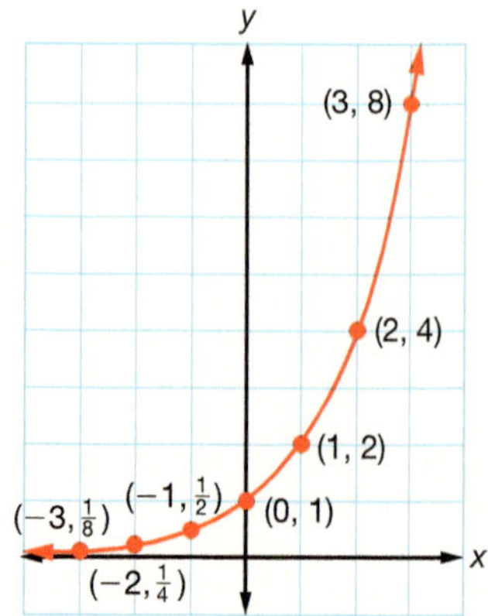

b.

x	$\left(\frac{1}{2}\right)^x$	Ordered Pairs
-3	$\left(\frac{1}{2}\right)^{-3} = 2^3 = 8$	$(-3, 8)$
-2	$\left(\frac{1}{2}\right)^{-2} = 2^2 = 4$	$(-2, 4)$
-1	$\left(\frac{1}{2}\right)^{-1} = 2^1 = 2$	$(-1, 2)$
0	$\left(\frac{1}{2}\right)^0 = 1$	(0, 1)
1	$\left(\frac{1}{2}\right)^1 = \frac{1}{2}$	$\left(1, \frac{1}{2}\right)$
2	$\left(\frac{1}{2}\right)^2 = \frac{1}{4}$	$\left(2, \frac{1}{4}\right)$
3	$\left(\frac{1}{2}\right)^3 = \frac{1}{8}$	$\left(3, \frac{1}{8}\right)$

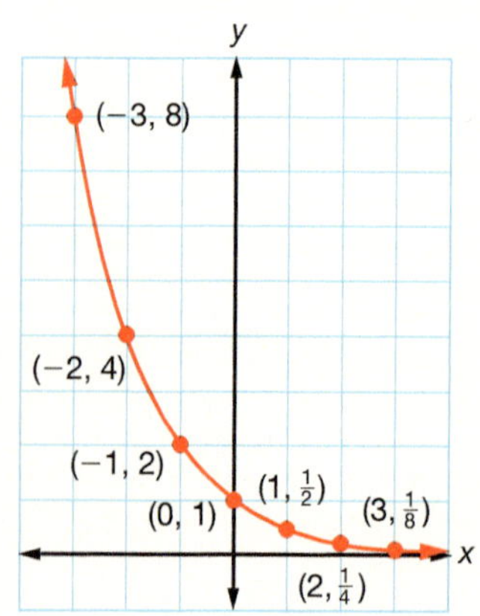

Notice the similarities and differences in these graphs. Both graphs have a y-intercept of (0, 1) and approach the x-axis but never intersect it. We call the line $y = 0$ (the x-axis) in these graphs an *asymptote*, a line that a graph approaches but never intersects. The function $y = 2^x$ continually increases from left to right, while $y = \left(\frac{1}{2}\right)^x$ decreases. The range of both functions is the set of positive real numbers, and their domain is the set of real numbers.

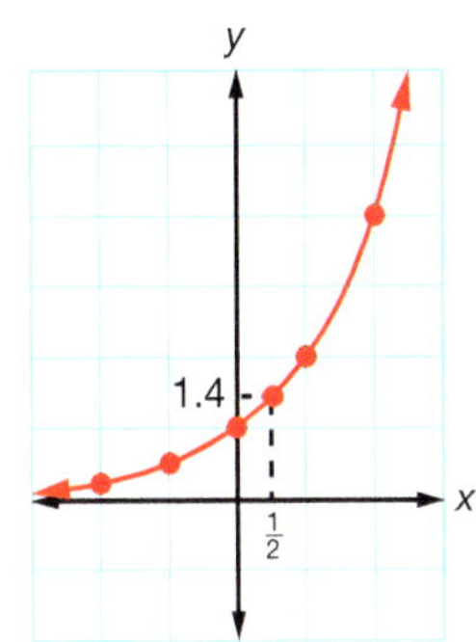

While the concept of rational exponents will not be covered in this course, the graph to the right can be used to estimate the value of $2^{\frac{1}{2}}$ or $2^{0.5}$ as 1.4. A calculator can confirm the values on the graph that are associated with non-integral exponents.

Example 2

Graph $y = 3 \cdot 2^x$ on the same coordinate plane as $y = 2^x$.

Answer

x	$3 \cdot 2^x$	Ordered Pairs
-3	$3 \cdot 2^{-3} = 3\left(\frac{1}{8}\right) = \frac{3}{8}$	$\left(-3, \frac{3}{8}\right)$
-2	$3 \cdot 2^{-2} = 3\left(\frac{1}{4}\right) = \frac{3}{4}$	$\left(-2, \frac{3}{4}\right)$
-1	$3 \cdot 2^{-1} = 3\left(\frac{1}{2}\right) = \frac{3}{2}$	$\left(-1, \frac{3}{2}\right)$
0	$3 \cdot 2^0 = 3(1) = 3$	(0, 3)
1	$3 \cdot 2^1 = 3(2) = 6$	(1, 6)
2	$3 \cdot 2^2 = 3(4) = 12$	(2, 12)
3	$3 \cdot 2^3 = 3(8) = 24$	(3, 24)

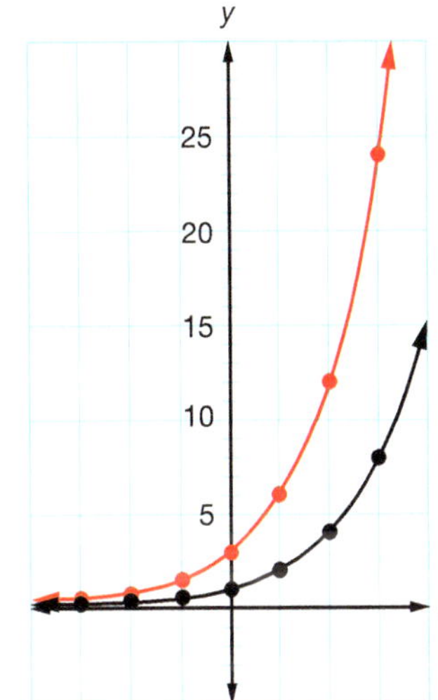

$y = 3 \cdot 2^x$

$y = 2^x$

How does the curve of $y = a \cdot 2^x$ differ from that of $y = 2^x$? For every value of x, the corresponding value of y will be a times its previous value. Since every graph of $y = b^x$ crosses the y-axis at (0, 1), the curve $y = ab^x$ will cross the axis at $(0, a)$.

Example 3

Photocopiers and graphics programs allow the user to enlarge or reduce an image using a percentage of the original, p. The equation $y = p^x$ can be used to find the effect of successive enlargements or reductions. What percent of the original is the image produced by each of the following? Round to the nearest tenth of a percent.

a. three consecutive reductions of 75%

b. four consecutive enlargements of 150%

Answer

a. $p = 0.75$; $y = 0.75^x$ — 1. Model the problem with an exponential function.

$y = 0.75^3 \approx 0.422$ — 2. Evaluate the function.

about 42.2% of the original — 3. Convert the decimal to a percent.

b. $p = 1.5$; $y = 1.5^x$ — 1. Model the problem with an exponential function.

$y = 1.5^4 \approx 5.063$ — 2. Evaluate the function.

about 506.3% of the original — 3. Convert the decimal to a percent.

You have studied problems involving simple interest in which there is a one-time payment of interest at the end of the investing period. Most investments involve compound interest, where interest is regularly credited to the account so that it can begin to earn interest along with the principal. Compound interest can be modeled with an exponential function.

Example 4

The total amount of a $10,000 investment account earning 5% annual interest compounded yearly can be modeled by the equation $y = 10{,}000(1.05)^n$, where n is the number of years the amount is invested. Assuming there are no withdrawals from the account, find the balance after the following numbers of years.

a. 1 yr **b.** 2 yr **c.** 5 yr
d. 10 yr **e.** 25 yr **f.** 50 yr

Answer

a. $y = 10{,}000(1.05)^1 = \$10{,}500.00$
b. $y = 10{,}000(1.05)^2 = 10{,}000(1.1025) = \$11{,}025.00$
c. $y = 10{,}000(1.05)^5 \approx \$12{,}762.82$
d. $y = 10{,}000(1.05)^{10} \approx \$16{,}288.95$
e. $y = 10{,}000(1.05)^{25} \approx \$33{,}863.55$
f. $y = 10{,}000(1.05)^{50} \approx \$114{,}674.00$

Use a calculator to evaluate the more complicated expressions.

Finding the earned interest, $\$114{,}674 - \$10{,}000 = \$104{,}674$, illustrates the power of compounded interest when compared to the simple interest of only $25,000 that would be earned over the same 50 yr period.

A. Exercises

Evaluate.

1. $y = \left(\frac{1}{2}\right)^x$ when $x = 5$

2. $y = 3^x$ when $x = -4$

3. $y = 3 \cdot 4^x$ when $x = 3$

4. $y = -3 \cdot 2^x$ when $x = -3$

Complete each table and then graph each exponential function.

5. $y = 3^x$

x	y
-2	
-1	
0	
1	
2	

6. $y = \left(\frac{1}{3}\right)^x$

x	y
-2	
-1	
0	
1	
2	

7. $y = \left(\frac{2}{3}\right)^x$

x	y
-3	
-2	
-1	
0	
1	
2	
3	

8. $y = \left(\frac{3}{2}\right)^x$

x	y
-3	
-2	
-1	
0	
1	
2	
3	

9. $y = 2 \cdot 3^x$

x	y
-2	
-1	
0	
1	
2	

10. $y = 2 \cdot \left(\frac{1}{3}\right)^x$

x	y
-2	
-1	
0	
1	
2	

11. $y = 3 \cdot \left(\frac{2}{3}\right)^x$

x	y
-3	
-2	
-1	
0	
1	
2	
3	

12. $y = 4 \cdot \left(\frac{3}{2}\right)^x$

x	y
-3	
-2	
-1	
0	
1	
2	
3	

13. Which of the following is the x-intercept of the graph of $y = ab^x$?

a. (1, 0) b. (0, 1) c. $(a, 0)$
d. $(0, a)$ e. There is none.

14. Which of the following is the y-intercept of the graph of $y = ab^x$?

a. (1, 0) b. (0, 1) c. $(a, 0)$
d. $(0, a)$ e. There is none.

15. Examine your graphs from exercises 5–12 and make a conjecture about all exponential functions $y = ab^x$ where $0 < b < 1$.

16. Examine your graphs from exercises 5–12 and make a conjecture about all exponential functions $y = ab^x$ where $b > 1$.

B. Exercises

Write an exponential function modeling consecutive enlargements or reductions by the given percentage. Then find the single setting that would accomplish the same result as the given number of consecutive enlargements or reductions. Round to the nearest whole percent.

17. 125%; 2
18. 60%; 3
19. 75%; 4
20. 125%; 3

21. The balance of an account earning 2% annual interest with an initial investment of $1000 can be modeled by the equation $y = 1000(1.02)^x$, where x is the number of years the money is invested. Assuming there are no withdrawals from the account, find the balance after the following numbers of years.

a. 1 yr b. 2 yr c. 10 yr
d. 25 yr e. 50 yr

22. Under ideal conditions, the population of bacteria in a petri dish doubles every hour. If the equation $y = 500 \cdot 2^h$ models the population of bacteria in the dish after h hours, how many bacteria are in the petri dish at each of the following times?

a. initially (at $h = 0$ hr) b. after 3 hr
c. after 12 hr d. after 1 day

Complete each table and then graph each exponential function.

23. $y = -3^x$

x	-2	-1	0	1	2
y					

24. $y = -\left(\frac{1}{3}\right)^x$

x	-2	-1	0	1	2
y					

25. $y = 3^{-x}$

x	-2	-1	0	1	2
y					

26. $y = \left(\frac{1}{3}\right)^{-x}$

x	-2	-1	0	1	2
y					

27. Compare the graphs of $y = 3^x$ and $y = -3^x$ (exercises 5 and 23) or the graphs of $y = \left(\frac{1}{3}\right)^x$ and $y = -\left(\frac{1}{3}\right)^x$ (exercises 6 and 24) to make a conjecture about the graphs of $y = b^x$ and $y = -b^x$.

28. Compare the graphs of $y = 3^x$ and $y = 3^{-x}$ (exercises 5 and 25) or the graphs of $y = \left(\frac{1}{3}\right)^x$ and $y = \left(\frac{1}{3}\right)^{-x}$ (exercises 6 and 26) to make a conjecture about the graphs of $y = b^x$ and $y = b^{-x}$.

29. Complete the table and then graph the function $y = 2^x + 3$.

x	-2	-1	0	1	2
y					

30. Compare the graphs of $y = 2^x$ and $y = 2^x + 3$ (Example 1*a* and exercise 29) to make a conjecture about the graphs of $y = b^x$ and $y = b^x + c$.

Match each equation to its graph.

31. $y = 2^x$ **32.** $y = \left(\frac{1}{2}\right)^x$ **33.** $y = -2^x$ **34.** $y = 2^x - 3$

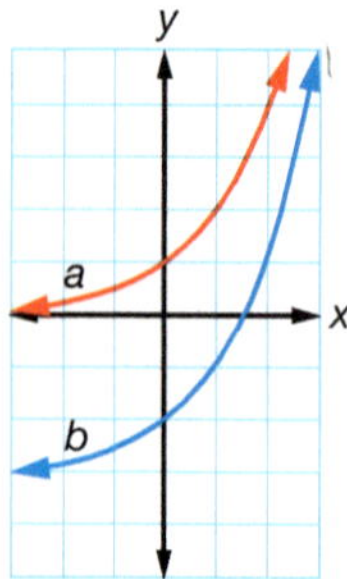

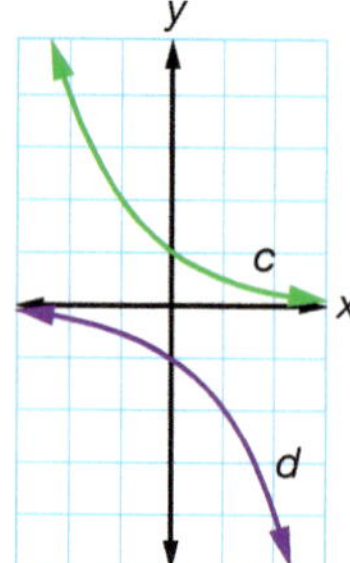

■ C. Exercises

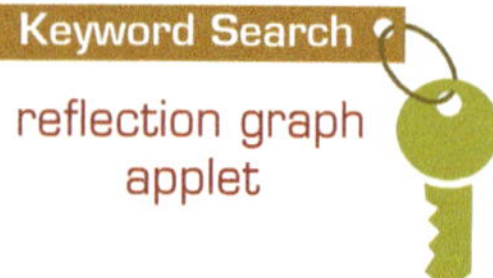

Your new boss proposes to pay you 2 pennies for your first day of work, 4 pennies for the second, 8 for the third, and so on until the thirtieth day of work.

35. When you model your daily pay, p, with the exponential function $p = b^d$, what is the value of b?

36. Use a calculator to determine your pay on the thirtieth day.

Answer the following exercises regarding the general exponential function $y = ab^x$.

37. The definition of the exponential function $y = ab^x$ states that $b \neq 1$. Write the simplified function rule and describe the graph of the function if $b = 1$.

38. If $a > 0$, will the function $y = ab^x$ always increase as x increases? Explain.

39. The graphs of $y = 2^x$ and $y = \left(\frac{1}{2}\right)^{-x}$ are identical. Show algebraically that these functions are equal.

Dominion Modeling

In Genesis 1:28 humans are given a mandate to subdue the earth. *Subdue* means to conquer or to bring into subjection; to have dominion over. The earth's vastness may have made this task seem impossible at one time, and the entrance of sin (Gen. 3:17–19) only complicated matters. But through God-given resources, abilities, and knowledge, people have overcome much of this great vastness through population growth and technological advances. The ultimate victory will be through Christ Himself (1 Cor. 15:57).

God did not command us to subdue the heavens, whose vastness stretches beyond human comprehension. God's awesome power is demonstrated by the inconceivable distances to even the closest objects in His created universe.

Express each number in scientific notation.

40. 1 billion

41. 1 millionth

42. The distance light travels in 1 sec is approximately 186,000 mi.

43. The distance light travels in 1 yr (1 light year) is about 6 trillion miles.

Use scientific notation to find the following values. Express each answer in scientific notation.

44. Find the diameter in miles of our galaxy, the Milky Way, which is 100,000 light years.

45. Find the distance in miles to our nearest neighboring galaxy, Andromeda, which is 2.2 million light years away.

CUMULATIVE REVIEW

Graph each inequality on its own coordinate plane. [6.7]

46. $y \le -2x + 3$

47. $y > \frac{3}{4}x$

Simplify. [8.1–8.2]

48. $(r^2)^{-2}s^4s^{-1}r^5$

49. $27m^{-3}n(-3mn^2)^{-2}$

50. $\frac{a^{-4}abc^{-2}}{c^{-1}c^{-2}b^3}$

51. $\frac{z^{-2}z^{-1}y^{-2}}{x^{-2}xy^3z^{-4}}$

Write and solve a system of equations for each word problem. [7.6–7.7]

52. On Saturday Tami pedaled to the state park at 21 mi/hr. After lunch she rode back at 14 mi/hr. If her total riding time was 5 hr, how far was it to the park?

53. A speed boat goes 15 mi down the river in 30 min and takes 45 min to return. Determine the rate of the boat in still water and the rate of the current.

54. How many milliliters of distilled water and concentrated (18%) acetic acid should be mixed to make 1 L of dilute (4%) acetic acid? Round to the nearest milliliter.

55. Mrs. Fields wants to split her $10,000 investment between a mutual fund that has averaged a 6% annual return and a more conservative bond fund that has averaged a 2% annual return. How much should she invest in each fund if she wants to earn $500 in interest during the next year?

TECHNOLOGY CORNER (TI-84+ Family)

The graphing calculator is especially helpful for comparing the graphs of several different functions. Use [Y=] to enter the illustrated exponential functions. The [^] key is used to indicate an exponent.

Select [GRAPH] and use [WINDOW] to adjust the portion of the graph that is illustrated.

Select [TRACE] and use [▲] and [▼] to move between the graphs of different functions. Use [◄] and [►] to reveal the coordinates of the points on a graph.

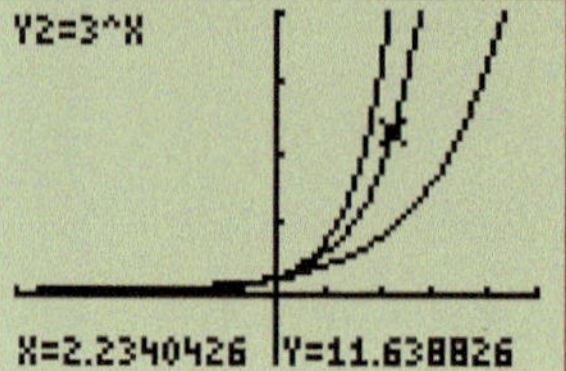

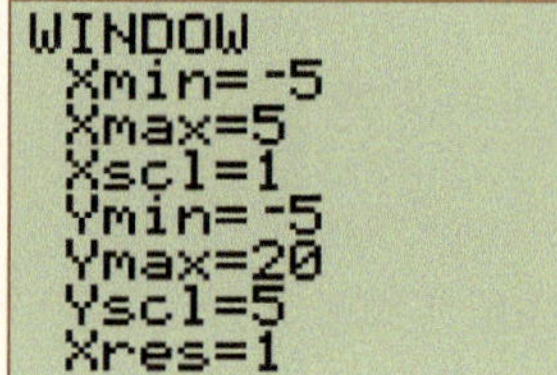

These graphs illustrate that, in functions of the form $y = b^x$, the function increases more rapidly in the first quadrant when b is larger.

Enter functions Y_4, Y_5, and Y_6 for comparison with the previous functions.

From these graphs it can be seen that the graph of $y = \left(\frac{1}{b}\right)^x$ is a mirror image or *reflection* of the graph of $y = b^x$, with the y-axis acting as the mirror.

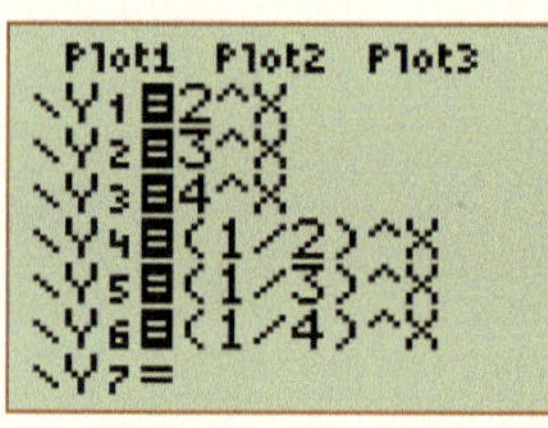

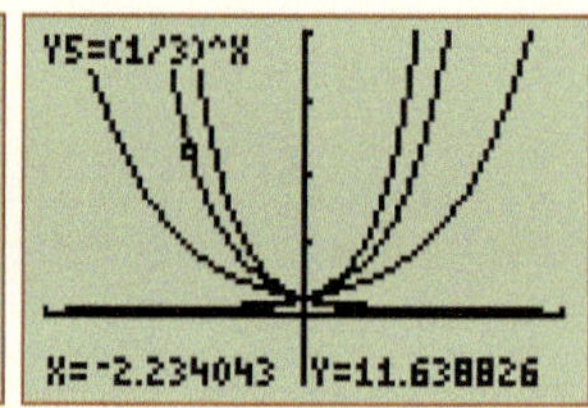

You can also explore what happens when the original graphs are modified to the form $y = b^{-x}$.

Use the [INS] function ([2nd] [DEL]) to insert a negative sign ([(-)]) in the first three functions.

When the functions are graphed, it appears as though there are only three functions graphed. Using the [TRACE] function will help you see that Y_2 and Y_5 have the same graph. Which other functions have the same graph?

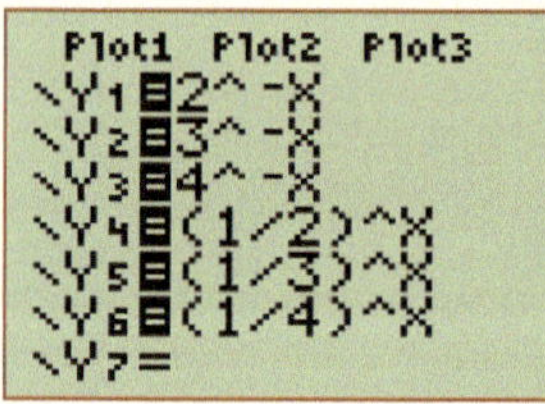

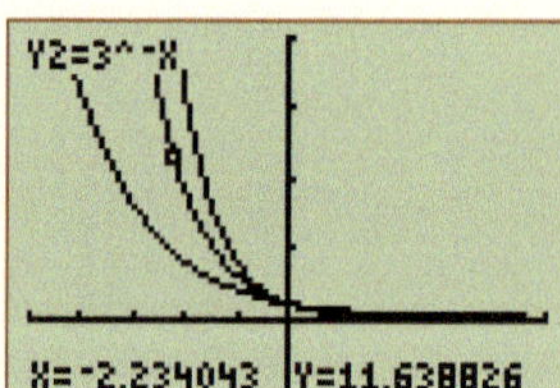

This can also be seen by turning off the graphs of the other functions. Moving the cursor to the highlighted = in a function and selecting [ENTER] removes the highlighting and turns the graph of that function off. (You can turn it back on by selecting [ENTER] again.)

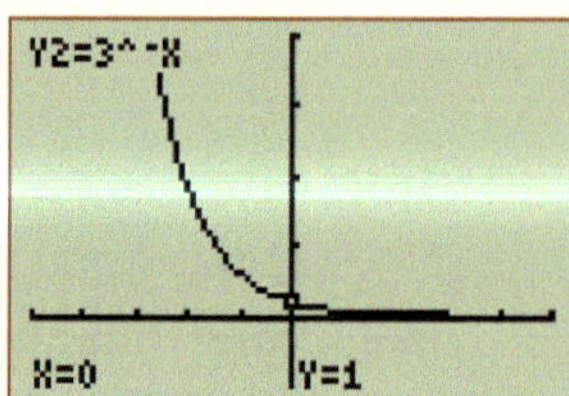

This exploration leads to the conjecture that $y = b^{-x}$ and $y = \left(\frac{1}{b}\right)^x$ are identical functions.

8.6 Exponential Growth and Decay

Wildlife management often involves restocking lakes and ponds to estabish desired fish populations.

Exponential growth and decay occur when the value of a function has the same percent increase or decrease during each unit of time.

Under ideal conditions, the number of fish in a pond initially stocked with 1000 fish increases by 50% each year. The next year's population is 150% of the current population. The table and graph illustrate this exponential growth.

Year	Population Increase	Population
0	1000	1000
1	1000×1.5	1500
2	$1500 \times 1.5 = 1000(1.5)^1 \times 1.5 = 1000(1.5)^2$	2250
3	$2250 \times 1.5 = 1000(1.5)^2 \times 1.5 = 1000(1.5)^3$	3375
4	$3375 \times 1.5 = 1000(1.5)^3 \times 1.5 = 1000(1.5)^4$	5063
5	$5063 \times 1.5 \approx 1000(1.5)^4 \times 1.5 = 1000(1.5)^5$	7594

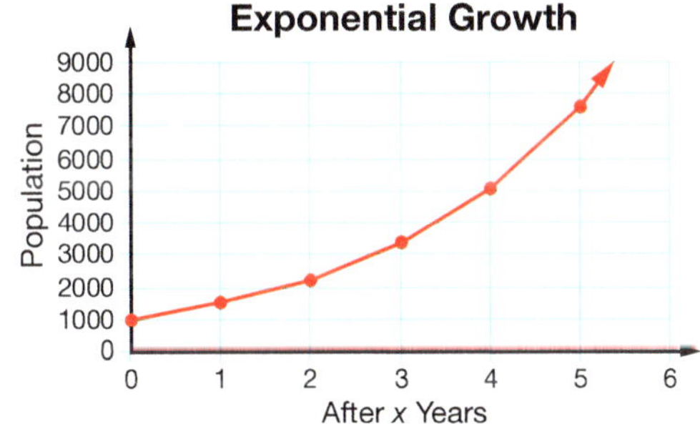

This type of growth is difficult to maintain for an extended period of time. At some point, the conditions necessary to sustain this growth cannot be maintained. Suppose the population decreases by 25% each year after the population reaches 10,000. Next year's population will be 75% of the current population. The table and graph illustrate this exponential decay.

Year	Population Decrease	Population
0	10,000	10,000
1	$10{,}000 \times 0.75$	7500
2	$7500 \times 0.75 = 10{,}000(0.75)^1 \times 0.75 = 10{,}000(0.75)^2$	5625
3	$5625 \times 0.75 = 10{,}000(0.75)^2 \times 0.75 = 10{,}000(0.75)^3$	4219
4	$4219 \times 0.75 \approx 10{,}000(0.75)^3 \times 0.75 = 10{,}000(0.75)^4$	3164
5	$3164 \times 0.75 \approx 10{,}000(0.75)^4 \times 0.75 = 10{,}000(0.75)^5$	2373

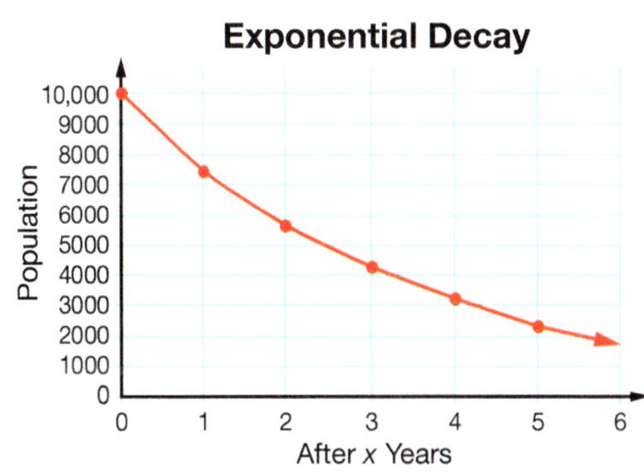

Definitions

Given an initial amount, A, and the **growth rate**, r, the exponential function $f(t) = A(1 + r)^t$ can be used to model both **exponential growth** ($r > 0$) and **exponential decay** ($-1 < r < 0$). The quantity $(1 + r)$ is called the **growth factor**.

Example 1

For each exponential function, identify the initial amount, the growth rate, and the growth factor. Then state whether the function models exponential growth or exponential decay.

a. $f(t) = 500(1 - 0.75)^t$ **b.** $y = 100(1 + 2.5)^x$

c. $f(d) = 2^d$ **d.** $y = 10{,}000\left(\frac{1}{2}\right)^x$

Answer

	Initial Amount	Growth Rate (r)	Growth Factor (b)	Type	
a.	500	-0.75 or -75%	0.25	decay	Notice that $r < 0$.
b.	100	2.5 or 250%	3.5	growth	Notice that $r > 0$.
c.	1	1 or 100%	2	growth	Rewrite as $f(d) = 1(1 + 1)^d$.
d.	10,000	$-\frac{1}{2}$ or -50%	$\frac{1}{2}$	decay	Rewrite as $y = 10{,}000\left(1 - \frac{1}{2}\right)^x$.

Example 2

In a single elimination tournament, half of the teams are eliminated in each round.

a. Write an exponential function representing the number of teams left after each round if the tournament begins with 64 teams.

b. Determine the number of teams left after three rounds.

c. Graph the function to find how many rounds it takes to determine a champion.

Answer

a. $f(n) = 64\left(1 - \frac{1}{2}\right)^n$ or $f(n) = 64\left(\frac{1}{2}\right)^n$ The rate of change is $r = -\frac{1}{2}$, and $A = 64$.

b. $f(3) = 64\left(\frac{1}{2}\right)^3 = 64\left(\frac{1}{8}\right) = 8$ teams Let $n = 3$ and evaluate.

c.

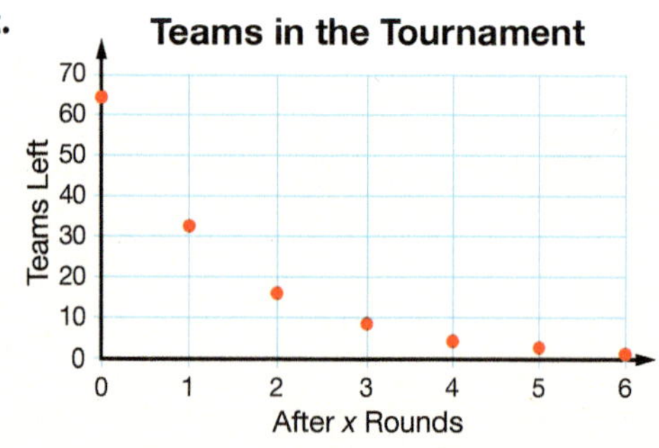

After 6 rounds, only the champion is left.

n	$f(n)$
0	$f(0) = 64\left(\frac{1}{2}\right)^0 = 64(1) = 64$
1	$f(1) = 64\left(\frac{1}{2}\right)^1 = 64\left(\frac{1}{2}\right) = 32$
2	$f(2) = 64\left(\frac{1}{2}\right)^2 = 64\left(\frac{1}{4}\right) = 16$
3	$f(3) = 64\left(\frac{1}{2}\right)^3 = 64\left(\frac{1}{8}\right) = 8$
4	$f(4) = 64\left(\frac{1}{2}\right)^4 = 64\left(\frac{1}{16}\right) = 4$
5	$f(5) = 64\left(\frac{1}{2}\right)^5 = 64\left(\frac{1}{32}\right) = 2$
6	$f(6) = 64\left(\frac{1}{2}\right)^6 = 64\left(\frac{1}{64}\right) = 1$

Example 3

A 5 g mold culture increases in size by 200% every day. Write an exponential function modeling this growth and find the size of the culture after one week.

Answer

$f(d) = 5(1 + 2)^d$ or $f(d) = 5(3)^d$

1. The rate of change is $r = 200\% = 2$, and $A = 5$.

$f(7) = 5(3)^7 = 5(2187) = 10{,}935$ g

2. Let $d = 7$ and evaluate.

Compound interest is another example of exponential growth. If interest is compounded annually, it is credited to the account at the end of the year. Other typical compounding periods are semiannually (twice a year), quarterly (four times a year), monthly (twelve times a year), and daily (365 times a year).

You must be careful to determine the interest rate per compounding period by dividing the annual percentage rate (APR) by the number of compounding periods in a year. An investment earning 6% APR compounded monthly will grow at a rate of $\frac{0.06}{12} = 0.005 = 0.5\%$ each month. Its growth would be modeled by the function

$$f(m) = A(1 + 0.005)^m = A(1.005)^m,$$

where m is the number of months the money is invested.

When determining the value of an account, be sure to use the number of compounding periods instead of the number of years. To find the value after 10 yr of a \$10,000 investment at 6% compounded monthly, use $10(12) = 120$ mo.

$$f(120) = 10{,}000(1.005)^{120} \approx \$18{,}193.97$$

Compound Interest Formula	
Compound interest is a special case of exponential growth.	
$A = P(1 + r)^x$, where	A = the account balance after t years, P = the principal amount, x = the number of periods, and r = the periodic interest rate.
If there are n compounding periods in each of t years, then	$r = \frac{APR}{n}$ and $x = nt$.

Example 4

Mr. Mendoza invests $6000 in a certificate of deposit (CD) earning 4% interest compounded quarterly.

a. Write an exponential function modeling the value of his investment.

b. What is the account balance after 10 yr?

c. Create a table of ordered pairs and graph the function.

Answer

a. $f(q) = 6000(1 + 0.01)^q = 6000(1.01)^q$

The initial amount, A, is $6000. The quarterly interest rate, r, is $\frac{0.04}{4} = 0.01$.

b. $f(40) = 6000(1.01)^{40} \approx \$8933.18$

Four interest payments per year for 10 yr is 40 compounding periods.

c.

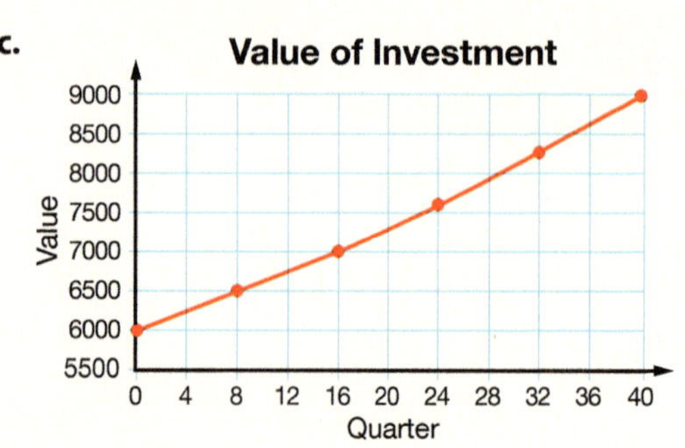

q	$f(q)$
0	$6000(1.01)^0 = \$6000$
8	$6000(1.01)^8 \approx \$6497$
16	$6000(1.01)^{16} \approx \$7035$
24	$6000(1.01)^{24} \approx \$7618$
32	$6000(1.01)^{32} \approx \$8250$
40	$6000(1.01)^{40} \approx \$8933$

The value of many items decreases by the same percent each year. This *depreciation* can be modeled by an exponential decay function.

Example 5

An automobile that sold for $30,000 depreciates about 12% each year.

a. Write an exponential function modeling the value of the car.

b. Determine the value (to the nearest hundred dollars) of the car after five, ten, and fifteen years.

c. Find the loss of value for the car during its first five years, its second five years, and its third five years.

Answer

a. $f(t) = 30{,}000(1 - 0.12)^t = 30{,}000(0.88)^t$

The initial amount, A, is $30,000. The growth rate is −12% or −0.12, and the growth factor is 0.88.

b. $f(5) = 30{,}000(0.88)^5 \approx \$15{,}800$

$f(10) = 30{,}000(0.88)^{10} \approx \$8400$

$f(15) = 30{,}000(0.88)^{15} \approx \$4400$

Evaluate the function at 5, 10, and 15 yr.

c. $30{,}000 - 15{,}800 = \$14{,}200$

$15{,}800 - 8400 = \$7400$

$8400 - 4400 = \$4000$

Find the difference of the values at the beginning and the end of each time period.

A. Exercises

For each exponential function, identify the initial amount, the growth rate (as a percent), and the growth factor. Then state whether the function models exponential growth or exponential decay.

1. $y = 35(1 + 0.05)^x$

2. $f(t) = 16\left(1 - \frac{1}{5}\right)^t$

3. $y = \left(\frac{7}{10}\right)^x$

4. $f(p) = 10{,}000(1.1)^p$

Write each described exponential function in the form $f(t) = ab^t$. Then state whether the function models exponential growth or exponential decay.

5. initial amount: 10; growth rate: 150%

6. initial amount: 0.5; growth rate: −30%

7. initial amount: 81; growth rate: $\frac{1}{2}$

8. initial amount: 125; growth factor: 0.2

9. initial amount: 75; growth factor: 0.5

10. initial amount: 60,000; growth factor: $\frac{3}{2}$

Write an exponential function modeling each description.

11. This year's annual sales of $800,000 are projected to decrease by 10% each year.

12. The school committee plans for the student body of 200 to grow by 3% each year.

13. A culture containing 1500 bacteria doubles in size each day.

14. A used car purchased for $4000 retains only three-fourths of its value each year.

$5000 is invested at 6% APR for 10 yr. Determine the number of compounding periods and the rate per compounding period when interest is compounded as follows.

15. annually

16. semiannually

17. quarterly

18. monthly

19. daily

B. Exercises

Write an exponential expression for the value of each investment. Do not evaluate.

20. $5000 compounded annually at 4% APR for 10 yr

21. $5000 compounded semiannually at 4% APR for 10 yr

22. $2000 compounded quarterly at 3% APR for 20 yr

23. $37,855 compounded monthly at 4% APR for 30 yr

24. $500 compounded daily at 6% APR for 10 yr

Write an exponential function and use it to find the amount in each account.

25. $10,000 compounded quarterly at 6% APR for 12 yr

26. $4500 compounded quarterly at 4.4% APR for 8 yr

27. $2000 compounded semiannually at 6% APR for 20 yr

28. $2000 compounded monthly at 6% APR for 20 yr

29. $5000 compounded annually at 4.5% APR for 6.5 yr

30. The population of Greenwood has increased by about 2% each year since 2000, when it had 25,000 residents.

  a. Write an exponential function, $f(x)$ (where x is the number of years after 2000), modeling the population of Greenwood.

  b. Use the function to approximate the population in 2010, 2020, 2030, and 2040.

  c. Graph the function, showing Greenwood's population through the year 2050.

31. A patient takes 500 mg of a medication, and the amount of medication in the patient's body decreases by 10% each hour.

  a. Write an exponential function, $f(x)$, modeling the amount of medication in the patient's body.

  b. How much medication (to the nearest milligram) is in the patient's body after 4 hr?

  c. Graph the function, showing the amount of medication in the patient's body for the first 8 hr.

32. A ball dropped from a height of 10 m bounces to a height that is $\frac{2}{5}$ of its previous height.

  a. Write an exponential function, $f(x)$, modeling the height of each bounce.

  b. How high is the fifth bounce of the ball?

  c. Graph the function, showing the first five bounces of the ball.

33. A piece of paper is 0.01 cm thick, and its thickness doubles each time it is folded in half.

  a. Write an exponential function, $f(n)$, modeling the thickness of the paper after n folds.

  b. How thick would the paper be after ten folds?

  c. Graph the function, showing the thickness of the paper for the first ten folds.

34. A culture of 500 bacteria doubles every 20 min.

  a. Write an exponential function, $f(p)$ (where p is the number of 20 min periods) modeling the number of bacteria.

  b. Determine the number of bacteria after 2 hr.

  c. Determine the approximate number of bacteria after one day.

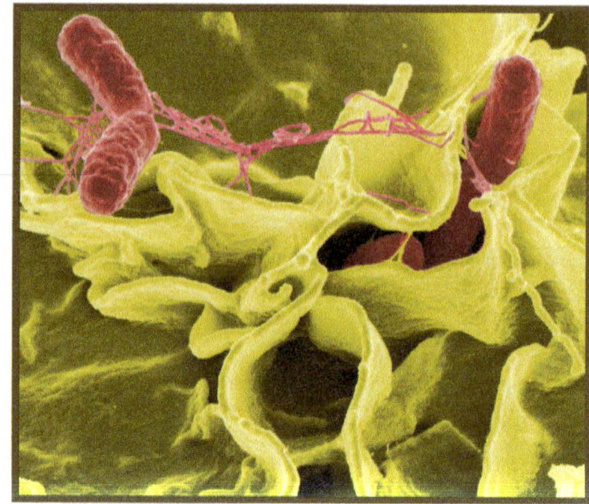

These rod-shaped Salmonella bacteria have a diameter from 0.7 to 1.5 μm and cause illnesses such as food poisoning and typhoid fever.

35. A pair of squirrels is introduced to an island, and their population doubles every 3 mo.

  a. Write an exponential function, $f(m)$, modeling the number of squirrels on the island after m months.

  b. Write an exponential function, $f(n)$, modeling the number of squirrels on the island after n years.

  c. Use either of your functions to determine the number of squirrels on the island after 3 yr. Then use the other function to check your work.

C. Exercises

Exponential functions model growth if $r > 0$ or decay if $-1 < r < 0$.

36. Describe the value of the growth factor when the function models exponential growth.

37. Describe the value of the growth factor when the function models exponential decay.

38. What values are not possible for the growth factor?

39. An investment of $10,000 earns 6% APR for 10 yr. To the nearest dollar, how much more is earned if the investment is compounded as follows?

a. semiannually instead of annually

b. quarterly instead of semiannually

c. monthly instead of quarterly

d. monthly instead of annually

Dominion Modeling

What people are saying and communicating on the Internet is referred to as Internet chatter. Most of the chatter is found in chatrooms, blogs, bulletin boards, social networks, micro-blogging services, and uploaded videos. The Central Intelligence Agency (CIA) "listens to" or tracks Internet chatter to better understand what is going on in the world. This tracking can be a powerful tool in preserving our freedoms and security. Health agencies track Internet chatter to help assess the magnitude of a viral outbreak. Businesses monitor chatter to obtain feedback on their products or services. Growing Internet chatter offers greater opportunities to make major decisions based on feedback from millions of consumers and public responses.

Meeting the spiritual needs of a lost world is a difficult task, but communicating over the Internet is an effective way to share our faith since there are many opportunities to present encouraging words and messages of hope (Philem. 6).

In 2010 about 75% of the 3.1×10^8 people in the United States used the Internet, compared to about 29% of the world's population of 6.9×10^9. At that time, approximately 2.4×10^{11} emails were sent each day, with more than 80% of those emails being classified as spam or unwanted.

Express each answer in both scientific and standard notation.

40. How many Internet users were there in the United States in 2010?

41. How many Internet users were there in the world in 2010?

42. What was the average number of daily emails per Internet user in the world (to the nearest whole number)?

43. A micro-blogging service sent 540 million micro-messages in just five days. What was the average number of messages sent per second through this service?

44. Explain one way that you can use this powerful tool to share your faith in Christ.

CUMULATIVE REVIEW

Determine whether the given ordered pair is a solution of the system $2x - 3y > 12$ and $y \leq -\frac{2}{3}x + 4$. [7.8]

45. $(2, 1)$

46. $(6, -5)$

Write the system represented by each graph. [7.8]

47.

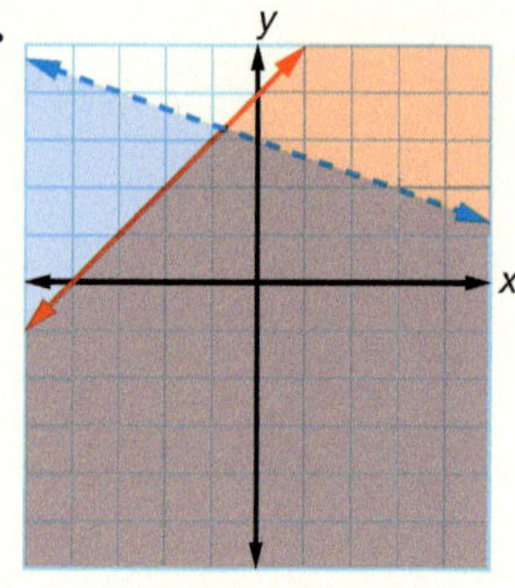

48.

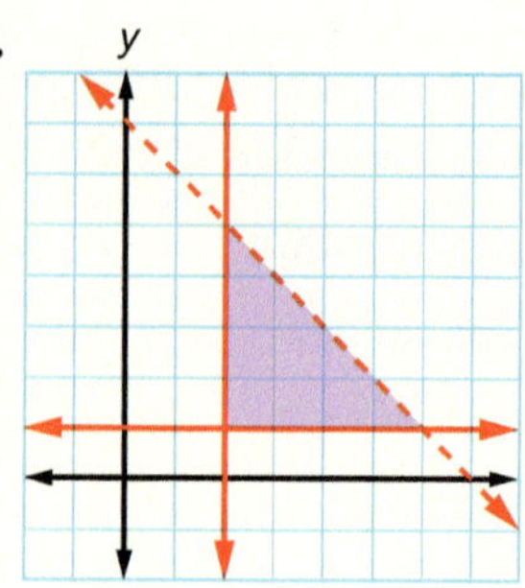

Solve each system. Then classify it as consistent independent, consistent dependent, or inconsistent. If the solution is an entire line, write its equation in slope-intercept form. [7.5]

49. $y = -2x + 3$
$4x + 2y = 6$

50. $5x + 2y = 8$
$x - 2y = 4$

51. $2x - 3y = 6$
$-4x + 6y = 12$

Simplify. [8.1–8.3]

52. $(12x^{-4}y^3)^2\left(-\frac{1}{4}x^3y^{-1}\right)^3$

53. $\left(\frac{3x}{y^4}\right)^2\left(\frac{y}{x^2}\right)^{-3}$

54. $(2.2 \times 10^{-8}) - (6 \times 10^{-9})$

SEQUENCES

Infinite Geometric Sums

Challenge

Can you state the first four partial sums for each of the following sequences? (Remember that $S_1 = A_1$, $S_2 = A_1 + A_2$, $S_3 = A_1 + A_2 + A_3$, $S_4 = A_1 + A_2 + A_3 + A_4$, ...)

a. 0.9, 0.09, 0.009, 0.0009, ...

b. $\frac{1}{2}, \frac{1}{4}, \frac{1}{8}, \frac{1}{16}, \ldots$

c. $4, -2, 1, -\frac{1}{2}, \ldots$

We have previously discussed partial sums for arithmetic sequences. The first four partial sums for the sequence 1, 3, 5, 7, ... would be 1, 1 + 3, 1 + 3 + 5, 1 + 3 + 5 + 7, which simplifies to 1, 4, 9, 16. We can also find partial sums for geometric sequences such as the three in the Challenge. In the first sequence, the first four partial sums are 0.9, 0.9 + 0.09, 0.9 + 0.09 + 0.009, 0.9 + 0.09 + 0.009 + 0.0009, which simplifies to 0.9, 0.99, 0.999, 0.9999. In the second sequence, the first four partial sums are $\frac{1}{2}, \frac{3}{4}, \frac{7}{8}, \frac{15}{16}$.

Incredibly, all the terms of an infinite sequence sometimes have a finite sum! Limits can be used to prove that the sum of the terms of an infinite geometric sequence $A_n = A_1 r^{n-1}$ with $|r| < 1$ is given by the formula $S = \frac{A_1}{1 - r}$. The sum of all the terms of the sequence $\frac{1}{2}, \frac{1}{4}, \frac{1}{8}, \frac{1}{16}, \ldots$ can be found using this formula. Since $A_1 = \frac{1}{2}$ and $r = \frac{1}{2}$, we have the sum $S = \frac{0.5}{1 - 0.5} = \frac{0.5}{0.5} = 1$. If an infinite geometric sequence diverges (i.e., $|r| > 1$), then it does not have a finite sum.

One application of this formula for an *infinite geometric sum* is to change a repeating decimal into fractional form. Consider $0.\overline{15} = 0.15 + 0.0015 + 0.000015 + \ldots$. Since $A_1 = 0.15$ and $r = 0.01$, we have the sum $S = \frac{0.15}{1 - 0.01} = \frac{15}{99} = \frac{5}{33}$.

Exercises

Can the formula $S = \frac{A_1}{1 - r}$ be used to find the sum of all the terms of each sequence? Explain why or why not.

1. $5, 1, \frac{1}{5}, \frac{1}{25}, \ldots$

2. $\frac{1}{32}, \frac{1}{16}, \frac{1}{8}, \frac{1}{4}, \ldots$

3. 7, 14, 21, 28, ...

4. $1, -\frac{3}{2}, \frac{9}{4}, -\frac{27}{8}, \ldots$

Find the sum of each infinite geometric sequence.

5. 0.9, 0.09, 0.009, 0.0009, ...

6. $4, -2, 1, -\frac{1}{2}, \ldots$

7. $1, \frac{1}{3}, \frac{1}{9}, \frac{1}{27}, \ldots$

8. $\frac{1}{81}, \frac{1}{27}, \frac{1}{9}, \frac{1}{3}, \ldots$

Use $S = \frac{A_1}{1 - r}$ to change each repeating decimal to a fraction.

9. $0.\overline{7}$

10. $0.\overline{36}$

CHAPTER 8 REVIEW

Simplify.

1. $(6m^2n^5)(-3mn^7)$
2. $(3g^2h)^3$
3. $81a^4b^3c^2 \div 9abc^2$
4. $\frac{4w^7}{z}\left(\frac{z^3}{2w^2}\right)$
5. $\frac{8m^4n^2p^5}{2m^2np}$
6. $\left(\frac{x^2}{y}\right)^2\left(\frac{y^3}{x}\right)^3$
7. $\frac{y^2}{x^{-3}y^2}$
8. $\frac{(y^{2a})^3}{y^{4a}}$
9. $2^5 \cdot 2^{x-5}$
10. $2j^{-2}k^4j^5 - 9kj^3k^3 - j^4k^9k^{-5}j^{-1}$

Simplify, leaving all variables in the numerator.

11. $\frac{-3m^{-3}r^7s^{-4}}{mr^2s^2}$
12. $2a^{-4}b^3c^{-2}\left(\frac{a^2b^2}{c^{-3}}\right)$
13. $\frac{-3x^2y^{-3}}{x^{-1}y}\left(-\frac{y}{2x^3}\right)^2$

Simplify, leaving each answer in positive exponential form.

14. $\frac{a^{-2}b^2}{a^{-1}b^{-2}}$
15. $5x^{-3}y^2z^4(-3x^{-3}yz^{-5})$
16. $\frac{x^{-5}y^3}{z^{-2}}\left(\frac{xz^3}{y}\right)^2$
17. $(-2a^3b^{-5})^{-2}$
18. Use the formula $V = \frac{1}{3}\pi r^2H$ to write a simplified algebraic expression for the volume of a cone with a radius of ab^3 and a height of $\frac{6b}{a}$.
19. Express 0.000319 in scientific notation.
20. Express 0.00032×10^8 in standard notation.

Simplify, expressing each answer in scientific notation.

21. $(9 \times 10^8) \div (3 \times 10^3)$
22. $(6 \times 10^{-2})(5 \times 10^3)$
23. $(5 \times 10^4)^3$
24. $(9 \times 10^3) - (7 \times 10^2)$
25. $(4 \times 10^{-3}) + (5 \times 10^{-2})$

Express each answer in scientific notation.

26. An angstrom, the unit used to express the size of atoms, is one ten-millionth of a millimeter. Write this in meters using scientific notation.
27. Light travels at 300 million meters per second. How many kilometers does it travel in 1 hr?

State the coordinates of the vertex or point of inflection of each function.

28. $f(x) = (x - 1)^2$
29. $f(x) = x^3 + 7$
30. $f(x) = (x + 3)^3 + 5$
31. $f(x) = (x - 2)^3 - 4$

Describe the translation(s) of the graph of $y = x^2$ or $y = x^3$ that produce the graph of each function. Then create a table of at least five ordered pairs and graph each function.

32. $y = x^3 - 2$
33. $y = (x - 3)^2 + 2$
34. Write a function rule that translates the function $y = x^2$ three units up and five units right.

Write the equation of each graph.

35.

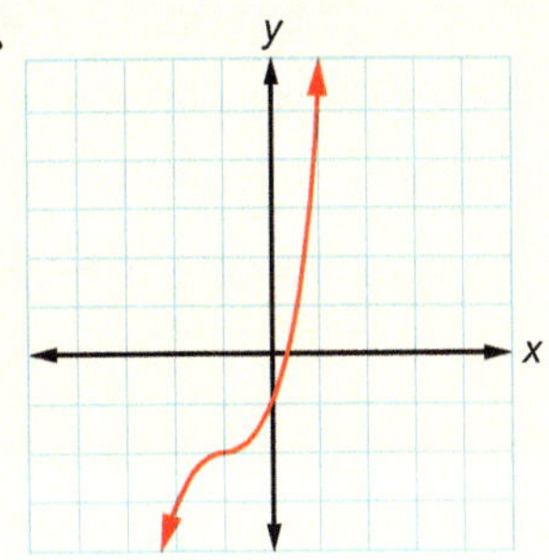

36.

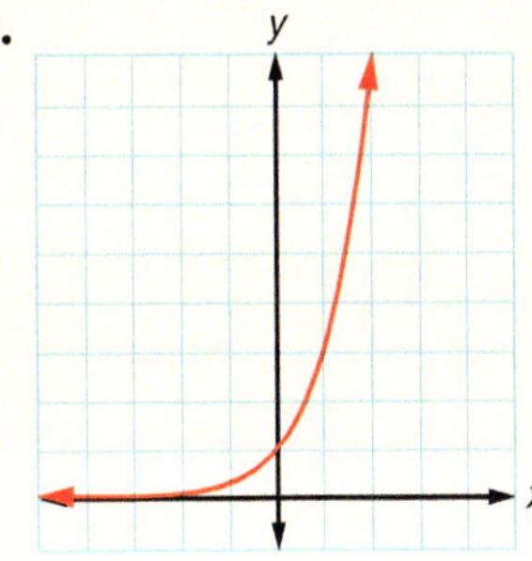

Evaluate.

37. $y = -2 \cdot 3^x$ when $x = -2$

38. $p = 100\left(\frac{1}{2}\right)^d$ when $d = 3$

Create a table of at least five ordered pairs and graph each function.

39. $y = \left(\frac{3}{5}\right)^x$

40. $y = 4\left(\frac{1}{2}\right)^x$

41. What is the y-intercept of the graph of functions of the form $y = b^x$ where $b > 0$?

42. Under what conditions is the graph of the exponential function $y = ab^x$ decreasing?

For each exponential function, identify the initial amount, the growth rate (as a percent), and the growth factor. Then state whether the function models exponential growth or exponential decay.

43. $y = 1.5^x$

44. $f(t) = 10{,}000\left(1 - \frac{2}{3}\right)^t$

Write an exponential function modeling consecutive enlargements or reductions by the given percentage. Then find the single setting that would accomplish the same result as the given number of consecutive enlargements or reductions.

45. 80%; 3

46. 120%; 2

Write an exponential function modeling each description and use it to answer each exercise.

47. Mr. Leeds projects that his current salary of \$50,000 will increase by 5% each year. What will his salary be 10 yr from now (to the nearest dollar)?

48. A mold culture covering 125 cm^2 shrinks to two-fifths of the previous day's size each day. How big is the mold culture after five days?

\$5000 is invested at 3% APR for 10 yr. Determine the number of compounding periods and the rate per compounding period when interest is compounded as follows. Then determine the ending balance.

49. quarterly

50. monthly

51. An initial population of 500 mosquitoes can triple each day.

a. State the growth rate (as a percent) and the growth factor for the mosquito population.

b. Write an exponential function, $f(d)$, modeling the number of mosquitoes.

c. Determine the number of mosquitoes after one week.

52. It takes about one week for half of the uranium-237 atoms present in a sample to undergo beta decay. A laboratory determines that a sample contains 16 g of U-237.

a. Write an exponential function, $f(w)$, modeling the amount of U-237 in the sample.

b. Approximately how much U-237 is left in the sample after five weeks?

c. Graph the function, showing the amount of U-237 in the sample during the first five weeks.

9 Polynomials

Mathematical modeling relies upon functions. For instance, population modeling uses formulas to predict animal and human population growth, enabling scientists to make proactive decisions in controlling or preparing for the dynamic changes in our world. Population models are used to understand the spread of parasites, viruses, and diseases. Our environmental health requires an understanding of the dynamics of populations.

In the Dominion Mandate God gave humans rule "over the fish of the sea, and over the fowl of the air, and over the cattle, and over all the earth, and over every creeping thing that creepeth upon the earth" (Gen. 1:26). Modeling interactions in our environment provides a means for responsibly managing our God-given resources.

The manipulation of formulas provides the necessary foundation to predicting, solving, and even preventing complex problems in the world in which we live. To whom much is given, much is required (Luke 12:48); and there is little doubt that we have been given responsibility over a large domain. Proper preparation includes developing proficiency in our math skills.

After this chapter you should be able to

1. classify and evaluate any given polynomial.
2. add and subtract polynomials.
3. multiply polynomials.
4. use the FOIL method to multiply binomials.
5. recognize special binomial products and use their patterns to compute products.
6. divide polynomials.

9.1 Classifying and Evaluating Polynomials

The energy of this wave at Waimea Bay, Hawaii, is modeled by the monomial $\frac{1}{8}\rho gH^2$, where ρ is the water density, g is the acceleration due to gravity, and H is the height of the wave.

Do you remember what terms are? A *term* was defined earlier as a constant, a variable, or the product or quotient of a constant and one or more variables. Remember that like terms contain the same variables with the same exponents and that only like terms can be added and subtracted.

The algebraic expression $2x^2 + 3x + 5y - 4x$ contains four terms: $2x^2$, $3x$, $5y$, and $-4x$. The like terms in this example are $3x$ and $-4x$. A polynomial is a special type of algebraic expression that can be written as a sum of terms without any variable divisors.

Definitions

A **polynomial** is an expression that can be written as a sum of terms consisting of constants, variables with nonnegative integral exponents, or products of a constant and one or more such variables.

A **monomial** is a polynomial with exactly one term.

A **binomial** is a polynomial with exactly two terms.

A **trinomial** is a polynomial with exactly three terms.

Examples of Polynomials			
Monomials	**Binomials**	**Trinomials**	**Other Polynomials**
$4y^3$	$2x^2 + 3x$	$5x^2 + 2x - 6$	$x^4 + 2x^3 - x^2 + 9$
$8ab^2c$	$5x^2 + y^3$	$2a^2b^5 + 8c^4 - 2a^5$	$x^3y + 5xy^2 + 3xy - 7y^5$

Notice that the expression $\frac{1}{2}x$ fulfills the definition of a polynomial because $\frac{1}{2}$ is a constant. An expression containing the term x^3y^{-2} or $\frac{x^3}{y^2}$ would not be a polynomial since the negative exponent indicates a variable divisor.

Polynomials can also be classified by their degree.

Definition

The **degree of a term** is the sum of the exponents of its variables.

The degree of the term $5x^3y^2z^4$ is 9, and the degree of the term $4x^2y$ is 3. Remember that a variable without a written exponent has a degree of one. The *degree of a constant* is zero because any constant can be written as a product including a variable with an exponent of zero. For example, $8 = 8(1) = 8x^0$.

Definition

The **degree of a polynomial** is the highest degree of any term in the polynomial.

Example 1

State the degree of the polynomial $3x^3y + 2x - 4xy^2$.

Answer $3x^3y + 2x - 4xy^2$

first term	$3x^3y^1$	$3 + 1 = 4$	1. Find the degree of each term.
second term	$2x^1$	1	
third term	$-4x^1y^2$	$1 + 2 = 3$	

The degree of the polynomial is 4.

2. The degree of the polynomial is equal to the highest degree of any of its individual terms.

It is customary to write polynomials in a *standard form,* where the terms are arranged in order of descending degrees of one of the variables. The polynomial

$$4x^2 + 3 + x + 6x^3$$

should be written with terms arranged in descending degrees of x:

$$6x^3 + 4x^2 + x + 3.$$

This allows polynomials with one variable to be quickly classified according to their degree.

Polynomial	Classification	
	Degree	Number of Terms
$6x^3 + 4x^2 + x + 3$	3	polynomial
$5x^2 - 2x - 3$	2	trinomial
$-2x^4$	4	monomial
$8x + 1$	1	binomial
-8	0	monomial

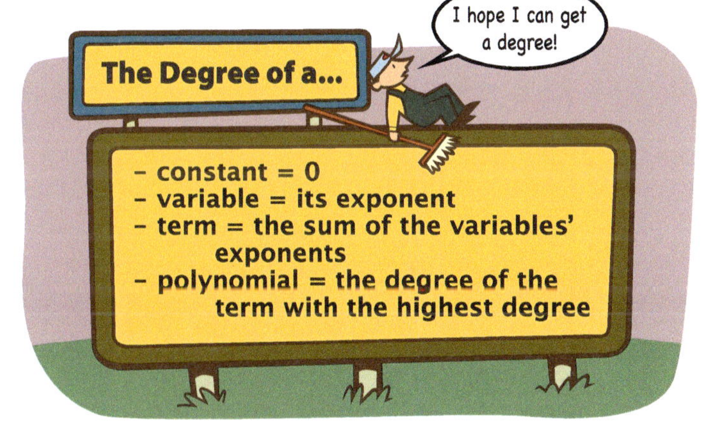

Polynomials are evaluated by substituting the given values and following the order of operations.

Example 2

Evaluate $2x^2 - 5x + 6$ when $x = 3$.

Answer $2(3)^2 - 5(3) + 6$ — 1. Substitute 3 for x in the trinomial.

$= 2(9) - 5(3) + 6$ — 2. Follow the order of operations and simplify.

$= 18 - 15 + 6 = 9$

Example 3

Evaluate $5x^3y + 2x - 3y + 5$ when $x = -2$ and $y = 1$.

Answer $5(-2)^3(1) + 2(-2) - 3(1) + 5$ — 1. Substitute -2 for x and 1 for y.

$= 5(-8)(1) + 2(-2) - 3(1) + 5$ — 2. Evaluate.

$= -40 - 4 - 3 + 5 = -42$

A. Exercises

State the degree of each term.

1. $4x^5y^2$
2. $-5x$
3. 1492
4. $5x^3y$

Classify each expression as a monomial, binomial, trinomial, other polynomial, or not a polynomial. Then state the degree of each polynomial.

5. $4x^3y^5 + 2xy$
6. $-5xy^3z^2$
7. $-4x^2y + 5xy + y - 8$
8. $3x^3y + 2x^2 - 5$
9. $\frac{4y^3}{x^2} + y^2 - 1$
10. $3x + 4y - 8z + 9$
11. $8x^3 - 2xyz$
12. $4a^{-3} - 1776$
13. $\frac{x^2y^3}{5} + \frac{xy}{2} - \frac{3}{8}$
14. 23

B. Exercises

Write each polynomial in standard form. Then classify it by its degree and its number of terms.

15. $3x + 5 - 2x^2$
16. $9 + 4x$
17. $\frac{1}{3}x^2 - 6x^5 + 12 - 4x$
18. $-12 + 6y + y^3$
19. $2xy - 7x^2y + 6x^4$
20. $-7x^2y^2 + x^4 + 5xy + 3x^3y + y^4$

Evaluate when $x = 3$, $y = -1$, and $z = 2$.

21. $x^2 + 4z - 8$
22. $5x^2 - 3y^2 - 4z$
23. $2x^3y + x^2y - 5xyz$
24. $5x^3y^2 + 8x^2y$
25. $5x^4y^2z^3$
26. $-3z^3 + 5z^2 - 4z + 7$

Evaluate.

27. $x^2 + x + 1$ when $x = 11$
28. $2z^5 + 5z^2$ when $z = 2$
29. $5m - 3$ when $m = 0$
30. $-x^2 - 4x - 7$ when $x = -1$
31. $y^5 - 2y^3 - 7$ when $y = -2$
32. $-2x^4 - x^3 + x^2 - x + 1$ when $x = -4$

C. Exercises

Evaluate.

33. $-4x^2y + 5xy + y - 8$ when $x = 5$ and $y = -3$
34. $3x^3y + 2x^2 - 5$ when $x = -2$ and $y = 4$
35. $x^2y - y^2$ when $x = \frac{2}{3}$ and $y = \frac{1}{2}$
36. $-a^2b + 2ab + 1$ when $a = 0.5$ and $b = 0.2$

Dominion Modeling

Population modeling uses mathematical information to inform and aid in the well-being of current and future generations, both in nature and in human populations. Food, water, energy, shelter, employment, education, transportation, natural resources, and ministries (including local churches, missions, and crisis centers) are just some of the areas in which human physical and spiritual needs can be better managed through population modeling using mathematical functions and formulas.

Population modeling offers a mathematical method of predicting changes, with the primary objective of managing resources and meeting basic needs. Management of wildlife populations focuses on population control, while the modeling of human populations concentrates on resource management.

37. Name three major areas of resource in which population modeling can help us prepare for critical human needs.

38. Name three factors that affect population growth through the rate of immigration (movement into a population) or emigration (movement out of a population).

39. Name three factors that may significantly affect the death rate.

40. The primary objective for weather modeling is prediction; for traffic modeling, it is regulation. What is the primary objective for human population modeling?

41. Name three areas of ministry that may represent a growing need in a growing population.

CUMULATIVE REVIEW

Match each term to its definition.

42. consistent [7.5]
43. coefficient [2.3]
44. conjunction [4.4]
45. disjunction [4.5]
46. domain [5.2]

a. the multiplier of the variable in a term
b. the set of x-coordinates in a relation
c. a system that has at least one solution
d. a compound statement using "and"
e. a compound statement using "or"

Write an equation and solve. [2.6–2.7]

47. Find three consecutive odd integers whose sum is 39.

48. Find three consecutive integers such that the difference of five times the third and twice the first equals seven more than the second.

Use the exponential function $y = 1.03^x$ for exercises 49–51. [8.6]

49. State the growth rate.

50. State the growth factor.

51. Does the function model exponential growth or exponential decay?

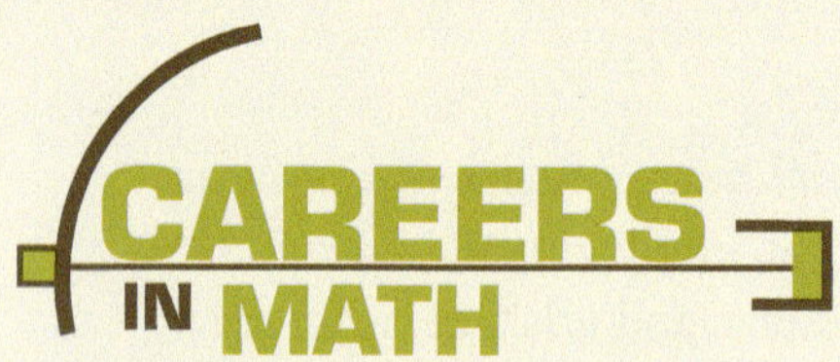

Engineer

Engineers apply math and science to solve technical problems. They design, build, test, and evaluate products. There are many different engineering fields, but all require an in-depth knowledge of mathematics. The following examples illustrate the creativity and diversity of engineering.

Civil engineers design and build roads and bridges. They calculate the direction and amount of stress on bridge girders in order to determine how much load a bridge will hold. They also determine the best and most economical route for a road, its curvature, and the banking on a curve.

Electrical engineers design and build power grids, transformers, and delivery systems, as well as electric motors. They must understand resistance and how to determine the size and type of wiring that should be used.

Aeronautical engineers design and build aircraft and the complex systems within them. These engineers must design to a set of specifications. For instance, how high must the plane fly? How fast must it be able to fly? What level of stress must it withstand?

Mechanical engineers work with machines that produce, transmit, and use power, such as commercial air-conditioning systems.

Chemical engineers convert raw materials into useful forms and develop new materials.

All of these fields deal with creating products that glorify our Lord through calling attention to His creation. The finished products then help us fulfill the Dominion Mandate as stewards of the earth.

9.2 Adding and Subtracting Polynomials

When adding polynomials, you are simply combining like terms. It may be helpful to align like terms for vertical computation.

Example 1

Add $(3a + 2b - 8c + 8)$ and $(2a - 4b + 6c + 9)$.

Answer

$$\begin{array}{r} 3a + 2b - 8c + 8 \\ +\ 2a - 4b + 6c + 9 \\ \hline 5a - 2b - 2c + 17 \end{array}$$

1. Arrange the polynomials so like terms are aligned vertically.
2. Add the columns.

Example 2

Simplify $(2x^2 + 3xy + 5y^2) + (8x^2 - 5xy + 3y) + (x^2 + 2y^2)$.

Answer

$$\begin{array}{l} 2x^2 + 3xy + 5y^2 \\ 8x^2 - 5xy \qquad\quad + 3y \\ +\ x^2 \qquad\quad + 2y^2 \\ \hline 11x^2 - 2xy + 7y^2 + 3y \end{array}$$

1. Arrange the polynomials so like terms are aligned vertically.
2. Add the columns.

A mechanic can calculate the horsepower of an n-cylinder engine with b-inch diameter bores from the formula $HP = \frac{nb^2}{2.5}$. Subtract monomials to find how much more horsepower a six-cylinder car has than a four-cylinder car with the same bore.

You can also add polynomials horizontally. Removing the parentheses is equivalent to distributing 1 to each of the terms.

Example 3

Simplify $(3x^2 + 2x - 5) + (8x^2 - 7x - 10)$.

Answer

$3x^2 + 2x - 5 + 8x^2 - 7x - 10$

$= 11x^2 - 5x - 15$

1. Remove the parentheses.
2. Combine like terms.

Example 4

Simplify $(15a^2 - 7ab - 12b + 3c^2) + (8ab - 3b + 12c^2)$.

Answer

$15a^2 - 7ab - 12b + 3c^2 + 8ab - 3b + 12c^2$

$= 15a^2 + ab - 15b + 15c^2$

1. Remove the parentheses.
2. Combine like terms.

Adding Polynomials	
Vertically	**Horizontally**
1. Arrange the polynomials so like terms are aligned vertically. 2. Add the columns.	1. Remove the parentheses. 2. Combine like terms.

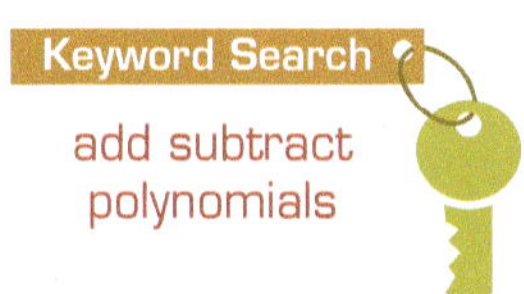

Recall that subtraction is defined as addition of the opposite. Therefore, subtraction of polynomials can be performed by adding the opposite of each term.

Example 5

Simplify $(3x^2 + 2x - 5) - (4x^2 - 5x + 2)$.

Answer $(3x^2 + 2x - 5) + (-1)(4x^2 - 5x + 2)$
$= 3x^2 + 2x - 5 - 4x^2 + 5x - 2$

1. Subtract by adding the opposite of each term in the polynomial being subtracted. (This can be thought of as distributing -1 to each term.)

$= -x^2 + 7x - 7$

2. Combine like terms.

Polynomials can also be subtracted vertically by adding the opposite. Be sure to write the opposite of each term and align like terms before adding.

Example 6

Simplify.
$$\begin{array}{r} 5a - 7b + 6c + 10 \\ \underline{-(3a + 4b - 8c - 9)} \end{array}$$

Answer
$$\begin{array}{r} 5a - 7b + 6c + 10 \\ \underline{+(-3a - 4b + 8c + 9)} \\ 2a - 11b + 14c + 19 \end{array}$$

1. Change the problem to an addition problem by finding the opposite (additive inverse) of each term in the second polynomial.
2. Add the columns.

Example 7

Simplify $(4x^2 + 3xy - y^2) - (5x^2 - 8xy + 4y^2)$.

Answer $4x^2 + 3xy - y^2 - 5x^2 + 8xy - 4y^2$

1. Remove the parentheses by adding the opposite of the second polynomial.

$= -x^2 + 11xy - 5y^2$

2. Combine like terms.

A. Exercises

Add using the vertical format.

1. $$\begin{array}{r} 3m + 4 \\ 7m + 2 \\ \underline{+ 9m - 12} \end{array}$$

2. $$\begin{array}{r} 5k^2 - 3k + 2 \\ \underline{+ 8k^2 + 9k - 8} \end{array}$$

3. $$\begin{array}{rrr} 4m^2 & + 2m & \\ & 8m & - 9 \\ \underline{+ 3m^2} & & \underline{+ 4} \end{array}$$

4. $$\begin{array}{rrr} 3ay & + 4y & - 5z \\ & 3y & + 10z \\ \underline{+ 7ay} & \underline{- 5y} & \end{array}$$

Subtract using the vertical format.

5. $$\begin{array}{r} 5a + 3 \\ \underline{-(2a - 5)} \end{array}$$

6. $$\begin{array}{r} 3a^2 + 2a - 4 \\ \underline{-(a^2 \quad\quad - 6)} \end{array}$$

7. $$\begin{array}{r} x + 2y - 3 \\ \underline{-(4x - 8y + 3)} \end{array}$$

8. $$\begin{array}{r} x^2 + 4xy - 10y^2 \\ \underline{-(4x^2 - 8xy + 3y^2)} \end{array}$$

Add.

9. $(5x + y) + (4x + 2y)$
10. $(2x^2 - 5x + 3) + (2x^2 + 7x + 5)$
11. $(3x^3 + 4x - y) + (9x^2 - 5x + 7) + (2x^3 + 3x + 8)$
12. $(4a + 9b - 10d) + (5a - 3b + 6d) + (18a - 2c)$

Subtract.

13. $(6x + y - 7) - (5x - y + 7)$
14. $(3x^2y - x + 5) - (3x^2y - x + 5)$
15. $(a^3 + 4b^2 - 3c) - (2a^3 - 12c)$
16. $(x^5 - x^4 + 8x - 7) - (x^4 + 2x - 2)$
17. $(y^2 + 3y + 8) - (9y^2 - 4y - 3)$
18. $(5x + 2y + 9) - (7x - 3y - 7)$

B. Exercises

Simplify.

19. $(-4x^3 + 3x^2 - 5x + 10) + (-6x^2 + 3x + 9) + (5x^2 - 10)$
20. $(8k^2 + 3km - 8m^2) + (3k^2 + 5m^2) + (2k^2 - 4km)$
21. $(-x^3 + x^2y + xy^2 + y^4) - (5x^3 + 2x^2y - y^3 + 3y^4)$
22. $(a^3b + a^2b^3 - 4ab) - (12a^2b^3 + 9)$
23. $(0.82a + 0.19b - 0.84) + (1.23a + 11.2b + 4.2) + (6.71a - 3.8b + 5.2)$
24. $\left(\frac{1}{2}x + \frac{1}{5}y\right) + \left(\frac{3}{2}x + \frac{8}{7}\right) + \left(\frac{1}{3}x - \frac{4}{5}y\right)$
25. $(4x - 2y + 8) + (3x + 9y - 4) - (2x - y + 4)$
26. $(3x^2 + 5) - (12x^2 - 7x) + (3x - 2)$
27. $(3x^2 - 10xy + 7y^2) - (2x^2 + 8xy + 8y^2) + (-9x^2 + 3y^2)$
28. $(12a^2 - 23ab - 6b^2) + (4a^2 - b^2) - (3a^2 + 2ab + b^2) - (a^2 + ab + b^2)$
29. Subtract $2x + 9$ from $3x - 7$.
30. Subtract $4x - 7y - 9$ from 0.

Write a polynomial expression for the perimeter of each figure.

31. 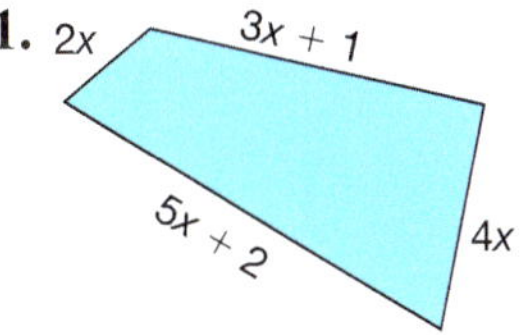

32. 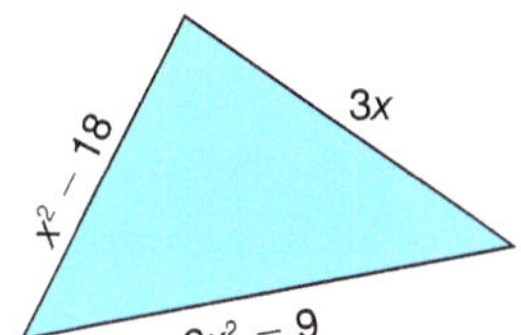

33. 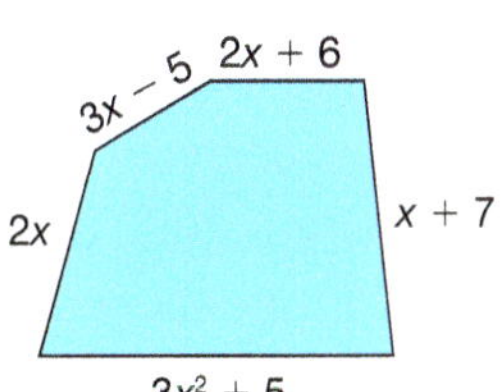

34. The length of a rectangle is $5n + 3$, and its width is $2n - 6$. Write a polynomial expression for its perimeter.
35. The perimeter of a rectangle is $6x + 6$, and its width is $x + 2$. Write a polynomial expression for its length.
36. Owen is x years old, Suzette is five years older than Owen, and Jane is three years younger than Owen. Write a polynomial expression representing their combined ages.

C. Exercises

Simplify.

37. $(-3x^2 + 2x - 5) - (x^3 - 2x^2 + 4x + 7) + (3x^3 + 5x^2 - 6x + 2)$

38. $\left(\frac{1}{2}x^2 + \frac{1}{3}xy + \frac{1}{4}y^2\right) + \left(\frac{1}{6}x^2 + \frac{1}{8}y^2\right) - \left(\frac{1}{2}x^2 + \frac{3}{2}xy\right)$

39. During the championship game, Tony made n points; Neal scored four more points than Tony; Cedric had twice as many points as Neal; Rob scored two more points than Cedric; and Moe had half as many points as Rob. What was the combined score of these five players?

40. Determine which has the greater perimeter: a square with a side equal to $3x + 6$, or a rectangle with a length of $4x + 8$ and a width of $2x + 3$ (where x is a positive number).

41. Write a polynomial expression for the sum of any two consecutive numbers and use the expression to explain why this sum must always be odd.

Dominion Modeling

Population growth can be modeled by the following exponential function:

$P(t) = A(1 + r)^t$, where A = the initial population,
r = the average periodic growth rate, and
t = the number of periods (most often in years).

A spreadsheet is a powerful mathematical tool that can be used to efficiently solve complex mathematical problems. Though the answers to the following problems can be found using a calculator, you will benefit from learning how to use a spreadsheet to quickly model population growth. While most spreadsheets follow the same basic principles, formula formats may vary slightly between software products.

42. In 2000 the population of Wilmington, NC, was 75,838. Use the exponential growth function to estimate Wilmington's 2008 population, assuming the city's annual growth rate was approximately 3.55%. Enter the following formula into cell C1 of a spreadsheet: =75838*(1.0355)^8.

Create a spreadsheet modeling the growth of Wilmington's population.

Enter 0 in cell A1 and enter 1 in cell A2. Highlight both cells as shown and place the cursor on the little square in the lower right-hand corner of the highlighted area. Drag the square down to row 51. If this procedure (referred to as *replicating* the formula) is done correctly, the cells will be filled with sequential numbers to 50. Next, enter the year 2000 in cell B1 and 2001 in cell B2. Highlight these two cells and drag to replicate the years to 2050.

	A	B
1	0	
2	1	
3		
4		

Change the formula in cell C1 to the following: =75838*(1.0355)^A1. Then highlight cell C1 and drag the little square in the lower right-hand corner down to row 51. Notice how the variable A1 is automatically changed as the formula is replicated down the column. The cell references are copied relative to each cell's position on the spreadsheet. The result represents a yearly estimate of Wilmington's population given a consistent growth rate of 3.55% over a period of 50 yr.

43. Predict Wilmington's 2050 population, assuming an annual growth rate of 3.55%.

44. Comparing Wilmington's 2000 and 2050 populations, name three areas of major concern for the city developers.

CUMULATIVE REVIEW

45. What is the additive inverse of -36? [1.3]

46. What is the additive identity? [1.3]

State each property.

47. Addition Property of Equality [2.4]

48. Addition Property of Inequality [4.2]

Use an example to illustrate each statement. [1.3–1.4]

49. Subtraction is not commutative.

50. Subtraction is not associative.

Simplify. [2.3, 8.1]

51. $(3x^2)(5x^3)$

52. $(-4a^2b)(3ab)^2$

53. $5(3x - 2y)$

54. $6x(xy + 5z)$

9.3 Multiplying Polynomials

Death Valley holds two records for the Western Hemisphere: highest temperature of 56.7°C and highest summer average temperature of 36.7°C. Use the binomial $\frac{9}{5}C + 32$ to convert these temperatures to Fahrenheit.

The Distributive Property is used to multiply a number by a sum. It can also be used to find the product of a monomial and a polynomial.

Finding the area of the rectangle illustrates this process.

$$2x(x + 2) = 2x^2 + 4x$$

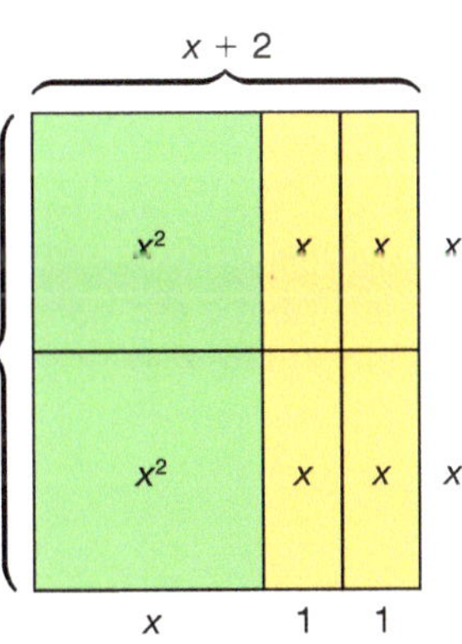

To multiply a monomial by a polynomial, use the Distributive Property to multiply each term of the polynomial by the monomial. This multiplication can also be arranged vertically using a format similar to that used for the multiplication of whole numbers.

$$\begin{array}{r} 512 \\ \times\ \ 4 \\ \hline 2048 \end{array} \qquad \begin{array}{r} x^2 - \ 3x + \ \ 8 \\ \times \qquad\qquad 4x \\ \hline 4x^3 - 12x^2 + 32x \end{array}$$

Example 1

Multiply $3x(x^2 - 7x + 8)$.

Answer

horizontal method

$3x(x^2 - 7x + 8)$

$= 3x(x^2) + 3x(-7x) + 3x(8)$

$= 3x^3 - 21x^2 + 24x$

vertical method

$$\begin{array}{r} x^2 - \ 7x + \ 8 \\ \times \qquad\quad 3x \\ \hline 3x^3 - 21x^2 + 24x \end{array}$$

Example 2

Multiply $2xy^2(3x - 8xy + 5y - 3z)$.

Answer $2xy^2(3x - 8xy + 5y - 3z)$

$= 2xy^2(3x) + 2xy^2(-8xy) + 2xy^2(5y) + 2xy^2(-3z)$
$= 6x^2y^2 - 16x^2y^3 + 10xy^3 - 6xy^2z$
$= 6x^2y^2 - 16x^2y^3 + 10xy^3 - 6xy^2z$

When two binomials are multiplied, the Distributive Property requires each term in the first binomial to be distributed to each term in the second binomial.

Finding the area of the rectangle illustrates this process.

$(2x + 1)(x + 2)$
$= 2x(x + 2) + 1(x + 2)$
$= 2x^2 + 4x + x + 2$
$= 2x^2 + 5x + 2$

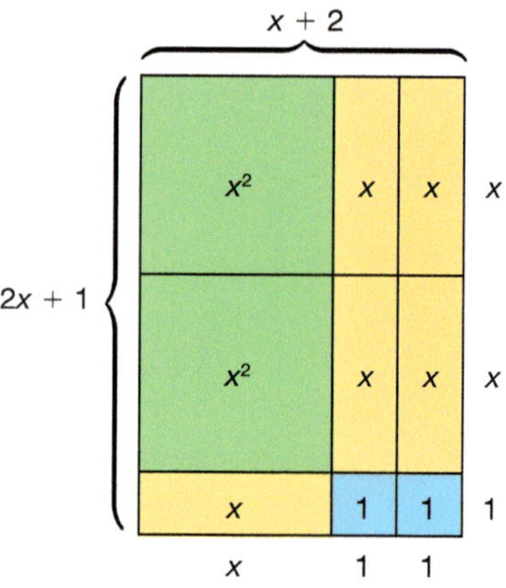

Example 3

Multiply $(3x - 7)(2x - 5)$.

Answer $(3x - 7)(2x - 5)$

$= 3x(2x - 5) - 7(2x - 5)$
$= 6x^2 - 15x - 14x + 35$
$= 6x^2 - 29x + 35$

When both polynomials have two or more terms, using the vertical multiplication format helps to align like terms to be added. Notice how each term in the bottom factor is multiplied by each term in the top factor.

$$\begin{array}{r} 512 \\ \times\ 24 \\ \hline 2048 \\ 1024 \\ \hline 12{,}288 \end{array} \qquad \begin{array}{r} x^2 - 3x + 8 \\ \times \quad 4x - 7 \\ \hline -7x^2 + 21x - 56 \\ 4x^3 - 12x^2 + 32x \\ \hline 4x^3 - 19x^2 + 53x - 56 \end{array}$$

Example 4

Multiply $(x - 2)(3x^2 + 4x + 9)$.

Answer

$$\begin{array}{r} 3x^2 + 4x + 9 \\ \times \qquad x - 2 \\ \hline -6x^2 - 8x - 18 \\ 3x^3 + 4x^2 + 9x \qquad \\ \hline 3x^3 - 2x^2 + x - 18 \end{array}$$

1. Write the factors (descending powers of x) in the vertical multiplication format.
2. Multiply each term in the top factor by each term in the bottom factor, aligning like terms in the same column.
3. Combine like terms.

By examining the distributive process for the general case below and the diagram, you can see why each term in the first factor is multiplied by each term in the second factor.

$$(a + b)(c + d + e) = a(c + d + e) + b(c + d + e)$$
$$= ac + ad + ae + bc + bd + be$$

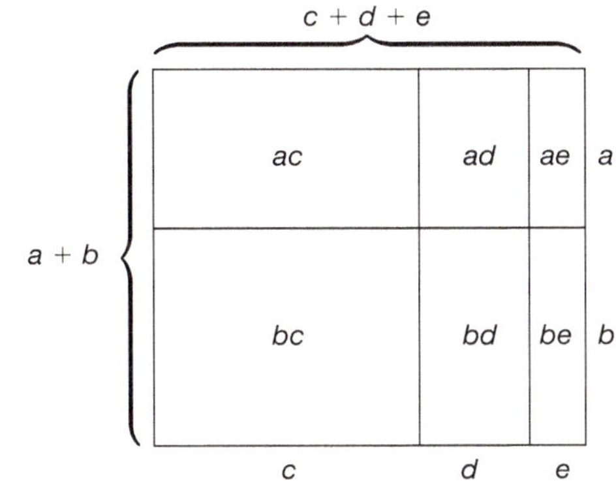

Example 5

Multiply $(2a + b)(7a + 4b - 5c)$.

Answer

$(2a + b)(7a + 4b - 5c)$

$= 2a(7a + 4b - 5c) + b(7a + 4b - 5c)$

$= 14a^2 + 8ab - 10ac + 7ab + 4b^2 - 5bc$

$= 14a^2 + 15ab - 10ac + 4b^2 - 5bc$

Multiplying Polynomials

1. Arrange both factors in descending powers of the same variable.
2. Multiply each term of the multiplier by each term of the multiplicand.
3. Combine like terms to obtain the final product.

Example 6

Multiply $(2x^3 - 3x + x^2 + 4)(5x - 2)$.

Answer

$$\begin{array}{r} 2x^3 + x^2 - 3x + 4 \\ \times \qquad\qquad 5x - 2 \\ \hline -4x^3 - 2x^2 + 6x - 8 \\ 10x^4 + 5x^3 - 15x^2 + 20x \qquad \\ \hline 10x^4 + x^3 - 17x^2 + 26x - 8 \end{array}$$

1. Arrange the terms in descending powers of x.
2. Place the problem in vertical format.
3. Multiply.
4. Combine like terms.

A. Exercises

Write the polynomial multiplication modeled by each diagram and simplify.

1. 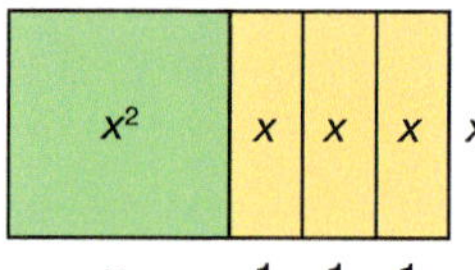

2. 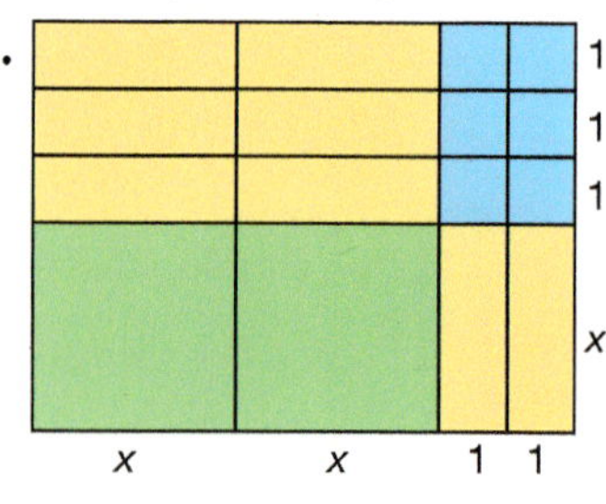

3. 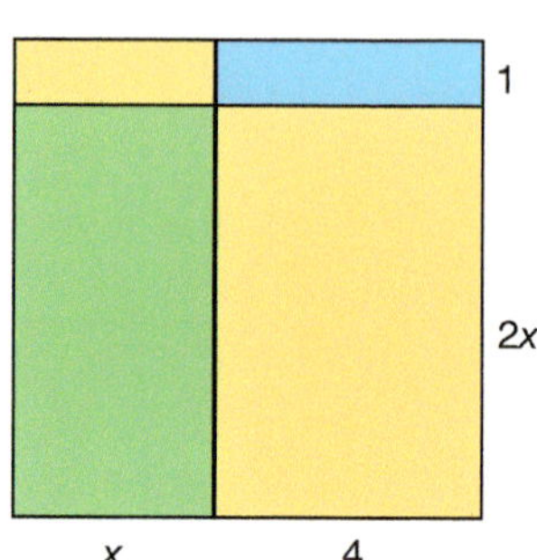

Multiply using the vertical format.

4. $\begin{array}{r} 2x^2 + 3x - 8 \\ \times \qquad\quad 7x \\ \hline \end{array}$

5. $\begin{array}{r} 5a + 10b \\ \times \quad -4b \\ \hline \end{array}$

6. $\begin{array}{r} 3x + 4 \\ \times\ 2x + 1 \\ \hline \end{array}$

7. $\begin{array}{r} 3x - 6 \\ \times\ 3x + 5 \\ \hline \end{array}$

8. $\begin{array}{r} 3a^2 - 2a - 4 \\ \times \qquad a - 7 \\ \hline \end{array}$

9. $\begin{array}{r} 6x^2 + 3xy - 9y^2 \\ \times \qquad 2x + y \\ \hline \end{array}$

Multiply.

10. $3a(k + 2m - p)$
11. $-7k(8k^2 + 9k + 6)$
12. $-2k(k^3 + 6k^2 + 3k - 9)$
13. $3a^2(4a^4 + 5a - 18)$
14. $(x + 7)(x + 5)$
15. $(y - 4)(y - 3)$
16. $(b + 2)(b - 1)$
17. $(c - 3)(c + 8)$
18. $(a^2 - 3a - 9)(a + 2)$
19. $(4x^2 - 3x - 6)(x + 2)$

B. Exercises

Multiply.

20. $6m(4m^2 - 3mn - 10)$
21. $-8x^3y(3x^2 + 7xy + 10)$
22. $\frac{1}{5}a^2(25a^3 + 15a^2 - 50)$
23. $16x\left(\frac{1}{4}x^2 + \frac{1}{8}xy + \frac{1}{2}\right)$
24. $(2a + 4b)(3a - 9)$
25. $(4b^2 - 8)(-2b^2 + 6b)$
26. $(2x^2 + 5)(3x^2 - 9)$
27. $(-3x^2 + 4y)(4xy + 3y^2)$
28. $(2x^2 - x + 4)(3x - 7)$
29. $(2x^2 - 5x + 8)(3x + 6)$
30. $(x + y)(x^2 + xy + y^2)$
31. $(x - 3)(x^2 + 3x + 9)$
32. $(a + b)(7a + 3b - 9c + d)$
33. $(x^2 - 3x - 9)(x^2 + 4x - 6)$
34. $(2x^2 - 5x + 7)(3x^2 - 7x - 4)$
35. $(x - 7)(x - 5)(x - 3)$

Write a polynomial expression for the area of each figure.

36. 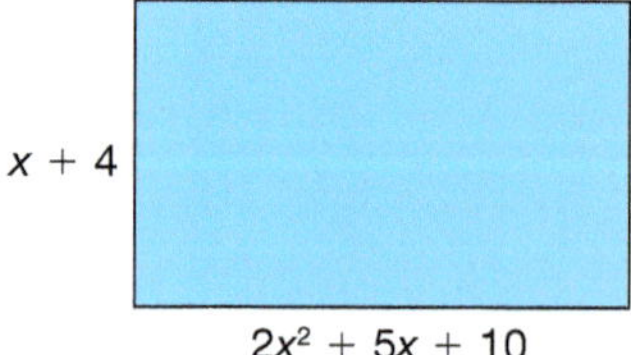

37. 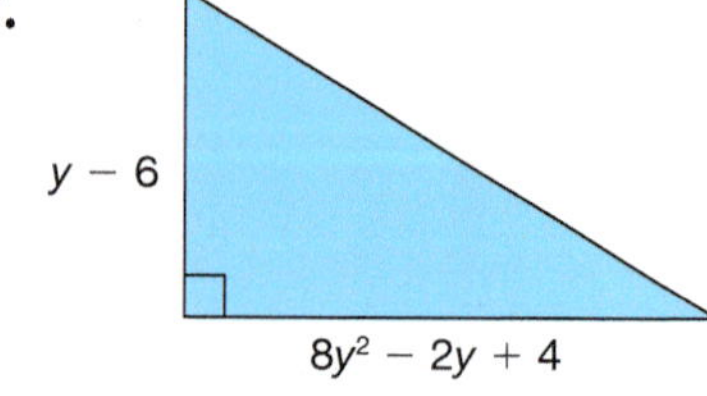

38. 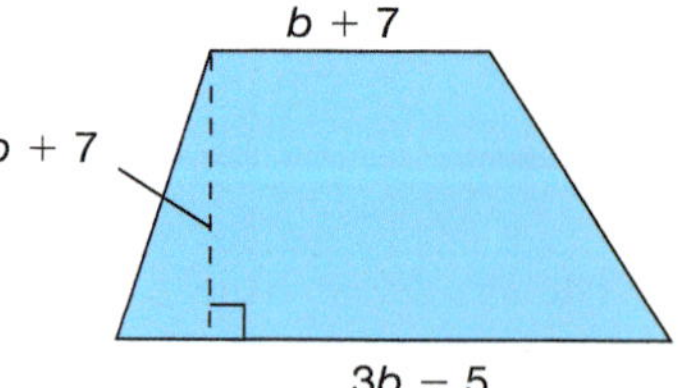

C. Exercises

Simplify.

39. $-3x(x + 4) - (x + 5)(x - 5)$

40. $(2x + 3)(x - 7) - (3x - 1)(x - 5)$

41. Write a polynomial expression for the product of two consecutive odd integers.

42. Write a polynomial expression for the product of three consecutive integers.

43. Draw a diagram illustrating the multiplication of $(x + 3)(2x + 1)$. State the product and explain how this answer is illustrated by your diagram.

44. Explain how the Distributive Property is used to multiply $(x + 2)(x^2 - 2x + 1)$.

Dominion Modeling

Wildlife and natural resources are a gift from God, and their proper management is our responsibility. Genesis 9:3 states, "Every moving thing that liveth shall be meat for you; even as the green herb have I given you all things." When we fail to manage the resources of this world, both people and nature suffer.

In the early 1900s the white-tailed deer population for the entire United States was 500,000; today it is about 20 million. Unmanaged, deer populations can double every two or three years. According to the National Highway Traffic Safety Administration, about 1.5 million deer-vehicle collisions occur yearly in the United States, resulting in about 150 deaths, 10,000 injuries, and $1.1 billion in property damage. Unmanaged wildlife can lead to overpopulation or extinction. Unrestrained population growth leaves deer prone to starvation during harsh winters, and malnutrition lowers their resistance to the spread of disease. Population modeling can help us make wise choices that benefit both wildlife and ourselves.

Create a spreadsheet modeling the growth of the deer population in the state of Wisconsin.

Enter 1,500,000 in cell A1 to represent an initial deer population. Enter a growth rate of 0.04 in cell A2. Enter 0 and 1 in cells B1 and B2 respectively. Replicate the values in column B to 20 by highlighting the two values and dragging the little square down.

Now enter the growth formula from Section 9.2, Dominion Modeling, in cell C1 as =$A$1*(1+$A$2)^B1. The $ signs signify an absolute reference. When you replicate this formula, the "A" and "1" in the formula will not change. Without the $ signs, these elements would change relative to the cell position. Now replicate the formula in C1 down the column to show 20 yr of population growth.

Use your spreadsheet to answer the following questions.

45. If Wisconsin's deer population is 1.5 million and the managed population growth is 4%, what is the estimated population in 10 yr?

46. If the deer population of 1.5 million goes unmanaged and the growth rate is 29%, what is the estimated population in 10 yr? (Change the growth rate in your spreadsheet.)

47. What is the predicted unmanaged deer population in 20 yr if the growth rate is 23%?

48. Name two factors that may reduce a deer population's unmanaged growth rate.

CUMULATIVE REVIEW

Simplify. [1.7, 8.2]

49. $3^{-1} - y^0 + 3^2$

50. $(3x^{-2}y)^{-3}$

Solve. [2.7, 4.6]

51. $5(x - 2) = 4 + 3x$

52. $|2x + 3| = 4$

Solve each system. [7.2, 7.4]

53. $2x - y = 7$
$3x + 2y = 0$

54. $5x + 4y = 0$
$y = -2x + 3$

Simplify, expressing each answer in scientific notation. [8.3]

55. $(4.4 \times 10^{-5}) + (3.8 \times 10^{-4})$

56. $(5.1 \times 10^{9}) \times (3 \times 10^{-4})$

Express each answer in scientific notation. [8.3]

57. Earth's average distance from the sun is 1.5×10^8 mi, while Neptune's is 4.5×10^9 mi. Neptune's average distance from the sun is how many times that of Earth?

58. Neptune's farthest distance from the sun, its aphelion, is 4.55 billion miles; and its closest distance from the sun, its perihelion, is 4.45 billion miles. What is the difference of these distances in miles?

TECHNOLOGY CORNER (TI-84+ Family)

The graphing calculator can be used to check the results of simplifying a polynomial expression. Use [Y=] to enter the polynomial expression $(x^2 + 3x + 2)(x - 1)$ as Y_1. Graph the function using the ZStandard option from the [ZOOM] menu.

Then use [Y=] to enter a possible answer, $x^3 + 2x^2 - x + 2$, as Y_2. Move the cursor onto the \ icon to the left of Y_2 and repeatedly press [ENTER] until the icon changes to -0.

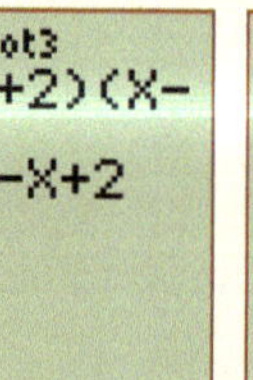

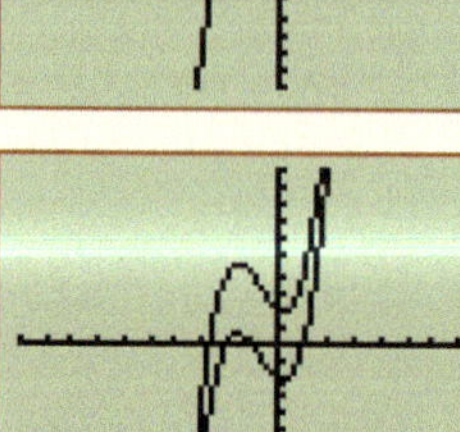

Be sure SEQUENTIAL is selected in the [MODE] menu, and select [GRAPH] to draw the graphs of Y_1 and Y_2. If the polynomials are equal, the graphs will coincide. Notice the effect of selecting the -0 icon before Y_2. These expressions cannot be equal since their graphs do not coincide.

Re-evaluate the product and edit Y_2 to match the new answer, $x^3 + 2x^2 - x - 2$. It appears that the graphs coincide and that the expressions are equal.

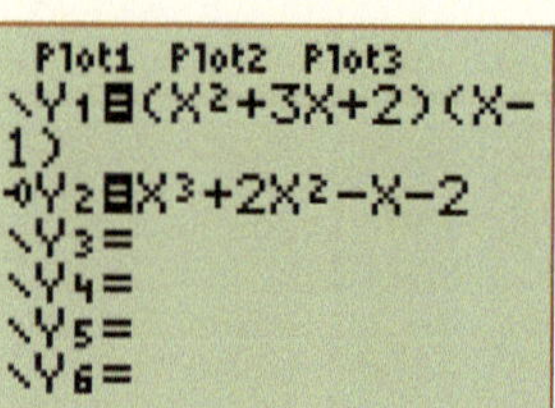

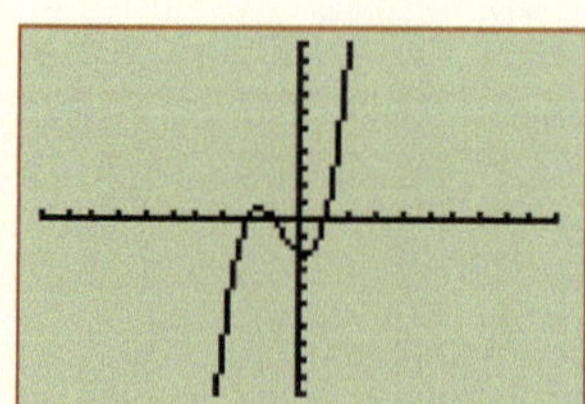

9.4 Multiplying Binomials Using FOIL

The Enigma was used by Germany in World War II to send secret messages. Complex polynomials are now used to encrypt data sent over the Internet.

While you can multiply binomials using the methods described in the previous section, you can also use an easily remembered pattern to mentally compute this frequently occurring product.

Consider the following binomial multiplication.

$$(a + b)(c + d)$$
$$= a(c + d) + b(c + d)$$
$$= ac + ad + bc + bd$$

The answer is the sum of the following four products, which can be easily remembered using the acronym FOIL.

First terms: a and c
Outer terms: a and d
Inner terms: b and c
Last terms: b and d

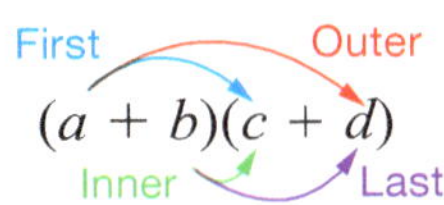

Example 1

Multiply $(x + 2)(x + 6)$.

Answer

$(x + 2)(x + 6)$
$= x^2 + 6x + 2x + 12$
$= x^2 + 8x + 12$

1. Find the sum of the products of the first, outer, inner, and last terms.
2. Combine like terms.

$(x + 2)(x + 6)$

When the binomials are written in descending degree, the products of the inner and outer terms are often like terms that need to be combined.

Example 2

Multiply $(y + 6)(4y - 9)$.

Answer

$(y + 6)(4y - 9) = 4y^2 - 9y + 24y - 54$
$= 4y^2 + 15y - 54$

1. Find the sum of the products of the first, outer, inner, and last terms.
2. Combine like terms.

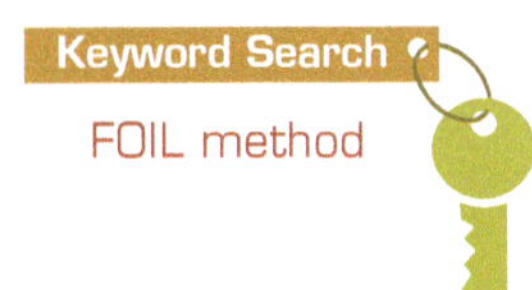

To mentally multiply binomials, use the *FOIL method* and combine any like terms.

Example 3

Multiply $(3a - 5)(a + 3)$ using the FOIL method.

Answer $(3a - 5)(a + 3) = 3a^2 + 9a - 5a - 15$ 1. Use FOIL.
$= 3a^2 + 4a - 15$ 2. Combine like terms.

Be careful when squaring a binomial. Many people mistakenly think that $(a + b)^2 = a^2 + b^2$. Notice that by applying the definition of an exponent, you can determine the correct answer using the FOIL method.

Example 4

Multiply $(2x - 5)^2$.

Answer $(2x - 5)^2 = (2x - 5)(2x - 5)$ 1. Square the quantity by multiplying it by itself.
$= 4x^2 - 10x - 10x + 25$ 2. Use FOIL.
$= 4x^2 - 20x + 25$ 3. Combine like terms.

Binomials and their products can be used to model real-life applications.

Example 5

A rectangular swimming pool has sides that are in a ratio of 3 : 2. If the pool is surrounded by a 5 ft wide concrete walkway, write a polynomial expression for the total area of the pool and walkway.

Answer

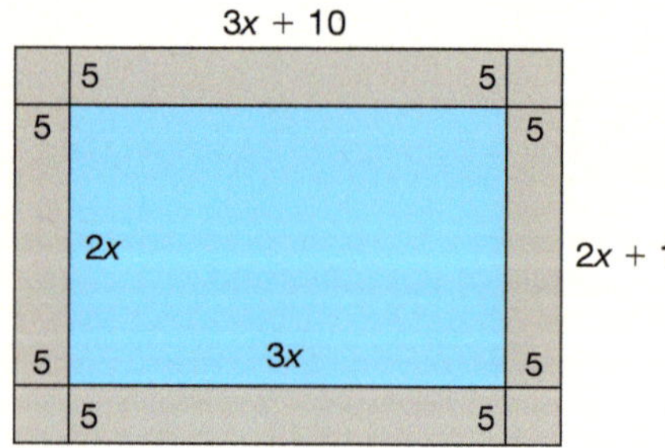

1. Draw a diagram to model the problem. Since the length and width of the pool are in a 3 : 2 ratio, represent them with $2x$ and $3x$. Add 10 to these distances.

area $= (2x + 10)(3x + 10)$ 2. Recall that area = length × width.
$= 6x^2 + 20x + 30x + 100$ 3. Use FOIL and combine like terms.
$= 6x^2 + 50x + 100$

A. Exercises

Multiply using the FOIL method.

1. $(x + 5)(x + 2)$ **2.** $(y - 7)(y + 9)$ **3.** $(a + 6)(a - 4)$
4. $(b - 3)(b - 4)$ **5.** $(x - 2)(x + 1)$ **6.** $(w - 6)(w + 8)$
7. $(v + 1)(v - 5)$ **8.** $(z - 3)(z - 8)$ **9.** $(c - 3)(c - 3)$
10. $(n - 4)(n - 4)$ **11.** $(x - 6)(x + 2)$ **12.** $(t - 6)(t - 7)$
13. $(s - 7)(s - 9)$ **14.** $(d + 9)(d - 8)$ **15.** $(y - 5)(y + 5)$
16. $(g + 3)(g - 3)$ **17.** $(a - 7)(a + 7)$ **18.** $(a - b)(a + b)$
19. $(x - 8)(x - 8)$ **20.** $(c - 1)(c - 1)$ **21.** $(a - b)(a - b)$
22. $(y + 9)(y + 9)$ **23.** $(x + 3)(x + 3)$ **24.** $(a + b)(a + b)$

B. Exercises

Multiply using the FOIL method.

25. $(3x + 2)(x + 4)$

26. $(2y - 6)(4y + 1)$

27. $(5z - 6)(3z + 2)$

28. $(2c - 9)(4c - 11)$

29. $(a + 6)^2$

30. $(3f - 2)^2$

31. $(-7g - 10)^2$

32. $(11x + 12)^2$

33. $(y + 7)(3y - 4)$

34. $(5z + 6)(2z - 7)$

35. $(3b + 8)(2b - 3)$

36. $(c + 4)(2c - 9)$

37. $(3h + 4)(3h - 4)$

38. $(2x - 13)(2x + 13)$

39. $(2v + 4)(3v - 6)$

40. $(5w + 6)(w - 4)$

41. $(3x^2 + 6)(x^2 - 1)$

42. $(2y^2 + 14)(3y^2 - 7)$

43. $(z^3 + 4)(2z^3 - 11)$

44. $(2b^3 + 3b)(3b^3 + 4b)$

45. $(a^2 - a)(a^3 + 1)$

46. $(4x^3 + 2)(5x^3 - 6x^2)$

47. Write a polynomial expression for the area of a rectangle with a length of $4x + 3$ and a width of $2x - 4$.

48. Write a polynomial expression for the area of a triangle if its base is $2x + 12$ and its height is $x + 7$.

49. The length of a soccer field is 40 yd longer than its width, and there is a 5 yd wide grass border around the entire field. Write a polynomial expression for the total area of the field and grass border.

50. The typical widescreen aspect ratio is 16 : 9. If a television has a 1 in. frame on each side and across the top and a 3 in. frame across the bottom, write a polynomial expression for the total area of the front of the television.

C. Exercises

51. If n represents the first of four consecutive odd integers, write a polynomial expression for the product of the third and fourth integers.

Simplify.

52. $(8x + 3)(3x - 2) + (x - 4)(x + 4)$

53. $(-2x - 5)(x - 12) - (3x + 2)(-x + 7)$

Write a polynomial expression for each shaded area.

54.

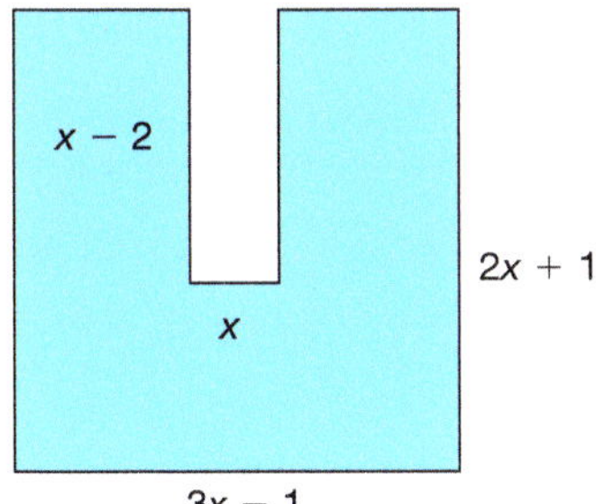

55.

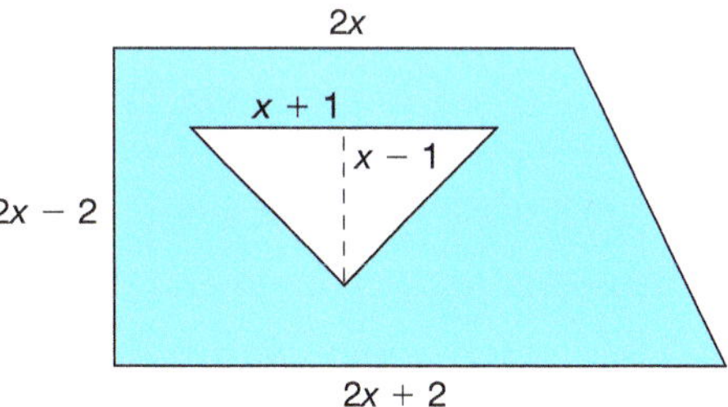

Dominion Modeling

The US population was estimated to be 310,000,000 in July 2010. The US population growth rate has been slowing in recent decades, from more than 2% in 1950 to less than 1% in 2009. It is predicted that it may slow to as little as 0.5% in the near future.

Create a spreadsheet modeling the population growth of the United States.

Enter the July 2010 US population in cell A1 and a growth rate of 0.008 in cell A2. Enter 0 and 1 in cells B1 and B2 respectively. Enter 2010 and 2011 in cells C1 and C2 respectively. Replicate the values in columns B and C to 40 yr and 2050 by highlighting these four cells and dragging the little square down.

Enter the following growth formula in cell D1: =$A$1*(1+$A$2)^B1. Recall that the $ signs signify absolute references, so those elements do not change when a formula is replicated. Replicate the formula in cell D1 down the column to model the predicted population until 2050.

Use your spreadsheet to answer the following questions.

56. What is the predicted US population for 2050 with a growth rate of 0.8%?

57. What would the 2050 US population be if the annual growth rate were 1.4%? (Change the growth rate in your spreadsheet.)

58. Assuming an annual 0.8% growth rate, predict the first year in which the US population would exceed 500 million people. (Be careful to replicate the data in all related columns.)

59. Name two major factors that would cause the prediction in the previous exercise to be significantly inaccurate.

CUMULATIVE REVIEW

State each property. [1.5, 8.1]

60. Commutative Property of Multiplication

61. Associative Property of Multiplication

62. Power Property of Exponents

63. Complete the rule for the power of a product: $(abc)^n =$ ___.

Solve. [2.5]

64. $2x + 7 = 37$

65. $\frac{2y}{3} - 12 = 8$

Find the missing value for each variation. [5.6]

66. If y is directly proportional to x and $y = 20$ when $x = 10$, find y when $x = 50$.

67. If y varies inversely with x and $y = 6$ when $x = 2$, find y when $x = 24$.

Simplify. [8.2]

68. $\frac{28a^4bc^5}{-4ac^2}$

69. $\frac{x^4y^3z^2}{-xy^2z}$

SEQUENCES

Sums of Perfect Squares and Cubes

Challenge Can you supply the next two numbers in each of the following sequences?

a. 1, 8, 27, 64, 125, …

b. 1, 9, 36, 100, 225, …

c. 1, 5, 14, 30, 55, …

The first Challenge problem is the sequence $A_n = n^3$. The second is the sequence of partial sums for $A_n = n^3$; the sixth partial sum, S_6, is $1 + 8 + 27 + 64 + 125 + 6^3 = 441$. The seventh partial sum, S_7, is $S_6 + A_7$, where $A_7 = 7^3$; so $S_7 = 441 + 343 = 784$. The third Challenge problem is the sequence of partial sums for $A_n = n^2$.

General-term formulas for the partial sums of $A_n = n^2$ and $A_n = n^3$ are polynomial functions that are too difficult to notice a pattern for. Since inductive reasoning fails in these cases, deductive reasoning is used in advanced math classes to develop these formulas.

The partial-sum formula for $A_n = n^2$ is $S_n = \frac{2n^3 + 3n^2 + n}{6}$. We can verify that this formula is true for $n = 5$ by substitution; $\frac{2 \cdot 5^3 + 3 \cdot 5^2 + 5}{6} = \frac{330}{6} = 55$, which is S_5 for the third sequence. The partial-sum formula for $A_n = n^3$ is $S_n = \frac{n^4 + 2n^3 + n^2}{4}$. If $n = 7$, we get $\frac{7^4 + 2 \cdot 7^3 + 7^2}{4} = \frac{3136}{4} = 784$ for S_7, which agrees with the previous result.

Exercises

Use substitution to verify that $S_n = \frac{2n^3 + 3n^2 + n}{6}$ for $A_n = n^2$ for each of the following values of n.

1. 1 **2.** 2 **3.** 3

Use substitution to verify that $S_n = \frac{n^4 + 2n^3 + n^2}{4}$ for $A_n = n^3$ for each of the following values of n.

4. 1 **5.** 2 **6.** 3

7. If we used the results of these first six exercises to say that the two formulas are always true, would we be using deductive or inductive reasoning?

Use the formulas to find the sum of each of the following.

8. the first 20 perfect squares **9.** the first 20 perfect cubes **10.** the first 50 perfect squares

9.5 Special Products

As the Suez Canal shortens the trip around Africa, so recognizing patterns in special products will shorten your work.

You may have recognized some patterns in the products computed in the previous section. Recognizing the patterns in special binomial products can save you time.

The first special product is the *square of a sum*. The area of the square illustrates this pattern.

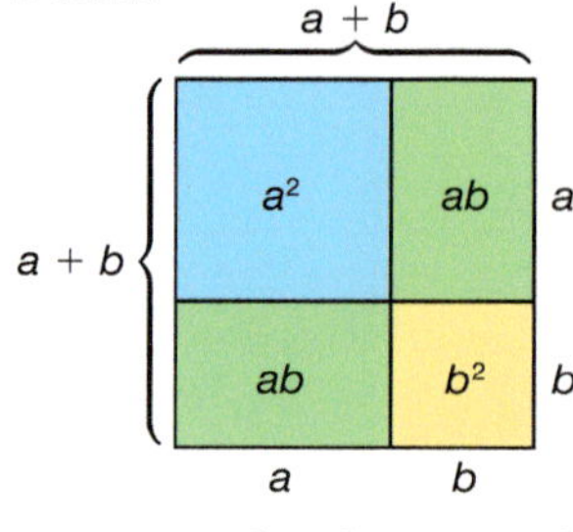

$(a + b)^2 = a^2 + 2ab + b^2$

Example 1

Simplify $(a + b)^2$.

Answer

$(a + b)^2 = (a + b)(a + b)$	1. Apply the definition of a power.
$= a^2 + ab + ab + b^2$	2. Use FOIL.
$= a^2 + 2ab + b^2$	3. Combine like terms.

Notice that the first and last terms of the trinomial are the squares of the first and last terms of the binomial. The trinomial's middle term is the sum of the two equal products of the inner and the outer terms (twice the product of the terms). This pattern can be used to quickly square a binomial sum.

$$(a + b)^2 = a^2 + 2ab + b^2$$

Example 2

Use the square of a sum pattern to expand the following expressions.

a. $(x + 2)^2$ **b.** $(3x + 4)^2$

Answer

a. $(x + 2)^2 = x^2 + 2(x)(2) + 2^2$	1. Use $(a + b)^2 = a^2 + 2ab + b^2$.
$= x^2 + 4x + 4$	2. Simplify.
b. $(3x + 4)^2 = (3x)^2 + 2(3x)(4) + 4^2$	1. $a = 3x$ and $b = 4$.
$= 9x^2 + 24x + 16$	2. Simplify.

A similar pattern can be found for the *square of a difference*.

$$(a - b)^2 = (a - b)(a - b)$$
$$= a^2 - ab - ab + b^2$$
$$= a^2 - 2ab + b^2$$

The two equal products of the inner and the outer terms are negative, so the middle term of the trinomial is $-2ab$. The last term of the trinomial is positive since it is the product of two factors of $-b$. This pattern can be used to quickly find the square of a difference.

$$(a - b)^2 = a^2 - 2ab + b^2$$

Example 3

Use the square of a difference pattern to expand the following expressions.

a. $(y - 3)^2$ **b.** $(2x - 5y)^2$

Answer

a. $(y - 3)^2 = y^2 - 2(y)(3) + 3^2$ — 1. Use $(a - b)^2 = a^2 - 2ab + b^2$.

$= y^2 - 6y + 9$ — 2. Simplify.

b. $(2x - 5y)^2 = (2x)^2 - 2(2x)(5y) + (5y)^2$ — 1. $a = 2x$ and $b = 5y$.

$= 4x^2 - 20xy + 25y^2$ — 2. Simplify.

Now consider the product of two binomials, one indicating the sum of two terms, $a + b$, and the other indicating the difference of the same terms, $a - b$. Two such binomial expressions are called *conjugates*.

Example 4

Multiply $(a + b)(a - b)$.

Answer

$(a + b)(a - b)$ — 1. Use FOIL.

$= a^2 - ab + ab - b^2$

$= a^2 - b^2$ — 2. The sum of the middle terms equals zero.

A *product of conjugates* is the difference of two squares.

$$(a + b)(a - b) = a^2 - b^2$$

Example 5

Use the product of conjugates pattern to expand the following expressions.

a. $(x - 4)(x + 4)$ **b.** $(3x + 7y)(3x - 7y)$

Answer

a. $(x - 4)(x + 4) = x^2 - 4^2$ — 1. Use $(a + b)(a - b) = a^2 - b^2$.

$= x^2 - 16$ — 2. Simplify.

b. $(3x + 7y)(3x - 7y) = (3x)^2 - (7y)^2$ — 1. $a = 3x$ and $b = 7y$.

$= 9x^2 - 49y^2$ — 2. Simplify.

With sufficient practice, you should be able to use the patterns to quickly compute the following products mentally.

Special Products of Binomials	
square of a sum	$(a + b)^2 = a^2 + 2ab + b^2$
square of a difference	$(a - b)^2 = a^2 - 2ab + b^2$
product of conjugates	$(a + b)(a - b) = a^2 - b^2$

These patterns can also be used to quickly compute some numerical products.

Example 6

Compute the following expressions using special product patterns.

a. 87×93 **b.** 72^2

Answer

a. $87 \times 93 = (90 - 3)(90 + 3)$ — 1. Write as the product of conjugates.

$= 90^2 - 3^2$ — 2. Use $(a + b)(a - b) = a^2 - b^2$.

$= 8100 - 9 = 8091$ — 3. Simplify.

b. $72^2 = (70 + 2)^2$ — 1. Write as the square of a sum.

$= 70^2 + 2(70)(2) + 2^2$ — 2. Use $(a + b)^2 = a^2 + 2ab + b^2$.

$= 4900 + 280 + 4 = 5184$ — 3. Simplify.

Patterns for higher powers of binomials can be developed using repeated multiplication.

Example 7

Expand $(a - b)^3$.

Answer

$(a - b)^3 = (a - b)(a - b)^2$ — 1. Rewrite as a product.

$= (a - b)(a^2 - 2ab + b^2)$ — 2. Expand $(a - b)^2$.

$= a^3 - 2a^2b + ab^2 - a^2b + 2ab^2 - b^3$ — 3. Use the Distributive Property.

$= a^3 - 3a^2b + 3ab^2 - b^3$ — 4. Simplify.

A. Exercises

Fill in the blanks in the following equations.

1. $(a + b)^2 = a^2 + \_\_\_ + b^2$ **2.** $(a - b)^2 = a^2 \_\_\_ 2ab \_\_\_ b^2$ **3.** $(a + b)(a - b) = a^2 \_\_\_ b^2$

Find each square.

4. $(x - 6)^2$ **5.** $(y - 16)^2$ **6.** $(2a + 3)^2$

7. $(6x - 12)^2$ **8.** $(x - 9y)^2$ **9.** $(a + 3b)^2$

Multiply.

10. $(y + 12)(y - 12)$ **11.** $(x - 1)(x + 1)$ **12.** $(3c + 4)(3c - 4)$

13. $(5z - 3)(5z + 3)$ **14.** $(x + 2y)(x - 2y)$ **15.** $(2d + 11)(2d - 11)$

B. Exercises

Express each of the following as the product of conjugates and then multiply mentally.

16. $15 \cdot 25$ **17.** $27 \cdot 33$ **18.** $36 \cdot 44$ **19.** $23 \cdot 37$

Express each of the following as the square of a sum or a difference and then multiply.

20. 29^2 **21.** 63^2 **22.** 52^2 **23.** 69^2

Simplify.

24. $(5x - 3)^2$
25. $(4x - 3y)(4x + 3y)$
26. $(y^3 - 8)(y^3 + 8)$
27. $(x^2 - 8y^3)^2$
28. $(a + 3)(a - 3) - (a + 2)^2$
29. $(b - 2)^2 - (b + 2)(b - 2)$
30. $(y - 6)(y + 2)(y + 6)$
31. $(z - 3)(z + 3)(z + 3)$
32. $(x - 5)^2 - (x + 5)^2$
33. $(a + b)^3$
34. $(x + 2)^3$
35. $(x - 1)^3$

C. Exercises

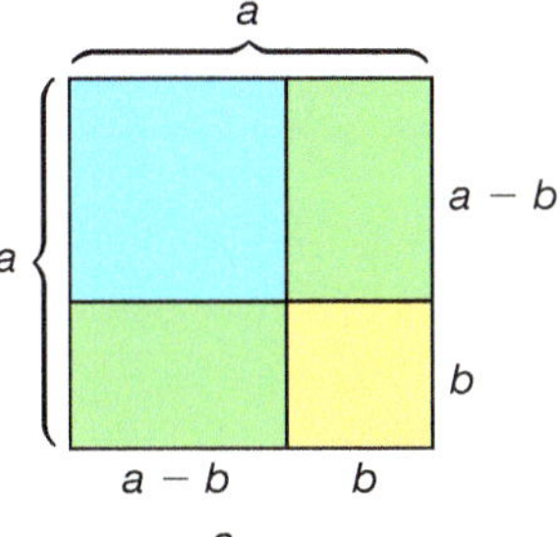

36. Show that the square of a difference pattern can be derived from the square of a sum pattern in which $(a - b)^2 = [a + (-b)]^2$.

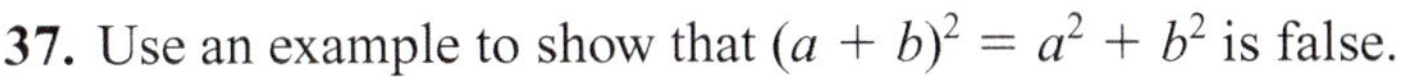

37. Use an example to show that $(a + b)^2 = a^2 + b^2$ is false.
38. Use the first diagram to the right to explain why $(a - b)^2 = a^2 - 2ab + b^2$.
39. Calculate the shaded area in the second diagram in two different ways to show that $(a - b)(a + b) = a^2 - b^2$.

a
a − b
a
b
a − b
b

Dominion Modeling

40. The world's population grew from about 6.456 billion to 6.531 billion from 2005 to 2006. Find the percent increase in the world's population during this year (to the nearest tenth of a percent).
41. The world's population was estimated to be about 6.880 billion in 2010 and increasing with an annual growth rate of about 1.1%. Use these estimates to predict the world's population in 2025 (to the nearest million).

It is estimated that over 55 million people die and over 130 million are born each year. Each newborn coming into the world is created in the image of God (Gen. 1:27). Each new individual and each new generation need to be reached with the gospel of Christ, so the ministry will never be finished until He returns.

CUMULATIVE REVIEW

42. What is the multiplicative identity? [1.5]
43. What is the multiplicative inverse of $-\frac{3}{7}$? [1.6]

Solve. [2.6–2.7, 4.3]

44. $4(2x - 1) = 21$
45. $3(x + 7) - (2x - 4) = 0$
46. $\frac{4(y + 1)}{5} + 6 = 27$
47. $-8y + 4 > -20$

Simplify. [8.2]

48. $\frac{a^4b^3c}{ab^2}$
49. $\frac{-42x^5y^4z^{10}}{-6xz^3}$

Use the exponential function $y = 3\left(\frac{1}{2}\right)^x$ for exercises 50–51. [8.5]

50. State the graph's y-intercept.
51. Is the graph increasing or decreasing?

9.6 Dividing Polynomials

The regular improvement of your math skills will not only increase your career choices but also provide greater opportunities to serve God and others. Jesus, in His humanity, "increased in wisdom and stature, and in favour with God and man" (Luke 2:52). Improving our minds through math and our hearts through God's Word provides greater opportunity for serving others.

The manipulation of polynomials improves your math skills, providing the steps necessary for expanding your understanding in new realms of mathematics, science, and technology. In science, a ratio of polynomials may be used to represent the relative change between two chemicals during a reaction. In software engineering, cyclic redundancy checking (CRC) is used to ensure the accurate transfer of data over the Internet. CRC is also known as polynomial code checksum because of the polynomial equations used in its error-checking algorithms.

Steel expands with an increase in temperature according to the formula $e = \frac{st}{1.1 \cdot 10^5}$, where s is the length of steel and t is the temperature change (in °F). Determine how much the 1595 ft Brooklyn Bridge expands between New York City's average temperature range of 26°F to 84°F.

When a polynomial is multiplied by a monomial, the monomial is distributed to each term. The division of a polynomial by a monomial is similar.

$$\frac{a+b+c}{d} = \frac{1}{d}(a+b+c) = \frac{a}{d} + \frac{b}{d} + \frac{c}{d}$$

To divide a polynomial by a monomial, divide each term of the polynomial by the monomial.

Example 1

Divide $(12x^2 + 20x)$ by $4x$.

Answer	$\frac{12x^2}{4x} + \frac{20x}{4x}$	1. Divide each term of the polynomial by the monomial.
	$= 3x + 5$	2. Simplify each term.
Check	$4x(3x + 5) = 12x^2 + 20x$	3. Multiplying the quotient by the divisor should produce the dividend.

Example 2

Divide $\frac{8x^3 + 16x^2 - 4x}{-4x}$.

Answer	$\frac{8x^3}{-4x} + \frac{16x^2}{-4x} + \frac{-4x}{-4x}$	1. Distribute the division to each term.
	$= -2x^2 - 4x + 1$	2. Simplify each term.
Check	$-4x(-2x^2 - 4x + 1) = 8x^3 + 16x^2 - 4x$	3. Check using multiplication.

Dividing by a monomial is fairly simple and, with practice, can often be done mentally. When the divisor is a polynomial, we can adopt the long division algorithm often used for dividing larger whole numbers. Review the division, multiplication, and subtraction steps used to divide 1635 by 32.

$$\begin{array}{r} 51 \\ 32\overline{)1635} \\ -160 \\ \hline 35 \\ -32 \\ \hline 3 \end{array}$$

$1635 \div 32 = 51\frac{3}{32}$

To check, multiply the divisor by the quotient to produce the dividend.

$$32\left(51 + \frac{3}{32}\right) = 32(51) + \cancel{32}\left(\frac{3}{\cancel{32}}\right) = 1632 + 3 = 1635$$

Now observe how the same long division algorithm can be used to divide polynomials.

Example 3

Divide $(x^2 + 6x - 9)$ by $(x + 2)$.

Answer

$x + 2\overline{)x^2 + 6x - 9}$

1. Write the problem using the long division format. Be sure that the terms of the dividend and the divisor are in order of descending degree.

$$\begin{array}{r} x \\ x + 2\overline{)x^2 + 6x - 9} \\ -(x^2 + 2x) \\ \hline 4x - 9 \end{array}$$

2. a. Divide the first term of the dividend by the first term of the divisor and write the result in the quotient.
   b. Multiply the entire divisor by x, aligning the product under the dividend.
   c. Subtract and bring down the next term to obtain a new dividend.

$$\begin{array}{r} x + 4 \\ x + 2\overline{)x^2 + 6x - 9} \\ -(x^2 + 2x) \\ \hline 4x - 9 \\ -(4x + 8) \\ \hline -17 \end{array}$$

3. a. Divide the first term of the new dividend by the first term of the divisor and write the result in the quotient.
   b. Multiply the entire divisor by 4, aligning the product under the dividend.
   c. Subtract to obtain a new dividend. Notice that there is no next term to bring down. This process is continued until the degree of the new dividend is less than the degree of the divisor.

$$\begin{array}{r} x + 4 - \frac{17}{x+2} \\ x + 2\overline{)x^2 + 6x - 9} \end{array}$$

4. Write the final quotient, placing the remainder over the divisor.

The division can be checked by multiplying the divisor by the quotient.

$$\begin{aligned} &(x + 2)\left(x + 4 - \frac{17}{x + 2}\right) \\ &= (x + 2)(x + 4) + \cancel{(x + 2)}\left(\frac{-17}{\cancel{x + 2}}\right) \\ &= x^2 + 6x + 8 - 17 \\ &= x^2 + 6x - 9 \end{aligned}$$

Dividing Polynomials
1. Arrange the terms in both the dividend and the divisor in descending degree.
2. Complete the divide, multiply, and subtract procedure. a. Divide the first term of the dividend by the first term of the divisor and write the result in the quotient. b. Multiply the result by the entire divisor and place the product under the dividend. c. Subtract this product from the dividend and bring down the next term to obtain a new dividend.
3. Repeat step 2 until the degree of the new dividend is less than the degree of the divisor. Place any remainder over the divisor and add the resulting rational expression to the quotient.
4. Check your results by multiplying your answer by the divisor.

Example 4

Divide $\frac{2x^4 - x^3 + 5x^2 + 9x - 6}{x - 1}$.

Answer

$$\begin{array}{r} 2x^3 + x^2 + 6x + 15 + \frac{9}{x-1} \\ x - 1 \overline{)2x^4 - x^3 + 5x^2 + 9x - 6} \\ \underline{-(2x^4 - 2x^3)} \qquad\qquad\qquad \\ x^3 + 5x^2 \qquad\quad \\ \underline{-(x^3 - x^2)} \qquad\quad \\ 6x^2 + 9x \quad \\ \underline{-(6x^2 - 6x)} \quad \\ 15x - 6 \\ \underline{-(15x - 15)} \\ 9 \end{array}$$

Follow the steps listed in the table.

The quotient is $2x^3 + x^2 + 6x + 15 + \frac{9}{x-1}$.

Check $(x - 1)\left(2x^3 + x^2 + 6x + 15 + \frac{9}{x-1}\right) = (x - 1)(2x^3 + x^2 + 6x + 15) + 9 = 2x^4 - x^3 + 5x^2 + 9x - 6$

Example 5

Divide $(x^3 - 8)$ by $(x - 2)$.

Answer

$$\begin{array}{r} x^2 + 2x + 4 \\ x - 2 \overline{)x^3 + 0x^2 + 0x - 8} \\ \underline{-(x^3 - 2x^2)} \qquad\qquad \\ 2x^2 + 0x \qquad \\ \underline{-(2x^2 - 4x)} \qquad \\ 4x - 8 \\ \underline{-(4x - 8)} \\ 0 \end{array}$$

Since there are no x^2 and x terms in the dividend, you will need to show placeholders for them in the long division. Insert zeros as coefficients of the missing terms.

The quotient is $x^2 + 2x + 4$.

Check $(x - 2)(x^2 + 2x + 4) = x^3 + 2x^2 + 4x - 2x^2 - 4x - 8 = x^3 - 8$

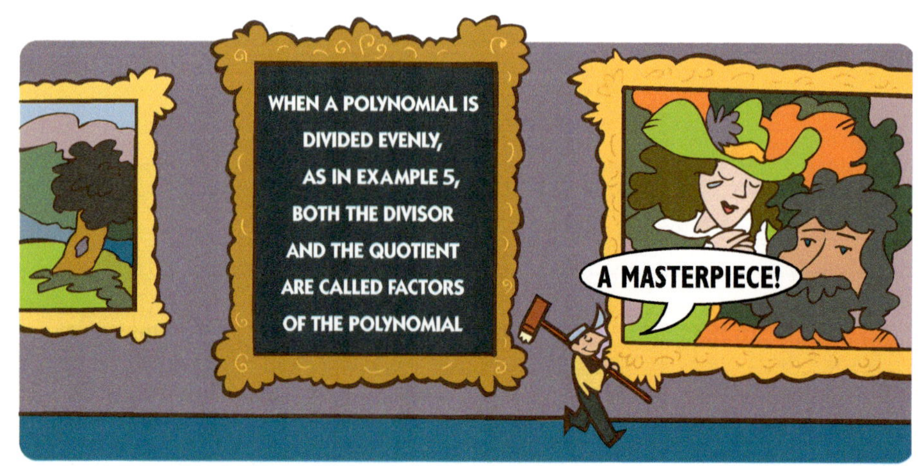

Example 6

Divide $(3a^3 + a - 5a^2 + 2)$ by $(1 + 3a)$.

Answer

$$\begin{array}{r} a^2 - 2a + 1 + \frac{1}{3a+1} \\ 3a+1 \overline{)\,3a^3 - 5a^2 + a + 2} \\ -(3a^3 + a^2) \\ \hline -6a^2 + a \\ -(-6a^2 - 2a) \\ \hline 3a + 2 \\ -(3a + 1) \\ \hline 1 \end{array}$$

First arrange each polynomial in descending powers of a; then divide.

The quotient is $a^2 - 2a + 1 + \frac{1}{3a + 1}$.

Keyword Search

polynomial long division

A. Exercises

Divide.

1. $(5x^2 + 25x)$ by $5x$
2. $(x^3 + x^2 - x)$ by x
3. $(8x^3 - 2x)$ by $2x$
4. $(abc + bcd)$ by bc
5. $\frac{3x^2 + 15}{3}$
6. $\frac{25ab^3 - 35a}{5a}$
7. $\frac{7a^2b - 42ab^2}{7ab}$
8. $\frac{16x^5y + 8xy}{8xy}$
9. $(24a^2b - 12ab + 9b^2) \div 3b$
10. $(81x^3 - 27x^2y + 9xy^3) \div 3x$
11. $(15x^3 - 3x^2 + 27x) \div (-2x)$
12. $(-4a^3b^2 + 32ab^3 - 18b^5) \div (-3b^2)$
13. $\frac{3n^4 + 5n^3 - n^2}{n^2}$
14. $\frac{34a^3b^2 - 10a^2b^3 + 2ab^5}{ab}$
15. $\frac{x^4 + 9x^3 - 5x}{3x}$
16. $\frac{7a^5b^2c^2 - 4a^3bc^4 + 3bc^8}{-2bc^2}$

B. Exercises

Divide.

17. $(x^2 + x - 6) \div (x + 3)$
18. $(x^2 - 7x + 12) \div (x - 4)$
19. $(3x^2 - 26x - 9) \div (x - 9)$
20. $(2x^2 + 10x + 12) \div (x + 2)$
21. $(a^2 + 4a + 12) \div (a - 6)$
22. $(5y^2 - 2y + 9) \div (y + 1)$
23. $(24x^2 + 13x - 9) \div (x + 1)$
24. $\frac{x^2 + 4x - 3}{x - 1}$
25. $\frac{6y^2 - 7y - 5}{2y + 1}$
26. $\frac{3x^2 + 11x - 15}{3x - 4}$

Divide. Be careful to use placeholders and descending order of powers.

27. $(2 + x^2 - x) \div (3 + x)$
28. $(7x^2 + 8 - 3x) \div (x + 4)$
29. $(x^3 - 2x^2 + 3x) \div (x - 4)$
30. $(4x^2 - 16) \div (x - 2)$
31. $(x^3 - 27) \div (x - 3)$
32. $(x^4 - 16) \div (x - 2)$
33. $\frac{x^4 - 8x^2 + 16}{x^2 + 4}$
34. $\frac{3x^3 - 5x^2 + 4x + 2}{x^2 - 2x + 2}$

Write a polynomial expression for each of the following.

35. the height of the parallelogram

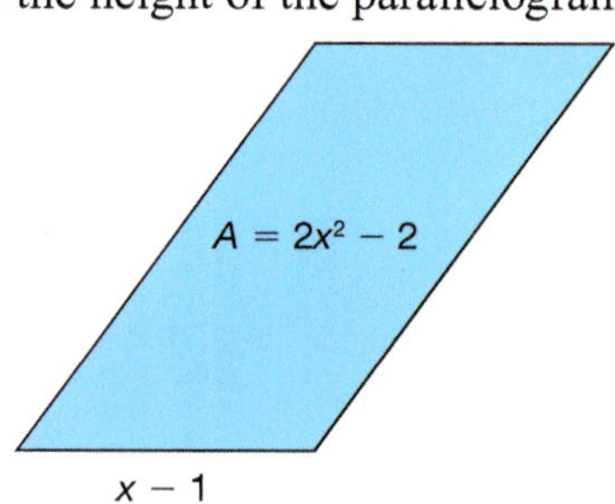

36. the width of the rectangle

$A = 2x^2 + 15x + 18$

$2x + 3$

C. Exercises

Divide.

37. $(x^5 - 2x^2 + 6) \div (x^2 + 4)$

38. $\left(-6m^3 - 2m^2 + \frac{7}{2}m + \frac{3}{2}\right) \div (2m + 1)$

Write a polynomial expression for each of the following.

39. the height of the triangle

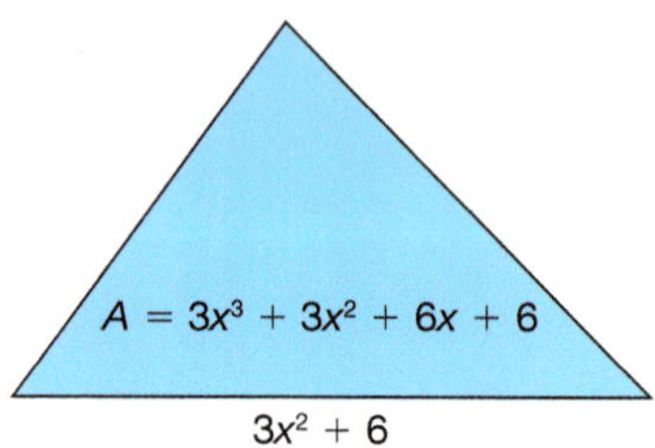

40. the height of the trapezoid

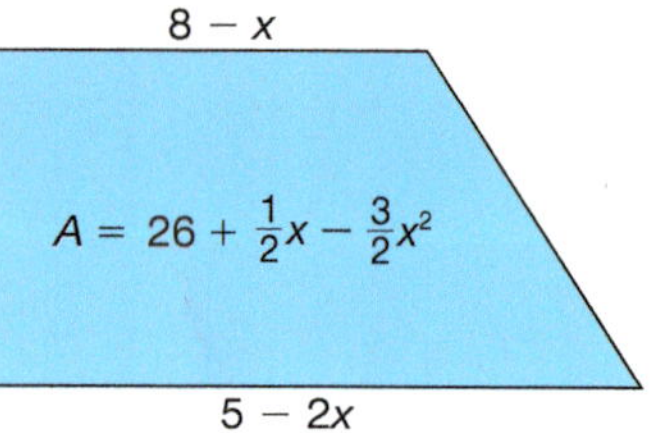

Dominion Modeling

Food and shelter are two of the most basic human needs, yet individuals in every population suffer from the lack of these necessities. Rescue missions in almost every major city provide food and shelter to the needy with a greater goal of meeting spiritual needs. Paul tells us that the entire law is fulfilled in the one commandment to love our neighbor as ourself (Gal. 5:14). Our neighbor certainly includes the needy in our communities.

In 2010 a rescue mission in a city of 150,000 served 64,675 meals and provided shelter for 2310 individuals during a given year. Annual food expenses were \$291,037.50, and shelter expenses for an individual were \$43.25 per day.

41. What was the average cost per meal?

42. What was the cost per person for three meals and shelter for one day?

43. What percent of the city's population was the number of individuals for whom shelter was provided (to the nearest hundredth of a percent)?

44. Use a spreadsheet, an expected population growth rate of 1.3%, and the answer to the previous exercise to predict the number of individuals for whom the mission should plan to provide shelter in 2050.

45. Name two factors that could cause this number to be much higher than expected.

46. Study the following Scripture passages and describe how a Christian worldview of ministering to the poor differs from a secular worldview: Genesis 1:27; Proverbs 14:31; 21:13; 28:27; 29:7; Matthew 25:35–40.

CUMULATIVE REVIEW

Simplify. [1.6]

47. $\dfrac{0}{5x^2y}$

48. $\dfrac{3x+2y}{0}$

Evaluate. [1.7]

49. $\dfrac{1}{2^{-1}}\left(\dfrac{3}{2}\right)^{-1}(-2)^{-2}$

50. $4^{-2}(-1^2)\left(\dfrac{1}{4}\right)^{-3}$

Write the equation of each line in slope-intercept form. [6.4–6.5]

51. a line passing through (4, 1) and (2, 6)

52. a line perpendicular to $y = -\frac{1}{2}x + 4$ and passing through $(-3, -6)$

Multiple choice: Choose the correct equation for each graph. [5.7, 8.4–8.5]

53. 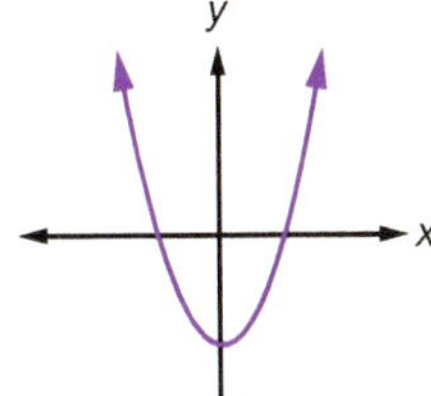

a. $y = (x - 3)^2$
b. $y = (x + 3)^2$
c. $y = x^2 - 3$
d. $y = x^2 + 3$

54. 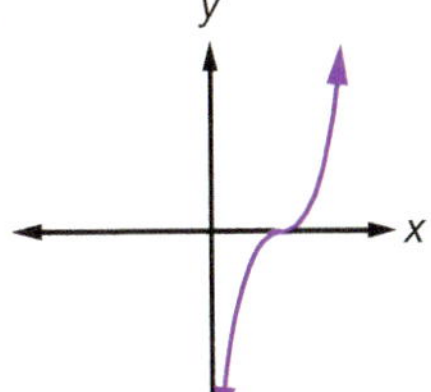

a. $y = (x - 2)^3$
b. $y = (x + 2)^3$
c. $y = x^3 - 2$
d. $y = x^3 + 2$

55. 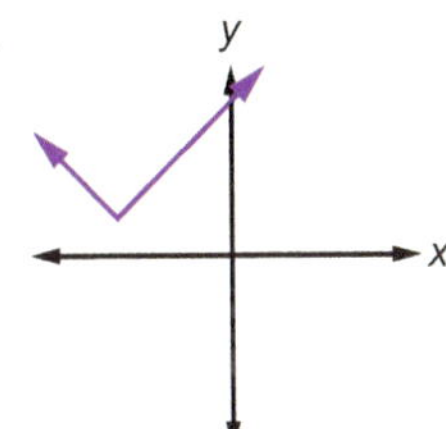

a. $y = |x - 3| - 1$
b. $y = |x + 3| + 1$
c. $y = |x - 3| + 1$
d. $y = |x + 3| - 1$

56. 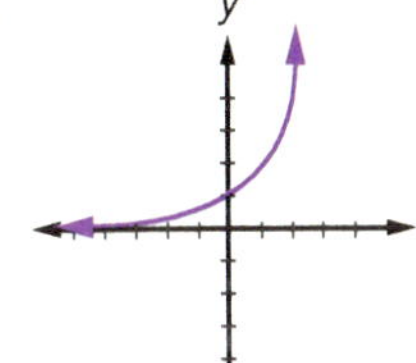

a. $y = 2^x$
b. $y = \left(\frac{1}{2}\right)^x$
c. $y = 3 \cdot 2^x$
d. $y = 3 \cdot \left(\frac{1}{2}\right)^x$

Find the digit that will properly fill in each box in the following division problems.

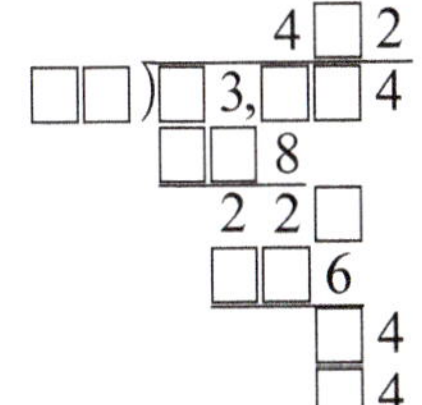

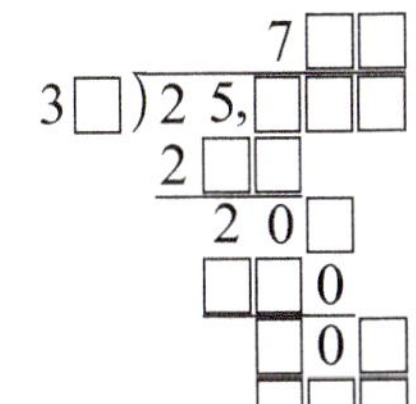

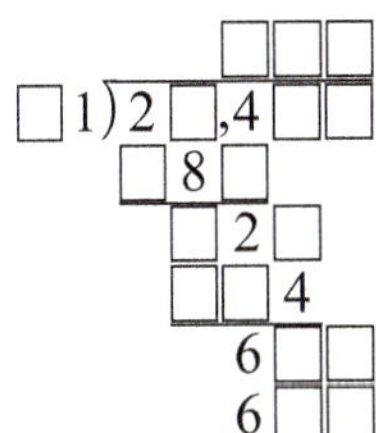

CHAPTER 9 REVIEW

Classify each expression as a monomial, binomial, trinomial, other polynomial, or not a polynomial. Then state the degree of each polynomial.

1. $x - 5$

2. 7

3. $3x^2y - 6xy + 4y$

4. $4x^{-3} + 3x^{-2} - x^{-1} + 5$

Evaluate when $x = 2, y = -1$, and $z = 0$.

5. $4x^3 + 2x - 5z$

6. $8y^4 + 7y^2 + 3y + 9$

7. $3x^2 - 5xz$

8. $6y^2 - 3y - 7x$

Simplify.

9. $(2x^2 + 3x - 5) + (7x^2 - 9)$

10. $(r^2 + 2t) + (r^2 + 6rt - 9t)$

11. $(5a^2 - ab - 3b^2) - (6a^2 - 7ab + 10b^2)$

12. $(15a^2 - 7ab - 12b + 3c^2) - (8ab - 3b + 12c^2)$

13. $(x^2 + y^2) - (x + y^2) + (12x^2 + y)$

14. $\left(\frac{1}{2}x^2 + \frac{1}{3}x\right) + \left(\frac{2}{3}x - \frac{1}{2}\right) - \left(\frac{1}{3}x^2 - \frac{1}{2}\right)$

15. $3x(x^2 - 5x + 4)$

16. $7xy^2(3x - 5xy^4)$

17. $-3a^4bc^2(2a^3b + 7abc^5)$

18. $(x^2 + 3x + 5)(x + 2)$

19. $(x + 2)(x^2 - 2x + 4)$

20. $(2b^2 - b)(b - 6)$

21. $(x + 4)(x - 6)$

22. $(a + 5)(a + 7)$

23. $(x - 9)(x - 11)$

24. $(a + 3b)(a - 7b)$

25. $(3x - 7)(2x + 5)$

26. $(x^3 - 2x)(2x^3 - 3x)$

27. $(y - 6)(y + 6)$

28. $(x + 3)^2$

29. $(x - 2y)^2$

30. $(2y - 3)^2$

31. $(4a - 5b)^2$

32. $(7xy + 3z^3)(7xy - 3z^3)$

33. $(x + 3)(x - 3)^2$

34. $(2c + d)^3$

35. $\frac{25x^2 - 35x}{5x}$

36. $\frac{21x^3y^4 + 7x^2y}{7xy}$

37. $(8x^2y^3z + 12xy^5) \div 2xy^2$

38. $(15x^3 + 27x^2 - 12x) \div 3x$

39. $(6x^2 + 21x + 9) \div (x + 3)$

40. $(x^2 + x + 1) \div (x - 1)$

41. $\frac{x^2 + 5}{x - 2}$

42. $\frac{8x^3 + 27}{3 + 2x}$

Write a polynomial expression for each of the following.

43. the width of the rectangle

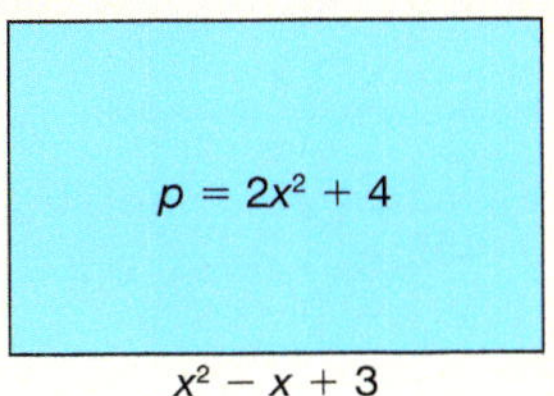

44. the perimeter of the triangle

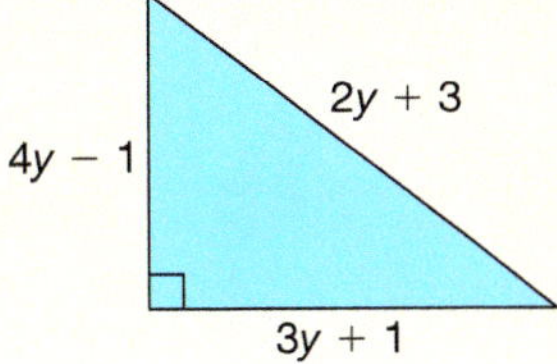

45. the area of the rectangle

46. the area of the triangle

47. the area of the shaded region

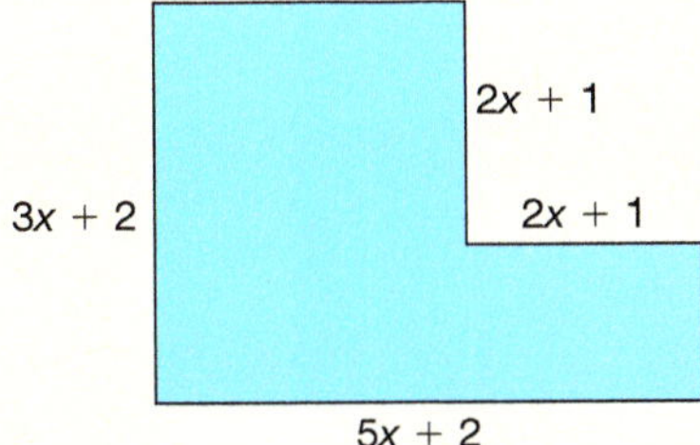

48. the area of the shaded region

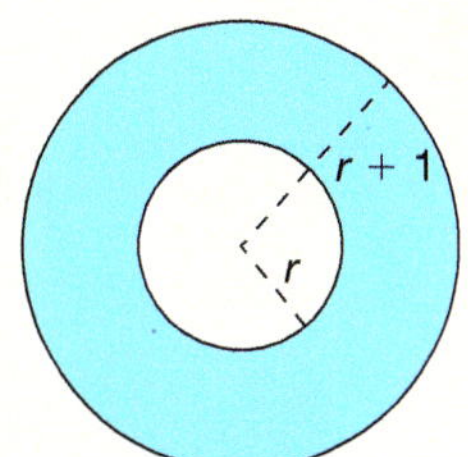

49. Corners with side length x are cut from an 8.5" × 11" piece of cardboard, and the resulting flaps are folded up and fastened to make a box. Write a polynomial expression in standard form for each of the following.

a. the length of the box

b. the width of the box

c. the area of the bottom of the resulting box

d. the volume of the resulting box

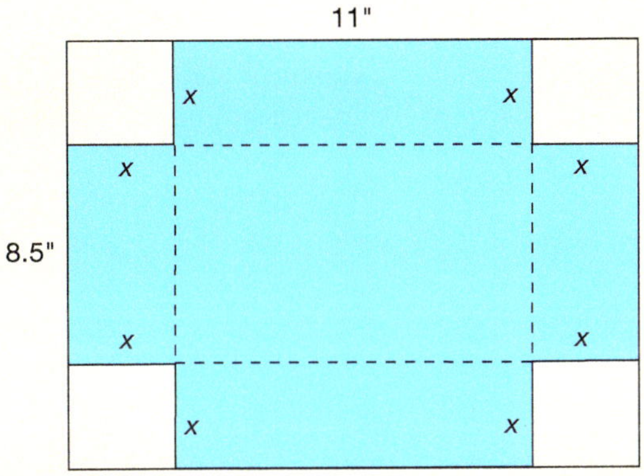

10 Factoring Polynomials

Mont Blanc, the tallest mountain in the Alps, is located on the French-Italian border and is home to several glaciers. At its base lies the French town of Chamonix, the location of the first Olympic winter games in 1924. The area remains a popular location for mountain climbers and skiers.

Tourists note the beauty of the mountain, but the locals know all too much about the potential dangers of the glaciers. In 1892 a glacial lake hidden beneath a thick layer of ice broke through Mont Blanc's glacier ice. Approximately 200,000 cubic meters of water cascaded down the mountain, flooding the village of Saint-Gervais-les-Bains and killing 175 people.

Since then, the community has worked hard to monitor the ice movement of the glaciers and has searched for ways to determine whether hidden pools of water are developing under the glaciers. A large tunnel system through the mountain has been used to facilitate this monitoring, but recently satellite imagery has proved a more valuable tool.

The local power plants use the melt water from the glaciers to produce electricity in hydroelectric power plants. Harnessing God-given natural resources and using them to benefit people is an example of how we can fulfill God's command in Genesis 1 to subdue the earth.

After this chapter you should be able to

1. factor common monomials from polynomials.
2. factor trinomials as the product of two binomials.
3. recognize prime polynomials.
4. factor a difference of squares and perfect square trinomials using special patterns.
5. factor by grouping.
6. factor polynomials completely over the set of integers.

10.1 Factoring Common Monomials

The monomial $\frac{1}{2}gt^2$ describes the distance that these skydivers fall.

A *factor* is one of several quantities multiplied to form a product. When you learned to find the greatest common factor (GCF) of two numbers, you factored each number into its prime factors. Factoring is the process of determining individual factors of a product.

Example 1

Find the GCF of the given monomials.

a. 108 and 72 **b.** $15x$ and $42x^2y$

Answer

a. $108 = 2 \cdot 2 \cdot 3 \cdot 3 \cdot 3$ — 1. Factor each number into its prime factors.

$72 = 2 \cdot 2 \cdot 2 \cdot 3 \cdot 3$

$\text{GCF} = 2 \cdot 2 \cdot 3 \cdot 3 = 36$ — 2. Find the GCF by finding the product of the shared factors.

b. $15x = 3 \cdot 5 \cdot x$ — 1. Factor each term into its prime factors.

$42x^2y = 2 \cdot 3 \cdot 7 \cdot x \cdot x \cdot y$

$\text{GCF} = 3 \cdot x = 3x$ — 2. Find the GCF by finding the product of the shared factors.

In this section you will learn to find a common monomial factor in a polynomial and then apply the Distributive Property in reverse.

Definitions

The **greatest common factor** of a polynomial is the product of all factors shared by every term of the polynomial.

Two numbers or terms are considered **relatively prime** if their greatest common factor is one.

Example 2

Factor $8x^2 - 16$.

Answer

$8x^2 = 2 \cdot 2 \cdot 2 \cdot x \cdot x$ — 1. Factor each term into its prime factors. Both terms contain three factors of 2.

$16 = 2 \cdot 2 \cdot 2 \cdot 2$

$\text{GCF} = 2 \cdot 2 \cdot 2 = 8$ — 2. Find the GCF.

$8x^2 - 16 = 8(x^2) - 8(2)$ — 3. Factor the GCF, 8, from each term of the polynomial.

$= 8(x^2 - 2)$

Check $8(x^2 - 2) = 8x^2 - 16$ — 4. Check using the Distributive Property.

Factoring Common Monomials from a Polynomial

1. Find the GCF of all terms in the polynomial.
2. Divide the GCF out of each term (applying the Distributive Property in reverse).
3. Write the factorization as the GCF times the sum of the quotients.
4. Check your solution by multiplying the two factors.

Example 3

Factor $42a^3 + 14a$.

Answer

$42a^3 = 2 \cdot 3 \cdot 7 \cdot a \cdot a \cdot a$

$14a = 2 \cdot 7 \cdot a$

$\text{GCF} = 2 \cdot 7 \cdot a = 14a$

1. Factor each term and determine the GCF. Both terms contain one factor each of 2, 7, and a.

$42a^3 + 14a$

$= 14a(3a^2) + 14a(1)$

2. Use the Commutative and Associative Properties to factor the GCF from each term.

$= 14a(3a^2 + 1)$

3. Apply the Distributive Property in reverse to obtain two factors.

Check $14a(3a^2 + 1) = 42a^3 + 14a$

4. Check using the Distributive Property.

Example 4

Factor $36x^2y + 9xy^3 - 27xy$.

Answer

$2^2 \cdot 3^2x^2y + 3^2xy^3 - 3^3xy$

$\text{GCF} = 9xy$

1. The GCF uses the lowest power of all common factors.

$9xy(2^2x) + 9xy(y^2) - 9xy(3)$

2. Factor the GCF from each term.

$= 9xy(4x + y^2 - 3)$

3. Write the factorization as the GCF times the sum of the quotients.

Check $9xy(4x + y^2 - 3) = 36x^2y + 9xy^3 - 27xy$

4. Check using the Distributive Property.

Not all polynomials can be factored using these methods. If no common factor other than 1 is found, then the polynomial is considered *prime*.

Example 5

Factor $51x^3 + 26y$.

Answer

$3(17)x^3 + 2(13)y$

1. After factoring each term, you can see that the only common factor is 1 and that the terms are relatively prime.

$51x^3 + 26y$

2. When the terms are relatively prime, the polynomial is prime and will not factor.

A polynomial can contain a common binomial factor, such as

$$3(x + y) - y(x + y).$$

You can factor out the common binomial by considering $(x + y)$ as a single quantity and applying the Distributive Property.

$$3(x + y) - y(x + y) = (x + y)(3 - y)$$

This is just like factoring the expression

$$3z - yz = z(3 - y),$$

where z represents the binomial.

Example 6

Factor $x(x + 1) + 2(x + 1)$.

Answer $x(x + 1) + 2(x + 1)$
$(x + 1)(x + 2)$

1. The binomial $x + 1$ is a factor of each term.
2. Apply the Distributive Property in reverse to write the sum as a product of polynomials.

A. Exercises

Find the GCF. If the GCF is one, write "relatively prime."

1. 45 and 75
2. 33 and 51
3. $24x$ and $32xy$
4. $7xz^2$ and $15y$
5. x^2y^3 and x^3y^2
6. $4ab^2$ and $10a^2b$

Factor. If the polynomial will not factor, write "prime."

7. $3x + 3y$
8. $ax + ay$
9. $8x - 32y$
10. $15x + 30$
11. $a^2 - a$
12. $-50x + 10$
13. $x^2y + wz$
14. $x^2 - 3x$
15. $7x^2 - 49xy + 28y^2$
16. $21x^3 - 6x^2 - 3$
17. $6a^2 + 48a - 24$
18. $5x + 7y - 11z$

B. Exercises

Factor. If the polynomial will not factor, write "prime."

19. $6a^2b + 3ab^2$
20. $10a^2bc^3 + 5ab^2c$
21. $91a^3b^2 + 13a^2b$
22. $300x^2 - 100x$
23. $14x^2y - 15z$
24. $-4\pi r + 2\pi x$
25. $12x^3 - 16xy^2$
26. $20a^3x + 50ax + 10x$
27. $56x^2 + 42xy - 28x$
28. $15x^2 - 16y + 19z^3$
29. $7a^5 - 84a^3 + 21a$
30. $x^5 + x^3 - x^2 + x$
31. $x(x - 2) + 8(x - 2)$
32. $4y(2y - 1) - 3(2y - 1)$
33. $5(b + c) + a(b + c)$
34. $6x(y - z) - 7(y - z)$

Use factoring to solve.

35. If the area of a rectangle is 42 in.^{2}, list all possible natural number dimensions of the rectangle.

36. Mrs. Smitherman is coordinating the third- and fourth-grade program. She wishes to arrange the children into the fewest number of rows that contain the same number of children. She also does not wish to mix the classes within a row. If there are 60 third graders and 48 fourth graders, how many children should be placed in each row? How many rows will be needed?

37. Taste So Good Cookie Company is repackaging their cookies into two different-sized boxes that will contain either 52 or 117 cookies each. If either box is to contain rows of the same number of cookies, how many cookies will be in each row?

38. The area of the rectangle to the right is $18x^2 - 6x$, and its width is $6x$. Find the length of the rectangle.

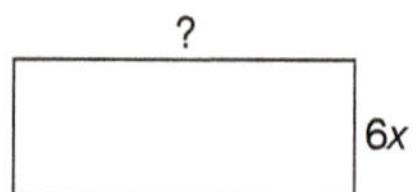

39. The circumference of a circle is $2x\pi + 2y\pi$. Write the factored form of this expression. What is the radius of the circle?

C. Exercises

Factor.

40. $6x^4y^5z^3 - 27x^3y^6z^4 + 18x^2y^4z^5$

41. $34a^7b^8c^2 + 51a^3b^5d^4 - 289a^2b^9c^{10}d$

The diagram to the right illustrates an insulated cylinder. The volume of the insulated region can be found using the following formula: $V = \pi r_1^2H - \pi r_2^2H$.

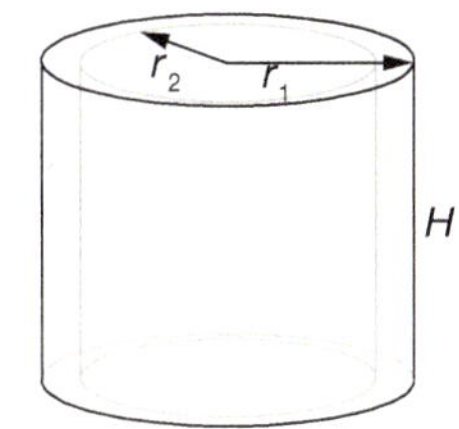

42. Factor the right side of the formula.

43. If the height of the cylinder is 12 in. with an outer radius of 5 in. and an inner radius of 4.5 in., find the volume of the region between the two cylinders in terms of π.

Dominion Modeling

As climates around the world go through cyclical weather patterns, glaciers can expand or shrink. Since glaciers are often a significant source of both water and electrical power, concern about glacial retreat is not unjustified. The hydroelectric plant at the base of Mont Blanc had to reposition its water intake as a result of glacial retreat.

Probing is a method used to measure snow and ice depth. Modern technology, such as satellite imagery, allows for more efficient measurements.

44. If the glacier Mer de Glace has lost 8.3% of its length during the last 130 years and the glacier is currently 11 km long, how long was the glacier 130 years ago? How much length has been lost?

45. The midsection of the glacier has lost 150 m, or 27%, of its thickness.

a. What was the glacier's thickness (to the nearest tenth of a meter) when first measured?

b. Convert this original measurement to the nearest foot.

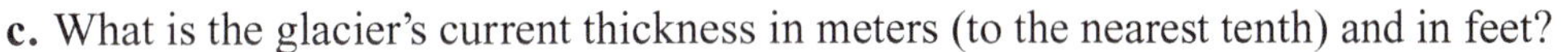

c. What is the glacier's current thickness in meters (to the nearest tenth) and in feet?

Interestingly, Mont Blanc has been called "the mountain that will not shrink." Due to less temperature variation as altitude increases, glacier thickness above the equilibrium line (snow line) has not shown significant change over the last one hundred years.

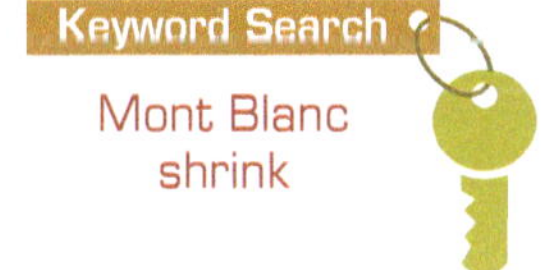

CUMULATIVE REVIEW

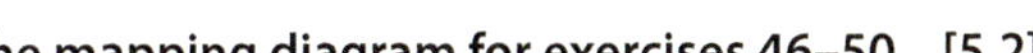

Use the mapping diagram for exercises 46–50. [5.2]

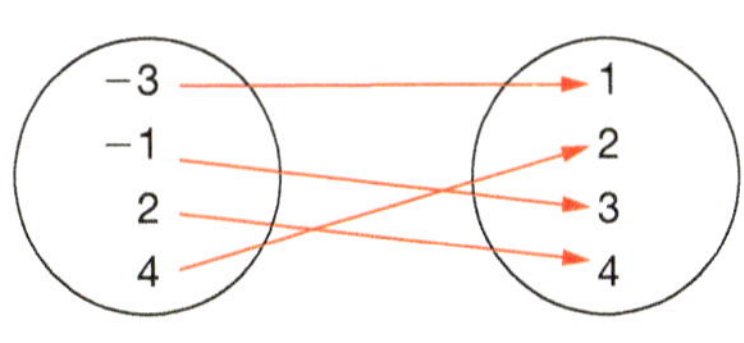

46. State the domain of the relation.

47. State the range of the relation.

48. Is the relation a function?

49. Write the relation as a set of ordered pairs.

50. Graph the relation.

Multiply. [9.4]

51. $(x + 5)(x - 8)$

52. $(x - 5)(x - 8)$

53. $(x - 5)(x + 8)$

54. $3(x - 2)(x + 7)$

55. $4y(x + 1)(x - 8)$

SEQUENCES

Factorial Sequences

Challenge **Can you supply the next two numbers in each of the following sequences?**

a. 1, 2, 6, 24, 120, …

b. $1, \frac{1}{2}, \frac{1}{6}, \frac{1}{24},$

c. $1, \frac{3}{2}, \frac{5}{3}, \ldots$

The first sequence above is the *factorial sequence* that is used in the study of probability. The sequence 1, 2, 6, 24, 120, … has the general-term formula $A_n = n!$, where $n!$ is defined as $n(n-1)(n-2)\ldots(3)(2)(1)$. Thus, A_6 in the first sequence is $6! = 6 \cdot 5 \cdot 4 \cdot 3 \cdot 2 \cdot 1 = 720$. By definition, $0! = 1$.

The second sequence involves the reciprocals of the terms of the factorial sequence and could be represented as $A_n = \frac{1}{n!}$. The third partial sum, S_3, for this sequence is $1 + \frac{1}{2} + \frac{1}{6} = \frac{10}{6} = \frac{5}{3}$, which leads to the conclusion that the third sequence is the partial-sum sequence for the second.

The sum of the reciprocals of the factorial sequence terms might seem unimportant, but one of the most important math constants in college mathematics, e, can be defined by this sum: $e = \frac{1}{0!} + \frac{1}{1!} + \frac{1}{2!} + \frac{1}{3!} + \ldots \approx 2.718$. You can verify this approximation for e using your calculator. Most scientific and graphing calculators have a key for e^x (associated with the LN key) and a factorial function "!" (often found in the PRB submenu under MATH).

Exercises

Write each factorial expression as a product of factors. Then evaluate the factorial.

1. 3!

2. 6!

Write each product of factors in factorial form.

3. $8 \cdot 7 \cdot 6 \cdot 5 \cdot 4 \cdot 3 \cdot 2 \cdot 1$

4. $1 \cdot 2 \cdot 3 \cdot 4 \cdot 5$

Write the partial sums for $A_n = \frac{1}{(n-1)!}$ to the given number of decimal places. (Hint: Since $S_3 = \frac{1}{0!} + \frac{1}{1!} + \frac{1}{2!} = 1 + 1 + \frac{1}{2} = 2.5$, enter 2.5 into your calculator, then add $\frac{1}{3!}$ to get S_4, and add $\frac{1}{4!}$ to get S_5, etc.)

5. S_4 through S_6; 3

6. S_7 through S_9; 5

7. S_{10} through S_{12}; 7

8. What is the value of e to nine decimal places?

10.2 Factoring Trinomials of the Form $x^2 + bx + c$

Remember that the FOIL method can be used to find the product of two binomial factors.

$$\begin{aligned}(x + 5)(x + 7) &= x^2 + 7x + 5x + (5 \cdot 7)\\ &= x^2 + (7 + 5)x + 35\\ &= x^2 + 12x + 35\end{aligned}$$

sum of the constants ↗ ↖ product of the constants

Notice that the constant term of the trinomial, 35, is the product of the binomial's constants, 5 and 7, and that the coefficient of the middle term, 12, is their sum. To factor a trinomial of the form $x^2 + bx + c$, you must find the two binomial factors that were multiplied to obtain the trinomial. The key is to identify the pair of factors of c, the constant term, whose sum is b, the coefficient of the middle term.

Example 1

Factor $x^2 - 5x + 6$.

Answer	$x^2 - 5x + 6$	
	$(x \quad)(x \quad)$	1. Write the factors of x^2 as the first terms in the binomials.
	$6 = -2(-3)$ and $-2 + (-3) = -5$	2. Find factors of $+6$ whose sum is -5. (A positive product implies that the factors have the same sign.)
	$(x - 2)(x - 3)$	3. Write these factors as the last terms in the binomials.
Check	$(x - 2)(x - 3)$ $= x^2 - 3x - 2x + 6$ $= x^2 - 5x + 6$	4. Multiply the binomials using the FOIL method.

Factoring Trinomials of the Form $x^2 + bx + c$

1. Write the factors of the first term.
2. Find factors of the constant term, c, whose sum is b. Write these factors as the last terms in the binomials.
3. Check the results by multiplying the binomials.

Example 2

Factor $y^2 + 5y - 6$.

Answer	$y^2 + 5y - 6$	
	$(y \quad)(y \quad)$	1. Write the factors of y^2 as the first terms in the binomials.
	$-6 = 6(-1)$ and $6 + (-1) = 5$	2. Find factors of -6 whose sum is $+5$. (A negative product implies that the factors have different signs.)
	$(y + 6)(y - 1)$	
Check	$(y + 6)(y - 1)$ $= y^2 - 1y + 6y - 6$ $= y^2 + 5y - 6$	3. Multiply the binomials using the FOIL method. Be sure that the sum of the inner and outer products produces the trinomial's middle term.

Determining the Signs of the Factors of c
1. If the sign of c is positive ($x^2 \pm bx + c$), a. the signs are the same and b. the factors will have the sign of the trinomial's middle term.
2. If the sign of c is negative ($x^2 \pm bx - c$), a. the signs are different and b. the larger factor will have the sign of the trinomial's middle term.

Example 3

Factor $x^2 + x - 72$.

Answer

$x^2 + 1x - 72$

$(x\quad)(x\quad)$ — 1. Factor x^2.

$-72 = 9(-8)$ and $9 + (-8) = 1$ — 2. Find factors of -72 whose sum is $+1$. (A negative product implies that the factors have different signs, with the larger factor being positive to match the sign of the trinomial's middle term.)

$(x + 9)(x - 8)$

Check

$(x + 9)(x - 8)$ — 3. Check using FOIL.

$= x^2 - 8x + 9x - 72$

$= x^2 + x - 72$

Example 4

Factor $x^2 + 3x + 6$.

Answer

$x^2 + 3x + 6$

$(x\quad)(x\quad)$ — 1. Factor x^2.

$6 = 1(6)$, but $1 + 6 \neq 3$ — 2. Find factors of $+6$ whose sum is $+3$. (Both signs must be positive.)

$6 = 2(3)$, but $2 + 3 \neq 3$

$(x + ?)(x + ?)$

$x^2 + 3x + 6$ — 3. Since there are no factors of $+6$ whose sum is $+3$, the trinomial is prime.

The first step in factoring a polynomial is to factor out any common monomial factors. When the constant term is a large number, a more organized approach may be required to find the correct pair of factors.

Example 5

Factor $2a^2 - 64a - 288$.

Answer $2(a^2 - 32a - 144)$

1. Factor out the common factor of 2.

$2(a \quad)(a \quad)$

2. Factor the first term of the trinomial, a^2.

factors	sum
1(−144)	−143
2(−72)	−70
3(−48)	−45
4(−36)	−32

3. Find factors of −144 whose sum is −32. (The signs must be different, with the larger factor being negative.) Use an organized list to find the correct factors. Note that the complete set of factors is not listed. Stop when you find the correct sum.

$2(a + 4)(a - 36)$

Check $2(a + 4)(a - 36)$
$= 2(a^2 - 36a + 4a - 144)$
$= 2(a^2 - 32a - 144)$
$= 2a^2 - 64a - 288$

4. Multiply the binomials using FOIL and then distribute the 2.

Some trinomials contain two variables. Consider the following product:

$$(x + 5y)(x - 4y) = x^2 - 4xy + 5xy - 20y^2$$
$$= x^2 + xy - 20y^2.$$

Notice that the factors of $-20y^2$ are $5y$ and $-4y$. When factoring a trinomial of the form $x^2 + bxy + cy^2$, be sure to include a factor of y in the second term of each binomial.

Example 6

Factor $x^2 + 11xy + 24y^2$.

Answer $x^2 + 11xy + 24y^2$

$(x \quad y)(x \quad y)$

1. Factor the first term, x^2. Factor y^2 from the last term.

$24 = 3(8)$ and $3 + 8 = 11$

2. Find factors of +24 whose sum is +11. (Both signs must be positive.)

$(x + 3y)(x + 8y)$

Check $(x + 3y)(x + 8y)$
$= x^2 + 8xy + 3xy + 24y^2$
$= x^2 + 11xy + 24y^2$

3. Check using FOIL.

A. Exercises

Factor. If the trinomial will not factor, write "prime."

1. $a^2 + 8a + 15$ **2.** $x^2 + 15x + 54$ **3.** $a^2 - 4a + 3$

4. $x^2 + 4x - 32$ **5.** $b^2 + b - 20$ **6.** $x^2 - 2x - 3$

7. $a^2 + 2a - 8$ **8.** $a^2 + 3a - 28$ **9.** $x^2 + 7x + 4$

10. $a^2 - 4a + 5$ **11.** $x^2 - 9x + 20$ **12.** $x^2 - 11x + 10$

13. $y^2 + 7y + 10$ **14.** $x^2 - 6x + 12$ **15.** $x^2 - 3x - 30$

16. $a^2 - 13a - 30$

B. Exercises

Factor. If the trinomial will not factor, write "prime."

17. $x^2 + 20xy + 100y^2$
18. $x^2 + 6xy + 9y^2$
19. $x^2 - 50xy + 96y^2$
20. $x^2 - 18xy + 45y^2$
21. $a^2 - 14ab - 72b^2$
22. $x^2 - 9xy - 24y^2$

Rearrange the terms in descending order; then factor.

23. $56 + 18x + x^2$
24. $-26y + y^2 + 48$
25. $15x + x^2 - 54$

Factor. Begin by factoring out the common monomial.

26. $6b^2 - 42b + 72$
27. $3x^2 - 24x - 27$
28. $ax^2 - 11ax - 12a$

Rearrange the terms in descending order and factor out the common monomial before factoring.

29. $24x + 36 + 3x^2$
30. $4x^2 + 96 - 44x$
31. $6x^3 - 16x^2 + x^4$

32. The area of a rectangle can be expressed as $x^2 - 18x + 56$. Factor to find binomial expressions representing the length and width. What is the difference between the length and the width?

33. The area of the shaded region is represented by $x^2 + 25x + 60$. Find a polynomial that represents the area of the larger rectangle, and then factor to find binomial expressions representing its length and width.

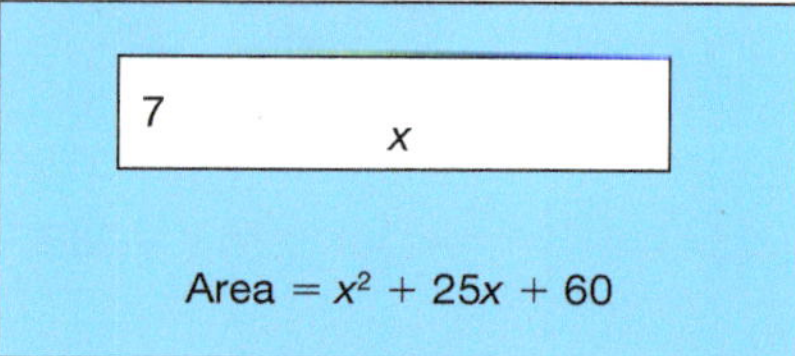

C. Exercises

Factor.

34. $-ax^3 + 45ax^2 + 144ax$
35. $4a^4b - 12a^3b^2 - 40a^2b^3$
36. $bx^4 - 19bx^2 - 120b$

Dominion Modeling

As the climate goes through different cycles, Christians understand that God controls everything about the climate, including the ice (Job 38:29–30). Is the earth warmer now than it was in the 1970s and 1980s? The answer is yes, but does this necessarily mean that it will continue to grow warmer? The answer to this question is known only by the Creator, Who controls the weather.

Some glaciers are obviously shrinking, and some are disappearing. The yearly glacier net balance (the difference between the amount of accumulated snow and the amount that melts or evaporates) of Washington's Southern Cascade Glacier was studied from 1953 to 2003. A sample of the data is given (in meters).

37. Make a scatterplot of the data. (Chapter 6, Technology Corner, shows how to do this on a graphing calculator.)

38. Since the points do not appear to be in a line, use a graphing calculator (or spreadsheet) to complete a quadratic regression on the data.

 a. State the function rule that models the data set.

 b. Sketch the graph of the function on your scatterplot.

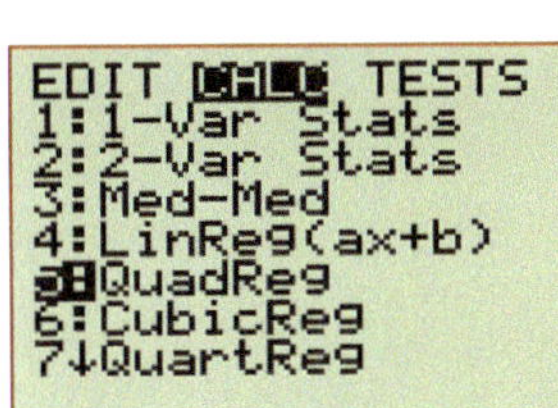

Year	Net Balance	Year	Net Balance
1970	−7.8	1982	−9.9
1972	−5.8	1984	−10.3
1974	−6.0	1986	−12.0
1976	−5.0	1988	−15.8
1978	−6.5	1990	−16.8
1980	−9.0	1992	−19.8

39. Including data from the entire fifty-year study presents a different picture. Create a scatterplot by entering the data set below.

Year	Net Balance	Year	Net Balance	Year	Net Balance	Year	Net Balance	Year	Net Balance
1953	0.0	1964	−4.2	1974	−6.0	1984	−10.3	1994	−21.8
1956	0.2	1966	−5.3	1976	−5.0	1986	−12.0	1996	−22.1
1958	−3.0	1968	−6.0	1978	−6.5	1988	−15.8	1998	−23.3
1960	−3.0	1970	−7.8	1980	−9.0	1990	−16.8	2000	−22.0
1962	−4.0	1972	−5.8	1982	−9.9	1992	−19.8	2002	−23.0

40. Draw and determine the equation of a trend line that models the data set.

CUMULATIVE REVIEW

Use the following polynomial for exercises 41–45: $6x^3y^3 - 27x^2y^4z + 8x^2y^3$.

41. Classify the polynomial by its number of terms. [9.1]

42. State the coefficient of the last term. [2.3]

43. State the degree of the first term. [9.1]

44. State the degree of the polynomial. [9.1]

45. Factor the polynomial. [10.1]

Multiply using FOIL. [9.4]

46. $(2x + 5)(3x + 7)$

47. $(2x + 5)(3x - 7)$

48. $(2x - 5)(3x - 7)$

Write an equation and solve. [3.7]

49. Josh leaves his home near Tallahassee and travels west toward Pensacola. Marcus leaves his home near Pensacola at the same time and travels east to meet Josh. If the distance between their homes is 175 mi, Marcus travels at 55 mi/hr, and the trip takes 90 min, find Josh's average speed (to the nearest mile per hour).

50. Arianna and Bethany plan to meet at the church, but Bethany has an extra 10 mi to travel. It took Arianna 12 min to travel 7 mi, and it took Bethany 25 min. What was the average rate for each driver (to the nearest mile per hour)?

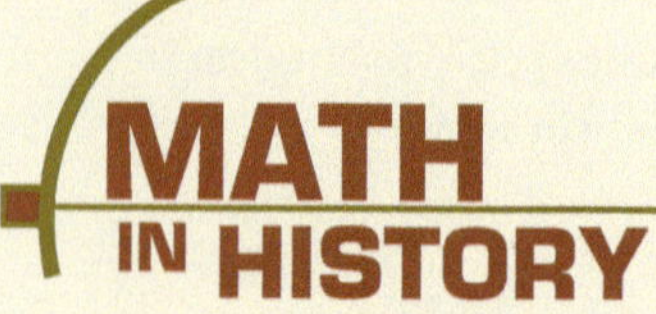

MATH IN HISTORY

Grace Hopper (1906–92)

Grace Hopper was a distinguished American naval officer, mathematician, and computer scientist. After studying math and physics at Vassar College, she attended Yale University, earning her PhD in 1934. She taught mathematics at Vassar from 1931 to 1943. During World War II she joined the Naval Reserve and was assigned to the Bureau of Ordnance Computation Project at Harvard University, where she soon began programming the Harvard Mark computers. After the war, she remained in the Naval Reserve; but she also began working as senior mathematician for a major computer corporation.

Hopper was a leader in the field of software development, seeking to make computers more programmer friendly and application friendly. In 1952 she invented the first computer compiler, which is an intermediate program that converts a programmer's instructions into the computer's machine language. Her compiler work for the UNIVAC computer led to the development of the popular computer business language COBOL in 1959.

She was recalled to active duty in 1967 and served until 1986, when she retired as Rear Admiral Hopper at the age of 80, the oldest active duty officer in the United States. At her retirement celebration, she was awarded the Defense Distinguished Service Medal. Hopper received 37 honorary degrees between 1972 and 1987 and was buried with full naval honors at Arlington National Cemetery.

Grace Hopper is a shining example of a fruitful life lived in the service and ideals of her country. She diligently developed her God-given talents to become a successful student, a distinguished educator, a business innovator, and a national hero. Her life story of diligence and service should be a challenge to us to develop our God-given abilities into a life of service and fruitfulness.

10.3 Factoring Trinomials of the Form $ax^2 + bx + c$

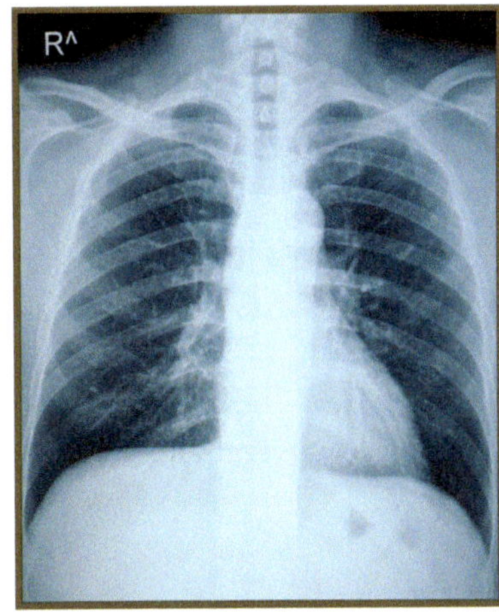

Just as the x-ray exposes the structure inside this person's chest, so factoring reveals the underlying structure of a polynomial.

To learn how to factor a trinomial of the form $ax^2 + bx + c$, observe the multiplication of two binomials to produce such a trinomial. Multiply the binomials $(3x - 2)(x + 5)$ using the FOIL method.

$$\begin{aligned}(3x - 2)(x + 5) &= 3x^2 + 15x - 2x - (2 \cdot 5)\\ &= 3x^2 + (15 - 2)x - 10\\ &= 3x^2 + 13x - 10\end{aligned}$$

Notice that the coefficients of the outer and inner products, 15 and -2, have a sum of $+13$ and that their product, -30, is the product of the leading coefficient and the constant term, $3(-10)$.

When factoring trinomials of the form $ax^2 + bx + c$, find factors of the product ac whose sum is b. Then rewrite the middle term as a sum of terms using these factors as coefficients.

Example 1

Factor $3x^2 + 7x + 2$.

Answer

$3x^2 + 7x + 2$	1. Notice that $ac = 6$ and $b = 7$.
$6 = 1(6)$ and $1 + 6 = 7$	2. Find factors of $+6$ whose sum is $+7$.
$3x^2 + 1x + 6x + 2$	3. Rewrite the middle term, $7x$, as a sum of terms with 1 and 6 as their coefficients.
$(3x^2 + 1x) + (6x + 2)$	4. Group the first two terms and the last two terms and factor the GCF from each grouping.
$x(3x + 1) + 2(3x + 1)$	a. Factor x from the first pair of terms and 2 from the second pair.
$(x + 2)(3x + 1)$	b. The polynomial now consists of two larger terms sharing a common binomial factor of $3x + 1$. Factor it from both terms.

Check

$$\begin{aligned}&(x + 2)(3x + 1)\\ &= 3x^2 + x + 6x + 2\\ &= 3x^2 + 7x + 2\end{aligned}$$

5. Check using FOIL. Notice that the four-term polynomial matches the polynomial in step 3.

Factoring Trinomials of the Form $ax^2 + bx + c$
1. Find factors of ac whose sum is b.
2. Rewrite the middle term as a sum of terms with these factors of ac as their coefficients.
3. Factor the four-term polynomial by grouping the terms in pairs. a. Factor the common monomial from each pair. b. Factor the common binomial from each new term.
4. Check your factorization by multiplying the factors.

Keyword Search

British factoring method

If the binomials that result from factoring the GCF from each grouping are not identical, you know that there is a mistake in your work.

Example 2

Factor $2x^2 - 19x + 24$.

Answer

$2x^2 - 19x + 24$

1. Notice that $ac = 48$ and $b = -19$.

$48 = -3(-16)$ and $-3 + (-16) = -19$

2. Find factors of $+48$ whose sum is -19. (Both factors must be negative.)

$2x^2 - 3x - 16x + 24$

3. Rewrite the middle term, $-19x$, as a sum of terms with -3 and -16 as their coefficients.

$(2x^2 - 3x) + (-16x + 24)$

4. Group the first two terms and the last two terms and factor the GCF from each grouping.

$x(2x - 3) - 8(2x - 3)$

   a. Factor x from the first pair of terms and -8 from the second pair.

$(x - 8)(2x - 3)$

   b. Factor the common binomial, $2x - 3$, from both terms.

Check

$(x - 8)(2x - 3)$
$= 2x^2 - 3x - 16x + 24$
$= 2x^2 - 19x + 24$

5. Check using FOIL.

Example 3

Factor $3x^2 - 20 - 7x$.

Answer

$3x^2 - 7x - 20$

1. Be sure to arrange the polynomial in standard form before factoring. Notice that $ac = -60$ and $b = -7$.

$-60 = -12(5)$ and $-12 + 5 = -7$

2. Find factors of -60 whose sum is -7. (The signs must be different, with the larger factor being negative.)

$3x^2 - 12x + 5x - 20$

3. Rewrite the middle term, $-7x$, as a sum of terms with -12 and 5 as their coefficients.

$(3x^2 - 12x) + (5x - 20)$

4. Group the first two terms and the last two terms and factor the GCF from each grouping.

$3x(x - 4) + 5(x - 4)$

   a. Factor $3x$ from the first pair of terms and 5 from the second pair.

$(3x + 5)(x - 4)$

   b. Factor the common binomial, $x - 4$, from both terms.

Check

$(3x + 5)(x - 4)$
$= 3x^2 - 12x + 5x - 20$
$= 3x^2 - 7x - 20$

5. Check using FOIL.

Remember that the first step in factoring a polynomial is to check to see whether you can factor out a common monomial factor before applying other methods.

Example 4

Factor $12x^3 - 26x^2 + 10x$.

Answer $2x(6x^2 - 13x + 5)$

1. Factor the common monomial $2x$ from each term. Then proceed to factor the remaining trinomial. Notice that $ac = 30$ and $b = -13$.

$30 = -10(-3)$ and $-10 + (-3) = -13$

2. Find factors of 30 whose sum is -13.

$2x(6x^2 - 10x - 3x + 5)$

3. Rewrite the middle term as a sum of terms with these factors as coefficients.

$2x[(6x^2 - 10x) + (-3x + 5)]$

4. Group the first two terms and the last two terms within brackets and factor the GCF from each grouping.

$2x[2x(3x - 5) + (-1)(3x - 5)]$

a. Factor $2x$ from the first pair of terms and -1 from the second pair.

$2x[(2x - 1)(3x - 5)]$
$= 2x(2x - 1)(3x - 5)$

b. Factor the common binomial, $3x - 5$, from both terms.

Check $2x(2x - 1)(3x - 5)$
$= 2x(6x^2 - 10x - 3x + 5)$
$= 2x(6x^2 - 13x + 5)$
$= 12x^3 - 26x^2 + 10x$

5. Multiply the binomials using FOIL and then distribute the $2x$ to produce the original trinomial.

Example 5

Factor $2x^2 + 3xy - 14y^2$.

Answer $2x^2 + 3xy - 14y^2$

1. Notice that $ac = -28$ and $b = 3$.

$-28 = 7(-4)$ and $7 + (-4) = 3$

2. Find factors of -28 whose sum is $+3$. (The signs must be different, with the larger factor being positive.)

$2x^2 + 7xy - 4xy - 14y^2$

3. Rewrite the middle term as a sum of terms with these factors as coefficients.

$(2x^2 + 7xy) + (-4xy - 14y^2)$

4. Group the first two terms and the last two terms and factor the GCF from each grouping.

$x(2x + 7y) - 2y(2x + 7y)$

a. Factor x from the first pair of terms and $-2y$ from the second pair.

$(x - 2y)(2x + 7y)$

b. Factor the common binomial, $2x + 7y$, from both terms.

Check $(x - 2y)(2x + 7y)$
$= 2x^2 + 7xy - 4xy - 14y^2$
$= 2x^2 + 3xy - 14y^2$

5. Check using FOIL.

Using the method described in the previous examples, you can follow the same procedure with each problem. The strategy taught in Section 10.2 can be modified slightly to factor trinomials of the form $ax^2 + bx + c$. If the coefficients of the first and last term of the trinomial have multiple pairs of factors, the process can be somewhat tedious; but with small coefficients the method can be quite efficient.

Example 6

Factor $2x^2 + 3x - 14$.

Answer	$2 = 2 \cdot 1$ and $14 = 1 \cdot 14$ $= 2 \cdot 7$		1. List all the factors of a and c.
	$2x^2 + 3x - 14$		2. Find the combination of first and third factors that produces a middle term of $3x$.
	combination	middle term	
	$(2x - 1)(x + 14)$	$28x - 1x = 27x$	a. Since the last term is negative, the signs in the binomials should be different.
	$(2x - 14)(x + 1)$	$2x - 14x = -12x$	
	$(2x - 2)(x + 7)$	$14x - 2x = 12x$	
	$(2x - 7)(x + 2)$	$4x - 7x = -3x$	b. If you obtain the opposite of the middle term, switch the signs in the binomials.
	$(2x + 7)(x - 2)$	$-4x + 7x = 3x$	
Check	$(2x + 7)(x - 2) = 2x^2 + 3x - 14$		3. Check using FOIL.

Notice that several combinations were tried before the correct factors were found. You cannot assume that the sign of the larger factor will match the sign of the middle term. Rather, in problems like this, it will be the sign of the larger product. With practice, you should be able to pick the correct combination on the first or second try.

A. Exercises

Factor.

1. $2x^2 + 11x + 12$
2. $4x^2 + 15x + 9$
3. $3x^2 - 14x + 8$
4. $9x^2 - 30x + 25$
5. $3x^2 + 5x - 28$
6. $2x^2 + x - 21$
7. $6x^2 + 5x - 4$
8. $6b^2 - 11b - 10$
9. $15c^2 + 19c + 6$
10. $12x^2 + 17x + 6$

B. Exercises

Factor.

11. $6a^2 - 5a - 4$
12. $9x^2 + 3x - 2$
13. $5x^2 - 14x - 3$
14. $2x^2 + 13x - 7$
15. $5x^2 + 8xy - 21y^2$
16. $6x^2 - 19xy - 20y^2$

Factor. Be sure to factor any common monomials first.

17. $6x^2 + 3x - 30$
18. $30x^2 - 100x - 250$
19. $16a^2 + 40ab - 24b^2$
20. $12x^2 - 8xy - 20y^2$
21. $m^2 - mn - 20n^2$
22. $a^2 + 3ab - 28b^2$
23. $x^2 + 3xy - 10y^2$
24. $x^2 - 7xy - 98y^2$
25. $2x^2 - 8x - 64$
26. $8a^2 - 104a - 112$
27. $12ab^3 + 9ab^2 - 21ab$
28. $51x^3y - 34x^2y - 85xy$
29. $45x^2 - 3x - 18$
30. $28x^2 - 26x - 24$
31. $2x^2 - 18xy - 104y^2$
32. $3x^2 + 9xy - 462y^2$

C. Exercises

Factor.

33. $144x^2 - 102x + 18$
34. $3a^4b + a^3b - 10a^2b$

The total area of a garden, including a 2 ft perimeter pathway, can be represented by the trinomial $15x^2 + 32x + 16$.

35. Factor the trinomial.

36. Sketch a diagram of the garden area, including the pathway. Label it with the path's dimensions and with the monomial expressions for the length and width of the interior area of the garden.

37. What is the length-to-width ratio of the garden's plantable area?

The width of a sidewalk surrounding a swimming pool is represented by x, and the total area of the pool and sidewalk is represented by $4x^2 + 104x + 651$.

38. Factor to find the dimensions of the pool.

39. Let $x = 5$ ft. Demonstrate that the area found using the binomials from factoring will result in the same area found using the original trinomial.

40. How much does the area change if the sidewalk is widened to 7 ft?

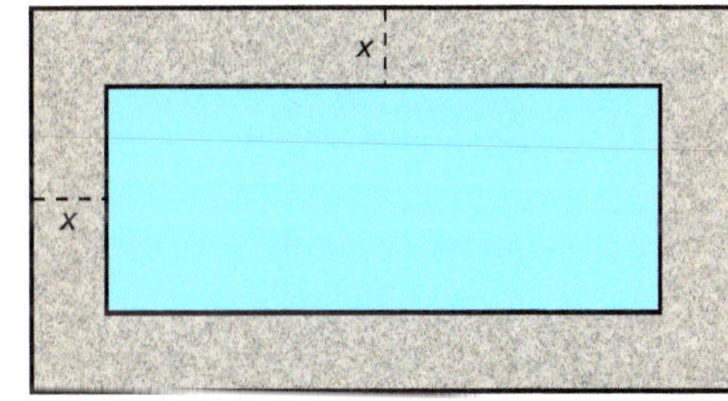

total area = $4x^2 + 104x + 651$

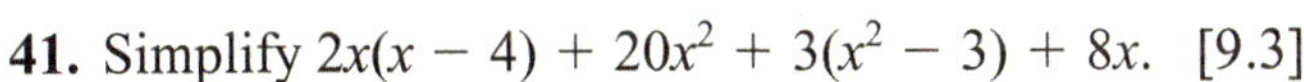

CUMULATIVE REVIEW

41. Simplify $2x(x - 4) + 20x^2 + 3(x^2 - 3) + 8x$. [9.3]

42. Solve $5x - 7 = 3x + 11$. [2.7]

43. Solve for x and y when $x = 8 - y$ and $3x - 8 = 5y$. [7.2]

44. Divide $\frac{2x^3 + x - 18}{x - 2}$. [9.6]

45. One factor of $12x^2 + 5x - 28$ is $4x + 7$. Find the other factor by dividing. [9.6]

46. One factor of $15x^2 - 4x - 4$ is $3x - 2$. Find the other factor by dividing. [9.6]

47. Write $5 > x$ using a less than sign. [4.1]

48. Graph $y < -3x + 5$. [6.7]

Write an equation and solve.

49. A bus leaves Regina for Raleigh, averaging 50 mi/hr. Two hours later another bus leaves, averaging 55 mi/hr. How long will it take the second bus to catch up with the first bus? [3.7, 7.6]

50. Lynda has \$5.90 in quarters and dimes. If she has four fewer dimes than quarters, how many of each are there? [2.8]

10.4 Special Patterns

The volume of metal used to make a washer is calculated with a binomial: $V = \pi R^2 t - \pi r^2 t$, where R and r are the inner and outer radii and t is the thickness.

Integers such as 36 and 121 are called *perfect squares* because they can be expressed as the square of another integer. Recall that the product of a pair of conjugates is a difference of two squares.

$$(x + y)(x - y) = x^2 - y^2$$

Since factoring is the opposite process of multiplication, factoring a *difference of squares* binomial produces a product of conjugates.

$$x^2 - y^2 = (x + y)(x - y)$$

Notice that both terms in the factorization are found by taking the square roots of terms in the original binomial.

Factoring the Difference of Squares
$a^2 - b^2 = (a + b)(a - b)$

Example 1

Factor $x^2 - 49$.

Answer		
	$x^2 - 7^2$	1. Rewrite the binomial as a difference of squares.
	$(x + 7)(x - 7)$	2. Apply the difference of squares pattern: $a^2 - b^2 = (a + b)(a - b)$.

When asked to factor a polynomial, always factor any common factors as the initial step.

Example 2

Factor $5 - 125x^2$.

Answer		
	$5(1 - 25x^2)$	1. Factor out the common factor of 5.
	$5[1^2 - (5x)^2]$	2. Rewrite the binomial as a difference of squares.
	$5(1 + 5x)(1 - 5x)$	3. Apply the difference of squares pattern.

Example 3

Factor $36x^6z^2 - 64y^2$.

Answer

$4(9x^6z^2 - 16y^2)$	1. Factor out the common factor of 4.
$4[(3x^3z)^2 - (4y)^2]$	2. Rewrite the binomial as a difference of squares.
$4(3x^3z + 4y)(3x^3z - 4y)$	3. Apply the difference of squares pattern.

A difference of squares, such as $x^2 - 25$, could be rewritten as the trinomial $x^2 + 0x - 25$ and factored by finding the factors of -25 whose sum is 0. This produces the same factors as those predicted by the pattern, $(x + 5)(x - 5)$. However, if you write $x^2 + 25$ as the trinomial $x^2 + 0x + 25$ and attempt to find two positive factors of 25 whose sum is 0, it quickly becomes apparent that the polynomial is prime. In fact, any binomial that is a sum of squares $(a^2 + b^2)$ cannot be factored using real numbers. Recognizing these patterns provides a much quicker solution.

You have already studied the patterns of two perfect squares:

$(x + y)^2 = x^2 + 2xy + y^2$ and
$(x - y)^2 = x^2 - 2xy + y^2$.

While these *perfect square trinomials* can be factored using the previous techniques, recognizing their patterns will enable you to write their factorizations quickly. Notice that the first and last terms of each trinomial are positive and are perfect squares. The middle term is twice the product of the square roots of the first and last terms.

Factoring Perfect Square Trinomials

$$a^2 + 2ab + b^2 = (a + b)^2$$
$$a^2 - 2ab + b^2 = (a - b)^2$$

Example 4

Factor $x^2 + 10x + 25$.

Answer

$x^2 + 2(x)(5) + 5^2$	1. The first and last terms are perfect squares. The middle term is twice the product of their square roots.
$(x + 5)^2$	2. Apply the square of a sum pattern: $a^2 + 2ab + b^2 = (a + b)^2$.

Example 5

Factor $12x^2 - 36x + 27$.

Answer

$3(4x^2 - 12x + 9)$	1. Factor out the common factor of 3.
$3[(2x)^2 - 2(2x)(3) + 3^2]$	2. The first and last terms are perfect squares. The middle term is the opposite of twice the product of their square roots.
$3(2x - 3)^2$	3. The negative middle term indicates that the square of a difference pattern should be used: $a^2 - 2ab + b^2 = (a - b)^2$.

Example 6

Factor $x^6y^4 - 8x^3y^2z + 16z^2$.

Answer $(x^3y^2)^2 - 2(x^3y^2)(4z) + (4z)^2$

1. The first and last terms are perfect squares. The middle term is the opposite of twice the product of their square roots.

$(x^3y^2 - 4z)^2$

2. Apply the square of a difference pattern.

The patterns in math reflect consistency and order. Christians have a much greater appreciation for the consistency found in math and science because they know that all things were created by Christ (Col. 1:16). Common patterns point to a single Creator of the universe and of life. The consistency and order discovered in the study of mathematics and science reveal God's power, order, and glory, "the invisible things of him from the creation of the world" (Rom. 1:20).

A. Exercises

Fill in the missing parts of each factorization.

1. $y^2 - 144 = (y \_\_\_ 12)(y \_\_\_ 12)$

2. $x^2 - y^2 = (x \_\_\_ y)(x \_\_\_ y)$

3. $49x^2 - 4 = (7x - \_\_\_)(7x + \_\_\_)$

4. $9x^2 - 100 = (3x - \_\_\_)(3x + \_\_\_)$

5. $36a^2 - 49 = (\_\_\_ - 7)(6a + \_\_\_)$

6. $25x^2 - 49 = (5x - \_\_\_)(\_\_\_ + 7)$

Identify the perfect square trinomials by writing "yes" or "no." Justify your answer.

7. $x^2 + 6x + 9$

8. $x^2 + 14x + 49$

9. $x^2y - 8xy + 16y^2$

10. $a^2b^2 - 12abc + 36c^2$

11. $4x^2 - 10x + 25$

12. $a^2 - 4ab - 4b^2$

Factor.

13. $a^2 - 225b^2$

14. $a^2x^2 - y^2$

15. $x^2 - 6x + 9$

16. $16a^2 + 8a + 1$

17. $x^2 - 16x + 64$

18. $m^2 + 24m + 144$

19. $a^2 - 10ab + 25b^2$

20. $a^2 - 14a + 49$

B. Exercises

Factor. Be sure to factor any common monomials first. If the polynomial will not factor, write "prime."

21. $2a^2 - 242$

22. $3x^2 - 147$

23. $9x^2 + 12x + 4$

24. $36x^2 + 12x + 1$

25. $a^2x^2 - 16x^2$

26. $a^4 - a^2b^2$

27. $9x^2 + 30xz + 25z^2$

28. $x^8 + 26x^5 + 169$

29. $16x^2 - 4$

30. $x^7 - x^5$

31. $xt^4 - 20xt^2 + 100$

32. $x^2 + 0.6x + 0.09$

33. $63x^2 - 175$

34. $147a^2 - 507$

C. Exercises

Factor.

35. $162a^2b^4c - 18b^4c$

36. $x^6 - x^2$

37. $27x^5y - 72x^3y^2 + 48xy^3$

Factor the polynomial; then write it as a division problem (e.g., $12 = 2 \cdot 6$ implies that $\frac{12}{2} = 2$ or $\frac{12}{6} = 2$).

38. $9x^2 + 48x + 64$

Dominion Modeling

The Les Bois hydroelectric power plant monitors the ice movement of the Mont Blanc glaciers using a system of tunnels that run under the glaciers. An instrument attached to a bicycle wheel was used to determine that the ice on the bottom layer of one of the glaciers moves down the mountain at an average of 1 in./hr.

39. Assuming that the ice continues to flow at the current average, how long would it take a portion of ice to move from the glacier's head down the length of the 6.8 mi glacier?

40. Eight stakes were evenly spaced along a horizontal line across the top of a glacier, and their movement (in meters) was recorded. Use a graphing calculator or a spreadsheet to make a scatterplot of the data.

Stake	Movement (meters)
1	230
2	235
3	250
4	265
5	270
6	245
7	230
8	210

41. Use a graphing calculator (or spreadsheet) to complete a quadratic regression on the data.

a. State the equation of a parabola that models the data set.

b. Sketch the graph of the function on your scatterplot.

42. Research glacial flow rates and explain the factor(s) that cause the stakes closest to the edges of the glacier to move more slowly.

CUMULATIVE REVIEW

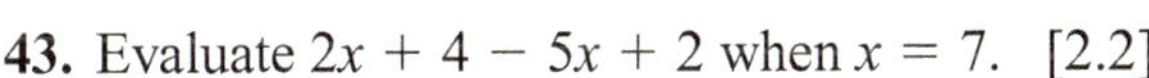

43. Evaluate $2x + 4 - 5x + 2$ when $x = 7$. [2.2]

44. Simplify $2x + 4 - 5x + 2$. [2.3]

45. Solve $2x + 4 - 5x + 2 = 0$. [2.6]

46. Graph $y = 2x + 4 - 5x + 2$. [6.3]

47. Simplify $23(-101)$. [1.5]

48. Factor 7373. [10.1]

49. Write exercises 47 and 48 as division problems. [1.6]

Factor each polynomial; then write it as a division problem. [10.1–10.2]

50. $2x + 4$

51. $-40xy^2 + 5xy + 20x$

52. $x^2 + 8x - 9$

A *pentamino* is a figure consisting of five adjoining congruent squares.

Determine whether each figure can be formed using the pentaminos pictured above without any overlaps or rotations.

1. 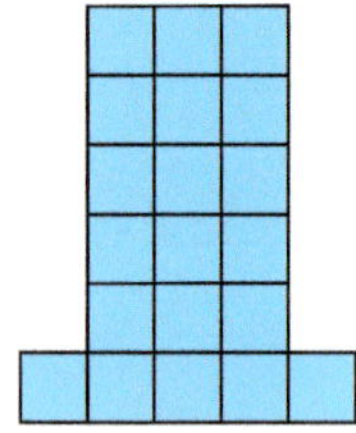

2. 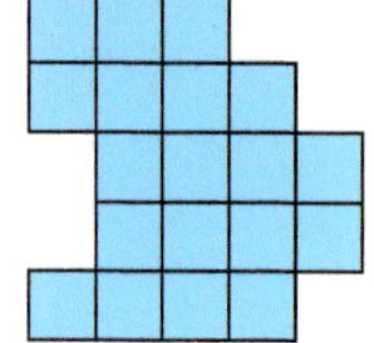

3. 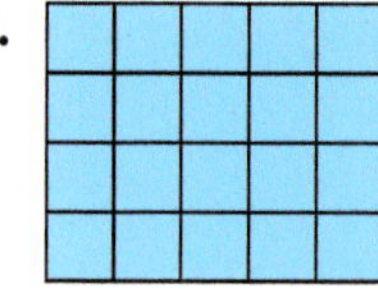

Determine whether each figure can be formed using the pentaminos pictured above if rotations and overlaps are allowed.

4. 

5. 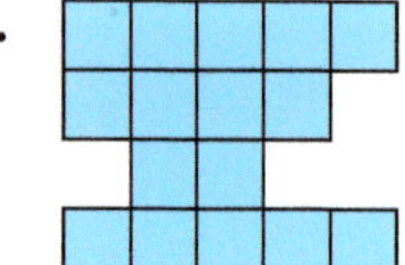

6. 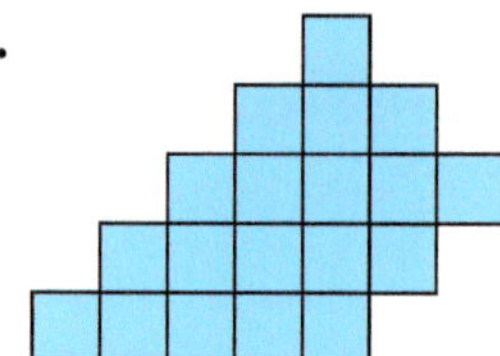

TECHNOLOGY CORNER (TI-84+ Family)

In Chapter 8 you graphed simple *quadratic* functions (containing an x^2 term) and *cubic* functions (containing an x^3 term). Looking at the graphs of these types of functions in factored form can be informative. Set your calculator's graphing window by pressing [ZOOM] [6]. Then press [Y=] and [CLEAR] any functions that have been entered previously, and make sure all plots are off.

Graph the function $y = (x - 4)(x + 1)$. Where does the parabola cross the x-axis?

Graph the function $y = x(x + 2)(x - 3)$. What are the three points where the cubic function crosses the x-axis?

Note that the x-coordinates of the x-intercepts (also called the *zeros* of the function) have the opposite signs as the constant terms in the binomials. This is similar to what we saw in the horizontal translations in Sections 5.7 and 8.4.

You can also check your factoring by graphing both the standard and factored forms of the expressions as functions. Their graphs should be identical.

10.5 Factoring Completely

All of the factoring you have done has been done over the set of integers. Consider $x^2 - 3$. Could this factor to $(x - \sqrt{3})(x + \sqrt{3})$? Technically, the answer is yes; but $\sqrt{3}$ is not an integer. If a polynomial cannot be factored over the set of integers, it can be factored over the set of real numbers or over the set of complex numbers.

In this book we will consider a polynomial completely factored when it cannot be factored further over the set of integers. Sometimes after you factor a polynomial, you can factor one of the resulting polynomial factors. In order for a polynomial to be completely factored, each polynomial factor must itself be completely factored.

Cumberland Falls, Kentucky, is 68 ft high. Its flow of 3217 ft^3/sec (cfs) makes it the most powerful waterfall in the South. Determine its energy (horsepower) using $E = 0.11345rh$, where r is the flow in cfs and h is the height in feet.

Example 1

Factor $5ax^5 - 5axy^4$.

Answer

$5ax(x^4 - y^4)$	1. Factor out the common monomial $5ax$.
$5ax(x^2 + y^2)(x^2 - y^2)$	2. Factor the difference of squares.
$5ax(x^2 + y^2)(x - y)(x + y)$	3. Factor the resulting difference of squares. Remember that a sum of squares, such as $x^2 + y^2$, cannot be factored using real numbers.

Check

$5ax(x^2 + y^2)(x - y)(x + y) = 5ax^5 - 5axy^4$	4. Multiplying the factors produces the original expression.

Example 2

Factor $x^4 - 9x^2 + 8$.

Answer

$x^4 - 9x^2 + 8$	1. There are no common monomials and no special patterns.
$(x^2 - 1)(x^2 - 8)$	2. Factor by finding the factors of $+8$ whose sum is -9.
$(x - 1)(x + 1)(x^2 - 8)$	3. Factor the first binomial again since it is a difference of squares.

In some expressions you can factor common factors from several groupings of the terms to leave a common binomial.

Example 3

Factor $2xy - 7x + 10y - 35$.

Answer

$(2xy - 7x) + (10y - 35)$	1. Group the first two and the last two terms since there are no common monomials and no special patterns.
$x(2y - 7) + 5(2y - 7)$	2. Factor a common monomial factor from each binomial.
$(x + 5)(2y - 7)$	3. Factor the common binomial, $2y - 7$, from both terms. This is the complete factorization.

Example 4

Factor $am - an - pm + pn$.

Answer	
$(am - an) + (-pm + pn)$	1. Group the terms with a common factor.
$a(m - n) - p(m - n)$	2. Factor a common monomial factor from each binomial.
$(a - p)(m - n)$	3. Factor the common binomial, $m - n$, from both terms.

Example 5

Factor $2y^3 + 5y^2 - 8y - 20$.

Answer	
$(2y^3 + 5y^2) + (-8y - 20)$	1. Group the first two and the last two terms since there are no common monomials and no special patterns.
$y^2(2y + 5) - 4(2y + 5)$	2. Factor a common monomial factor from each binomial.
$(y^2 - 4)(2y + 5)$	3. Factor the common binomial, $2y + 5$, from both terms.
$(y - 2)(y + 2)(2y + 5)$	4. Factor $y^2 - 4$ to complete the factorization.

Strategies for Factoring Completely over the Set of Integers

1. Factor out common monomials. (Section 10.1)

2. Consider the type of polynomial and specific factoring strategies.

Binomials	**Trinomials**	**Polynomials**
Difference of squares $a^2 - b^2 = (a - b)(a + b)$ (Section 10.4)	a. **Perfect square trinomial** $a^2 \pm 2ab + b^2 = (a \pm b)^2$ (Section 10.4) b. $\boldsymbol{x^2 + bx + c}$ or $\boldsymbol{x^2 + bxy + cy^2}$ Find factors of c that sum to b. (Section 10.2) c. $\boldsymbol{ax^2 + bx + c}$ or $\boldsymbol{ax^2 + bxy + cy^2}$ Rewrite the middle term using factors of ac whose sum is b as coefficients, and factor by grouping. (Section 10.3)	Factor by grouping. (Section 10.5)

3. Continue factoring until each factor is no longer factorable.

4. Check.

A. Exercises

Factor completely.

1. $5a^2 + 10a + 5$
2. $x^3 - 2x^2 + x$
3. $200 - 2a^2$
4. $8a^8 - 2a^6$
5. $3x^2 - 18x + 24$
6. $3a^5 - 4a^3$
7. $5x^2 - 70x + 245$
8. $m^{10} + 2m^5 + 1$

Factor by grouping.

9. $x(2x - y) + y(2x - y)$
10. $4(b + c) - m(b + c)$
11. $(at - aw) + (bt - bw)$
12. $(xt - 3xw) + (2t - 6w)$
13. $a(m + n) + b(m + n) - 2(m + n)$
14. $m(2a + b) - n(2a + b) + 2n(2a + b)$

B. Exercises

Factor completely.

15. $8a^2 - 48ab^2 + 72b^4$

16. $3x^3 - 33x^2 + 84x$

17. $x^4 - 16$

18. $a^4x^4 - 81$

19. $18x^2 - 3x - 36$

20. $30y^2 - 35y - 100$

21. $40w^2 - 62w + 24$

22. $70x^2 - 32x - 24$

23. $wy + wz + xy + xz$

24. $xy + 5x + 3y + 15$

25. $xy - 6x + 7y - 42$

26. $x^2 - 2x - 3xy + 6y$

27. $6x^2 + xy^2 + 24x + 4y^2$

28. $(2ax^2 + bx^2) - (2ay^2 + by^2)$

29. $y^3 - 3y^2 - 4y + 12$

30. $a^3 + 2a^2 - 9a - 18$

31. $5y^3 - 15y^2 - 5y + 15$

32. $3x^3 - 6x^2 - 48x + 96$

C. Exercises

Given one factor of the polynomial, factor the polynomial completely.

33. $(x - 2)$ is a factor of $x^3 - 3x^2 - 10x + 24$.

34. $(x + 2)$ is a factor of $x^3 + x^2 - 8x - 12$.

35. $(y + 1)$ is a factor of $y^3 - 9y^2 + 4y + 15$.

36. $(y + 4)$ is a factor of $3xy^2 + 18xy + 24x - y^2 - 6y - 8$.

Factor completely.

37. $x^2 + 2xy + y^2 - 9$

38. $x^2 + 2xy + y^2 + 2x + 2y + 1$

Dominion Modeling

As a result of a flood in 1892 caused by a hidden glacial lake, the French villages surrounding Mount Blanc began diligently monitoring its glaciers. In 2010 scientists used satellite imagery to locate a similar hidden lake estimated to contain 65,000 m^3 of water. The lake was found to be at an altitude of approximately 3200 m, making it a challenge for the engineers hired to drain it.

39. Use unit analysis to convert the following measurements.

a. altitude: 3200 m to feet

b. volume: 65,000 m^3 to cubic feet

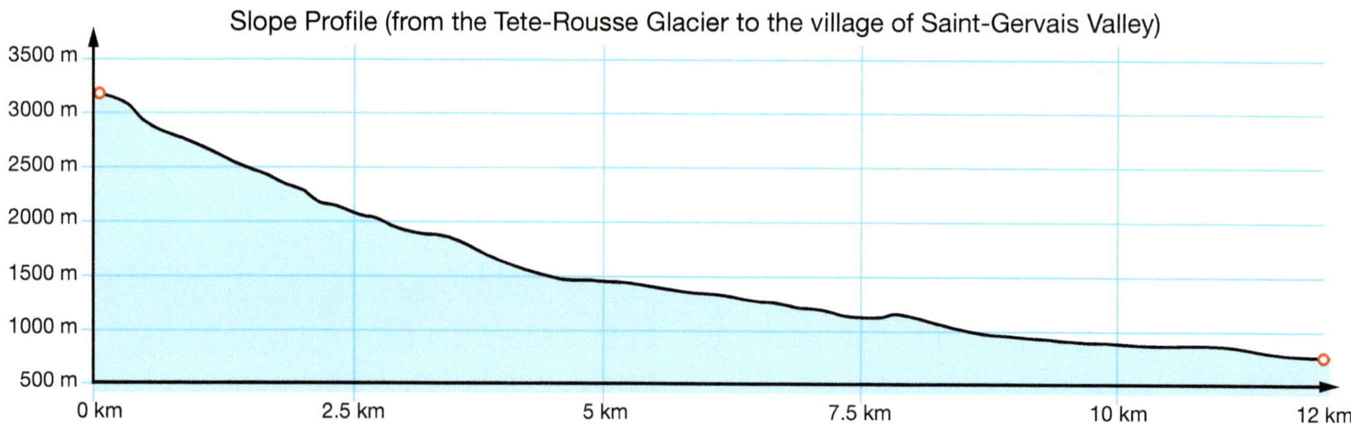

40. Given the elevation profile for the path from the glacier (the red circle on the left) to the village of Saint-Gervais-les-Bains (the red circle on the right), estimate each of the following slopes as a decimal (to the nearest hundredth) and then convert to a percent grade.

a. the average slope for the first 2.5 km (measured horizontally)

b. the average slope from the glacier to the village

41. It has been estimated that it would take 15 min for the water from the lake to reach the village of Saint-Gervais-les-Bains, which is about 12 km from the glacier.

a. Determine the speed of the water in kilometers per minute.

b. What is the rate in miles per hour (to the nearest whole number)?

CUMULATIVE REVIEW

Solve each literal equation for q. [3.1]

42. $q + 12p = 3t$

43. $Aq - 5 = 7n$

44. $\frac{q}{6} + \frac{3q}{4} = 2m$

45. $5q + xq = 7$

Write an equation and solve. [2.8]

46. Wesley has four more quarters than nickels and two more dimes than quarters. If the coins' total value is $4.40, how many of each does he have?

47. Cassie has 42 coins. She has twice as many dimes as nickels, and the rest are quarters. If she has a total of $7, how many of each does she have?

Write and solve a system of equations for each word problem. [7.6–7.7]

48. Geoff flew his twin-engine plane from Cincinnati to St. Louis. He made the 350 mi flight against a sustained wind in 2 hr. If flying conditions were the same for the return trip, which took 10 min less, what was the average speed of the plane with no wind and what was the wind speed (to the nearest mile per hour)?

49. Cameron and his friend took a canoe trip 12 mi up a lazy river in 3 hr and returned in 2 hr 24 min. What was the rate of the current, and what would be the boys' paddling rate in still water?

50. Vanessa needs a 25% saline solution. If she has a 40% solution on hand, how much water and how much 40% solution should she mix to obtain 3 gal of a 25% saline solution?

51. A metalsmith is working to create an alloy of copper and nickel. If he currently has an alloy that is 20% copper, how much of the 20% alloy and how much copper should he use to produce 10 lb of an alloy that is 60% copper?

CHAPTER 10 REVIEW

Factor out the common monomial.

1. $3r^2 + 5r$ **2.** $8x^2y^3 + 6xy^2 - 2x$ **3.** $4x^3 - 8x^2$ **4.** $x^4 + 3x^3y + 7x^2$

Factor.

5. $y^2 - 15y + 56$ **6.** $x^2 + 3x - 54$ **7.** $2d^2 + 9d + 7$ **8.** $9t^2 + 21t - 8$

Rearrange the terms in descending order; then factor.

9. $5x - 14 + x^2$ **10.** $a^2 - 84 - 5a$ **11.** $-6 - b + 15b^2$ **12.** $7 + 2y^2 + 15y$

Use special patterns to factor.

13. $a^2 - 9$ **14.** $25x^2 - 144y^2$ **15.** $a^2 + 14a + 49$ **16.** $9x^2 - 30x + 25$

Factor. Begin by factoring out the common monomial.

17. $x^2y + xy - 56y$ **18.** $7y^2 - 21y + 14$

19. $18x^2 - 100x + 50$ **20.** $12x^2 - 68x + 80$

Factor completely. If the polynomial will not factor over the set of integers, write "prime."

21. $2x^2 - 30x + 108$ **22.** $a^2 - 22a + 121$

23. $z^2 - 13z + 48$ **24.** $9x^2 - 25x$

25. $18y^2 - 50$ **26.** $b^2 + 4$

27. $24x^2 - 96y^2$ **28.** $5b^2 - 50b + 125$

29. $9b^2 - 12bc + 4c^2$ **30.** $2a^2 + 9a - 18$

31. $15x^2 + xy - 2y^2$ **32.** $3y^2 + 17y + 20$

33. $b^2 - 15b - 36$ **34.** $12x^2 + 78xy + 90y^2$

35. $6a^2 - 20a + 14$ **36.** $15x^2 + 14y$

37. $12y^2 + 22y - 70$ **38.** $9r^2 + 9r + 2rs + 2s$

39. $10x^3 + 14x^2 - 15x - 21$ **40.** $x^4 + 6x^2 + 9$

41. $c^4 + 3c^2 + 2$ **42.** $(x + y)^2 - 9$

43. $256a^8 - 1$ **44.** $9x^4 - 288x^2 + 2304$

45. $x^2y + x^2 - xy - 1$

46. Rewrite the formula for the surface area of a cone, $S = \pi rl + \pi r^2$, by factoring the right side of the equation.

47. Use the fact that $x^3 - y^3 = (x - y)(x^2 + xy + y^2)$ to simplify $\frac{x^3 - y^3}{x - y}$.

The total area of both the picture and its frame, in square inches, is represented by the polynomial $4w^2 + 24w + 35$.

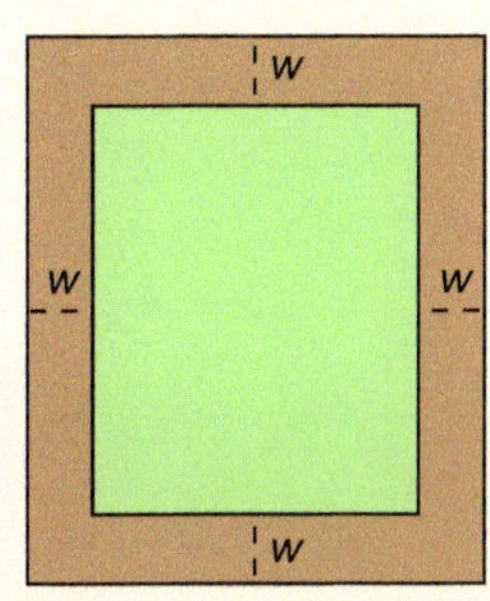

48. Factor the polynomial to find binomial expressions representing the height and width of the outside of the frame.

49. Using the factors found in the previous exercise, find the area of the picture alone.

50. Assume that the frame is 2 in. wide.

a. What is the total area of the picture and the frame?

b. What is the area of the frame?

11 RADICALS

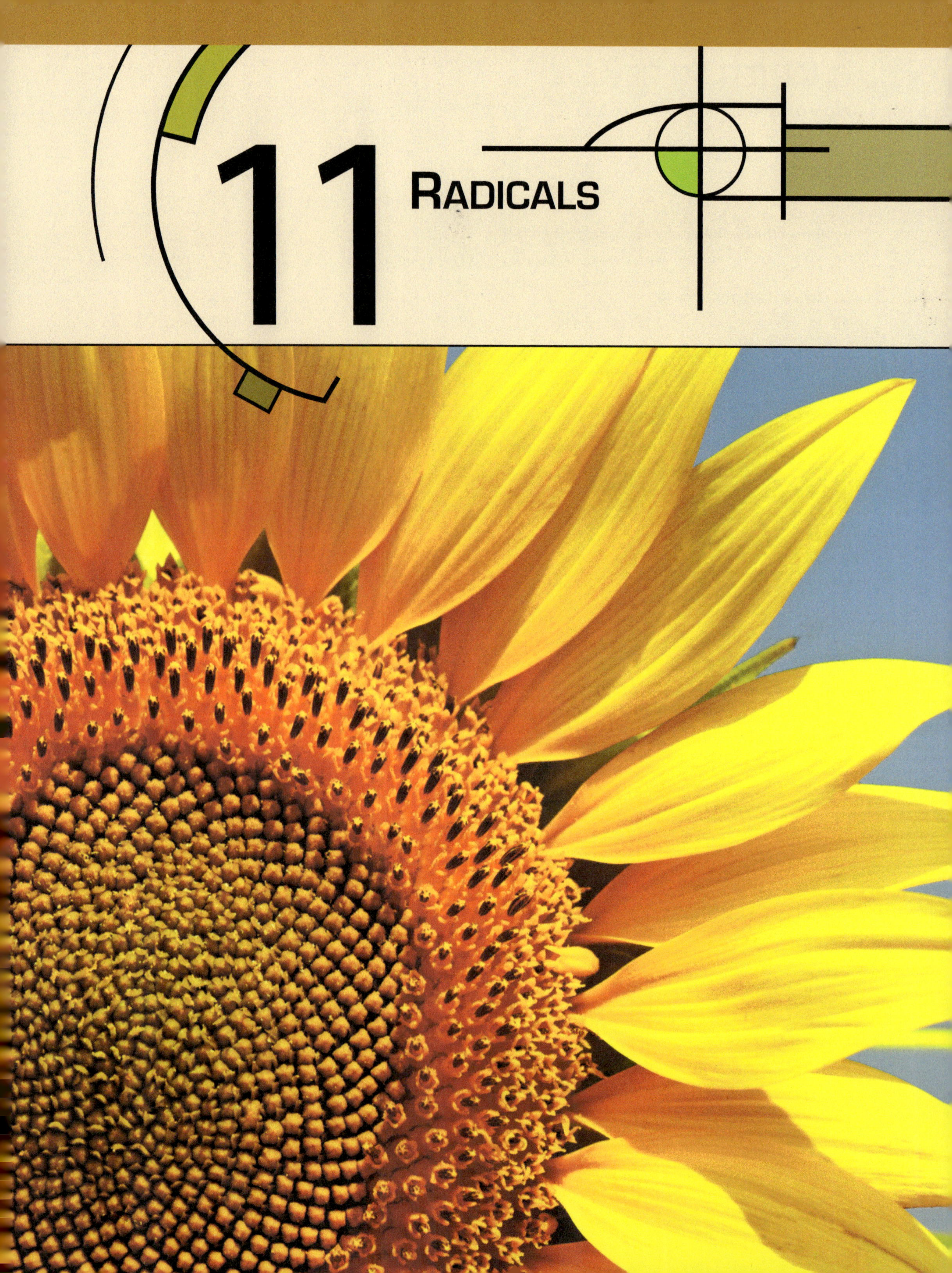

Most students learn the same mathematical tools. Although the Christian uses them for many of the same purposes as does the unbeliever, he uses them with a different view of life and the world. Viewing our world as the creation of God and viewing ourselves as a creation in the image of God (Gen. 1:26–28) reveals truth concerning our existence. Knowing the truth revealed in God's Word provides value and purpose for life as well as unique insight into the world around us. This Christian worldview affects how we live and the choices we make. It keeps us from living foolishly, wastefully, or shamefully (Prov 15:14). Additionally, a Christian worldview provides an accurate understanding of our spiritual condition and the condition of our physical world.

God's creation reveals some interesting mathematical patterns. The golden ratio, often represented by the lowercase Greek letter phi (ϕ), is a number that is frequently found in nature. In the Dominion Modeling exercises throughout this chapter, you will discover interesting mathematical characteristics of the golden ratio and its relationship to nature, art, graphic design, architecture, and science.

After this chapter you should be able to

1. approximate the root of a number using the two nearest integers and using a calculator.
2. convert between radical form and exponential form.
3. simplify radical expressions.
4. multiply and divide radical expressions.
5. add and subtract radical expressions.
6. solve problems using the Pythagorean theorem.
7. develop and apply the distance formula.
8. develop and apply the formula for the midpoint of a segment.
9. solve radical equations.
10. graph simple radical functions.

11.1 Expressing Roots

Reception from a TV tower depends on the height of the tower. The approximate effective radius is $d = \sqrt{1.5h}$.

Squaring a number and taking the square root of the result are similar to inverse operations, such as addition and subtraction.

$3^2 = 9$ and $(-3)^2 = 9$,

so the square roots of 9 are 3 and -3.

Definition

A **square root** is one of a number's two equal factors. The symbol for a square root is the *radical sign*, $\sqrt{\ }$.

The expression under the radical sign is called the *radicand*. For example, in $\sqrt{9}$, 9 is the radicand. The square root of 9 has two values: ± 3. To avoid confusion, mathematicians have agreed to use the radical sign to indicate the positive square root, called the *principal root*. If the negative square root is desired, a negative sign is placed in front of the radical sign.

$\sqrt{9} = 3$ and $-\sqrt{9} = -3$

Example 1

Find the indicated roots.

a. $\sqrt{4}$ **b.** $-\sqrt{16}$

Answer **a.** $\sqrt{4} = \sqrt{2^2} = 2$ **b.** $-\sqrt{16} = -\sqrt{4^2} = -4$

To indicate roots other than square roots, an *index* is added to the radical sign.

Definitions

A **cube root**, $\sqrt[3]{\ }$, is one of a number's three equal factors.

An ***n*th root**, $\sqrt[n]{\ }$, is one of a number's n equal factors.

If $2^3 = 8$,
then $\sqrt[3]{8} = 2$.

Example 2

Find the indicated roots.

a. $\sqrt[3]{125}$ **b.** $\sqrt[4]{81}$

Answer **a.** $\sqrt[3]{125} = \sqrt[3]{5^3} = 5$ **b.** $\sqrt[4]{81} = \sqrt[4]{3^4} = 3$

The even root of a negative number is not a real number since any number to an even power is positive. However, the odd root of a negative number is a negative real number.

Example 3

Find the indicated roots.

a. $\sqrt{-4}$ **b.** $\sqrt[3]{-\frac{27}{125}}$ **c.** $\sqrt[5]{-32}$

Answer **a.** $\sqrt{-4}$ is not a real number.

b. $\sqrt[3]{-\frac{27}{125}} = \sqrt[3]{\left(-\frac{3}{5}\right)^3} = -\frac{3}{5}$

c. $\sqrt[5]{-32} = \sqrt[5]{(-2)^5} = -2$

While some radicals such as $\sqrt[3]{-125}$ and $\sqrt{0.25}$ are rational numbers, many radicals are irrational numbers. These nonrepeating, nonterminating decimals can be either stated exactly as radicals or approximated by rounding to a designated decimal place.

Knowing the perfect squares up to 400 will help you to quickly approximate many square roots to the nearest whole number. Determine the perfect squares that are just less than and just greater than the radicand. For $\sqrt{7}$,

$$4 < 7 < 9,$$
$$\sqrt{4} < \sqrt{7} < \sqrt{9}, \text{ and}$$
$$2 < \sqrt{7} < 3.$$

Since 7 is closer to 9 than it is to 4, $\sqrt{7}$ is closer to 3 than it is to 2. To express $\sqrt{7}$ to the nearest hundredth, round a calculator's approximation of 2.645751311 to 2.65.

Example 4

Mentally estimate $\sqrt{129}$ to the nearest integer. Then use a calculator to find $\sqrt{129}$ to the nearest hundredth.

Answer

$121 < 129 < 144$ — 1. $11^2 = 121$ and $12^2 = 144$.

$\sqrt{121} < \sqrt{129} < \sqrt{144}$

$11 < \sqrt{129} < 12$

$\sqrt{129} \approx 11$ — 2. 129 is closer to 121 than to 144.

$\sqrt{129} \approx 11.36$ — 3. Round a calculator's approximation of 11.35781669 to the nearest hundredth.

Example 5

Mentally estimate $\sqrt[3]{50}$ to the nearest integer. Then use a calculator to find $\sqrt[3]{50}$ to the nearest thousandth.

Answer

$27 < 50 < 64$ — 1. $3^3 = 27$ and $4^3 = 64$.

$\sqrt[3]{27} < \sqrt[3]{50} < \sqrt[3]{64}$

$3 < \sqrt[3]{50} < 4$

$\sqrt[3]{50} \approx 4$ — 2. 50 is closer to 64 than to 27.

$\sqrt[3]{50} \approx 3.684$ — 3. Round a calculator's approximation of 3.684031499 to the nearest thousandth.

The definitions of roots and powers allow you to convert an equation from exponential to radical form and vice versa. In general, $b^n = a$ is equivalent to $\sqrt[n]{a} = b$. The radicand is the value obtained when b is raised to the nth power.

Up to this point, the definition of exponents has been limited to integers. Radical expressions are used to give meaning to rational exponents.

Compare the following computations to identify a pattern for converting rational exponents to radical form.

$$\sqrt{4} \cdot \sqrt{4} = 2 \cdot 2 = 4 \text{ and } 4^{\frac{1}{2}} \cdot 4^{\frac{1}{2}} = 4^{\left(\frac{1}{2} + \frac{1}{2}\right)} = 4^1 = 4$$

Both expressions are equivalent to 4, resulting in the following conclusion.

$$\sqrt{4} \cdot \sqrt{4} = 4^{\frac{1}{2}} \cdot 4^{\frac{1}{2}}$$

$$(\sqrt{4})^2 = \left(4^{\frac{1}{2}}\right)^2$$

Therefore, $\sqrt{4} = 4^{\frac{1}{2}}$.

You can convert between the radical form and the exponential form of any expression using the following definition.

Definition

The exponential expression $x^{\frac{a}{b}}$ is equivalent to the radical $\sqrt[b]{x^a}$ or $\left(\sqrt[b]{x}\right)^a$.

$$8^{\frac{2}{3}} = \sqrt[3]{8^2} = (\sqrt[3]{8})^2$$

In a fractional exponent, the denominator becomes the index, and the numerator becomes the power of the radicand.

Example 6

a. Convert $5^{\frac{2}{3}}$ to radical form.

b. Convert $\sqrt[5]{x^3}$ to exponential form.

Answer

a. $5^{\frac{2}{3}} = \sqrt[3]{5^2} = \sqrt[3]{25}$ — The denominator becomes the index, and the numerator becomes the power of the radicand.

b. $\sqrt[5]{x^3} = x^{\frac{3}{5}}$ — The index of 5 becomes the denominator, and the power of the radicand becomes the numerator.

Example 7

a. Write $3^{\frac{1}{2}}x^{\frac{3}{4}}$ as a single radical.

b. Convert $\sqrt[4]{7x^3y^2}$ to exponential form.

Answer

a. $3^{\frac{1}{2}}x^{\frac{3}{4}} = 3^{\frac{2}{4}}x^{\frac{3}{4}}$

1. Rename fractional exponents using a common denominator so that their bases can be placed under the same radical.

$= \sqrt[4]{3^2x^3} = \sqrt[4]{9x^3}$

2. Assign the numerators as powers of the factors in the radicand. The common denominator is the index of the radical.

b. $\sqrt[4]{7x^3y^2} = 7^{\frac{1}{4}}x^{\frac{3}{4}}y^{\frac{2}{4}}$

1. The index becomes the denominator of the exponents. The power of 7 is 1, so its exponent is $\frac{1}{4}$.

$= 7^{\frac{1}{4}}x^{\frac{3}{4}}y^{\frac{1}{2}}$

2. Reduce exponents where possible.

A. Exercises

Find the indicated roots. If the root is not a real number, state "not real."

1. $\sqrt{49}$
2. $\sqrt{81}$
3. $-\sqrt{36}$
4. $-\sqrt{121}$
5. $\sqrt{\frac{49}{16}}$
6. $\sqrt{\frac{9}{25}}$
7. $\sqrt{0.01}$
8. $-\sqrt{2.25}$
9. $\sqrt{-196}$
10. $\sqrt[3]{216}$
11. $\sqrt[4]{16}$
12. $\sqrt[4]{-1}$
13. $\sqrt[3]{-27}$
14. $-\sqrt[3]{-8}$
15. $\sqrt[3]{\frac{125}{64}}$
16. $\sqrt[3]{-0.125}$
17. $\sqrt{1492^2}$
18. $(\sqrt{1492})^2$
19. $\sqrt[3]{1776^3}$
20. $\sqrt{(-17)^2}$

B. Exercises

Mentally estimate each root to the nearest integer.

21. $\sqrt{18}$
22. $\sqrt{75}$
23. $\sqrt{300}$
24. $\sqrt[3]{29}$

Use a calculator to approximate each radical.

25. $\sqrt{12}$ to the nearest tenth
26. $\sqrt{95}$ to the nearest hundredth
27. $\sqrt{43}$ to the nearest thousandth
28. $\sqrt[3]{20}$ to the nearest hundredth
29. $\sqrt[3]{572{,}000}$ to the nearest tenth

Convert each radical to exponential form.

30. $\sqrt{5}$
31. $\sqrt{3x^3}$
32. $\sqrt[3]{2y^2}$
33. $\sqrt[5]{3x^2y^3}$
34. $\sqrt[8]{7^2ab^4}$

Convert each exponential expression to radical form.

35. $3^{\frac{1}{2}}y^{\frac{1}{2}}$
36. $5^{\frac{1}{3}}x^{\frac{2}{3}}$
37. $7^{\frac{1}{6}}a^{\frac{1}{2}}b^{\frac{1}{3}}$
38. $2^{\frac{4}{9}}y^{\frac{1}{3}}z^{\frac{2}{3}}$
39. $3^{\frac{3}{8}}r^{\frac{1}{4}}s^{\frac{1}{2}}$

C. Exercises

40. When is the root of a negative number equal to a real number value?

41. Use a counterexample to demonstrate that $\sqrt{x^2} = x$ is not always true. Then explain why $\sqrt{x^2} = |x|$.

Simplify.

42. $8^{\frac{2}{3}}$
43. $125^{\frac{4}{3}}$
44. $\sqrt{x^2 + 6x + 9}$

Dominion Modeling

In Chapter 3 a *Fibonacci sequence* was defined recursively as $A_n = A_{n-1} + A_{n-2}$, which implies that each term is the sum of the two preceding terms. In the most common Fibonacci sequence, $A_1 = 1$ and $A_2 = 1$, yielding 1, 1, 2, 3, 5, 8, …. Consecutive numbers of this Fibonacci sequence often appear in nature, such as in the spirals of pinecones, pineapples, or seeds in a sunflower.

The *golden ratio sequence* can be generated by dividing each Fibonacci term by its previous term. It can be defined recursively as $B_n = \frac{A_{n+1}}{A_n}$, where A_n is a Fibonacci sequence and the first golden ratio sequence term, B_1, is $\frac{A_2}{A_1}$. As n gets larger, the ratio $\frac{A_{n+1}}{A_n}$ gets closer to the *golden ratio*, which is usually represented by the Greek letter ϕ (phi) and is approximately 1.618. The following exercises illustrate how a golden ratio sequence can be used to approximate the value of ϕ.

45. In a spreadsheet, enter a value of 1 in both cell A1 and cell A2. Enter the formula =A1+A2 in cell A3; then replicate this formula down through cell A26. Enter the formula =A2/A1 in cell B1; then replicate this formula down through cell B25. To the nearest millionth, what value does the golden ratio sequence approach?

46. Change the values in cell A1 and cell A2 to two other numbers. What value does the sequence approach?

CUMULATIVE REVIEW

Solve. [4.3]

47. $3(x + 4) \leq 9$

48. $2y - 5 \leq 3(y + 2)$

Evaluate. [5.3]

49. If $f(x) = 5x - 4$, find $f(2)$.

50. If $f(x) = -2x^2 - 4$, find $f(3)$.

Simplify. [8.1]

51. $2abc(-5a^2b^3c^5)$

52. $-3x^2y^4z(-2xy^5z)$

53. $(2x^2)^3(2x)^2$

54. $3x^4y^2(2x^5y)^3$

Factor. [10.4]

55. $4x^2 - 289$

56. $7x^2 - 112$

TECHNOLOGY CORNER (TI-84+ Family)

Decimal approximations of roots can be found easily on your calculator. The $\sqrt{\ }$ function ([2nd] [x^2]) automatically inserts a beginning parenthesis, which requires no matching ending parenthesis to accurately complete simple calculations such as $\sqrt{9 + 16}$. To evaluate expressions such as $\sqrt{9} + 16$, insert an ending parenthesis to end the radicand and then enter the rest of the expression.

√(9+16
5
√(9)+16
19

The [MATH] menu allows you to enter cube roots and other roots.

To evaluate $\sqrt[3]{-8}$, enter [MATH] [4] [(-)] 8.

To evaluate $\sqrt[4]{81}$, enter [4] [MATH] [5] 81.

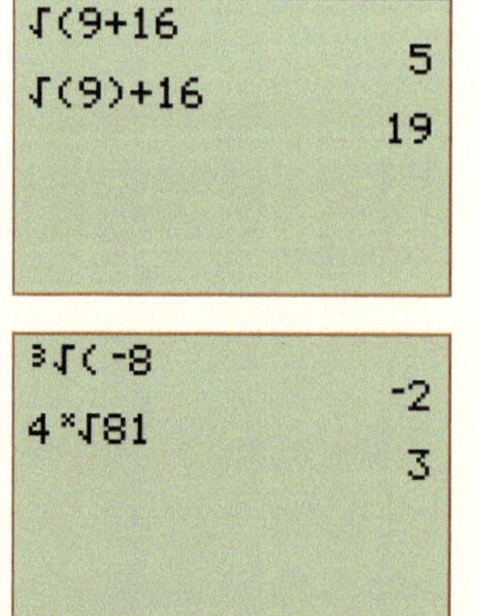

Rational exponents can be entered using the [^] key followed by the fraction within parentheses.

To evaluate $4^{\frac{3}{2}}$, enter 4 [^] [(] 3 [÷] 2 [)].

The reciprocal key, [x^{-1}], can also be used to enter fractional exponents.

To evaluate $(-8)^{\frac{1}{3}}$, enter [(-)] 8 [^] 3 [x^{-1}].

4^(3/2)
8
-8^3-1
-2

11.2 Simplifying Radicals

The water flow g (in gal/min) through a hose with a nozzle diameter d (in inches) and a static pressure p (in psi) is determined using the formula $g = 29.7d^2\sqrt{p}$.

In mathematics, answers are generally expected to be stated in simplest form. When a radical is in simplest form, its radicand contains no perfect power factors. The Product Property of Radicals is used to simplify radicals.

Product Property of Radicals

For $x \geq 0$ and $y \geq 0$, $\sqrt[n]{xy} = \sqrt[n]{x} \cdot \sqrt[n]{y}$.

To simplify a radical, apply this property after writing the radicand as the product of its prime factors.

Example 1

Simplify.

a. $\sqrt{24}$ **b.** $\sqrt{75}$ **c.** $\sqrt[3]{40}$

Answer

a. $\sqrt{24} = \sqrt{2 \cdot 2 \cdot 2 \cdot 3}$ — 1. Write the radicand as the product of its prime factors.
$= \sqrt{2^2} \cdot \sqrt{6}$ — 2. Apply the Product Property.
$= 2\sqrt{6}$ — 3. Simplify.

b. $\sqrt{75} = \sqrt{3 \cdot 5 \cdot 5}$ — 1. Write the radicand as the product of its prime factors.
$= \sqrt{5^2} \cdot \sqrt{3}$ — 2. Apply the Product Property.
$= 5\sqrt{3}$ — 3. Simplify.

c. $\sqrt[3]{40} = \sqrt[3]{2 \cdot 2 \cdot 2 \cdot 5}$ — 1. Write the radicand as the product of its prime factors.
$= \sqrt[3]{2^3} \cdot \sqrt[3]{5}$ — 2. Apply the Product Property.
$= 2\sqrt[3]{5}$ — 3. Simplify.

The steps outlined below can be used to simplify radical expressions containing numerical or variable factors.

Simplifying Radicals

1. Write the radicand as the product of its prime factors.
2. Apply the Product Property to the perfect powers.
3. Simplify the roots of the perfect powers.
4. Multiply the factors inside and outside the radical.

Example 2

Simplify.

a. $\sqrt{648}$ **b.** $\sqrt[3]{14{,}580}$

Answer

a. $\sqrt{648} = \sqrt{2 \cdot 2 \cdot 2 \cdot 3 \cdot 3 \cdot 3 \cdot 3}$

$= \sqrt{2^2} \cdot \sqrt{3^2} \cdot \sqrt{3^2} \cdot \sqrt{2}$

$= 2 \cdot 3 \cdot 3 \cdot \sqrt{2} = 18\sqrt{2}$

1. Write the radicand as the product of its prime factors.
2. Apply the Product Property to the perfect squares.
3. Simplify.

b. $\sqrt[3]{14{,}580} = \sqrt[3]{2 \cdot 2 \cdot 3 \cdot 3 \cdot 3 \cdot 3 \cdot 3 \cdot 3 \cdot 5}$

$= \sqrt[3]{3^3} \cdot \sqrt[3]{3^3} \cdot \sqrt[3]{2 \cdot 2 \cdot 5}$

$= 3 \cdot 3 \cdot \sqrt[3]{20} = 9\sqrt[3]{20}$

1. Write the radicand as the product of its prime factors.
2. Apply the Product Property to the perfect cubes.
3. Simplify.

Care must be taken when the Product Property is used to simplify radicals containing variables. The claim that $\sqrt{x^2} = x$ is true only when x is not negative.

$$\sqrt{2^2} = \sqrt{4} = 2, \text{ but } \sqrt{(-2)^2} = \sqrt{4} = 2 \text{ as well.}$$

If x is negative, then $\sqrt{x^2} = -x$ (the opposite of x), which is a positive number. This can be simplified by using the definition of absolute value to state

$$\sqrt{x^2} = |x|.$$

Consider the following cube roots:

$$\sqrt[3]{2^3} = \sqrt[3]{8} = 2 \text{ and } \sqrt[3]{(-2)^3} = \sqrt[3]{-8} = -2.$$

Notice that in this case, $\sqrt[3]{x^3} = x$ is true for all values of x.

When you are finding even roots, the absolute value is used to ensure a positive result.

Definition

The ***n*th root of x^n** or $\sqrt[n]{x^n} = \begin{cases} |x| & \text{if } n \text{ is even.} \\ x & \text{if } n \text{ is odd.} \end{cases}$

You may be able to save time by factoring the radicand into a product with perfect power factors.

Example 3

Simplify $\sqrt{32x^3y^2}$.

Answer

$\sqrt{16 \cdot 2x^2xy^2}$

$= \sqrt{16} \cdot \sqrt{x^2} \cdot \sqrt{y^2} \cdot \sqrt{2x}$

$= 4|x||y|\sqrt{2x}$

$= 4x|y|\sqrt{2x}$

1. Write the radicand as a product with perfect square factors.
2. Apply the Product Property and simplify. Use the absolute value to ensure a positive result.
3. If x were negative, the original expression would not be a real number; therefore, x must be positive and the absolute value is not needed.

Example 4

Simplify $\sqrt[3]{16x^2y^4z^6}$.

Answer $\sqrt[3]{8 \cdot 2x^2y^3yz^3z^3}$

$= \sqrt[3]{8} \cdot \sqrt[3]{y^3} \cdot \sqrt[3]{z^3} \cdot \sqrt[3]{z^3} \cdot \sqrt[3]{2x^2y}$

$= 2yzz\sqrt[3]{2x^2y} = 2yz^2\sqrt[3]{2x^2y}$

1. Write the radicand as a product with perfect cube factors.
2. Apply the Product Property and simplify. Notice that the absolute value is not used for cube roots.

Example 5

Simplify.

a. $\sqrt[3]{(x+2)^3}$ **b.** $\sqrt{(x+2)^2}$

Answer **a.** $\sqrt[3]{(x+2)^3} = x + 2$ **b.** $\sqrt{(x+2)^2} = |x+2|$

Note that $\sqrt{x^4} = \sqrt{x^2} \cdot \sqrt{x^2} = |x| \cdot |x| = |x|^2 = x^2$. The absolute value is not necessary since x^2 is always positive.

The use of absolute value to guarantee a nonnegative result is necessary only when the simplification of an even root of a variable raised to an even power produces a variable with an odd exponent.

$\sqrt{x^2} = |x|$ $\sqrt{x^3} = x\sqrt{x}$ $\sqrt{x^4} = x^2$

$\sqrt{x^5} = x^2\sqrt{x}$ $\sqrt{x^6} = |x|^3$ $\sqrt{x^7} = x^3\sqrt{x}$

It would be impossible for us to perform all our responsibilities to subdue the earth (Gen. 1:28) without learning and applying mathematics. Even the communication of the gospel is aided through the mathematical engineering of high-tech communication systems. Yet the power of the gospel is found not in human reasoning, mathematics, or technology, but in God (Rom. 1:16).

A. Exercises

Simplify each radical that can be simplified.

1. $\sqrt{12}$ **2.** $\sqrt{18}$ **3.** $\sqrt{27}$ **4.** $\sqrt{32}$

5. $\sqrt{50}$ **6.** $\sqrt{15}$ **7.** $\sqrt{192}$ **8.** $\sqrt{99}$

9. $\sqrt{70}$ **10.** $\sqrt{200}$ **11.** $\sqrt[3]{54}$ **12.** $\sqrt[3]{320}$

13. $\sqrt[3]{160}$ **14.** $\sqrt[4]{32}$ **15.** $\sqrt[4]{1000}$ **16.** $\sqrt[7]{512}$

B. Exercises

Simplify each radical that can be simplified.

17. $\sqrt{x^2}$ **18.** $\sqrt{x^8}$ **19.** $\sqrt{y^5}$ **20.** $\sqrt{y^{10}}$

21. $\sqrt{16x^2}$ **22.** $\sqrt{8x^6}$ **23.** $\sqrt{18x^2y}$ **24.** $\sqrt{4a^6b^4}$

25. $\sqrt{50x^3y^3}$ **26.** $\sqrt{20m^5n^5}$ **27.** $\sqrt{18x^7y^9}$ **28.** $\sqrt{9x^4y^2z^6}$

29. $\sqrt{75a^3bc^8}$ **30.** $\sqrt{90m^9n^{15}}$ **31.** $\sqrt{\frac{x^6y^6}{100}}$ **32.** $\sqrt[3]{8x^4y}$

33. $\sqrt[3]{15x^2yz^2}$ **34.** $\sqrt[4]{10x^5y^6z^7}$ **35.** $\sqrt[5]{32a^5b^{15}}$

Tsunamis are large ocean waves caused by earthquakes, volcanoes, or landslides on the ocean floor. They can reach speeds as high as 600 mi/hr in deep water and produce waves as tall as 120 ft when they hit the shore. The wave's speed, V, in miles per hour can be modeled using the formula $V = \frac{15}{22}\sqrt{gd}$, where g is the gravitational acceleration of 32 ft/sec^2 and d is the depth of the water in feet.

These before and after satellite photos demonstrate the destructive power of a tsunami.

Use the formula to determine the speed of a wave at the following depths.

36. 242 ft

37. 3872 ft

Skid marks can be used to estimate the speed a car was going before it skidded to a stop. The speed, S, in miles per hour can be modeled using the formula $S = \sqrt{30Df}$, where D is the length of the skid marks in feet and f is the coefficient of friction between the tires and the road.

Use the formula to determine the speed a car was going before it skidded to a stop.

38. skid marks: 40 ft; $f = 0.75$

39. skid marks: 160 ft; $f = 0.75$

40. skid marks: 160 ft; $f = \frac{1}{3}$

C. Exercises

Simplify each radical that can be simplified.

41. $\sqrt{x^2 - 9}$

42. $\sqrt[3]{432r^{17}s^{24}t^{19}}$

43. $3x^2y\sqrt[3]{405x^4y^2z^5}$

44. $19ab^4c^2\sqrt{1152a^9b^{11}c^4}$

45. $11xyz\sqrt[4]{2^5 \cdot 3^4 \cdot 5x^{12}y^6z^9}$

Dominion Modeling

The sequences studied in Section 11.1, Dominion Modeling, use a Fibonacci sequence to find decimal approximations of the golden ratio. The golden ratio is an irrational number with an exact value of $\phi = \frac{1 + \sqrt{5}}{2}$.

The golden ratio can be defined geometrically using the ratio of the lengths of line segments. If the ratio of the longer part of a segment to its shorter part equals the ratio of the total length of the segment to its longer part, the ratio is the golden ratio.

$$\phi = \frac{a}{b} = \frac{a + b}{a}$$

46. The golden ratio is an irrational number. What is an irrational number?

47. Use a calculator to approximate $\frac{1 + \sqrt{5}}{2}$ to the nearest millionth.

48. A line segment is divided so that $a = 144$ and $b = 89$.

a. Find the decimal equivalents of $\frac{a}{b}$ and $\frac{a + b}{a}$ to the nearest thousandth.

b. Is the segment divided into a golden ratio?

49. A line segment is divided so that $a = 3$ and $b = 2$.

a. Find the decimal equivalents of $\frac{a}{b}$ and $\frac{a + b}{a}$ to the nearest thousandth.

b. Is the segment divided into a golden ratio?

CUMULATIVE REVIEW

Evaluate. [2.2]

50. $3a^2 - 2b + 2c^3$ when $a = \frac{1}{2}$, $b = 2$, and $c = -\frac{1}{2}$

51. $x^2 + 5x + \sqrt{x} - 6$ when $x = 9$

State the degree of each term. [9.1]

52. $5xy^3$

53. $-7a^6b^3$

Factor. [10.2–10.3, 10.5]

54. $x^2 - 2x - 63$

55. $a^2 + 4a - 32$

56. $a^2 + 2ab + b^2$

57. $m^2 + 28m + 196$

58. $5x^2 + 10x + 5$

59. $12b^8 - 3b^6$

11.3 Multiplying Radicals

The Product Property of Radicals is also used to multiply radicals. Note that radicals must have the same index before they can be multiplied.

$$\sqrt[n]{x} \cdot \sqrt[n]{y} = \sqrt[n]{xy}$$

Example 1

Simplify $\sqrt{2} \cdot \sqrt{6}$.

Answer

$\sqrt{2} \cdot \sqrt{6} = \sqrt{2(2 \cdot 3)}$ — 1. Apply the Product Property, writing the radicand as the product of its prime factors.

$= \sqrt{2^2} \cdot \sqrt{3} = 2\sqrt{3}$ — 2. Simplify the square root of the perfect square.

Example 2

Simplify $-2\sqrt{18} \cdot 5\sqrt{50}$.

Answer

$-2\sqrt{18} \cdot 5\sqrt{50} = -2 \cdot 5\sqrt{18} \cdot \sqrt{50}$ — 1. Use the Commutative and Associative Properties to rearrange the factors.

$= -10\sqrt{(2 \cdot 3^2)(2 \cdot 5^2)}$ — 2. Apply the Product Property, writing the radicand as the product of its prime factors.

$= -10 \cdot 2 \cdot 3 \cdot 5 = -300$ — 3. Simplify the square root of each perfect square.

Unless otherwise stated, from here on we will assume that all variables in the radicand represent nonnegative values. Therefore, it will not be necessary to use absolute values to ensure that even roots of variables are positive.

Example 3

Simplify $\sqrt{6ab} \cdot \sqrt{2bc}$.

Answer

$\sqrt{6ab} \cdot \sqrt{2bc} = \sqrt{(3 \cdot 2)2ab^2c}$ — 1. Apply the Product Property, writing the radicand as the product of its prime factors.

$= 2b\sqrt{3ac}$ — 2. Simplify the square root of each perfect square.

Multiplying Radicals

1. Apply the Product Property, writing the radicand as the product of its prime factors.
2. Simplify the root of each perfect power.
3. Simplify the resulting expression.

Example 4

Simplify $\sqrt[3]{225} \cdot \sqrt[3]{315}$.

Answer

$\sqrt[3]{225} \cdot \sqrt[3]{315} = \sqrt[3]{(3^2 \cdot 5^2)(3^2 \cdot 5 \cdot 7)}$
$= \sqrt[3]{3^3 \cdot 3 \cdot 5^3 \cdot 7}$ — 1. Apply the Product Property, writing the radicand as the product of its prime factors.

$= 3 \cdot 5\sqrt[3]{3 \cdot 7}$ — 2. Simplify the cube root of each perfect cube.

$= 15\sqrt[3]{21}$ — 3. Simplify.

As you become more familiar with factors and perfect squares, you may recognize factors that are perfect squares under the radical. Simplifying each radical before multiplying can save you time and effort.

Example 5

Simplify $\sqrt{50} \cdot \sqrt{32}$.

Answer

$\sqrt{25 \cdot 2} \cdot \sqrt{16 \cdot 2}$ — 1. Apply the Product Property, writing the radicand as a product with perfect square factors.

$= 5\sqrt{2} \cdot 4\sqrt{2}$ — 2. Simplify the square root of each perfect square.

$= 20 \cdot 2 = 40$ — 3. Simplify.

Example 6

Simplify $\sqrt[3]{4x^2y^5z} \cdot \sqrt[3]{32xy}$.

Answer

$\sqrt[3]{4x^2y^5z} \cdot \sqrt[3]{32xy} = \sqrt[3]{2^2 \cdot 2^5x^3y^6z}$ — 1. Apply the Product Property, writing the radicand as the product of its prime factors.

$= \sqrt[3]{2^6 \cdot 2x^3y^6z}$ — 2. Simplify the cube root of each perfect cube.

$= 2^{\frac{6}{3}}x^{\frac{3}{3}}y^{\frac{6}{3}}\sqrt[3]{2z}$ — a. Group factors so the exponents are multiples of three (perfect powers).

$= 2^2xy^2\sqrt[3]{2z}$ — b. Divide the exponents of the perfect powers by the index, 3. Leave all other factors in the radicand.

$= 4xy^2\sqrt[3]{2z}$ — 3. Simplify.

More complicated expressions involving radicals can also be simplified using the Product Property by applying other previously established principles.

Example 7

Simplify.

a. $\sqrt{5}(\sqrt{3} - 2)$ **b.** $2\sqrt{6}(\sqrt{15} - 3\sqrt{10})$

Answer

a. $\sqrt{5}(\sqrt{3} - 2) = \sqrt{15} - 2\sqrt{5}$ — Apply the Distributive Property.

b. $2\sqrt{6}(\sqrt{15} - 3\sqrt{10})$

$= 2\sqrt{(2 \cdot 3)(3 \cdot 5)} - 6\sqrt{(2 \cdot 3)(2 \cdot 5)}$ — 1. Apply the Distributive Property, writing each radicand as the product of its prime factors.

$= 2 \cdot 3\sqrt{2 \cdot 5} - 6 \cdot 2\sqrt{3 \cdot 5}$ — 2. Simplify the square root of each perfect square.

$= 6\sqrt{10} - 12\sqrt{15}$ — 3. Simplify.

Heron's formula, named after the Greek mathematician Heron of Alexandria, can be used to calculate the area of a triangle directly from the lengths of its three sides and its semiperimeter, s:

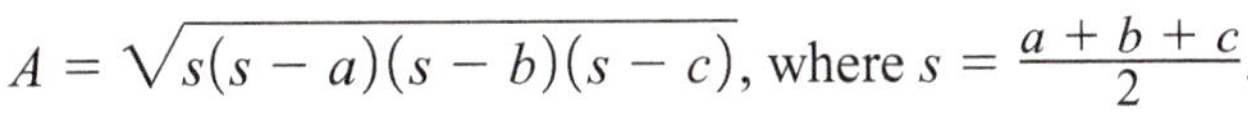

$$A = \sqrt{s(s-a)(s-b)(s-c)}, \text{ where } s = \frac{a+b+c}{2}.$$

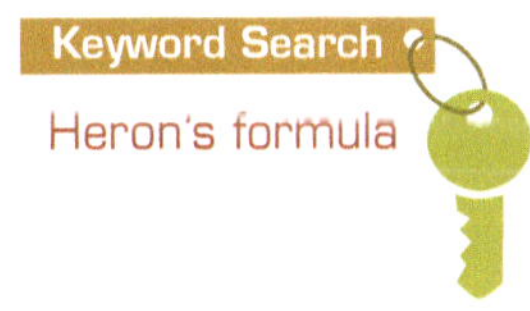

Example 8

The lengths of the sides of a triangle are 6 ft, 8 ft, and 10 ft. Using Heron's formula, find the area of the triangle.

Answer

$s = \frac{6 + 8 + 10}{2} = \frac{24}{2} = 12 \text{ ft}$ — 1. Calculate the triangle's semiperimeter.

$A = \sqrt{12(12-6)(12-8)(12-10)}$ — 2. Substitute known values into Heron's formula.

$= \sqrt{12(6)(4)(2)}$ — 3. Simplify.

$= \sqrt{(2^2 \cdot 3)(3 \cdot 2)(2^2)(2)}$

$= 2 \cdot 3 \cdot 2 \cdot 2 = 24 \text{ ft}^2$

A. Exercises

Simplify.

1. $\sqrt{5} \cdot \sqrt{3}$
2. $\sqrt{7} \cdot \sqrt{5}$
3. $\sqrt{8} \cdot \sqrt{3}$
4. $\sqrt{18} \cdot \sqrt{3}$
5. $\sqrt{12} \cdot \sqrt{5}$
6. $\sqrt{15} \cdot \sqrt{6}$
7. $\sqrt{5} \cdot \sqrt{10}$
8. $\sqrt{33} \cdot \sqrt{55}$
9. $\sqrt{28} \cdot \sqrt{35}$
10. $\sqrt{42} \cdot \sqrt{28}$
11. $\sqrt{5} \cdot \sqrt{118}$
12. $\sqrt{54} \cdot \sqrt{48}$
13. $\sqrt{3}(\sqrt{2} - 1)$
14. $\sqrt{5}(5 + \sqrt{3})$
15. $\sqrt{7}(\sqrt{21} + \sqrt{2})$
16. $\sqrt{6}(\sqrt{18} + \sqrt{2})$

B. Exercises

Simplify. Assume that all variables in the radicand represent nonnegative values.

17. $\sqrt[3]{40} \cdot \sqrt[3]{54}$
18. $\sqrt[3]{81} \cdot \sqrt[3]{64}$
19. $\sqrt[3]{56} \cdot \sqrt[3]{98}$
20. $\sqrt[4]{32} \cdot \sqrt[4]{64}$
21. $\sqrt{4ab} \cdot \sqrt{5b}$
22. $\sqrt{3x^2y^3} \cdot \sqrt{2xy}$
23. $\sqrt{2w^{15}z^3} \cdot \sqrt{8wz}$
24. $\sqrt{x^2y} \cdot \sqrt{xy^2} \cdot \sqrt{xy}$
25. $4\sqrt{8a^5c^7} \cdot \sqrt{6a^4c}$
26. $2x\sqrt{15x^3y} \cdot \sqrt{21x^4y^3}$
27. $\sqrt[3]{4x^5y^7} \cdot \sqrt[3]{2x^6y}$
28. $\sqrt[3]{25a^4bc^6} \cdot \sqrt[3]{10abc^3}$
29. $\sqrt[3]{10a^4b^3c^9} \cdot \sqrt[3]{50a^4b^2c}$
30. $\sqrt[4]{32x^3yz^6} \cdot \sqrt[4]{x^2y^2z^2}$
31. The side of a square measures $3\sqrt{5}$ in. What is the area of the square?
32. The width of a rectangle is $2\sqrt{7}$ ft, and its length is $5\sqrt{14}$ ft. Express the area of the rectangle as a simplified radical.
33. The width of a rectangle is $\sqrt[3]{x^4}$ units, and its length is $\sqrt[3]{x^5}$ units. Express the area of the rectangle in simplest terms.
34. The length of an edge of a cube is $\sqrt[3]{3x^5}$ units. Express the volume of the cube in simplest terms.
35. The lengths of the sides of a triangle are 3 m, 4 m, and 5 m. Calculate the area of the triangle using Heron's formula.

C. Exercises

36. Use Heron's formula to calculate the area of an equilateral triangle whose sides measure $2\sqrt{3}$ units.

Simplify. State each product in simplest form.

37. $\sqrt[3]{3249} \cdot \sqrt[3]{171}$
38. $\sqrt[4]{8000x^{13}y^{12}} \cdot \sqrt[4]{540x^{30}y^{22}}$

Radicals with different indices but the same radicand can be multiplied after they are converted to exponential form. Complete the following steps to multiply $\sqrt{x} \cdot \sqrt[3]{x}$.

39. Convert each radical to exponential form.
40. Multiply the exponential form of the expressions.
41. Express the product as a radical.

Dominion Modeling

Mathematicians, artists, scientists, engineers, and even philosophers have long been intrigued by the many interesting definitions and unique characteristics of the golden ratio. By expressing the golden ratio as a continued fraction, we can illustrate one of its unique characteristics.

It can be shown that $\phi = 1 + \cfrac{1}{1 + \cfrac{1}{1 + \cfrac{1}{1 + \cfrac{1}{1 + \dots}}}}$.

Because this fraction is infinite, the denominator of its second term is equal to ϕ and we can substitute ϕ for this denominator.

$$\phi = 1 + \frac{1}{\phi}$$

This can be rearranged to state that

$$\frac{1}{\phi} = \phi - 1.$$

In other words, the reciprocal of the golden ratio can be obtained by simply subtracting one.

42. Simplify each expression, writing your answer as an improper fraction. Then use a calculator to find each decimal approximation to the nearest thousandth.

a. $1 + \frac{1}{2}$

b. $1 + \cfrac{1}{1 + \frac{1}{2}}$

c. $1 + \cfrac{1}{1 + \cfrac{1}{1 + \frac{1}{2}}}$

d. $1 + \cfrac{1}{1 + \cfrac{1}{1 + \cfrac{1}{1 + \frac{1}{2}}}}$

43. Predict the next three improper fractions in the sequence of expressions suggested by the answers to the previous exercise. Then find their decimal approximations to the nearest thousandth.

44. Use a calculator to evaluate $\frac{1}{\phi} = \frac{2}{1 + \sqrt{5}}$. Round to the nearest millionth.

45. Use a calculator to evaluate $\frac{1 + \sqrt{5}}{2} - \frac{2}{1 + \sqrt{5}}$.

CUMULATIVE REVIEW

Solve. [2.7]

46. $5x - 11 = x + 7$

47. $3(2n - 6) - 2n(3 + 4) = 4n - 6$

Simplify. [8.2, 9.6]

48. $\frac{21x^7}{3x^3}$

49. $\frac{-2r^3s^5t}{14rs^4t}$

50. $\frac{24a^5b^{-3}c}{12a^{-2}b^4c^{-1}}$

51. $\left(\frac{2x^{-3}}{x^2}\right)^{-2}$

52. $\frac{8x^4y + 16xy}{4xy}$

53. $\frac{n^5 + 5n^3 - n^2}{n^2}$

Convert each radical to exponential form. [11.1]

54. $\sqrt[4]{x^2y^3z^4}$

55. $\sqrt[8]{a^7b^4c^2}$

SEQUENCES

Summation (Sigma) Notation

Challenge **Can you find the given partial sum for each sequence A_n?**

a. S_5 for $A_n = 3n - 1$

b. S_4 for $A_n = 2^{n-1}$

c. S_3 for $A_n = n^2$

The Greek letter sigma (Σ) can be used instead of S_n to represent the sum of the terms of a sequence. In general, $\sum_{k=1}^{n} A_k = S_n$, where A_k is the general-term formula for any sequence and k is a variable used to count from 1 to n. The three Challenge sequences can be restated with this new notation as follows.

a. $\sum_{k=1}^{5}(3k - 1)$ **b.** $\sum_{k=1}^{4} 2^{k-1}$ **c.** $\sum_{k=1}^{3} k^2$

When n is small, evaluate the sum by substituting into the sequence formula. The fifth partial sum $S_5 = \sum_{k=1}^{5}(3k - 1) = 2 + 5 + 8 + 11 + 14 = 40$. When n is large, use a general partial-sum formula if one exists. To find $\sum_{k=1}^{25}(3k - 1)$, use the arithmetic partial-sum formula $S_n = \frac{n(A_1 + A_n)}{2}$ to get $S_{25} = \frac{25(2 + 74)}{2} = 950$.

Finding $\sum_{k=1}^{4} 2^{k-1} = 1 + 2 + 4 + 8 = 15$ involves four substitutions into the geometric sequence general-term formula. The *geometric partial-sum formula* $S_n = \frac{A_1(1 - r^n)}{1 - r}$ is developed in upper-level math classes using deductive reasoning. To find $S_{14} = \sum_{k=1}^{14} 2^{k-1}$, note that $r = 2$ and apply the formula: $S_{14} = \frac{1(1 - 2^{14})}{1 - 2} = 16{,}383$. If $|r| < 1$, find the infinite geometric sum using the formula $S_\infty = \frac{A_1}{1 - r}$. To find $\sum_{k=1}^{\infty}\left(\frac{1}{2}\right)^{k-1} = 1 + \frac{1}{2} + \frac{1}{4} + \frac{1}{8} + \ldots$, note that $A_1 = \left(\frac{1}{2}\right)^{1-1} = 1$ and $r = \frac{1}{2}$ and apply the formula: $S_\infty = \frac{1}{1 - 0.5} = 2$.

Exercises

Find the given partial sum by substituting into the sequence general-term formula.

1. $\sum_{k=1}^{2}(2k + 1)$ **2.** $\sum_{k=1}^{3}\left(\frac{1}{2}\right)^k$ **3.** $\sum_{k=1}^{4} k^3$

Find the given partial sum by using the appropriate sum formula.

4. $\sum_{k=1}^{40}(2k + 1)$ **5.** $\sum_{k=1}^{11} 4^k$ **6.** $\sum_{k=1}^{\infty}\left(-\frac{1}{2}\right)^k$

Consider the partial sum $\sum_{k=1}^{n}\frac{1}{k(k + 1)}$.

7. Find A_1, A_2, A_3, and A_4. **8.** Find S_1, S_2, S_3, and S_4.

9. State the general partial-sum formula (notice the pattern in exercise 8). Then find S_{20}.

11.4 Dividing Radicals

A farmer determines the sprinkler radius for circular irrigation as $r = \sqrt{\frac{A}{\pi}}$. Rationalize the denominator.

The Product Property of Radicals, $\sqrt[n]{xy} = \sqrt[n]{x} \cdot \sqrt[n]{y}$, was useful for multiplying and simplifying radicals. A similar property applies to division.

Quotient Property of Radicals

For $x \geq 0$ and $y > 0$, $\sqrt[n]{\frac{x}{y}} = \frac{\sqrt[n]{x}}{\sqrt[n]{y}}$.

The numerical example below demonstrates how a square root can be found with and without the use of the Quotient Property.

$$\sqrt{\frac{16}{25}} = \frac{4}{5} \text{ and } \frac{\sqrt{16}}{\sqrt{25}} = \frac{4}{5}, \text{ so } \sqrt{\frac{16}{25}} = \frac{\sqrt{16}}{\sqrt{25}}.$$

Division problems are often expressed as fractions, the most useful form for working with radicals.

Example 1

Simplify $\sqrt{40} \div \sqrt{5}$.

Answer

$\frac{\sqrt{40}}{\sqrt{5}} = \sqrt{\frac{40}{5}}$ — 1. Write the division as a fraction and apply the Quotient Property of Radicals.

$= \sqrt{8}$ — 2. Simplify the radicand.

$= 2\sqrt{2}$ — 3. Simplify the radical.

Example 2

Simplify $\frac{\sqrt{15}}{\sqrt{12}}$.

Answer

$\frac{\sqrt{15}}{\sqrt{12}} = \sqrt{\frac{15}{12}} = \sqrt{\frac{5}{4}}$ — 1. Apply the Quotient Property and reduce the radicand.

$= \frac{\sqrt{5}}{\sqrt{4}} = \frac{\sqrt{5}}{2}$ — 2. Apply the Quotient Property to simplify the denominator.

Fractions with radicals in the denominator or radicals with fractions in the radicand are not considered to be simplified. Neither the fraction $\frac{\sqrt{5}}{\sqrt{7}}$ nor the radical $\sqrt{\frac{5}{7}}$ is simplified. These expressions cannot be simplified using the Quotient Property, as $\frac{\sqrt{40}}{\sqrt{5}}$ and $\frac{\sqrt{15}}{\sqrt{12}}$ were in the previous examples.

Definition

The process of removing radicals from the denominator of a fraction is called **rationalizing the denominator**.

During this process, the radical is multiplied by an expression equal to one that changes the denominator from an irrational radical to an integer.

Example 3

Simplify $\frac{\sqrt{5}}{\sqrt{7}}$.

Answer $\frac{\sqrt{5}}{\sqrt{7}} = \frac{\sqrt{5}}{\sqrt{7}} \cdot \frac{\sqrt{7}}{\sqrt{7}}$ 1. Multiply by one in the form of $\frac{\sqrt{7}}{\sqrt{7}}$.

$= \frac{\sqrt{35}}{(\sqrt{7})^2} = \frac{\sqrt{35}}{7}$ 2. Simplify the denominator.

Example 4

Simplify $\sqrt{3} \div \sqrt{6}$.

Answer $\frac{\sqrt{3}}{\sqrt{6}} = \sqrt{\frac{3}{6}} = \sqrt{\frac{1}{2}}$ 1. Apply the Quotient Property and reduce the radicand.

$= \frac{\sqrt{1}}{\sqrt{2}} \cdot \frac{\sqrt{2}}{\sqrt{2}} = \frac{\sqrt{2}}{2}$ 2. Apply the Quotient Property again to rewrite the fraction; then multiply by one in the form of $\frac{\sqrt{2}}{\sqrt{2}}$.

If you chose to rationalize the denominator first in Example 4, you would need to simplify the resulting radical and then reduce the fraction.

Example 5

Simplify $\frac{\sqrt{2}}{\sqrt{75}}$.

Answer $\frac{\sqrt{2}}{\sqrt{75}} = \frac{\sqrt{2}}{5\sqrt{3}}$ 1. Simplify radicals where possible: $\sqrt{75} = \sqrt{3 \cdot 5^2} = 5\sqrt{3}$.

$= \frac{\sqrt{2}}{5\sqrt{3}} \cdot \frac{\sqrt{3}}{\sqrt{3}} = \frac{\sqrt{6}}{5 \cdot 3} = \frac{\sqrt{6}}{15}$ 2. Rationalize the denominator.

Example 6

Simplify $\frac{25\sqrt{3}}{9\sqrt{15}}$.

Answer $\frac{25\sqrt{3}}{9\sqrt{15}} = \frac{25\cancel{\sqrt{3}}}{9\cancel{\sqrt{3}}\sqrt{5}} = \frac{25}{9\sqrt{5}}$ 1. Reduce the fraction by canceling the common radical factor.

$= \frac{25}{9\sqrt{5}} \cdot \frac{\sqrt{5}}{\sqrt{5}} = \frac{\overset{5}{\cancel{25}}\sqrt{5}}{9 \cdot \cancel{5}} = \frac{5\sqrt{5}}{9}$ 2. Rationalize the denominator and reduce the resulting fraction.

Expressions with roots other than square roots are rationalized using similar methods. However, it will not help to change the radicand in the denominator into a perfect square. If you are taking a cube root, the radicand must be a perfect cube. For a fourth root, the radicand must be a perfect fourth power, and so on.

Rationalizing the Denominator

1. Write the radicals in fractional form, $\frac{\sqrt[n]{a}}{\sqrt[n]{b}}$.
2. Simplify both the numerator and the denominator as much as possible.
3. If a radical is left in the denominator, multiply by the form of one that makes the radicand in the denominator a perfect power.
4. Simplify both the numerator and the denominator and reduce the fraction if necessary.

Example 7

Simplify $\frac{\sqrt[3]{8}}{\sqrt[3]{16}}$.

Answer

$$\frac{\sqrt[3]{8}}{\sqrt[3]{16}} = \sqrt[3]{\frac{8}{16}} = \sqrt[3]{\frac{1}{2}} = \frac{1}{\sqrt[3]{2}}$$

1. Apply the Quotient Property and reduce the radicand; then apply the Quotient Property again to rewrite the fraction.

$$= \frac{1}{\sqrt[3]{2}} \cdot \frac{\sqrt[3]{2^2}}{\sqrt[3]{2^2}} = \frac{\sqrt[3]{2^2}}{\sqrt[3]{2^3}}$$

2. Since the radical is a cube root, you need three factors of 2 to get a perfect cube.

$$= \frac{\sqrt[3]{4}}{2}$$

3. Simplify.

Example 8

Simplify $\sqrt{18a^2bc^2} \div \sqrt{9ab^3c^3}$.

Answer

$$\frac{\sqrt{18a^2bc^2}}{\sqrt{9ab^3c^3}} = \sqrt{\frac{18a^2bc^2}{9ab^3c^3}} = \sqrt{\frac{2a}{b^2c}}$$

1. Apply the Quotient Property and reduce the radicand.

$$= \frac{\sqrt{2a}}{b\sqrt{c}}$$

2. Apply the Quotient Property again and simplify the radical in the denominator.

$$= \frac{\sqrt{2a}}{b\sqrt{c}} \cdot \frac{\sqrt{c}}{\sqrt{c}} = \frac{\sqrt{2ac}}{bc}$$

3. Rationalize the denominator.

A. Exercises

Simplify.

1. $\frac{\sqrt{4}}{\sqrt{25}}$
2. $\frac{\sqrt{80}}{\sqrt{20}}$
3. $\sqrt{5} \div \sqrt{4}$
4. $\sqrt{15} \div \sqrt{5}$
5. $\frac{\sqrt{8}}{\sqrt{4}}$
6. $\frac{6\sqrt{3}}{\sqrt{9}}$
7. $\frac{\sqrt{10}}{3\sqrt{5}}$
8. $\frac{2}{\sqrt{3}}$
9. $\frac{3}{\sqrt{13}}$
10. $\frac{3}{\sqrt{5}}$
11. $\frac{1}{\sqrt{6}}$
12. $\sqrt{3} \div \sqrt{2}$
13. $\sqrt{14} \div \sqrt{21}$
14. $\frac{\sqrt{114}}{\sqrt{32}}$
15. $\frac{3\sqrt{5}}{\sqrt{15}}$
16. $\frac{3\sqrt{4}}{\sqrt{32}}$

B. Exercises

Simplify.

17. $\sqrt[3]{\frac{1}{2}}$
18. $\sqrt[3]{\frac{1}{9}}$
19. $\frac{\sqrt[3]{9}}{\sqrt[3]{36}}$
20. $\frac{\sqrt[3]{7}}{\sqrt[3]{4}}$
21. $\frac{\sqrt[3]{5}}{\sqrt[3]{25}}$
22. $\frac{\sqrt[3]{5}}{\sqrt[3]{12}}$

Perform the indicated operations and express each answer in simplest form.

23. $\sqrt{8x^4y^3} \div \sqrt{14x^3y^5}$
24. $\sqrt{5xy} \div \sqrt{10x^2y^3}$
25. $\sqrt{r^3s^2t} \div \sqrt{rst}$
26. $\sqrt{6mn} \div \sqrt{3mn^2}$
27. $(\sqrt{2tx^2z^3} \cdot \sqrt{7txz^2}) \div \sqrt{10t^3x^2}$
28. $\sqrt{15x^4y^2} \div (\sqrt{xyz} \cdot \sqrt{3y^2})$
29. $\sqrt[3]{15x^2y^4} \div \sqrt[3]{5x^3y}$
30. $\sqrt[3]{9a^3b^4c} \div \sqrt[3]{27ab^5c}$
31. The area of a rectangle is $36\sqrt{5}$ in.^{2}, and its length is $4\sqrt{15}$ in. Express the width of the rectangle as a simplified radical.
32. The area of a triangle is $15\sqrt{2}$ cm^2, and its base is $2\sqrt{5}$ cm. Express the height of the triangle as a simplified radical.

The body surface area (BSA) of an adult is used to determine the dosage of some medicines. The Mosteller formula, BSA $= \sqrt{\frac{HW}{3600}}$, gives the BSA in square meters, where H is the person's height in centimeters and W is the person's mass in kilograms.

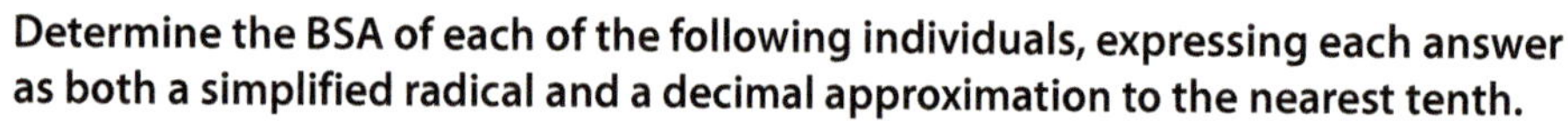

Determine the BSA of each of the following individuals, expressing each answer as both a simplified radical and a decimal approximation to the nearest tenth.

33. 180 cm; 60 kg
34. 175 cm; 72 kg

C. Exercises

Simplify.

35. $(\sqrt{18x^2y^4} \cdot \sqrt{9xy^5}) \div (\sqrt{15x^4y^2} \cdot \sqrt{6xy^9})$
36. $(2\sqrt{8a^3b^2} \cdot 3\sqrt{4ab^3}) \div (\sqrt{24a^5b} \cdot \sqrt{6a^3b^9})$
37. $\sqrt[4]{4096} \div \sqrt[3]{512}$

Simplify by converting to exponential form and expressing rational exponents with a denominator of four.

38. Express $\sqrt{180a^3b^7}$ as a fourth root. Then simplify the fourth root.
39. $\sqrt[4]{56x^3y^4} \div \sqrt{18x^3y^2}$

Dominion Modeling

In Section 11.3, Dominion Modeling, you saw how an interesting definition of ϕ led to one of the golden ratio's unique characteristics, that its reciprocal can be obtained by subtracting one. Another interesting definition of the golden ratio uses an infinite nested radical.

$$\phi = \sqrt{1 + \sqrt{1 + \sqrt{1 + \sqrt{1 + \ldots}}}}$$

Squaring both sides produces

$$\phi^2 = 1 + \sqrt{1 + \sqrt{1 + \sqrt{1 + \ldots}}}.$$

Since $\phi = \sqrt{1 + \sqrt{1 + \sqrt{1 + \ldots}}}$, we can substitute to produce

$$\phi^2 = 1 + \phi.$$

A second unique characteristic of ϕ is that its square can be obtained by simply adding one.

40. Use a calculator to find the decimal approximation of $\sqrt{1 + \sqrt{1 + \sqrt{1 + \sqrt{1 + \sqrt{1 + \sqrt{1}}}}}}$ to the nearest thousandth.

41. Use a calculator to evaluate $\left(\frac{1 + \sqrt{5}}{2}\right)^2$. Round to the nearest millionth.

42. Use a calculator to evaluate $\left(\frac{1 + \sqrt{5}}{2}\right)^2 - \frac{1 + \sqrt{5}}{2}$.

CUMULATIVE REVIEW

State the slope of each line. [6.5]

43. a line perpendicular to the graph of $y = 3x - 7$

44. a line parallel to the graph of $y = -\frac{2}{3}x + 4$

Graph each inequality. [6.7]

45. $y > -x + 3$

46. $2x + 3y \le 12$

Simplify. [9.2]

47. $7x^2 - 4xy - 5y - 3x^2 + 6y$

48. $(a^3 + 4b^2 - 3c) + (2a^3 - 10c)$

49. $(2x^3y + x^2y^3 - 4xy) - (12x^2y^3 + 15)$

Factor. [10.3, 10.5]

50. $3x^2 + 16x + 5$

51. $2b^2 - 28b + 98$

52. $9x^4 + 6x^2y^2 + y^4$

Solve each equation for the variable. Write the solution in the appropriate squares in the grid.

■	1	2	3	■	4	5		■
6	■	7		■	8		■	9
10	11	■	12			■	13	
14		15		■	16	17		
■	18		■	■	■	19		■
20			21	■	22			23
24		■	25			■	26	
	■	27		■	28	29	■	
■	30			■	31			■

ACROSS

1. $5x + 35 = 3160$
4. $3y - 105 = 342$
7. $6x^2 - 672 = x^2 + 32(979)$
8. $3x - (8 - x) = 2x(5) - 242$
10. $9x - 16 = 227$
12. $\frac{x}{33} + \frac{x}{9} = 42$
13. $x - 100 = -30$
14. $\frac{x}{28} = 157$
16. $\frac{x}{6} + 27x = 263{,}245$
18. $\frac{x}{2} + 10(x - 3) = 568.5$
19. $x^2 = 144$
20. $13x - 15{,}079 = 52{,}755$
22. $2y - 583 = 4293$
24. $\sqrt{4} - 222 = -5x$
25. $\frac{x}{5} + \frac{x}{25} = 54$
26. $z = \sqrt{625} + 2\sqrt{100}$
27. $2^7 = k + 29$
28. $x^2 - 168 = 3313$
30. $16(2\sqrt[3]{27}) = x - 28$
31. $x = 8(100) + 3$

DOWN

2. $2x = 7 \cdot 8$
3. $x + 599 = 75^2$
4. $10\sqrt{4900} = x - 26^2 - 3$
5. $x - 1 = \sqrt{64} \cdot 6$
6. $25^2 - x = 1$
9. $x^2 = 160{,}000$
11. $x \cdot 56^0 = 73{,}524$
13. $40^3 + 25^3 = x + 19^2 + 30$
15. $3x = 2913$
17. $\frac{1}{4}x = \frac{921}{6}$
20. $24^2 = x + 3^3$
21. $t - 669 = 85^2 + 20^2$
22. $50^2 = g - 58$
23. $\frac{6}{z} = \frac{1}{71} \cdot \frac{1}{2}$
27. $10^2 - 2^3 = e$
29. $2 \cdot 3^2 \cdot 5 = z$

11.5 Adding and Subtracting Radicals

The time, t, required to free-fall d feet is $t = \frac{\sqrt{d}}{4}$. How much longer does it take to fall 40 ft than to fall 10 ft?

When adding or subtracting terms that contain radicals, you must be sure they are like radicals. Adding and subtracting like radicals is quite similar to adding and subtracting variable expressions. The Distributive Property is used to add two expressions, such as $2x + 3x$.

$$2x + 3x = (2 + 3)x = 5x$$

Similarly, like radicals are added or subtracted using the Distributive Property.

$$2\sqrt{5} + 3\sqrt{5} = (2 + 3)\sqrt{5} = 5\sqrt{5}$$

Definition

Like radicals are radicals that have the same radicand and the same index.

Adding and Subtracting Radicals

1. Simplify each radical.
2. Combine like radicals by adding or subtracting their rational factors.

Example 1

Simplify $4\sqrt{3} + 2\sqrt{3}$.

Answer $4\sqrt{3} + 2\sqrt{3}$
$= (4 + 2)\sqrt{3} = 6\sqrt{3}$

Since each radical is already simplified, combine the like radicals by adding their rational factors.

Example 2

Simplify $\sqrt{45} + \sqrt{18} - \sqrt{5}$.

Answer $\sqrt{3^2 \cdot 5} + \sqrt{3^2 \cdot 2} - \sqrt{5}$ — 1. Simplify each radical that can be simplified.
$= 3\sqrt{5} + 3\sqrt{2} - \sqrt{5}$
$= 2\sqrt{5} + 3\sqrt{2}$ — 2. Combine like radicals. Remember that $3\sqrt{5} - \sqrt{5} = (3 - 1)\sqrt{5}$.

Can $\sqrt{5} + \sqrt{2}$ be combined? Are they like radicals? Notice that $\sqrt{9} + \sqrt{16} = 3 + 4 = 7$, but $\sqrt{25} = 5$. Therefore, $\sqrt{9} + \sqrt{16} \neq \sqrt{25}$.
In general, $\sqrt{x} + \sqrt{y} \neq \sqrt{x + y}$.

Example 3

Simplify $-8\sqrt[3]{5} + 9\sqrt[3]{40} - 2\sqrt[3]{135}$.

Answer

$-8\sqrt[3]{5} + 9\sqrt[3]{2^3 \cdot 5} - 2\sqrt[3]{3^3 \cdot 5}$ — 1. Simplify each radical.

$= -8\sqrt[3]{5} + 18\sqrt[3]{5} - 6\sqrt[3]{5}$

$= 4\sqrt[3]{5}$ — 2. Combine like radicals.

Example 4

Simplify $2x\sqrt{50x} - 5\sqrt{8x^3}$.

Answer

$2x\sqrt{5^2 \cdot 2x} - 5\sqrt{2^2 \cdot 2x^2x}$ — 1. Simplify each radical.

$= 2x \cdot 5\sqrt{2x} - 5 \cdot 2x\sqrt{2x}$

$= 10x\sqrt{2x} - 10x\sqrt{2x}$

$= 0$ — 2. Combine like radicals.

Remember that only like terms, those that have identical variable and radical factors, can be combined.

Example 5

Simplify $\sqrt{2x^2} + \sqrt{75x^4} - 2x^2\sqrt{3} - x\sqrt{18}$.

Answer

$\sqrt{2x^2} + \sqrt{3 \cdot 25x^4} - 2x^2\sqrt{3} - x\sqrt{2 \cdot 9}$ — 1. Simplify each radical. You may recognize factors that are perfect squares.

$= x\sqrt{2} + 5x^2\sqrt{3} - 2x^2\sqrt{3} - 3x\sqrt{2}$

$= -2x\sqrt{2} + 3x^2\sqrt{3}$ — 2. Combine like terms.

A. Exercises

Simplify.

1. $\sqrt{3} + 5\sqrt{3}$
2. $2\sqrt{6} + 3\sqrt{6}$
3. $\sqrt{3} + 4\sqrt{2} + \sqrt{3}$
4. $2\sqrt{5} - 6\sqrt{5} + 3\sqrt{7}$
5. $\sqrt{8} + 3\sqrt{2}$
6. $\sqrt{20} + 4\sqrt{5}$
7. $5\sqrt{12} - 6\sqrt{3}$
8. $\sqrt{12} - \sqrt{48}$
9. $\sqrt{8} + 3\sqrt{2} - 4\sqrt{2}$
10. $-3\sqrt{3} + \sqrt{18} + \sqrt{27}$
11. $\sqrt[3]{54} + 5\sqrt[3]{2}$
12. $\sqrt[3]{24} - \sqrt[3]{375}$

Complete the following table.

	x	y	$\sqrt{x}$	$\sqrt{y}$	$\sqrt{x} + \sqrt{y}$	$\sqrt{x + y}$
13.	25	144				
14.	27	12				
15.	45	75				

16. Based on the results of exercises 13–15, what conclusion can you draw about the expressions $\sqrt{x} + \sqrt{y}$ and $\sqrt{x + y}$?

B. Exercises

Simplify.

17. $\sqrt{5} + \sqrt{20} - \sqrt{72}$

18. $4\sqrt{2} - 6\sqrt{72} + 8\sqrt{32}$

19. $\sqrt[3]{16} + \sqrt[3]{18}$

20. $\sqrt[3]{40} - \sqrt[3]{875}$

21. $\sqrt[3]{16} + \sqrt[3]{81} + \sqrt[3]{250}$

22. $3\sqrt[3]{108} - 2\sqrt[3]{32} + 4\sqrt[3]{64}$

23. $\sqrt{x} + 3\sqrt{xy} + 5\sqrt{x}$

24. $4\sqrt{x} + 3\sqrt{y} - 2\sqrt{x}$

25. $a\sqrt{3} - 3a\sqrt{3} + \sqrt{2}$

26. $3\sqrt{ab} + 4\sqrt{a} - 5\sqrt{b} + 5\sqrt{b}$

27. $\sqrt{x^2y} - x\sqrt{y}$

28. $\sqrt{32a^3} + \sqrt{2a^3} - 3\sqrt{6}$

29. $\sqrt{\frac{x^2}{16}} + \sqrt{\frac{x^2}{9}}$

30. $\sqrt{\frac{x^2}{16} + \frac{x^2}{9}}$

31. The width of a rectangle is $2\sqrt{3}$ units, and its length is three times longer than its width. Express the perimeter of the rectangle as a simplified radical.

32. Express the perimeter of the polygon to the right as a simplified radical expression.

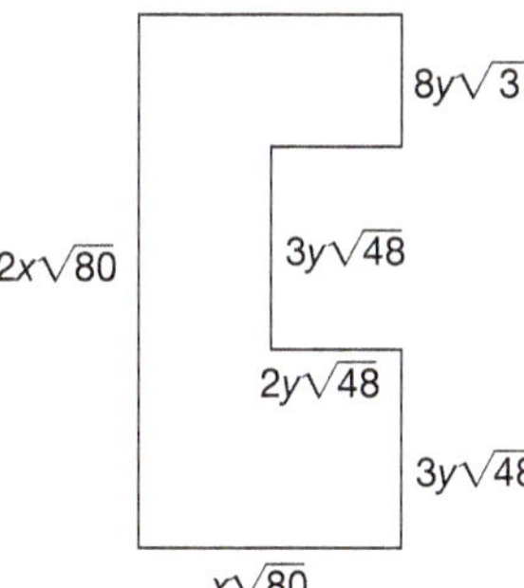

C. Exercises

33. Use Heron's formula (p. 443) to calculate the area of a triangle whose sides are $4\sqrt{6}$ cm, $3\sqrt{6}$ cm, and $6\sqrt{6}$ cm.

34. Express the area of the rectangle to the right as a simplified radical expression.

Simplify.

35. $2\sqrt[3]{16x^6y^4} - xy\sqrt[3]{2x^3y}$

36. $2y\sqrt[3]{72x^5y^2} - 3x\sqrt[3]{243x^2y^5}$

37. $\sqrt{x^3 - 8x^2 + 16x} + \sqrt{64x^3} + \sqrt{x^2 + 4x + 4}$

Simplify by converting to exponential form with a common denominator. Express each answer as a simplified radical.

38. $\sqrt{a^3b} + a\sqrt[6]{a^3b^3}$

39. $5y\sqrt[3]{x^5y} - 3\sqrt[6]{x^{10}y^8}$

Dominion Modeling

The Fibonacci numbers, the golden ratio, and the golden spiral all appear in God's creation, but people have also used these mathematical relationships to create works of art and beautiful architectural structures. The Fibonacci sequence can be used to approximate both the golden ratio and a golden spiral.

The chambered shell of Nautilus pompilius *is well known for its spiral shape.*

40. To create a figure that approximates a golden spiral, draw a series of squares on graph paper using the numbers of the Fibonacci sequence as their side lengths, as illustrated in the first diagram to the right. Then draw a quarter-circle in each successive square to create a Fibonacci spiral that approximates the golden spiral.

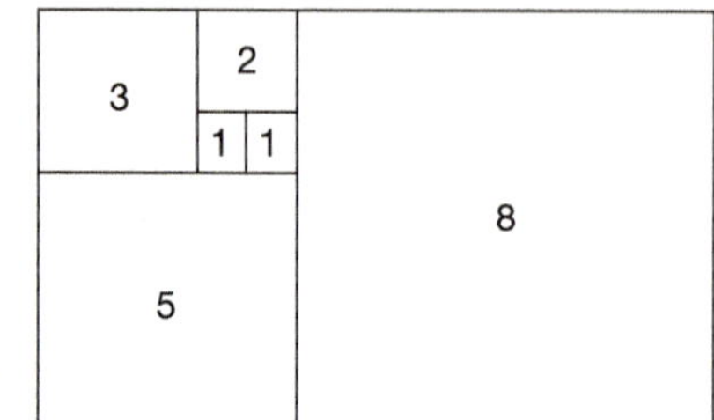

41. Express the length-to-width ratio of the large rectangle containing the spiral as an improper fraction. Then find its decimal equivalent.

42. Add another square to the figure from exercise 40 and extend the spiral through this square. Express the length-to-width ratio of this new largest rectangle as both an improper fraction and a decimal approximation to the nearest thousandth.

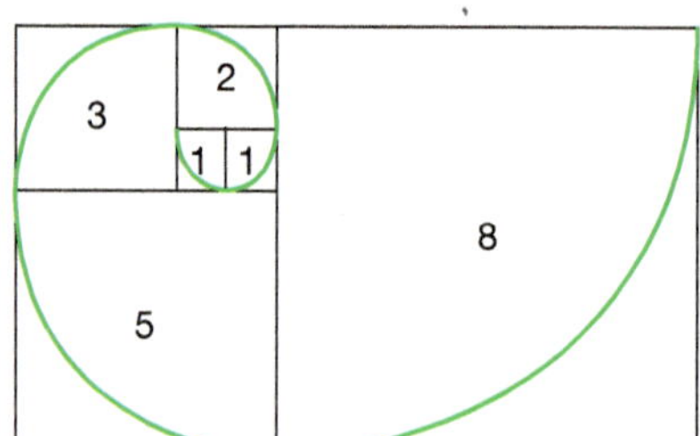

43. Add the next square in the sequence and extend the spiral through this square. Express the length-to-width ratio of this new largest rectangle as both an improper fraction and a decimal approximation to the nearest thousandth.

These Fibonacci rectangles can be used to approximate a *golden rectangle*, which has a length-to-width ratio equal to ϕ, the golden ratio. If a square is added to the width or cut from the length of a golden rectangle, the result is another golden rectangle.

CUMULATIVE REVIEW

Evaluate. [1.8]

44. $3 + 5(2 + 6)$

45. $-9 - [-8 + (3 \cdot 2 + 6)] + 4 + 2 \cdot 5$

Solve. [2.7]

46. $3x - \sqrt{18} = x + \sqrt{50}$

47. $5(x - 8) = 2x - 5x$

A segment has endpoints with the given coordinates. Find the length of the segment and the midpoint of the segment. [1.2]

48. 3 and 12

49. -8 and 14

Factor. [10.4]

50. $x^2 - 32x + 256$

51. $4x^2 - 16y^2$

Divide. [9.6]

52. $(x^2 - 2x - 35) \div (x + 5)$

53. $(6x^2 + 11x - 31) \div (3x - 5)$

11.6 The Pythagorean Theorem

Ancient Egyptian farmers divided their farmland into plots with square corners. Each year when the Nile River flooded, new boundary lines had to be marked and these right angles had to be found again. The Egyptians discovered that they could form a right angle by making a triangle whose sides measured 3, 4, and 5 units. The angle opposite the side 5 units long was always a right angle.

Definitions

A **right angle** is an angle whose measure is 90°.

A **right triangle** is a triangle that has a right angle.

A group of Greek philosophers and mathematicians led by Pythagoras (582–500 BC) studied the Egyptians' use of triangles. Pythagoras and his followers believed that, in any right triangle, the sum of the squares of the lengths of the *legs* (the two shorter sides) is equal to the square of the length of the *hypotenuse* (the side opposite the right angle). Their work popularized this property of right triangles, which is now called the Pythagorean theorem.

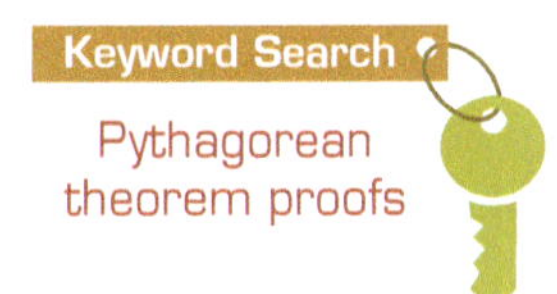

Definitions

A **theorem** is a statement that has been proved to be always true.

Pythagorean theorem: The sum of the squares of the lengths of the legs of a right triangle is equal to the square of the length of the hypotenuse. In the illustrated right triangle, $a^2 + b^2 = c^2$.

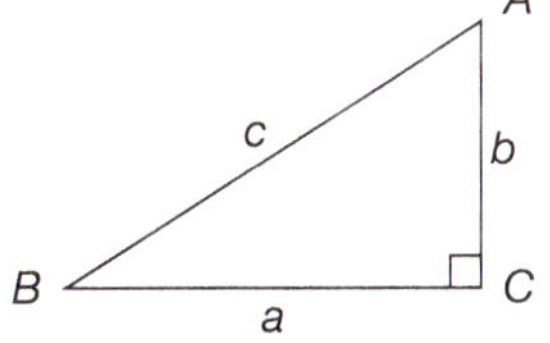

Example 1

Find the length of side b in right triangle ABC.

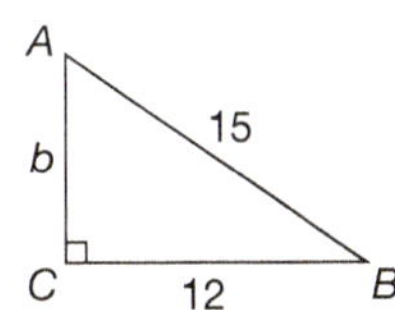

Answer

b = the unknown leg length; $c = 15$, the hypotenuse; and $a = 12$	1. Identify the known lengths of the sides and assign variables, making sure that the hypotenuse is c.
$a^2 + b^2 = c^2$ $12^2 + b^2 = 15^2$	2. Substitute values into the Pythagorean theorem.
$144 + b^2 = 225$ $b^2 = 81$	3. Solve for b^2.
$\lvert b \rvert = 9$ $b = \pm 9$	4. Take the square root of each side.
Therefore, $b = 9$ units.	5. Since b represents a length, use only the positive solution.

Solving an Equation of the Form $x^2 = c$

To solve an equation of the form $x^2 = c$, take the square root of both sides of the equation: $x = \pm\sqrt{c}$.

Example 2

Find the length of the hypotenuse in right triangle ABC.

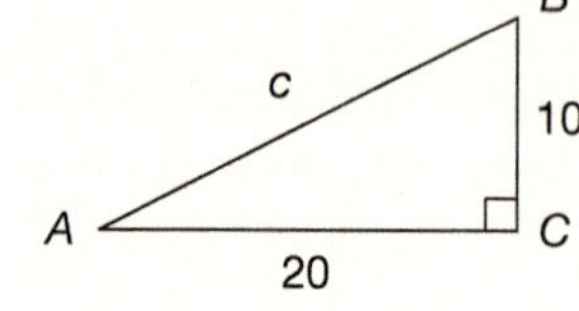

Answer

$a = 10, b = 20$, and
$c =$ the length of the hypotenuse

1. Identify the known lengths and assign variables.

$$a^2 + b^2 = c^2$$
$$10^2 + 20^2 = c^2$$
$$100 + 400 = c^2$$
$$c^2 = 500$$

2. Substitute values into the Pythagorean theorem and solve for c^2.

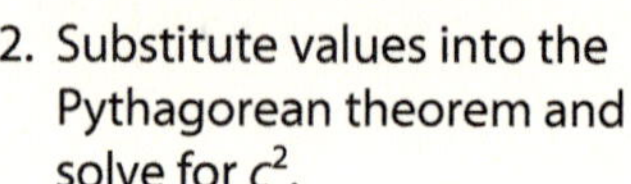

$$c = \pm\sqrt{500} = \pm 10\sqrt{5}$$

3. Take the square root of each side and simplify.

Therefore, $c = 10\sqrt{5}$ units.

4. Since c represents a length, use only the positive solution.

Example 3

Use the Pythagorean theorem to find the distance between $P_1\,(-3, 1)$ and $P_2\,(5, 4)$.

Answer

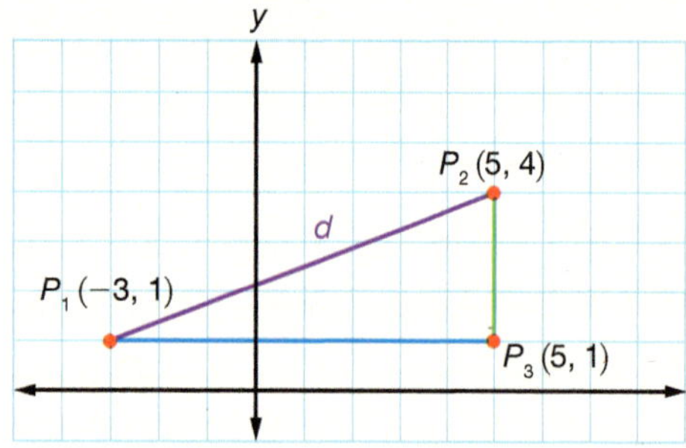

1. Graph the points and draw the segment connecting them to represent the unknown distance, d.

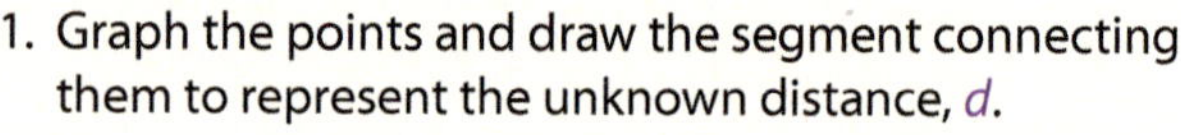

2. Form a right triangle by drawing a vertical line from P_2 and a horizontal line from P_1. These lines form a right angle at their intersection, $P_3\,(5, 1)$.

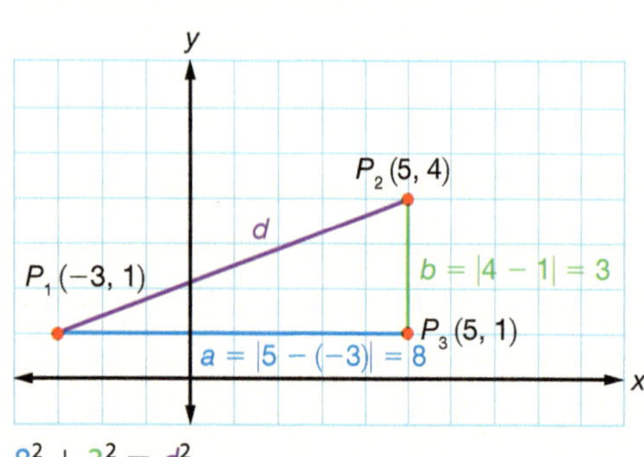

3. Determine the lengths of the legs.
   a. Since P_3 and P_1 have the same y-coordinate, a is the distance between their x-coordinates.
   b. Since P_3 and P_2 have the same x-coordinate, b is the distance between their y-coordinates.

$$8^2 + 3^2 = d^2$$
$$64 + 9 = d^2$$
$$d^2 = 73$$
$$d = \pm\sqrt{73}$$

4. Substitute into the Pythagorean theorem and solve for d. Note that $\sqrt{73}$ cannot be simplified.

The distance between P_1 and P_2 is $\sqrt{73} \approx 8.5$ units.

While the procedure from Example 3 could be used to find the distance between any two points on the Cartesian coordinate plane, a formula can be developed by applying this procedure to the generic points $P_1\ (x_1, y_1)$ and $P_2\ (x_2, y_2)$.

The length of the horizontal leg is the distance between the x-coordinates,

$$a = |x_2 - x_1|;$$

and the length of the vertical leg is the distance between the y-coordinates,

$$b = |y_2 - y_1|.$$

Using the Pythagorean theorem produces

$$d^2 = |x_2 - x_1|^2 + |y_2 - y_1|^2.$$

Taking the square root of both sides and dropping the absolute value signs (since squaring always produces a positive result) produces the following formula.

Distance Formula

The distance, d, between any two points $P_1\ (x_1, y_1)$ and $P_2\ (x_2, y_2)$ is given by the formula $d = \sqrt{(x_2 - x_1)^2 + (y_2 - y_1)^2}$.

Example 4

Use the distance formula to find the distance between (2, 1) and (5, 1).

Answer

Let $(x_1, y_1) = (2, 1)$ and $(x_2, y_2) = (5, 1)$. — 1. Assign variables to the given points.

$d = \sqrt{(x_2 - x_1)^2 + (y_2 - y_1)^2}$ — 2. Substitute into the distance formula and simplify.

$= \sqrt{(5 - 2)^2 + (1 - 1)^2}$

$= \sqrt{3^2 + 0^2} = \sqrt{9} = 3$

The distance is 3 units.

Example 5

Find the distance between $(-2, -9)$ and $(6, -5)$.

Answer Let $(x_1, y_1) = (-2, -9)$ and $(x_2, y_2) = (6, -5)$. — 1. Assign variables to the given points.

$d = \sqrt{(x_2 - x_1)^2 + (y_2 - y_1)^2}$ — 2. Substitute into the distance formula and simplify.

$= \sqrt{[6 - (-2)]^2 + [-5 - (-9)]^2}$

$= \sqrt{8^2 + 4^2} = \sqrt{64 + 16} = \sqrt{80} = 4\sqrt{5}$

The distance is $4\sqrt{5}$, or almost 9 units.

The distance formula extends the formula for the length of a segment on a number line, $l = |b - a|$, to any two points in the plane. Likewise, the formula for the midpoint of a segment on a number line, $m = \frac{a + b}{2}$, can be extended to any segment in the coordinate plane.

Midpoint Formula

The midpoint of a segment with endpoints $P_1\ (x_1, y_1)$ and $P_2\ (x_2, y_2)$ is $M\left(\frac{x_1 + x_2}{2}, \frac{y_1 + y_2}{2}\right)$.

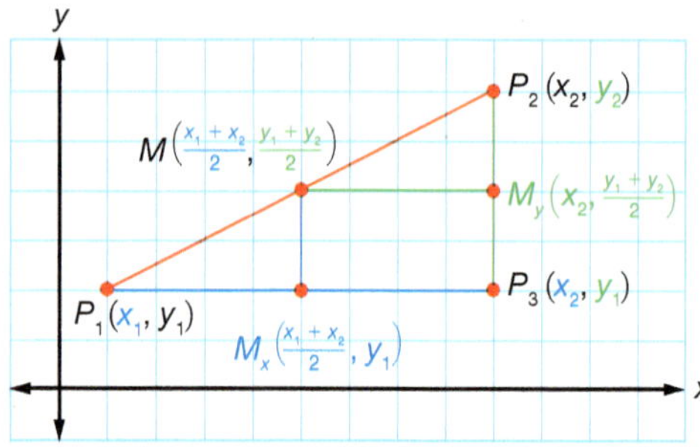

Notice that each coordinate of the midpoint is the average of the corresponding coordinates of the endpoints.

Example 6

Find the midpoint of the segment with endpoints at $(-3, 1)$ and $(5, 4)$.

Answer Let $(x_1, y_1) = (-3, 1)$ and $(x_2, y_2) = (5, 4)$. — 1. Assign variables to the given points.

$\left(\frac{x_1 + x_2}{2}, \frac{y_1 + y_2}{2}\right) = \left(\frac{-3 + 5}{2}, \frac{1 + 4}{2}\right) = \left(\frac{2}{2}, \frac{5}{2}\right)$ — 2. Substitute into the midpoint formula and simplify.

The midpoint is $\left(1, 2\frac{1}{2}\right)$.

A. Exercises

Find the length of the unknown side of each right triangle.

	a	b	c
1.	3	4	
2.		12	13
3.	4	5	
4.	3	3	
5.	2		8
6.		6	12
7.		7	9
8.	2		5

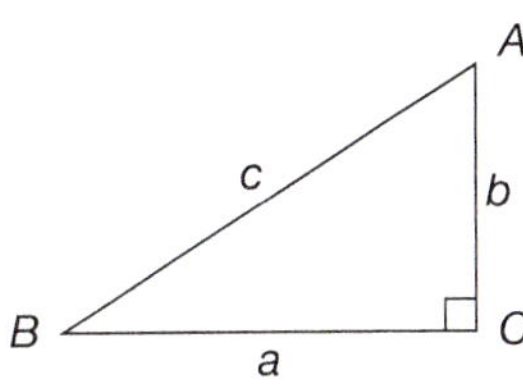

The converse of the Pythagorean theorem states, "If a triangle has sides of lengths a, b, and c, where c is the longest side and $a^2 + b^2 = c^2$, then the triangle is a right triangle." Use this theorem to determine whether each triangle is a right triangle.

9.

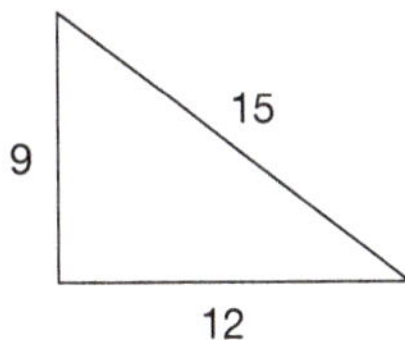

10.

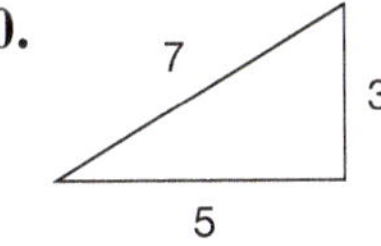

11.

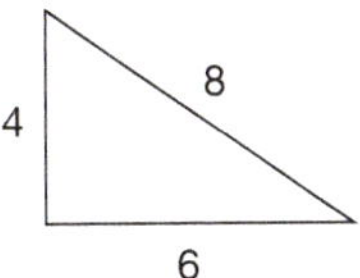

12.

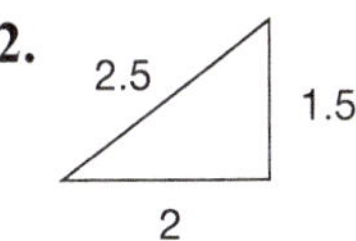

Given the lengths of the three sides of a triangle, determine whether it is a right triangle.

13. 10, 8, and 6 **14.** 7, 14, and 17 **15.** 4, $\sqrt{3}$, and 1 **16.** $4\sqrt{2}$, $3\sqrt{2}$, and $5\sqrt{2}$

Find the midpoint of a segment with the given endpoints.

17. (2, 4) and (6, 6)

18. (−6, 8) and (8, −4)

19. (−5, 0) and (−10, 14)

20. (9, −12) and (7, −4)

B. Exercises

Find the distance between the given pairs of points. Express each answer as a simplified radical.

21. (3, 4) and (0, 0) **22.** (3, 2) and (9, 10) **23.** (6, 0) and (−1, 1)

24. (5, 6) and (−4, −6) **25.** (−3, −1) and (2, −13) **26.** (10, 4) and (0, 12)

27. (1, 8) and (4, −1) **28.** (−7, 3) and (1, 8) **29.** (−2, −6) and (2, 1)

30. (1, 0) and (4, 2) **31.** $(2\sqrt{7}, -3)$ and $(\sqrt{7}, 5)$ **32.** $(5\sqrt{3}, \sqrt{24})$ and $(-\sqrt{12}, \sqrt{54})$

Find the length of the third side of each right triangle in which c is the hypotenuse.

33. $a = 3$; $b = \sqrt{7}$ **34.** $b = \sqrt{51}$; $c = 10$ **35.** $a = \sqrt{11}$; $b = \sqrt{13}$

Find the midpoint of a segment with the given endpoints.

36. $(2\sqrt{3}, -6)$ and $(4\sqrt{3}, 8)$

37. $(5\sqrt{2}, 3\sqrt{5})$ and $(\sqrt{2}, -7\sqrt{5})$

38. $(\sqrt{27}, -\sqrt{5})$ and $(-\sqrt{3}, \sqrt{5})$

Find the perimeter of each figure.

39.

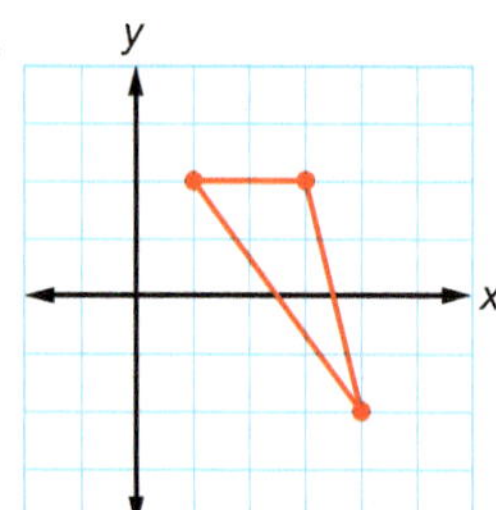

40.

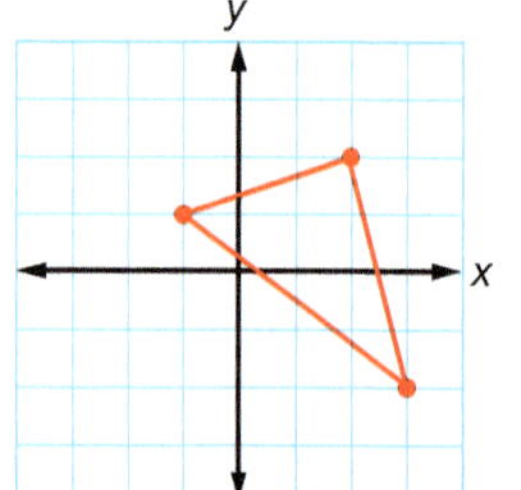

41.

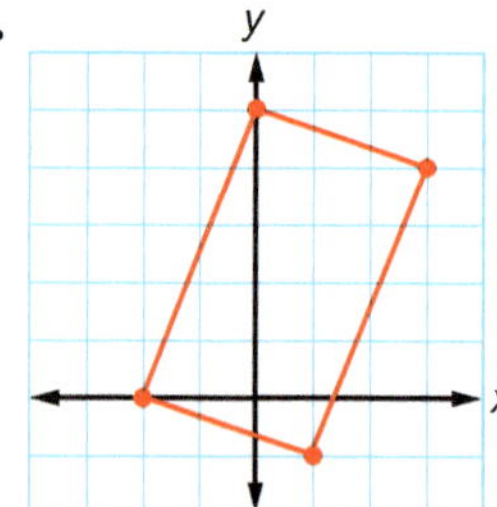

42.

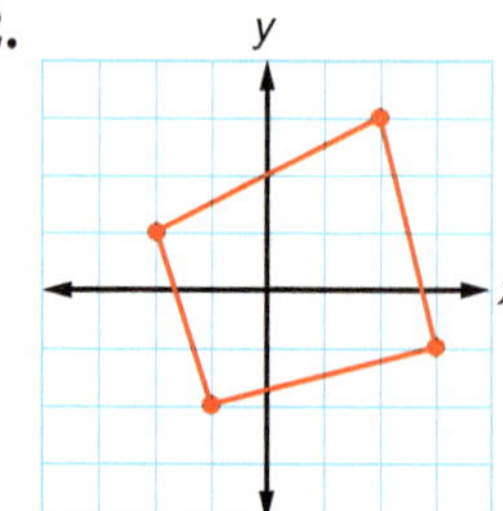

43. A surveyor is measuring between property stakes across a pond. He needs to find the distance from A to B. He knows that the distance from A to C is 0.3 mi and from C to B is 0.6 mi. He also knows that $\overline{AC}$ and $\overline{CB}$ form a right angle. What is the distance from A to B?

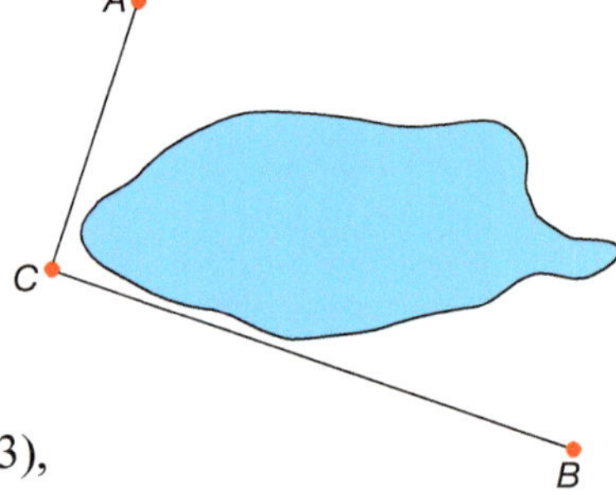

44. The bases are each 90 ft apart on a baseball diamond. Find the distance a player has to throw the ball to make an out at first base if he is standing at third base.

45. A destroyer is located at point (4, 9) and an aircraft carrier is at point (15, 3), where the numbers represent distances in nautical miles. What is the distance (to the nearest tenth of a nautical mile) between the two vessels?

46. The M. T. Container Manufacturing Company is planning to make a new box 3 in. high and 10 in. long. If the M. T. box measures 14 in. diagonally across the bottom, what is the width of the box?

47. A construction worker needs to pour concrete slab for a garage. His carpenter's square (a right-angle tool) is much too small to ensure perfect right angles, so he depends on the Pythagorean theorem. He measures along the concrete forms as illustrated in the diagram and adjusts the forms to create a perfect right angle. What must measurement c in the diagram equal to create a right angle?

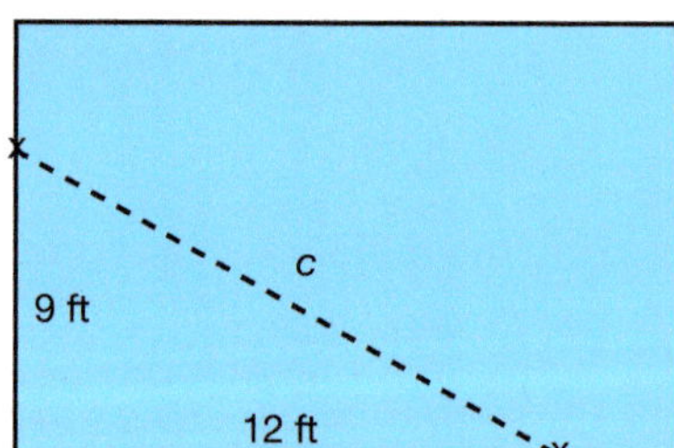

C. Exercises

48. The dimensions of a rectangular box are length = 15 in., width = 12 in., and height = 9 in. Use the Pythagorean theorem to find the following.

a. the length of a diagonal of the bottom of the box

b. the length of a diagonal of the box, d

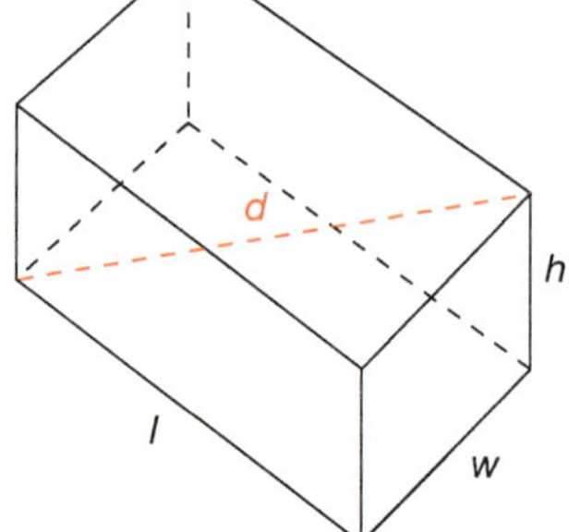

49. The dimensions of a rectangular box are l, w, and h. Use the Pythagorean theorem to derive expressions for the following.

a. the length of a diagonal of the bottom of the box

b. the length of a diagonal of the box, d

50. Mr. Ferguson is going to construct a ramp to his barn. The barn loft is 48 in. above ground level, and the base of the ramp will need to be 20 ft from the barn.

a. How long (in feet and inches) must the ramp be (to the nearest inch)?

b. What is the slope of the ramp? Express your answer as both a fraction and a percent.

A Pythagorean triple consists of three positive integers a, b, and c, such that $a^2 + b^2 = c^2$. Commonly used Pythagorean triples include (3, 4, 5) and (6, 8, 10). If m and n are positive integers with $m > n$, then $(m^2 - n^2, 2mn, m^2 + n^2)$ is a Pythagorean triple.

51. Given $m = 12$ and $n = 10$, find the related Pythagorean triple and verify that it is a Pythagorean triple.

52. Show algebraically that $(m^2 - n^2, 2mn, m^2 + n^2)$ is a Pythagorean triple.

Dominion Modeling

The golden ratio can be found in many famous paintings and architectural designs. Objects in art and architecture with golden ratio dimensions are commonly considered to be of "perfect proportion" and more pleasing to the eye than objects with other proportions. For example, studies have shown that the majority of people who are shown three rectangles will choose a golden rectangle over other rectangles as the best looking. Because of its inherent aesthetic value and its frequent presence in nature, this ratio is often called "golden" or "divine."

Wounded Christ Between Angels
Bartolommeo Vivarini
From the Bob Jones University Collection

53. Perform an Internet search using the keywords "golden ratio art." List three other famous artists and the name of one painting by each in which the golden ratio is incorporated.

54. Perform an Internet search using the keywords "golden ratio architecture." List three famous architectural designs in which the golden ratio is incorporated.

CUMULATIVE REVIEW

Translate each word phrase into an algebraic expression. [2.1]

55. three less than twice the square root of a number

56. three less than the square root of twice a number

57. twice the square root of three less than a number

58. the square root of the quantity three less than twice a number

Express each number in scientific notation. [8.3]

59. 0.000015

60. $15{,}000 \times 10^{-2}$

State the conjugate of each binomial. [9.5]

61. $\sqrt{x} + 2$

62. $2x - \sqrt{3}$

Multiply. [9.4–9.5]

63. $(5x - 3)(3x + 2)$

64. $(4x - 7)^2$

11.7 Multiplying and Dividing Radical Expressions

You have learned how to multiply and divide polynomials and how to multiply and divide radicals. By combining these skills, you will be able to multiply and divide more complicated radical expressions.

Definition

A **radical expression** is an algebraic expression containing at least one radical.

Some examples of radical expressions are $\sqrt{3} + 4$, $\sqrt{6} - \sqrt{3}$, $5 + \sqrt[3]{7}$, and $\sqrt[3]{2x^2y}$. Radical expressions are not polynomials, but the same principles apply when you are multiplying radical expressions with more than one term.

Example 1

Simplify $(\sqrt{7} + \sqrt{2})(\sqrt{3} - 6)$.

Answer $\sqrt{7} \cdot \sqrt{3} - 6\sqrt{7} + \sqrt{2} \cdot \sqrt{3} - 6\sqrt{2}$

$= \sqrt{21} - 6\sqrt{7} + \sqrt{6} - 6\sqrt{2}$

1. Multiply using the FOIL method.
2. Since no terms in this radical expression can be simplified or combined, this is the final answer.

Example 2

Simplify $(\sqrt{5} - 7)(\sqrt{5} + 7)$.

Answer $(\sqrt{5})^2 - 7^2$

$= 5 - 49 = -44$

1. Recognize and apply the product of conjugates pattern: $(a + b)(a - b) = a^2 - b^2$.
2. Simplify.

Example 3

Simplify $(\sqrt{3} - \sqrt{2})^2$.

Answer $(\sqrt{3})^2 - 2(\sqrt{3})(\sqrt{2}) + (\sqrt{2})^2$

$= 3 - 2\sqrt{6} + 2 = 5 - 2\sqrt{6}$

1. Recognize and apply the square of a difference pattern: $(a - b)^2 = a^2 - 2ab + b^2$.
2. Simplify.

Multiplying Radical Expressions

1. Follow the same rules that you follow for multiplying polynomials.
2. After multiplying, simplify radicals and combine like terms.

Example 4

Simplify $(\sqrt{3} - 2)(\sqrt{3} + 2)(\sqrt{6} - 4)$.

Answer

$[(\sqrt{3})^2 - 2^2](\sqrt{6} - 4)$

$= (3 - 4)(\sqrt{6} - 4)$

$= -1(\sqrt{6} - 4)$

$= -\sqrt{6} + 4 = 4 - \sqrt{6}$

1. Recognize and apply the product of conjugates pattern to multiply the first two radical expressions.
2. Simplify the product. The Commutative Property of Addition can be used to rewrite the answer.

Remember that expressions with radicals in the denominator are not simplified. To simplify $\frac{3}{\sqrt{5}}$, you multiplied by one in the form of $\frac{\sqrt{5}}{\sqrt{5}}$ to eliminate $\sqrt{5}$ from the denominator.

$$\frac{3}{\sqrt{5}} \cdot \frac{\sqrt{5}}{\sqrt{5}} = \frac{3\sqrt{5}}{5}$$

When a radical expression appears in the denominator of a fraction, as in $\frac{4}{\sqrt{5} - 2}$, you must eliminate the radical from the denominator. However, simply multiplying by the radical will not eliminate it. Instead, you must use the conjugate of the denominator to eliminate the radical. Recall that $a - b$ and $a + b$ are *conjugates* and that their product is the difference of their squares, $a^2 - b^2$. Multiplying the numerator and the denominator by the conjugate of $\sqrt{5} - 2$, which is $\sqrt{5} + 2$, produces an equivalent expression without radicals in the denominator.

$$\frac{4}{\sqrt{5} - 2} \cdot \frac{\sqrt{5} + 2}{\sqrt{5} + 2} = \frac{4(\sqrt{5} + 2)}{(\sqrt{5})^2 - 2^2} = \frac{4\sqrt{5} + 8}{5 - 4} = \frac{4\sqrt{5} + 8}{1} = 4\sqrt{5} + 8$$

Example 5

Simplify $\frac{4 + \sqrt{3}}{5 + \sqrt{3}}$.

Answer

$\frac{4 + \sqrt{3}}{5 + \sqrt{3}} \cdot \frac{5 - \sqrt{3}}{5 - \sqrt{3}}$

$= \frac{20 - 4\sqrt{3} + 5\sqrt{3} - 3}{25 - 3}$

$= \frac{17 + \sqrt{3}}{22}$

1. Multiply the numerator and the denominator by the conjugate of the denominator.
2. Multiply the numerator using the FOIL method. The product of conjugates in the denominator equals the difference of their squares.
3. Simplify.

Dividing Radical Expressions

1. Place the dividend over the divisor.
2. Multiply the numerator and the denominator by the conjugate of the denominator.
3. Simplify.

Example 6

Simplify $(\sqrt{m} - \sqrt{n}) \div (\sqrt{m} + \sqrt{n})$.

Answer

$$\frac{\sqrt{m} - \sqrt{n}}{\sqrt{m} + \sqrt{n}}$$

1. Place the dividend over the divisor.

$$= \frac{\sqrt{m} - \sqrt{n}}{\sqrt{m} + \sqrt{n}} \cdot \frac{\sqrt{m} - \sqrt{n}}{\sqrt{m} - \sqrt{n}}$$

2. Multiply the numerator and the denominator by the conjugate of the denominator.

$$= \frac{(\sqrt{m})^2 - 2(\sqrt{m})(\sqrt{n}) + (\sqrt{n})^2}{(\sqrt{m})^2 - (\sqrt{n})^2}$$

3. Recognize and apply the square of a difference pattern in the numerator. The product of conjugates in the denominator equals the difference of their squares.

$$= \frac{m - 2\sqrt{mn} + n}{m - n}$$

4. Simplify.

Example 7

Simplify $\frac{3\sqrt{2} + 4}{2\sqrt{3} + 5}$.

Answer

$$\frac{3\sqrt{2} + 4}{2\sqrt{3} + 5} \cdot \frac{2\sqrt{3} - 5}{2\sqrt{3} - 5}$$

1. Multiply the numerator and the denominator by the conjugate of the denominator.

$$= \frac{6\sqrt{6} - 15\sqrt{2} + 8\sqrt{3} - 20}{4(3) - 25}$$

2. Multiply the numerator using the FOIL method. The product of conjugates in the denominator equals the difference of their squares.

$$= \frac{6\sqrt{6} - 15\sqrt{2} + 8\sqrt{3} - 20}{-13}$$

$$= \frac{-6\sqrt{6} + 15\sqrt{2} - 8\sqrt{3} + 20}{13}$$

3. Simplify. Distribute the negative sign in the denominator across the numerator.

A. Exercises

Simplify.

1. $(2 - \sqrt{3})(2 + \sqrt{3})$

2. $(3 + \sqrt{7})(3 - \sqrt{7})$

3. $(\sqrt{3} - \sqrt{11})(\sqrt{3} + \sqrt{11})$

4. $(\sqrt{7} - \sqrt{10})^2$

5. $(\sqrt{5} - \sqrt{3})^2$

6. $(\sqrt{7} + 8)^2$

7. $(7 + \sqrt{3})(4 - \sqrt{11})$

8. $(3 - \sqrt{10})(4 + \sqrt{20})$

9. $(\sqrt{6} - \sqrt{2})(\sqrt{8} + 4)$

10. $(\sqrt{12} - \sqrt{3})(\sqrt{5} + \sqrt{2})$

State the conjugate of each radical expression.

11. $-3 + \sqrt{5}$

12. $\sqrt{7} - \sqrt{11}$

Simplify.

13. $\frac{1}{2 + \sqrt{3}}$

14. $\frac{2 + \sqrt{6}}{3 - \sqrt{6}}$

15. $\frac{\sqrt{11}}{\sqrt{3} + \sqrt{2}}$

16. $\frac{9}{3 + \sqrt{2}}$

17. $\frac{1}{\sqrt{2}}$

18. $\frac{7\sqrt{3}}{2\sqrt{4} + 1}$

B. Exercises

Simplify.

19. $(\sqrt{5} - 3)(\sqrt{2} + 1)(\sqrt{5} + 3)$

20. $(3 - \sqrt{2})^2(\sqrt{5} + 3)$

21. $\sqrt{5}(\sqrt{35} + 5)(\sqrt{15} + 2)$

22. $(x\sqrt{3} - 5)(x\sqrt{6} - 3)$

23. $\sqrt{x}(\sqrt{3} - 5)(\sqrt{x^3} + 2)$

24. $\sqrt{xy^2}(\sqrt{x} - \sqrt{y})(\sqrt{x} + \sqrt{y})$

25. $\sqrt{5}(\sqrt{3} - \sqrt{2})(\sqrt{5} - 6)$

26. $\sqrt{x}(x + \sqrt{8})^2$

27. $3\sqrt{5}(\sqrt{7} - \sqrt{3})(\sqrt{7} + \sqrt{3})$

28. $\sqrt{7}(\sqrt{3} + \sqrt{21})^2$

29. $\dfrac{2}{\sqrt{3} - 1}$

30. $\dfrac{\sqrt{5} - 3}{\sqrt{5} + 3}$

31. $\dfrac{5}{\sqrt{2} - \sqrt{6}}$

32. $\dfrac{\sqrt{5} + 1}{\sqrt{7} - 2}$

33. $\dfrac{16}{\sqrt{7} + \sqrt{3}}$

34. $\dfrac{\sqrt{8}}{3\sqrt{2} - \sqrt{3}}$

35. Express the area of the rectangle as a simplified radical expression.

$15\sqrt{2} - \sqrt{6}$

$3\sqrt{5} + \sqrt{15}$

C. Exercises

Simplify.

36. $\dfrac{\sqrt{2} + \sqrt{5}}{\sqrt{3} - \sqrt{10}}$

37. $\dfrac{3\sqrt{2} - 8\sqrt{7}}{4\sqrt{3} + \sqrt{5}}$

38. $(\sqrt{x} - 4)^2(\sqrt{x} - 7)^2$

Simplify by converting to exponential form with a common denominator. Express each answer as a simplified radical.

39. $\sqrt[3]{2} \cdot \sqrt{2}$

40. $\sqrt[3]{4} \cdot \sqrt[4]{4}$

Dominion Modeling

Why do some webpages look so much better than others? What makes a good design layout? For some people, design comes naturally and intuition results in a layout that is visually pleasing to most other people. While there are many factors that contribute to a well-designed page, knowing and applying mathematics can enhance the layout and make it more pleasing to the eye.

The rule of thirds is one mathematical guide that can lead to a better layout. Dividing a picture into thirds creates nine rectangular areas and four points of intersection. Aligning primary objects with these points of intersection often creates a more pleasing balance and layout than simply centering the objects.

Many designers have applied the golden ratio, ϕ, and golden rectangles (which have a length-to-width ratio equal to the golden ratio) to create more pleasing layouts.

Use $\phi \approx 1.618$ and the definition of a golden rectangle, $\frac{l}{w} = \phi$, to complete the following exercises.

41. Find the width of a golden rectangle whose length is 11 cm (to the nearest tenth).

42. Find the length of a golden rectangle whose width is 8.5 cm (to the nearest tenth).

43. A designer is creating a webpage that is to be 1024 pixels wide with a wider primary content column on the left and a secondary content column on the right. She wants the ratio of the page width to the width of the primary content column to be as close to the golden ratio as possible.

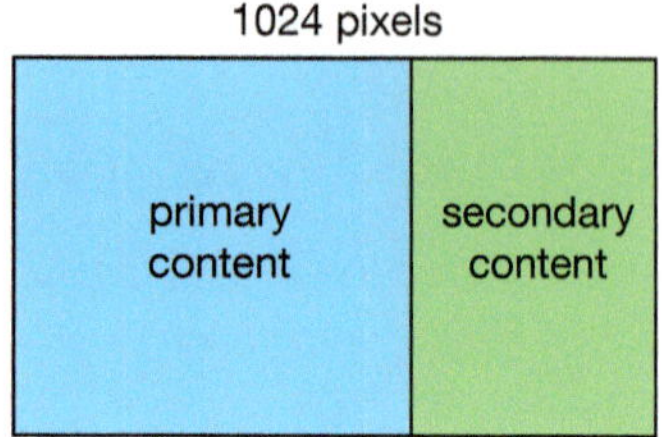

a. Find the width of the primary content column (to the nearest pixel).

b. Find the width of the secondary content column.

c. Find the ratio of the widths of the primary and secondary columns (to the nearest thousandth).

CUMULATIVE REVIEW

Solve. [2.5, 4.3]

44. $2x + \sqrt{5} = \sqrt{17}$

45. $-3x + 2 > 23$

46. Solve the system by graphing. [7.1]

$2x - y = -1$
$x + 2y = 7$

47. Divide $(x^2 - 5x + 3)$ by $(x^2 - 3x + 2)$. [9.6]

Factor. [10.2–10.3]

48. $x^2 + 4xy - 45y^2$

49. $2a^2 - 18ab - 44b^2$

Simplify. [11.4]

50. $\frac{15\sqrt{x}}{\sqrt{3y}}$

51. $\frac{8}{\sqrt[3]{5}}$

52. Find the length of the hypotenuse of a right triangle with legs of length $2 + \sqrt{15}$ units and $2\sqrt{3} - \sqrt{5}$ units. [11.6]

53. Find x. [11.6]

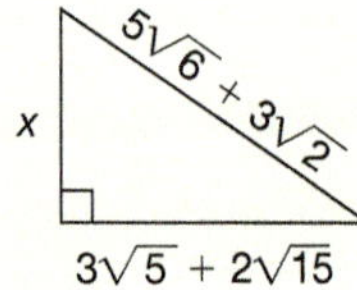

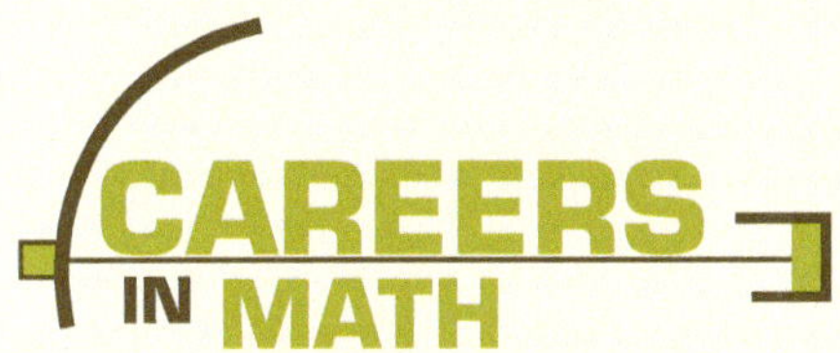

Chemist/Chemical Engineer

Chemists can do amazing things with the raw materials found in nature. The microchips in your computer, your cell phone, and even the family car are products of chemical engineering. The silicon in microchips is made from natural sand, or silica. The process involves the removal of oxygen, accomplished by heating silica and carbon to temperatures over 2000°C. Such processes require the scientific and mathematical knowledge and skills of a chemist.

The career of a Christian chemist offers not only the means to make a good living, but also many opportunities for demonstrating care and love for others. For instance, chemists work on cures for serious diseases or medicines for those suffering from physical infirmities, as well as the development of new materials for warmth or stronger materials for structural safety. Their efforts help others live more safely and comfortably through the wise use of the natural materials provided in God's creation.

Chemists develop new materials to improve existing products or make new ones. Analytical chemists determine the composition of a substance by identifying its elements and compounds and their relative quantities. Chemical engineers work together with industrial chemists to develop the processes needed to make new products.

God has created an astonishing variety of compounds with many differing chemical properties. A chemist can use his knowledge and skills to harness the potential of God's creation for His glory.

11.8 Radical Equations

Radical equations are used during accident investigations.

Equations such as $\sqrt{x} = 10^2$, $3 + \sqrt{x} = 24$, and $6\sqrt{x} = 9$ are called *radical equations* because they contain a variable in the radicand. Equations such as $x + \sqrt{3} = 0$ and $5x - \sqrt{13} = 25$ are not radical equations because the variable is not in the radicand.

Definition

A **radical equation** is an equation that contains a variable in a radicand.

Example 1

Solve $\sqrt{x} = 10$.

Answer $(\sqrt{x})^2 = 10^2$ — 1. Square both sides of the equation to eliminate the square root.

$x = 100$ — 2. When you square a square root, you get the radicand. That is, $(\sqrt{x})^2 = x$.

Check $\sqrt{100} = 10$ — 3. Check your solution in the original equation.

After both sides of an equation are squared, any resulting solutions must be checked in the original equation. If a derived solution does not make the original equation true, then it is invalid and is called an *extraneous* solution.

How do we get an invalid answer? Consider the following.

$-2 = 2$ a false statement

$(-2)^2 = 2^2$ square both sides

$4 = 4$ a true statement

Squaring both sides of the equation did not result in both sides being multiplied by the same number, so a false statement was changed into a true statement. When squaring both sides of an equation, you must check the solution to determine its validity.

Solving Radical Equations
1. Isolate the radical on one side of the equation.
2. Square both sides of the equation.
3. Solve the resulting equation.
4. Check any solutions in the original equation.

Example 2

Solve $8 + \sqrt{x - 3} = 4$.

Answer

$$\sqrt{x - 3} = -4$$
$$(\sqrt{x - 3})^2 = (-4)^2$$
$$x - 3 = 16$$
$$x = 19$$

1. Isolate the radical.
2. Square both sides.
3. Solve.

Check

$$8 + \sqrt{19 - 3} = 4$$
$$8 + 4 = 4;\text{ false}$$

There is no solution.

4. Check your solution in the original equation. Since $12 \neq 4$, this solution does not check.
5. The derived solution, 19, is extraneous.

You may have noticed that when the radical was isolated, the resulting equation set the principal square root equal to a negative number. Since this is not possible, you could confidently state at that point that there is no solution.

Example 3

Solve $\sqrt{2x + 5} = 8$.

Answer

$$\sqrt{2x + 5} = 8$$
$$(\sqrt{2x + 5})^2 = 8^2$$
$$2x + 5 = 64$$
$$2x = 59$$
$$x = \frac{59}{2}$$

1. The radical is already isolated.
2. Square both sides.
3. Solve.

Check

$$\sqrt{2\left(\frac{59}{2}\right) + 5} = 8$$
$$\sqrt{64} = 8$$

$\frac{59}{2}$ is the solution.

4. Check your solution in the original equation.
5. The solution makes the original equation true.

Example 4

Solve $10 - 2\sqrt{x} = 0$.

Answer	$-2\sqrt{x} = -10$	1. Isolate the radical.
	$\sqrt{x} = 5$	
	$(\sqrt{x})^2 = 5^2$	2. Square both sides.
	$x = 25$	3. The equation is solved.
Check	$10 - 2\sqrt{25} = 0$	4. Check your solution in the original equation.
	$10 - 2(5) = 0$	
	25 is the solution.	5. The solution makes the original equation true.

A. Exercises

Solve and check your solution.

1. $\sqrt{x} = 4$
2. $\sqrt{y} = 5$
3. $\sqrt{x + 2} = 8$
4. $\sqrt{y - 5} = 2$
5. $3 + \sqrt{x} = 9$
6. $\sqrt{x} = 3.2$
7. $\sqrt{x + 5} = 0$
8. $3\sqrt{2r} = 6$
9. $\sqrt{x + 5} + 5 = 7$
10. $3 - \sqrt{x - 5} = 4$
11. $2 - \sqrt{x} = -8$
12. $\sqrt{4a - 3} + 5 = 8$

B. Exercises

Solve and check your solution.

13. $7 + 3\sqrt{x - 3} = -14$
14. $3 + 4\sqrt{x + 3} = 11$
15. $5 + 2\sqrt{y - 3} = 27$
16. $5 + \sqrt{x + 1} = 7$
17. $14 - \sqrt{x - 3} = 10$
18. $8 + 2\sqrt{x} = 18$
19. $12 - 4\sqrt{x - 3} = 7$
20. $5\sqrt{x} - 13 = 2\sqrt{x} + 26$

Write an equation and solve.

21. The square root of twice a number is 12. Find the number.

22. Five more than twice the square root of a number is 17. Find the number.

Use the formula $S = \sqrt{30Df}$ to determine the distance required for a car to skid to a stop. (S is the car's speed in miles per hour, D is the length of the skid marks in feet, and f is the coefficient of friction between the tires and the road.)

23. $f = \frac{2}{3}$; $S = 60$ mi/hr
24. $f = 0.8$; $S = 45$ mi/hr

25. Solve the equation $S = \sqrt{30Df}$ for D.

The formula $T = 2\pi\sqrt{\frac{l}{g}}$ relates a pendulum's period, T, in seconds (the time it takes to swing back and forth) to its length, l, in centimeters using g, the gravitational acceleration of 981 cm/sec^2.

How long would the pendulum have to be (to the nearest tenth of a centimeter) to make a pendulum with the given period?

26. 1 sec
27. 2 sec

28. Solve the equation $T = 2\pi\sqrt{\frac{l}{g}}$ for l.

Use the formula $V = \frac{15}{22}\sqrt{gd}$ to determine the depth of the water for tsunami waves traveling at the given speeds. (V is the wave's speed in miles per hour, g is the gravitational acceleration of 32 ft/sec^2, and d is the depth of the water in feet.)

29. 120 mi/hr

30. 180 mi/hr

31. Solve the equation $V = \frac{15}{22}\sqrt{gd}$ for d.

C. Exercises

Solve.

32. $\sqrt{5x + 1} - \sqrt{x + 3} = 0$

33. $\sqrt{3x - 3} + \sqrt{x + 1} = 2\sqrt{x + 1}$

34. $\sqrt{x} + 2 = \sqrt{9x}$

35. The instantaneous velocity of a free-falling object can be calculated using the formula $v = \sqrt{2gd}$, where g is the gravitational acceleration of 9.8 m/sec^2 and d is the distance the object has traveled in meters. Find the distance a rock must fall to reach a velocity of 28 m/sec.

Beginning in cell A1 of a spreadsheet, enter distance values from 1 to 100 in column A. (This can be quickly accomplished by entering 1 and 2 in the first two cells, highlighting both cells, and then dragging the corner handle down to replicate the numbers.) Enter the formula =SQRT(2*9.8*A1) in cell B1. Replicate this formula down to row 100 by highlighting the cell and dragging the corner handle.

36. Using the results on the spreadsheet, state the only three distances less than 100 m that produce an integral value for the instantaneous velocity of a falling object.

Dominion Modeling

Understanding truth provides great insight into the world around us, which in turn results in a more abundant life with greater joy than could be experienced without this understanding (cf. Jer. 9:24). Viewing mathematical patterns in nature can be interesting, but knowing that God has created all things for His glory and that it is by His power that they continue to exist (Col. 1:16–17) is exciting and provides greater purpose.

Count the number of spirals in each direction.

The golden ratio is all around us. The Fibonacci sequence can be seen in the spiraling bracts of a pinecone, the seeds of a sunflower, or the scales of a pineapple. The numbers of clockwise and counterclockwise spirals are consecutive Fibonacci sequence numbers. Consecutive leaves of the sunflower plant are staggered by 222.5°, allowing a maximum amount of sunlight and rain to reach each leaf.

37. Find the ratio of the number of degrees in a circle to the number of degrees by which sunflower leaves are staggered (to the nearest thousandth).

Many have claimed that several proportions in the human body are approximately equal to the golden ratio. When repairing or replacing damaged teeth, dentists have found that making the ratio of the width of the middle two front teeth to their visible length equal to the golden ratio typically produces the aesthetically ideal length. The ratio of the length of your forearm to the length of your hand and the ratio of your height to the height of your navel are both approximately ϕ.

38. Record the following measurements in a table similar to the one below: forearm, hand, total height, and height of navel. Then determine each of your personal ratios (to the nearest thousandth).

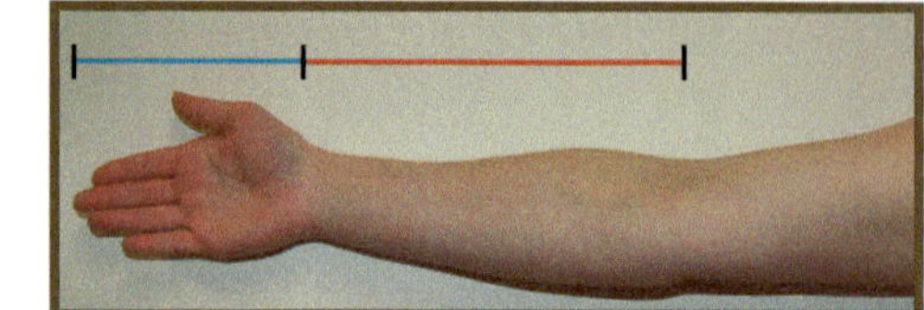

39. Using your values as the experimental values and ϕ as the know value, calculate the percent error of each ratio. Record your results in the table.

Measurement		Personal Ratio		Percent Error
forearm		$\frac{\text{forearm}}{\text{hand}}$		
hand				
total height		$\frac{\text{total height}}{\text{height of navel}}$		
height of navel				

40. Obtain and record the forearm-to-hand length ratios and total-to-navel height ratios of four other people. For each set of five measurements, find the average ratio and its percent error using ϕ as the known value.

41. Do the results of the previous exercises confirm or refute the claim that the golden ratio is found in the design of the human body? Explain your reasoning.

"I will praise thee; for I am fearfully and wonderfully made: marvellous are thy works; and that my soul knoweth right well" (Ps. 139:14).

"Many, O Lord my God, are thy wonderful works which thou hast done, and thy thoughts which are to us-ward: they cannot be reckoned up in order unto thee: if I would declare and speak of them, they are more than can be numbered" (Ps. 40:5).

CUMULATIVE REVIEW

42. Graph $y = \frac{1}{2}x + 2$. [6.3]

43. State the slope and y-intercept of $y = \frac{1}{2}x + 2$. [6.3]

Factor. [10.4]

44. $x^2 - 9$

45. $x^2 - \frac{1}{9}$

Divide. [9.6]

46. $\frac{x^3 - x^2 - 18x + 8}{x + 4}$

47. $\frac{x^2 - 26x + 169}{x - 13}$

Simplify. [11.5]

48. $\frac{\sqrt{3}}{2} + \frac{\sqrt{3}}{5}$

49. $\frac{4\sqrt{20}}{5} - \frac{5\sqrt{5}}{2}$

Write the function rule for each translation of the base function $y = x^2$. [8.4]

50. 3 units left and 2 units up

51. 4 units right and 1 unit down

11.9 Radical Functions

Radical functions are functions with the independent variable in the radicand, such as $y = \sqrt{x}$. Cube root and higher root functions are also radical functions, but this section will focus on square root functions.

Example 1

Graph the radical function $y = \sqrt{x}$.

Answer

x	$y = \sqrt{x}$
0	0
1	1
4	2
9	3
−4	undefined

1. Choose values that make the radicand a perfect square. Notice that the function is undefined when x is a negative number.

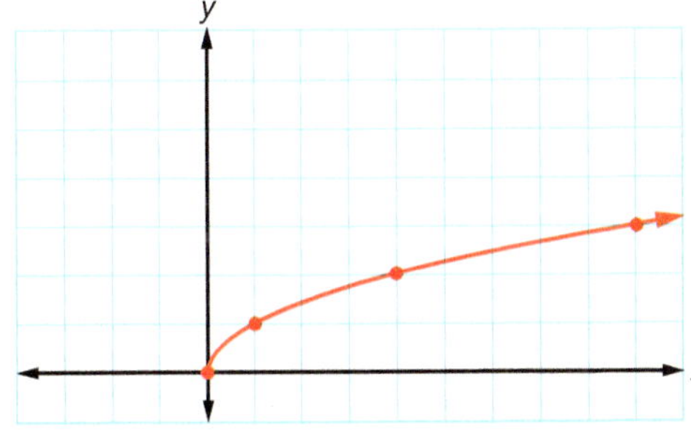

2. Graph the function.

One of the most important characteristics of a radical function is that it has a limited domain. In a square root function, the radicand cannot be negative; that is, the square root of a negative number is not defined for real numbers. Therefore, when a square root function is graphed, the value of the radicand must be greater than or equal to zero.

Example 2

Graph $y = \sqrt{x - 3}$.

Answer

$x - 3 \geq 0$
$x \geq 3$

1. Determine the domain.

x	$y = \sqrt{x - 3}$
3	0
4	1
7	2
12	3

2. Using the domain for x, choose values that make the radicand a perfect square. It is helpful to begin with the value that makes the radicand equal to zero.

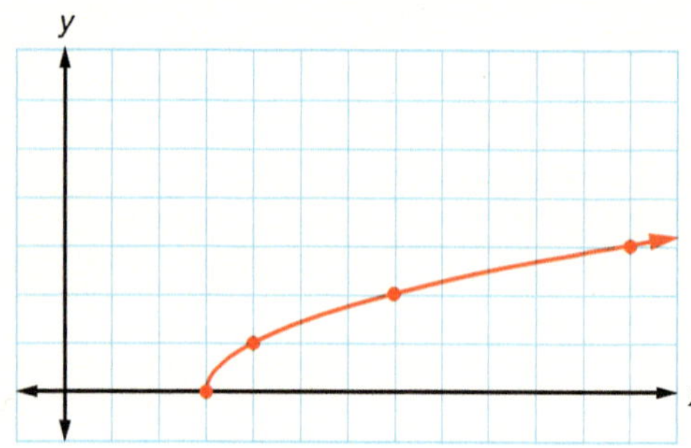

3. Graph the function.

Recall that, as in Example 2, subtracting from the input of a function shifts the graph of the function to the right. Adding to the input will shift the graph to the left. Also, adding to or subtracting from the output of a function will shift the graph up or down respectively.

Example 3

Graph $y = \sqrt{x + 4} - 1$.

Answer

$x + 4 \geq 0$
$x \geq -4$

1. Determine the domain.

x	$y = \sqrt{x + 4} - 1$
−4	−1
−3	0
0	1
5	2

2. Using the domain for x, choose values that make the radicand a perfect square. Begin with the value that makes the radicand equal to zero.

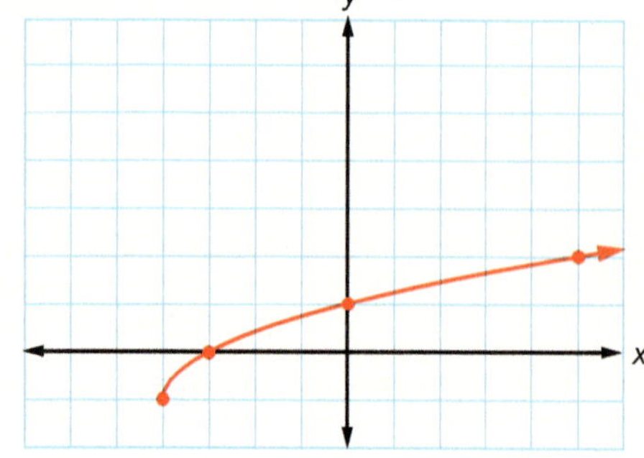

3. Graph the function.

Example 4

Graph $y = \sqrt{-x}$.

Answer

$-x \geq 0$

$x \leq 0$

1. Determine the domain.

x	$y = \sqrt{-x}$
0	0
−1	1
−4	2
−9	3

2. Using the domain for x, choose values that make the radicand a perfect square.

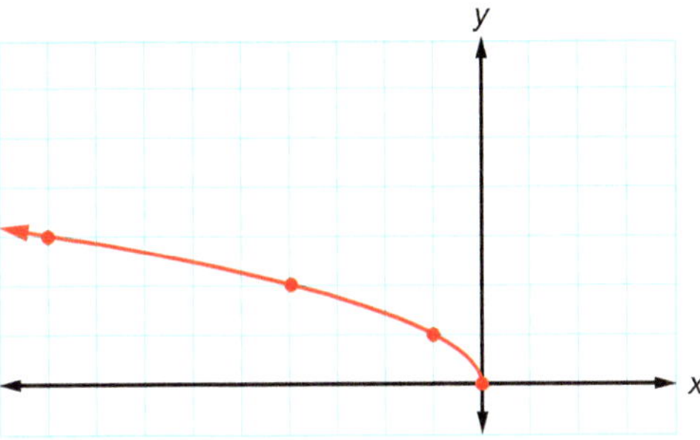

3. Graph the function.

Example 4 illustrates a reflection. The graph of $y = \sqrt{-x}$ is a reflection of the graph of $y = \sqrt{x}$ across the y-axis.

Example 5

Graph $y = -\sqrt{x}$.

Answer

$x \geq 0$

1. Determine the domain.

x	$y = -\sqrt{x}$
0	0
1	−1
4	−2
9	−3

2. Using the domain for x, choose values that make the radicand a perfect square.

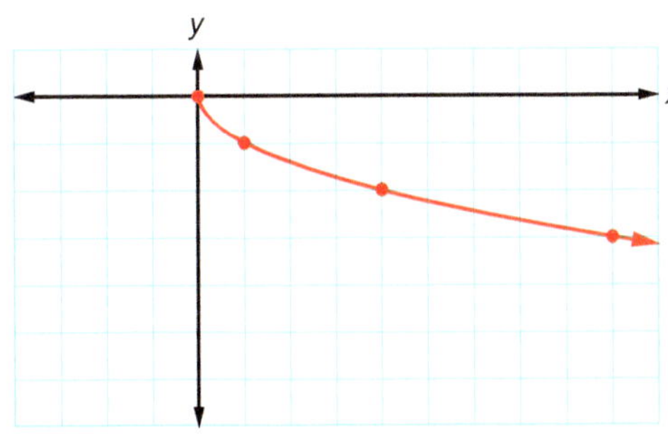

3. Graph the function.

Example 5 illustrates another reflection. The graph of $y = -\sqrt{x}$ is a reflection of the graph of $y = \sqrt{x}$ across the x-axis.

A. Exercises

Determine the domain of each radical function.

1. $y = \sqrt{x}$ **2.** $y = \sqrt{x + 3}$ **3.** $y = \sqrt{x} - 2$

4. $y = \sqrt{x} + 7$ **5.** $y = \sqrt{x - 5}$ **6.** $y = \sqrt{x - 3} + 4$

7. $y = \sqrt{x + 2} - 3$ **8.** $y = \sqrt{-x}$

Make a table of ordered pairs and graph each function.

9. $y = \sqrt{x}$ **10.** $y = \sqrt{x + 3}$ **11.** $y = \sqrt{x} - 2$

12. $y = \sqrt{x} + 7$ **13.** $y = \sqrt{x - 5}$ **14.** $y = \sqrt{x - 3} + 4$

15. $y = \sqrt{x + 2} - 3$ **16.** $y = \sqrt{-x}$

B. Exercises

Match each function with its graph.

17. $f(x) = \sqrt{x}$

18. $f(x) = \sqrt{x} + 2$

19. $f(x) = \sqrt{x + 2}$

20. $f(x) = \sqrt{-x}$

21. $f(x) = -\sqrt{x}$

22. $f(x) = \sqrt{x - 3} - 1$

a.

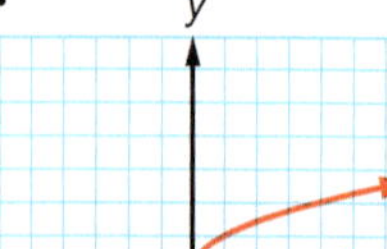

b.

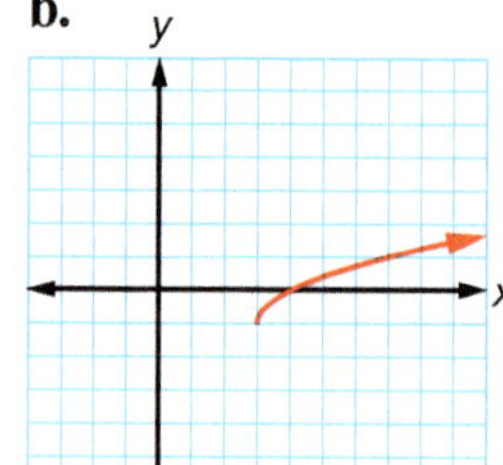

c.

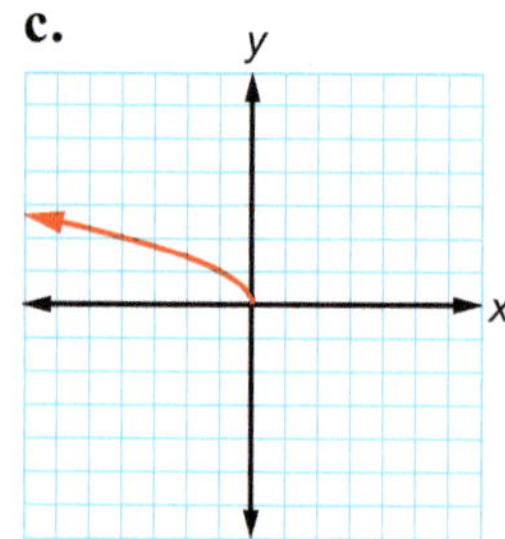

d.

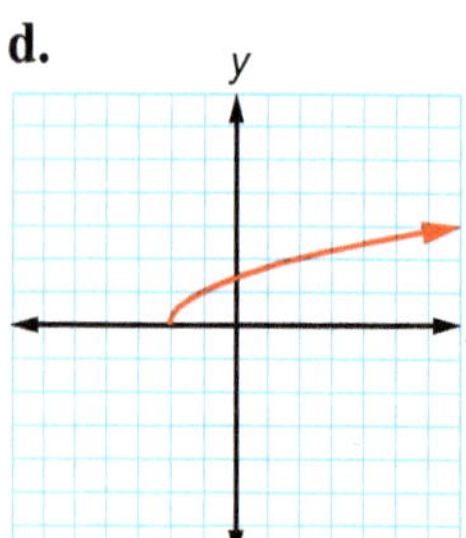

e.

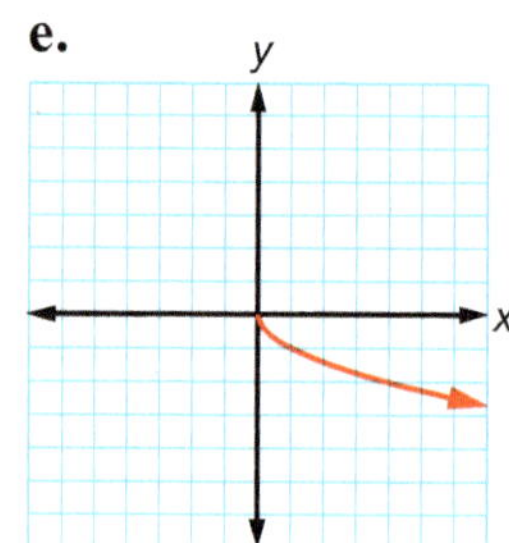

f.

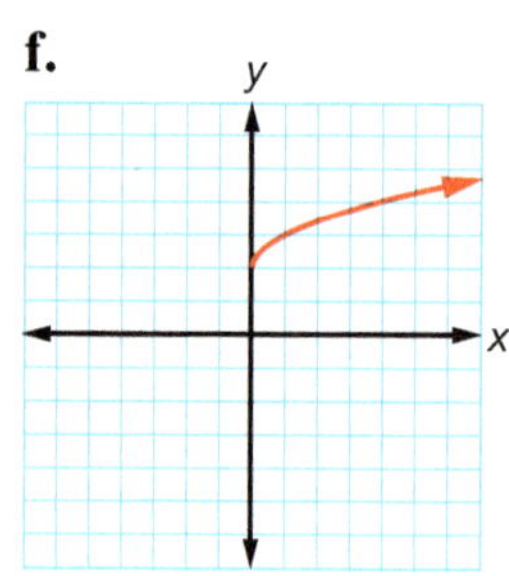

Write the function rule for each translation or reflection of the base function $f(x) = \sqrt{x}$.

23. translated 1 unit up

24. translated 3 units left

25. translated 2 units right and 2 units down

26. translated 5 units left and 3 units up

27. reflected across the y-axis

28. reflected across the x-axis

Sketch each translation of $y = \sqrt{x}$ without making a table of ordered pairs.

29. $y = \sqrt{x} - 3$ **30.** $y = \sqrt{x + 2}$ **31.** $y = \sqrt{x + 3} - 2$

32. $y = \sqrt{x - 2} + 2$ **33.** $y = -\sqrt{x} - 3$ **34.** $y = \sqrt{-x} - 4$

C. Exercises

35. Without graphing, describe the graph of $y = -\sqrt{-x}$.
36. What is the domain of $y = \sqrt[3]{x}$?
37. Make a table of ordered pairs and graph $y = \sqrt[3]{x}$.
38. What is the domain of $y = \sqrt[4]{x}$?
39. Make a table of ordered pairs and graph $y = \sqrt[4]{x}$.
40. Describe the difference between the domain of $y = \sqrt[n]{x}$ when n is odd and when n is even.
41. Describe what happens to the graph of $y = \sqrt[n]{x}$ as n increases.

CUMULATIVE REVIEW

Find each missing length, expressing each answer as both a simplified radical and estimated to the nearest integer. [11.1, 11.6]

42. 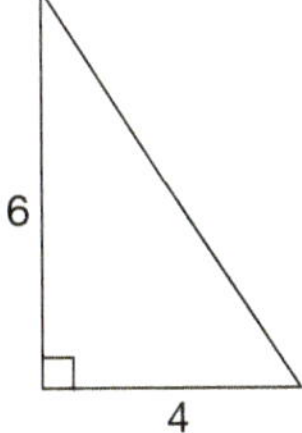

43. 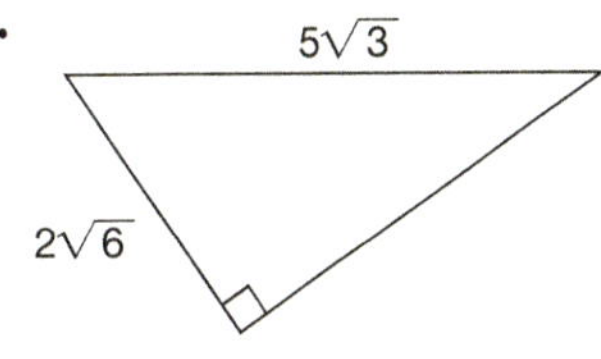

Find the distance between the given pairs of points. Then find the midpoint of each segment. [11.6]

44. $(5, -3)$ and $(9, -1)$
45. $(-2, -4)$ and $(7, 8)$
46. Write the equation of the line passing through $(5, -3)$ and $(9, -1)$. [6.4]
47. Write the equation of the line passing through the origin and perpendicular to the line containing the points $(-2, -4)$ and $(7, 8)$. [6.5]

Simplify. [11.3–11.4, 11.7]

48. $\sqrt[3]{54} \cdot \sqrt[3]{36}$
49. $\sqrt{12a^7b} \div \sqrt{18ab^3}$
50. $\dfrac{\sqrt{2}}{1 - \sqrt{2}}$
51. $\dfrac{2 - \sqrt{2}}{2 + \sqrt{2}}$

CHAPTER 11 REVIEW

Mentally estimate each root to the nearest integer.

1. $\sqrt{127}$ **2.** $\sqrt[3]{50}$

Use a calculator to approximate each radical to the nearest hundredth.

3. $\sqrt{0.63}$ **4.** $\sqrt[3]{619}$

Convert each radical to exponential form.

5. $\sqrt{2x^3}$ **6.** $\sqrt[4]{3y^2}$

Convert each exponential expression to simplified radical form.

7. $3^{\frac{2}{3}}a^{\frac{5}{3}}$ **8.** $(16xy)^{\frac{1}{3}}z^{\frac{4}{3}}$

Simplify. If the expression is not a real number, state "not real." Do not assume that variables in the radicand are nonnegative.

9. $\sqrt[3]{-8}$ **10.** $\sqrt{-25}$ **11.** $\sqrt{84}$

12. $\sqrt[5]{1728}$ **13.** $\sqrt{12a^3b^4}$ **14.** $\sqrt[3]{81x^2y^4}$

Simplify. Assume that all variables in the radicand represent nonnegative values.

15. $\sqrt[3]{21} \cdot \sqrt[3]{9}$ **16.** $\sqrt{8} \cdot \sqrt{18}$ **17.** $\sqrt{20ab} \cdot \sqrt{63a^2b^3}$

18. $\sqrt[3]{9a^5b^2} \cdot \sqrt[3]{3ab^4}$ **19.** $\sqrt{35}(\sqrt{7} + \sqrt{5})$ **20.** $\sqrt{10g}(\sqrt{2g} - 3\sqrt{15})$

21. $\frac{5}{\sqrt{6}}$ **22.** $\frac{\sqrt[3]{3}}{\sqrt[3]{27}}$ **23.** $\frac{\sqrt{7x}}{\sqrt{27x^2y}}$

24. $\sqrt{2x^{12}y^3} \div \sqrt{8x^4y^9}$ **25.** $\sqrt{6} - \sqrt{24} + 5\sqrt{6}$ **26.** $\sqrt{18} + \sqrt{12} + \sqrt{8}$

27. $\sqrt[3]{7} + \sqrt[3]{56} - \sqrt[3]{189}$ **28.** $a\sqrt{32a} + \sqrt{8a^3} - \sqrt{18a^5}$ **29.** $(\sqrt{2} + \sqrt{3})(\sqrt{2} - \sqrt{3})$

30. $(5\sqrt{2} - \sqrt{3})^2$ **31.** $\frac{6}{3 + \sqrt{2}}$ **32.** $\frac{3\sqrt{3}}{\sqrt{3} - 3\sqrt{2}}$

Find each missing length, expressing each answer as both a simplified radical and a decimal approximation to the nearest hundredth.

33.

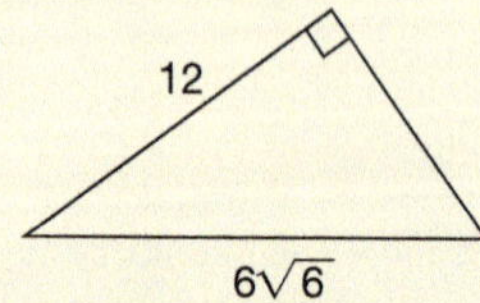

34.

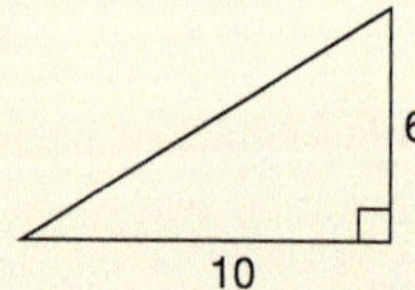

Find the distance between the given pairs of points. Express each answer as a simplified radical.

35. (1, 7) and (−3, 9) **36.** (4, −5) and (−5, 4)

Find the midpoint of a segment with the given endpoints.

37. (1, 7) and (−3, 9) **38.** (−3, 2) and (−1, −5)

Solve and check your solution.

39. $\sqrt{x} = 16$ **40.** $\sqrt{x + 5} = 25$

41. $\sqrt{x - 2} + 8 = 5$ **42.** $10 - 3\sqrt{x + 7} = 4$

The length of each edge of a cube is $2\sqrt[3]{5}$ cm.

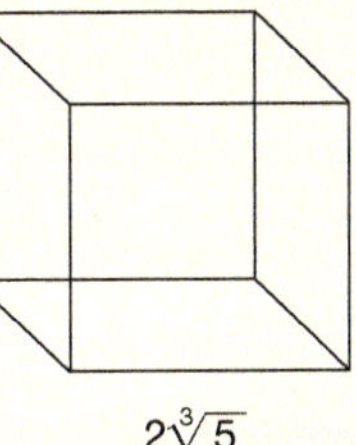

$2\sqrt[3]{5}$

43. What is the area of a side of the cube?

44. What is the volume of the cube?

Use the formula $T = 2\pi\sqrt{\frac{l}{g}}$ to answer the following exercises. (T is a pendulum's period in seconds, l is its length in feet, and g is the gravitational acceleration of 32 ft/sec^2.)

45. On March 31, 1851, Léon Foucault used a 220 ft long pendulum hung in the Panthéon of Paris to demonstrate the rotation of the earth to Napoleon Bonaparte. How long did it take for the pendulum to swing back and forth? Express your answer as both a simplified radical expression and a decimal approximation to the nearest tenth of a second.

46. If the Foucault pendulum in Chicago's Museum of Science and Industry takes 9 sec to swing back and forth, how long is this pendulum (to the nearest tenth of a foot)?

47. The aspect ratio of many widescreen televisions is 16 : 9. If the advertised 42 in. screen of a television is its diagonal measurement, determine the width and the height of the screen (to the nearest tenth of an inch).

42 in.
$9x$
$16x$

Determine the domain and make a table of ordered pairs; then graph each function.

48. $y = \sqrt{x} - 3$

49. $y = \sqrt{x - 2}$

50. $y = \sqrt{-x}$

51. Which of the following is the function rule for the given graph?

a. $f(x) = \sqrt{x - 1} + 3$ **b.** $f(x) = \sqrt{x + 1} + 3$

c. $f(x) = \sqrt{x + 1} - 3$ **d.** $f(x) = \sqrt{x - 3} + 1$

e. $f(x) = \sqrt{x + 3} + 1$ **f.** $f(x) = \sqrt{x + 3} - 1$

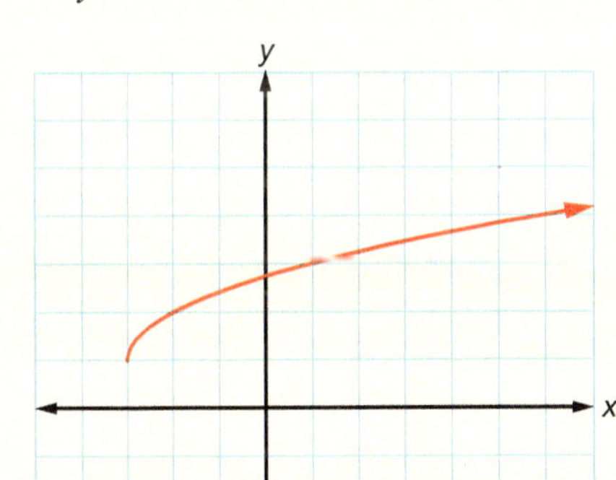

52. Write the function rule for the translation of $y = \sqrt{x}$ three units down and five units left.

12 Quadratic Functions

What comes to your mind when you see a waterfall or a fountain? Waterfalls such as Niagara Falls in North America and Angel Falls in Venezuela inspire thoughts of the power of God and the beauty of His creation. When Christian mathematicians look at a waterfall and consider the laws of physics in action, they can also see the consistency of God's creation and be reminded of His faithfulness to us.

Man-made fountains are also fascinating to study. The path the water follows as it is projected from the fountainhead can be mathematically modeled by a quadratic equation, whose familiar parabolic form illustrates the trajectory of many projectiles. Although fountains are often designed solely for beauty, some are also used for water-cooled air-conditioning systems. Sprinklers create variations of fountains that are used to keep our lawns looking nice and, more importantly, to irrigate crops.

After this chapter you should be able to

1. solve quadratic equations by factoring, taking roots, completing the square, and using the quadratic formula.
2. express quadratic functions in both standard form and vertex form.
3. graph quadratic functions.
4. find the zeros of a quadratic function.
5. use quadratic equations and functions to solve problems.

12.1 Solving Quadratic Equations by Factoring

It is said that Galileo dropped balls from the Leaning Tower of Pisa and showed that the falling of objects can be described by a quadratic relationship.

The first equations you learned to solve were linear equations. A linear equation in one variable can be put into the general form $ax + b = 0$, where $a \neq 0$. The left side of the equation is a binomial with a degree of 1. For this reason, linear equations are also called first-degree equations. Second-degree equations are called *quadratic equations*. "Quadratic" comes from the word "quadrate," meaning square or squared.

Definition

A **quadratic equation** is an equation of the second degree. Its *standard form* is $ax^2 + bx + c = 0$, where a, b, and c are real numbers and $a \neq 0$.

What is the value of x in the equation $3x = 0$? If $3x = 0$, then $x = 0$. In the equation $xy = 0$, either $x = 0$ or $y = 0$, or possibly both equal zero. The *Zero Product Property* is used to solve quadratic equations. It states that if the product of two or more factors is zero, then at least one of the factors is zero.

Zero Product Property

If $pq = 0$, then $p = 0$ or $q = 0$.

Example 1

Solve $(x - 5)(x + 2) = 0$.

Answer

$x - 5 = 0$	or $x + 2 = 0$	1. Set each factor equal to zero using the Zero Product Property.
$x = 5$	$x = -2$	2. Solve each linear equation.

There are two solutions: $x = 5, -2$.

Many quadratic equations will need to be factored before you can apply the Zero Product Property. Remember to look for common factors as your first step in factoring.

Example 2

Solve $x^2 + 8x + 12 = 0$.

Answer	$(x + 2)(x + 6) = 0$	1. Factor the trinomial.
	$x + 2 = 0$ or $x + 6 = 0$	2. Apply the Zero Product Property.
	$x = -2$ $\quad$ $x = -6$	3. Solve each equation.
Check	$(-2)^2 + 8(-2) + 12 = 4 - 16 + 12 = 0$	4. Substitute each solution into the original equation.
	$(-6)^2 + 8(-6) + 12 = 36 - 48 + 12 = 0$	

Example 3

Solve $8x^2 - 16x = 0$.

Answer	$8x(x - 2) = 0$	1. Factor $8x$ from each term.
	$8x = 0$ or $x - 2 = 0$	2. Apply the Zero Product Property.
	$x = 0$ $\quad$ $x = 2$	3. Solve each equation.
Check	$8(0)^2 - 16(0) = 0$	4. Substitute each solution into the original equation.
	$8(2)^2 - 16(2) = 32 - 32 = 0$	

When solving a quadratic equation by factoring, be sure that the equation is in standard form first.

Solving Quadratic Equations by Factoring

1. Write the equation in standard form: $ax^2 + bx + c = 0$.
2. Factor the polynomial side of the equation.
3. Set each factor that contains a variable equal to zero using the Zero Product Property.
4. Solve the resulting equations.
5. Check your solutions in the original equation.

Example 4

Solve $2x^2 + 8x - 42 = 0$.

Answer	$2(x^2 + 4x - 21) = 0$	1. Factor the trinomial.
	$2(x + 7)(x - 3) = 0$	
	$2 \neq 0$ or $x + 7 = 0$ or $x - 3 = 0$	2. Apply the Zero Product Property.
	$x = -7$ $\quad$ $x = 3$	3. Solve each equation.
Check	$2(-7)^2 + 8(-7) - 42 = 98 - 56 - 42 = 0$	4. Substitute each solution into the original equation.
	$2(3)^2 + 8(3) - 42 = 18 + 24 - 42 = 0$	

When an equation has more than one solution, those solutions are often listed as a set. The solution set for Example 4 is $\{-7, 3\}$.

Example 5

Solve $3x^2 = 147$ and state the solution set.

Answer	$3x^2 - 147 = 0$	1. Write the equation in standard form.
	$3(x^2 - 49) = 0$	2. Factor 3 from each term.
	$3(x - 7)(x + 7) = 0$	3. Factor the binomial.
	$3 \neq 0$ or $x - 7 = 0$ or $x + 7 = 0$	4. Apply the Zero Product Property.
	$x = 7$ $\quad$ $x = -7$	5. Solve each equation.
	The solution set is $\{\pm 7\}$.	6. Check mentally and state the solution set.

Quadratic equations will have at most two real number solutions. The Zero Product Property can be applied to equations with higher degrees after they have been factored. An equation such as $4x^3 = 14x^2 + 8x$ can have up to three real number solutions.

Example 6

Find the solution set for the equation $4x^3 = 14x^2 + 8x$.

Answer

$4x^3 - 14x^2 - 8x = 0$	1. Set the equation equal to zero.
$2x(2x^2 - 7x - 4) = 0$	2. Factor $2x$ from each term.
$2x(2x^2 - 8x + 1x - 4) = 0$ $2x[2x(x - 4) + 1(x - 4)] = 0$ $2x(2x + 1)(x - 4) = 0$	3. Factor the trinomial using the technique you learned in Section 10.3.
$2x = 0$ or $2x + 1 = 0$ or $x - 4 = 0$	4. Apply the Zero Product Property.
$x = 0$ $2x = -1$ $x = 4$ $x = -\frac{1}{2}$	5. Solve each equation.
The solution set is $\left\{-\frac{1}{2}, 0, 4\right\}$.	6. State the solution set.

Check

$4(0)^3 - 14(0)^2 - 8(0) = 0$ $4\left(-\frac{1}{2}\right)^3 - 14\left(-\frac{1}{2}\right)^2 - 8\left(-\frac{1}{2}\right) = -\frac{1}{2} - \frac{7}{2} + 4 = 0$ $4(4)^3 - 14(4)^2 - 8(4) = 256 - 224 - 32 = 0$	7. Substitute each solution into the original equation.

Dividing both sides of an equation by a variable expression causes a solution to be lost. Dividing both sides of the original equation in Example 6 by x would eliminate the solution of zero.

Some word problems require you to solve a quadratic equation. Be sure to set the equation equal to zero before factoring and applying the Zero Product Property.

Example 7

An architect wishes to design a room for a home addition in which the length is 5 ft longer than the width. If the room needs to be 204 ft², find the dimensions.

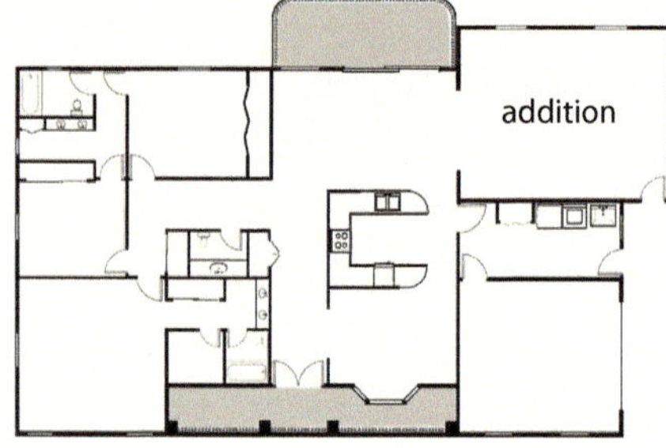

Answer

Let w = the width and $w + 5$ = the length.	1. Assign variable expressions to the unknowns.
$(w + 5)w = 204$	2. Use the formula $A = lw$ to write an equation modeling the problem.
	3. Solve the resulting quadratic equation.
$w^2 + 5w = 204$ $w^2 + 5w - 204 = 0$	a. Write the equation in standard form.
$(w + 17)(w - 12) = 0$	b. Factor the trinomial.
$w + 17 = 0$ or $w - 12 = 0$	c. Apply the Zero Product Property.
$w = -17$ $w = 12$	d. Solve each equation.
dimensions: 12 ft × 17 ft	4. Interpret the solutions to your equation. Since −17 is not a reasonable length, the width must be 12 ft and the length is 12 + 5 = 17 ft.

Check

area = 12 ft × 17 ft = 204 ft²	5. Check your solution in the original problem.

A. Exercises

Solve.

1. $(x - 2)(x + 4) = 0$
2. $(y + 5)(y + 2) = 0$
3. $2z(z + 7) = 0$
4. $3x(x + 12) = 0$
5. $8x(x + 1)(x - 10) = 0$
6. $-5a(a - 1)(3a + 2) = 0$
7. $(4b + 1)(3b + 2) = 0$
8. $(2x - 11)(4x - 5) = 0$

Solve by factoring.

9. $y^2 + 6y + 9 = 0$
10. $a^2 - 8a + 15 = 0$
11. $x^2 - 9 = 0$
12. $y^2 - 225 = 0$
13. $a^2 + 5a - 36 = 0$
14. $x^2 - x - 56 = 0$

B. Exercises

Solve by factoring and state the solution set.

15. $x^2 - x = 0$
16. $x^2 + 3x = 0$
17. $c^2 = 7c$
18. $d^2 = 5d$
19. $x^2 - 10x = -16$
20. $m^2 - 2m = 35$
21. $p^2 = 2p + 24$
22. $x^2 - 12 = -4x$
23. $y^2 = 169$
24. $16y^2 = 9$
25. $4x^2 = 9$
26. $2a^2 + 100 = -30a$
27. $3x^2 - 216 = 63x$
28. $8x^2 + 10x = -3$
29. $10t^2 = -7t + 12$
30. $14x^2 + 12 = 29x$
31. $5r^3 + 5r^2 = 10r$
32. $4s^3 = 16s$

Write a quadratic equation and solve.

33. Two consecutive integers have a product of 240. Find the integers.
34. Two consecutive integers have a product that is 14 less than ten times the smaller integer. Find the integers.
35. Brady wants to lay a sidewalk within his flower garden, which is 52 ft long and 44 ft wide. How wide should the sidewalk be if he desires to keep 1748 ft^2 for the garden area?
36. The difference of two numbers is 6 and the difference of their squares is 60. Find the numbers.

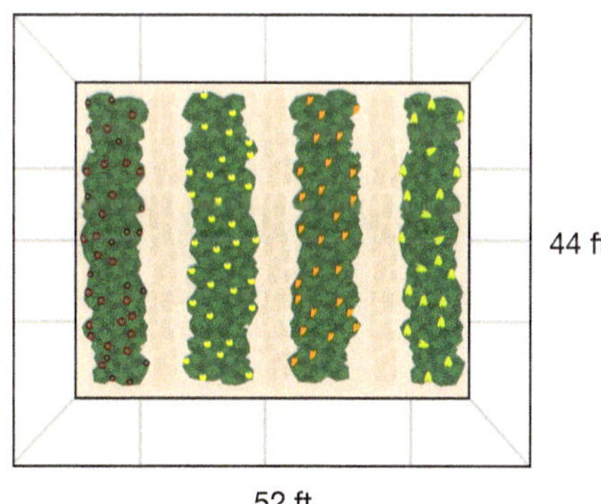

C. Exercises

Write a quadratic equation and solve.

37. Javier said, "If five times my age in eight years is subtracted from the square of my present age, the result is 86." Find Javier's present age.
38. A gardener currently has a vegetable garden 12 ft wide and 20 ft long. He wishes to increase the area to 560 ft^2 by adding the same amount to the garden's length and width. How many feet should be added? What are the new dimensions?
39. Find the solution set for the equation $15x^3 = 17x^2 + 42x$.
40. Use the fact that $x - 2$ is a factor of $x^3 + 9x^2 + 2x - 48$ to find the solution set for the equation $x^3 + 9x^2 + 2x - 48 = 0$.

Dominion Modeling

Water wheels, invented by the ancient Greeks, are often located close to a waterfall to take advantage of the water's potential energy. Water wheels were used into the twentieth century to operate mills. Similar technology is used in modern hydroelectric power plants.

The vertical velocity of falling water, v, in feet per second can be calculated using the formula $v = \sqrt{2gd}$, where g is the gravitational acceleration of 32 ft/sec^2 and d is the distance the water has fallen in feet.

41. Find the velocity of the water if it has fallen 1.5 ft.
42. Derive the quadratic equation describing the distance, d, that the water must fall to attain a given vertical velocity, v, by solving the equation $v = \sqrt{2gd}$ for d.

CUMULATIVE REVIEW

Solve. Write ∅ if no solution exists. [4.6]

43. $|x - 4| = 2$
44. $|x + 8| = -2$
45. $\frac{1}{2}|x + 7| = 1$

Find all positive values for k such that the trinomial will factor over the set of integers. [10.2]

46. $x^2 + kx + 21$
47. $x^2 - kx + 12$
48. $x^2 + kx - 12$

Simplify. [11.2]

49. $\sqrt{108}$
50. $\sqrt{(-7)^2 - 4(1)(-2)}$
51. $\sqrt{12^2 - 4(-3)(-1)}$
52. $\sqrt{22}(4\sqrt{20} - 3\sqrt{55})$

MIND OVER MATH

How quickly can you find each product? Would it be faster with a calculator? Why or why not?

1. $\frac{583}{12} \cdot 8 \cdot 134 \cdot \frac{52}{77} \cdot \frac{6}{901} \cdot 596 \cdot 0$
2. $\frac{7}{19} \cdot \frac{19}{5} \cdot \frac{5}{41} \cdot \frac{41}{37} \cdot \frac{37}{8} \cdot \frac{8}{3} \cdot \frac{3}{7}$
3. $\frac{19}{23} \cdot \frac{14}{15} \cdot \frac{37}{40} \cdot \frac{15}{19} \cdot \frac{23}{27} \cdot \frac{11}{14} \cdot \frac{27}{37}$

12.2 Solving Quadratic Equations by Taking Roots

Equations of the form $x^2 - c = 0$ in which c is a perfect square can be solved using the factoring method from the previous section.

$$x^2 - 64 = 0$$
$$(x - 8)(x + 8) = 0$$
$$x - 8 = 0 \text{ or } x + 8 = 0$$
$$x = 8 \text{ or } x = -8$$

Another method of solving equations of this form is to isolate the squared term and then take the square root of each side of the equation.

Example 1

Solve $x^2 - 64 = 0$.

Answer

$x^2 = 64$	1. Isolate the squared term.
$\sqrt{x^2} = \sqrt{64}$	2. Take the square root of each side.
$\lvert x \rvert = 8$	3. Remember that $\sqrt{x^2} = \lvert x \rvert$.
$x = \pm 8$	4. Solve.

Consider $x^2 = 0$. How many solutions are there for x? Is there a value of x such that $x^2 = -16$?

Square Root Property

1. If $n \geq 0$ and $x^2 = n$, then $x = \pm\sqrt{n}$.
   a. If $n > 0$, then x has two real solutions.
   b. If $n = 0$, then x has one real solution.
2. If $n < 0$ and $x^2 = n$, then x has no real solutions.

Example 2

Solve $2x^2 - 24 = 0$.

Answer

$2x^2 - 24 = 0$	1. Isolate the squared variable.
$2x^2 = 24$	a. Isolate the term containing x^2.
$x^2 = 12$	b. Isolate x^2 by dividing both sides by 2.
$x = \pm\sqrt{12}$	2. Apply the Square Root Property.
$x = \pm 2\sqrt{3}$	3. Simplify.

Check

$2(2\sqrt{3})^2 - 24 = 2(12) - 24 = 0$	4. Substitute each solution into the original equation.
$2(-2\sqrt{3})^2 - 24 = 2(12) - 24 = 0$	

The same procedure can be applied to equations containing a polynomial that is squared.

Example 3

Solve $(x + 2)^2 = 36$.

Answer	$(x + 2)^2 = 36$	1. The left side is already a perfect square.
	$x + 2 = \pm 6$	2. Apply the Square Root Property.
	$x = -2 \pm 6$	3. Solve by adding -2 to both sides. Note that -2 ± 6 means $-2 + 6$ or $-2 - 6$.
	$x = 4, -8$	
Check	$(4 + 2)^2 = 36$ and $(-8 + 2)^2 = 36$	4. Both answers are solutions.

Solving Equations of the Form $ax^2 - c = 0$

1. Isolate the squared expression.
2. Apply the Square Root Property.
3. Simplify the radical and solve each equation.

Example 4

Solve $(x - 5)^2 = 21$. State your solutions as simplified radical expressions and as a solution set containing decimal approximations rounded to the nearest hundredth.

Answer	$(x - 5)^2 = 21$	1. The left side is already a perfect square.
	$x - 5 = \pm\sqrt{21}$	2. Apply the Square Root Property.
	$x = 5 \pm \sqrt{21}$	3. Solve and state the solutions as simplified radical expressions.
	$x \approx 5 \pm 4.58$	4. Use a calculator to find the decimal approximations.
	The solution set is {0.42, 9.58}.	
Check	$(0.42 - 5)^2 = (-4.58)^2 = 20.9764 \approx 21$ $(9.58 - 5)^2 = 4.58^2 = 20.9764 \approx 21$	5. Use the decimal approximations and a calculator to check the solutions.

Example 5

Solve $7x^2 + 6 = 2x^2 + 8$.

Answer	$5x^2 + 6 = 8$	1. Isolate x^2.
	$5x^2 = 2$	
	$x^2 = \frac{2}{5}$	
	$x = \pm\sqrt{\frac{2}{5}}$	2. Apply the Square Root Property.
	$x = \pm\frac{\sqrt{2}}{\sqrt{5}} \cdot \frac{\sqrt{5}}{\sqrt{5}} = \pm\frac{\sqrt{10}}{5}$	3. Rationalize the denominator.

The equation $x^2 + 7 = 0$ cannot be solved using the factoring method from the previous section. Solving for x^2 and applying the Square Root Property produces $x = \pm\sqrt{-7}$. Since we cannot square a real number to produce -7, these solutions are complex numbers (which will be discussed in later courses). For now, solve the problem, leaving the negative number under the radical; and state that there are no real solutions.

Assuming negligible air resistance, the height of a dropped object can be modeled by the function $h(t) = -16t^2 + h_i$, where t is the time in seconds and h_i is the initial height in feet.

Example 6

If Jed is 150 ft high when his keys fall from his pocket, how long does it take for his keys to reach the ground?

Answer

$h(t) = -16t^2 + h_i$ $0 = -16t^2 + 150$	1. Substitute the initial height of 150 ft and find t when the keys are at a height of 0 ft.
$16t^2 = 150$ $t^2 = \frac{150}{16}$	2. Isolate t^2.
$t = \pm\sqrt{\frac{150}{16}} = \pm\frac{5\sqrt{6}}{4}$	3. Apply the Square Root Property and simplify.
$t \approx 3.06$ sec	4. Calculate the decimal approximation of the answer. Only the positive answer is a reasonable solution.

A. Exercises

Solve using the Square Root Property.

1. $x^2 = 25$ **2.** $x^2 = 121$ **3.** $x^2 = 6$

4. $x^2 = 15$ **5.** $x^2 - 100 = 0$ **6.** $x^2 - 16 = 0$

7. $x^2 - 5 = 31$ **8.** $x^2 + 3 = 147$ **9.** $x^2 + 5 = 0$

10. $3x^2 = 15$ **11.** $3x^2 = 51$ **12.** $-4x^2 = 64$

B. Exercises

Solve by taking roots.

13. $3x^2 = 81$ **14.** $3x^2 = 72$ **15.** $4x^2 = 50$

16. $6x^2 = 40$ **17.** $4x^2 - 16 = 3x^2 + 9$ **18.** $2x^2 - 17 = x^2 + 3$

19. $3x^2 + 10 = x^2 + 24$ **20.** $(x - 6)^2 = 13$ **21.** $(x + 2)^2 = 12$

22. $(x + 4)^2 = 25$ **23.** $(x + 3)^2 = 121$ **24.** $(x - 5)^2 + 17 = 0$

25. $4(x + 2)^2 = 0$ **26.** $5x^2 + x - 2 = x^2 + x + 7$ **27.** $(4x + 1)^2 = 9$

28. $(9x - 2)^2 = 16$ **29.** $4(x + 1)^2 = 9$ **30.** $9(x - 2)^2 = 16$

Write a quadratic equation and solve. Use a calculator and the decimal approximation of 3.14 for π.

31. A circular sprinkler is advertised to have coverage of about 450 ft^2. Find its range (radius) to the nearest foot.

32. The volume of the circular cylinder to the right is 45,600 ft^3. Find its radius to the nearest foot.

30 ft

Write a quadratic equation and solve. Assume that air resistance is negligible.

33. Dave dropped an object partway up his rock climb. If the object took exactly 2 sec to hit the ground, how high had he climbed?

34. A construction worker dropped a bolt from the twelfth floor, a height of 120 ft. How long (to the nearest tenth of a second) did it take to reach the ground?

35. How many seconds will it take for an object to fall 200 ft? 400 ft? Explain why the time does not double.

C. Exercises

Solve by taking roots.

36. $2(5x - 3)^2 = 7$

37. $4x^2 - 20x + 25 = 32$

38. An architect plans to center a square building that will be 150 ft on each side within a 40,000 ft^2 square lot. How far will the resulting courtyard extend in each direction from the building?

39. Another common formula that models objects in free fall is $d = \frac{1}{2}gt^2$, where d is the distance fallen, g is the gravitational acceleration, and t is the time.

a. Use the distance formula $d = |b - a|$ to write an expression for the distance an object falls from an initial height of h_i to its current height, h_c.

b. Substitute the expression for distance into the equation $d = \frac{1}{2}gt^2$ and solve for h_c. Be sure to consider the relative magnitudes of h_c and h_i when interpreting the absolute value.

c. The gravitational acceleration near the surface of the earth is approximately 32 ft/sec^2. Substitute this value into the derived formula for h_c to verify the function rule for free fall used in Example 6.

d. Convert 32 ft/sec^2 to meters per second squared (to the nearest tenth of a meter). Then state the function rule for the height of an object in free fall for t seconds with an initial height of h_i meters.

Dominion Modeling

40. Water droplets coming off a waterfall will fall at the same rate as any other falling object. The function $h(t) = -16t^2 + 65$ models the height of a water droplet at time t.

a. Use the function to determine the height of the waterfall.

b. How long (to the nearest tenth of a second) does it take for a droplet to complete its fall?

41. The path of a water droplet can be modeled by the function $h(x) = ax^2 + 65$, where $h(x)$ is the height of the water and x is its horizontal distance from the top of the waterfall. Substitute each of the following ordered pairs into the equation and solve for a. Then write the function rule that models the path of the droplet.

a. (4, 0)

b. (5.5, 0)

c. (4.75, 0)

d. Model the waterfall by plotting all three functions on your graphing calculator. State the size of an appropriate viewing window.

© Odell Garrison

Hikers to Dry Falls near Highlands, NC, can pass under the falls and stay dry.

Keyword Search
quadratic equation uses

CUMULATIVE REVIEW

42. What is the easiest way to find half of any fraction with an even numerator? [1.5]

43. What is the easiest way to find half of any fraction with an odd numerator? [1.5]

Translate each phrase into a numerical expression. Then evaluate your expression. [1.5, 1.7]

44. half of $\frac{4}{5}$, squared

45. half of $\frac{3}{8}$, squared

Factor and solve. [10.5, 12.1]

46. $x^2 - 14x - 51 = 0$

47. $9x^2 - 64 = 0$

48. $x^4 - 13x^2 + 36 = 0$

49. $2x^3 + x^2 - 8x - 4 = 0$

Write an equation and solve. Use a table when needed.

50. Jesse has $30 more in an account earning 6% than he has in another account earning 5%. If his total annual interest will be $100, how much is in each account? [3.6]

51. Shay has a total of $8.20. If she has three more dimes than nickels and five fewer quarters than dimes, how many of each coin does she have? [2.8]

TECHNOLOGY CORNER (TI-84+ Family)

Your graphing calculator can be used to find decimal equivalents of answers in radical form and to check your answers in the original equation. The [ANS] function ([2nd] [(-)]) is used to insert the previous answer into the expression to be calculated. By repeatedly selecting the [ENTRY] function ([2nd] [ENTER]), you can scroll through recently entered expressions in order to find and edit a similar expression without reentering the entire expression.

The answers to quadratic equations often come in conjugate pairs, such as $\frac{3 \pm \sqrt{5}}{7}$, the solutions to $49x^2 - 42x + 4 = 0$. To find the decimal equivalent of $\frac{3 + \sqrt{5}}{7}$, press [(] 3 [+] [2nd] [x^2] 5 [)] [)] [÷] 7 [ENTER]. To find the decimal equivalent of the second solution, $\frac{3 - \sqrt{5}}{7}$, press [ENTRY] and use [◄] to place the cursor over the [+] sign. Change it to [−] and press [ENTER].

```
(3+√(5))/7
           .7480097111
(3-√(5))/7
           .1091331461
```

The second solution can be checked in the equation $49x^2 - 42x + 4 = 0$ by entering 49 [ANS] [x^2] [−] 42 [ANS] [+] 4 [ENTER]. Since the expression evalutes to 0, your solution is correct.

```
(3+√(5))/7
           .7480097111
(3-√(5))/7
           .1091331461
49Ans²-42Ans+4
                     0
■
```

To check the first solution, press [ENTRY] three times or until the entry (3+√(5))/7 is displayed at the bottom of the screen. Press [ENTER] to display the decimal solution. Press [ENTRY] twice to display 49Ans²−42Ans+4 at the bottom of the screen; then press [ENTER]. Again, you are looking for a result of 0.

```
           .1091331461
49Ans²-42Ans+4
                     0
(3+√(5))/7
           .7480097111
49Ans²-42Ans+4
                     0
```

John von Neumann (1903–57)

John von Neumann, a Hungarian-American mathematician, is well known in a variety of mathematical fields, especially computer science, game theory, and quantum mechanics. His discoveries helped bring modern society into both the atomic age and the computer age. Von Neumann was a child prodigy with a photographic memory. By age 10 he had completed college calculus. He earned a PhD in math at age 23 from the University of Budapest and had written ten major professional papers by age 25.

After teaching at two universities in Germany, von Neumann came to the United States as a visiting professor at Princeton. In 1933 he and Albert Einstein became two of the original faculty members of the Institute for Advanced Study at Princeton.

In 1937 von Neumann became an American citizen. During World War II he worked with the US War Department, studying submarine warfare, economic intelligence, and ordnance problems. His most famous work originated in secret meetings at Los Alamos, New Mexico, where scientists and mathematicians gathered to develop the atomic bomb. By creating a means of detonation by an inward blast, von Neumann expedited the development of the atomic bomb by a year.

After the war ended in 1945, he became the director of Princeton Institute's Electronic Computer Project, where he and his group developed new high-speed computers and where he continued working with computers until his death. His work in mathematics and computer technology confirmed his sense that God does indeed exist. He once said, "There is no comparison between the human nervous system and the most complicated machine that human intelligence has ever devised, or can devise. No man can tell me that behind the complications of the human nervous system there is no such thing as a greater intelligence. For me, that other intelligence is God."

12.3 Completing the Square

Some quadratic equations, such as $a^2 + 6a + 2 = 0$, cannot be solved by factoring. The method discussed in the previous section cannot be applied since isolating the squared expression, a^2, leaves a variable in the other side of the equation, $-6a - 2$.

Adding the right number, 7, to both sides produces an equivalent equation, $a^2 + 6a + 9 = 7$. The left side is now a perfect square trinomial that can be factored. The equation $(a + 3)^2 = 7$ can be solved by taking the square root of each side.

Determining exactly which number should be added to both sides of the equation $a^2 + 6a + 2 = 0$ is simpler when the constant term is moved to the right side of the equation: $a^2 + 6a = -2$. Looking at a geometric representation of $a^2 + 6a$ can help you determine which number should be added to both sides in order to create a perfect square trinomial.

1. Create a geometric representation for $a^2 + 6a$.

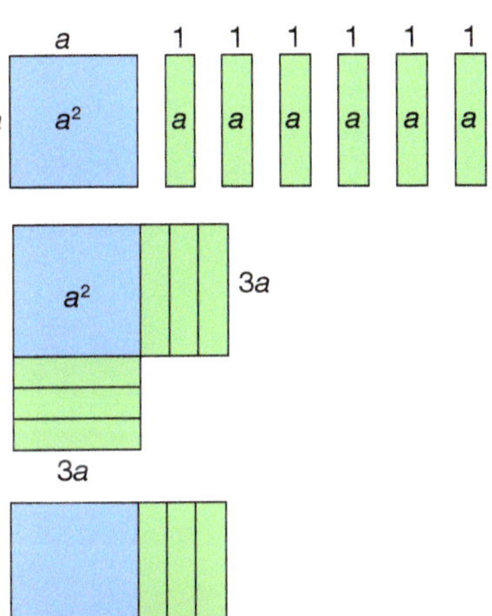

2. Arrange the tiles to form the upper and left edges of a square.

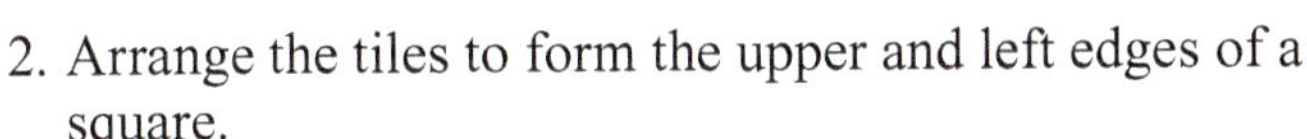

3. Complete the square. Note that half of the six strips were used on each side of the square, resulting in the addition of nine unit squares to complete the square's shape.

When the square of one-half of the coefficient of the a term, $\left(\frac{6}{2}\right)^2 = 3^2 = 9$, is added to both sides of $a^2 + 6a = -2$, the left side can be factored as a perfect square trinomial. This allows you to solve the equation by taking roots.

$$a^2 + 6a + 9 = -2 + 9$$
$$(a + 3)^2 = 7$$
$$a + 3 = \pm\sqrt{7}$$
$$a = -3 \pm \sqrt{7}$$

The process of adding a number to the expression $x^2 + bx$ to produce a perfect square trinomial is called *completing the square.*

Completing the Square

To complete the square of an expression in the form of $x^2 + bx$, add the square of half the coefficient of x: $x^2 + bx + \left(\frac{b}{2}\right)^2$.

Example 1

For each expression, complete the square and factor the resulting trinomial.

a. $x^2 - 16x$ **b.** $x^2 + \frac{3}{5}x$

Answer

a. $\left(\frac{-16}{2}\right)^2 = (-8)^2 = 64$

$x^2 - 16x + 64$

$(x - 8)^2$

b. $\left(\frac{1}{2} \cdot \frac{3}{5}\right)^2 = \left(\frac{3}{10}\right)^2 = \frac{9}{100}$

$x^2 + \frac{3}{5}x + \frac{9}{100}$

$\left(x + \frac{3}{10}\right)^2$

1. Find the square of half the coefficient of x.
2. Add the result to the binomial.
3. Factor the perfect square trinomial.

You can use this method to solve any quadratic equation of the form $x^2 + bx + c = 0$. Remember that if you add a number to one side of an equation, you must add the same number to the other side.

Example 2

Solve $x^2 + 8x = 9$.

Answer

$x^2 + 8x + 16 = 9 + 16$ — 1. Complete the square by adding $\left(\frac{8}{2}\right)^2 = 16$ to both sides.

$(x + 4)^2 = 25$ — 2. Factor the left side of the equation and simplify the right side.

$x + 4 = \pm 5$ — 3. Apply the Square Root Property.

$x = -4 \pm 5$ — 4. Solve.

$x = 1, -9$

Check

$1^2 + 8(1) = 1 + 8 = 9$

$(-9)^2 + 8(-9) = 81 - 72 = 9$

Solving Quadratic Equations by Completing the Square

1. Write the equation in the form $x^2 + bx = c$.
2. Complete the square by adding $\left(\frac{b}{2}\right)^2$ to both sides of the equation.
3. Factor the trinomial.
4. Apply the Square Root Property: If $x^2 = n$, then $x = \pm\sqrt{n}$.
5. Solve the resulting equations.

Example 3

Solve $a^2 - 7a - 3 = 0$.

Answer

$a^2 - 7a = 3$ — 1. Write the equation in $x^2 + bx = c$ form by adding 3 to both sides.

$a^2 - 7a + \left(\frac{7}{2}\right)^2 = 3 + \frac{49}{4}$ — 2. Complete the square by adding $\left(\frac{-7}{2}\right)^2 = \left(\frac{7}{2}\right)^2 = \frac{49}{4}$ to both sides.

$\left(a - \frac{7}{2}\right)^2 = \frac{61}{4}$ — 3. Factor the left side and simplify the right side.

$a - \frac{7}{2} = \pm\frac{\sqrt{61}}{2}$ — 4. Apply the Square Root Property.

$a = \frac{7 \pm \sqrt{61}}{2}$ — 5. Solve.

The solution set is $\left\{\frac{7 - \sqrt{61}}{2}, \frac{7 + \sqrt{61}}{2}\right\} \approx \{-0.41, 7.41\}$.

Example 3 illustrates that the sign of b can be ignored when $\frac{b}{2}$ is squared. Leaving the expression as $\left(\frac{b}{2}\right)^2$ may make the factoring of the perfect square trinomial easier.

Example 4

Cassia increased the size of her square garden to 550 ft^2 by adding 4 ft to its width and 7 ft to its length. Find the dimensions of the original garden.

Answer

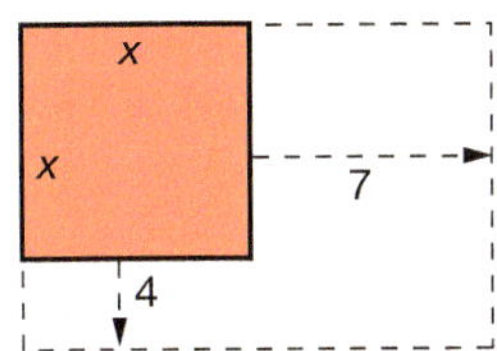

1. Draw a diagram of the problem and assign variable expressions to the unknowns.

Let x = the original length and width,
$x + 4$ = the new width, and
$x + 7$ = the new length.

$$(x + 7)(x + 4) = 550$$

2. Use the area formula to write an equation for the new area.

$$x^2 + 11x + 28 = 550$$
$$x^2 + 11x = 522$$

3. Write the equation in $x^2 + bx = c$ form.

$$x^2 + 11x + \left(\frac{11}{2}\right)^2 = 522 + \frac{121}{4}$$

4. Complete the square by adding $\left(\frac{11}{2}\right)^2 = \frac{121}{4}$ to both sides.

$$\left(x + \frac{11}{2}\right)^2 = \frac{2088}{4} + \frac{121}{4} = \frac{2209}{4}$$

5. Factor the left side and simplify the right side.

$$x + \frac{11}{2} = \pm\sqrt{\frac{2209}{4}} = \pm\frac{47}{2}$$

6. Apply the Square Root Property.

$$x = \frac{-11 \pm 47}{2}$$

7. Solve. Note that a negative answer is not reasonable.

$$x = \frac{-11 + 47}{2} = 18 \text{ or } x = \frac{-11 - 47}{2} = -29$$

The original garden was 18 ft on each side.

While the integral answers in Example 4 indicate that the equation could have been solved by factoring after converting to standard form, finding the correct factoring of $x^2 + 11x - 522$ could be challenging.

A. Exercises

What number must be added to complete each square?

1. $x^2 + 8x$ **2.** $x^2 - 12x$ **3.** $x^2 + 16x$ **4.** $x^2 - 10x$

5. $x^2 - 2x$ **6.** $x^2 + 52x$ **7.** $x^2 + 3x$ **8.** $x^2 + 5x$

9. $x^2 - 11x$ **10.** $x^2 - 9x$ **11.** $x^2 + \frac{3}{2}x$ **12.** $x^2 + \frac{5}{6}x$

13. $x^2 - \frac{10}{11}x$ **14.** $x^2 - \frac{4}{5}x$

Solve by completing the square.

15. $a^2 + 2a = 15$ **16.** $x^2 + 6x = -5$ **17.** $x^2 + 4x = 5$ **18.** $y^2 - 6y = 7$

B. Exercises

Solve by completing the square and state the solution set.

19. $x^2 - 10x - 24 = 0$ **20.** $a^2 + 6a - 55 = 0$ **21.** $x^2 - 2x - 99 = 0$ **22.** $m^2 - 14m + 40 = 0$

Solve by completing the square. State irrational solutions as both simplified radical expressions and as decimal approximations rounded to the nearest hundredth.

23. $m^2 - \frac{4}{3}m = 0$

24. $x^2 + \frac{5}{6}x = 0$

25. $x^2 + x = 2$

26. $a^2 - 7a = -10$

27. $x^2 + 20x + 15 = 0$

28. $x^2 - 12x - 5 = 0$

29. $x^2 - 18x + 9 = 0$

30. $x^2 - 8x - 2 = 0$

31. $x^2 = 8x - 4$

32. $x^2 = 4 + 12x$

33. $4x = x^2 - 3$

34. $10x = x^2 + 7$

35. Increasing the width of a square by 2 in. and the length by 7 in. causes the area of the resulting rectangle to be 204 in.^{2}. What was the size of the original square?

36. Find two integers such that one is 6 more than the other and their product is 216.

C. Exercises

Solve by completing the square.

37. $x^2 + 3x + 1 = 0$

38. $3m^2 + 15m - 36 = 0$

Dominion Modeling

Circular sprinkler systems are used in both agricultural and residential settings. Consider a field whose length is 10 ft longer than its width and whose area is 2376 ft^2.

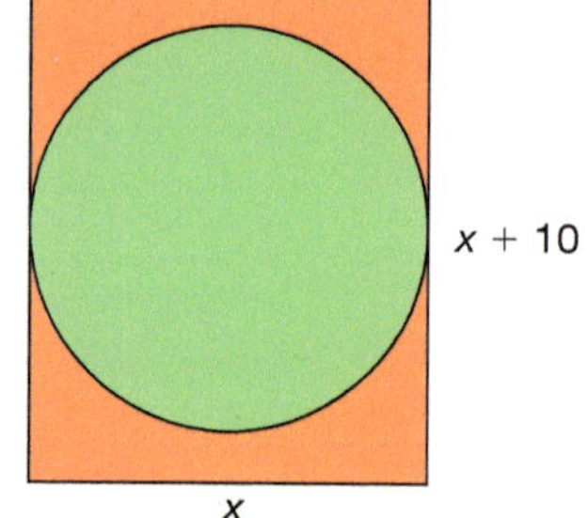

39. Find the dimensions of the field.

40. A single sprinkler head that rotates 360° is positioned in the middle of the field and adjusted to water the entire width of the field.

a. What is the area that this sprinkler waters (to the nearest square foot)?

b. What percent of the field is not watered?

41. By placing two sprinkler heads in opposite corners to rotate 90° with a radius the same as the width of the field, the farmer notices that practically all of the field gets watered.

a. What is the area that each sprinkler waters (to the nearest square foot)?

b. Assuming the unwatered areas are negligible, what percent of the field is watered twice?

CUMULATIVE REVIEW

Simplify. [1.3, 9.6]

42. $\frac{3}{4} + \frac{25}{16}$

43. $-\frac{6}{7} + \frac{16}{49}$

44. $\frac{4x^2 + 12x - 20}{4}$

45. $\frac{3x^2 + 5x - 15}{3}$

46. $\dfrac{\frac{1}{2}x^2 - 3x + 1}{\frac{1}{2}}$

47. $2\left(\frac{1}{2}x^2 - 3x + 1\right)$

Solve. [11.8]

48. $\sqrt{x} = 9$

49. $\sqrt{x} + 10 = 3$

50. $2\sqrt{x} = 5$

51. $-3\sqrt{x - 4} + 9 = 0$

12.4 Completing the Square with Leading Coefficients

The method for completing the square used in the previous section works only when the leading coefficient is one. Given $ax^2 + bx + c = 0$, where $a \neq 1$, you will need to first divide both sides of the equation by a, the leading coefficient.

Example 1

Solve $4x^2 + 16x - 8 = 0$.

Answer

$\frac{4x^2 + 16x - 8}{4} = \frac{0}{4}$	1. Divide both sides of the equation by the leading coefficient, 4.
$x^2 + 4x - 2 = 0$	
$x^2 + 4x = 2$	2. Write the equation in $x^2 + bx = c$ form.
$x^2 + 4x + 4 = 2 + 4$	3. Complete the square by adding $\left(\frac{4}{2}\right)^2 = 4$ to both sides.
$(x + 2)^2 = 6$	4. Factor the trinomial and simplify the right side.
$x + 2 = \pm\sqrt{6}$	5. Apply the Square Root Property.
$x = -2 \pm \sqrt{6}$	6. Solve.

Example 2

Solve $3x^2 - 2x - 1 = 0$.

Answer

$\frac{3x^2 - 2x - 1}{3} = \frac{0}{3}$	1. Divide both sides of the equation by the leading coefficient, 3.
$x^2 - \frac{2}{3}x - \frac{1}{3} = 0$	
$x^2 - \frac{2}{3}x = \frac{1}{3}$	2. Write the equation in $x^2 + bx = c$ form.
$x^2 - \frac{2}{3}x + \left(\frac{1}{3}\right)^2 = \frac{1}{3} + \frac{1}{9}$	3. Complete the square by adding $\left(\frac{1}{2} \cdot \frac{2}{3}\right)^2 = \left(\frac{1}{3}\right)^2 = \frac{1}{9}$ to both sides. Remember that the sign of b can be ignored because $\frac{b}{2}$ is squared.
$\left(x - \frac{1}{3}\right)^2 = \frac{4}{9}$	4. Factor the trinomial and simplify the right side.
$x - \frac{1}{3} = \pm\frac{2}{3}$	5. Apply the Square Root Property.
$x = \frac{1}{3} \pm \frac{2}{3}$	6. Solve.
$x = 1, -\frac{1}{3}$	

Example 3

Solve $2x^2 = 10x + 3$.

Answer

$2x^2 - 10x = 3$

$\frac{2x^2 - 10x}{2} = \frac{3}{2}$

$x^2 - 5x = \frac{3}{2}$

1. Write the equation in $x^2 + bx = c$ form by subtracting $10x$ from both sides and then dividing both sides by 2.

$x^2 - 5x + \left(\frac{5}{2}\right)^2 = \frac{3}{2} + \frac{25}{4}$

2. Complete the square by adding $\left(\frac{5}{2}\right)^2 = \frac{25}{4}$ to both sides.

$\left(x - \frac{5}{2}\right)^2 = \frac{31}{4}$

3. Factor the trinomial and simplify the right side.

$x - \frac{5}{2} = \pm\frac{\sqrt{31}}{2}$

4. Apply the Square Root Property.

$x = \frac{5}{2} \pm \frac{\sqrt{31}}{2}$

$x = \frac{5 \pm \sqrt{31}}{2}$

5. Solve.

Remember that if a radicand is negative, there are no real number solutions.

In previous sections we have modeled the fall of objects, such as dropped keys, whose initial vertical velocity is zero. If a projectile has an initial vertical velocity v_i (measured in ft/sec) and h_i is the initial height, then the function that models the height at time t is $h(t) = -16t^2 + v_it + h_i$.

Example 4

A player pops a baseball straight up from an initial height of 3 ft with an initial velocity of 88 ft/sec. How long after the batter hits the ball is the ball 15 ft high?

Answer

$h(t) = -16t^2 + v_it + h_i$

$15 = -16t^2 + 88t + 3$

1. Substitute the given information into the formula.

$16t^2 - 88t = -12$

$\frac{16t^2 - 88t}{16} = -\frac{12}{16}$

$t^2 - \frac{11}{2}t = -\frac{3}{4}$

2. Write the equation in $ax^2 + bx = c$ form. Then divide both sides by 16 so that $a = 1$.

$t^2 - \frac{11}{2}t + \left(\frac{11}{4}\right)^2 = -\frac{3}{4} + \frac{121}{16}$

3. Complete the square by adding $\left(\frac{1}{2} \cdot \frac{11}{2}\right)^2 = \left(\frac{11}{4}\right)^2 = \frac{121}{16}$ to both sides.

$\left(t - \frac{11}{4}\right)^2 = \frac{109}{16}$

4. Factor the trinomial and simplify the right side.

$t - \frac{11}{4} = \pm\frac{\sqrt{109}}{4}$

5. Apply the Square Root Property.

$t = \frac{11 \pm \sqrt{109}}{4}$

6. Solve.

$t \approx 5.36, 0.14$ sec

7. The baseball is 15 ft high after 0.14 sec and after 5.36 sec.

Keyword Search

completing the square

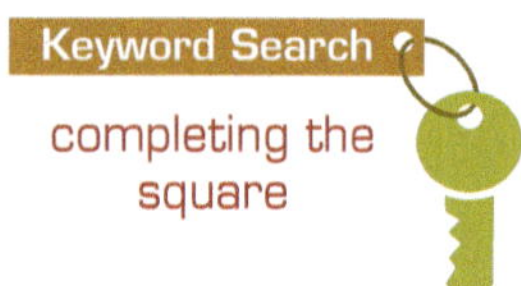

A. Exercises

Solve by completing the square.

1. $2x^2 + 12x = 4$
2. $3y^2 - 12y = -3$
3. $3a^2 - 24a = 0$
4. $5b^2 + 60b = -55$
5. $2y^2 + 16y = 6$
6. $6x^2 + 24x = -12$
7. $2s^2 + 12s + 28 = 0$
8. $3r^2 + 72r = -24$
9. $2x^2 - 16x = 8$
10. $5y^2 - 40y + 100 = 0$

B. Exercises

Solve by completing the square.

11. $2y^2 + y = 3$
12. $2x^2 - x = 10$
13. $4c^2 - 8c - 5 = 0$
14. $9d^2 + 36d - 13 = 0$
15. $2x^2 - x = 1$
16. $3y^2 - y = 2$
17. $2g^2 - g = 2$
18. $2h^2 - 6h = -1$
19. $2y^2 - 3y = -35$
20. $2x^2 - 10x = 3$
21. $3k^2 - 2k - 8 = 0$
22. $3y^2 = -8y - 5$
23. $2x^2 = -3x - 1$
24. $6n^2 = 9n + 27$
25. $4m^2 - 4m = 11$
26. $3x^2 + 18x = 2$
27. $2y^2 + 24y + 1 = 0$
28. $3t^2 + 24t + 1 = 0$
29. Find two integers such that the first is 1 less than twice the second and their product is 231.
30. Find two integers such that the first is 5 more than twice the second and the sum of their squares is 850.
31. Find the length and width of a one-story rectangular house if its length is 15 ft less than twice its width and its area is 1568 ft^2.
32. A ball is thrown upward at an initial velocity of 26 ft/sec^2 from a height of 3 ft. Using the function from Example 4, find the approximate length of time (to the nearest tenth of a second) it will take the ball to return to a height of 5 ft.

C. Exercises

33. Find two integers such that the first is 1 less than twice the second and the difference of their squares is 456.

The height (in feet) of a football kicked from the 30 yd line can be modeled by the equation $h = -0.02x^2 + 1.2x$, where x is the horizontal distance traveled in yards. Use a calculator to answer the following questions.

34. On what yard line will the ball hit the ground?
35. At what approximate yardage will it return to a height of 4 ft?
36. Will the football reach a height of 20 ft?

Solve by completing the square.

37. $5x^2 + 24x + 1 = 0$
38. $\frac{1}{2}x^2 + 3x = 17$
39. The equation $x^4 + 3x^2 - 2 = 0$ can be solved by completing the square. Let $x^2 = y$ and substitute to get a quadratic equation in y; then solve for y. Approximate y to two decimal places. Then find x from y.

CUMULATIVE REVIEW

Evaluate. [1.7–1.8]

40. $3[2 \cdot 3 - 4 + (5 - 8) \div (-3)] - 4$

41. $4^3 - (3 - 8)^2 + 6 \div 2$

42. $\left(\frac{1}{2}\right)^{-3} + (2^{-2})^3 + \frac{2}{3^{-4}}$

43. $\left(\frac{1}{2} - 2\right)^{-2} + \left(\frac{1}{3}\right)^3 - 5^0$

Simplify. [8.1, 9.5, 11.2, 11.7]

44. $(x^{-3})^{-2}$

45. $(2x + 5)^2 + \frac{1}{(2x + 5)^{-1}}$

46. $\sqrt[3]{378}$

47. $\sqrt[4]{1536}$

48. $\frac{6 + \sqrt{12}}{2}$

49. $\frac{10 + \sqrt{20}}{4}$

12.5 The Quadratic Formula

Any quadratic equation can be solved by completing the square. By completing the square on the standard form of a quadratic equation, $ax^2 + bx + c = 0$, you can derive a formula that expresses the roots of any quadratic equation in terms of its coefficients, a, b, and c.

The welder's sparks trace curves called parabolas, which are described by quadratic equations.

Example 1

Solve $ax^2 + bx + c = 0$ $(a \neq 0)$ for x by completing the square.

Answer

$$ax^2 + bx + c = 0$$

1. Write the equation in $x^2 + bx = c$ form.

$$ax^2 + bx = -c$$

a. Subtract c from both sides.

$$x^2 + \frac{b}{a}x = -\frac{c}{a}$$

b. Divide both sides by a.

$$x^2 + \frac{b}{a}x + \left(\frac{b}{2a}\right)^2 = -\frac{c}{a} + \frac{b^2}{4a^2}$$

2. Complete the square by adding $\left(\frac{1}{2} \cdot \frac{b}{a}\right)^2 = \left(\frac{b}{2a}\right)^2 = \frac{b^2}{4a^2}$ to both sides.

$$\left(x + \frac{b}{2a}\right)^2 = -\frac{4ac}{4a^2} + \frac{b^2}{4a^2}$$

$$\left(x + \frac{b}{2a}\right)^2 = \frac{-4ac + b^2}{4a^2}$$

3. Factor the trinomial and combine the terms on the right side using the common denominator $4a^2$: $-\frac{c}{a} \cdot \frac{4a}{4a} = -\frac{4ac}{4a^2}$.

$$x + \frac{b}{2a} = \pm\sqrt{\frac{-4ac + b^2}{4a^2}}$$

4. Apply the Square Root Property.

$$x + \frac{b}{2a} = \frac{\pm\sqrt{b^2 - 4ac}}{2a}$$

5. Simplify the right side by reordering the terms in the numerator and finding the square root of the denominator.

$$x = -\frac{b}{2a} \pm \frac{\sqrt{b^2 - 4ac}}{2a}$$

6. Solve by adding $-\frac{b}{2a}$ to both sides.

$$x = \frac{-b \pm \sqrt{b^2 - 4ac}}{2a}$$

Theorem

If $ax^2 + bx + c = 0$, where $a \neq 0$, then $x = \dfrac{-b \pm \sqrt{b^2 - 4ac}}{2a}$.

The steps in Example 1 constitute a proof of this theorem, and the resulting equation is called the *quadratic formula*. Be sure to write an equation in the standard form $ax^2 + bx + c = 0$ before determining the values of a, b, and c.

Example 2

Use the quadratic formula to find the solution set for $x^2 + 5x - 6 = 0$.

Answer

$1x^2 + 5x - 6 = 0$

$a = 1$, $b = 5$, and $c = -6$

1. Identify the values of a, b, and c from the standard form of the quadratic equation, $ax^2 + bx + c = 0$.

$$x = \frac{-5 \pm \sqrt{5^2 - 4(1)(-6)}}{2(1)}$$

2. Substitute these values into the quadratic formula, $x = \dfrac{-b \pm \sqrt{b^2 - 4ac}}{2a}$.

$$= \frac{-5 \pm \sqrt{25 - (-24)}}{2} = \frac{-5 \pm \sqrt{49}}{2} = \frac{-5 \pm 7}{2}$$

3. Simplify the radical.

$$x = \frac{-5 + 7}{2} = \frac{2}{2} = 1 \text{ or } x = \frac{-5 - 7}{2} = \frac{-12}{2} = -6$$

4. Simplify the two expressions separately.

The solution set is $\{-6, 1\}$.

The equation in the previous example can also be solved by factoring or by completing the square. The advantage of using the quadratic formula is that it can be used to solve any quadratic equation.

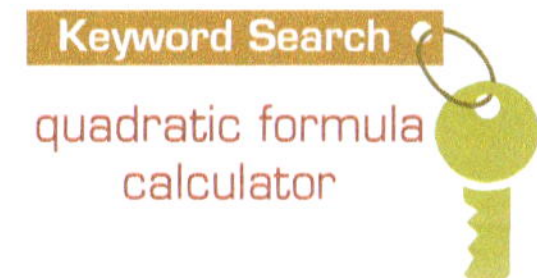

Solving Quadratic Equations by the Quadratic Formula

1. Write the equation in standard form: $ax^2 + bx + c = 0$.
2. Identify the values of a, b, and c.
3. Substitute these values into the quadratic formula and simplify.

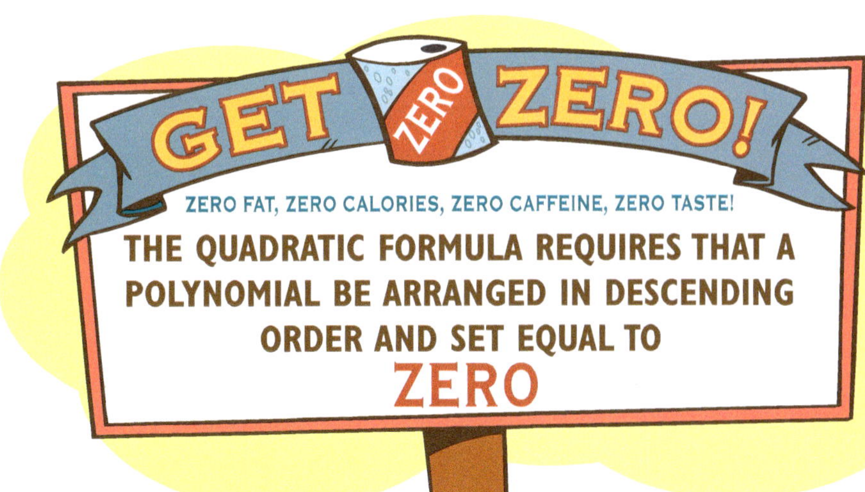

Example 3

Use the quadratic formula to solve $3y^2 - 4y = 8$.

Answer

$3y^2 - 4y - 8 = 0$ — 1. Write the equation in standard form.

$a = 3, b = -4, \text{ and } c = -8$ — 2. Identify the values of a, b, and c.

$$y = \frac{-(-4) \pm \sqrt{(-4)^2 - 4(3)(-8)}}{2(3)}$$

3. Substitute these values into the quadratic formula.

$$= \frac{4 \pm \sqrt{16 + 96}}{6} = \frac{4 \pm \sqrt{112}}{6} = \frac{4 \pm 4\sqrt{7}}{6}$$

4. Simplify the radical.

$$= \frac{\cancel{2}(2 \pm 2\sqrt{7})}{\cancel{2}(3)} = \frac{2 \pm 2\sqrt{7}}{3}$$

5. Simplify the expression.

Example 4

A box (without a lid) is to be made from a sheet of cardboard 24 in. long and 18 in. wide by cutting squares from each corner and then folding the cardboard up. To the nearest tenth of an inch, what is the length of each side of the squares that should be cut out of each corner if the bottom of the box is to have an area of 265 in.^{2}?

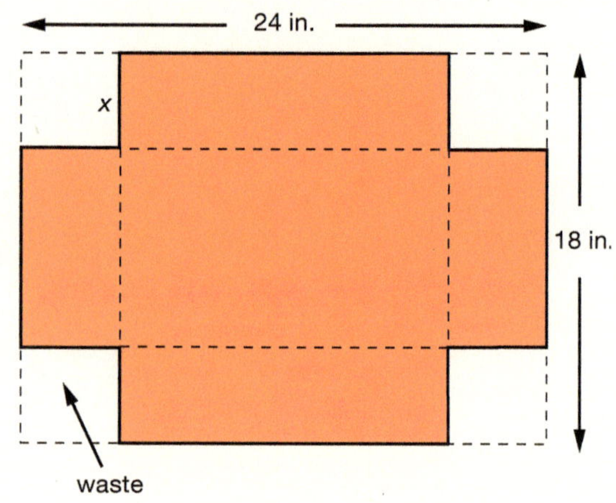

Answer

Let x = the length of each square to be cut out, $24 - 2x$ = the length of the bottom of the box, and $18 - 2x$ = the width of the bottom of the box. — 1. Assign expressions to the unknowns.

$265 = (24 - 2x)(18 - 2x)$ — 2. Use the formula $A = lw$ to write an equation modeling the problem.

3. Write the equation in standard form.
   a. Use FOIL to multiply the right side.
   b. Subtract 265 from both sides.

$265 = 432 - 84x + 4x^2$

$0 = 4x^2 - 84x + 167$

$a = 4, b = -84, \text{ and } c = 167$ — 4. Identify the values of a, b, and c.

$$x = \frac{-(-84) \pm \sqrt{(-84)^2 - 4(4)(167)}}{2(4)}$$

$$= \frac{84 \pm \sqrt{7056 - 2672}}{8} = \frac{84 \pm \sqrt{4384}}{8}$$

5. Substitute these values into the quadratic formula.

$$x = \frac{84 + \sqrt{4384}}{8} \approx 18.8 \text{ in. or } x = \frac{84 - \sqrt{4384}}{8} \approx 2.2 \text{ in.}$$

6. Calculate the decimal approximations of the possible solutions.

The length of each square is ≈ 2.2 in.

7. Since 18.8 in. is larger than the width of the cardboard, 2.2 in. is the only reasonable solution.

A. Exercises

State the values of *a*, *b*, and *c* when the equation is written in standard form. Do not solve.

1. $x^2 + 6x + 5 = 0$

2. $2x^2 - 6x + 1 = 0$

3. $x^2 + 5x = -6$

4. $3y^2 - 5 = 2y$

5. $w^2 - 6 = 0$

6. $z^2 = 7z$

Use the quadratic formula to find the solution set.

7. $x^2 + 6x + 8 = 0$

8. $x^2 + 4x + 3 = 0$

9. $x^2 + 8x + 15 = 0$

10. $x^2 - 9x + 20 = 0$

11. $a^2 - 9 = 0$

12. $b^2 - 9b = -8$

13. $2c^2 - c - 3 = 0$

14. $d^2 + 3d = 28$

15. $y^2 - 6y = -9$

16. $4x^2 - 12x + 9 = 0$

B. Exercises

Use the quadratic formula to solve.

17. $2x^2 - 9x + 4 = 0$

18. $3x^2 = 2x + 5$

19. $2y^2 - 5y = -1$

20. $3w^2 + 3 = 7w$

21. $3x^2 = 7$

22. $4x^2 = 5x$

23. $2x^2 - 5x - 10 = 0$

24. $4y^2 - 3y = 8$

25. $7m^2 + 9m = -2$

26. $15x^2 + 22x + 8 = 0$

27. $-3k^2 + 2k + 10 = 0$

28. $4x^2 + 4x + 3 = 0$

29. $8x^2 = 26x + 7$

30. $5r^2 - 7 = 2r$

Use the quadratic formula and a calculator to solve. Round decimal approximations to the nearest hundredth.

31. $3x^2 - 8x + 2 = 0$

32. $3y^2 - 2 = 3y$

33. Which property justifies step 1*b* in Example 1, in which both sides are divided by *a*?

34. Which property justifies step 5 in Example 1, in which the terms are reordered in the numerator?

35. Find two integers whose sum is 30 and whose product is 216.

36. Find two integers such that the larger is 2 less than eight times the smaller and the difference of their squares is 884.

37. A box is to be made from a sheet of cardboard 30 in. long by 24 in. wide by cutting squares from each corner and then folding the cardboard up. How tall will the box be if the bottom of the box is to have an area of 310 in.^{2}?

38. A rectangular piece of cardboard is made into a box by cutting 3 in. squares from each corner and then folding the cardboard up. The length of the cardboard box is 3 in. more than its width. If the volume of the box is 540 in.^{3}, what are the dimensions of the original piece of cardboard?

C. Exercises

Use the quadratic formula and a calculator to solve. Round decimal approximations to the nearest hundredth.

39. $-2.5x^2 + 7.4x - 3.8 = 0$

40. $\frac{2}{3}x^2 - \frac{4}{5}x - \frac{1}{2} = 0$

41. Show how to solve $ry^2 + sy + t = 0$ for y by completing the square. (Assume that $r \neq 0$.)

A baseball is thrown from third to first base. The height of the ball is modeled by the equation $h = -0.0034x^2 + 0.41x + 5.5$, where h is the height in feet and x is the horizontal distance in feet.

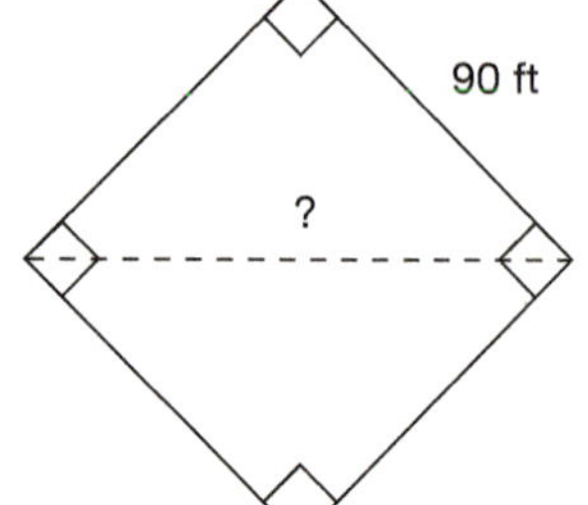

42. How far (to the nearest tenth of a foot) does the ball travel before it returns to a height of 10 ft?

43. Should the first baseman be able to catch the ball? Explain your reasoning.

44. Will the ball reach a height of 17 ft? If so, at what distance(s) from third base?

Dominion Modeling

A landscape designer is planning a fountain and positions two fountainheads so that their two parabolic streams of water will intersect. These streams can be modeled by the equations $y = -0.14x^2 + 1.7x$ and $y = -0.14x^2 + 3.9x - 22$, where y is the height of the water in feet and x is its distance from the first fountainhead in feet.

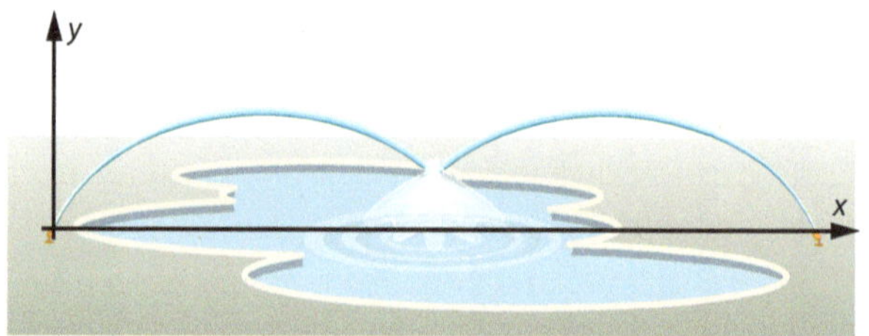

45. Using the elimination method of solving systems of equations (described in Section 7.4) will eliminate the y and x^2 terms. Write the linear equation that results.

46. Solve for x and then for y.

47. Use these values to describe the location of the intersection of the streams relative to the first fountainhead.

CUMULATIVE REVIEW

Classify each number as rational, irrational, or not real. [1.1, 12.2]

48. 12

49. $\sqrt{5}$

50. $\frac{\sqrt{169}}{2}$

51. $\frac{-17}{3}$

52. $\sqrt{-17}$

53. $\frac{\sqrt{160}}{4}$

Write an equation and solve. [3.6]

When Matthew purchased a baseball glove, he saved \$28 off the original price and paid \$44.94, which included the 7% sales tax.

54. What was the original price?

55. What percent did he save?

Sandy bought a dress for \$42 (before sales tax) after obtaining a 25% discount.

56. How much did the dress originally cost?

57. The following day she went back to the same store, and there was now a sign up saying to take an additional 30% off. What total percent could she have saved by waiting a day?

12.6 More Quadratic Equations

When solving a quadratic equation, choose the method that is most appropriate for the given problem.

General Method of Solving Quadratic Equations
Write the quadratic equation in the form $ax^2 + bx + c = 0$. 1. If $b = 0$, isolate x^2 and apply the Square Root Property. 2. If the polynomial factors easily, solve the equation by factoring. 3. Otherwise, use the quadratic formula: $x = \frac{-b \pm \sqrt{b^2 - 4ac}}{2a}$.

Quadratic equations are usually not solved by completing the square since one of the three methods listed above is generally easier. However, knowing how to complete the square is essential to the development of the quadratic formula. This method is also used in graphing quadratic functions and in other applications that will be studied in later courses.

When the quadratic formula is used, the value of the radicand determines the number and type of solutions to the equation. This value, $D = b^2 - 4ac$, is called the *discriminant*. The following table describes the number and type of real number solutions to $ax^2 + bx + c = 0$ when a, b, and c are rational numbers.

Jørn Utzon designed the "sails" on Australia's Sydney Opera House using a pair of shells, each represented by a quadratic equation.

Determining the Number and Type of Solutions

$D = b^2 - 4ac$	Solutions	Example
$D > 0$ and a perfect square	two rational	$x = \frac{1 \pm \sqrt{25}}{4} = \frac{3}{2}, -1$
$D > 0$ but not a perfect square	two irrational	$x = \frac{1 \pm \sqrt{13}}{4} \approx 1.15, -0.65$
$D = 0$	one rational	$x = \frac{1 \pm \sqrt{0}}{4} = \frac{1}{4}$
$D < 0$	no real	$x = \frac{1 \pm \sqrt{-9}}{4}$; no real solution

Example 1

Evaluate the discriminant; then describe the solution(s) of each equation.

a. $x^2 + 5x - 7 = 0$ **b.** $x^2 = 156 - x$ **c.** $3x^2 + 5x = -9$

Answer

a. $x^2 + 5x - 7 = 0$

$a = 1, b = 5$, and $c = -7$ — 1. Identify the values of *a*, *b*, and *c* when the equation is in standard form, $ax^2 + bx + c = 0$.

$D = b^2 - 4ac$ — 2. Evaluate the discriminant.

$= 5^2 - 4(1)(-7) = 25 + 28 = 53$

There are two irrational solutions. — 3. $D > 0$ and is not a perfect square.

b. $x^2 = 156 - x$

$x^2 + x - 156 = 0$ — 1. Write the equation in standard form.

$D = b^2 - 4ac$ — 2. Evaluate the discriminant.

$= 1^2 - 4(1)(-156) = 1 + 624 = 625$

There are two rational solutions. — 3. $D > 0$ and is a perfect square.

c. $3x^2 + 5x = -9$

$3x^2 + 5x + 9 = 0$ — 1. Write the equation in standard form.

$D = b^2 - 4ac$ — 2. Evaluate the discriminant.

$= 5^2 - 4(3)(9) = 25 - 108 = -83$

There are no real solutions. — 3. $D < 0$.

To find a quadratic equation that has stated values as its solutions, reverse the steps used to solve a quadratic equation by factoring.

Example 2

Write a quadratic equation in standard form with the given solutions.

a. $x = 5, -7$ **b.** $x = \frac{2}{3}, -\frac{1}{4}$

Answer

a. $x = 5$ or $x = -7$ — 1. Write each solution as a linear equation in the form $ax + b = 0$.

$x - 5 = 0$ $\quad x + 7 = 0$

$(x - 5)(x + 7) = 0 \cdot 0$ — 2. Use the Multiplication Property of Equality to form a quadratic equation.

$x^2 + 7x - 5x - 35 = 0$ — 3. Use FOIL on the left side to produce an equation in standard form.

$x^2 + 2x - 35 = 0$

b. $x = \frac{2}{3}$ or $x = -\frac{1}{4}$ — 1. Write each solution as a linear equation in the form $ax + b = 0$.

$3x = 3\left(\frac{2}{3}\right)$ $\quad 4x = 4\left(-\frac{1}{4}\right)$ — a. Clear the fractions.

$3x = 2$ $\quad 4x = -1$

$3x - 2 = 0$ $\quad 4x + 1 = 0$ — b. Set each equation equal to zero.

$(3x - 2)(4x + 1) = 0$ — 2. Use the Multiplication Property of Equality to form a quadratic equation.

$12x^2 + 3x - 8x - 2 = 0$ — 3. Use FOIL on the left side to produce an equation in standard form.

$12x^2 - 5x - 2 = 0$

Example 3

A carpenter wants to fit three boards together to form a right triangle to use as a brace. The hypotenuse must be 3 ft longer than one leg and 5 ft longer than the other leg. Find the measurements for the sides of the triangle needed by the carpenter.

Answer

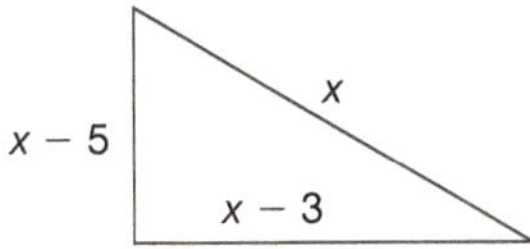

1. Sketch a right triangle. Let x = the hypotenuse, $x - 3$ = the length of one leg, and $x - 5$ = the length of the other leg.

$a^2 + b^2 = c^2$

$(x - 5)^2 + (x - 3)^2 = x^2$

2. Substitute the expressions into the Pythagorean theorem.

$x^2 - 10x + 25 + x^2 - 6x + 9 = x^2$

$x^2 - 16x + 34 = 0$

3. Simplify and write the equation in standard form.

$D = b^2 - 4ac = (-16)^2 - 4(1)(34) = 120$

4. Evaluate the discriminant. Since $D > 0$ and is not a perfect square, the solutions will be irrational.

$x = \frac{-(-16) \pm \sqrt{120}}{2(1)} = \frac{16 \pm 2\sqrt{30}}{2} = 8 \pm \sqrt{30}$

$x = 8 + \sqrt{30} \approx 13.48$ or $x = 8 - \sqrt{30} \approx 2.52$

5. Use the quadratic formula to solve, substituting known values into the equation.

hypotenuse ≈ 13.48 ft

$\text{leg}_1 = x - 5 \approx 13.48 - 5 = 8.48$ ft

$\text{leg}_2 = x - 3 \approx 13.48 - 3 = 10.48$ ft

6. Interpret the results. 2.52 ft would give negative lengths for both legs, so 13.48 ft is the only reasonable length for the hypotenuse.

A. Exercises

Matching: Determine the easiest method for solving. Do not solve.

a. factoring **b.** taking roots **c.** completing the square **d.** quadratic formula

1. $a^2 - 2a - 8 = 0$ **2.** $x^2 - 3x - 5 = 0$ **3.** $x^2 - 7 = 0$

4. $x^2 + 3x = 0$ **5.** $7x^2 - 15x + 54 = 0$ **6.** $(x - 21)^2 = 4$

Evaluate the discriminant; then select the choice describing the solution(s) of each equation.

a. two rational **b.** two irrational **c.** one rational **d.** no real

7. $x^2 + 20x + 51 = 0$ **8.** $x^2 + 11 = 0$ **9.** $x^2 - 6x + 9 = 0$

10. $x^2 = 2x$ **11.** $-3x^2 + x = -8$ **12.** $x^2 + 2x = -9$

13. $2x^2 + 5x + 1 = 0$ **14.** $1 + 10x = -25x^2$

Solve using the method of your choice and describe the solution(s).

15. $x^2 + 12x - 45 = 0$ **16.** $q^2 + 9q - 4 = 0$

17. $a^2 + 3a = -1$ **18.** $x^2 + 2x + 9 = 0$

19. $2x^2 - 7x + 2 = 0$ **20.** $4x^2 - 5x - 9 = 0$

B. Exercises

Write a quadratic equation in standard form with the given solutions.

21. $x = -8, 4$

22. $x = 0, 5$

23. $x = \frac{1}{2}, \frac{3}{4}$

24. $x = -\frac{1}{2}, \frac{2}{3}$

Solve.

25. $6y^2 = 13y - 2$

26. $3a^2 - 5 = -2a$

27. $9a^2 + 24a = -16$

28. $4r^2 - 5 = r$

29. $x^2 + 8x - 2 = 0$

30. $-t^2 + 4 = 0$

31. $a^2 + 44a + 2 = 0$

32. $x^2 - 4x + 2 = 0$

33. $(3x + 4)(2x - 5) = 4$

34. $(x + 5)(2x - 7) = 9$

35. $(3x - 1)^2 + 6 = 0$

36. $x(5x - 9) + 6 = 0$

37. What are the dimensions of a rectangle whose perimeter is 54 in. and whose area is 180 in.^{2}?

38. The hypotenuse of a right triangle measures 8 in. more than one leg and 3 in. more than the other leg. Find the lengths of the sides of the triangle (to the nearest tenth of an inch).

C. Exercises

Solve. State irrational solutions as decimal approximations rounded to the nearest thousandth.

39. $(y + 4)^2 - y(y + 2) = 9$

40. $(x - 5)^2 + (x - 5) - 2 = 0$

41. $(x - 6)(x + 2) - (x - 6) = 9$

42. $(x - 1)^2 - (3x + 2)(x - 5) = 0$

Write a quadratic equation in standard form with the given solutions.

43. $x = \frac{-3 \pm \sqrt{5}}{3}$

44. $x = \frac{-2 \pm \sqrt{3}}{6}$

CUMULATIVE REVIEW

Solve. [11.8]

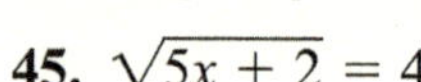

45. $\sqrt{5x + 2} = 4$

46. $\sqrt{3x + 1} = \sqrt{2x - 9}$

Graph. [6.3]

47. $y = -\frac{5}{3}x + 7$

48. $y = 4x - 9$

Graph the following equations on the same coordinate plane. [5.7, 11.9]

49. $y = |x|$

50. $y = 3|x|$

51. $y = \frac{1}{2}|x|$

52. $y = -3|x|$

53. Mr. Sutton is taking the scouts camping and has a rope that is to reach from a stake in the ground to the top of a tent pole. If the pole is 8 ft high and the rope is 12 ft long, how far from the pole should he place the stake to make the rope taut? [11.6]

54. If one side of a square is increased by 3 in. and the adjacent side is decreased by 1 in., the rectangle formed has an area of 77 in.^{2}. Find the length of a side of the square. [12.3]

12.7 Quadratic Functions: $f(x) = ax^2 + c$

Polynomial functions of the second degree are called *quadratic functions*.

Definition

A **quadratic function** is a function that can be written in the form $f(x) = ax^2 + bx + c$, where a, b, and c are real numbers and $a \neq 0$. This form is called the *standard form* of a quadratic function.

In Section 8.4 you were introduced to the parabolic graph of the basic quadratic function $f(x) = x^2$. The graph of any quadratic function is also a *parabola*.

Example 1

Graph $f(x) = x^2$ and $g(x) = 3x^2$ and compare their graphs.

Answer

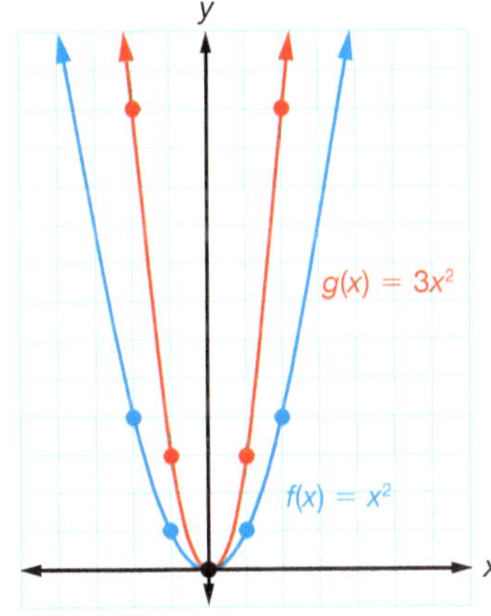

1. Make a table for each function, using at least five values for x.

x	$f(x) = x^2$	$g(x) = 3x^2$
0	0	0
1	1	3
−1	1	3
2	4	12
−2	4	12

2. For each function, plot the ordered pairs and draw the parabola.
3. Each point on $g(x)$ is three times higher than its corresponding point on $f(x)$. This makes its graph appear "steeper" or "narrower."

Graphing Quadratic Functions

1. Find at least five ordered pairs that satisfy the function.
2. Plot the ordered pairs on a Cartesian coordinate plane.
3. Connect the ordered pairs with a smooth curve.

Example 2

Graph $p(x) = -x^2$ and $q(x) = \frac{1}{2}x^2$ and compare their graphs to the graph of $f(x) = x^2$.

Answer

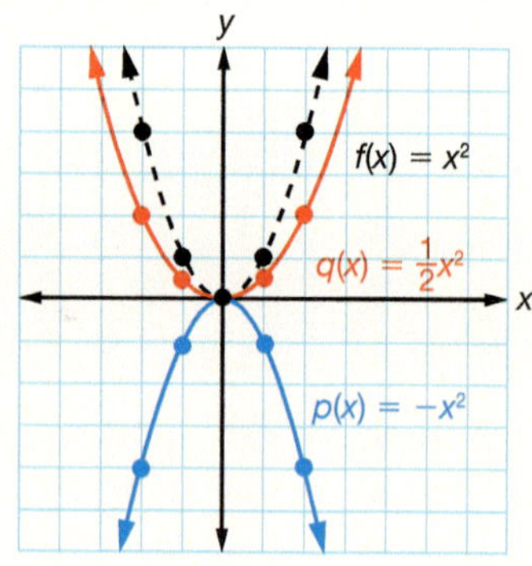

1. Make a table for each function.

x	$p(x) = -x^2$	$q(x) = \frac{1}{2}x^2$
0	0	0
1	−1	$\frac{1}{2}$
−1	−1	$\frac{1}{2}$
2	−4	2
−2	−4	2

2. For each function, plot the ordered pairs and draw the parabola.
3. a. Each point on $p(x)$ is the located the same distance below the x-axis as its corresponding point on $f(x)$ is above the x-axis. Since $p(x)$ has the same shape as $f(x)$, it is said to be the *reflection* of $f(x)$ across the x-axis.

   b. Each point on $q(x)$ is half the height of its corresponding point on $f(x)$. This makes its graph appear "flatter" or "wider."

Recall that the "turning point" of a parabola is called its *vertex*. Since the vertex is either the lowest or the highest point on the graph, it is called a *minimum* or a *maximum*. The vertices of all the parabolas in Examples 1 and 2 are at the origin, (0, 0). These parabolas are also symmetric across the y-axis. The vertex falls on this "fold line," called the *line of symmetry*.

The Effect of a on the Graph of $y = ax^2 + bx + c$			
If $a > 0$, the graph opens upward.	If $a < 0$, the graph opens downward.	If $\lvert a \rvert > 1$, the graph is "steeper."	If $\lvert a \rvert < 1$, the graph is "flatter."
line of symmetry; V; minimum	V; maximum; line of symmetry	line of symmetry; V	line of symmetry; V

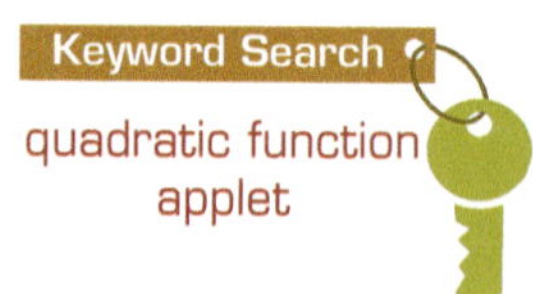

Example 3

Graph $r(x) = x^2 + 2$ and compare its graph to the graph of $f(x) = x^2$.

Answer

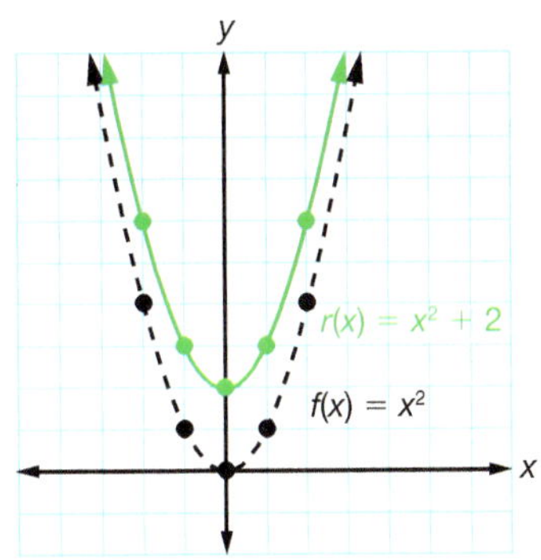

1. Make a table for each function. Note that each value for $r(x)$ is 2 more than the corresponding value for $f(x)$.

x	$f(x) = x^2$	$r(x) = x^2 + 2$
−2	4	6
−1	1	3
0	0	2
1	1	3
2	4	6

2. Plot the ordered pairs and draw the parabola.
3. Each point on $r(x)$ is 2 units above its corresponding point on $f(x)$. The graph of $r(x)$ is the result of translating the graph of $f(x)$ up 2 units.

Example 4

Graph $s(x) = 3x^2 - 4$ and compare its graph to the graph of $g(x) = 3x^2$.

Answer

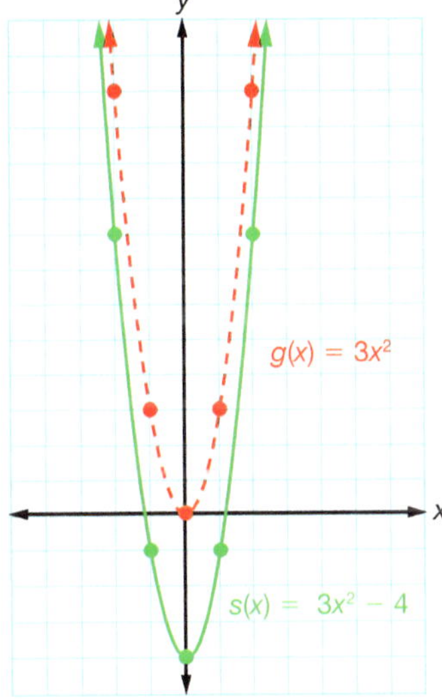

1. Make a table for each function. Note that each value for $s(x)$ is 4 less than the corresponding value for $g(x)$.

x	$g(x) = 3x^2$	$s(x) = 3x^2 - 4$
−2	12	8
−1	3	−1
0	0	−4
1	3	−1
2	12	8

2. Plot the ordered pairs and draw the parabola.
3. Each point on $s(x)$ is 4 units below its corresponding point on $g(x)$. The graph of $s(x)$ is the result of translating the graph of $g(x)$ down 4 units.

Notice the effect that adding a constant, c, has on the graph of $f(x) = ax^2$. If c is positive, the graph slides up c units; if c is negative, the graph slides down c units. This causes the vertex of the graph of $f(x) = ax^2 + c$ to be at $(0, c)$.

A. Exercises

For each function, find $f(0)$, $f(2)$, $f(-1)$, and $f(4)$.

1. $f(x) = x^2$

2. $f(x) = -x^2$

3. $f(x) = x^2 - 4$

4. $f(x) = x^2 + 5$

5. $f(x) = -2x^2$

6. $f(x) = 3x^2$

7. $f(x) = 3x^2 - 7$

8. $f(x) = -2x^2 + 8$

Complete the following table.

	Function	Direction of Opening	Vertex		Comparison of Shape to $f(x) = x^2$
			At	Minimum or Maximum	
9.	$f(x) = 4x^2$				
10.	$f(x) = -\frac{1}{3}x^2$				
11.	$f(x) = -x^2 + 4$				
12.	$f(x) = \frac{1}{5}x^2 - 7$				

Graph each quadratic function.

13. $f(x) = x^2$
14. $f(x) = 2x^2$
15. $f(x) = x^2 + 2$
16. $f(x) = x^2 - 7$
17. $f(x) = x^2 - 1$
18. $f(x) = x^2 + 3$
19. $f(x) = -x^2 + 2$
20. $f(x) = -x^2 + 5$
21. $f(x) = 2x^2 - 3$
22. $f(x) = 4x^2 - 5$

B. Exercises

Graph each quadratic function.

23. $f(x) = -3x^2$
24. $f(x) = \frac{1}{3}x^2$
25. $f(x) = -\frac{1}{3}x^2$
26. $f(x) = \frac{1}{4}x^2$
27. $f(x) = 2x^2 + 3$
28. $f(x) = \frac{1}{2}x^2 - 5$
29. $f(x) = -2x^2 + 7$
30. $f(x) = 7x^2 - 10$

31. The path of a projectile recorded in the table below can be modeled by a quadratic function.

Horizontal Distance (ft)	x	0	25	50	75	100				
Approximate Height (ft)	y	1	263	450	563	600	563	450	263	1

a. Plot the first five ordered pairs from the table.

b. Use the fact that parabolas are symmetric and that the maximum height of the projectile is 600 ft to fill in the missing distances and to finish the graph.

c. Approximate the horizontal distance traveled by the projectile before it hits the ground.

32. Where is the vertex of any quadratic function of the form $f(x) = ax^2$?

33. What is the x-intercept for graphs of the form $f(x) = ax^2$? What is the y-intercept?

34. Where is the line of symmetry for any function of the form $f(x) = ax^2 + c$? What is the equation for this line?

35. State the coordinates of the vertex of any quadratic function of the form $f(x) = ax^2 + c$.

36. The graph of $f(x) = \frac{1}{2}x^2 + 5$ is contained in which quadrants of the coordinate plane?

37. The graph of $f(x) = \frac{1}{2}x^2 - 5$ is contained in which quadrants of the coordinate plane?

C. Exercises

38. Given $f(x) = 0.2x^2 + 0.3$, do the following.

a. Find $f(x)$ for $x = 0, \pm1, \pm2$, and ±3.

b. Graph the function.

c. State the coordinates of the vertex and whether it is a minimum or maximum point.

d. State the equation of the line of symmetry.

Dominion Modeling

Use your knowledge of symmetry and your graphing calculator to complete a quadratic regression, given the following information. (See Chapter 6, Technology Corner; but use 5:QuadReg.)

39. Water flowing from a fountain follows a parabolic path. If the water reaches a height of 9 ft and the distance between the fountainhead [use point (0, 0)] and where the water hits the pool is 5 ft, find a quadratic function that models the path of the water.

40. Find the height of the water 1 ft, 2.5 ft, 3 ft, and 3.5 ft away from the fountainhead.

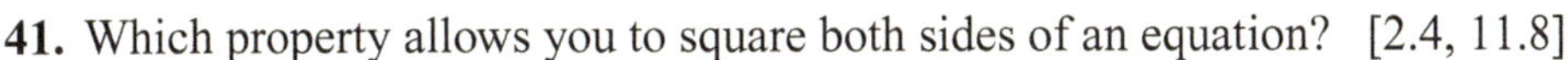

CUMULATIVE REVIEW

41. Which property allows you to square both sides of an equation? [2.4, 11.8]

42. Which property allows you to complete the square on a quadratic equation? [2.4, 12.3]

Simplify.

43. $52^{-\frac{1}{2}}$ [1.7, 11.4]

44. $2\sqrt{3} + 5\sqrt{12} + 7\sqrt{2} + \sqrt{27}$ [11.5]

45. $-4 - \{-[-3 + (-7)] + [-5(-9 - 1)]\} \div (-5)$ [1.8]

Solve. [2.6, 2.8, 4.3]

46. $3 - (-x + 2) = 10 \div 2$

47. $5 - x > 8$

48. The sum of three consecutive odd numbers is 333. Find the numbers.

49. Ray has $3.41 in pennies and dimes. How many of each are there if he has five more than six times as many pennies as dimes?

50. If 97 is at least 3 more than twice a number, find the possible solutions.

12.8 Quadratic Functions: $f(x) = a(x - h)^2 + k$

Model rocketry provides exciting applications of mathematics.

The previous section illustrated how the constant term c in the function $f(x) = ax^2 + c$ translates the graph of $f(x) = ax^2$ vertically. If $c > 0$, the parabola slides up c units; and if $c < 0$, the parabola slides down c units. The effect of adding a middle term, bx, is more easily seen when the equation is written in the form $f(x) = a(x - h)^2 + k$.

In this form, a constant h is subtracted from the independent variable before the expression is squared. How does this affect the graph? Note that the x-coordinate of the parabola's vertex can be determined by setting the expression being squared equal to zero.

Example 1

Graph $g(x) = (x - 3)^2$ and compare its graph to the graph of $f(x) = x^2$.

Answer $x - 3 = 0$ when $x = 3$

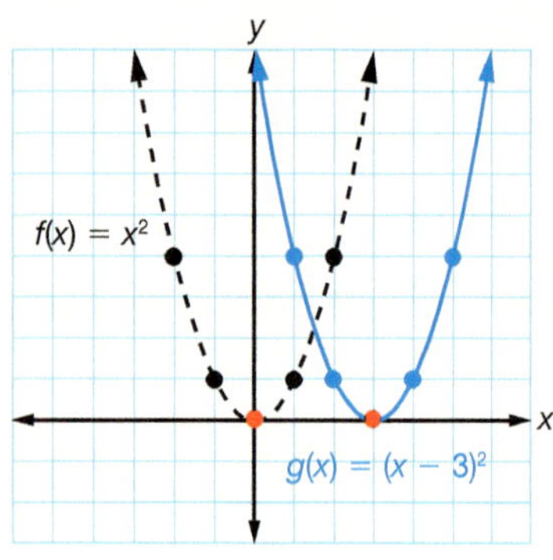

1. Find the x-coordinate of the parabola's vertex.
2. Make a table of ordered pairs, including the vertex and several points on either side of it.

x	$(x - 3)^2$	$g(x)$	
1	$(1 - 3)^2 = (-2)^2$	4	
2	$(2 - 3)^2 = (-1)^2$	1	
3	$(3 - 3)^2 = 0^2$	0	(vertex)
4	$(4 - 3)^2 = 1^2$	1	
5	$(5 - 3)^2 = 2^2$	4	

3. Plot the points and draw the parabola.
4. Each point on $g(x)$ is 3 units to the right of its corresponding point on $f(x)$.

Example 2

Graph $h(x) = (x + 1)^2$ and compare its graph to the graph of $f(x) = x^2$.

Answer $x + 1 = 0$ when $x = -1$

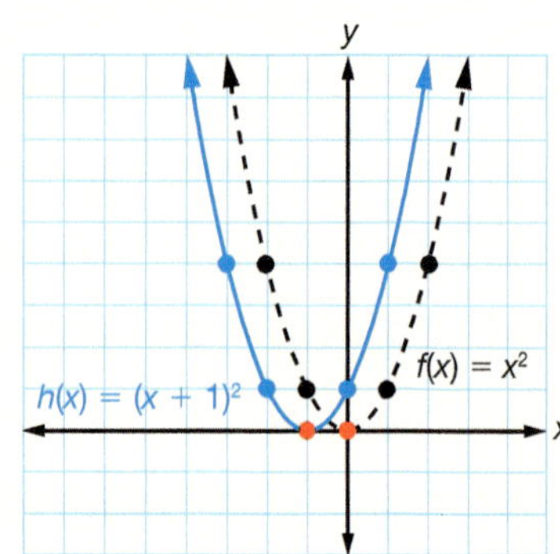

1. Find the x-coordinate of the parabola's vertex.
2. Make a table of ordered pairs, including the vertex and several points on either side of it.

x	$(x + 1)^2$	$h(x)$	
−3	$(-3 + 1)^2 = (-2)^2$	4	
−2	$(-2 + 1)^2 = (-1)^2$	1	
−1	$(-1 + 1)^2 = 0^2$	0	(vertex)
0	$(0 + 1)^2 = 1^2$	1	
1	$(1 + 1)^2 = 2^2$	4	

3. Plot the points and draw the parabola.
4. Each point on $h(x)$ is 1 unit to the left of its corresponding point on $f(x)$.

The examples show that the graph of $f(x) = ax^2$ is translated $|h|$ units horizontally when a constant h is subtracted from the variable before the expression is squared. The graph of $f(x) = a(x - h)^2$ slides to the right if $h > 0$ and slides to the left if $h < 0$.

The graph of $f(x) = a(x - h)^2 + k$ is the result of combining horizontal and vertical translations of $f(x) = ax^2$. Initially finding the x-coordinate of the vertex by setting the expression within the parentheses equal to zero will help you determine which values of x to use in your table and will eliminate any confusion about the direction of the translations.

Example 3

Graph $p(x) = (x - 2)^2 - 1$.

Answer $x - 2 = 0$ when $x = 2$

1. Find the x-coordinate of the parabola's vertex.

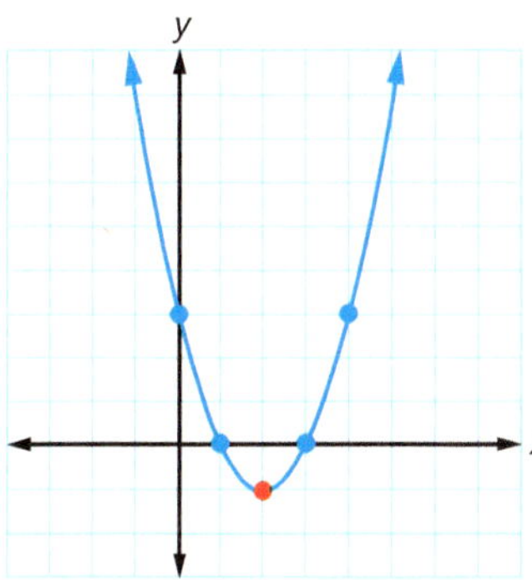

2. Make a table of ordered pairs, including the vertex and several points on either side of it.

x	$(x - 2)^2 - 1$	$p(x)$	
0	$(0 - 2)^2 - 1 = (-2)^2 - 1$	3	
1	$(1 - 2)^2 - 1 = (-1)^2 - 1$	0	
2	$(2 - 2)^2 - 1 = 0^2 - 1$	-1	(vertex)
3	$(3 - 2)^2 - 1 = 1^2 - 1$	0	
4	$(4 - 2)^2 - 1 = 2^2 - 1$	3	

3. Plot the points and draw the parabola.

Notice the result of letting $x = h$ in $f(x) = a(x - h)^2 + k$.

$$f(h) = a(h - h)^2 + k$$
$$f(h) = a(0) + k$$
$$f(h) = k$$

Therefore, the vertex of the parabola is at (h, k). Since the parabola's line of symmetry is a vertical line through the vertex, its equation is $x = h$.

Example 4

Graph $q(x) = -2(x - 3)^2 + 5$.

Answer $x - 3 = 0$ when $x = 3$;
vertex: $(3, 5)$
line of symmetry: $x = 3$

1. Find the vertex and the equation for the parabola's line of symmetry.

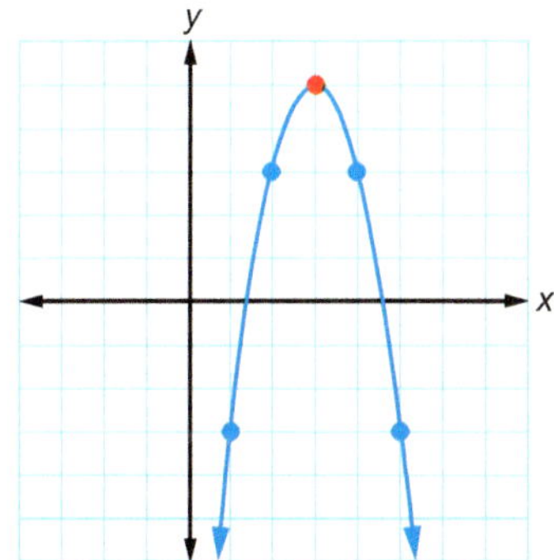

2. Make a table of ordered pairs, including the vertex and several points on either side of it.

x	$-2(x - 3)^2 + 5$	$q(x)$	
1	$-2(1 - 3)^2 + 5 = -2(-2)^2 + 5 = -8 + 5$	-3	
2	$-2(2 - 3)^2 + 5 = -2(-1)^2 + 5 = -2 + 5$	3	
3	$-2(3 - 3)^2 + 5 = -2(0)^2 + 5 = 0 + 5$	5	(vertex)
4	$-2(4 - 3)^2 + 5 = -2(1)^2 + 5 = -2 + 5$	3	
5	$-2(5 - 3)^2 + 5 = -2(2)^2 + 5 = -8 + 5$	-3	

3. Plot the points and draw the parabola.

The $f(x) = a(x - h)^2 + k$ form of a quadratic function can be called the *vertex form* or the *graphing form* because it displays so much information about the graph of a quadratic function.

Function	$f(x) = a(x - h)^2 + k$	Vertex	Direction of Opening	Comparison of Shape to $f(x) = x^2$
$f(x) = -x^2$	$f(x) = -1(x - 0)^2 + 0$	$(0, 0)$	down $(a < 0)$	same $(\|a\| = 1)$
$f(x) = 3x^2 - 4$	$f(x) = 3(x - 0)^2 - 4$	$(0, -4)$	up $(a > 0)$	steeper $(\|a\| > 1)$
$f(x) = (x + 1)^2$	$f(x) = 1[x - (-1)]^2 + 0$	$(-1, 0)$	up $(a > 0)$	same $(\|a\| = 1)$
$f(x) = (x - 2)^2 - 1$	$f(x) = 1(x - 2)^2 - 1$	$(2, -1)$	up $(a > 0)$	same $(\|a\| = 1)$
$f(x) = -\frac{1}{3}(x + 7)^2 + 9$	$f(x) = -\frac{1}{3}[x - (-7)]^2 + 9$	$(-7, 9)$	down $(a < 0)$	flatter $(\|a\| < 1)$

If a quadratic function is in standard form, $f(x) = ax^2 + bx + c$, you can change it to the equivalent vertex form $f(x) = a(x - h)^2 + k$ using a variation of the procedure outlined in Section 12.3 for completing the square.

Example 5

Write $f(x) = x^2 - 6x + 10$ in graphing form. Identify the vertex and direction of opening.

Answer

$f(x) = (x^2 - 6x) + 10$

1. Group the $ax^2 + bx$ terms.

$f(x) = (x^2 - 6x + 9) + 10 - 9$

2. Complete the square by adding $\left(\frac{6}{2}\right)^2 = 9$ inside the parentheses. Subtract 9 outside the parentheses so the value of the right side of the equation will not change.

$f(x) = (x - 3)^2 + 1$

3. Factor the trinomial inside the parentheses and simplify the constants outside the parentheses.

$a = 1, h = 3, k = 1$

4. Identify a, h, and k and use them to describe the graph.

vertex: $(3, 1)$

a. The vertex is at (h, k).

The graph opens upward and has the same shape as $y = x^2$.

b. Since $a = 1$, the graph opens upward.

If the function has a leading coefficient other than one, completing the square is more complicated. By completing the square on $f(x) = ax^2 + bx + c$, we can develop formulas for h and k.

$f(x) = (ax^2 + bx) + c$

$f(x) = a\left(x^2 + \frac{b}{a}x\right) + c$

1. Group the $ax^2 + bx$ terms and then factor out the leading coefficient, a.

$f(x) = a\left(x^2 + \frac{b}{a}x + \frac{b^2}{4a^2}\right) + c - \frac{b^2}{4a}$

2. Complete the square by adding $\left(\frac{1}{2} \cdot \frac{b}{a}\right)^2 = \frac{b^2}{4a^2}$ inside the parentheses. Because of the factor of a outside the parentheses, the value added to the equation is actually $a\left(\frac{b^2}{4a^2}\right) = \frac{b^2}{4a}$. Subtract this value outside the parentheses.

$f(x) = a\left(x + \frac{b}{2a}\right)^2 + \left(c - \frac{b^2}{4a}\right)$

3. Factor the trinomial.

$h = -\frac{b}{2a}$ and $k = \left(c - \frac{b^2}{4a}\right)$

4. Comparing the resulting function to $f(x) = a(x - h)^2 + k$ produces general expressions for h and k.

We will use this formula for h, but it is typically easier to find the value for k by substituting the value of h into $f(x)$.

Example 6

State the vertex of $f(x) = -2x^2 - 12x - 19$ and then graph the function.

Answer

$f(x) = -2x^2 - 12x - 19$

1. Find the vertex.

$h = -\frac{-12}{2(-2)} = -\frac{12}{4} = -3$

a. Use $h = -\frac{b}{2a}$.

$k = f(-3) = -2(-3)^2 - 12(-3) - 19$
$= -18 + 36 - 19 = -1$

b. Use $k = f(h)$.

vertex: $(-3, -1)$

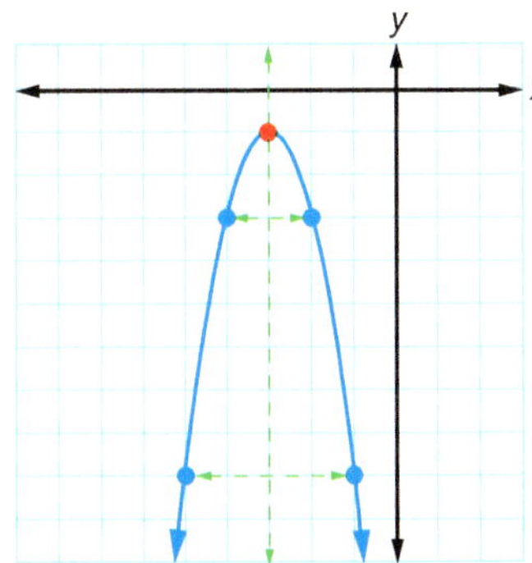

2. Note that $a = -2$, which indicates that the graph will open downward and will be stretched by a factor of 2.
3. Make a table of ordered pairs, including two points to the right of the vertex.

x	$-2x^2 - 12x - 19$	$f(x)$
-3	vertex	-1
-2	$-2(-2)^2 - 12(-2) - 19$ $= -8 + 24 - 19$	-3
-1	$-2(-1)^2 - 12(-1) - 19$ $= -2 + 12 - 19$	-9

4. Graph the three points and use the line of symmetry, $x = -3$, to plot the points to the left of the vertex. Then draw the parabola.

A. Exercises

Complete the following table.

	Function	Vertex (h, k)	Line of Symmetry	Value of a	Direction of Opening	Comparison of Shape to $f(x) = x^2$
1.	$f(x) = (x - 2)^2 + 9$					
2.	$f(x) = -(x + 5)^2 + 1$					
3.	$f(x) = 3(x - 2)^2 - 4$					
4.	$f(x) = \frac{1}{4}x^2 - 6$					
5.	$f(x) = -5(x + 1)^2$					
6.	$f(x) = \frac{2}{3}(x - 8)^2$					
7.	$f(x) = -\frac{7}{4}x^2 + 7$					
8.	$f(x) = -6x^2$					
9.	$f(x) = x^2 - 8x + 9$					
10.	$f(x) = x^2 - 10x - 7$					

Graph each quadratic function.

11. $f(x) = (x - 4)^2 + 2$
12. $f(x) = (x - 2)^2$
13. $f(x) = (x - 1)^2 - 3$
14. $f(x) = -2(x - 2)^2 + 5$
15. $f(x) = (x + 3)^2$
16. $f(x) = (x + 1)^2 - 2$
17. $f(x) = -(x + 4)^2 + 1$
18. $f(x) = (x - 3)^2 - 8$
19. $f(x) = 3(x + 1)^2 + 4$
20. $f(x) = -\frac{1}{2}(x - 2)^2 + 3$

B. Exercises

Complete the square to write each quadratic function in graphing form. State the vertex, but do not graph.

21. $f(x) = x^2 + 4x - 12$
22. $f(x) = x^2 - 6x + 8$
23. $f(x) = x^2 + 14x + 40$
24. $f(x) = x^2 - 8x + 15$

For each quadratic function, state the vertex and then graph the function.

25. $f(x) = x^2 + 10x + 24$
26. $f(x) = x^2 - 4x + 5$
27. $f(x) = -x^2 - 8x - 12$
28. $f(x) = -2x^2 + 8x - 7$
29. $f(x) = 3x^2 - 18x + 20$
30. $f(x) = 5x^2 - 10x - 1$
31. The profit resulting from manufacturing and selling a product is represented by the function $P(x) = -\frac{1}{4}(x - 200)^2 + 750$, where x is the number of items manufactured.
    a. How much income is gained (or lost) by manufacturing and selling 100 items? 250 items?
    b. How many items should be produced and sold for maximum profit?
    c. What is the maximum profit?
32. The cost equation used by the R & E Manufacturing Co. is $C(h) = 2h^2 - 8h + 12$, where h is the number of hours it takes to produce a particular product and $C(h)$ is the cost of production in hundreds of dollars. What is the minimum cost possible?
33. The height of a punted football can be modeled by the function $f(t) = -16t^2 + 66t + 6$, where $f(t)$ is the height in feet and t is the time in seconds. Find the maximum height of the football and the length of time it would take to reach that height. Do not round.
34. Julie wants to place 44 ft of picket fencing around a garden so as to create the largest rectangular area possible.
    a. If x represents the width, write an algebraic expression that represents the length.
    b. Write a function rule, $f(x)$, that models the area of the garden.
    c. What is the maximum point on the parabola?
    d. Find the dimensions of the garden with the maximum area.
    e. What is the maximum area?
35. Pastor Keifer wants to fence in a rectangular play area next to the church, using fencing for the remaining three sides. The church has 90 ft of fencing budgeted for the project.
    a. If x represents the width, write an algebraic expression that represents the length.
    b. Write a function rule, $f(x)$, that models the area of the play area.
    c. What is the maximum point on the parabola?
    d. Find the dimensions of the maximum area.
    e. What is the maximum area?

C. Exercises

For each quadratic function, state the vertex and then graph the function.

36. $f(x) = \frac{1}{4}x^2 - x - 2$

37. $f(x) = -\frac{1}{2}x^2 - x - 2$

38. The altitude of a volcanic bomb (a mass of molten rock spewn from a volcano) can be modeled by the function $h(t) = -4.9t^2 + 80t + 1000$, where $h(t)$ is the altitude of the bomb and t is the time in seconds. Find the maximum height of the projectile (to the nearest meter). How long after the eruption does the bomb reach its maximum height (to the nearest second)?

Dominion Modeling

A football is thrown from a height of 6 ft on the 30 yd line. The ball reaches a maximum height of 12 ft on the 50 yd line.

39. Using the 30 yd line as the origin of a coordinate plane, state the coordinates of the point at which the ball is thrown and the coordinates of the vertex of the ball's parabolic path.

40. Substitute the values from the vertex and the other known point on the parabola into the formula $y = a(x - h)^2 + k$ and solve the resulting equation to determine the value of a.

41. Write a quadratic function rule in vertex form that models the height of the ball with respect to the horizontal distance. Then write an equivalent function rule in standard form.

Graph the function found in exercise 41 on your graphing calculator and use the TRACE function to answer the following questions.

42. Approximately where will the ball bounce?

43. Between which yard lines should the receiver position himself if the ball is considered catchable between the heights of 6 ft and 3 ft?

CUMULATIVE REVIEW

Simplify. [9.3, 9.6]

44. $(x - 4)(x^4 + 4x^3 + 16x^2 + 64x + 256)$

45. $(x^2 + 5x - 7) \div (x + 1)$

Solve each system. [7.2]

46. $4x + 3y = 8$
$2x - y = -1$

47. Find two numbers such that one is 32 more than twice the other and their difference is 103.

Solve. [2.7, 12.5]

48. $3x + 8 = 5x + 9$

49. $x^2 + 8x - 105 = 0$

50. $x^2 = 7x + 144$

51. $2x^2 - 3x + 1 = 0$

52. $18x^3 + 33x^2 - 105x = 0$

53. Sales tax varies directly with the cost of an item. If a \$527 item is taxed \$36.89, how much tax will you pay on an item costing \$297? [5.6]

12.9 Zeros of a Function

We have discussed falling and projected objects, including balls, keys, and water droplets. The consistency displayed by these physical phenomena results from God's providential preservation of His creation (Col. 1:16–17; Heb. 1:2–3). In His goodness, God the Creator has allowed us to use mathematics to model His creation and to use these models to help those around us. We can see the consistency in His handiwork and marvel at His greatness and faithfulness as demonstrated in His creation (Ps. 19:1). No wonder Paul could write that "the invisible things of him from the creation of the world are clearly seen" (Rom. 1:20).

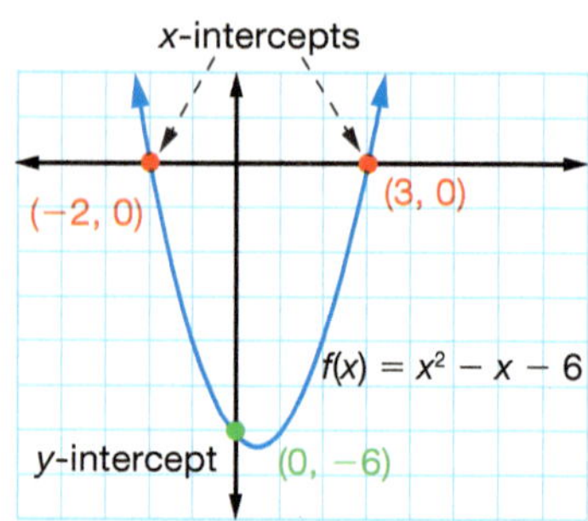

We have seen how to determine the vertex of a parabola and whether it is a minimum or maximum point. It is also helpful to know where a graph crosses the x- and y-axes. Since the y-intercept always has an x value of zero, the y-intercept can be found by letting $x = 0$ and solving for $y = f(0)$.

Example 1

Find the y-intercept of each function algebraically.

a. $g(x) = 2x^2 - 3x + 5$ **b.** $h(x) = (x - 1)^2 + 2$

Answer

a. $g(0) = 2(0)^2 - 3(0) + 5 = 5$

b. $h(0) = (0 - 1)^2 + 2 = (-1)^2 + 2 = 3$

1. Evaluate each function when $x = 0$.

a. (0, 5) **b.** (0, 3)

2. State the y-intercept.

The x-intercepts of a graph always have y values of zero. Therefore, the x-intercepts can be found by letting $y = 0$ or $f(x) = 0$ and solving for x. The x-intercepts of $f(x) = x^2 - x - 6$ are found by setting the equation equal to zero and solving the resulting quadratic equation, $x^2 - x - 6 = 0$. Since $f(x) = 0$ for these values of x, they are called the *zeros* or *roots* of the function.

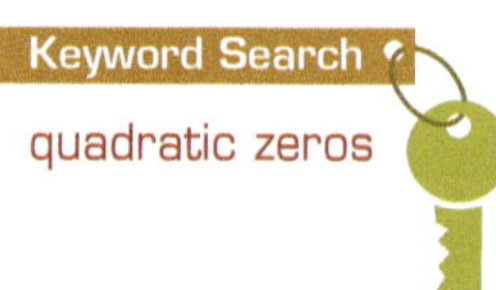

Definition

A **zero** (or **root**) **of a function** is a value of x at which the graph of the function crosses the x-axis. It is the x-coordinate of an x-intercept of the graph.

Example 2

Find the zero(s) of each function algebraically.

a. $r(x) = x^2 - 4x - 5$ **b.** $s(x) = -x^2 + 4x - 4$

Answer

a.
$x^2 - 4x - 5 = 0$
$(x - 5)(x + 1) = 0$
$x - 5 = 0$ or $x + 1 = 0$
$x = 5 \qquad x = -1$

b.
$-x^2 + 4x - 4 = 0$
$-1(x^2 - 4x + 4) = 0$
$-1(x - 2)(x - 2) = 0$
$x - 2 = 0$ or $x - 2 = 0$
$x = 2 \qquad x = 2$

1. Set each function equal to zero.
2. Solve by factoring.

a. $x = 5, -1$ **b.** $x = 2$

3. State the zero(s).

The zeros of the functions in Example 2 are easily confirmed by their graphs below. The single root in the second parabola is often called a *double root* since two factors produce the same root. In this case, the parabola's vertex is located at that x-intercept.

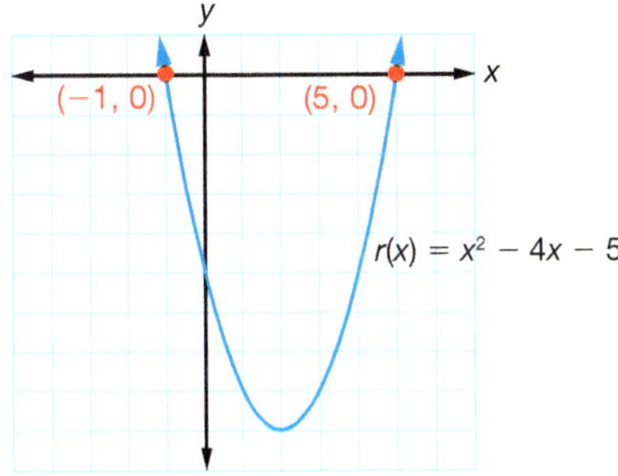

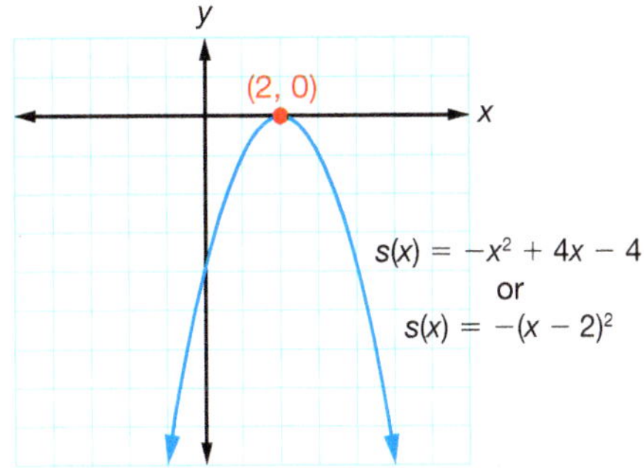

Example 3

Find the zero(s) of each function (to the nearest hundredth).

a. $p(x) = (x - 1)^2 - 6$ **b.** $q(x) = x^2 - x + 4$

Answer

a. $(x - 1)^2 - 6 = 0$ 1. Set the function equal to zero.

$(x - 1)^2 = 6$ 2. Solve by isolating the squared expression and applying the Square Root Property.

$x - 1 = \pm\sqrt{6}$

$x = 1 \pm \sqrt{6}$

$x \approx 3.45, -1.45$ 3. Use a calculator to find the decimal approximations.

b. $x^2 - x + 4 = 0$ 1. Set the function equal to zero.

$$x = \frac{-(-1) \pm \sqrt{(-1)^2 - 4(1)(4)}}{2(1)} = \frac{1 \pm \sqrt{-15}}{2}$$ 2. Use the quadratic formula to solve.

There are no real zeros. 3. The discriminant is negative.

The graph of $p(x) = (x - 1)^2 - 6$ can be used to approximate its roots at -1.5 and 3.5, but an algebraic solution is needed to get exact values. The fact that the graph of $q(x) = x^2 - x + 4$ does not cross the x-axis illustrates why there are no real solutions when the discriminant ($D = b^2 - 4ac$) is negative.

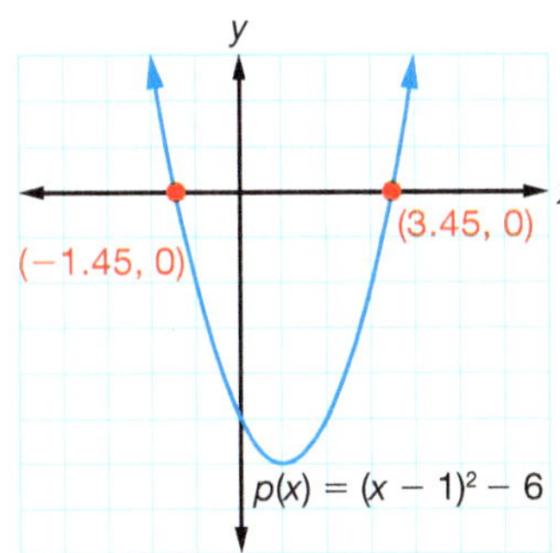

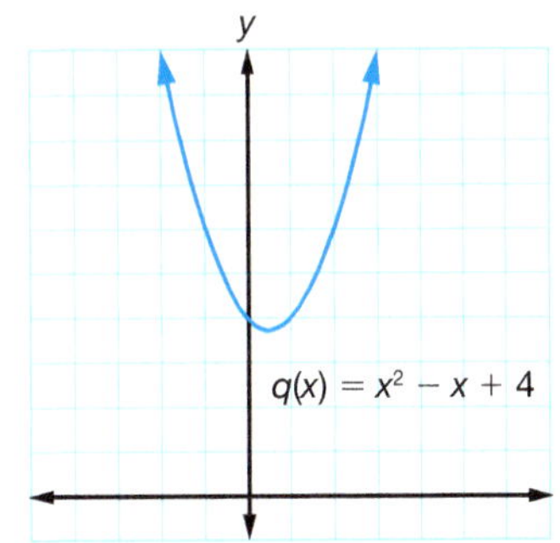

Example 4

Given the quadratic function $f(x) = 3x^2 + 8x - 3$, find its y-intercept, zeros, and vertex; then graph the function.

Answer

$f(x) = 3x^2 + 8x - 3$	1. Find the y-intercept.
$f(0) = 3(0)^2 + 8(0) - 3 = -3$	a. Find $f(0)$.
y-intercept: $(0, -3)$	b. State the y-intercept.
	2. Find the zeros.
$3x^2 + 8x - 3 = 0$	a. Set the function equal to zero.
$(3x - 1)(x + 3) = 0$	b. Solve by factoring.
$3x - 1 = 0$ or $x + 3 = 0$ $3x = 1$ $x = -3$ $x = \frac{1}{3}$	
$x = \frac{1}{3}, -3$	c. State the zeros.
$a = 3, b = 8$, and $c = -3$	3. Find the vertex.
$h = -\frac{8}{2(3)} = -\frac{4}{3}$ or $-1\frac{1}{3}$	a. Use $h = -\frac{b}{2a}$.
$k = f\left(-\frac{4}{3}\right) = 3\left(-\frac{4}{3}\right)^2 + 8\left(-\frac{4}{3}\right) - 3$ $= \frac{16}{3} - \frac{32}{3} - \frac{9}{3} = -\frac{25}{3}$ or $-8\frac{1}{3}$	b. Use $k = f(h)$.
$\left(-1\frac{1}{3}, -8\frac{1}{3}\right)$	c. State the vertex.

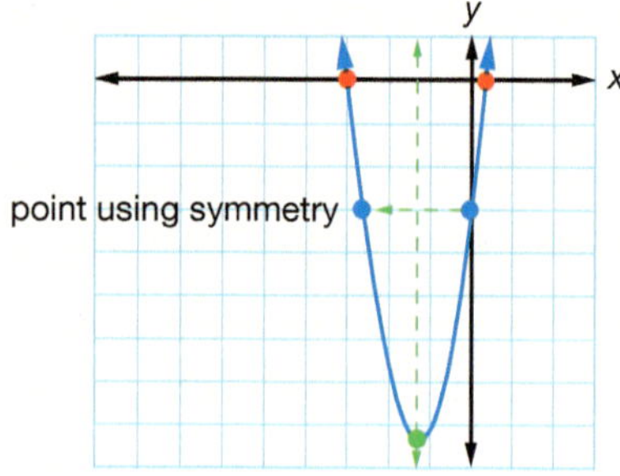

4. Graph.
   a. Plot the vertex, noting that it is a minimum since $a > 0$.
   b. Plot the zeros.
   c. Plot the y-intercept and use symmetry to plot a fifth point. Note that the y-intercept is $1\frac{1}{3}$ units to the right of the line of symmetry.
   d. Note that the parabola is "steeper" than the graph of $y = x^2$, as predicted by $|a| > 1$.

A. Exercises

Find the y-intercept of each function algebraically.

1. $f(x) = \frac{1}{4}x^2$

2. $f(x) = x^2 + 8$

3. $f(x) = 3x^2 - 5$

4. $f(x) = 4x^2 - 5x$

5. $f(x) = 3x^2 - 2x + 1$

6. $f(x) = \frac{1}{2}x^2 - 2x - 3$

7. $f(x) = (x - 2)^2 + 4$

8. $f(x) = (x + 3)^2 - 5$

9. $f(x) = 2(x - 1)^2 - 7$

10. $f(x) = \frac{1}{2}(x - 4)^2 - 11$

Find the zero(s) of each function algebraically.

11. $f(x) = x^2 - 2x - 8$

12. $f(x) = x^2 + 6x + 8$

13. $f(x) = (x - 3)^2$

14. $f(x) = -x^2 - 6x - 9$

15. $f(x) = -2(x + 3)^2 - 2$

16. $f(x) = x^2 + 3$

17. $f(x) = 2(x + 1)^2 - 8$

18. $f(x) = (x - 1)^2 - 8$

19. $f(x) = x^2 + 6x - 8$

20. $f(x) = 2x^2 - 5x - 8$

B. Exercises

21. What is the y-intercept of any function in the form $f(x) = ax^2 + bx + c$?
22. Which quadrants contain the graph of any function $f(x) = ax^2 + bx + c$ if $a < 0$ and the discriminant $D < 0$?
23. Which line contains the vertex of any quadratic function with one real root?
24. How many zeros are there for any quadratic function whose discriminant is zero?

For each quadratic function, state the *y*-intercept, zero(s), and vertex; then graph the function.

25. $f(x) = 4x^2$
26. $f(x) = x^2 - 4$
27. $f(x) = x^2 + 3x$
28. $f(x) = x^2 - 6x + 8$
29. $f(x) = x^2 - 5x + 6$
30. $f(x) = (x + 2)^2 - 3$
31. $f(x) = 4x^2 - 12x + 9$
32. $f(x) = x^2 - 4x - 4$
33. $f(x) = -x^2 - 2x - 3$
34. $f(x) = \frac{1}{4}(x - 5)^2 + 3$
35. The accountants at an appliance manufacturing corporation model their profits from a particular model of refrigerator with the function $P(x) = -x^2 + 400x - 6000$, where x is the number of refrigerators produced weekly.
    - **a.** How many refrigerators should the company produce weekly to make the maximum profit?
    - **b.** What is the maximum profit?
    - **c.** Use a compound inequality to state the number of refrigerators that need to be produced weekly for the company to make a profit.
36. The B & J Co. wants to produce items at a minimum cost to the company. The cost of the items is modeled by the function $C(x) = x^2 - 140x + 5700$, where x is the number of items produced.
    - **a.** How many items should the company produce to keep the cost to a minimum?
    - **b.** What is the minimum cost?
    - **c.** When would the company have a cost of $0?

C. Exercises

37. Write a general expression for the y value of the y-intercept of a graph in the form $f(x) = a(x - h)^2 + k$.
38. Complete the following steps in order to determine whether x feet of fencing will enclose more area when arranged as a square or as a circle.
    - **a.** Express the length of a side of a square (s) and of the radius of a circle (r) in terms of x.
    - **b.** Express the area of the square and of the circle in terms of x.
    - **c.** Which area is larger? Justify your answer.
39. Given $f(x) = -1.5x^2 - 7x + 12$, describe its graph as follows.
    - **a.** Find the vertex.
    - **b.** State the equation of the line of symmetry.
    - **c.** Which way does the parabola open?
    - **d.** Is the parabola "steeper" or "flatter" than $y = x^2$?
    - **e.** Find its zeros.

Dominion Modeling

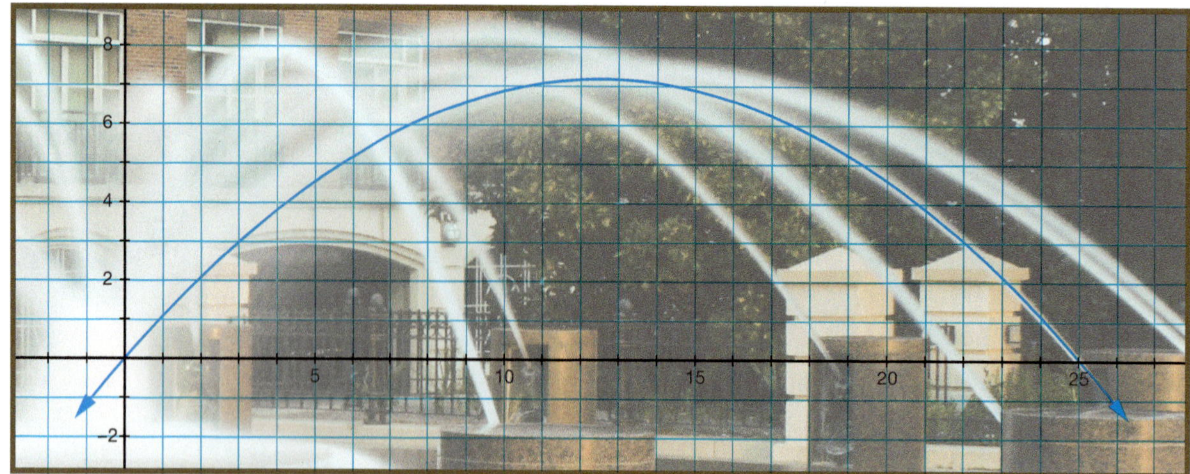

40. Make a table of six ordered pairs by estimating the location of the following points on the parabola outlined in the photo above: the vertex, the zeros, and three additional points on the fountainhead side of the parabola.

41. Complete the following steps to write a function that models the parabolic path of the stream of water.

a. Substitute the values from the vertex into the formula $f(x) = a(x - h)^2 + k$.

b. Determine the value of a (to the nearest thousandth) by substituting into the formula any other point on the parabola. You may want to use one of the zeros.

c. Write the function for the parabola.

42. Plot the points from your table and graph your function using a graphing calculator, a spreadsheet, or interactive geometry software. Check that your function's graph models the points from the table.

CUMULATIVE REVIEW

Find the slope, *y*-intercept, and zero of each linear equation. [6.4, 12.9]

43. $2x - y = 7$

44. $5x + 3y = 6$

Find the distance between the given pairs of points. [11.6]

45. $(2, 5)$ and $(-3, 7)$

46. $(-3, 6)$ and $(2, -8)$

Graph. [6.7]

47. $y < x - 2$

48. $3x + 2y > 4$

Solve. [12.1–12.2, 12.5]

49. $-4 = x^2 + 4x$

50. $3x^2 + 7x - 5 = 0$

51. $x^2 + 3x = 2x^2 - 5x + 4$

52. $(x - 5)^2 = 7$

SEQUENCES

Quadratic Arithmetic Sums

Challenge Can you state the first five partial sums and the partial-sum formula S_n for each of the following sequences?

a. 1, 2, 3, 4, 5, …

b. 2, 4, 6, 8, 10, …

c. 1, 3, 5, 7, 9, …

Since all three of the Challenge sequences are arithmetic, we could use the arithmetic partial-sum formula $S_n = \frac{n(A_1 + A_n)}{2}$ in each case. In the first sequence, $A_n = n$, so we have $S_n = \frac{n(1 + n)}{2} = \frac{n^2 + n}{2}$. However, noticing that each term in the second sequence is double the corresponding term in the first sequence leads to $S_n = 2\left(\frac{n^2 + n}{2}\right) = n^2 + n$. For the third sequence, recognizing the pattern in the first five partial sums results in $S_n = n^2$.

The general-term formula A_n for an arithmetic sequence will always be linear (first degree), and the partial-sum formula will always be quadratic (second degree). We can prove these facts with deductive reasoning.

First, manipulate the arithmetic general-term formula as follows: $A_n = A_1 + (n - 1)d$ $= A_1 + dn - d = dn + (A_1 - d)$. Then represent d as m, n as x, and $(A_1 - d)$ as b, resulting in $A_n = mx + b$, which is a linear function.

To prove that S_n must be quadratic, use $A_n = dn + c$. (Since A_1 and d are both constants, $A_1 - d$ is also a constant and could be represented with b as above or with c as here.)

$$S_n = \frac{n(A_1 + A_n)}{2} = \frac{n(A_1 + dn + c)}{2} = \frac{A_1n + dn^2 + cn}{2} = \frac{dn^2 + A_1n + cn}{2} = \frac{dn^2 + (A_1 + c)n}{2},$$

which can be written as $\frac{d}{2}n^2 + \frac{(A_1 + c)}{2}n$, a quadratic expression.

Exercises

Classify each reasoning example as deductive or inductive.

1. Since the terms of $A_n = 1, 4, 9, 16, 25, \ldots$ are perfect squares, $A_n = n^2$.
2. $(n - 1)d = nd - d$

State the real number property or principle that justifies each deductive reasoning step.

3. If $A_n = dn + c$, then $n(A_1 + A_n) = n(A_1 + dn + c)$.
4. $n(A_1 + dn + c) = A_1n + dn^2 + cn$
5. $A_1n + dn^2 + cn = dn^2 + A_1n + cn$
6. $dn^2 + A_1n + cn = dn^2 + (A_1 + c)n$

Given $S_n = n^2 + n$ for $A_n = 2, 4, 6, 8, 10, \ldots$, find S_n for each of the following sequences.

7. $A_n = 6, 12, 18, 24, 30, \ldots$
8. $A_n = 3, 6, 9, 12, 15, \ldots$

Given $A_n = 3, 8, 13, 18, \ldots$, find each general formula.

9. A_n
10. S_n

CHAPTER 12 REVIEW

Solve.

1. $(x - 2)(x + 3) = 0$
2. $(x + 6)(x - 10) = 0$
3. $2a(4a - 1)(3a + 9) = 0$

Solve by factoring.

4. $x^2 + 3x - 10 = 0$
5. $x^2 - 6x - 7 = 0$
6. $8y^3 - 2y = 0$

Solve by taking roots.

7. $x^2 = 6$
8. $3y^2 - 7 = 5$
9. $(x - 4)^2 = 11$
10. $25(x - 1)^2 = 12$

Solve by completing the square.

11. $x^2 + 8x + 7 = 0$
12. $x^2 + 5x = 3$
13. $2x^2 - 20x - 5 = 0$
14. $3t^2 + 18t - 3 = 0$

State the values of *a*, *b*, and *c* when the equation is written in standard form. Do not solve.

15. $x^2 + 4x - 1 = 0$
16. $3x^2 = 2 + x$

Evaluate the discriminant; then state the number and nature of the root(s).

17. $x^2 - 2x + 1 = 0$
18. $x^2 + 4x - 5 = 0$
19. $x^2 + 4x + 5 = 0$
20. $4x^2 + 3x = 3$

Use the quadratic formula to solve.

21. $x^2 - 6x + 2 = 0$
22. $x^2 - 4x - 21 = 0$
23. $3x^2 + 2x - 5 = 0$
24. $4x^2 - 5 = 2x + 1$

Solve and state the solution set.

25. $(x - 7)(x + 2) = 0$
26. $a^2 = 16$
27. $x^2 - 9 = 0$
28. $2x^2 + 8x + 9 = 0$
29. $a + 3 = a^2 - 2$
30. $5x^2 = 8x$
31. $x + 7 = x^2 + 10x + 25$
32. $x^2 + 6x + 9 = 0$

Write a quadratic equation and solve.

33. Find two numbers whose difference is 19 and whose product is -84.
34. Find two integers such that the first is 2 more than twice the second and their product is 112.
35. The area of a rectangular piece of concrete is 120 ft^2, and its perimeter is 46 ft. What are the dimensions of the rectangle?
36. Mr. Franks runs a hot dog distribution warehouse that distributes hot dogs to local grocery stores. The amount of profit he makes depends on the number of boxes of hot dogs he sells. The profit follows the equation $P(x) = -x^2 + 140x - 3940$. What is his maximum profit, and how many boxes must he sell to make the maximum profit?

Given $f(x) = x^2 + 5x - 7$ and $g(x) = 2x^2 - 7x + 1$, find the following.

37. $f(-1)$
38. $g(3)$

Use the graph of $p(x)$ for exercises 39–42.

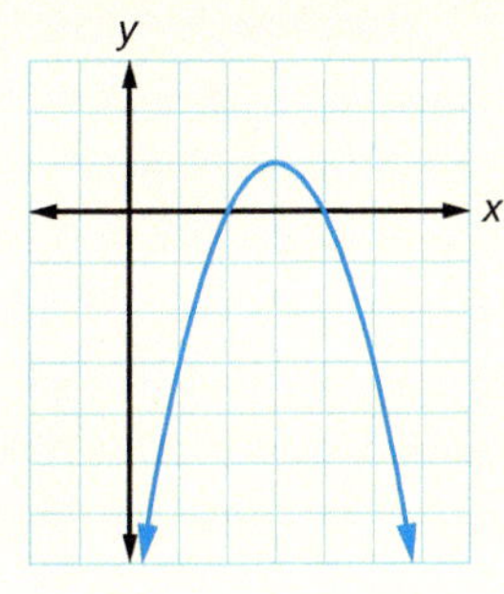

39. Find $p(5)$.

40. Find the zeros.

41. What is the maximum value of the function? Which value of the independent variable produces this maximum value?

42. Find the equation of the line of symmetry.

For each quadratic function, state the vertex and zero(s); then graph the function.

43. $f(x) = \frac{1}{2}x^2$

44. $f(x) = x^2 - 6$

45. $f(x) = -3(x - 2)^2$

46. $f(x) = (x + 6)^2 + 2$

47. $f(x) = x^2 + 6x + 1$

48. $f(x) = 3 + 2x - x^2$

Given $q(x) = 2x^2 - 8x + 6$, find the following without graphing.

49. Find the y-intercept.

50. Find the zeros.

51. Find the vertex. Is it a minimum or a maximum?

52. Will the graph cross the line $y = 30$? Why or why not?

13
RATIONAL
EXPRESSIONS

Among the discoveries of Pythagoras and his students are the three classical means. The arithmetic mean is used to find average test scores or the average time it takes to run a mile. The arithmetic mean of n numbers is found by dividing their sum by n.

Less familiar are the harmonic and geometric means. The harmonic mean of n numbers is found by dividing the number of values by the sum of their reciprocals. It is used to find the average speed directly from several differing speeds when the distances traveled at each speed are the same. The geometric mean of n numbers is found by taking the nth root of their product. The geometric mean has applications in right triangles and is also used to find average growth rates. The Dominion Modeling exercises in this chapter will focus on the difference between arithmetic and harmonic means.

You may wonder whether harmonic means are found in music. Pythagoras is credited for relating arithmetic and harmonic means to the intervals that create harmonies that are pleasing to the ear. Tradition suggests that upon hearing the hammers of a blacksmith, Pythagoras became curious about their harmonious tones. The story suggests that the hammers had relative weights of 6, 8, 9, and 12. Inspired by the intervals he heard, Pythagoras applied the arithmetic and harmonic means to a string on a musical instrument by subdividing it into the same proportions as the hammers' weights.

What accounts for the ability of mathematics to model the workings of the physical universe? While some would describe this ability as mysterious, incredible, or even unreasonably effective, the Christian sees mathematics as an attempt to discover and describe the order established by God. We would expect the mathematical structure found in human reasoning to mirror God's universe since they are both reflections of their Creator.

After this chapter you should be able to

1. determine values for which a rational expression is undefined.
2. simplify rational expressions.
3. add, subtract, multiply, and divide rational expressions.
4. simplify complex and mixed rational expressions.
5. solve rational equations.
6. solve word problems using rational equations.
7. graph rational functions of the form $f(x) = \frac{a}{x - h} + k$ and identify the graph's center and asymptotes.

13.1 Simplifying Rational Expressions

When a larger tube tapers into a smaller one, the taper per inch, T, is found by the formula $T = \frac{D - d}{L}$, where L is the length of the pipe and D and d are the larger and smaller diameters, respectively.

Recall that a rational number is defined as a ratio of integers whose denominator is not zero. You have already learned how to simplify, add, subtract, multiply, and divide rational numbers. In this chapter, you will apply the same principles to rational expressions.

Definition

A **rational expression** is a ratio of polynomials whose denominator is not zero.

The following are all examples of rational expressions.

1. $\frac{x}{x + 2}$ **2.** $\frac{x - 4}{5}$ **3.** $\frac{x - 6}{x - 9}$

4. $\frac{2x^2 + 3x + 6}{x^2 - 8x + 7}$ **5.** $\frac{x + 7}{x^2 + 3x - 9}$ **6.** $\frac{3x^2 + 4x - 5}{x^2 + 1}$

Suppose $x = -2$ in the first expression. Substituting -2 for x would give us $\frac{-2}{0}$. Therefore, the rational expression is undefined when $x = -2$, and -2 is called an *excluded value*.

To find the excluded values for a rational expression, determine which value(s) of x will make the denominator zero. In the second expression, the denominator has no variables. Since $5 \neq 0$, the rational expression is always defined.

Example 1

Determine the excluded value(s) for each expression.

a. $\frac{x - 6}{x - 9}$ **b.** $\frac{2x^2 + 3x + 6}{x^2 - 8x + 7}$ **c.** $\frac{x + 7}{x^2 + 3x - 9}$

Answer

a. $\frac{x - 6}{x - 9}$

$x - 9 \neq 0$ 1. The denominator cannot equal zero.

$x \neq 9$ 2. Solve.

$\frac{x - 6}{x - 9}$ is undefined when $x = 9$.

b. $\frac{2x^2 + 3x + 6}{x^2 - 8x + 7}$

$x^2 - 8x + 7 \neq 0$ 1. The denominator cannot equal zero.

$(x - 7)(x - 1) \neq 0$ 2. Solve by factoring.

$x - 7 \neq 0$ or $x - 1 \neq 0$

$x \neq 7, 1$

$\frac{2x^2 + 3x + 6}{x^2 - 8x + 7}$ is undefined when $x = 7, 1$.

Example 1, continued

c. $\frac{x+7}{x^2+3x-9}$

$x^2 + 3x - 9 \neq 0$

$x \neq \frac{-3 \pm \sqrt{3^2 - 4(1)(-9)}}{2(1)} \neq \frac{-3 \pm \sqrt{45}}{2} \neq \frac{-3 \pm 3\sqrt{5}}{2}$

$\frac{x+7}{x^2+3x-9}$ is undefined when $x = \frac{-3 \pm 3\sqrt{5}}{2} \approx 1.85, -4.85$.

1. The denominator cannot equal zero.
2. Since the trinomial cannot be factored, use the quadratic formula to solve.

The rational expression $\frac{3x^2 + 4x - 5}{x^2 + 1}$ has no excluded values because there are no real number solutions to $x^2 + 1 = 0$.

Remember that fractions such as $\frac{12}{15}$ are reduced to lowest terms by canceling all common factors. The simplification of any rational expression is based on one of the fundamental principles of fractions: $\frac{ac}{bc} = \frac{a}{b}$ when $b \neq 0$ and $c \neq 0$.

Example 2

Simplify.

a. $\frac{4a^3}{8a}$ **b.** $\frac{180m^3n}{27mn^2}$

Answer **a.** $\frac{\cancel{4}\cancel{a}a^2}{\cancel{4} \cdot 2\cancel{a}} = \frac{a^2}{2}$ **b.** $\frac{\cancel{9} \cdot 20\cancel{m}m^2\cancel{n}}{\cancel{9} \cdot 3\cancel{m}\cancel{n}n} = \frac{20m^2}{3n}$ Factor and cancel common factors.

Example 3

Simplify.

a. $\frac{3(x-1)}{3}$ **b.** $\frac{4(a+3)}{a+3}$ **c.** $\frac{3x+6}{3}$

Answer

a. $\frac{\cancel{3}(x-1)}{\cancel{3}} = x - 1$ Cancel the common factor, 3.

b. $\frac{4\cancel{(a+3)}}{\cancel{a+3}} = \frac{4}{1} = 4$ Cancel the common binomial factor, $a + 3$.

c. $\frac{3x+6}{3} = \frac{3(x+2)}{3}$ 1. Factor the numerator.

$= \frac{\cancel{3}(x+2)}{\cancel{3}} = x + 2$ 2. Cancel the common factor, 3.

Remember that only common factors can be canceled. Common terms cannot be canceled.

$\frac{x+2}{2} \neq \frac{x+\cancel{2}}{\cancel{2}} \neq x + 1$

By substituting 4 for x, you can see the error in this simplification.

$\frac{x+2}{2} = \frac{4+2}{2} = \frac{4+\cancel{2}}{\cancel{2}} = 4 + 1$ incorrect

$\frac{x+2}{2} = \frac{4+2}{2} = \frac{6}{2} = 3$ correct

Example 4

Simplify $\frac{x^2 + 4x + 4}{x^2 - 4}$.

Answer

$\frac{(x + 2)(x + 2)}{(x - 2)(x + 2)}$ — 1. Factor each polynomial completely.

$= \frac{(x + 2)\cancel{(x + 2)}}{(x - 2)\cancel{(x + 2)}} = \frac{x + 2}{x - 2}$ — 2. Cancel the common factor.

In Example 4 the original expression has excluded values of ± 2, but the only excluded value for the simplified expression is 2. For this reason, excluded values for the original expression are often listed with the simplified rational expression.

$$\frac{x^2 + 4x + 4}{x^2 - 4} = \frac{x + 2}{x - 2};\ x \neq \pm 2$$

Factors such as $x - 3$ and $3 - x$ are opposites of each other. When two such opposite factors appear in the numerator and the denominator, factoring -1 from one of the factors allows you to cancel a common factor.

$$\frac{3 - x}{x - 3} = \frac{-1(-3 + x)}{x - 3} = \frac{-1\cancel{(x - 3)}}{\cancel{x - 3}} = -1$$

Example 5

Simplify $\frac{4x^2 + 4x - 48}{3 - x}$, leaving the answer in factored form. Include any excluded values.

Answer

$x \neq 3$ — 1. State any excluded values for the original expression.

$\frac{4(x^2 + x - 12)}{3 - x} = \frac{4(x - 3)(x + 4)}{3 - x}$ — 2. Factor the polynomial completely.

$= \frac{4\cancel{(x - 3)}(x + 4)}{-1\cancel{(x - 3)}}$ — 3. Recognize that $x - 3$ and $3 - x$ are opposites and factor -1 out of one of those factors so they can be canceled.

$= -4(x + 4);\ x \neq 3$ — 4. Simplify the result and state the excluded value.

Example 6

Write a rational expression representing the portion of the square that is shaded. Then simplify and state the result as a percent (to the nearest tenth).

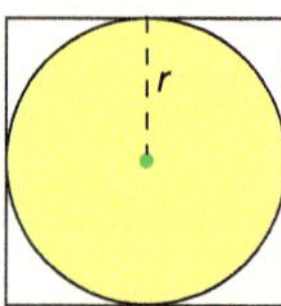

Answer

$A_{circle} = \pi r^2$

$A_{square} = s^2 = (2r)^2 = 4r^2$ — 1. Write expressions for the area of the circle and the square.

$\frac{A_{circle}}{A_{square}} = \frac{\pi r^2}{4r^2}$ — 2. Write the rational expression.

$= \frac{\pi \cancel{r^2}}{4\cancel{r^2}} = \frac{\pi}{4}$ — 3. Cancel common factors.

$\approx 0.785 = 78.5\%$ — 4. Complete the division; then convert the result to a percent.

A. Exercises

State any excluded values for each rational expression.

1. $\frac{3x + 4}{x}$
2. $\frac{x^2 - 9}{x + 4}$
3. $\frac{a + 7}{a + 2}$
4. $\frac{b + 8}{b - 5}$
5. $\frac{x^2 + 3x - 2}{x^2 + 5x - 24}$
6. $\frac{y - 7}{y^2 + 7y - 18}$
7. $\frac{c + 4}{c^2 - 10c + 24}$
8. $\frac{z - 5}{z^3 - 3z^2 - 4z}$

Simplify, leaving each answer in factored form.

9. $\frac{3x^2}{15x}$
10. $\frac{24x^2y^3}{8xy}$
11. $\frac{3(a + 7)}{3}$
12. $\frac{2(y - 5)}{y - 5}$
13. $\frac{6x - 6}{6}$
14. $\frac{x^2 + x}{x}$
15. $\frac{4a^2 + 2a}{2a}$
16. $\frac{3c}{3c - 6c^2}$

B. Exercises

Simplify, leaving each answer in factored form. State any excluded values.

17. $\frac{x^2 - 4x - 45}{x^2 - 10x + 9}$
18. $\frac{3x^2 - 20x + 12}{3x^2 + x - 2}$
19. $\frac{x^2 - y^2}{x + y}$
20. $\frac{y^2 - x^2}{x^2 - y^2}$
21. $\frac{3x^2 + 3xy}{x^2 - y^2}$
22. $\frac{x^7y^2 + x^6y^2 - 6x^5y^2}{x^5 - 9x^3}$
23. $\frac{x^3 - 3x^2 - 10x}{x^3 + 2x^2}$
24. $\frac{3x^5y - 3x^4y - 60x^3y}{9x^4 + 36x^3}$

Write a rational expression representing the portion of each figure that is shaded. State the result as a simplified expression and as a percent (to the nearest tenth).

25. 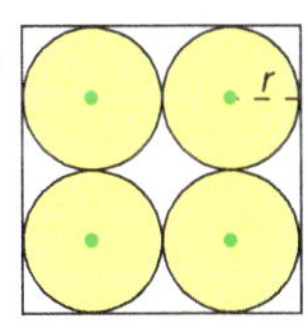

26. 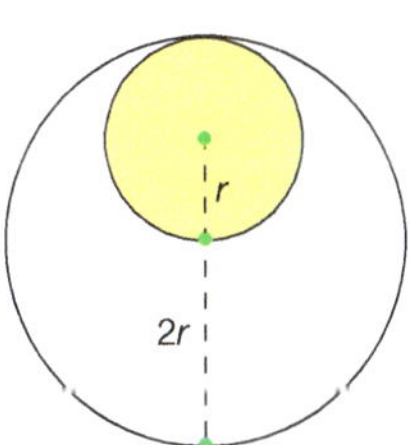

27. 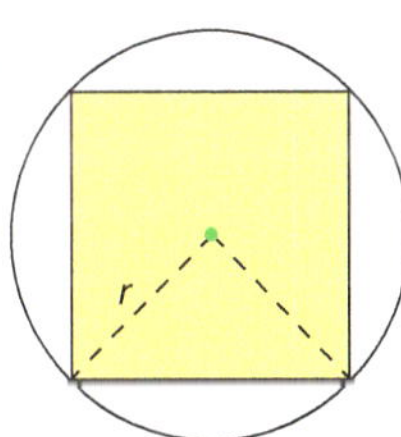

28. 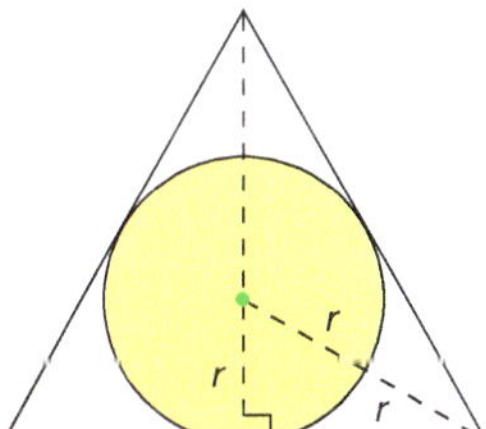

The ideal mechanical advantage (IMA) of a lever is given by the formula IMA = $\frac{l_e}{l_r}$, where l_e is the distance from the fulcrum to the effort and l_r is the distance from the fulcrum to the resistance.

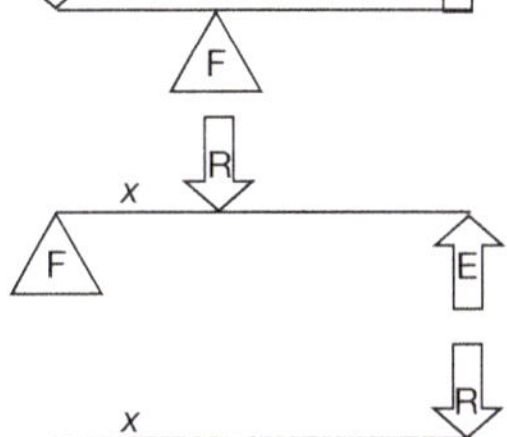

29. Write a rational expression representing the IMA of a 5 ft first-class lever if the fulcrum is placed x ft from the resistance.
30. Write a rational expression representing the IMA of a 5 ft second-class lever if the resistance is placed x ft from the fulcrum.
31. Write a rational expression representing the IMA of a 5 ft third-class lever if the effort is applied x ft from the fulcrum.

C. Exercises

Simplify, leaving each answer in factored form. State any excluded values.

32. $\frac{48x^2 + 34x - 5}{1 - 8x}$

33. $\frac{5xz + 5z}{10x^2 - 10}$

34. $\frac{x^5 + x^4 - 4x^3 - 4x^2}{x^5 + 4x^4 + 4x^3}$

35. Explain why the expression $\frac{9x^2 - 16}{3x^2 + 4x}$ is not identical to the simplified expression $\frac{3x - 4}{x}$.

36. Write a rational expression representing the greatest portion of a cube that can be filled by a sphere. State the result as a simplified expression and as a percent (to the nearest tenth).

Dominion Modeling

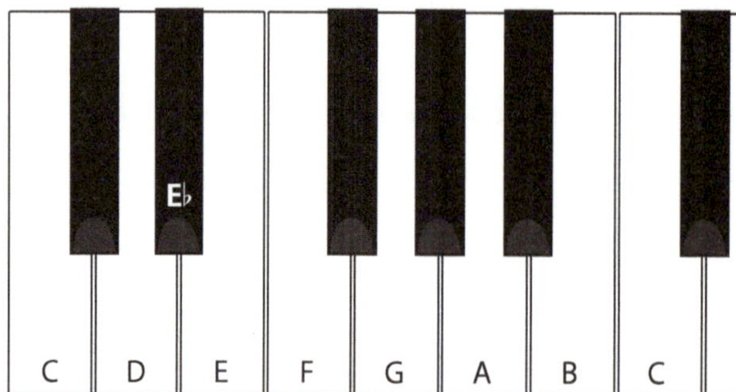

A *musical interval* is the distance in pitch between two notes. A scale in the key of C major is played on the white keys of a piano in succession from left to right. An interval of a 2nd can be heard by playing the first and second notes in the scale (C and D), an interval of a 3rd by playing the first and third notes (C and E), and so on up to an octave (equivalent to an interval of an eighth), which can be heard by playing two consecutive Cs.

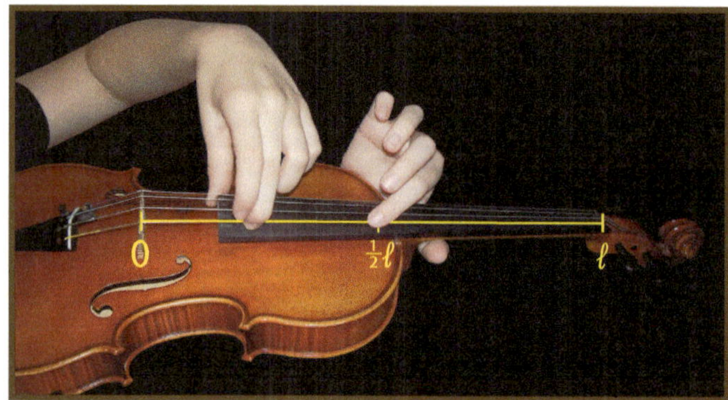

Pythagoras found that when a string was pressed at its midpoint, the plucked string produced a tone one octave higher than the tone produced by plucking the string without pressing it down. The arithmetic mean, A, of 0 and l (the original string length) predicts the length of string needed to produce a tone one octave higher.

$$A = \frac{a + b}{2} = \frac{0 + l}{2} = \frac{l}{2} \text{ or } \frac{1}{2}l$$

37. Where should a string be pressed (in terms of l) to produce a tone that is one octave higher than the tone produced when the string is pressed at $\frac{1}{2}l$?

38. Describe how to produce a tone three octaves higher than the tone produced by the entire length of the string.

CUMULATIVE REVIEW

Solve each literal equation for the indicated variable. [3.1]

39. $mp - r = 0$ for m

40. $ns^2 + ms + q = 0$ for s

Simplify. [1.7, 11.4]

41. $3^{-1} + 5^{-1}$

42. $\dfrac{3x^2y\sqrt{20}}{9xy\sqrt{6}}$

Use the linear equation $3x - 5y = 11$ for exercises 43–44. [6.1, 6.3]

43. Find the x- and y-intercepts.

44. Write the equation in slope-intercept form; then state the slope.

State whether each relationship is a direct or an inverse variation. [5.6]

45. The product of the variables is a constant.

46. The ratio of the variables is a constant.

State whether each table represents a direct or an inverse variation. Write the function rule for the variation. [5.6]

47.

x	2	4	5	7
y	6	12	15	21

48.

x	1	2	3	4
y	24	12	8	6

MIND OVER MATH

One benefit of studying mathematics is the development of correct reasoning skills. Can you detect the faulty reasoning in the following proofs?

Proof that 1 = 2

$a = b$	1. Let $a = b$.
$ab = b^2$	2. Multiply both sides by b.
$ab - a^2 = b^2 - a^2$	3. Subtract a^2 from both sides.
$a(b - a) = (b + a)(b - a)$	4. Factor both sides.
$a = b + a$	5. Divide both sides by $(b - a)$.
$a = a + a$	6. Since $a = b$, substitute a for b.
$a = 2a$	7. Simplify the right side by adding like terms.
$1 = 2$	8. Divide both sides by a.

Proof that 4 = 8

$16 - 48 = 64 - 96$	1. Both sides of the equation equal -32.
$16 - 48 + 36 = 64 - 96 + 36$	2. Add 36 to both sides.
$4^2 - 2(4)(6) + 6^2 = 8^2 - 2(8)(6) + 6^2$	3. Factor the perfect square trinomials.
$(4 - 6)^2 = (8 - 6)^2$	
$4 - 6 = 8 - 6$	4. Take the square root of both sides.
$4 = 8$	5. Add 6 to both sides.

13.2 Multiplying and Dividing Rational Expressions

The basic rule for multiplying fractions is to place the product of the numerators over the product of the denominators. You can multiply and then cancel or cancel and then multiply. Canceling first is usually easier.

A player's batting average is .314, which is calculated using the rational expression $A = \frac{H}{B}$, where H is the number of hits and B is the number of at bats.

Example 1

Multiply $\frac{5}{9} \cdot \frac{27}{40}$.

Answer $\frac{\cancel{5}}{\cancel{9}} \cdot \frac{\cancel{27}^{3}}{\cancel{40}_{8}} = \frac{3}{8}$

The multiplication of rational expressions is done in the same way. As with rational numbers, it is best to cancel first.

Example 2

Simplify $\frac{3a^2c^2}{2b} \cdot \frac{10a}{9bc}$.

Answer $\frac{\cancel{3}a^2\cancel{c}c}{\cancel{2}b} \cdot \frac{\cancel{10}^{5}a}{\cancel{9}_{3}b\cancel{c}}$ 1. Cancel common factors.

$= \frac{5a^3c}{3b^2}$ 2. Multiply.

In Example 2 we have assumed that $b \neq 0$ and $c \neq 0$ since the original expression would be undefined if $b = 0$ or $c = 0$. For the rest of this chapter, you may assume that no denominator of a rational expression has a value of zero.

Multiplying Rational Expressions
1. Factor all algebraic quantities in the numerators and the denominators.
2. Cancel common factors appearing in both a numerator and a denominator.
3. Express the result as the product of the numerators divided by the product of the denominators.

Example 3

Simplify $\frac{x^2 + 2xy + y^2}{2x + 2y} \cdot \frac{18}{x^2 - y^2}$.

Answer $\frac{(x + y)(x + y)}{2(x + y)} \cdot \frac{18}{(x + y)(x - y)}$ 1. Factor the numerators and the denominators.

$= \frac{\cancel{(x + y)}\cancel{(x + y)}}{\cancel{2}\cancel{(x + y)}} \cdot \frac{\cancel{18}^{9}}{\cancel{(x + y)}(x - y)}$ 2. Simplify by canceling common factors.

$= \frac{9}{x - y}$ 3. The result is the product of the numerators divided by the product of the denominators.

Example 4

Simplify $\frac{x^3 - x^2 - 72x}{3x^2} \cdot \frac{12x}{x^2 + x - 56}$, leaving the answer in factored form.

Answer

$$\frac{x(x-9)(x+8)}{3x^2} \cdot \frac{12x}{(x+8)(x-7)}$$

1. Factor the numerators and the denominators.

$$= \frac{\cancel{x}(x-9)\cancel{(x+8)}}{\cancel{3}\cancel{x}\cancel{x}} \cdot \frac{\overset{4}{\cancel{12}}\cancel{x}}{\cancel{(x+8)}(x-7)}$$

2. Simplify by canceling common factors.

$$= \frac{4(x-9)}{x-7}$$

3. The result is the product of the numerators divided by the product of the denominators.

To divide by a fraction, multiply by the reciprocal of the divisor.

$$\frac{1}{2} \div \frac{7}{4} = \frac{1}{\cancel{2}} \cdot \frac{\overset{2}{\cancel{4}}}{7} = \frac{2}{7}$$

This same procedure is used for dividing rational expressions.

Example 5

Divide $\frac{x^2}{3y^2}$ by $\frac{x}{9y}$.

Answer

$$\frac{x^2}{3y^2} \div \frac{x}{9y} = \frac{x^2}{3y^2} \cdot \frac{9y}{x}$$

Invert the divisor and multiply.

$$= \frac{x\cancel{x}}{\cancel{3}\cancel{y}y} \cdot \frac{\overset{3}{\cancel{9}}\cancel{y}}{\cancel{x}} = \frac{3x}{y}$$

Dividing Rational Expressions

Multiply by the reciprocal of the divisor.

Example 6

Simplify $\frac{x^2 + 3x + 2}{x^2 + x - 2} \div \frac{x^2 + 2x + 1}{x^2 - 1}$.

Answer

$$\frac{x^2 + 3x + 2}{x^2 + x - 2} \cdot \frac{x^2 - 1}{x^2 + 2x + 1}$$

1. Multiply by the reciprocal of the divisor.

$$= \frac{(x+2)(x+1)}{(x+2)(x-1)} \cdot \frac{(x-1)(x+1)}{(x+1)(x+1)}$$

2. Factor the numerators and the denominators.

$$= \frac{\cancel{(x+2)}\cancel{(x+1)}}{\cancel{(x+2)}\cancel{(x-1)}} \cdot \frac{\cancel{(x-1)}\cancel{(x+1)}}{\cancel{(x+1)}\cancel{(x+1)}} = 1$$

3. Simplify by canceling common factors. Remember that $\frac{\cancel{a}}{\cancel{a}} = 1$.

These principles can also be used to convert unit rates such as miles per hour or meters per second. Treat each unit as a factor and arrange the unit multipliers to cancel the original units.

Example 7

How many feet does a car travel each second if it is traveling at 60 mi/hr?

Answer

$$\frac{60\ \cancel{mi}}{\cancel{hr}} \cdot \frac{5280\text{ ft}}{1\ \cancel{mi}} \cdot \frac{1\ \cancel{hr}}{60\ \cancel{min}} \cdot \frac{1\ \cancel{min}}{60\text{ sec}} = \frac{\cancel{60}(5280)\text{ ft}}{\cancel{60}(60)\text{ sec}} = 88\text{ ft/sec}$$

A. Exercises

Multiply.

1. $\frac{2}{5} \cdot \frac{19}{28}$

2. $\frac{16}{9} \cdot \frac{27}{8}$

3. $\frac{3a^2}{17b} \cdot \frac{34b^2}{6ab}$

4. $\frac{5x^2y}{18xy} \cdot \frac{3y^3}{9y}$

5. $\frac{x+2}{5} \cdot \frac{25}{x+2}$

6. $\frac{(x-3)(x+4)}{x-1} \cdot \frac{x-1}{(x-4)(x-3)}$

State the reciprocal of each expression.

7. $-\frac{2}{3}$

8. $\frac{32a^2b}{xy^3}$

9. $\frac{a^2-b^2}{a^2b}$

10. $x^2 - 2x - 3$

Divide.

11. $\frac{4}{5} \div \frac{8}{35}$

12. $\frac{14}{9} \div \frac{28}{27}$

13. $\frac{8x^4y^2}{x^3y} \div \frac{24x}{7y}$

14. $\frac{a^2b^3}{x^5y^2} \div \frac{a^4b}{xy^3}$

15. $\frac{(x+y)^2}{x-y} \div \frac{(x+y)(x-y)}{(x-y)^2}$

16. $\frac{4(x+1)}{5(x-5)} \div \frac{3(x+1)}{x-5}$

B. Exercises

Simplify, leaving each answer in factored form.

17. $\frac{x^2-5x-14}{8} \cdot \frac{4}{x-7}$

18. $\frac{x^2+3x-18}{x^2-x} \cdot \frac{x^3y}{xy-3y}$

19. $\frac{x^2+10x+16}{x-4} \cdot \frac{x^2-x-12}{x+2}$

20. $\frac{x^2-x-20}{x^2-6x+5} \cdot \frac{x+3}{x^2+7x+12}$

21. $\frac{2x^2-xy-y^2}{x^2-y^2} \cdot \frac{x^2+2xy+y^2}{3x^2-xy-4y^2}$

22. $\frac{x-2}{x^3-5x^2+6x} \cdot \frac{xy^2}{xy+4y}$

23. $\frac{x^2-y^2}{(x-y)^2} \div \frac{x^2-2xy+y^2}{x-y}$

24. $\frac{x^2+2x}{3x-9} \div \frac{x^3-9x}{x-3}$

25. $(x+1) \div \frac{x^2+2x+1}{x}$

26. $\frac{x^2-11x+18}{x-3} \div \frac{x^2+6x+8}{x+2}$

27. $\frac{x^2-xy-2y^2}{x^2-y^2} \div \frac{x^2-3xy+2y^2}{3x^2+4xy+y^2}$

28. $\frac{x^2+3x-54}{x^2+5x+6} \div \frac{x^2-12x+36}{x^2-3x-18}$

29. $\frac{a^2+ab-2b^2}{a^2-b^2} \cdot \frac{a^2-2ab-3b^2}{a^2-3ab}$

30. $\frac{3x^2-11xy-4y^2}{x^3y-4x^2y^2} \cdot \frac{x^2y-3xy^2}{3x+y}$

31. $\frac{a^2+7ab+10b^2}{a^2+ab-2b^2} \cdot \frac{3a^2-7ab+4b^2}{3a^2+11ab-20b^2}$

32. $\frac{3m^2-2mn-n^2}{m^3n^2} \div (m^2-2mn+n^2)$

33. $\frac{x^2-7x-18}{x+4} \cdot \frac{x^2-5x+6}{x-9} \cdot \frac{x+4}{x-3}$

34. $\frac{1}{2x^2+xy-y^2} \div \frac{x^2+2xy+y^2}{2x^2-3xy+y^2}$

35. $\frac{12x^2-29xy+14y^2}{x^2-xy-12y^2} \div \frac{3x^2+xy-2y^2}{x^2-2xy-8y^2}$

36. Convert 125 cm/sec to kilometers per hour.

37. Convert a flow rate of 4 fl oz/sec to gallons per minute.

38. Brock ran 440 yd in 1 min. What was his average speed in miles per hour?

39. A typist averages 45 words/min. The chapters in the book she is working on average 30 pages each with an average of 325 words per page.
    a. Express her rate of typing in chapters per hour.
    b. How long does it take her to type a chapter?

C. Exercises

Simplify, leaving each answer in factored form.

40. $\frac{x^4 + x^3 - 6x^2}{xy^3 - 2y^3} \cdot \frac{x^2 + 2x - 15}{x^2 + 8x + 15} \div \frac{x^2}{y^3}$

41. $\frac{4ab^4}{a^2 + 8ab + 15b^2} \cdot \frac{a^2 + 3ab - 10b^2}{a + 7b} \div \frac{a^2b^3}{a^2 + 2ab - 3b^2}$

42. $\frac{x^2 - 3x - 18}{x^2 - 2x - 48} \div \frac{x^2 - x - 12}{x^3 - 12x^2 + 36x} \cdot \frac{3x^2 - 6x - 144}{x^2 - 4x}$

43. $\frac{(3x + 1)(x - 1) + (5x - 3)(x + 2) - (19x - 12)}{x^2 - 1} \div \frac{4x^2 + 7x - 15}{x^2 + 4x + 3}$

CUMULATIVE REVIEW

Simplify. [1.4, 11.7]

44. $\frac{11}{25} - \frac{6}{25}$

45. $(\sqrt{26} - 4\sqrt{2})(2\sqrt{13} + 5)$

Graph. [4.4, 6.1]

46. $2 < x \le 8$ on a number line

47. $2x + 3y = 15$ on a coordinate plane

Solve. [2.8, 12.6]

48. $\frac{24}{25}x - 7 = x - \frac{3}{10}$

49. $3(2x - 5) + 2(5x + 1) = 7 - 3x(x - 5)$

50. Ten less than the square root of 1 more than twice a number is 1. Find the number. [11.8]

51. Find the slope of the line parallel to $7x = 4 - 5y$. [6.5]

Given the two points (6, −3) and (−2, −4), find the following. [6.2, 11.6]

52. the slope of the line passing through the points

53. the distance between the points

SEQUENCES

Harmonic Sequences

Challenge **Can you state the general-term formula A_n for each of the following sequences?**

a. $\frac{1}{2}, \frac{1}{4}, \frac{1}{6}, \frac{1}{8}, \frac{1}{10}, \ldots$

b. $\frac{1}{3}, \frac{1}{5}, \frac{1}{7}, \frac{1}{9}, \frac{1}{11}, \ldots$

c. $\frac{1}{2}, \frac{1}{5}, \frac{1}{8}, \frac{1}{11}, \frac{1}{14}, \ldots$

A *harmonic sequence* is a sequence whose terms are the reciprocals of the terms of an arithmetic sequence. Its name is derived from overtones (or harmonics) in music. Overtones are wavelengths from a vibrating string or column of air that are higher than the fundamental wavelength or frequency in sound. Overtone wavelengths of a vibrating string are $\frac{1}{2}, \frac{1}{3}, \frac{1}{4}, \frac{1}{5}, \frac{1}{6}, \ldots$.

Identifying the general-term formula for a harmonic sequence involves finding the arithmetic general-term formula that describes the denominators. By noticing the pattern of the given terms, we can find the general-term formula by inductive reasoning. The pattern of the denominators in the first sequence in the Challenge is 2, 4, 6, 8, 10, …, $2n$, so $A_n = \frac{1}{2n}$. Each denominator in the second sequence is one more than the corresponding denominator in the first sequence. Therefore, the general-term formula for the second sequence is $A_n = \frac{1}{2n + 1}$.

If the pattern is more difficult to recognize, as in the third sequence, we can use deductive reasoning. Using $A_n = A_1 + (n - 1)d$ and substituting 2 for A_1 and 3 for d produces $A_n = 2 + (n - 1)3 = 2 + 3n - 3 = 3n - 1$. Thus, the general-term formula for the third sequence is $A_n = \frac{1}{3n - 1}$. This is deductive reasoning since we reasoned from the general arithmetic formula $A_n = A_1 + (n - 1)d$ to the specific harmonic formula $A_n = \frac{1}{3n - 1}$.

An infinite harmonic sequence is unusual in that, even though its limit is zero and the sequence converges, the sum of all its terms is infinite.

Exercises

Are the following sequences harmonic? Explain your reasoning.

1. $1, \frac{1}{5}, \frac{1}{9}, \frac{1}{13}, \frac{1}{17}, \ldots$

2. $\frac{1}{2}, \frac{1}{4}, \frac{1}{8}, \frac{1}{16}, \frac{1}{32}, \ldots$

3. $1, \frac{1}{4}, \frac{1}{9}, \frac{1}{16}, \frac{1}{25}, \ldots$

4. $-\frac{1}{7}, -\frac{1}{3}, 1, \frac{1}{5}, \frac{1}{9}, \ldots$

State the general-term formula for each of the following harmonic sequences.

5. $1, \frac{1}{3}, \frac{1}{5}, \frac{1}{7}, \frac{1}{9}, \ldots$

6. $\frac{1}{4}, \frac{1}{8}, \frac{1}{12}, \frac{1}{16}, \frac{1}{20}, \ldots$

7. $-\frac{1}{2}, \frac{1}{3}, \frac{1}{8}, \frac{1}{13}, \frac{1}{18}, \ldots$

8. $\frac{1}{9}, \frac{1}{3}, -\frac{1}{3}, -\frac{1}{9}, -\frac{1}{15}, \ldots$

Classify each reasoning process as deductive or inductive.

9. reasoning by recognizing a pattern

10. reasoning from general to specific

13.3 Adding and Subtracting Expressions with Common Denominators

Operations with rational expressions follow the same rules as operations with rational numbers. Expressions with a common denominator are added by writing the sum of the numerators over the common denominator. Be sure to simplify the resulting expression.

Example 1

Simplify.

a. $\frac{4}{x} + \frac{9}{x}$ **b.** $\frac{2a}{15} + \frac{8a}{15}$

Answer **a.** $\frac{4}{x} + \frac{9}{x} = \frac{13}{x}$

b. $\frac{2a}{15} + \frac{8a}{15} = \frac{10a}{15}$

$= \frac{\overset{2}{\cancel{10}}a}{\underset{3}{\cancel{15}}} = \frac{2a}{3}$

1. Write the sum of the numerators over the common denominator.
2. If possible, simplify by canceling common factors.

Example 2

Simplify $\frac{2x + 9}{x - 4} + \frac{3x + 6}{x - 4}$.

Answer $\frac{2x + 9 + 3x + 6}{x - 4} = \frac{5x + 15}{x - 4}$

$= \frac{5(x + 3)}{x - 4}$

1. Add the numerators by combining like terms.
2. Factoring shows that the expression cannot be simplified.

Example 3

Simplify $\frac{3x^2 + 4x - 8}{x + 2} + \frac{x^2 - 3x - 6}{x + 2}$.

Answer $\frac{3x^2 + 4x - 8 + x^2 - 3x - 6}{x + 2} = \frac{4x^2 + x - 14}{x + 2}$

$= \frac{(4x - 7)\cancel{(x + 2)}}{\cancel{x + 2}} = 4x - 7$

1. Add the numerators by combining like terms.
2. Factor and simplify by canceling common factors.

To subtract rational expressions, follow the procedure for subtracting rational numbers. When the expressions have a common denominator, subtract the numerators and write the result over the common denominator. Then factor and simplify.

Example 4

Simplify $\frac{2+x}{8x} - \frac{x-8}{8x}$.

Answer

$\frac{(2+x)-(x-8)}{8x} = \frac{2+x-x+8}{8x} = \frac{10}{8x}$ — 1. Subtract the numerators by adding the opposite.

$= \frac{\cancel{10}^{5}}{\cancel{8}_{4}x} = \frac{5}{4x}$ — 2. Simplify by canceling common factors.

Example 5

Simplify $\frac{3x^2+3x-20}{x^2+9x+20} - \frac{x^2+x+4}{x^2+9x+20}$.

Answer

$\frac{(3x^2+3x-20)-(x^2+x+4)}{x^2+9x+20}$ — 1. Subtract the numerators by adding the opposite.

$= \frac{3x^2+3x-20-x^2-x-4}{x^2+9x+20} = \frac{2x^2+2x-24}{x^2+9x+20}$

$= \frac{2(x+4)(x-3)}{(x+5)(x+4)}$ — 2. Factor the numerator and the denominator.

$= \frac{2\cancel{(x+4)}(x-3)}{(x+5)\cancel{(x+4)}} = \frac{2(x-3)}{x+5}$ — 3. Simplify by canceling common factors.

Adding and Subtracting Rational Expressions with Common Denominators

1. Add or subtract the numerators.
2. Simplify the resulting rational expression.

A rational number can be thought of as having three signs: one in the numerator, one in the denominator, and one preceding the fraction. The rational number $-\frac{4}{5}$ can be expressed as any one of these three equivalent expressions:

$$-\frac{4}{5} = \frac{-4}{5} = \frac{4}{-5}.$$

It is easier to simplify a problem like $\frac{3}{5} + \frac{2}{-5}$ if you use another form of the rational number.

$$\frac{3}{5} + \frac{2}{-5} = \frac{3}{5} - \frac{2}{5} = \frac{1}{5}$$

This concept is especially helpful for simplifying rational expressions with denominators that have opposite binomial factors, such as $x - 4$ and $4 - x$. Factoring -1 from one of the binomials and using an equivalent rational expression allows you to combine expressions with opposite binomials in their denominators.

$$\frac{3}{x-4} + \frac{2}{4-x}$$

$$= \frac{3}{x-4} + \frac{2}{-(x-4)}$$

$$= \frac{3}{x-4} - \frac{2}{x-4} = \frac{1}{x-4}$$

Example 6

Simplify $\frac{4x}{x-7} - \frac{3x}{7-x}$.

Answer

$\frac{4x}{x-7} - \frac{3x}{-(x-7)}$

$= \frac{4x}{x-7} - \left(-\frac{3x}{x-7}\right) = \frac{4x}{x-7} + \frac{3x}{x-7}$

$= \frac{7x}{x-7}$

1. Recognize that the denominators are opposite binomials and factor -1 out of one of the denominators.
2. Use an equivalent expression to rewrite the subtraction as the addition of the opposite.
3. The result cannot be simplified.

A. Exercises

Perform the indicated operations and simplify.

1. $\frac{8}{x} + \frac{6}{x}$
2. $\frac{5}{x+2} + \frac{7}{x+2}$
3. $\frac{4x}{25} - \frac{9x}{25}$
4. $\frac{3x}{y^2} - \frac{5x}{y^2}$
5. $\frac{7}{2c} - \frac{9}{2c}$
6. $\frac{3x}{8} + \frac{5x}{8}$
7. $\frac{2x}{x-7} - \frac{14}{x-7}$
8. $\frac{8}{3x+2} + \frac{12x}{3x+2}$
9. $\frac{5x}{x^2+y} + \frac{4x}{x^2+y}$
10. $\frac{x-5}{x+2} - \frac{x+7}{x+2}$
11. $\frac{x+4}{7} - \frac{x-3}{7}$
12. $\frac{4a}{a-b} - \frac{4b}{a-b}$
13. $\frac{9x}{yz} + \frac{3y}{yz}$
14. $\frac{2x}{x+y} - \frac{2y}{x+y}$

B. Exercises

Multiple choice: List all correct answers.

15. Which of the following are equivalent expressions to $\frac{3a}{a-b}$?
 a. $\frac{3a}{b-a}$ b. $-\frac{3a}{b-a}$ c. $-\frac{-3a}{b-a}$ d. $\frac{-3a}{b-a}$
16. Which of the following are equivalent expressions to $\frac{y}{-x-y}$?
 a. $\frac{-y}{x-y}$ b. $-\frac{y}{x+y}$ c. $\frac{-y}{y+x}$ d. $\frac{y}{y+x}$

Simplify the following expressions with opposite denominators.

17. $\frac{3}{m-4} + \frac{6}{4-m}$
18. $\frac{4}{x-2} - \frac{10}{2-x}$
19. $\frac{x}{x-y} + \frac{y}{y-x}$
20. $\frac{a}{b-a} - \frac{b}{a-b}$

Perform the indicated operations and simplify.

21. $\frac{6}{a+b} - \frac{3a}{a+b}$
22. $\frac{3a}{a^2-b^2} - \frac{3b}{a^2-b^2}$
23. $\frac{5}{a^2+3a-2} - \frac{7}{a^2+3a-2}$
24. $\frac{x+2}{x^2-3x-9} - \frac{x-4}{x^2-3x-9}$
25. $\frac{x+2}{x^2-4x-7} + \frac{x-9}{x^2-4x-7}$
26. $\frac{a-2}{a^2-3a+9} + \frac{a+4}{a^2-3a+9}$
27. $\frac{2m^2+4mn+n^2}{m^2+mn-2n^2} + \frac{m^2+3mn+n^2}{m^2+mn-2n^2}$
28. $\frac{5x^2+3x-9}{x^2+7x+10} + \frac{2x^2+8x+3}{x^2+7x+10}$
29. $\frac{3x^2+4x-10}{x^2+5x-36} - \frac{2x^2+x+18}{x^2+5x-36}$
30. $\frac{a^3}{a-1} - \frac{a}{a-1}$
31. $\frac{2x^2-6x+4}{x^2-8x+15} + \frac{x^2-5x-19}{x^2-8x+15}$
32. $\frac{x^2-8x+13}{x^2-4x-12} - \frac{-x^2-4x-1}{x^2-4x-12}$
33. $\frac{a}{a^2-b^2} - \frac{b}{b^2-a^2}$
34. $\frac{y}{x^2-y^2} + \frac{x}{y^2-x^2}$

C. Exercises

Given $f(x) = \frac{x^2 - 4}{x}$ and $g(x) = \frac{x + 2}{x}$, find the simplified expression for each of the following.

35. $f(x) + g(x)$ **36.** $f(x) - g(x)$ **37.** $f(x) \cdot g(x)$ **38.** $f(x) \div g(x)$

Perform the indicated operations and simplify.

39. $\frac{-3x - 3}{2x^2 - 5x - 12} - \frac{5x^2 - 2x + 1}{2x^2 - 5x - 12} - \frac{x^2 + 6x - 7}{2x^2 - 5x - 12}$

40. Naming the appropriate properties, explain why $\frac{4x + 8}{2x + 6}$ can be reduced but $\frac{4x + 7}{2x + 6}$ cannot.

Dominion Modeling

In Section 13.1, Dominion Modeling, you saw how octaves can be produced using the arithmetic mean, $A = \frac{a + b}{2}$. The string lengths that produce the intervals of a 4th and a 5th can be determined using the arithmetic and harmonic means of $\frac{1}{2}l$ and l. The harmonic mean of two numbers, a and b, can be found using the formula $H = \frac{2ab}{a + b}$.

41. Find the arithmetic mean of the string lengths for two notes one octave apart, $\frac{1}{2}l$ and l.

42. Find the harmonic mean of the string lengths for two notes one octave apart, $\frac{1}{2}l$ and l.

43. Copy the number line below, which models a string of length l.

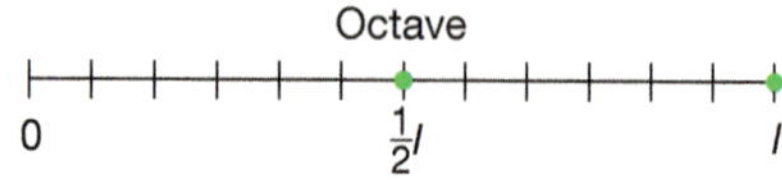

a. On the number line, graph points representing the arithmetic and harmonic means found in exercises 41–42.

b. Using the fact that an interval of a 5th is one step higher than an interval of a 4th, label the two points on your number line to show which length produces which interval.

44. Compare the hammer weights of 6, 8, and 9 to the heaviest weight of 12 by writing and simplifying three ratios. Label each ratio with the interval Pythagoras heard from each pair of hammers.

CUMULATIVE REVIEW

Explain why each expression is not simplified. Then simplify. [2.3, 11.2, 11.7, 13.1]

45. $\frac{4\sqrt{7}}{5 - \sqrt{2}}$ **46.** $\sqrt[3]{40}$ **47.** $3x - 5 + 2x + 1$

48. $\frac{9 \pm 3\sqrt{7}}{6}$ **49.** $\frac{3x^2 - 15x}{x - 5}$

Simplify. [1.3–1.4]

50. $\frac{4}{15} + \frac{3}{5}$ **51.** $\frac{13}{9} - \frac{8}{18}$ **52.** $\frac{3x}{4} + \frac{5x}{8}$

53. $\frac{2x + y}{3} - \frac{7x + 2y}{3}$ **54.** $\frac{6x - 2y}{5} + \frac{x + 5y}{3}$

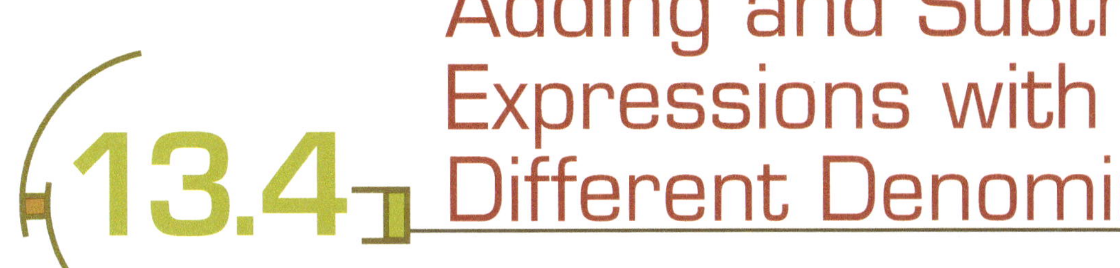

13.4 Adding and Subtracting Expressions with Different Denominators

To add or subtract rational expressions with different denominators, you must first find a common denominator. The *least common denominator* (LCD) is the least common multiple (LCM) of the denominators. To find the LCM of two expressions, factor each expression completely and then find the product of the highest power of each factor.

The total time for a roundtrip hike is the sum of two rational expressions: $t = \frac{d_{uphill}}{r_{uphill}} + \frac{d_{downhill}}{r_{downhill}}$.

LCM of Whole Numbers	
12	15
$2^2 \cdot 3$	$3 \cdot 5$
LCM $= 2^2 \cdot 3 \cdot 5 = 60$	

LCM of Algebraic Expressions	
$9x^3y$	$3y^2z$
$3^2 \cdot x^3 \cdot y$	$3 \cdot y^2 \cdot z$
LCM $= 3^2 \cdot x^3 \cdot y^2 \cdot z = 9x^3y^2z$	

Example 1

Find the LCM of $x^4 - 3x^3 - 10x^2$ and $3x^3 - 30x^2 + 75x$.

Answer

$x^4 - 3x^3 - 10x^2$ | $3x^3 - 30x^2 + 75x$

$= x^2(x^2 - 3x - 10)$ | $= 3x(x^2 - 10x + 25)$

$= x^2(x - 5)(x + 2)$ | $= 3x(x - 5)^2$

1. Factor each polynomial.

LCM $= 3x^2(x - 5)^2(x + 2)$

2. The LCM is the product of the highest power of each factor.

Expressions can be added or subtracted after they have been renamed as equivalent rational expressions with the LCD.

Example 2

Simplify $\frac{3}{2a} + \frac{5}{8}$.

Answer

$2a = 2 \cdot a;\ 8 = 2^3$

$\text{LCD} = 2^3 \cdot a = 8a$

1. Find the LCD by determining the LCM of $2a$ and 8.

$\frac{4 \cdot 3}{4 \cdot 2a} + \frac{5a}{8a}$

2. Multiply the numerator and the denominator of each fraction by the factor that makes each denominator equal to the LCD.

$= \frac{12}{8a} + \frac{5a}{8a} = \frac{5a + 12}{8a}$

3. Write the sum of the numerators over the LCD.

Example 3

Simplify $\frac{2x - 5}{3x} + \frac{x - 1}{2x} + \frac{5x + 6}{12x^2}$.

Answer

$3x = 3 \cdot x;\ 2x = 2 \cdot x;\ 12x^2 = 2^2 \cdot 3 \cdot x^2$

$\text{LCD} = 2^2 \cdot 3 \cdot x^2 = 12x^2$

1. Find the LCD.

$\frac{4x(2x - 5)}{4x(3x)} + \frac{6x(x - 1)}{6x(2x)} + \frac{5x + 6}{12x^2}$

$= \frac{8x^2 - 20x}{12x^2} + \frac{6x^2 - 6x}{12x^2} + \frac{5x + 6}{12x^2}$

2. Multiply the numerator and the denominator of each fraction by the factors that make each denominator equal to the LCD.

$= \frac{8x^2 - 20x + 6x^2 - 6x + 5x + 6}{12x^2}$

3. Add the numerators by combining like terms.

$= \frac{14x^2 - 21x + 6}{12x^2}$

4. The numerator cannot be factored, so the expression cannot be simplified.

When polynomials are subtracted, care must be taken with signs. You may find it helpful to rewrite subtraction as the addition of the opposite, being sure to change all the signs in the polynomial being subtracted.

Example 4

Simplify $\frac{x - 3}{x + 1} - \frac{x - 4}{x + 2}$.

Answer

$x + 1$ and $x + 2$ have no common factors.

$\text{LCD} = (x + 1)(x + 2)$

1. Find the LCD.

$\frac{(x - 3)(x + 2)}{(x + 1)(x + 2)} - \frac{(x - 4)(x + 1)}{(x + 2)(x + 1)}$

$= \frac{x^2 - x - 6}{(x + 1)(x + 2)} - \frac{x^2 - 3x - 4}{(x + 1)(x + 2)}$

2. Multiply the numerator and the denominator of each fraction by the factor that makes each denominator equal to the LCD.

$= \frac{x^2 - x - 6 - x^2 + 3x + 4}{(x + 1)(x + 2)}$

3. Subtract the second numerator by adding the opposite of each of its terms.

$= \frac{2x - 2}{(x + 1)(x + 2)}$

$= \frac{2(x - 1)}{(x + 1)(x + 2)}$

4. Factoring the numerator shows that the expression cannot be simplified.

Example 5

Simplify $\frac{b}{a^2 + ab} - \frac{a}{ab + b^2}$.

Answer

$\frac{b}{a(a+b)} - \frac{a}{b(a+b)}$

1. Factor the denominators to find the LCD, $ab(a + b)$.

$= \frac{bb}{a(a+b)b} - \frac{aa}{b(a+b)a}$

$= \frac{b^2}{ab(a+b)} - \frac{a^2}{ab(a+b)}$

2. Multiply the numerator and the denominator of each fraction by the factor that makes each denominator equal to the LCD.

$= \frac{b^2 - a^2}{ab(a+b)}$

3. Subtract.

$= \frac{(b-a)\cancel{(b+a)}}{ab\cancel{(a+b)}} = \frac{b-a}{ab}$

4. Factor and simplify by canceling the common factors $a + b$ and $b + a$.

Adding and Subtracting Rational Expressions with Different Denominators

1. Find the LCD of the rational expressions.
2. Make equivalent rational expressions having the LCD.
3. Combine the numerators and place the result over the common denominator.
4. Simplify the resulting rational expression.

Example 6

Simplify $\frac{x+1}{2x-4} + \frac{x+1}{2x-x^2}$.

Answer

$\frac{x+1}{2(x-2)} + \frac{x+1}{x(2-x)}$

1. Factor the denominators to find the LCD.

$= \frac{x+1}{2(x-2)} + \frac{x+1}{-x(x-2)}$

$= \frac{x+1}{2(x-2)} - \frac{x+1}{x(x-2)}$

2. Since $x - 2$ and $2 - x$ are opposites, factor -1 out of the factor $2 - x$ and restate as a subtraction problem.

$= \frac{(x+1)x}{2(x-2)x} - \frac{2(x+1)}{2x(x-2)}$

$= \frac{x^2 + x}{2x(x-2)} - \frac{2x+2}{2x(x-2)}$

3. Multiply the numerator and the denominator of each fraction by the factor that makes each denominator equal to the LCD, $2x(x - 2)$.

$= \frac{x^2 - x - 2}{2x(x-2)}$

4. Subtract.

$= \frac{\cancel{(x-2)}(x+1)}{2x\cancel{(x-2)}} = \frac{x+1}{2x}$

5. Factor and simplify by canceling common factors.

A. Exercises

Find the LCM of each pair of expressions. Leave polynomials in factored form.

1. $6x$ and $8x^3$
2. $4ab^2$ and $12bc$
3. $2r + 12$ and $3r + 18$
4. $3x - 1$ and $x - 3$
5. $a^2 - b^2$ and $a^2 - 2ab + b^2$
6. $t^2 - 3t + 2$ and $t^2 - t - 2$

Perform the indicated operations and simplify.

7. $\frac{3x}{4} + \frac{9}{16}$
8. $\frac{x}{9} - \frac{4x}{7}$
9. $\frac{x + y}{5} - \frac{x}{15}$
10. $\frac{y - 3}{18} + \frac{y + 7}{4}$
11. $\frac{a^2 + 3a}{5} + \frac{a^2 - 9}{2}$
12. $\frac{4}{3x} - \frac{7}{21x^3}$
13. $\frac{3}{c} - \frac{4}{c + 1}$
14. $\frac{9}{3m^3} + \frac{14}{8m^2}$
15. $\frac{5}{2x^3y^2} + \frac{10}{xy}$
16. $\frac{x^2 - 4}{x^2} + \frac{x - y}{x}$
17. $\frac{2b}{b + 5} - \frac{3}{b - 4}$
18. $\frac{g}{(g - 2)^2} - \frac{3}{g - 2}$

B. Exercises

Perform the indicated operations and simplify.

19. $\frac{2y}{(y - 7)(y + 4)} - \frac{9}{(y - 7)^2}$
20. $\frac{7}{(x + 2)(x - 5)} + \frac{9}{(x + 2)(x + 3)}$
21. $\frac{x - 3}{(x + 2)(4 - x)} - \frac{x + 9}{(x + 2)(x - 4)}$
22. $\frac{2b - 10}{(b - 5)^2} + \frac{3b - 5}{5 - b}$
23. $\frac{-d}{2d + 8} + \frac{2d}{3d - 12}$
24. $\frac{1}{2x + 8} - \frac{x + 3}{3x + 12}$
25. $\frac{1}{3x - 9} + \frac{1}{4x + 16}$
26. $\frac{5a}{a - 1} - \frac{6a^2}{a^2 - 1}$
27. $\frac{3}{x^2 - y^2} + \frac{5}{x + y}$
28. $\frac{10}{x^2 - 6x - 27} + \frac{8}{x^2 - 3x - 18}$
29. $\frac{3y}{y^2 + 7y + 10} - \frac{2y^2}{y^2 + 10y + 25}$
30. $\frac{m}{m^2 - n^2} - \frac{n}{n^2 - m^2}$
31. $\frac{x - 3}{x + 5} + \frac{2x - 7}{x + 3} + \frac{26}{x^2 + 8x + 15}$
32. $\frac{x}{x + 3} + x - \frac{x}{x + 5}$

C. Exercises

Perform the indicated operations and simplify.

33. $\frac{a + 3}{a + 5} - \frac{a + 2}{a - 5} + \frac{a}{a^2 - 25}$
34. $\frac{x + 3}{x^2 + 6x - 16} + \frac{x - 2}{x^2 - 6x + 8}$
35. $\frac{3 - x}{x^2 - 9} + \frac{x + 2}{x^2 + 5x + 6} - \frac{x + 2}{4 - x^2}$
36. $\frac{2}{x + 1} + \frac{1}{x - 1} + \frac{2}{1 - x^2}$
37. Explain the error in the work below and then complete the problem correctly.
$$\frac{3x + 5}{(x - 2)(x - 5)} - \frac{4x + 1}{(2 - x)(5 - x)} = \frac{3x + 5}{(x - 2)(x - 5)} + \frac{4x + 1}{(x - 2)(x - 5)}$$
38. Simplify $\frac{x - 1}{(x - 3)(x - 4)} - \frac{x - 9}{(x - 3)(4 - x)} - \frac{x - 3}{(3 - x)(4 - x)} + \frac{2 - x}{(3 - x)(x - 4)}$.

Dominion Modeling

Mathematicians have continued to study musical intervals and have discovered mathematical relationships between several major and minor intervals. Just as the intervals of a 4th and a 5th are the arithmetic and harmonic means of two notes one octave apart, so the intervals of a major 3rd and a minor 3rd are the arithmetic and harmonic means of two notes a 5th apart. Section 13.1, Dominion Modeling, mentioned that the interval of a major 3rd can be heard by playing the first and third notes of the C major scale (C and E). In a minor 3rd the second note is a half step lower, so an E-flat (E♭) would follow C instead of an E.

39. Find the arithmetic and harmonic means of the string lengths for two notes a 5th apart, $\frac{2}{3}l$ and l.

40. Make a number line modeling a string of length l. On the number line, graph and label points representing the intervals of a major 3rd, minor 3rd, 4th, 5th, and octave.

41. Which mean (arithmetic or harmonic) produces a major 3rd? Which produces a minor 3rd?

CUMULATIVE REVIEW

Simplify. [8.1–8.2, 13.2]

42. $(5x^{-3}y^2)^2(4x^2y^3)^{-1}$

43. $(-2)^{-3}(2a^{-2}b^2)^{-1}$

44. $\frac{x^2 - 1}{12} \cdot \frac{4}{x - 1}$

45. $\frac{4a^2 - 4b^2}{a - b} \cdot \frac{1}{4}$

Factor. [10.5]

46. $4x^2 - 100$

47. $3a^3 + 21a^2 + 36a$

Solve. [12.6]

48. $y^2 - 10y = -16$

49. $3x^3 - 3x^2 = 18x$

50. One angle of a triangle is 10° more than another. The third angle is 6° less than twice the first. Find the measure of each angle. [2.6]

51. A ladder must reach 20 ft up a telephone pole when it is positioned 6 ft from the base of the pole. How long does the ladder need to be? [11.6]

13.5 Complex and Mixed Expressions

Recall that a mixed number, such as $3\frac{1}{4}$, is actually the sum of a whole number and a rational number. By completing the addition, you can convert the number to an improper fraction.

$$3 + \frac{1}{4} = \frac{3}{1} \cdot \frac{4}{4} + \frac{1}{4} = \frac{12}{4} + \frac{1}{4} = \frac{13}{4}$$

The sum of a polynomial and a rational expression is a *mixed rational expression*. By adding the terms, you can write the sum as a single rational expression.

The amount of a loan payment, A, is determined using a formula with a complex rational expression, $A = P\frac{r(1 + r)^n}{(1 + r)^n - 1}$.

Example 1

Simplify $4y + \frac{3}{y - 2}$.

Answer

$\frac{4y}{1} \cdot \frac{y - 2}{y - 2} + \frac{3}{y - 2}$	1. Multiply the numerator and the denominator of the monomial by $y - 2$.
$= \frac{4y^2 - 8y}{y - 2} + \frac{3}{y - 2}$	2. Distribute.
$= \frac{4y^2 - 8y + 3}{y - 2}$	3. Add the numerators.

The division $\frac{1}{2} \div \frac{3}{4}$ can also be written as a fraction, $\frac{\frac{1}{2}}{\frac{3}{4}}$.

A fraction with a fraction in the numerator or denominator is called a *complex fraction*. A complex rational expression is a complex fraction that contains variables.

Definition

A **complex rational expression** is an algebraic expression of the form $\frac{n}{d}$ in which the numerator, n, the denominator, d, or both are rational expressions.

The first complex rational expression below has a rational expression in the numerator; the second has rational expressions in both the numerator and the denominator; the third has a rational expression only in the denominator.

$$\frac{\frac{x}{y}}{4x} \qquad \frac{\frac{x + 2}{x^2}}{\frac{x + 2}{x}} \qquad \frac{x^2 + 3x - 9}{\frac{x + 2}{x^3}}$$

Expressions containing fractions in the numerator or the denominator should always be simplified. Remembering that fractions are another way of writing a division problem can be helpful in this simplification.

Example 2

Simplify $\dfrac{\frac{x}{y}}{4}$.

Answer $\dfrac{\frac{x}{y}}{4} = \frac{x}{y} \div 4$ 1. Think of the complex fraction as a division problem.

$= \frac{x}{y} \cdot \frac{1}{4}$ 2. Multiply by the reciprocal of the denominator.

$= \frac{x}{4y}$ 3. Simplify.

Example 3

Simplify $\dfrac{\frac{x+2}{x^2}}{\frac{x+2}{x}}$.

Answer $\dfrac{x+2}{x^2} \cdot \dfrac{x}{x+2}$ 1. Express the division indicated by the complex fraction as multiplication of the numerator by the reciprocal of the denominator.

$= \dfrac{\cancel{x+2}}{\cancel{x}x} \cdot \dfrac{\cancel{x}}{\cancel{x+2}} = \dfrac{1}{x}$ 2. Simplify.

Complex fractions can also be simplified by multiplying the numerator and the denominator by the LCD of all of the fractions in the numerator and the denominator. Each of the following examples shows two different methods of simplifying a complex expression. Choose whichever method seems easier to you.

Simplifying Complex Rational Expressions	
Method 1	Multiply the numerator by the reciprocal of the denominator.
Method 2	Multiply the numerator and the denominator by their LCD.

Example 4

Simplify $\dfrac{\frac{x^2-9}{4x}}{\frac{x+3}{8x^2}}$.

Answer

Method 1

Multiply the numerator by the reciprocal of the denominator.

$\dfrac{x^2-9}{4x} \cdot \dfrac{8x^2}{x+3} = \dfrac{(x-3)\cancel{(x+3)}}{\cancel{4x}} \cdot \dfrac{\cancel{4x} \cdot 2x}{\cancel{x+3}}$

$= 2x(x-3) = 2x^2 - 6x$

Method 2

Multiply the numerator and the denominator by their LCD, $8x^2$.

$\dfrac{\frac{x^2-9}{4x} \cdot 8x^2}{\frac{x+3}{8x^2} \cdot 8x^2} = \dfrac{(x^2-9)2x}{x+3} = \dfrac{(x-3)\cancel{(x+3)}2x}{\cancel{x+3}}$

$= 2x(x-3) = 2x^2 - 6x$

The second method may be easier when the numerator or the denominator contains more than one term.

Example 5

Simplify $\frac{1 + \frac{a^2}{b^2}}{1 - \frac{a}{b}}$, leaving the answer in factored form.

Answer

Method 1

$$\frac{\frac{b^2}{b^2} + \frac{a^2}{b^2}}{\frac{b}{b} - \frac{a}{b}} = \frac{\frac{b^2 + a^2}{b^2}}{\frac{b-a}{b}}$$

1. Since the fraction bar also acts as a grouping symbol, first combine the terms in the numerator and the denominator.

$$= \frac{b^2 + a^2}{b\cancel{b}} \cdot \frac{\cancel{b}}{b - a}$$

$$= \frac{b^2 + a^2}{b(b - a)}$$

2. Multiply by the reciprocal of the denominator.

Method 2

$$\frac{\left(1 + \frac{a^2}{b^2}\right)b^2}{\left(1 - \frac{a}{b}\right)b^2}$$

1. Multiply the numerator and the denominator by their LCD, b^2.

$$= \frac{b^2 + \frac{a^2}{\cancel{b^2}} \cdot \cancel{b^2}}{b^2 - \frac{a}{\cancel{b}} \cdot \cancel{b}b}$$

2. Distribute and simplify.

$$= \frac{b^2 + a^2}{b^2 - ab}$$

$$= \frac{b^2 + a^2}{b(b - a)}$$

3. Factoring shows that the expression cannot be simplified.

A. Exercises

Simplify each mixed rational expression.

1. $2 + \frac{1}{x}$

2. $y + \frac{1}{2}$

3. $2y + \frac{1}{y + 1}$

4. $3x - \frac{1}{x - 3}$

5. $2a - \frac{a}{b}$

6. $3b + \frac{a^2}{b - a}$

Simplify each complex rational expression.

7. $\frac{\frac{1}{4}}{\frac{5}{8}}$

8. $\frac{8\frac{4}{5}}{3\frac{3}{7}}$

9. $\frac{\frac{a}{b}}{\frac{c}{d}}$

10. $\frac{\frac{x}{y}}{z}$

11. $\frac{x}{\frac{y}{z}}$

12. $\frac{\frac{1}{r}}{\frac{1}{s}}$

13. $\frac{\frac{a^2}{d}}{\frac{ad}{c^3}}$

14. $\frac{x^2y^3}{\frac{x}{y}}$

15. $\frac{\frac{y}{c}}{c + y}$

16. $\frac{5}{\frac{x + y}{x - y}}$

B. Exercises

Simplify each mixed rational expression.

17. $x + 2 + \frac{1}{x}$

18. $y - 1 + \frac{1}{y - 2}$

19. $2y + 3 - \frac{1}{y + 1}$

20. $x + 3 - \frac{8x}{x - 3}$

21. Given the general expression for a mixed number, $A\frac{b}{c}$, meaning $A + \frac{b}{c}$, write a general expression for the equivalent improper fraction.

22. Given the general expression for a mixed expression, $p(x) + \frac{n(x)}{d(x)}$, where $p(x)$, $n(x)$, and $d(x)$ are polynomials, write a general expression for the equivalent rational expression.

Simplify each complex rational expression, leaving the answer in factored form.

23. $\dfrac{x}{1 + \frac{1}{x}}$

24. $\dfrac{\frac{1}{m} + m}{\frac{1}{m} - m}$

25. $\dfrac{\frac{x}{y} + 1}{\frac{x}{y} - \frac{y}{x}}$

26. $\dfrac{\frac{1}{a + b}}{\frac{a + b}{a^2 - b^2}}$

27. $\dfrac{\frac{x + y}{x - y}}{x^2 - y^2}$

28. $\dfrac{x^2 + 2xy + y^2}{\frac{x + y}{x - y}}$

29. $\dfrac{\frac{3a}{2a^2 + 4a}}{\frac{6a}{a^2 + 2a}}$

30. $\dfrac{\frac{m + n}{m - n}}{\frac{1}{m^2 - n^2}}$

31. $\dfrac{\frac{4x^3 - 36x}{2x^3 - 2x^2 - 12x}}{x^2 + 5x + 6}$

32. $\dfrac{4x^3 - 36x}{\frac{2x^3 - 2x^2 - 12x}{x^2 + 5x + 6}}$

33. $\dfrac{\frac{3 - x}{x^2 + 7x + 10}}{\frac{x^2 - 9}{x^2 - 4x - 12}}$

Keyword Search

Doppler effect

The pitch of a car's horn as it is approaching a listener is higher than when the car is stationary. As the car moves away from the listener, the pitch is lower. This *Doppler effect* can be modeled by the following equation for the perceived frequency: $h = \left(\dfrac{1}{1 - \frac{v}{s}}\right)f$, where v is the speed at which the sound's source is approaching, s is the speed of sound, and f is the frequency in hertz (cycles per second) of the sound emitted by the source.

When a plane breaks through the wall of compressed sound waves that have accumulated in front of the plane, a sonic boom is produced. This is an extreme example of the Doppler effect.

34. Write the expression $\left(\dfrac{1}{1 - \frac{v}{s}}\right)f$ as a simplified rational expression.

35. Determine the frequency of the sound heard by a listener if a car's horn emits a frequency of 400 Hz and the car is approaching at 85 km/hr. Assume the speed of sound to be 1245 km/hr.

36. Determine the frequency of the sound heard by the listener as the car moves away at the same speed. (Hint: Use a negative value for speed.)

C. Exercises

Simplify.

37. $\dfrac{\frac{1}{2x}}{\frac{2x}{3}} \div \dfrac{\frac{3x}{4}}{\frac{4}{3} - x}$

38. $\dfrac{\frac{x}{y} - \frac{x - y}{x + y}}{\frac{y}{x} + \frac{x - y}{x + y}}$

39. Mr. Sung commutes d miles to and from work each day. His average speed for his drive to work early in the morning is 60 mi/hr, but during his drive home in rush-hour traffic his average speed is only 40 mi/hr. Using the fact that the average speed is the total distance divided by the total time, demonstrate that the average speed for his entire commute is 48 mi/hr, not 50 mi/hr.

40. Write a general expression for the average speed of a roundtrip in which the average speed for the drive to a destination is r_1 and the average speed for the return drive is r_2.

CUMULATIVE REVIEW

Explain why each expression is not simplified. Then simplify. [8.2, 11.2, 11.4, 13.1]

41. $\frac{3}{x^{-2}}$

42. $\frac{x^2 - 1}{x + 1}$

43. $\sqrt{50}$

44. $\sqrt{\frac{2}{5}}$

Solve. [3.2]

45. $\frac{4}{9} = \frac{x}{2}$

46. $\frac{7}{2y} = \frac{35}{20}$

47. $\frac{2a - 1}{2} = \frac{3a + 2}{4}$

48. $\frac{2d + 3}{3d - 5} = -\frac{1}{4}$

49. Find the LCM of $24xyz^2$, $2x^2y$, and $6x^3y^2z^5$. [13.4]

50. Simplify $\frac{2x - 4}{2 - x} - \frac{7x - 14}{x - 2}$. [13.3]

13.6 Solving Rational Equations

The total resistance of a circuit with parallel resistors is a complex rational expression.

Definition

A **rational equation** is an equation containing rational expressions.

You have already solved rational equations containing fractions in which the variables were in the numerators. You learned to clear the equation of fractions by multiplying both sides of the equation by the LCD of all the terms and then solving the resulting equation.

Example 1

Solve $\frac{x - 7}{7} - \frac{x + 2}{3} = 1$.

Answer

$$21\left(\frac{x - 7}{7} - \frac{x + 2}{3}\right) = 21(1)$$

1. Multiply both sides by the LCD, 21, to clear the fractions.

$$21\left(\frac{x - 7}{7}\right) - 21\left(\frac{x + 2}{3}\right) = 21(1)$$

$$3(x - 7) - 7(x + 2) = 21$$

2. Solve.

$$3x - 21 - 7x - 14 = 21$$

$$-4x - 35 = 21$$

$$-4x = 56$$

$$x = -14$$

To solve rational equations in which variables appear in the denominator, apply the same principles.

Example 2

Solve $\frac{1}{2x} + \frac{5}{12} = -\frac{1}{3x}$.

Answer

$$12x\left(\frac{1}{2x} + \frac{5}{12}\right) = 12x\left(-\frac{1}{3x}\right)$$

1. Multiply both sides by the LCD, $12x$, to clear the fractions.

$$\overset{6}{\cancel{12x}}\left(\frac{1}{\cancel{2x}}\right) + \cancel{12}x\left(\frac{5}{\cancel{12}}\right) = \overset{4}{\cancel{12x}}\left(-\frac{1}{\cancel{3x}}\right)$$

$$6 + 5x = -4$$

$$5x = -10$$

$$x = -2$$

2. Solve.

Check

$$\frac{1}{2(-2)} + \frac{5}{12} = -\frac{1}{3(-2)}$$

$$-\frac{1}{4} + \frac{5}{12} = \frac{1}{6}$$

$$-\frac{3}{12} + \frac{5}{12} = \frac{2}{12}$$

3. Check your solution in the original equation.

Remember that rational expressions are undefined for any values of the variable that make the denominator equal to zero. When you multiply by a variable expression to clear the denominators, the new equation may introduce solutions for which the original equation is undefined. You must check that your solution is not an excluded value for any of the expressions in the original rational equation. Any solution that does not check is extraneous.

Example 3

Solve $\frac{5}{x+1} = \frac{-2}{x-6}$.

Answer

$$\cancel{(x+1)}(x-6)\frac{5}{\cancel{x+1}} = (x+1)\cancel{(x-6)}\frac{-2}{\cancel{x-6}}$$

1. Multiply both sides by the LCD, $(x+1)(x-6)$, to clear the fractions.

$$5(x-6) = -2(x+1)$$

$$5x - 30 = -2x - 2$$

$$7x = 28$$

$$x = 4$$

2. Solve.

Check

Since $x \neq 1$ or -6, the solution is $x = 4$.

3. Check that your solution is not an excluded value in the original equation.

When the rational equation is a proportion, as in Example 3, multiplying both sides by the LCD produces the same result as setting the product of the extremes equal to the product of the means.

$$\frac{5}{x+1} = \frac{-2}{x-6}$$

$$5(x-6) = -2(x+1)$$

While this approach quickly produces the same equation as multiplying both sides by the LCD, remember that it can be applied only when you are solving a proportion.

Example 4

Solve $\frac{x - 12}{x - 10} = \frac{2}{x} - \frac{20}{x^2 - 10x}$.

Answer

$$\frac{x - 12}{x - 10} = \frac{2}{x} - \frac{20}{x(x - 10)}$$

1. Factor the denominators to identify the LCD as $x(x - 10)$.

$$x(x - 10)\left(\frac{x - 12}{x - 10}\right) = x(x - 10)\left(\frac{2}{x} - \frac{20}{x(x - 10)}\right)$$

$$x\cancel{(x - 10)}\left(\frac{x - 12}{\cancel{x - 10}}\right) = \cancel{x}(x - 10)\left(\frac{2}{\cancel{x}}\right) - \cancel{x(x - 10)}\left(\frac{20}{\cancel{x(x - 10)}}\right)$$

2. Multiply both sides by the LCD to clear the fractions.

$$x(x - 12) = 2(x - 10) - 20$$
$$x^2 - 12x = 2x - 40$$
$$x^2 - 14x + 40 = 0$$
$$(x - 10)(x - 4) = 0$$
$$x - 10 = 0 \quad \text{or} \quad x - 4 = 0$$
$$x = 10 \qquad\qquad x = 4$$

3. Solve.

Check Since $x \neq 0$ or 10, $x = 10$ is an extraneous solution. Therefore, $x = 4$ is the only solution to the equation.

4. Check that your solutions are not excluded values in the original equation.

Example 5

What number must be added to both the numerator and the denominator of $\frac{3}{5}$ to produce a fraction equivalent to $\frac{6}{7}$?

Answer

$$\frac{3 + n}{5 + n} = \frac{6}{7}$$

1. Write the appropriate equation, using n to represent the number added to the numerator and the denominator.

$$7\cancel{(5 + n)}\left(\frac{3 + n}{\cancel{5 + n}}\right) = \cancel{7}(5 + n)\left(\frac{6}{\cancel{7}}\right)$$

2. Multiply both sides by the LCD to clear the fractions.

$$7(3 + n) = 6(5 + n)$$
$$21 + 7n = 30 + 6n$$
$$n = 9$$

3. Solve.

Check $\frac{3 + 9}{5 + 9} = \frac{12}{14} = \frac{6}{7}$

4. Check your solution in the original equation.

A. Exercises

Solve.

1. $\frac{m}{3} = 9$

2. $\frac{x + 1}{3} = 4$

3. $\frac{2}{x} = 8$

4. $\frac{8}{y} = 16$

5. $\frac{8}{x + 2} = 3$

6. $\frac{10}{a - 5} = 7$

7. $\frac{x + 3}{x - 2} = 1$

8. $\frac{x - 5}{x - 4} = 2$

9. $\frac{a - 9}{15} = \frac{a - 6}{5}$

10. $\frac{8}{2y} = \frac{4}{y - 2}$

11. $\frac{1}{m - 9} = \frac{2}{m + 4}$

12. $\frac{2}{a - 7} = \frac{9}{a + 3}$

B. Exercises

Solve.

13. $\frac{a + 4}{11} - \frac{a + 2}{3} = 7$

14. $\frac{x - 12}{2} + \frac{x + 7}{5} = \frac{1}{10}$

15. $\frac{4}{3x} + 7 = \frac{1}{6x}$

16. $\frac{x + 5}{2x} - \frac{8x + 3}{9x} = \frac{5}{18}$

17. $\frac{5}{x+9} - \frac{7}{x-8} = 0$

18. $\frac{a-3}{4} - \frac{5}{a+2} = 1$

19. $\frac{4}{a+3} + \frac{2}{a} = \frac{6}{a^2+3a}$

20. $\frac{1}{c+1} - \frac{2}{c^2-1} = \frac{1}{1-c}$

21. $\frac{x+3}{x-4} = \frac{4}{x} + \frac{28}{x^2-4x}$

22. $x - \frac{1}{1-x} = \frac{1}{x-1}$

23. $\frac{m+4}{m-9} = \frac{m-3}{m-6}$

24. $\frac{x+12}{x-2} - \frac{4}{x} = \frac{-4}{x^2-2x}$

25. The width of a rectangular picture is one-half its length. If its area is 648 in.^{2}, what are the dimensions of the picture?

26. Jack and Erin spent $\frac{1}{4}$ of their money on rides at the fair. They paid \$20 for food and transportation and returned with $\frac{4}{7}$ of their money. How much money did they take to the fair?

27. What number must be added to both the numerator and the denominator of $\frac{5}{11}$ to produce a fraction equivalent to $\frac{3}{4}$?

28. What number must be added to the numerator and subtracted from the denominator of $\frac{1}{2}$ to produce a fraction equivalent to $-\frac{8}{7}$?

29. One number is 6 less than another number. The quotient of the larger divided by the smaller is $\frac{5}{2}$. What are the two numbers?

30. Divide 84 into two parts that are in the ratio of 3 : 4.

31. Jenny has made 35 of 50 free throw attempts in practice.
    a. What is the least number of free throws she can make in a row before her percentage of free throws made is at least 80%?
    b. What is the greatest number of free throws she can miss in a row before her percentage of free throws made falls below 60%?

C. Exercises

Solve.

32. $\frac{x^2}{9} = \frac{x}{3} + 4$

33. $\frac{x^2}{4} - \frac{3x}{8} = 2$

34. $\frac{4}{x^2-8x+15} = \frac{1}{x^2-3x-10}$

35. $\frac{1}{x^2-2x-8} - \frac{4}{x^2-x-12} = \frac{2}{x^2-2x-8}$

Dominion Modeling

It can be shown that $a : A = H : b$, where A is the arithmetic mean and H is the harmonic mean of two numbers, a and b (Howard Eves, *An Introduction to the History of Mathematics*, 6th ed.).

36. In Section 13.3, Dominion Modeling, we saw how the intervals of a 4th and a 5th can be produced using $\frac{3}{4}l$ and $\frac{2}{3}l$, the arithmetic and harmonic means, respectively, of $\frac{1}{2}l$ and l. Verify the proportion $a : A = H : b$ for these values.

37. Using the formulas $A = \frac{a+b}{2}$ and $H = \frac{2ab}{a+b}$, verify the proportion $a : A = H : b$.

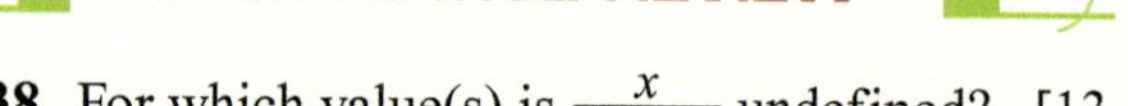

CUMULATIVE REVIEW

38. For which value(s) is $\frac{x}{3x + 5}$ undefined? [13.1]

Simplify. [13.2, 13.4]

39. $\frac{x + 3}{7} \div \frac{x + 5}{21}$

40. $\frac{x + 3}{7} - \frac{x + 5}{21}$

41. $\frac{8}{x} - \frac{5}{x + 2}$

Solve each literal equation for the indicated variable. [3.1]

42. $rt = d$ for r

43. $\frac{6yp - t}{q} = n$ for y

Solve each system. [7.5]

44. $4x + 3y = 9$
$3x - 7y = 16$

45. $y = \frac{3}{4}x - 3$
$3x - 4y = 12$

Graph the system of inequalities. [7.8]

46. $y \leq 2x + 3$
$y \geq x - 2$

47. The hypotenuse of a right triangle is 3 cm less than twice the length of one side. The other side is 2 cm less than the length of the hypotenuse. Find the lengths of the sides. [11.6, 12.6]

13.7 Applying Rational Equations

Do you recall the basic equation used to solve motion problems? You have used tables to organize the different rates, times, and distances in these problems. After filling in two of the columns with variable expressions and information from the problem, use $r \cdot t = d$ to derive the expressions in the third column. Then write an equation relating the quantities in the table and solve the problem.

Example 1

Troy and Philip decided to spend the day canoeing on a river with a 3 mi/hr current. They began paddling up the river at 9:00 AM. After stopping on an island for 1 hr to eat lunch, they paddled back down the river and arrived back at their starting point at 3:00 PM. If they typically paddle at 5 mi/hr in still water, how far up the river did they travel? When did they stop for lunch?

Answer Let d = the distance traveled up and down the river.

	r	t	d
Upstream	$5 - 3 = 2$	$\frac{d}{2}$	d
Downstream	$5 + 3 = 8$	$\frac{d}{8}$	d

1. **Read** the problem carefully and assign a variable to represent the main unknown.
2. **Plan** your solution by making a table.
   a. Fill in the distance and rate columns using the information in the problem and the variable expression.
   b. Determine expressions for each time by solving $r \cdot t = d$ for $t = \frac{d}{r}$.

Example 1, continued

$$\frac{d}{2} + \frac{d}{8} = 5$$
$$8\left(\frac{d}{2}\right) + 8\left(\frac{d}{8}\right) = 8(5)$$
$$4d + d = 40$$
$$5d = 40$$
$$d = 8 \text{ mi up the river}$$

$\frac{d}{2} = \frac{8}{2} = 4$ hr up the river, so they stopped for lunch at 1:00 PM.

3. **Solve** an equation relating the times of the trips.
   a. The total time spent traveling up and down the river was 5 hr.
   b. Multiply each term by the LCD, 8, to clear the fractions.
   c. Use the expression in the table to determine the time spent traveling up the river and the time at which they stopped for lunch.

Check

	r	t	d
Upstream	2	$\frac{8}{2} = 4$	8
Downstream	8	$\frac{8}{8} = 1$	8

4. **Check.** Determining the time spent traveling down the river to be 1 hr and adding the hour they spent eating lunch puts their return time at 3:00 PM.

These problems differ from those worked earlier only in that they often contain rational expressions.

Example 2

When Lisa flies home for Christmas, her flight time is $12\frac{1}{2}$ hr shorter than the time it would take her to drive the 780 mi. If the plane's average speed is six times the average speed of her car, find both speeds.

Answer

Let r = the speed of the car and $6r$ = the speed of the plane.

	r	t	d
By Car	r	$\frac{780}{r}$	780
By Plane	$6r$	$\frac{780}{6r}$	780

1. **Read** the problem carefully and assign variable expressions to represent the main unknowns.
2. **Plan** your solution by making a table.
   a. Fill in the distance and rate columns using the information in the problem and the variable expressions.
   b. Determine expressions for each time by solving $r \cdot t = d$ for $t = \frac{d}{r}$.

$$\frac{780}{r} = \frac{780}{6r} + \frac{25}{2}$$
$$6r\left(\frac{780}{r}\right) = 6r\left(\frac{780}{6r}\right) + 6r\left(\frac{25}{2}\right)$$
$$4680 = 780 + 75r$$
$$3900 = 75r$$
$$r = 52 \text{ mi/hr}$$
$$6r = 6(52) = 312 \text{ mi/hr}$$

3. **Solve** an equation relating the times of the trips.
   a. The time by car is $12\frac{1}{2}$ hr more than the time by plane.
   b. Multiply each term by the LCD, $6r$, to clear the fractions.
   c. Use the expression in the table to find the speed of the plane.

Check

	r	t	d
By Car	52	$\frac{780}{52} = 15$	780
By Plane	312	$\frac{780}{312} = \frac{5}{2}$	780

4. **Check.** Filling in the values in your table shows that the 15 hr it takes to drive is $12\frac{1}{2}$ hr longer than the $2\frac{1}{2}$ hr it takes to fly.

Work is a God-given responsibility (Gen. 2:15; Exod. 20:9; 1 Thess. 4:11). Paul and his associates worked "night and day" so they would not be a burden to others. Paul reprimanded lazy people: "If any would not work, neither should he eat" (2 Thess. 3:8–10). Every Christian has a responsibility to work hard because he serves not just a teacher, employer, or parent, but the Lord Jesus Himself (Rom. 12:11; Eph. 6:1–8; Col. 3:20–24).

A manager's ability to accurately predict the amount of time needed for his associates to complete a task allows him to maximize their efficiency without burdening them with unrealistic expectations. The work problems in this section can help you analyze some of the everyday jobs in which you might be involved.

Rates of work will typically be determined from the information in the problem. If Joe completes a project in 8 hr, his rate of work is $\frac{1 \text{ project}}{8 \text{ hr}} = \frac{1}{8}$ of the project per hour.

Example 3

The yearbook advisor knows that Allison can proofread a chapter in 60 min. Sean can proofread a similar chapter in 75 min. How long will it take them to proofread other similar chapters if they work together?

Answer Let m = the number of minutes it should take for them to do the job together.

1. **Read** the problem carefully and assign a variable to represent the main unknown.
2. **Plan** by making a table listing each rate of work.

	Time (minutes)	Rate of Work
Allison	60	$\frac{1 \text{ chapter}}{60 \text{ min}}$
Sean	75	$\frac{1 \text{ chapter}}{75 \text{ min}}$
Together	m	$\frac{1 \text{ chapter}}{m \text{ min}}$

3. Write and **solve** an equation relating the work rates.
   a. The rate of working together is the sum of the individual rates.
   b. Multiply each term by the LCD, $300m$, to clear the fractions.

$$\frac{1}{60} + \frac{1}{75} = \frac{1}{m}$$

$$300m\left(\frac{1}{60}\right) + 300m\left(\frac{1}{75}\right) = 300m\left(\frac{1}{m}\right)$$

$$5m + 4m = 300$$

$$9m = 300$$

$$m = \frac{100}{3} = 33\tfrac{1}{3} \text{ min}$$

Check

4. **Check** your answer in the original equation.

$$\frac{1}{60} + \frac{1}{75} = \frac{1}{\frac{100}{3}}$$

$$\overset{5}{\cancel{300}}\left(\frac{1}{\cancel{60}}\right) + \overset{4}{\cancel{300}}\left(\frac{1}{\cancel{75}}\right) = \overset{3}{\cancel{300}}\left(\frac{3}{\cancel{100}}\right)$$

$$5 + 4 = 9$$

You should always check that your solutions are reasonable. A time of $33\frac{1}{3}$ min to complete the task together is between the 30 min and $37\frac{1}{2}$ min it would take Allison and Sean to do half the job individually. Notice also that the answer is not the average of these times.

Example 4

Mr. Mathews owns a landscaping company and schedules the jobs for his company. He can complete the weekly maintenance for one of his corporate clients in 10 hr by himself or in 4 hr with his assistant, Ned. How long should he schedule for Ned to complete the job by himself?

Answer Let h = the number of hours it should take for Ned to complete the job.

1. **Read** the problem carefully and assign a variable to represent the main unknown.
2. **Plan** by making a table listing each rate of work.

	Time (hours)	Rate of Work
Mr. Mathews	10	$\frac{1 \text{ job}}{10 \text{ hr}}$
Ned	h	$\frac{1 \text{ job}}{h \text{ hr}}$
Together	4	$\frac{1 \text{ job}}{4 \text{ hr}}$

3. Write and **solve** an equation relating the work rates.
   a. The rate of working together is the sum of the individual rates.
   b. Multiply each term by the LCD, $20h$, to clear the fractions.

$$\frac{1}{10} + \frac{1}{h} = \frac{1}{4}$$

$$20h\left(\frac{1}{10}\right) + 20h\left(\frac{1}{h}\right) = 20h\left(\frac{1}{4}\right)$$

$$2h + 20 = 5h$$

$$20 = 3h$$

$$h = \frac{20}{3} = 6\frac{2}{3} \text{ hr}$$

Check

4. **Check** your answer in the original equation.

$$\frac{1}{10} + \frac{1}{\frac{20}{3}} = \frac{1}{4}$$

$$\overset{2}{\cancel{20}}\left(\frac{1}{\cancel{10}}\right) + \cancel{20}\left(\frac{3}{\cancel{20}}\right) = \overset{5}{\cancel{20}}\left(\frac{1}{\cancel{4}}\right)$$

$$2 + 3 = 5$$

Can you see how knowing the time it will take for workers to finish a project is helpful to a company or business trying to get a particular job done? When the labor costs can be estimated, the marketing manager can determine how to price and promote products and services.

A. Exercises

Make a table for each problem. Do not solve.

1. Mrs. Beck drove the 200 mi to her aunt's house at a speed 3 mi/hr faster than the speed at which she returned.
2. Bryce biked against a 5 mi/hr wind for 50 mi and then returned with the wind.
3. Pete takes 3 hr to mow the school's ball fields, and Adam takes 6 hr. How long will it take them working together?
4. Sheila and Keisha can each wash their parents' cars in 1 hr. How long will it take them working together?

Make a table and solve. Round to the nearest tenth if necessary.

5. At 7:00 AM Aiden started walking to school at a rate of 2.5 mi/hr. At 7:45 his brother William started riding a bicycle to school. If William rode at a rate of 10 mi/hr, how far did Aiden walk before William caught up with him?
6. The second bus for the senior trip left the school $\frac{1}{2}$ hr later than the first bus, which traveled at an average speed of 45 mi/hr. When it finally departed, the second bus averaged 50 mi/hr. How long did it take the second bus to catch up with the first bus?
7. Mr. O'Brian can plant his acreage in 74 hr. His neighbor can plant the same amount of acreage with his larger equipment in 50 hr. How long will it take to plant Mr. O'Brian's field if they work together?
8. Darla figured that making puppets for vacation Bible school would take her 4 hr. If she worked with her sister (who would take 6 hr to do the job alone), she could finish faster. If the girls work together, how long will it take them to finish the puppets?

9. The junior class is getting ready for the spring play. Landon is in charge of painting the sets, which will take 52 hr if he paints them by himself. Landon knows that Ethan is a good artist and could paint the sets in 36 hr. If he decides to ask Ethan to help him, how long will it take the two to paint the sets?
10. It takes 9 hr for one water pipe to fill a swimming pool. If another pipe is added and both pipes work together, the pool fills in only 3 hr. How long would it take the second pipe to fill the pool if it were the only one used?

B. Exercises

Solve. Round to the nearest tenth if necessary.

11. Two flights left Karr Airport at the same time. One plane flew 250 mi/hr. The other plane flew 375 mi/hr. The faster plane went 750 mi farther than the slower plane and traveled 1 hr longer. How far did each plane travel?
12. Mr. Truman drove to his cottage in the mountains at a rate of 45 mi/hr. On the way back he drove at a rate of 50 mi/hr but took a route 15 mi shorter than the route by which he traveled to the cottage. How many total miles did he travel if it took him $\frac{1}{2}$ hr longer to go than to return?
13. Logan and Carter walked 30 min to the bicycle repair shop. They rode home in 45 min, taking a route that was 6 mi longer than their walk. If they rode 7 mi/hr faster than they walked, how fast did they walk and ride?
14. Gabriel walks $\frac{1}{10}$ as fast as a train travels. It takes 1.5 hr longer for him to walk 12 mi than for the train to travel 60 mi. Find the speed of each.
15. An airplane flew 700 mi with the wind in the same amount of time that it flew 625 mi against the wind. The average speed of the plane with no wind is 265 mi/hr. What was the wind speed?
16. Donna and Evan canoed a total of 14 mi upstream and back. The roundtrip took 8 hr. If they row at a rate of 4 mi/hr in still water, how fast was the current?
17. Phil can load a truck in 2 hr. With Henry's help, he can finish loading a truck in 45 min. How long does it take if Henry is working alone?

18. A certain computer can grade 2000 objective-answer exams in 10 sec. A new computer can grade the 2000 exams in less time. If the two computers together can grade 2000 exams in 3 sec, how long does it take the new computer alone to do the grading?

19. If Lucas, who takes twice as long as Liam to remove the tassels from an acre of corn, works with Liam, he and Liam can do an acre in 50 min. How long does it take each one to do the job alone?

20. Mr. Endicott, Mr. Bookout, and Mr. Piatt are going to paint a house together. Mr. Endicott can paint one side of the house in 4 hr. To paint an equal area, Mr. Bookout takes 3 hr and Mr. Piatt takes 2 hr. If the men work together, how long will it take them to paint one side of the house?

21. Rachel takes 3 hr longer than Fran to clean the house. If the girls work together, they can get the work done in 4 hr. How long does it take each girl working alone to clean the house?

22. Mr. Jansen can shingle a roof in 10 hr, and Mr. Doyle can do it in 8 hr. Mr. Davies has never done the job alone; but last time he helped with a similar job, it took the three men $3\frac{1}{3}$ hr working together. How long would it take Mr. Davies alone?

23. The Schultzes biked 15 mi from Brattleboro to Keene with a tailwind of 2 mi/hr. If the return trip against the wind took 2 hr longer, how fast would the family bike on a calm day?

24. A bicycle club took a weekend trip to a national forest and rode 60 mi at a certain average rate. On their return trip over the same route, they traveled 4 mi/hr faster and took 4 hr less. What were their average rates going and returning?

25. How long will it take the Swansons to weed their vegetable garden if all four family members work together? They estimate that working alone, it would take Mr. Swanson 4 hr, Mrs. Swanson 4 hr, Brenda 6 hr, and Glen 7 hr.

26. At 1:00 PM Mrs. Fanelli tells Joey and Marcy that they can play miniature golf if they get the chores done (properly!). Joey usually takes 5 hr to do the chores, and Marcy does them in 4 hr. If they work together, how long will it take them to do the job? If the course closes at 5:00 PM and it takes 14 min to drive there, how long will they have to play?

C. Exercises

Solve. Round to the nearest tenth if necessary.

27. A touring bus traveling to and from a historical site covered 200 mi in $4\frac{1}{2}$ hr. If the rate returning was 10 mi/hr faster than the rate going, find the rate each way. Also find the amount of time it took to reach the site.

28. A machine at a light bulb factory produces metal bases for light bulbs. The company has a rush order that needs to be filled in 24 hr. After 7 hr the machine breaks down, and another machine is used to complete the job in 5 hr. It would have taken 15 hr to do the job using the first machine alone. How long would it have taken to do the job using only the second machine?

29. Pamela takes 2 hr longer than Carla to make a dress. Together they can finish the dress in 4 hr. How long does each girl take working alone?

30. Mr. Johns drives a semitrailer. On one trip to a city 275 mi from his home, he traveled 30 min longer going than he did returning along the same route because of construction work. How long did it take him to make the trip each way if his rate going was 5 mi/hr slower than his rate returning? How fast did he travel each way?

Dominion Modeling

The average of several speeds can be found using the arithmetic mean if the time traveled at each rate is the same. If the distance traveled at each rate is the same, then the harmonic mean of the individual rates provides the average rate of speed.

A missionary travels 50 mi to and from a village each week. He drives his vehicle part of the way, but he must hike the rest of the way into the mountains. It takes $2\frac{1}{2}$ hr to get to the village, but he returns home in just $1\frac{2}{3}$ hr.

31. To find the missing rates, fill in the table below using the formula $rt = d$.

	r	t	d
Going to the Village			
Returning Home			

32. Find the arithmetic mean of the missionary's rates going to the village and returning.

33. Find the average rate using $rt = d$ and the total distance and time traveled.

34. The answers to exercises 32–33 are not the same. Use $H = \frac{2ab}{a + b}$ to find the harmonic mean of his going and returning rates. Verify that this answer gives the average rate for the entire trip.

CUMULATIVE REVIEW

Graph each quadratic function. [12.7–12.9]

35. $y = x^2 + 3$

36. $y = 2(x - 2)^2$

37. $y = x^2 - 4x + 7$

State the coordinates of the vertex of each parabola. [8.4, 12.8]

38. $f(x) = (x - 12)^2 + 6$

39. $f(x) = -3(x + 8)^2 - 10$

40. Write a function rule that translates the function $y = x^2$ two units left and one unit down. [8.4]

41. Divide $x^2 + x + 4$ by $x - 1$. [9.6]

42. Find two numbers whose difference is 808 and whose sum is 950. [7.4]

43. The poll projecting that Senator Brown would receive 48% of the vote claimed a 3% margin of error. Write an absolute value inequality that models this claim. Then solve it to find the range of outcomes actually predicted by the poll. [4.7]

44. How many milliliters of a 28% potassium iodide solution and how many milliliters of a 60% potassium iodide solution should be combined by a chemist who needs 800 mL of a 40% solution? [3.8]

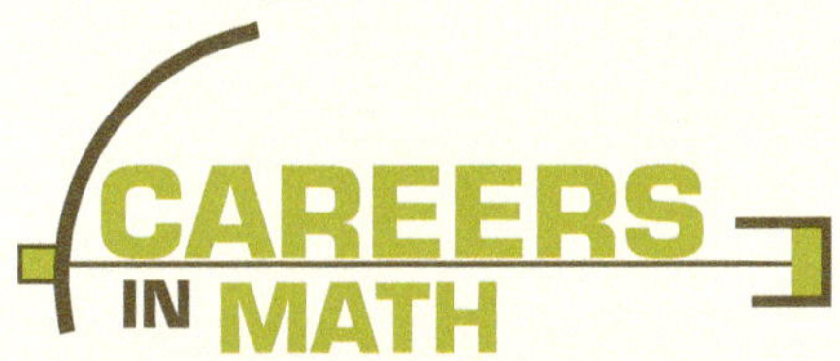

Operations Research Analyst

Operations research analysts study processes by performing statistical analysis and mathematical modeling of step-by-step procedures. Their goal is to maximize some desirable objective, such as profit, or minimize some undesirable objective, such as man-hours expended. The subject of their study can be as simple as an assembly line, where the maximum amount of a product is to be produced or the number of man-hours expended is to be minimized. It can be as complicated as using computers to schedule flights and maintenance for all the aircraft on a carrier at sea for several months.

Their first task is to write an objective function. Variables such as the number of workers with different specialties, the amount of time for each repair, and the operation of supply lines to guarantee the availability of repair parts must be managed efficiently.

Operations research is used to model entire management information systems rather than individual elements. Analysts usually work in a team with other specialists, such as mathematicians, physicists, engineers, computer programmers, or specialists from the field of work in which they are engaged.

Many of these positions are federal government jobs and require a college degree in one of the specialties listed above. These jobs originated with the military during World War II and were very beneficial in making decisions that helped the war effort.

Christians can be excited about an occupation that has a primary goal of determining new procedures to improve efficiency in completing tasks. This work is in harmony with God's command to subdue the earth by organizing something that is complex into something that is efficient and useful. It focuses on serving others in helping them achieve their goals.

13.8 Graphing Rational Functions

Recall that a rational number is a number that can be written as a ratio of integers. Similarly, a *rational function* is a function that can be expressed as a ratio, or quotient, of polynomials.

Definition

A **rational function** is a function expressed as a ratio of polynomials whose denominator is not zero.

The functions in this section differ from other functions you have studied in that they have a variable in the denominator. Rational functions are undefined for any value of the variable that causes the denominator to be equal to zero.

You have already seen how the inverse variation $xy = 2$ can be written as $y = \frac{2}{x}$ or $f(x) = \frac{2}{x}$. This is an example of one of the simplest forms of a rational function, $f(x) = \frac{a}{x}$.

Example 1

Graph $f(x) = \frac{2}{x}$.

Answer $x \neq 0$

1. Determine the value for which the function is undefined.
2. Make a table of ordered pairs.

x	$f(x) = \frac{2}{x}$
-4	$-\frac{1}{2}$
-2	-1
-1	-2
$-\frac{1}{2}$	-4
0	undefined
$\frac{1}{2}$	4
1	2
2	1
4	$\frac{1}{2}$

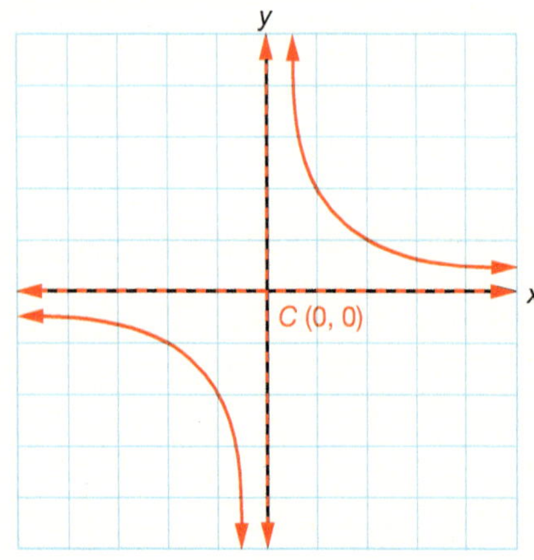

3. Graph the function.

This graph is called a *hyperbola*. The ends of each branch of the hyperbola in Example 1 approach but never reach the x- and y-axes. Therefore, the x-axis is the horizontal asymptote and the y-axis is the vertical asymptote in this graph. These asymptotes are indicated by dashed lines on the graph. The center of the hyperbola is at the intersection of its asymptotes, (0, 0).

Now consider a rational function of the form $f(x) = \frac{a}{x} + k$.

Example 2

Graph $f(x) = \frac{2}{x} + 3$.

Answer Since the function is undefined when $x = 0$, $x = 0$ is the equation of a vertical asymptote.

1. Determine the value for which the function is undefined.

2. Make a table of ordered pairs.

x	$\frac{2}{x} + 3$	$f(x)$
-4	$-\frac{1}{2} + 3$	$2\frac{1}{2}$
-2	$-1 + 3$	2
-1	$-2 + 3$	1
$-\frac{1}{2}$	$-4 + 3$	-1
0	undefined	
$\frac{1}{2}$	$4 + 3$	7
1	$2 + 3$	5
2	$1 + 3$	4
4	$\frac{1}{2} + 3$	$3\frac{1}{2}$

3. Graph the function.

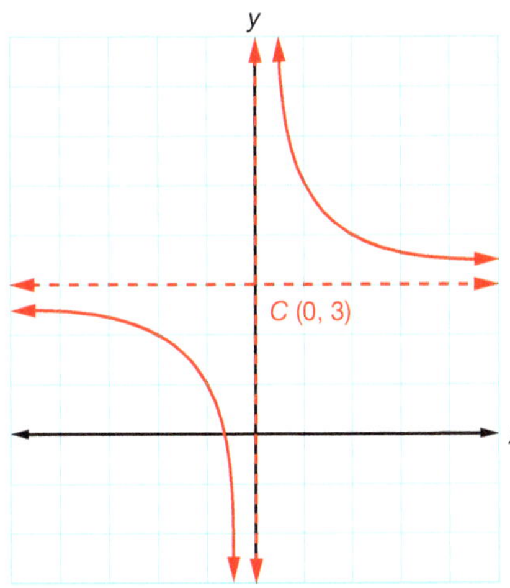

Observe that the graph of $f(x) = \frac{2}{x} + 3$ is the result of translating the graph of $f(x) = \frac{2}{x}$ three units up. The graph's center and horizontal asymptote have been moved up to (0, 3) and $y = 3$, while its vertical asymptote is still the y-axis.

In general, the graph of $f(x) = \frac{a}{x} + k$ is the result of translating the graph of $f(x) = \frac{a}{x}$ k units up if k is positive and k units down if k is negative.

Example 3

Graph $g(x) = \frac{-6}{x}$.

Answer Since the function is undefined when $x = 0$, $x = 0$ is the equation of a vertical asymptote.

1. Determine the value for which the function is undefined.

2. Make a table of ordered pairs.

x	$g(x) = \frac{-6}{x}$
-6	1
-4	$\frac{3}{2}$
-2	3
-1	6
0	undefined
1	-6
2	-3
4	$-\frac{3}{2}$
6	-1

3. Graph the function.

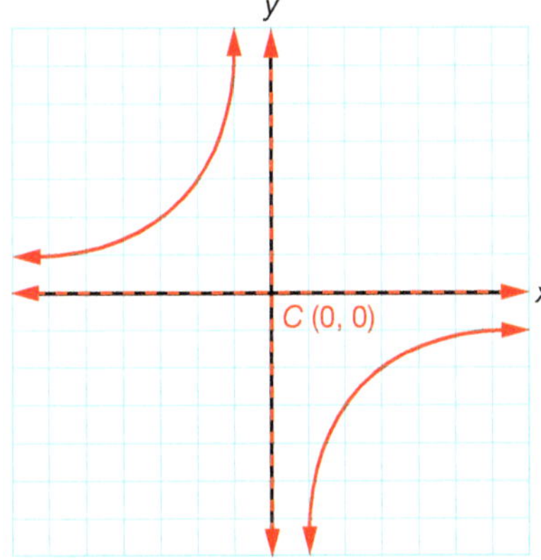

In Example 3, when x is a very large positive value, $g(x)$ is negative and very close to zero. Likewise, for a very large negative value of x, $g(x)$ is positive and very close to zero. This indicates that the line $y = 0$ (the x-axis) is the horizontal asymptote. Since the vertical asymptote is $x = 0$, the graph's center is at (0, 0). Notice that the negative value of a causes the graph to be located in the upper left and lower right quadrants formed by the asymptotes.

Example 4

Graph $h(x) = \frac{-6}{x - 2}$ and compare its graph to the graph of $g(x) = \frac{-6}{x}$ in Example 3.

Answer Since the function is undefined when $x = 2$, $x = 2$ is the equation of a vertical asymptote.

1. Determine the value for which the function is undefined.
2. Make a table of ordered pairs.

x	$\frac{-6}{x-2}$	$h(x)$
−4	$\frac{-6}{-4-2}$	1
−2	$\frac{-6}{-2-2}$	$\frac{3}{2}$
0	$\frac{-6}{0-2}$	3
1	$\frac{-6}{1-2}$	6
2	undefined	
3	$\frac{-6}{3-2}$	−6
4	$\frac{-6}{4-2}$	−3
6	$\frac{-6}{6-2}$	$-\frac{3}{2}$
8	$\frac{-6}{8-2}$	−1

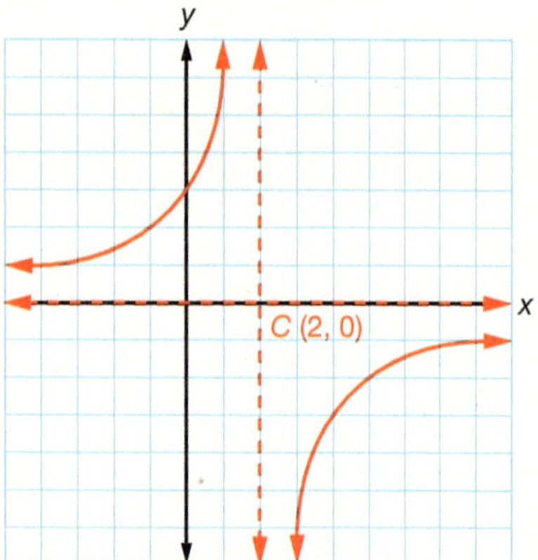

3. Graph the function.

The graph of $h(x)$ is the result of translating the graph of $g(x)$ 2 units right.

4. Compare the graphs in Examples 3 and 4.

The graph of a function of the form $y = \frac{a}{x - h} + k$ is a translation of the base function $y = \frac{a}{x}$. It is a hyperbola with its center at (h, k), a vertical asymptote of $x = h$, and a horizontal asymptote of $y = k$. You will need to choose several points on either side of the hyperbola's center to determine its exact shape.

Example 5

Given $f(x) = \frac{2}{x + 3} - 1$, identify the center of the hyperbola and both asymptotes. Then graph the function.

Answer $f(x) = \frac{2}{x - (-3)} + (-1)$

$(h, k) = (-3, -1)$

vertical asymptote: $x = -3$

horizontal asymptote: $y = -1$

1. The graph of a function of the form $f(x) = \frac{a}{x - h} + k$ has its center at (h, k), a vertical asymptote of $x = h$, and a horizontal asymptote of $y = k$.

Example 5, continued

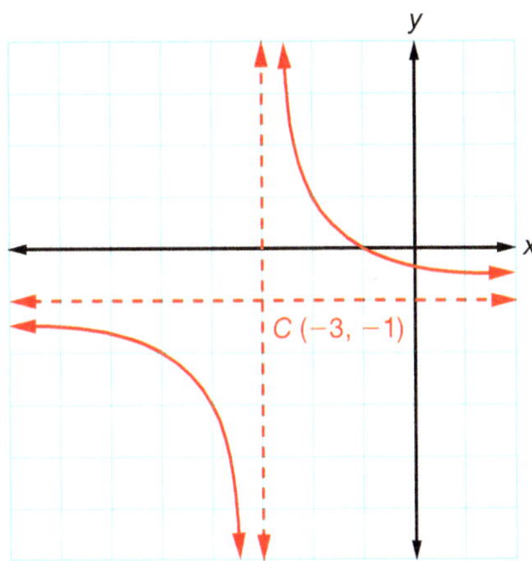

2. Draw the asymptotes and label the center.
3. Make a table of several convenient points on either side of the hyperbola's center to establish the general shape of the graph.

x	$\frac{2}{x+3} - 1$	$f(x)$
-7	$\frac{2}{-7+3} - 1$	$-1\frac{1}{2}$
-5	$\frac{2}{-5+3} - 1$	-2
-4	$\frac{2}{-4+3} - 1$	-3
-3	undefined	
-2	$\frac{2}{-2+3} - 1$	1
-1	$\frac{2}{-1+3} - 1$	0
1	$\frac{2}{1+3} - 1$	$-\frac{1}{2}$

4. Graph the function.

The graph of $f(x) = \frac{2}{x+3} - 1$ is the result of translating the graph of $f(x) = \frac{2}{x}$ three units left and one unit down. Notice the similarity to the translations of the absolute value, power, radical, and quadratic functions studied earlier.

The table below summarizes the families of functions studied in this course and their key characteristics.

Function Type	General Form	Variable	Example	Characteristics
Absolute Value (Section 5.7)	$f(x) = \lvert x - h \rvert + k$	within absolute value symbols	$f(x) = \lvert x - 3 \rvert + 4$	vertex: (h, k)
Linear (Section 6.3)	$f(x) = mx + b$	degree of 1	$f(x) = \frac{1}{2}x + 3$	y-intercept: $(0, b)$ slope: m
Power (Section 8.4)	$f(x) = ax^n$	the base of a power	$f(x) = 3(x - 2)^3 - 5$	a changes the graph's "steepness"
Exponential (Section 8.5)	$f(x) = ab^x$	the exponent of a power	$f(x) = 3 \cdot 2^x$	growth when $b > 1$ decay when $0 < b < 1$
Radical (Section 11.9)	$f(x) = \sqrt{x - h} + k$	in the radicand	$f(x) = \sqrt{x - 2} + 3$	"begins" at (h, k)
Quadratic (Section 12.7)	$f(x) = a(x - h)^2 + k$ $= ax^2 + bx + c$	degree of 2	$f(x) = (x - 5)^2 + 2$	vertex: (h, k)
Rational (Section 13.8)	$f(x) = \frac{a}{x - h} + k$	in the denominator	$f(x) = \frac{2}{x - 1} + 5$	center: (h, k) vertical asymptote: $x = h$ horizontal asymptote: $y = k$

A. Exercises

Multiple choice: Classify each function.

a. absolute value **b.** linear **c.** exponential

d. radical **e.** quadratic **f.** rational

1. $f(x) = 2x + 17$ **2.** $f(x) = (x + 12)^2 - 6$ **3.** $f(x) = |x + 7| - 21$

4. $f(x) = \sqrt{x + 7}$ **5.** $f(x) = 5^x$ **6.** $f(x) = \frac{1}{x}$

7. $f(x) - \frac{2}{3}x$ **8.** $f(x) = 5x^{\ 1}$ **9.** $f(x) = x^2 + 3x - 9$

10. $f(x) = x + |-5|$

State the value of *x* for which each function is undefined.

11. $f(x) = \frac{2}{x - 5}$ **12.** $f(x) = \frac{2}{x + 3}$

State the center and asymptotes for each function's graph.

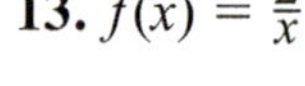

13. $f(x) = \frac{2}{x}$

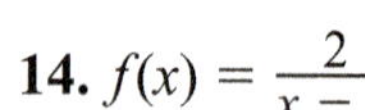

14. $f(x) = \frac{2}{x - 4}$

15. $f(x) = \frac{2}{x - 1} + 7$

16. $f(x) = \frac{2}{x - 5} - 1$

17. $f(x) = \frac{2}{2x - 1} + 1$

18. $f(x) = \frac{2}{x} - 9$

19.

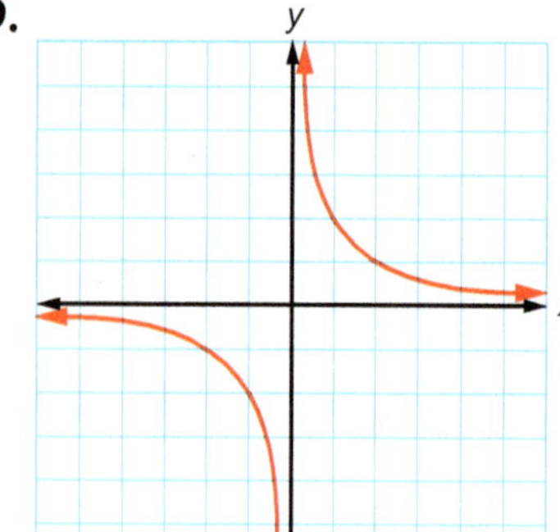

20.

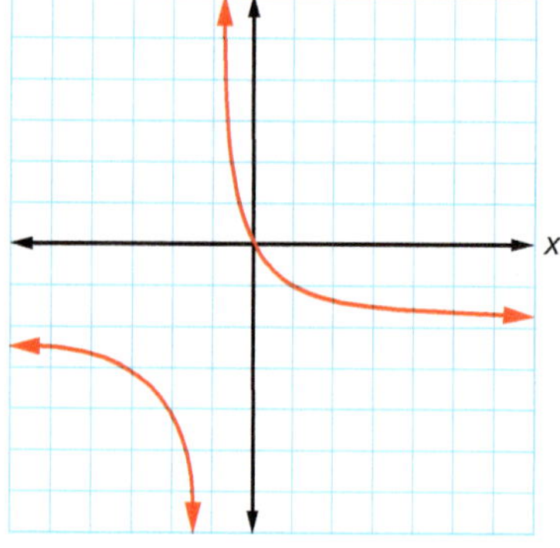

B. Exercises

Graph each rational function. Be sure to include its asymptotes.

21. $f(x) = \frac{3}{x}$ **22.** $f(x) = \frac{3}{x} + 3$ **23.** $f(x) = \frac{-2}{x + 2}$

24. $f(x) = -\frac{1}{x} + 1$ **25.** $f(x) = \frac{3}{x + 2} + 2$ **26.** $f(x) = \frac{-1}{x - 4} - 2$

Describe the translation(s) of $f(x) = \frac{1}{x}$ that produce the graph of each function.

27. $f(x) = \frac{1}{x - 10}$ **28.** $f(x) = \frac{1}{x + 14}$ **29.** $f(x) = \frac{1}{x} + 7$

30. $f(x) = \frac{1}{x} - 3$ **31.** $f(x) = \frac{1}{x + 4} - 9$ **32.** $f(x) = \frac{1}{x - 4} + 2$

Ohm's law states that the current in a circuit is directly proportional to the voltage and inversely proportional to the resistance in the circuit. This relationship can be modeled using the formula $I = \frac{V}{R}$, where I is the current in amps, V is the voltage in volts, and R is the resistance in ohms. If the voltage of a circuit is 120 V, the current is a rational function of the resistance in the circuit, $I = \frac{120}{R}$. Electrical engineers increase or decrease resistance in a circuit to maintain desirable current ranges.

33. Graph the portion of the rational function $I = \frac{120}{R}$ that gives reasonable values for I and R.

Use the graph from exercise 33 to answer the following questions.

34. What happens in the circuit as the resistance increases?

35. What happens in the circuit as the resistance decreases?

36. What type of variation is illustrated by the graph?

C. Exercises

Use long division to express each rational function in $f(x) = \frac{a}{x - h} + k$ form; then graph the function.

37. $f(x) = \frac{2x - 1}{x - 1}$

38. $f(x) = \frac{3x + 2}{x + 2}$

39. Why is the vertical asymptote not part of the solution to a rational function?

40. Why are points on the horizontal asymptote not solutions to a rational function $f(x) = \frac{a}{x - h} + k$?

41. Graph $f(x) = \frac{1}{x^2}$. Describe the characteristics of the graph compared to the graph of $f(x) = \frac{1}{x}$.

Dominion Modeling

In order to consider the reasonableness of building an airstrip for the delivery of supplies at a missionary compound, the mission board studies how long it takes the missionaries to travel to and from the nearest city. Since the trips have the same distance but different times, the average rate can be found using the harmonic mean of the individual rates. The harmonic mean of n numbers is found by dividing n by the sum of the numbers' reciprocals. Expressed as a formula, $H = \frac{n}{\frac{1}{x_1} + \frac{1}{x_2} + \ldots + \frac{1}{x_n}}$, where each variable x_i is one of the numbers.

42. Use $rt = d$ to find the average rate for each trip.

	r	t	d
Trip 1		5 hr	120 mi
Return 1		$6\frac{2}{3}$ hr	120 mi
Trip 2		$7\frac{1}{2}$ hr	120 mi
Return 2		6 hr	120 mi

43. Show that $\frac{2}{\frac{1}{a} + \frac{1}{b}}$ is equivalent to $\frac{2ab}{a + b}$.

a. Starting with the first expression, find the LCD of the fractions in the denominator.

b. Multiply the numerator and the denominator of the complex fraction by the LCD and simplify.

44. Find the harmonic mean of the four rates in the table above.

a. Substitute the individual rates into the formula for the harmonic mean.

b. Find the LCD of the fractions in the denominator.

c. Multiply the numerator and the denominator of the complex fraction by the LCD and simplify. State the rate to the nearest tenth.

CUMULATIVE REVIEW

Solve each compound inequality. [4.4–4.5]

45. $7 + 5x \leq 9 \vee 5x + 3 > 7$

46. $2x - 3 > 5 \wedge 5x - 1 < 9$

47. Solve $\frac{5x - 9}{x - 2} - 1 = \frac{13}{3} + \frac{1}{x + 2}$. [13.6]

Factor. [10.5]

48. $7x^4 + 11x^2 + 4$

49. $3x^4 - 31x^2 + 36$

50. $x^8 - y^8$

Given the two points (1, 3) and (−2, 5), find the following. [6.2, 11.6]

51. the slope of the line passing through the points

52. the distance between the points

53. A gardener has 80 lb of a lime/fertilizer mixture that is 25% fertilizer. How much lime should she add to produce a mixture that is 20% fertilizer? [3.8]

54. The square of a number reduced by twice the number is 15. Find the number. [12.6]

TECHNOLOGY CORNER (TI-84+ Family)

While a graphing calculator can be used to draw the graph of many functions, you need to be aware of the limitations of this technology. Graphing the function $f(x) = \frac{3x^2}{x^2 - 4}$ using ZStandard from the [ZOOM] menu when in CONNECTED mode produces the graph to the right.

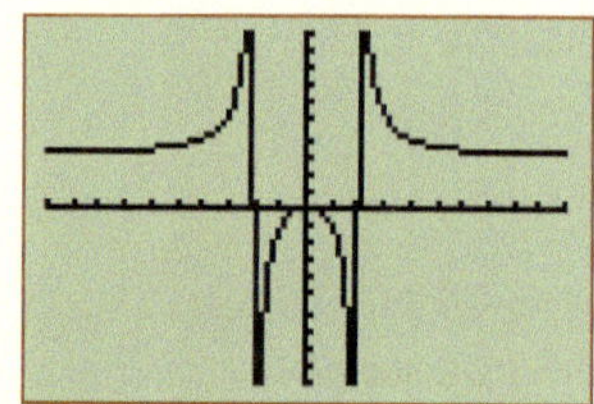

In CONNECTED mode, the calculator actually connects plotted points in an attempt to produce a smooth curve. When it connects the points just to the left and right of the undefined values $x = \pm 2$, the nearly vertical solid lines appear to be asymptotes, but they are actually an inaccurate drawing of the graph. Use the [TRACE] function to see how the y value switches between very large negative and positive values on either side of the undefined x value. Notice that no y value is given if you enter one of the undefined values of x when using the [TRACE] function.

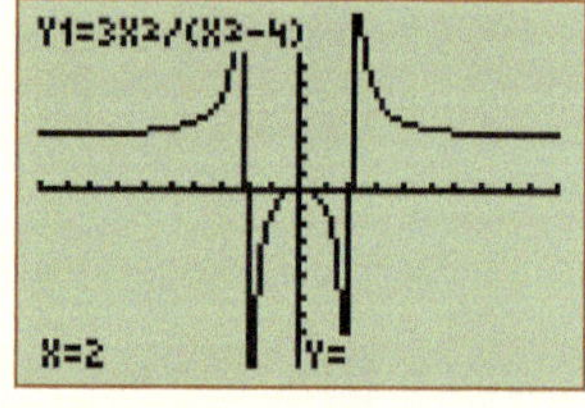

Switching to DOT mode causes the calculator to graph only the plotted points. This corrects the inaccurate drawing of the graph at the asymptotes, but the overall shape of the graph may be harder to see. By converting the function to a form that we are more familiar with, $f(x) = \frac{12}{x^2 - 4} + 3$, we can predict the vertical asymptotes at $x = \pm 2$ and the horizontal asymptote at $y = 3$.

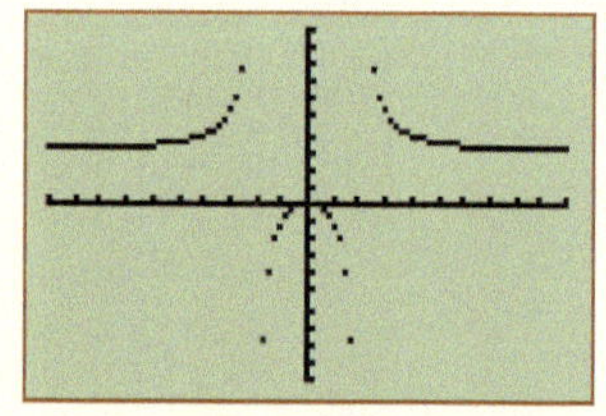

You can use your graphing calculator to investigate properties of other rational functions, such as $f(x) = \frac{2x^2 + 5x + 2}{x + 1}$. Using polynomial division to express the function as $f(x) = 2x + 3 - \frac{1}{x + 1}$ will help you recognize the vertical asymptote. While there is no horizontal asymptote, you can discover an interesting characteristic of the function by comparing its graph to the graph of $g(x) = 2x + 3$.

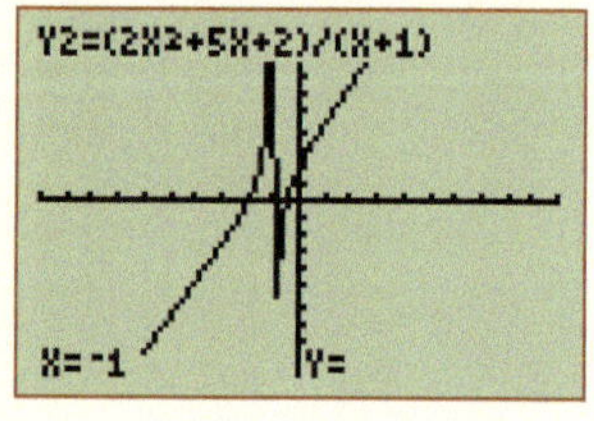

CHAPTER 13 REVIEW

State any excluded values for each rational expression.

1. $\frac{5}{x - 3}$

2. $\frac{x - 8}{x^2 - 4x - 12}$

3. $\frac{x - 6}{x^3 - x}$

Simplify.

4. $\frac{2x + 8}{2}$

5. $\frac{12x + 3}{8x + 2}$

6. $\frac{3x + 4}{6x^2 - 7x - 20}$

7. $\frac{x^2 + 7x + 10}{x^2 - x - 6}$

8. $\frac{2x - 14}{2x^2 - 22x + 56}$

Find the LCM of each pair of polynomials.

9. $8x^4$ and $28x^3y$

10. $4x - 20$ and $x^2 - 25$

Simplify, leaving each answer in factored form.

11. $\frac{a}{\frac{b}{c}}$

12. $\frac{4x^3}{(x - 5)^2} \cdot \frac{x - 5}{2x}$

13. $\frac{2x + 6}{3x + 5} + \frac{x + 4}{3x + 5}$

14. $2c + \frac{d}{c}$

15. $\frac{56x^2 - 18}{7x - 5} - \frac{7x^2 + 7}{7x - 5}$

16. $\frac{(9x + 2)(x - 3)}{(2x - 1)(x + 4)} \div \frac{(9x + 2)(x - 6)}{(x - 6)(2x - 1)}$

17. $\frac{x^2 + 13x + 22}{3} \cdot \frac{9}{x + 2}$

18. $x + y - \frac{1}{x - y}$

19. $\frac{x}{x^2 - y^2} + \frac{y}{x^2 - y^2}$

20. $\frac{5x + 2}{x + 2} - \frac{x - 1}{x + 3}$

21. $\frac{\frac{x}{y} - 1}{\frac{y}{x} - 1}$

22. $\frac{5x^3}{3x^2 + x - 10} \div \frac{15x}{6x^2 + 13x + 2}$

23. $\frac{3}{1 \quad x} + \frac{7}{x - 1}$

24. $\frac{\frac{2x}{y} + 1}{\frac{2x + y}{y^3}}$

25. $\frac{xy}{5xy^3 - 11x^2} - \frac{y^3}{5y^5 - 11xy^2}$

26. $\frac{x^2 - 5x - 14}{2x + 3} \cdot \frac{2x^2 + x - 3}{x^2 - 7x}$

27. $\frac{x - 1}{x^2 + 9x + 18} \div \frac{x^2}{x^2 + x - 30}$

28. $\frac{4x^2 + 6x}{6x^2 - x - 15} - \frac{6x + 9}{15 + x - 6x^2}$

29. $\frac{\frac{x^2 - 4}{x}}{x + 2}$

30. $\frac{x + 2y}{27y} + \frac{x}{3}$

31. $\frac{x^3 + 4x}{3x^2 + 23x + 14} \div \frac{x^4 - 16}{x^2 + 5x - 14}$

32. $\frac{5x^2 - 40x}{3x^2 + 7x + 2} \cdot \frac{3x + 1}{10x^2 - 160x + 640}$

33. $\frac{2}{xy} + \frac{3}{x^2 - 3xy} - \frac{1}{xy - 3y^2}$

34. $\frac{x + y}{x^2 + 2xy + y^2} + \frac{3}{x - y} - \frac{1}{x + y}$

Solve.

35. $\frac{x^2}{4} = \frac{13x}{8} + 3$

36. $\frac{6}{x - 3} = 3$

37. $\frac{x + 8}{2x - 6} = \frac{17}{23}$

38. $\frac{12x - 3}{x^2 + 4x - 5} - \frac{5}{x + 5} = \frac{6}{x - 1}$

39. $\frac{3}{8x} + \frac{1}{4} = \frac{5}{6x}$

40. $\frac{-5x + 8}{x^2 - 9} + \frac{5}{x + 3} = \frac{7x}{2x - 6}$

41. $\frac{2x - 2}{x^2 - 1} = \frac{x - 5}{x + 1}$

42. $\frac{6}{x^2 - x - 2} = \frac{-1}{x^2 + x - 6} - \frac{4}{x^2 + 4x + 3}$

Complete each table and write an equation modeling the data. Do not solve.

43.

	r	t	d
Going	$x - 5$	7	
Returning	$x + 5$	4	

44. total time = 7 hr

	r	t	d
Day 1	$x + 3$		40
Day 2	x		50

45.

	Time (hours)	Rate of Work
First	5	
Second	2	
Together	x	

46.

	Time (hours)	Rate of Work
Seth	x	
Molly	$2x$	
Dominic	$x + 5$	
Together	8	

Solve.

47. Dana has decorated the church for vacation Bible school in 5 hr and Violet has done it in 3 hr. If they work together, how long will it take them?

48. Joan can wallpaper a living room in 9 hr. Kelly is inexperienced at wallpapering, but she wants to learn. If they work together on wallpapering the living room, it takes 6 hr. How long would it take Kelly to wallpaper the living room by herself?

49. Micah bikes 40 mi in the same time that Alexis drives 100 mi. If Alexis travels 12 mi/hr more than twice Micah's rate, how fast does each travel?

50. Gavin left early to drive to his parents' house and was able to drive 300 mi before lunch. Due to increased traffic after lunch, his average speed for the remaining 105 mi was 15 mi/hr less than his average speed before lunch. If he drove for a total of 9 hr, determine his average speed before lunch and after lunch. What was his average speed for the trip?

51. Write a rational expression representing the portion of the regular hexagon that is shaded. State the result as a simplified expression and as a percent (to the nearest tenth).

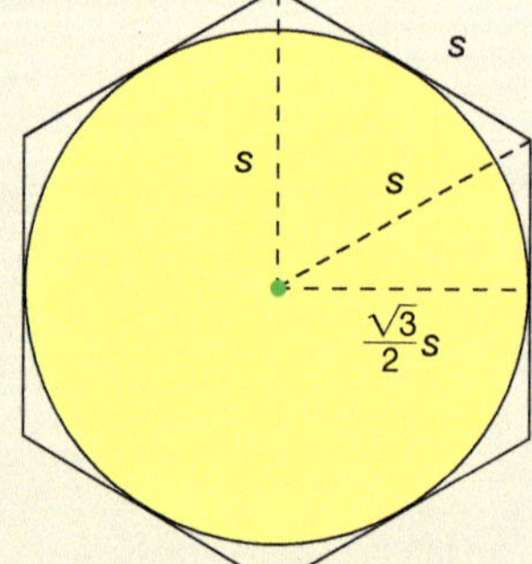

52. What number must be added to both the numerator and the denominator of $\frac{7}{8}$ to produce a fraction equivalent to $\frac{12}{13}$?

State the center and asymptotes for each function's graph.

53. $f(x) = \frac{12}{x + 3}$ **54.** $f(x) = \frac{2}{x - 3} + 4$

Graph each rational function. Be sure to include its asymptotes.

55. $f(x) = -\frac{2}{x} + 1$ **56.** $f(x) = \frac{6}{x + 2} - 3$

Multiple choice: Classify each function.

a. absolute value **b.** linear
c. exponential **d.** radical
e. quadratic **f.** rational

57. $f(x) = x + 3^2$ **58.** $f(x) = x^2 + \sqrt{7}$

SYMBOLS

Symbol	Meaning
$+$	addition, positive
$-$	subtraction, negative, opposite
$\times, \cdot$	multiplication
$\div$	division
$\frac{p}{q}$	division, fraction
$\pm$	plus or minus
$\lvert x \rvert$	absolute value of x
$^\circ$	degrees
$\%$	percent
$0.\overline{3}$	repeating decimal
$f(x)$	function of x
$\sqrt{\ }$	square root
$\sqrt[3]{\ }$	cube root
$\sqrt[n]{\ }$	nth root
$\angle$	angle
$\triangle$	triangle
$'$	prime
$\mathbb{N}$	natural numbers
$\mathbb{W}$	whole numbers
$\mathbb{Z}$	integers
$\mathbb{Q}$	rational numbers
$\mathbb{Q}'$	irrational numbers
$\mathbb{R}$	real numbers
π	pi
Δ	delta
ϕ	phi
Σ	sigma, sum (summation notation)
$=$	equal to
$\neq$	not equal to
$\approx$	approximately equal to
$\cong$	congruent to
$>$	greater than
$\ngtr$	not greater than
$<$	less than
$\nless$	not less than
$\geq$	greater than or equal to
$\ngeq$	not greater than or equal to
$\leq$	less than or equal to
$\nleq$	not less than or equal to
$\ldots$	ellipses (and so on)
$\{\ \}$	set braces
$\{x \mid \ldots\}$	the set of x such that ...
$\in$	element of
$\notin$	not an element of
$\subseteq$	subset of
$\not\subset$	not a subset of
$\varnothing$	empty (null) set
$\cup$	union
$\cap$	intersection
$\vee$	or
$\wedge$	and
∞	infinity
$!$	factorial

SELECTED ODD ANSWERS

Chapter 1—Real Number Operations

1.1 Sets of Numbers

1–9. odd 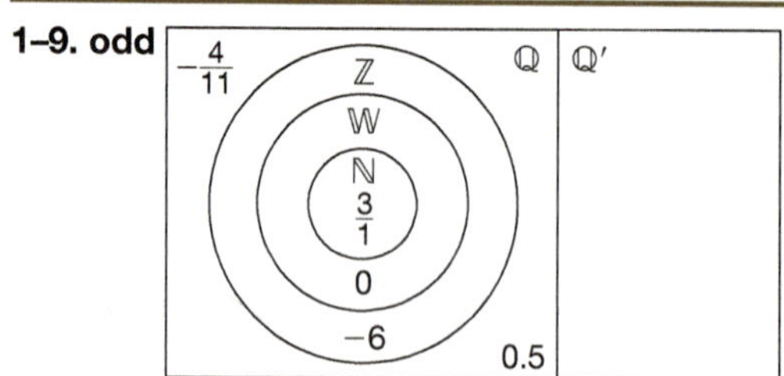

11. true **13.** false **15.** false **17.** true **19.** false **21.** false **23.** false **25.** false **27.** false **29.** $\{1, 3\}$ **31.** $\{-1\}$ **33.** $\{-1, 5\}$ **35.** {MN, WI, IA, IL, MO, KY, TN, AR, MS, LA, ME, MA, MD, MI, MT} **37.** {TX, LA, MS, AL, FL, NM, ND, SD, WV, NC, SC, NJ, NY, RI, NH} **39.** $\varnothing$ **41.** {ND, NY, NH} **43.** $\{0, 1, 2, 3, \ldots\}$

45. 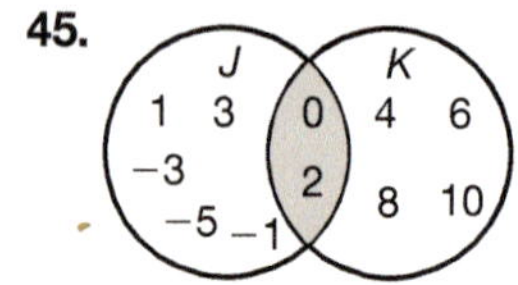

47. $\{5, 7\} \subseteq \{1, 3, 5, \ldots\}$ **49.** finite: $\{5, 7\}$; infinite: $\{1, 3, 5, \ldots\}$ **51.** $\frac{4}{3}$ units

53. 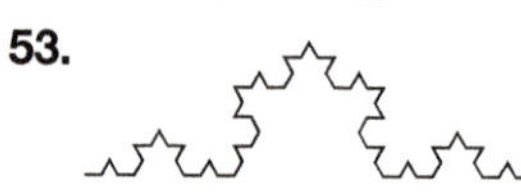

53a. 64 **53b.** $\frac{1}{27}$ unit **53c.** $\frac{64}{27}$ units

55. GCF = 8, LCM = 720 **57.** $\frac{37}{24}$ or $1\frac{13}{24}$ **59.** $\frac{5}{24}$ **61.** $\frac{7}{12}$ **63.** $\frac{21}{16}$ or $1\frac{5}{16}$

1.2 Number Lines, Opposites, and Absolute Values

1.

3. 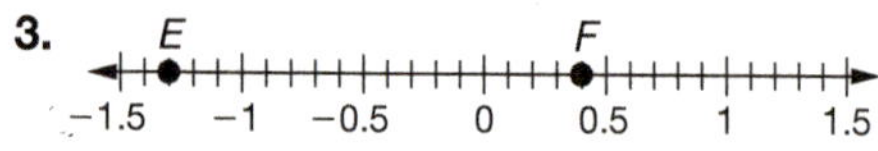

5. **7.** 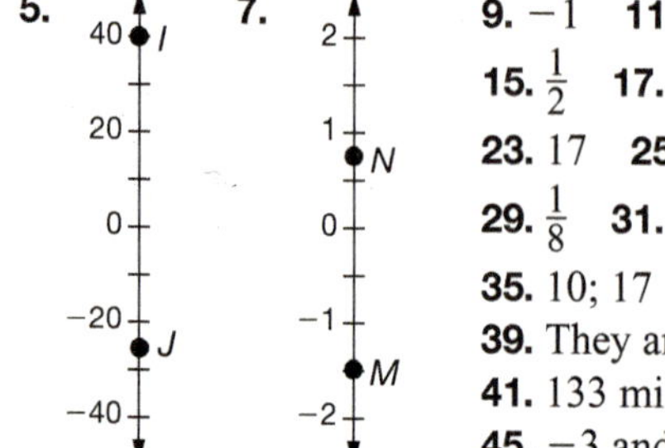

9. −1 **11.** 3 **13.** −15 **15.** $\frac{1}{2}$ **17.** $\frac{5}{4}$ **19.** 9 **21.** $-\frac{1}{2}$ **23.** 17 **25.** 2.8 **27.** 315 **29.** $\frac{1}{8}$ **31.** 12 **33.** 1.6 **35.** 10; 17 **37.** 13; 18.5 **39.** They are equal. **41.** 133 mi **43.** 0 **45.** −3 and 1

47a. There are four times as many segments. **47b.** It is $\frac{1}{3}$ the length. **47c.** It is $\frac{4}{3}$ the length. **49.** $\{h, m\}$ **51.** true

53. 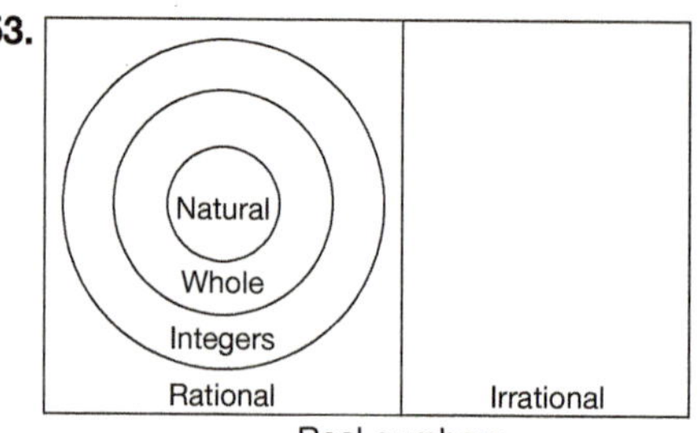

Sequences Evens, Odds, and Multiples

1. 4, 8, 12, 16, 20 **3.** 11, 22, 33, 44, 55 **5.** 20 **7.** 39

1.3 Adding Rational Numbers

1. −3 **3.** −4 **5.** 24 **7.** −2 **9.** −91 **11.** −21 **13.** 0 **15.** −617.1 **17.** $\frac{103}{35}$ **19.** $\frac{7}{15}$ **21.** $-\frac{17}{88}$ **23.** 0 **25.** −2 **27.** 10 **29.** 28 **31.** −12 **33.** −4 **35.** 5 + 2; 7 **37.** 2 + 7; 9 **39.** −6 + 5; −1 **41.** 12 knots **43.** Associative Property **45a.** $\left(\frac{1}{3}\right)^n$ or $\frac{1}{3^n}$ **45b.** $\left(\frac{4}{3}\right)^n$ **47.** true **49.** false **51.** false **53.** 5 **55.** −7

1.4 Subtracting Rational Numbers

1. 5 **3.** −17 **5.** −24 **7.** −2 **9.** −130 **11.** 14 **13.** 621.17 **15.** $\frac{7}{9}$ **17.** $-\frac{29}{10}$ **19.** $\frac{5}{36}$ **21.** 2 **23.** 0 **25.** −9 **27.** 6 − 2; 4 **29.** 8 − 3; 5 **31.** 16 − 5; 11 **33.** $15.92 **35.** 4 **37.** 28 **39.** $\frac{65}{84}$

41a. would continue to increase forever
41b. would get closer and closer to zero
41c. would continue to increase forever **43.** 19
45. No; the preceding two exercises illustrate that this is not always true.
47. Associative Property of Addition **49.** 8 **51.** 10

1.5 Multiplying Rational Numbers

1. −6 **3.** 24 **5.** 3 **7.** 36 **9.** 287 **11.** −546 **13.** −78.96 **15.** −126 **17.** 0 **19.** $-\frac{7}{33}$ **21.** $\frac{3}{8}$ **23.** $-\frac{2}{3}$ **25.** $\frac{49}{8}$ **27.** 5(4); 20 **29.** −2(9); −18 **31.** 6(8); 48 **33.** $1\frac{1}{5}$ gal **35.** 29.4 yd **37.** 33.02 cm **39.** $-a = -1 \cdot a$ **41.** $-3 + (-3) + (-3) + (-3) + (-3) + (-3) + (-3) + (-3)$ **45.** 44 **47.** −64 **49.** $-\frac{11}{12}$ **51.** $\frac{7}{6}$ **53.** −3.11

1.6 Dividing Rational Numbers

1. 9 **3.** undefined **5.** -9 **7.** -7 **9.** -9 **11.** undefined **13.** 8 **15.** 5 **17.** $-\frac{11}{18}$ **19.** $-\frac{10}{3}$ **21.** $\frac{21}{19}$ **23.** $-\frac{9}{40}$ **25.** $\frac{8}{15}$ **27.** $-\frac{42}{5}$ **29.** $\frac{81}{-3}$; -27 **31.** $\frac{9}{2}$; 4.5 **33.** $\frac{11}{4}$; 2.75 **35.** $\frac{4}{5}$ ft **37.** $\frac{0}{a} = 0$ **39.** 0.073
41. No; $6 \div 3 = 2$, but $3 \div 6 = \frac{1}{2}$.
45. Additive Identity Property
47. Associative Property of Addition
49. 5.7 **51.** $\frac{3}{4}$ **53.** $-\frac{3}{14}$

1.7 Exponents

1. $4 \cdot 4 \cdot 4 = 64$ **3.** $-3(-3)(-3)(-3)(-3) = -243$ **5.** 1 **7.** $\frac{1}{4 \times 4} = \frac{1}{16}$ **9.** $-\frac{1}{2 \times 2 \times 2 \times 2} = -\frac{1}{16}$ **11.** $-3(-3)(-3)(-3) = 81$ **13.** 3^5 **15.** 8^{15} **17.** 12^{-5} **19.** 10^{15} **21.** 3^{-5} **23.** 9^2 **25.** $\frac{1}{7^{11}}$ **27.** 17^{54} **29.** 28^8 **31.** 2^2 **33.** $\frac{1}{4^7}$ **35.** 1 **37.** b, c, and d **39.** positive **41.** 9^4; 6561 **43.** 8^2; 64 **45.** $\frac{2^6}{5}$ **47.** $\frac{1}{81x^{12}}$ **49a.** 3^n **49b.** $\left(\frac{1}{4}\right)^n$ or $\frac{1}{4^n}$ **49c.** $\left(\frac{3}{4}\right)^n$ **51.** -48 **53.** -2214 **55.** -416 **57.** 1992 **59.** 6.25

1.8 Order of Operations

1. 17 **3.** 10 **5.** -11 **7.** 14 **9.** -13 **11.** 21 **13.** -1 **15.** -6 **17.** 41 **19.** $\frac{3}{2}$ **21.** 11 **23.** 2 **25.** $2\sqrt{4} + 5$; 9
27. $(20 - 13)^2$; 49 **29.** $(2 + 5)6$; 42
31. $5 \cdot 6 + 20 \div 5$; 34 **33.** $9 - |2 + (-15)|$; -4
35. $-\frac{19}{24}$ **37.** 5.5 **39.** -2.5 **41.** 4 **43.** 21 **45.** $\mathbb{N}$ **47.** $\mathbb{Q}'$

Chapter 1 Review

1–3. odd

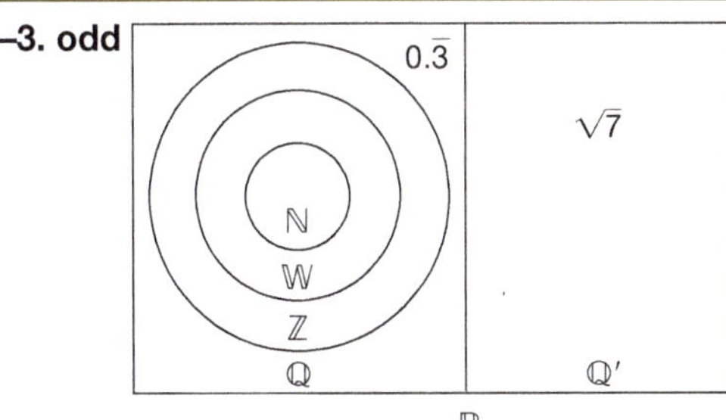

5. true **7.** $\{2\}$ **9.** $\{-1, 2\}$ **11.** -3.5 **13.** 1.5 **15.** 7 **17.** 9; 12.5 **19.** -4 **21.** -20 **23.** 4 **25.** 1 **27.** 4.44 **29.** $\frac{32}{21}$ **31.** $-\frac{11}{72}$ **33.** 8^{15} **35.** 3^{24} **37.** -3 **39.** 15 **41.** 93 **43.** -29 **45.** $5 - 8$; -3 **47.** $3^3 + 4$; 31 **49.** f **51.** h **53.** g **55.** c

Chapter 2—Variables and Equations

2.1 Variables and Algebraic Expressions

1. x; 3 and 6 **3.** t; 4 and 6 **5.** $-8, -3, -1, 4$
7. $-12, 3, 9, 24$ **9.** $-2, \frac{1}{2}, \frac{3}{2}, 4$ **11.** 0, 5, 21, 96 **13.** $8n$
15. $n - 10$ **17.** $5n + 4$ **19.** $n - 2n$ **21.** q^5 **23.** $325n$
25. $3x + 4$ (if x = the number of math books)
27. $25x + 90y$ **29.** $h^3 + h^2$ **31.** $\frac{2n}{n + 7}$
33. $n + (n + 2) + (n + 4)$ **35.** $n(n + 2) - 3$
37. $(90 - a) + (180 - a)$ **39.** $6d + 12h$ **41.** $3x + 2y + z$
43. $4x - 2y + 4z$ **47.** \$0.01/day ≈ \$0.43/mo ≈ \$5.26/yr
49. \$0.10/day ≈ \$2.85/mo ≈ \$34.69/yr **51.** -24 **53.** 108
55. $x^a x^b = x^{a+b}$ **57.** $(x^a)^b = x^{ab}$ **59.** x^{10}

2.2 Evaluating Algebraic Expressions

1. -32 **3.** -2 **5.** $-\frac{14}{5}$ **7.** $63\frac{61}{125} = \frac{7936}{125}$ **9.** -59
11. 13.856 psi **13.** 160 ft **15.** 12.56 ft **17.** 60 in.^{3}
19. -57.76 **21.** a^4 **23.** 183 **25.** $-\frac{29}{12}$ **27.** $\frac{7}{144}$
29. 94 mi **31.** \$22.50 **33.** \$715 **35.** \$155 **37.** \$403.75
39. $\frac{d}{r} = t$ **41.** $d = 2r$ **43.** Pi is not equal to 3.14 ($\pi \approx 3.14$). **45.** $\frac{25}{4}$ **47.** c is 1.5 times as large; A is 2.25 times as large. **49.** \$22.34/yr **51.** Negative one is not an element of the natural numbers. **53.** Associative Property of Multiplication **55.** parentheses or grouping symbols, exponents, multiplication and division left to right, addition and subtraction left to right **57.** y^7 **59.** $\frac{m + n}{2}$

2.3 Using the Distributive Property

1. 3, 4, -5; $4y$, $-5y$ **3.** 1, -2, -9; xy, $-2xy$, $-9xy$
5. $4x + 8y$ **7.** $5x^2 + 5xy$ **9.** $10a$ **11.** $16xy$
13. $-3d + 10f$ **15.** $3y^2 - 8y$ **17.** $6a + 9$ **19.** $x - 8y$
21. $2x$ **23.** $8x - 5y$ **25.** $5a - 4b$ **27.** $b - 18$
29. $-5.94m + 1.6n$ **31.** $2x$ **33.** $x^2 - y^2$ **35.** $-9p - 17$
37. $-x^3 - 13x^2 + 12x$ **39.** $-66x + 72$ **41.** $4x^2 + 2x - 15$
43. $A = \frac{1}{2}hb_1 + \frac{1}{2}hb_2$ **45.** $Ft = m(v_1 - v_2)$
47. payback period = initial investment ÷ savings per period
49a. \$90 **49b.** ≈ 6.11 yr or 73.3 mo **51.** $t - 5$ **53.** $x + 6$
55. $x^2 - 5x + 4$; -2 **57.** Multiplicative Identity Property
59. Distributive Property

2.4 Solving One-Step Equations

1. false **3.** true **5.** no **7.** yes
9. subtract 7; Addition Property of Equality
11. divide by 4; Multiplication Property of Equality
13. $x = 9$ **15.** $b = 76$ **17.** $x = \frac{5}{2}$ **19.** $c = 72$
21. Multiplication Property of Equality
23. Multiplicative Identity Property **25.** $x = \frac{1}{4}$ **27.** $y = 30$
29. $b = 4.38$ **31.** $d = 1.612$ **33.** $w = \frac{1}{18}$ **35.** $y = 20.5$

37. $4n = 20$; $n = 5$ **39.** $6r = 37.80$; $r = \$6.30$
41. $a + 115 = 180$; $a = 65°$ **43.** $x = d + b$ **45.** $x = nq$
47. ROI = (savings ÷ initial investment) × 100%
49. 16.4% **51.**

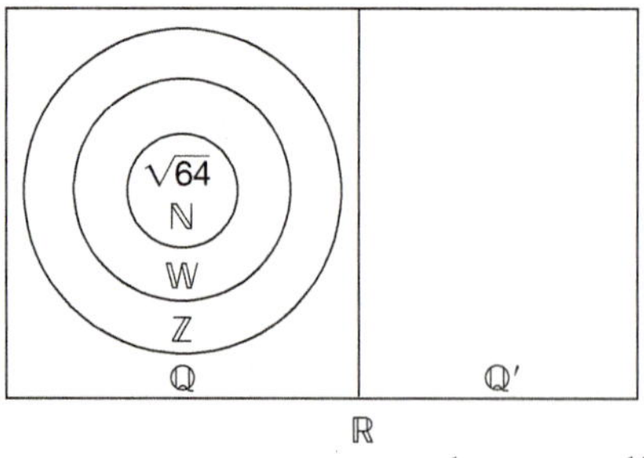

53. 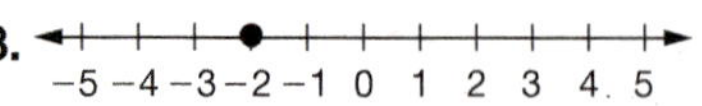

55. $\frac{1}{a^9b^2}$ **57.** $-\frac{17}{18}$
59. $xy^2 + x + 1$; 30.38

2.5 Solving Two-Step Equations

1. $x = 8$ **3.** $z = 5$ **5.** $c = -8$ **7.** $x = -355$ **9.** $p = 96$
11. $r = 3$ **13.** −41 **15.** −83 **17.** 14 **19.** 81 stamps
21. $w = 30$ **23.** $x = -1.9$ **25.** $y = -4.81$ **27.** $z = -\frac{39}{20}$
29. $x = 23$ **31.** 7 **33.** 13 nickels **35.** 47 min **37.** 24 ft
39. 7.5 sec **41.** 36 in first hour; 24 in second hour; 12 in third hour **43.** take the square root of both sides
45. No; no; neither $(-5)^2$ nor 5^2 equals −25. **47.** $t = \frac{v_f - v_i}{a}$
49. \$612.96/yr **51.** 0.61 yr ≈ 7.3 mo **53.** −19
55. −144 **57.** −13 **59.** $6xy - 12y^2$ **61.** $-4x + 10$

2.6 Simplifying Equations

1. $a = 1$ **3.** $c = -6$ **5.** $w = 9$ **7.** $y = -14$ **9.** $m = 4$
11. $p = -55$ **13.** $s = 2$ **15.** $k = 5$ **17.** $\frac{5}{33}$ **19.** $\frac{1784}{999}$
21. $n + (n + 23) = 83$; 30 and 53
23. $n + (n + 1) + (n + 2) = 84$; 27, 28, and 29
25. $g + (g + 11) = 63$; 26 girls and 37 boys
27. $3b + b + (b + 120) = 180$; $m\angle A = 36°$; $m\angle B = 12°$; $m\angle C = 132°$ **29.** $m + (m - 12) = 87$; camera: \$49.50; calculator: \$37.50
31. $(6s + 4) + s + (2s - 9) = 40$; 34, 5, and 1
33. $x + (x - 29) = 85$ **35.** 28 and 57 **37.** $x = \frac{m + n}{2}$
39. \$1700 **43.** {10, 20}
45. {..., −5, 0, 5, 10, 15, 20, 30, 40, ...} **47.** 135
49. $\frac{9}{2} = 4.5$ **51.** 5

Sequences Arithmetic Sequences

1. 2, 5, 8, 11, 14 **3.** −12, −19, −26, −33, −40 **5.** −260
7. $A_n = A_{n-1} - 5$ **9.** $A_n = 10n + 4$

2.7 Solving Multi-Step Equations

1. $x = 2$ **3.** $x = 3$ **5.** $x = 1$ **7.** no solution **9.** $x = 1$
11. ℝ; identity **13.** $y = 13$ **15.** $a = -9$ **17.** $m = -2$
19. $x = -22.18$ **21.** $x = -22$ **23.** ℝ; identity
25. $x = \frac{1040}{67}$ **27.** Commutative Property of Addition
29. Addition Property of Equality
31. Multiplication Property of Equality
33. $2n - 3 = 3n + 31$; −34
35. $4n = 3(n + 2) + 4$; 10 and 12
37. $2(3w) + 2w = 3w + 85$; 51 units by 17 units
39. r = ribbon length; $r + 2 = 2r - 6$; $r = 8$; desired length: 10 in. **41.** 26 items **43a.** $4l + 400$
43b. $27.5l$ **43c.** $27.5l = 4l + 400$; $l \approx 17$ lawns **45.** 1
47. $-\frac{80}{9} = -8\frac{8}{9}$ **49.** \$1.20 **51.** $f + 24(3) = 89$; 17 ft
53. $\frac{14}{9}$

2.8 Eliminating Fractions and Decimals

1. $x = 12$ **3.** $x = 21$ **5.** $y = \frac{50}{3}$ **7.** $b = -4$ **9.** $y = 30$
11. $b = 5$ **13.** $y = 1$ **15.** $x = 14.77$ **17a.** 70 **17b.** 70
17c. 350 **19.** Associative Property of Multiplication
21. $a = 22$ **23.** $m = 7$ **25.** $x = -\frac{155}{128} \approx -1.21$
27. $a = 4.375$ **29.** $x = 2$ **31.** 3 tens, 4 fives, and 8 ones
33. 66 dimes, 13 nickels, and 52 pennies
35. 7 quarters and 11 dimes **37.** 6 rolls of nickels, 12 rolls of dimes, and 18 rolls of quarters **39.** $x = \frac{4bd - dpq}{2}$
41. $b = 7500$ bikes **43.** $k \approx 6579$ bikes **45.** $6a - 9b$
47. $x = \frac{26}{21}$ **49.** $x = -2.8$ **51.** $p = -\frac{1}{2}$ **53.** $b = -\frac{1}{5}$

Chapter 2 Review

1. x; 3 and −4 **3.** A constant is any fixed number, while a numerical coefficient is a constant factor in a term.
5. $(90 - a) + (180 - a)$ **7.** 23, −1, 2, 47 **9.** −18 **11.** 58
13. 24 ft **15.** $-9m + 3n$ **17.** $-8a - 3b$ **19.** $5x - 8y$
21. yes **23.** $y = 8.32$ **25.** $a = 24$ **27.** $k = \frac{9}{2}$
29. $d = -\frac{2}{5}$ **31.** $v = -3$ **33.** $x = -1$
35. Multiplication Property of Equality
37. Associative Property of Addition
39. Addition Property of Equality **41.** $\frac{1}{7}r = 13$; 91 papers
43. $15d + 5 = 350$; 23 CDs **45.** $n + 4n = 155$; 31 and 124
47. $25m = 5m + 300$; 15 lawns
49. 8 nickels and 12 quarters

Chapter 3—Using Equations

3.1 Solving Literal Equations

1. $x = \frac{a}{b}$ **3.** $x = r^2$ **5.** $b = y - mx$ **7.** $t = \frac{I}{Pr}$ **9.** $t = \frac{d}{r}$
11. $s = \frac{p}{3}$ **13.** $P = \frac{14r}{3x}$ **15.** $x = \frac{y - b}{m}$
17. $x_1 = \frac{mx_2 + y_1 - y_2}{m}$ **19.** $H = \frac{S - 2\pi r^2}{2\pi r}$ **21.** $x = -\frac{b}{2}$
23. $n = \frac{bx}{x - b}$ **25.** $v = \frac{d_f - d_i}{t}$; $v = 88$ ft/sec
27. $h = \frac{S - cr}{c}$; $h = 9$ in. **29.** $x = \frac{bn}{n - b}$
31. $b = \frac{c}{c + 1}$ **33.** $T_2 = \frac{P_2V_2T_1}{P_1V_1}$; $T_2 = 171$ K

35.

work van	1500	346 gal
SUV	1154	
compact car	500	250 gal
hybrid	250	

37. An expression represents a single quantity, while an equation states that two different expressions are equal. **39.** $I = R - E$ **41.** $y = 12$ **43.** $s = 42$ **45.** $c^{-10}d^6f^{-3}$

3.2 Ratios and Proportions

1. 4 to 5 **3.** 1 to 3 **5.** 17 points/game **7.** $9.25/hr **9.** $18.25/yr **11.** $80.\overline{6}$ ft/sec **13.** $x = 56$ **15.** $z = 1000$ **17.** $b = \frac{7}{8} = 0.875$ **19.** $w = 2$ **21.** $d = 9$ **23.** $n = 9$ **25.** $q = \frac{1}{2} = 0.5$ **27.** $1.33/roll **29.** $18/shirt **31.** $190 **33.** 47 oz **35.** 18 volleyball players **37a.** $\frac{b}{s - b}$ **37b.** $\frac{s - b}{s}$ **39.** 0.347 cm^3 **41a.** 346 gal **41b.** 250 gal **43.** $\frac{4}{9}$ **45.** $\frac{25}{49}$ **47.** $a = 135$ **49.** $h = \frac{2w + 440}{11}$ **51.** $-1, 1, 3$

Sequences Fibonacci and Golden Ratio Sequences

1. 7, 12, 19, 31, 50 **3.** $-1, 2, 1, 3, 4$ **5.** $1, 2, \frac{3}{2}, \frac{5}{3}, \frac{8}{5}$ **7.** 0.618 (not 1.618)

3.3 Similar Figures and Scale Models

1. $\angle R \cong \angle Z$, $\angle S \cong \angle Q$, and $\angle B \cong \angle M$; $\frac{RS}{ZQ} = \frac{SB}{QM} = \frac{BR}{MZ}$ **3.** $x = 17.5$ units; $y = 20$ units **5.** $\frac{p_{DEF}}{21} = \frac{15}{6}$; $p_{DEF} = 52.5$ units **7.** 70 mi **9.** 1 in. : 20 mi **11.** 140 ft × 230 ft **13.** 12 in. × 8 in. **15.** 2.4 in. **17.** $MN = 12$ units; $TS = 6$ units **19.** $\frac{A_{MNO}}{A_{RTS}} = \frac{9}{4}$; $A_{RTS} = 24$ units^2 **21.** $h = 20$ units **23.** $\frac{V_S}{V_L} = \frac{8}{125}$; $V_L = 3000$ units^3 **25.** 21.6 units and 26.4 units; 60 units **27.** 4 : 25; 8 : 125 **29.** 147 ft^2 **31.** 160 ft **33.** 1029π units^3 **35.** 78 in. by 72 in. **37.** 30 in. **39.** 4.0 units **41a.** $\frac{3}{130}$ **41b.** $\frac{1}{60}$ **41c.** $\frac{1}{99}$ **43.** $\mathbb{Q}$ **45.** false **47.** 34; -5 **49.** $-\frac{7}{12}$ **51.** $\frac{47}{60}$

3.4 The Percent Equation

1. 0.69, $\frac{69}{100}$ **3.** 0.75, $\frac{3}{4}$ **5.** 25%, $\frac{1}{4}$ **7.** 350%, $\frac{7}{2}$ **9.** 7.5, 750% **11.** 126 **13.** 82.76 **15.** $\frac{3}{8}$ **17.** 33,075 **19.** 500 **21.** 40%; 60% **23.** 236 **25.** $8.57/hr **27.** 85% **29.** $20.6 million **31.** 33–46 lb **33.** 167.7 lb **35.** 16% protein, 28% fat, and 56% carbohydrates; healthy diet **37.** 662,080 vehicles **39.** 9,895,000 people **41.** 97.5 mi/gal **43.** a) 11.1 mi/gal; b) 43.0 mi/gal **45.** 5 **47.** 0.39 **49.** $x = \frac{28}{3} = 9.\overline{3}$ **51.** 99 in.^{2} **53.** $x = \frac{3c + 2b}{ac}$

3.5 Percent Change and Error

1. 25% **3.** 650% **5.** 980 **7.** 270 **9.** 1230 **11.** 15; 10% **13.** 22.2% **15.** 22.1% increase; 83 **17.** 115; 17.4% increase **19.** subtract 39; 31.2% decrease **21.** add 145.2; 419.2 **23.** 1400; subtract 1050 **25.** 1250; 1225 **27.** 3.3% **29.** 0.13 g/mL; 13% **31.** 0.7% **33a.** 24 **33b.** 16 **35.** -30 (if the number represents a quantity that can be negative; otherwise, it is not possible) **37a.** 25% decrease **37b.** 25% decrease **39.** 20% **41.** Answers are in terms of g, the current price per gallon. **41a.** $104.17g$ **41b.** $28.85g$ **41c.** $20.83g$ **41d.** $12.63g$ **43.** $y = 7$ **45.** 150% **47.** 150 **49.** $7\frac{1}{3}$" × 11" **51.** $x = 7200$

3.6 Money

1. $4.00 **3.** $13.00 **5.** $4123 **7.** 7% **9.** $12,579 **11.** $15; $90 **13.** $120; $138 **15.** $19.95; 24% **17.** $6.75; 15% **19.** $749; $501.83 **21.** $300 **23.** $5000 **25.** $10.07; $66.00 **27.** $851 **29.** 16% **31.** $67.50; $33.75 **33.** $131.25; 81.25% **35.** $2500 at 5% and $7500 at 6% **37.** $1800 for 9 mo and $2500 for 18 mo **39.** $10,080 for 2.5 yr and $7000 for 3 yr **41.** $55\frac{5}{9}$% **43.** -6 **45.** $\frac{197}{48}$ **47.** $2x^2 - 3xy + 2y^2$ **49.** 486 in.^{2} **51.** 6.25%

3.7 Motion

1. 54 **3.** $\frac{d}{19}$ **5.** 20 **7.** r **9.** $7ab$ **11.** 125 mi **13.** 400 mi/hr **15.** $\frac{3}{4}$ hr **17.** 1 hr **19.** 6 hr **21.** up: 10 ft/sec; down: 80 ft/sec **23.** 144 mi **25.** 1 hr **27.** 8 mi by canoe; 30 mi by bicycle **29.** $5000; ≈ 96 mo or 8 yr; 12.5% **31.** $13,500; ≈ 259 mo or 21.6 yr; 4.6% **33.** $13,000; ≈ 50 mo or 4.2 yr; 24.0% **35.** $w = 4$ **37.** $y = 9$ **39.** $t = \frac{v_f - v_i}{a}$; $t = 15$ sec **41.** $25,500 **43.** 52%

3.8 Mixtures

1. 14 gal **3.** 7.2 oz **5.** 80 L **7.** $\$5.31n$ **9.** $0.5x$ **11.** 18 gal; 7.6 gal; 42% salt **13.** 42 gal; 33% acid **15.** 4 lb of butterscotch and 16 lb of cinnamon balls **17.** 21 lb of Chinese and 14 lb of Indian **19.** 2 gal **21.** 2.5 L **23.** 11% **25.** 3 cans of grapefruit and 4 cans of pineapple **27.** 2.4 oz **29.** Car A: 16%, 34%, 20%, 16%, 11%; Car B: 23%, 16%, 23%, 3%, 9%, 13% **31.** 24 in. **33.** 8 mi/hr **35.** 12 games **37.** The school exceeded its goal by 5 students. **39.** $0

Chapter 3 Review

1. $m = dv$ **3.** $r = \frac{S - ch}{c}$ **5.** $A = \frac{SR}{3 + S}$
7. 275 words/page **9.** 60 mi/hr **11.** $y = \frac{125}{3} \approx 41.6$
13. 10 g of hydrogen and 80 g of oxygen **15.** 3 in.
17. $x = 13.6$ cm **19.** 134.4 cm^2 **21.** 50%, $\frac{1}{2}$ **23.** 150
25. 4% increase; 936 **27.** 175; 140
29. 120 people; 280 people **31.** 47.7 mi/hr **33.** 92.2%
35. 3.6% **37.** 38.5% **39.** $15,000
41. $5000 at 4% and $10,000 at 6% **43.** 5760 yd ≈ 3.3 mi
45. 30 lb of Colombian and 20 lb of Brazilian **47.** $6.\overline{6}$ L
49. 4 L

Chapter 4—Solving Inequalities

4.1 Inequalities

1. d **3.** c **5.** b **7.** a **9.** $x \leq -4$ **11.** $z < 21$
13a. yes **13b.** no **13c.** no **15a.** yes **15b.** yes
15c. no **17.** $x > 3; x \not\leq 3$ **19.** $x \geq 18; x \not< 18$
21. 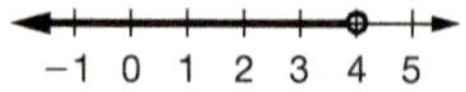 −1 0 1 2 3 4 5
23. 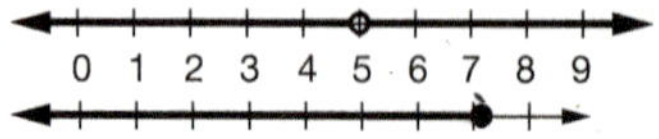 0 1 2 3 4 5 6 7 8 9
25. 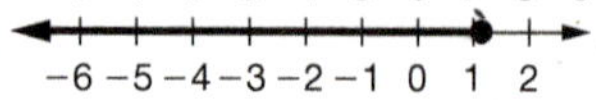 −6 −5 −4 −3 −2 −1 0 1 2
27. 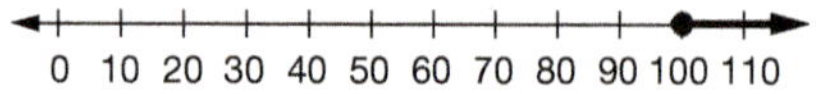 0 10 20 30 40 50 60 70 80 90 100 110
29. 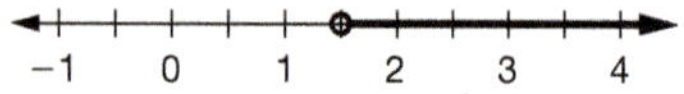 −1 0 1 2 3 4
31. 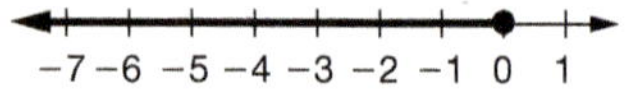 −7 −6 −5 −4 −3 −2 −1 0 1
33. v = voting age; $v \geq 18$ **35.** w = wattage; $w \leq 60$
37. x = daily vitamin C intake in milligrams; $x \geq 75$
39. 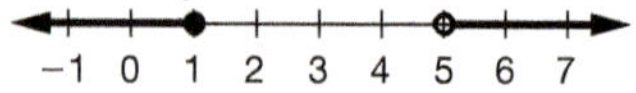 −1 0 1 2 3 4 5 6 7
41. 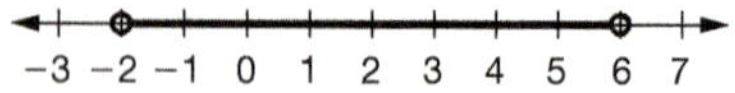 −3 −2 −1 0 1 2 3 4 5 6 7
43. = and ≠; < and ≥; > and ≤ **45.** $\frac{15}{28}$
47. Commutative Property of Addition
49. Addition Property of Equality
51. Additive Identity Property **53.** $x = 10$

4.2 Properties of Inequality

1. $x > -7$ 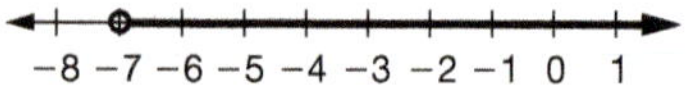 −8 −7 −6 −5 −4 −3 −2 −1 0 1
3. $x < -3$ −7 −6 −5 −4 −3 −2 −1 0 1 2
5. $x \neq -3$ −6 −5 −4 −3 −2 −1 0 1 2 3 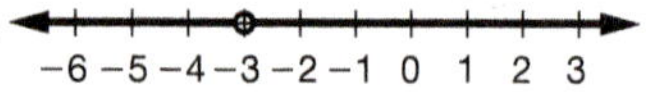
7. $x \leq 2$ 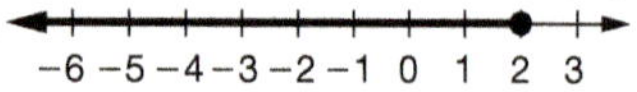 −6 −5 −4 −3 −2 −1 0 1 2 3
9. $x \leq -11$ **11.** $z \neq 31$ **13.** $b > -5$ **15.** $x \geq 24$
17. $r < -6$ **19.** $x \leq -\frac{4}{5}$ **21.** $m \leq -2$ **23.** $k \geq \frac{3}{2}$
25. any number less than 57 **27.** at most 5 teenagers
29. a maximum of 45.59 sec **31.** 3 cans **33.** $x < 0$
35. $x > -12$ **37.** ocean: at least 4%; Dead Sea: at least 32%
39. interest: $600; amount to repay: $5600; 24 monthly payments of $233.33 **41.** rational **43.** $x = \frac{4}{3}$ **45.** $z = 5$
47. $a = 3$ **49.** $x = c - a - b$

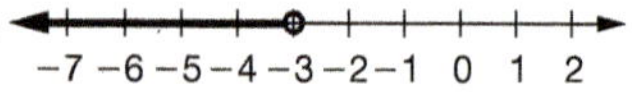

4.3 Solving Inequalities

1. $x \leq -2$ 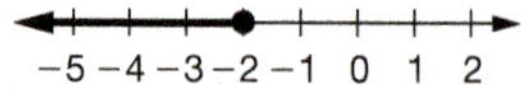 −5 −4 −3 −2 −1 0 1 2
3. $z \geq 4$ 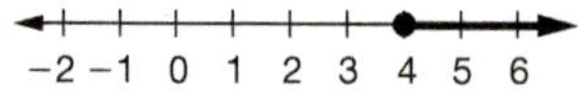 −2 −1 0 1 2 3 4 5 6
5. $y > -7$ 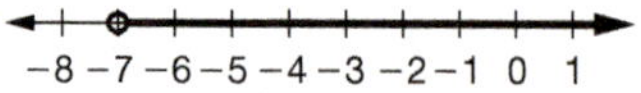 −8 −7 −6 −5 −4 −3 −2 −1 0 1 **7.** $z \neq 30$
9. $c < -130$ **11.** $m > -18$ **13.** $r \leq 3$ **15.** $y \leq 4$
17. $z \geq -3$ **19.** $y \geq -\frac{26}{33}$ **21.** $c \leq \frac{38}{5}$ **23.** $d < -\frac{1}{2}$
25. 14 **27.** above 15°C **29.** at least $3000
31. Jed: at least 24 mi/hr; brother: at least 9 mi/hr
33. $x \neq -7$ **35.** $x \geq 2$ **37.** 9.4% **39.** false **41.** −53
43. Multiplication Property of Equality
45. Addition Property of Equality **47.** $x = \frac{7}{4}$

Sequences Arithmetic Sums

1. 1, 4, 9, 16, 25 **3.** −10, −10, 0, 20, 50 **5.** 65, 335
7. 10,000

4.4 Conjunctions

1. 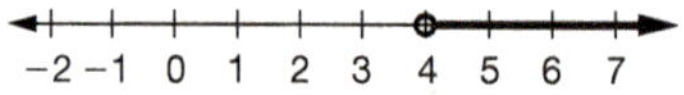 −2 −1 0 1 2 3 4 5 6 7
3. 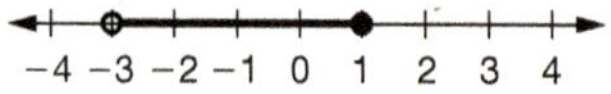 −4 −3 −2 −1 0 1 2 3 4
5. −3 −2 −1 0 1 2 3 4 5 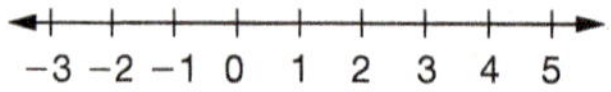
7. $-3 < x \leq 1$ **9.** $x < 4$ **11.** $-2 \leq x < 6$
13. $-1 \leq x < 6$ **15.** $-2 < x < 3$ **17.** $x < -12$ **19.** ∅
21. $x \leq 5$, but $x \neq 4$ **23.** $-6 < x < 3$ **25.** $4 < x < 7$
27. $3 < x \leq 5$ **29.** 3300 ft $< f <$ 13,000 ft
31. 3 ft $\leq s \leq$ 306 ft **33.** $30° < a < 45°$
35. 16 in. $< d <$ 24 in. **37.** $6 < x < 15$ **39.** $x = -4$
41. more than 35 of faucet B **43.** 3.9% **45.** $\frac{65}{64}$ **47.** $6x^5y$
49. $x = 31.5$ **51.** $I = \frac{L}{\pi r}$ **53.** $\frac{17}{6}$

4.5 Disjunctions

1. and **3.** union **5.** disjunction
7. 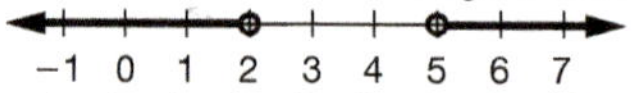 −1 0 1 2 3 4 5 6 7
9. 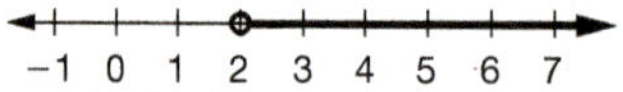 −1 0 1 2 3 4 5 6 7
11. 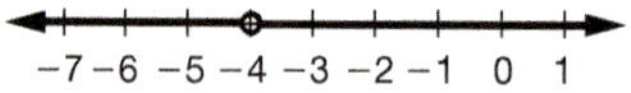 −7 −6 −5 −4 −3 −2 −1 0 1
13. ℝ **15.** $y \leq 0 \vee y > 5$ **17.** $b > 3$ **19.** $x < 6$
21. $x < \frac{5}{2}$ **23.** ∅ **25.** ℝ **27.** $z \leq -\frac{5}{2} \vee z > 3$
29. $3 \cdot 10^{16}$ Hz $< f < 3 \cdot 10^{19}$ Hz
31. $w < 1 \cdot 10^{-8}$ m or $w > 3.80 \cdot 10^{-7}$ m
33. $x < -3$ or $x > 1$ **35.** $-5 < x < 0$ **37.** $x \neq \frac{7}{3}$
39. $x \geq \frac{5}{3}$ **41.** $243.76 **43.** More than $5500 is paid for the loan ($5500 + interest paid) while less than $5500 is invested in the savings ($5500 − interest earned).
45. $\frac{y^6}{27x^3z^9}$ **47.** $-8x + 33$ **49.** 784 **51.** −16
53. 21; 6.5

4.6 Absolute Value Equations

1. $x = \pm 3$ **3.** $x = \pm 5$ **5.** $x = \pm\frac{8}{3}$ **7.** $\varnothing$ **9.** $y = \pm 43$
11. $\varnothing$ **13.** $x = -1, 2$ **15.** $x = -7, 5$ **17.** $y = -14, 19$
19. $y = -\frac{36}{5}, \frac{12}{5}$ **21.** $x = -\frac{7}{6}, 2$ **23.** $z = -\frac{8}{7}, 4$
25. $x = -\frac{41}{7}, 5$ **27.** $|x - 2| = 4; x = -2, 6$
29. $|x + 4| = 8; x = -12, 4$ **31.** $|x - 5| = 15; x = -10, 20$
33. $|x - 10| = 3$ **35.** $|x + 6| = 9$
37. 6%; 5.09%; 6% simple
39. 5.75%; 5.65%; 5.6% monthly **41.** 95% **43.** 22%
45. \$375 **47.** 16 in. **49.** about 1.5 hr

4.7 Absolute Value Inequalities

1. $x < -3$ or $x > 3$
3. $-5 \le x \le 5$
5. $x < -8$ or $x > 8$
7. $-7 \le x \le 13$
9. $x \le -10$ or $x \ge 4$ **11.** $x < -6$ or $x > -\frac{5}{2}$
13. $x < -\frac{99}{5}$ or $x > -\frac{9}{5}$ **15.** $x \le -\frac{11}{2}$ or $x \ge 9$
17. $-\frac{8}{3} \le x \le \frac{4}{3}$ **19.** $x < -17$ or $x > -1$ **21.** $|x - 5| \le 2$
23. $|l - 3| \le 0.02$; $2.98 \text{ cm} \le l \le 3.02 \text{ cm}$
25. $|t - 90| \le 50$ **27.** $|t - 30| \le 3$; $27 \text{ min} \le t \le 33 \text{ min}$
29. $\varnothing$ **31.** $x = -7$ **33.** $x \ne \frac{1}{3}$ **35.** $x = \frac{2c - b}{4}$ **37.** 200
39. 1 cm : 25 km **41.** 33%
43. \$7500 in the stock fund and \$2500 in bonds

Chapter 4 Review

1. $>$ **3.** $y < 3$ **5.**
7. $x < 11$ **9.** $z > -\frac{21}{5} = -4.2$ **11.** $b \le 4$ **13.** $x < \frac{17}{8}$
15. any number greater than 18
17. any number greater than $\frac{7}{6}$; 2 **19.** 94 **21.** disjunction
23. $-2 < x \le 4$ **25.** $\mathbb{R}$ **27.** $x \ge 10$ **29.** $3 < x < 4$
31. $x < 5$ **33.** $-\frac{5}{3} < x < 1$ **35.** $\mathbb{R}$ **37.** $\varnothing$ **39.** $z = -\frac{8}{7}, 4$
41. $x < -10$ or $x > 4$ **43.** $-7 < x < 3$ **45.** $|x - 4| = 3$
47. $|p - 73| > 14$; $p < \$59$ or $p > \$87$

Chapter 5—Relations and Functions

5.1 Points in the Coordinate Plane

1. Springfield **3.** A-6 **5.** $(-3, -1)$ **7.** $(3, 5)$
9. $(-4, 1)$ **11.** $(-5, -4)$ **13.** $(-1, 2)$

15–23. odd

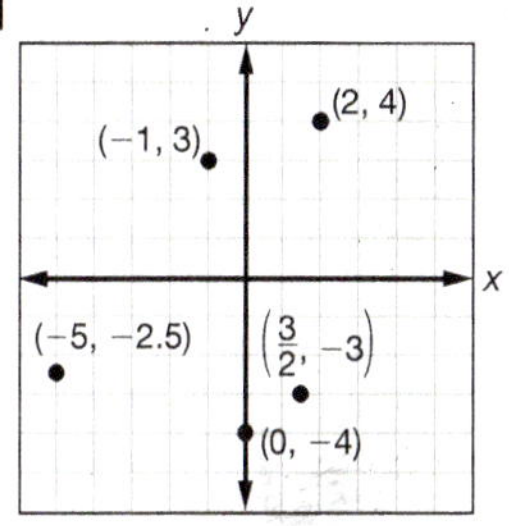

25. $(0, 0)$

27–29. odd

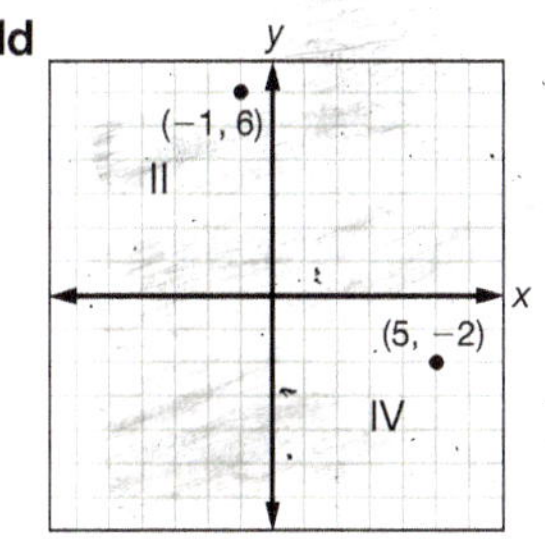

31. y-axis
33. x-axis
35. I
37. IV
39. $B(7, 0)$, $C(7, 4)$, $D(0, 4)$

41.

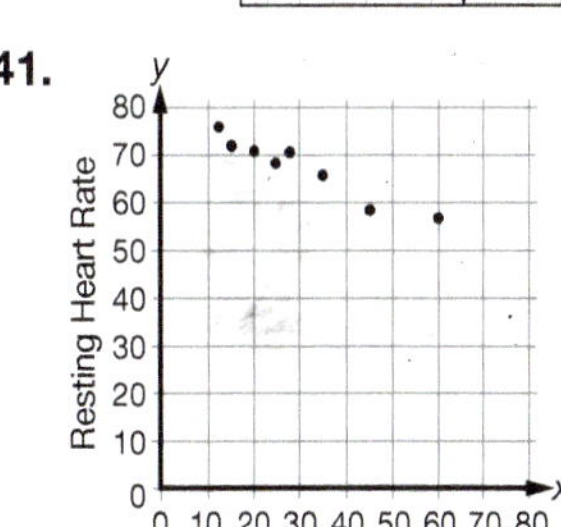

The resting heart rate decreases as exercise increases.

43. Cyprus
45. Dead Sea
47. 32.8° N, 35.6° E
49. 26°13.4′ S, 157°10.25′ E
51. 41°45′4″ N, 88°2′45″ W
53. 308%
55. 5.5%
57. \$7000 at 6% and \$3000 at 7.5%
59. $x \not> 4$
61. $a + b > c$

5.2 Relations and Functions

1. yes **3.** no **5.** no

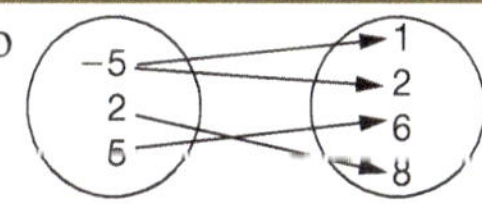

7. yes

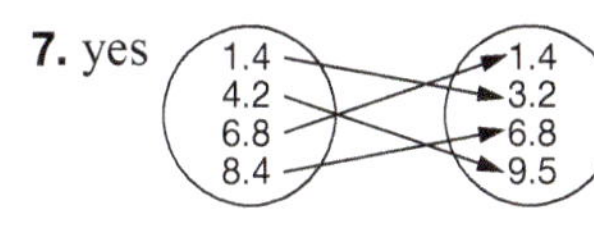

9. $D = \{-9, 4, 6\}$; $R = \{2, 8, 10, 14\}$; no
11. $D = \{1.3, 1.4, 1.5, 1.6\}$; $R = \{3.7\}$; yes
13. a **15.** c **17.** b

19. yes

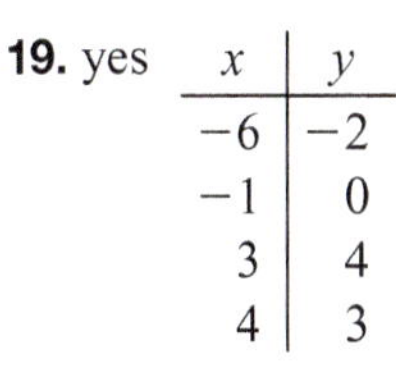

x	y
−6	−2
−1	0
3	4
4	3

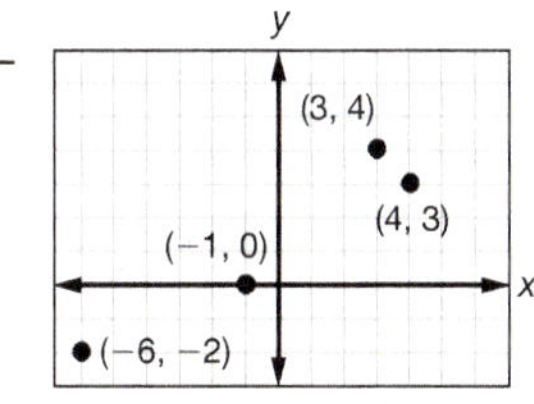

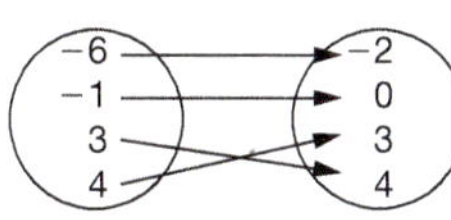

21. no
$\{(0, 400), (1, 100), (2, 200), (2, 300)\}$

x	y
0	400
1	100
2	200
2	300

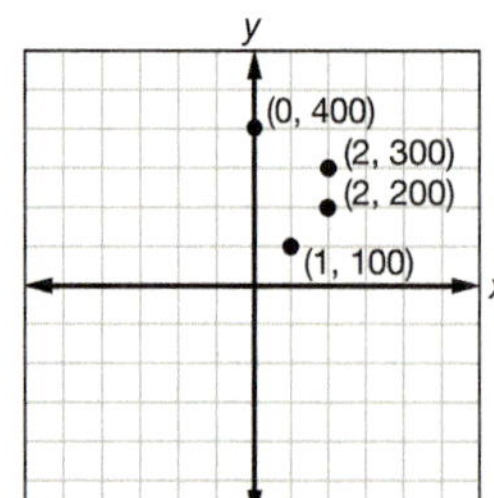

23. -4 **25.** 28 **27.** time; distance **29.** age; height **35.** o **37.** m **39.** 4, 5, 5, 6, -1, -3, -6, 0, -1 **41.** south **43.** northeast **45.** 1225 **47.** Addition Property of Equality **49.** $x = 4$ **51.** $x = 0.7$ **53.** $\frac{43}{90}$

5.3 Graphs of Relations and Functions

1. yes **3.** no **5.** yes
7. $D = \{-4, -3, -1, 0\}$; $R = \{-3, -2, -1, 0\}$; no
9. $D = \{-4, -1, 0, 1, 2, 3, 4\}$; $R = \{1, 2\}$; yes
11. $D = \{x \mid -2 \leq x \leq 2\}$; $R = \{y \mid -4 \leq y \leq 4\}$; yes
13. $\{(-1, -3), (0, 0), (1, 3)\}$ **15.** $\{(-2, -1), (0, 3), \left(\frac{1}{2}, 4\right), (1, 5)\}$

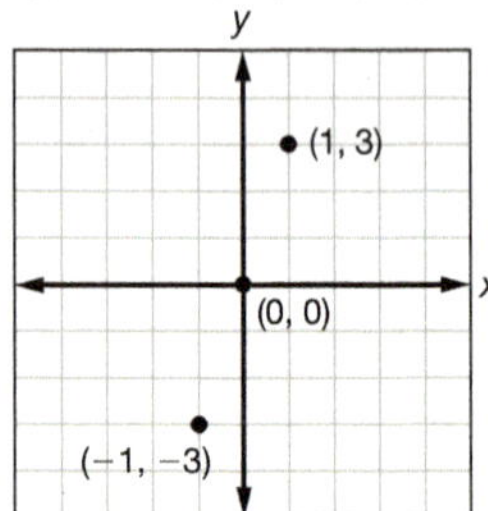

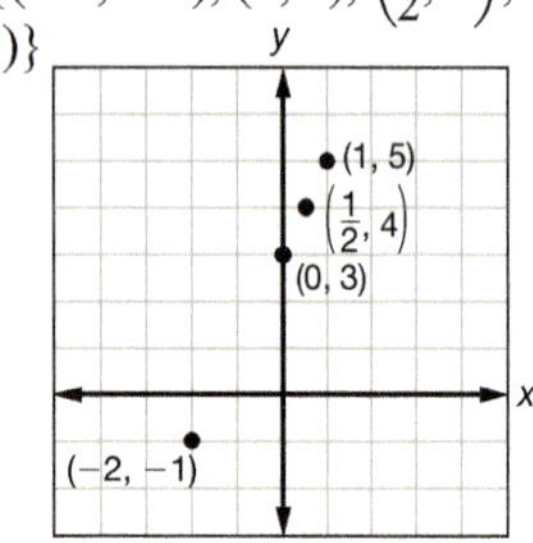

17. Table values will vary.

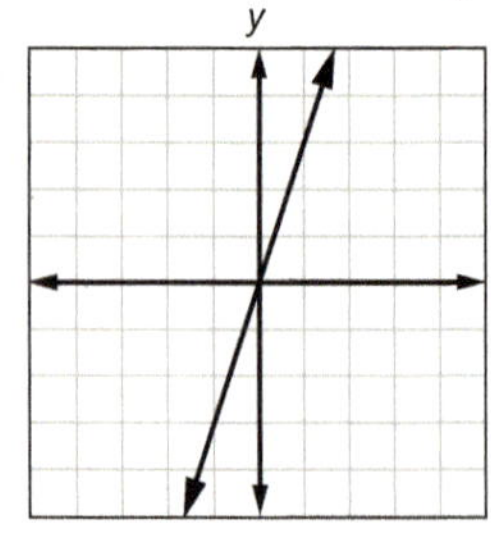

$D = \mathbb{R}$; $R = \mathbb{R}$

19. Table values will vary.

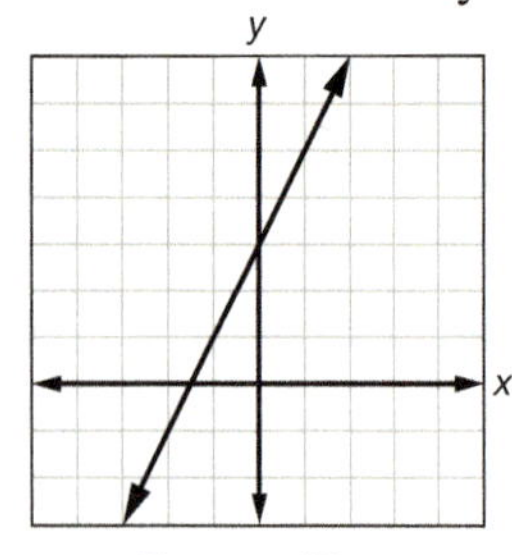

$D = \mathbb{R}$; $R = \mathbb{R}$

21. no

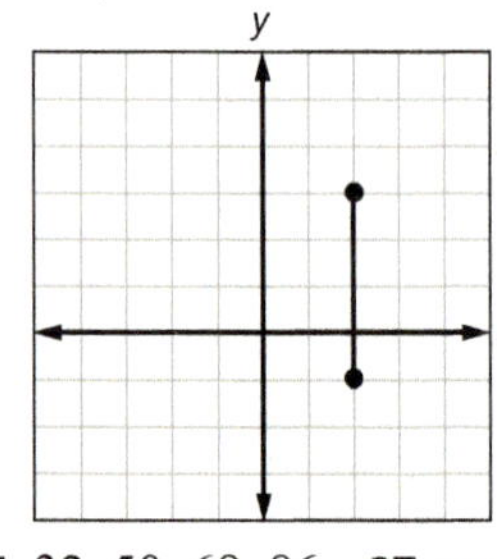

23. yes

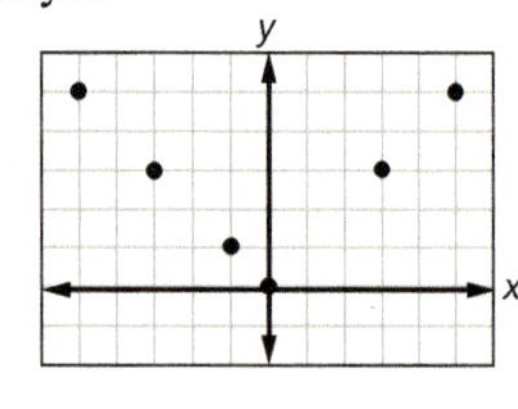

25. 32, 50, 68, 86 **27.** yes **29.** $D = \{s \mid s \geq 0\}$ **31.** 19 m

33.

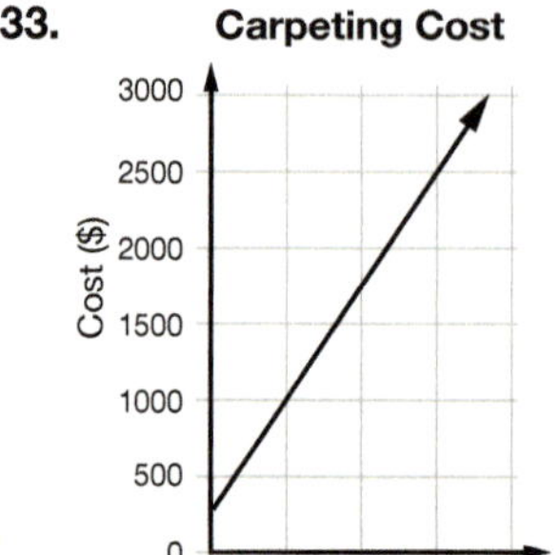

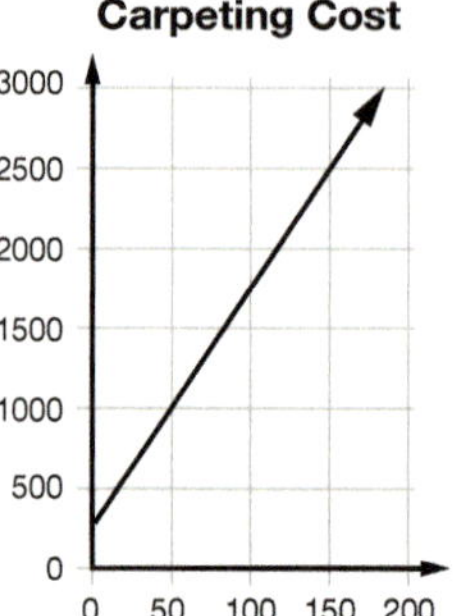

35. c
37. no

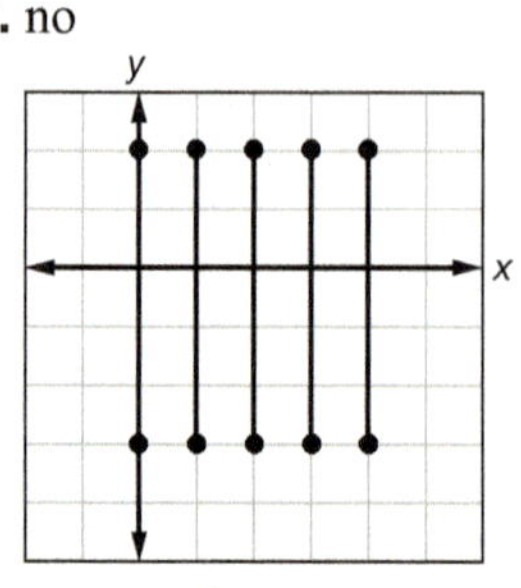

39.

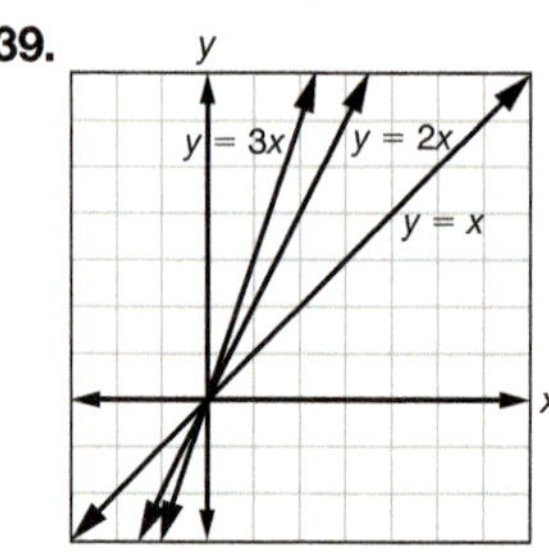

41. The larger the number, the steeper the graph.
43. 130.3 nautical miles
45. $x = \frac{114}{7}$
47. $x = 1.5$
49. $x = \frac{d - c}{3ab}$

51. −4 −3 −2 −1 0 1 2 3 4 5 6 **53.** $3 < x < 9$

5.4 Using Graphs

5a. II **5b.** III **5c.** I **7.**

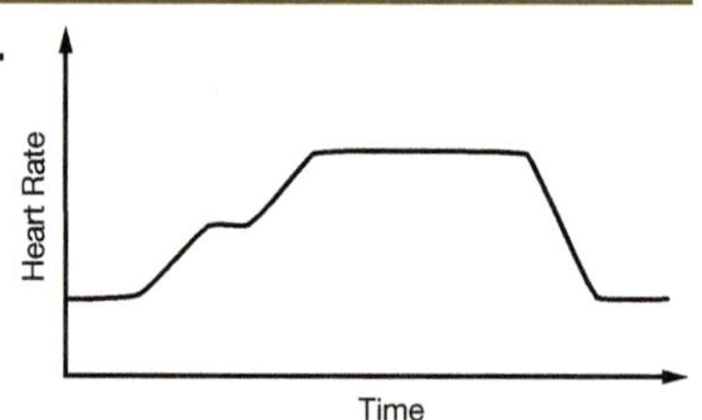

9.

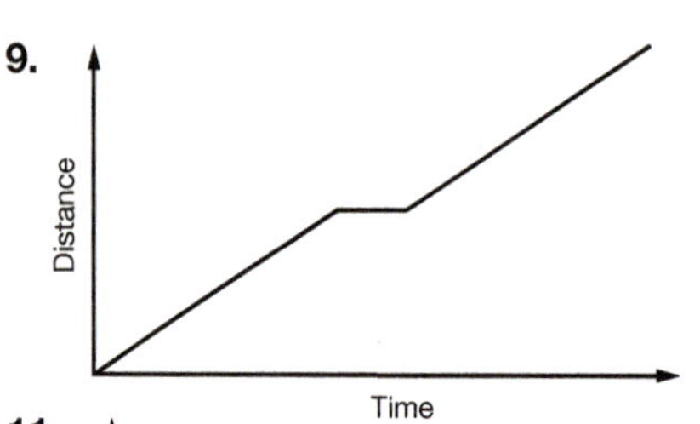

11.

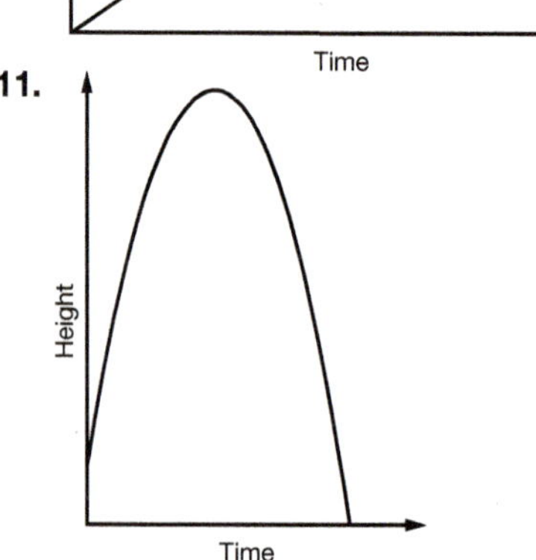

13.

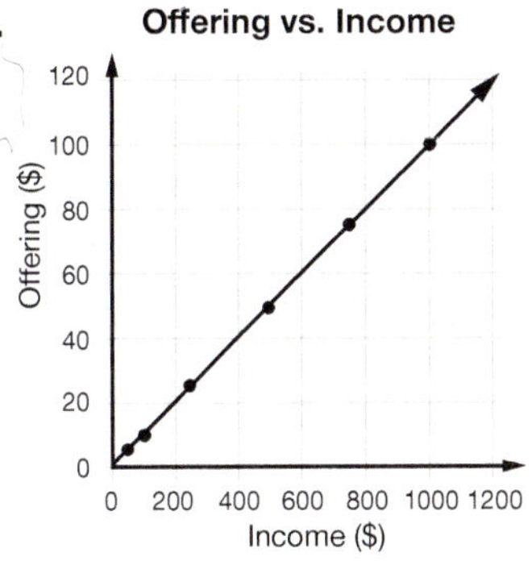

15. $30

17.

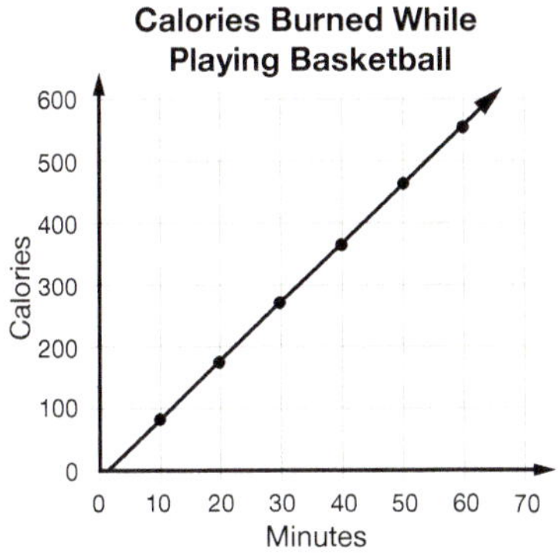

19. 16 min
21. elevation
23. As the elevation increases, the boiling point of water decreases quickly at first and then not so quickly.

25.

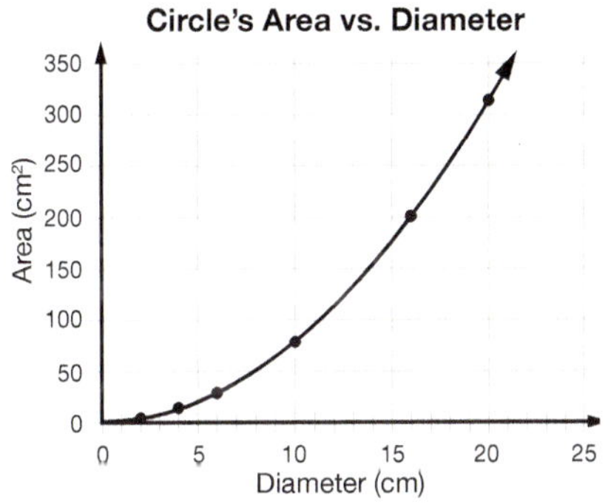

27. $\approx 20 \text{ cm}^2$

29.

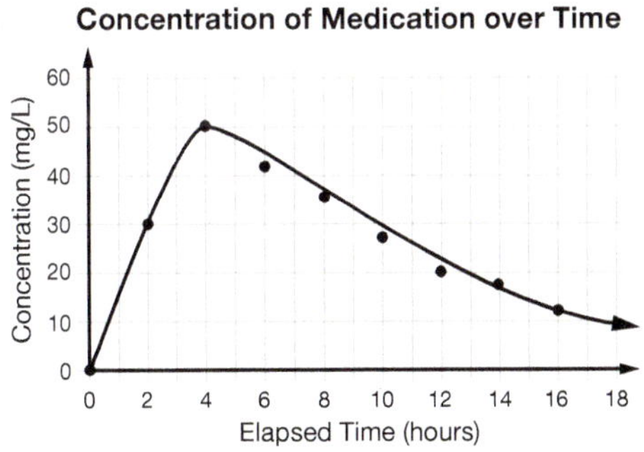

31. ≈ 31 mg/L **33.** The student stands 3 ft from the back wall for 3 sec and then moves back toward the wall for 3 sec until he is 1 ft from the wall. For the next 4 sec he moves forward until he is 9 ft from the wall, where he stands for the remaining 3 sec.
35. The ball is dropped from a height of 10 m and takes almost 1.3 sec to hit the ground. It rebounds to half its original height, and each successive bounce requires shorter time intervals until it comes to rest after its seventh bounce.

37.

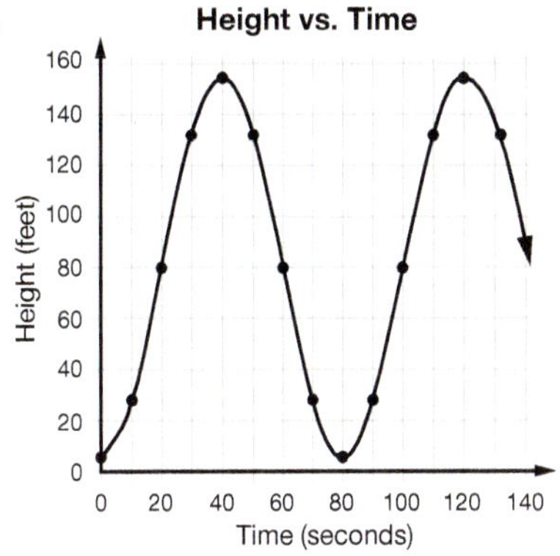

39. ≈ 50 ft **41.** a minute or nautical mile
43. No; other lines of latitude have a smaller circumference.
45. 5591 mi **47.** true **49.** $20 - 3x$ **51.** $x = -11, 6$
53. $-2 < x < 10$

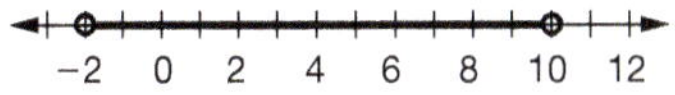

55. $x < -20$ or $x > 18.5$

5.5 Function Rules

1. $f = \{(-2, -6), (0, -4), (3, -1)\}$

3.

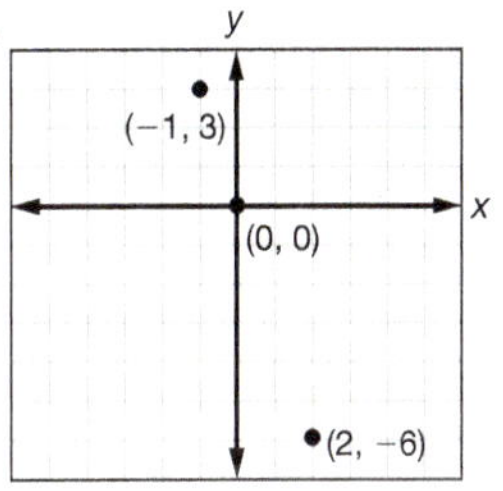

5.

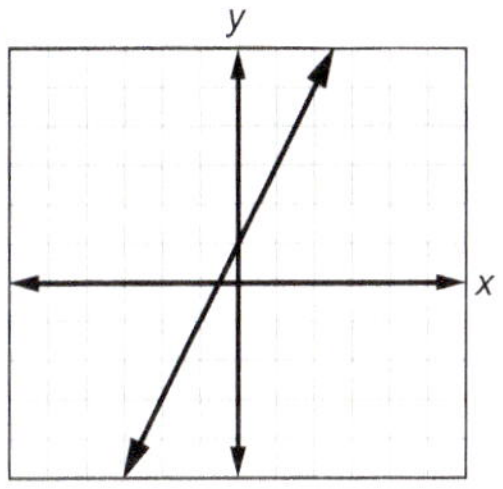

7. $f(x) = 4x$ **9.** $f(x) = x - 2$ **11.** $f(x) = 3x$
13. $f(x) = -x + 2$ **15.** $f(x) = x + 2$ **17.** $f(x) = \frac{1}{3}x$
19. $f(x) = 2x + 1$ **21.** $r(x) = 5x + 2$ **23.** $f(x) = 2x - 1$
25. $f(d) = 0.25d$ **27.** $N(a) = a + 2$ **29.** $P(c) = 0.25c$
31. $t(d) = \frac{d}{50}$ **33.** $A(r) = \frac{1}{2}\pi r^2$
35. $R(x) = 1.05x$;
$C(x) = 1200 + 0.48x$;
≈ 2100 balls per month

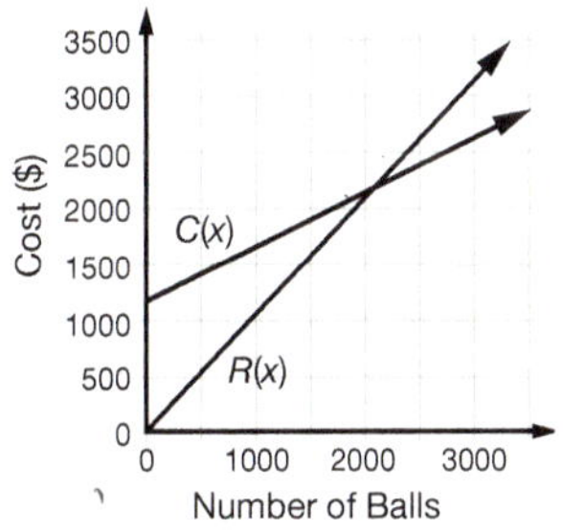

37. 425 min; $485,000; no

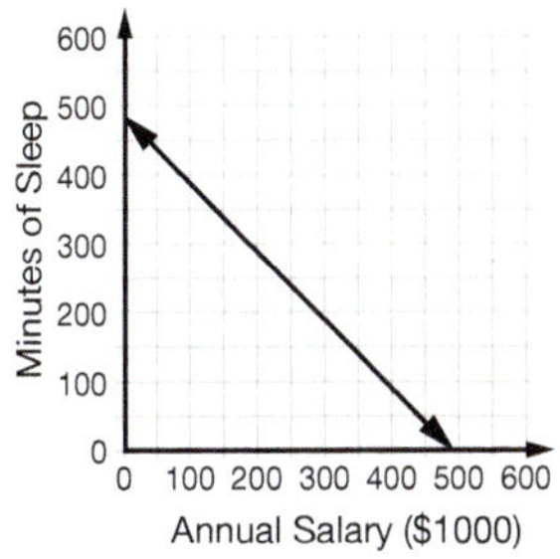

39. 2072 mi **41.** $\frac{9}{8}$ or $1\frac{1}{8}$ **43.** 42,000 **45.** $\frac{y^6}{x^2}$ **47.** $2x + 5$
49. $-\frac{16}{3}$

5.6 Direct and Inverse Variations

1. direct **3.** inverse **5.** inverse **7.** neither

9. $k = 2; y = 2x$ **11.** $k = 36; m = \frac{36}{a}$ **13.** $k = 5; r = \frac{5}{s}$

15. direct; $y = 6x$

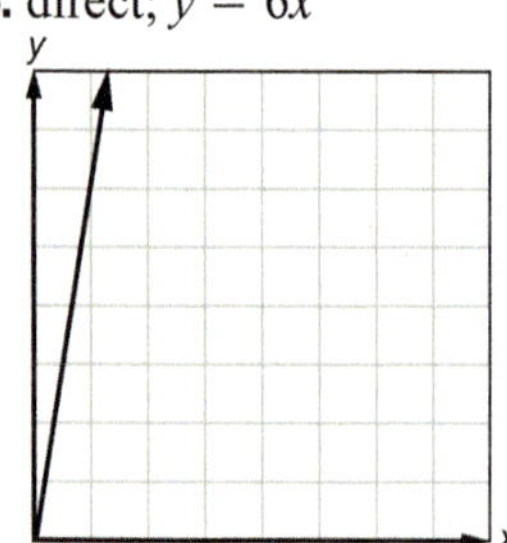

17. inverse; $y = \frac{18}{x}$

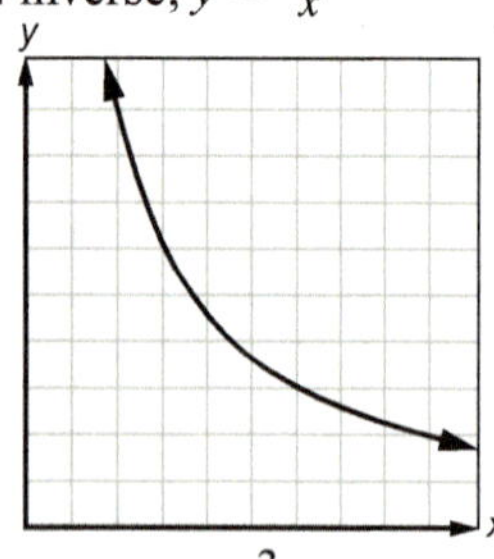

19. inverse; $y = \frac{10}{x}$

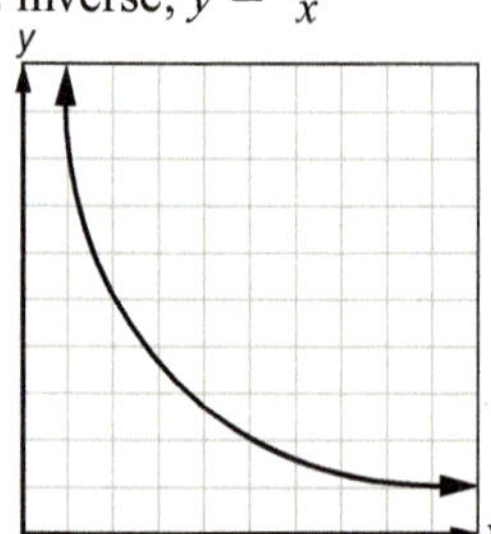

21. direct; $y = \frac{3}{2}x$

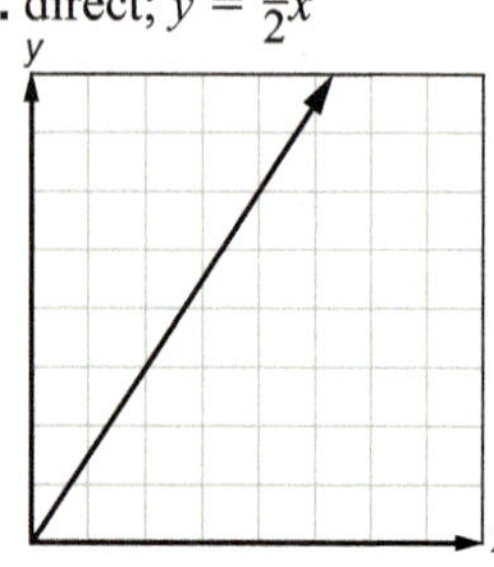

23. neither **25.** 40 **27.** 0.1 **29.** 6 **31.** direct; $k = \pi$
33. inverse; $k = 4$ **35.** 130 mi **37.** 2.5 units
39. 52,000 people **41.** neither

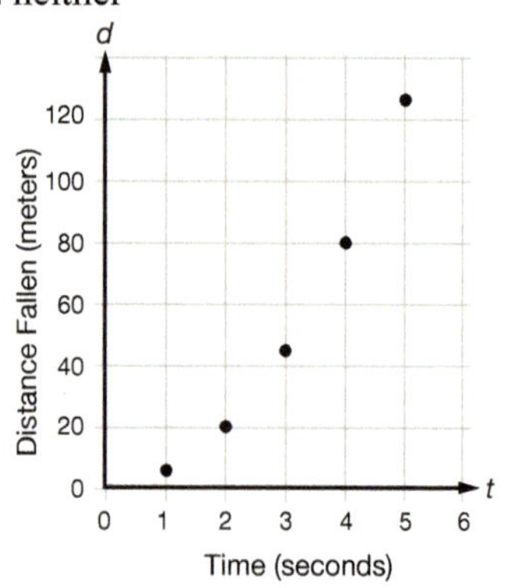

43. $d = 5t^2$ **43a.** 281.25 m **43b.** 10 sec **45.** 113 in.^{3}
47. $x = 82.69$ **49.** $x = 25$ **51.** $a = -14$ **53.** 16.7%
55. just under 6.6 hr (≈ 6 hr 34 min)

Sequences Geometric Sequences

1. 6, 18, 54, 162, 486 **3.** $\frac{1}{4}, -\frac{1}{2}, 1, -2, 4$ **5.** 19,531,250
7. $A_n = -3A_{n-1}$ **9.** $A_n = \frac{1}{25}(-5)^{n-1}$

5.7 Graphing Absolute Value Functions

1. Table values will vary.

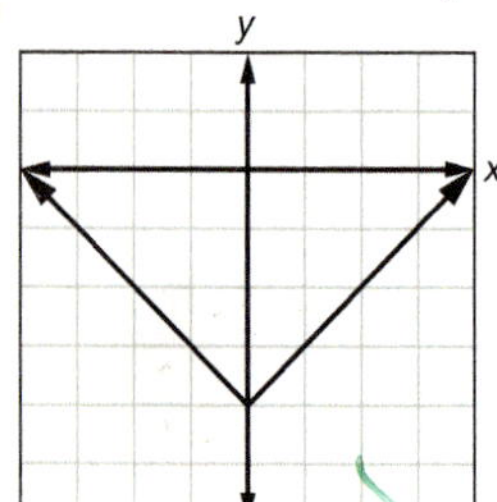

3. Table values will vary.

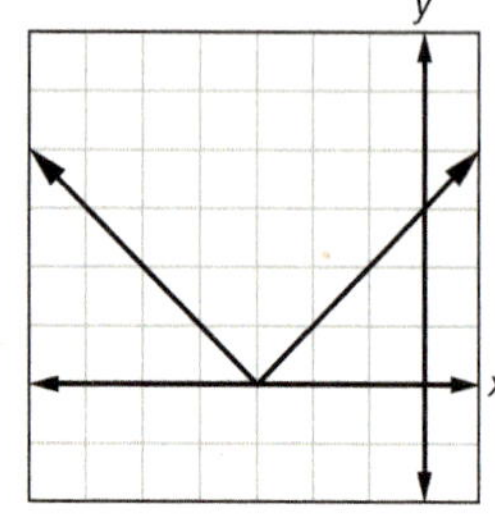

5. Table values will vary.

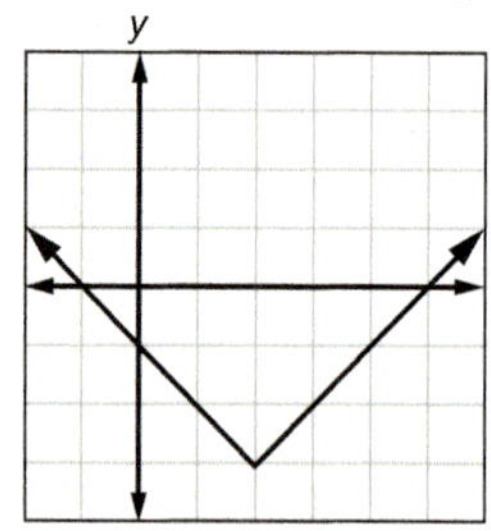

7. 4 units up **9.** 10 units right
11. $\frac{1}{2}$ unit up
13. $f(x) = |x| - 2.5$
15. $f(x) = |x + 4|$
17. 3 units right, 4 units down
19. 7 units left, 1 unit up
21. $(-5, 3)$ **23.** (a, b)
25. $f(x) = |x + 2| + 6$
27. $f(x) = |x - 6.23| + 5.9$

29.

31.

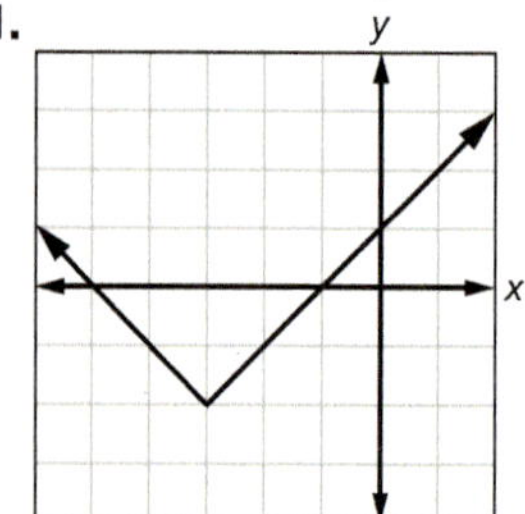

33.

35. Table values will vary.

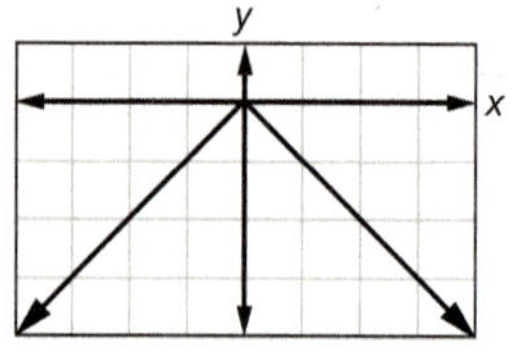

37. Table values will vary.

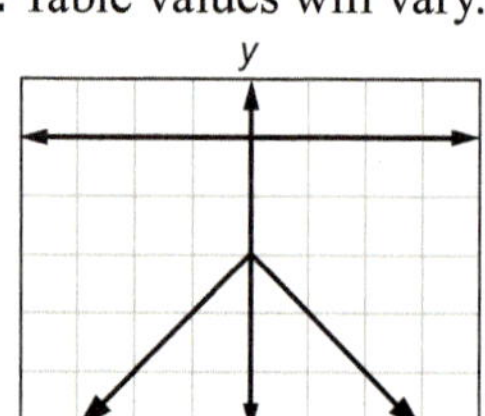

39. Table values will vary.

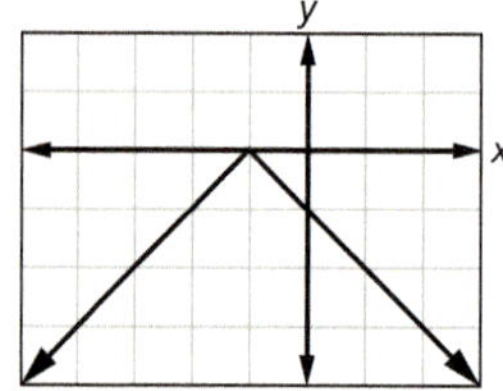

41. It turns the graph upside down. **43.** $\mathbb{R}$ **45.** $\leq$
47. $x > -17$ **49.** $x < -\frac{4}{7}$ **51.** $x < 10$

Chapter 5 Review

1. $A\ (4, 3), B\ (-3, 2), C\ (-5, -2), D\ (0, -4)$
3. W: III, X: II, Y: I, Z: y-axis
5. **Book Cost vs. Page Count**

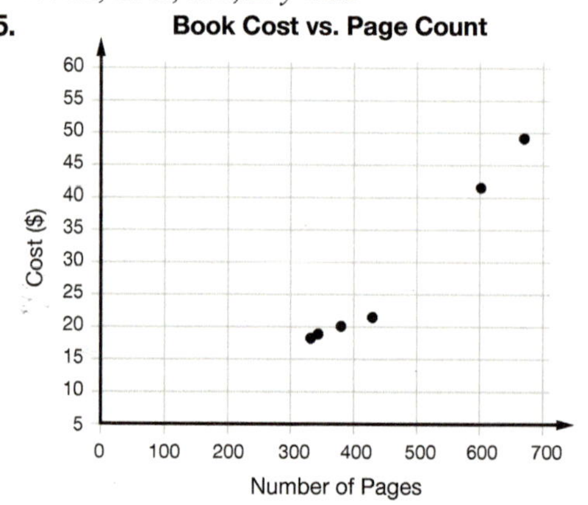

7. $D = \{-7, -3, -2, 1, 3, 5, 8\}; R = \{3, 4, 5, 6, 7, 8\}$
9. yes **11.** Table values will vary; yes.
13. the real numbers
15. $D = \{x \mid -4 < x < 3\}; R = \{y \mid -3 < y < 4\}$; yes

17. yes

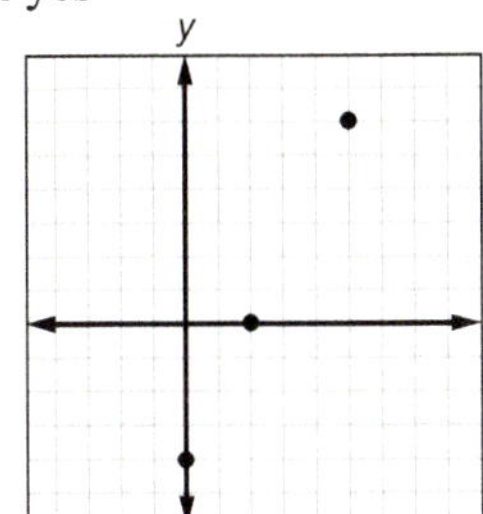

19. A: temperature slowly rises for 10 min; B: temperature is constant at 0°C for 20 min; C: temperature rises for 20 min; D: temperature is constant at 100°C for 30 min; E: temperature quickly rises in the last 10 min

21.

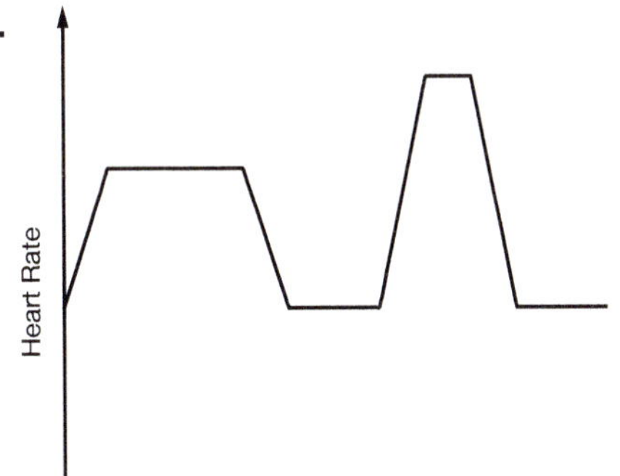

23. 9 gal

25.

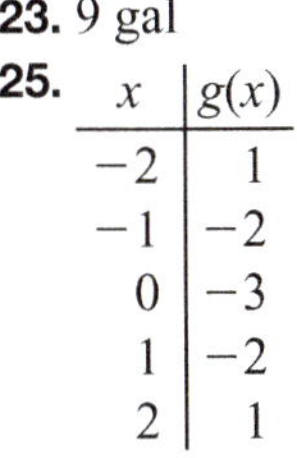

x	$g(x)$
-2	1
-1	-2
0	-3
1	-2
2	1

27.

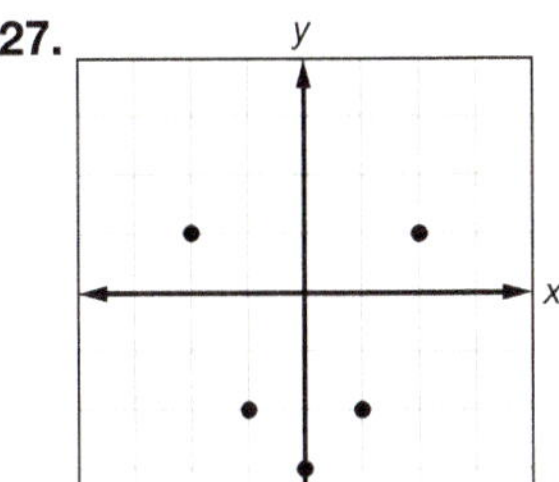

29. $f(x) = -\frac{1}{2}x$

31. direct; $y = 4.5x$

33. $k = 1.6$; $y = 1.6x$

35. 1.92 atm

37. -4, -2, 0, 2, 4 → -2, 0, 2

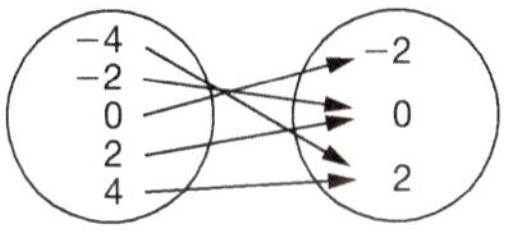

39.

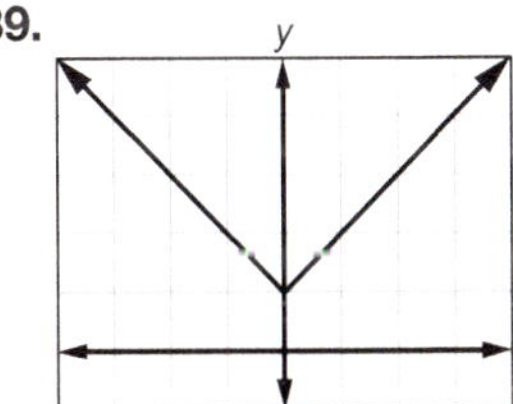

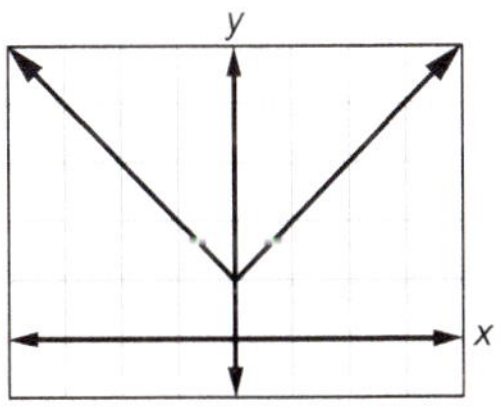

41.

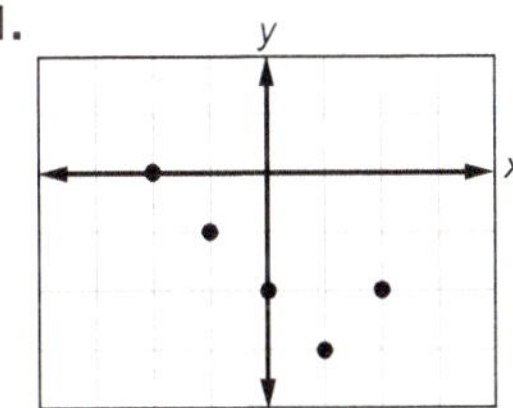

Chapter 6—Linear Functions

6.1 Graphing Lines

1. Table values will vary.

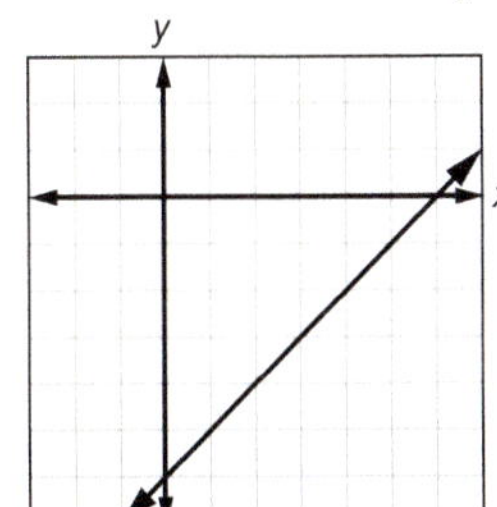

3. Table values will vary.

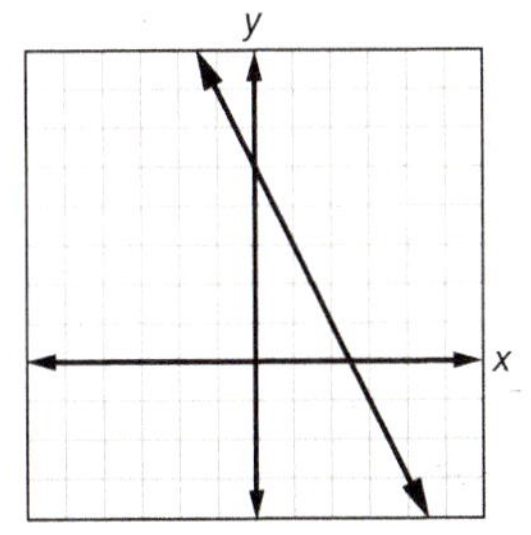

5. Table values will vary.

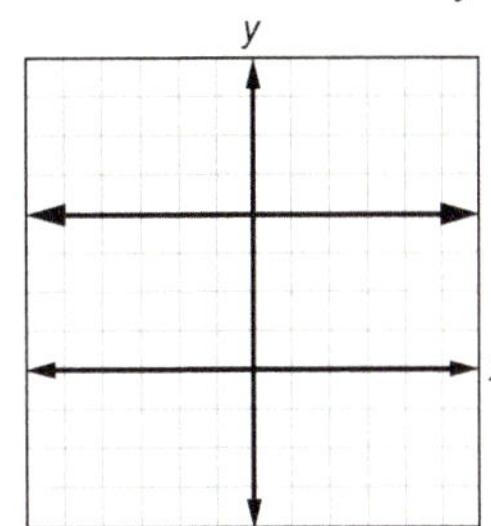

7. Table values will vary.

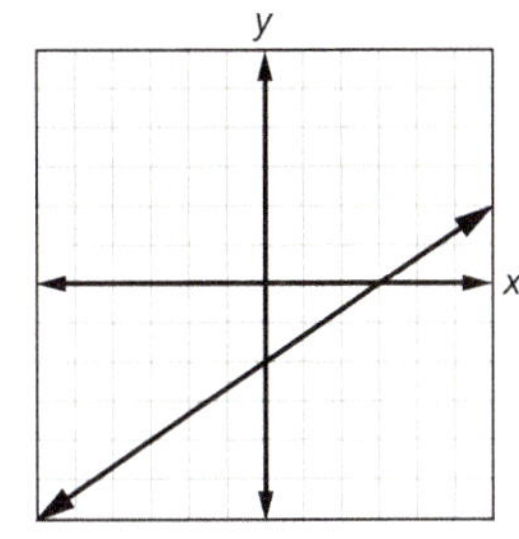

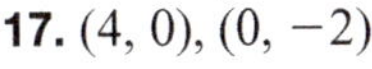

9. $y = -5x + 20$ **11.** $y = 4x - 7$ **13.** $x + y = 9$

15. $4x - 7y = 2$

17. $(4, 0), (0, -2)$

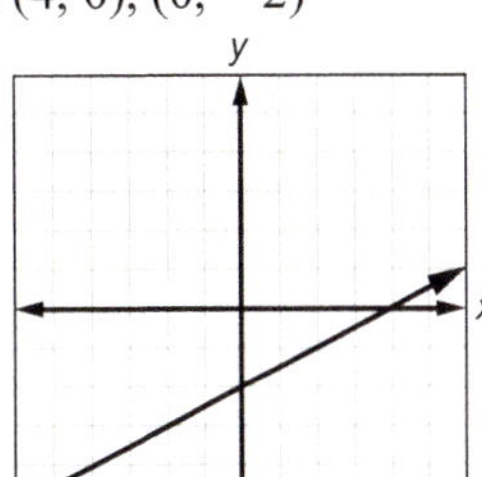

19. $(4, 0), (0, 4)$

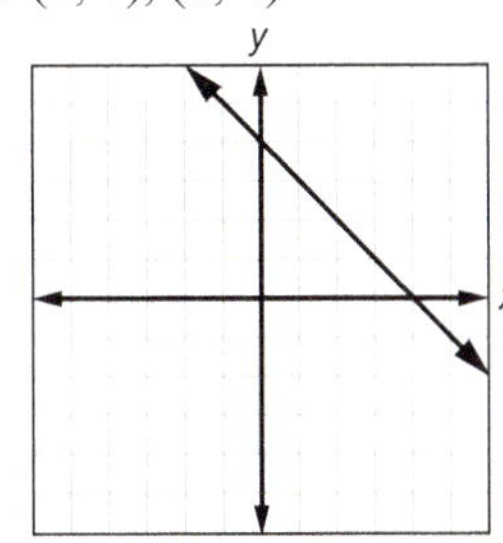

21. $(-3, 0), \left(0, \frac{3}{2}\right)$

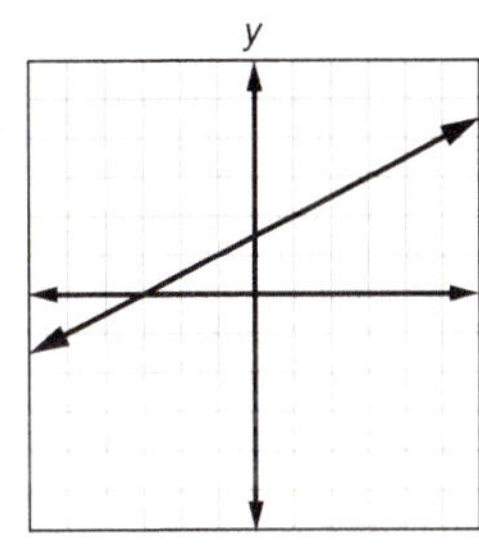

23. $(2, 0)$, none

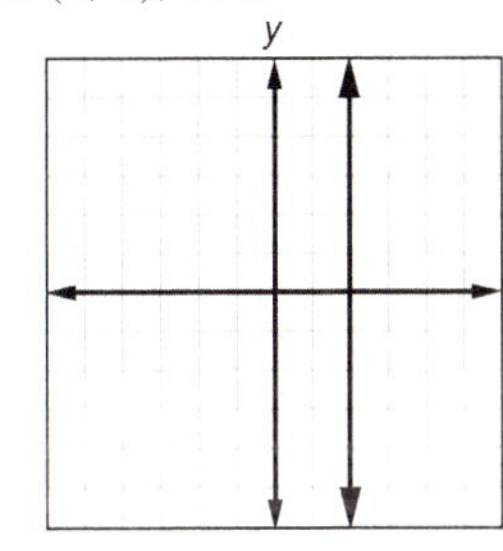

25.

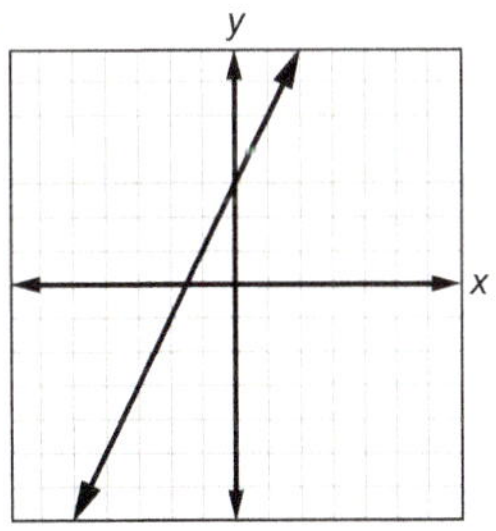

27.

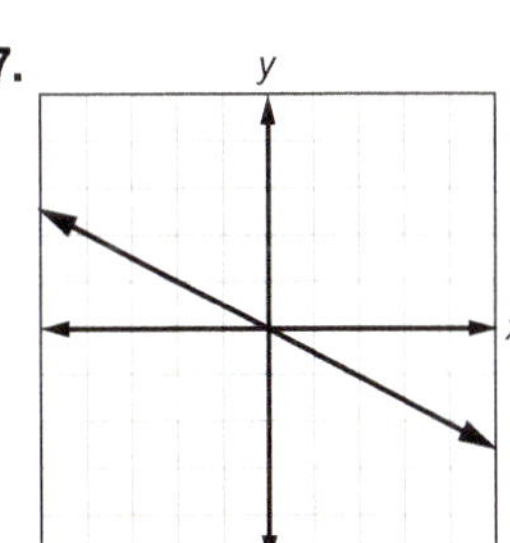

29.

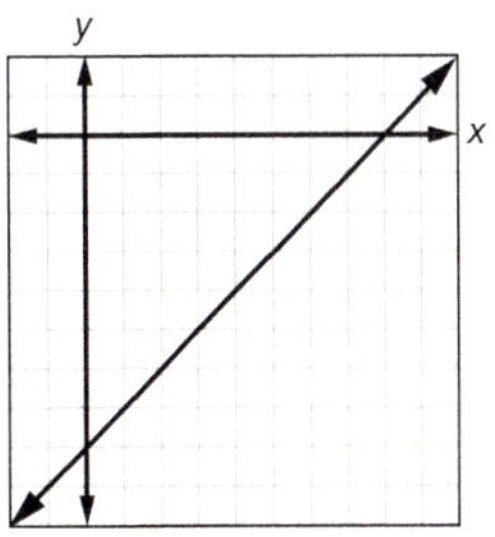

31.

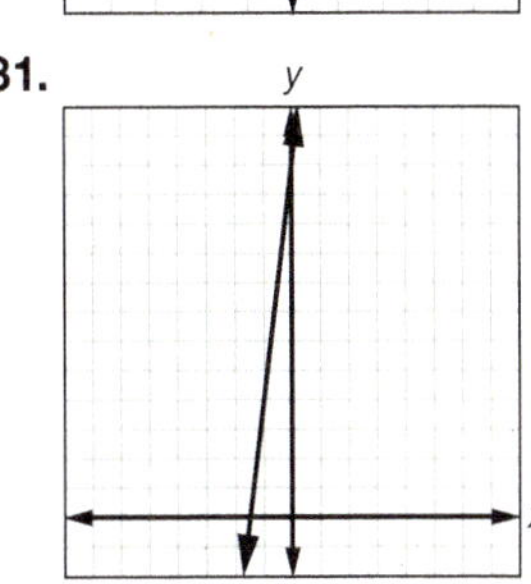

33. 0 **35.** horizontal **37.** $\left(\frac{C}{A}, 0\right), \left(0, \frac{C}{B}\right)$

39. yes; no; Two y-intercepts would be on the same vertical line.

41.

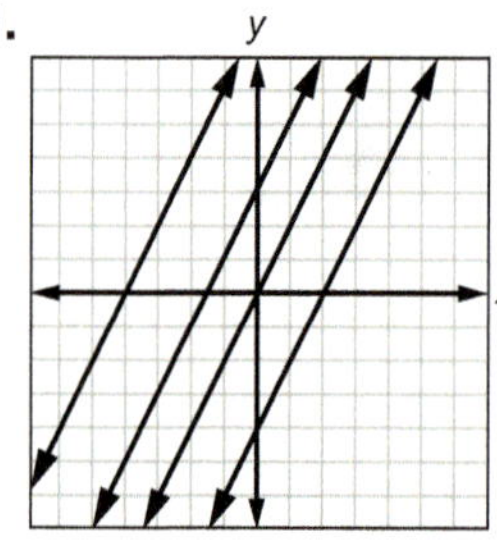

The equations have the same coefficient of x. The graphs are parallel.

43. $x = -\frac{3}{2}$

45. $x = \frac{15}{31}$

47. $l = \frac{S - b^2}{2b}$

49. 7.5 in. and 9 in.

51. 54

6.2 Slope

1. *a*, *d* 3. *e* 5. *d* 7. $m = -\frac{2}{3}$ 9. $m = \frac{1}{3}$ 11. undefined
13. $m = 4$ 15. $m = 0$ 17. $m = 1$ 19. $m = \frac{5}{2}$
21. undefined 23. $m = 2.8$

25.

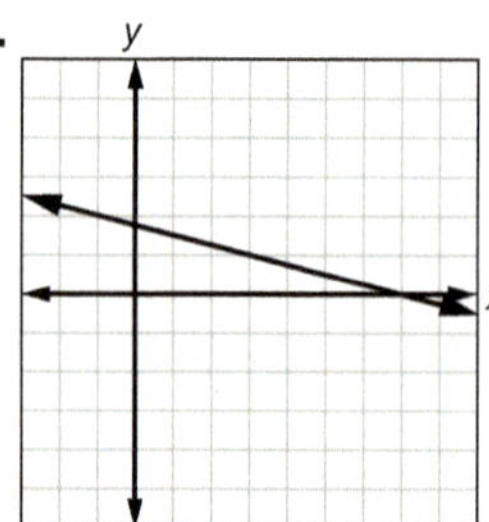

27.

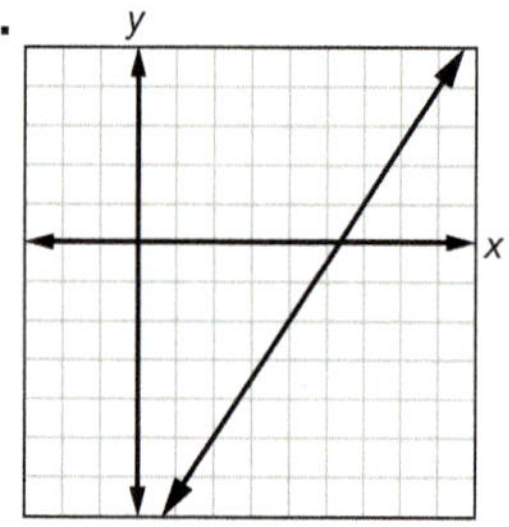

29a. 12 mi/hr 29b. 20 mi/hr 29c. 0 mi/hr 29d. stationary
31a. 18 cm/yr 31b. 8 cm/yr 31c. 5 cm/yr 31d. 1 cm/yr
33. 950.4 ft 35. 5.8 mi 37. 63°; difficult
39. $x = -2.5, 7.5$ 41. 0.76 43. \$70 45. $\frac{5}{6}$ qt $= 0.8\overline{3}$ qt
47. 1

6.3 Slope-Intercept Form of a Line

1. $m = 3$; (0, 4) 3. $m = -1$; (0, 0) 5. c 7. b

9.

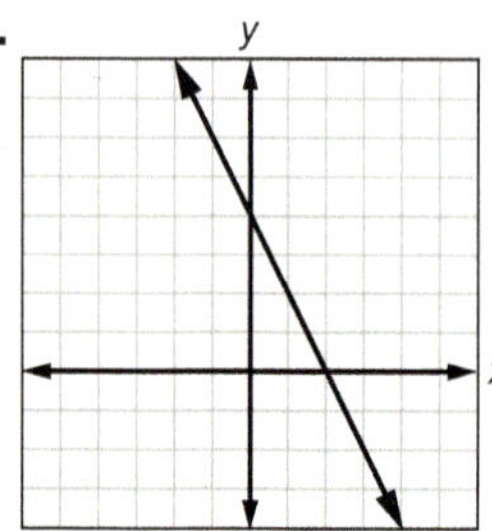

11.

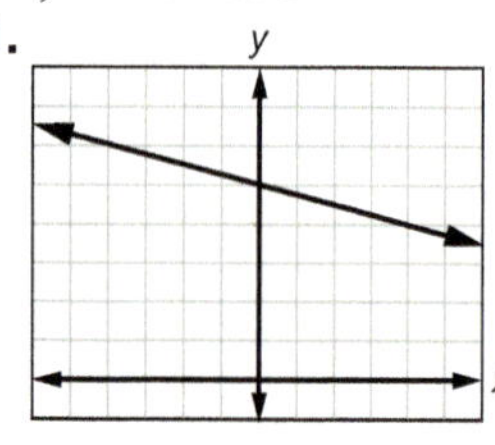

13.

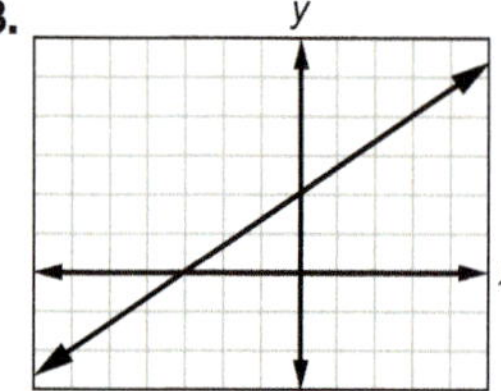

15. $y = -3x + 5$; $m = -3$; (0, 5)
17. $y = -\frac{1}{7}x + 2$; $m = -\frac{1}{7}$; (0, 2)
19. $y = \frac{8}{3}x - 3$; $m = \frac{8}{3}$; (0, −3)
21. $y = -dx + k$; $m = -d$; (0, k)

23. $y = -2x + 5$

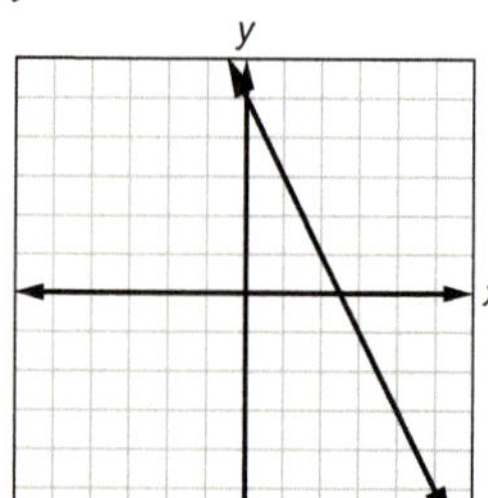

25. $y = -\frac{1}{2}x + 4$

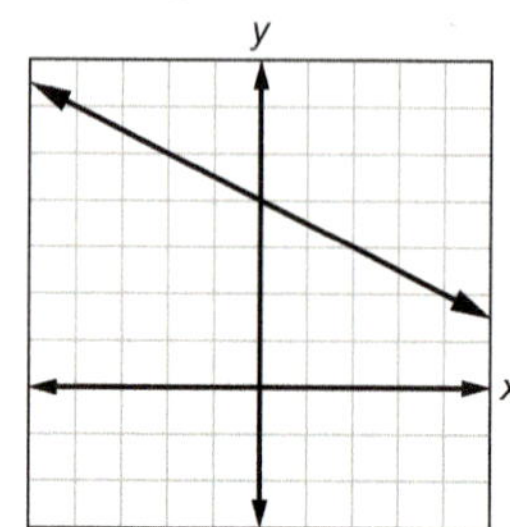

27. $y = -\frac{3}{4}x + 2$

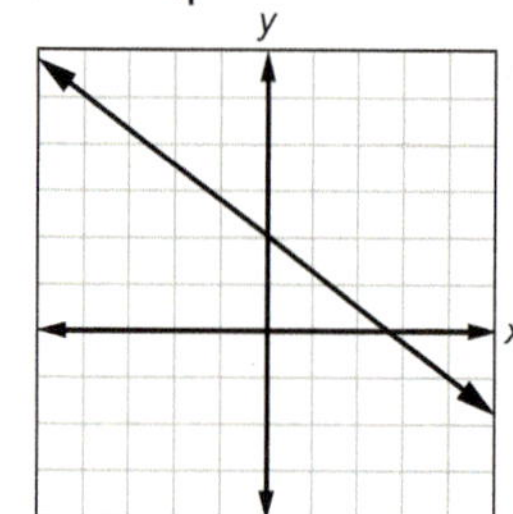

29. $y = \frac{1}{3}x + 3$

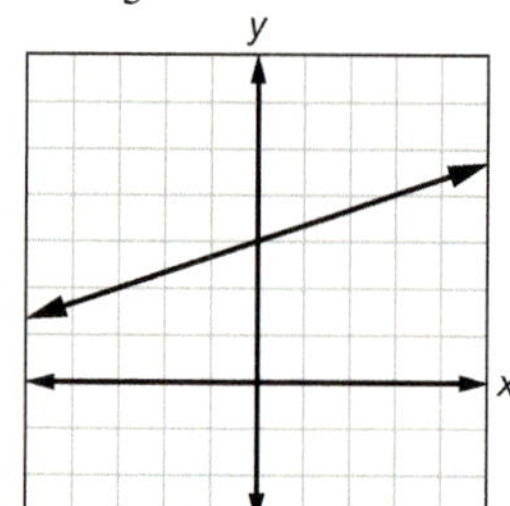

31. $y = 2x + 5$

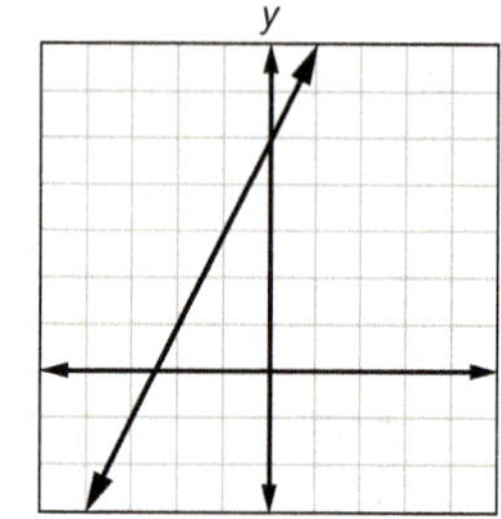

33. $f(x) = 18x + 226$; 586 students
35. $\frac{1}{45}$ gal/mi or $0.0\overline{2}$ gal/mi; $g(m) = -\frac{1}{45}m + 4.5$ or $g(m) = -0.0\overline{2}m + 4.5$; 0.5 gal
37. $m = -\frac{A}{B} = 6$; $b = \frac{C}{B} = -5$; $y = 6x - 5$
39. $y = -3x$

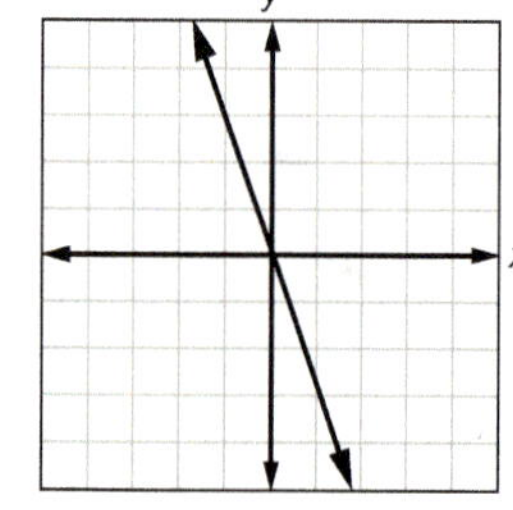

41. 16 ft 43. The ladder could fall backwards.
45. 2.5335 47. $-15\frac{1}{2}$
49. 1 51. $\frac{351}{152}$ or $2\frac{47}{152}$
53. 8.04 or $8\frac{1}{25}$

6.4 Writing Linear Equations

1. $y = -5x + 10$ 3. $y = -\frac{1}{3}x + 3$ 5. $y = 7$
7. $y = \frac{3}{2}x - 2$ 9. $y = 3x - 1$ 11. $y = -3x + 1$
13. $y = \frac{5}{3}x + 4$ 15. $y = -\frac{1}{4}x + \frac{25}{4}$ 17. $y = \frac{3}{5}x + 4$
19. $y = -x + 6$ 21. $y = 4x$ 23. $y = \frac{1}{11}x + \frac{26}{11}$
25. $y = -4$ 27. $y = 0.4x - 2.6$ or $y = \frac{2}{5}x - \frac{13}{5}$
29. $y = -\frac{5}{4}x + \frac{5}{2}$ or $y = -1.25x + 2.5$

31. $f(x) = 5000x + 190{,}000$; \$290,000
33. $f(x) = 1.25x - 231$; 499 hot lunches
35. $m = -0.3$ shirts/\$; $s(c) = -0.3c + 38$; \$126.67
37. $9x - 4y = 27$ **39.** $21x + 35y = 30$ **41.** 28 in.
43. 7/12 **45.** $-3x - 5$ **47.** $x < -65.25$ **49.** $x = 2, 10$
51. 17 quarters and 14 dimes **53.** 75%

6.5 Parallel and Perpendicular Lines

1. $-\frac{1}{7}$ **3.** $\frac{5}{3}$ **5.** $-\frac{7}{4}$ **7.** $m = \frac{2}{3}$ **9.** $m = \frac{8}{5}$ **11.** $m = -5$
13. $m = -\frac{1}{4}$ **15.** $m = -\frac{9}{5}$ **17.** $y = -x + 7$
19. $y = \frac{1}{9}x - \frac{71}{9}$ **21.** $y = x + 4$ **23.** $y = 5$
25. $y = \frac{7}{5}x - \frac{21}{5}$ **27.** neither **29.** perpendicular
31. $y = -x + 5$; $y = \frac{5}{3}x - 1$; neither
33. $y = 6x + 15$; $y = -\frac{1}{6}x + \frac{8}{3}$; perpendicular
35. $9x - 12y = -4$ **37.** $\frac{6}{7}$
39. $m_{AB} = m_{CD} = \frac{4}{3}$; $m_{BC} = m_{DA} = -\frac{3}{4}$ **41.** no **43.** 16
45. 9.375 in. or $9\frac{3}{8}$ in.; yes **47.** -252 **49.** $34x - 36y$
51. $x < 3$ or $x \geq 7$
53. $x \leq -7.5$ or $x \geq 4.5$

−8 −6 −4 −2 0 2 4 6

55. Joshua: 53 mi/hr; Savannah: 63 mi/hr

Sequences Converging and Diverging Sequences

1. 0 **3.** $\frac{1}{2}$ **5.** diverging **7.** diverging **9.** 1

6.6 Trend Lines and Correlation

1. e **3.** c **5.** a

7.

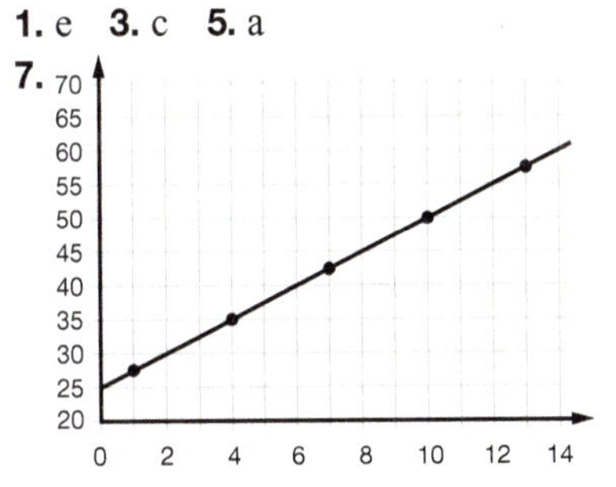

$y = 2.5x + 25$

9. $y = 75$

11.

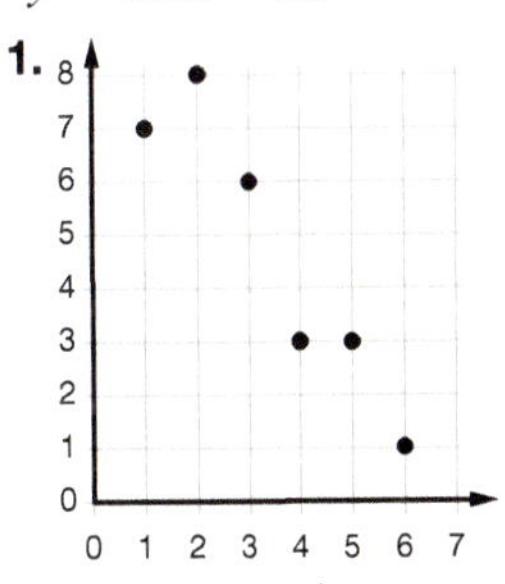

strong negative

13. $y = 4.7$
15. yes
17. $y = 0.82x + 16.93$
19. 81

21.

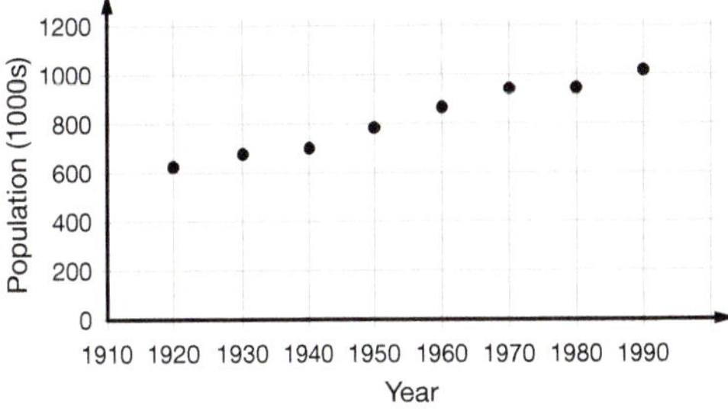

strong positive

23. 1,192,600 **25.** Extrapolating over a relatively short period of time to predict the population in 2020 is much more likely to be accurate than extrapolating over a long period of time.

27.

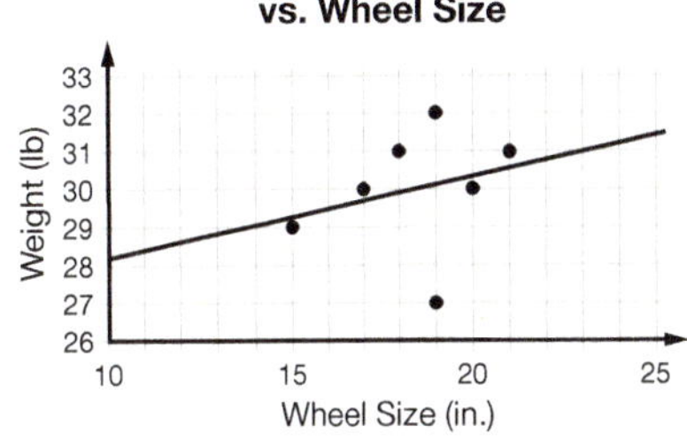

$y = 0.21x + 26.11$

29. 23 in. **31.** speed; stopping distance

33.

v^2	d
100	16
625	59
900	80
2025	154
2500	183
3025	216
3600	251

Stopping Distance vs. Speed Squared

Stopping Distance (ft): 0 50 100 150 200 250 300

Speed Squared (mi/hr)^2: 0 500 1000 1500 2000 2500 3000 3500 4000

35. 122 ft; 341 ft **37.** W **39.** $x = -\frac{31}{7}$ **41.** $x = 30$
43. $x = 4$ **45.** direct

6.7 Graphing Linear Inequalities

1. solid **3.** dashed **5.** D **7.** M **9.** b **11.** a **13.** b **15.** a

17.

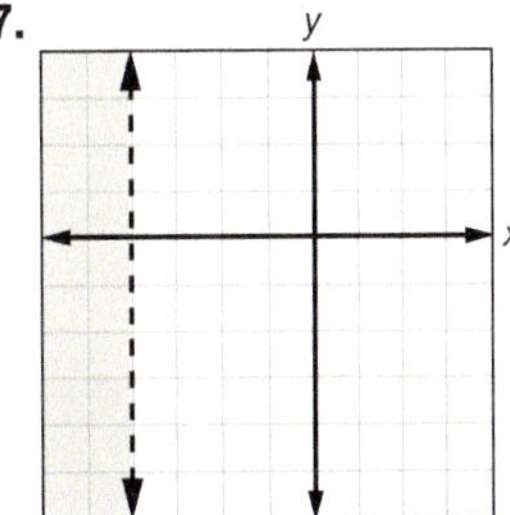

19.

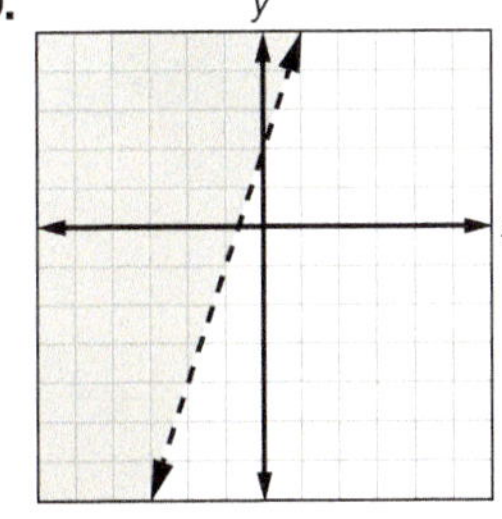

21.

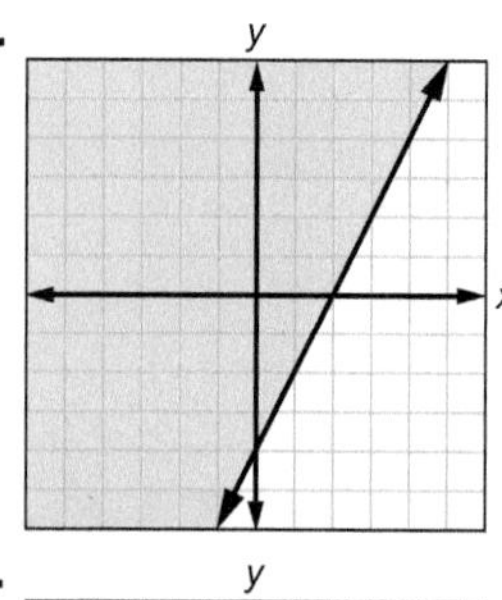

23.

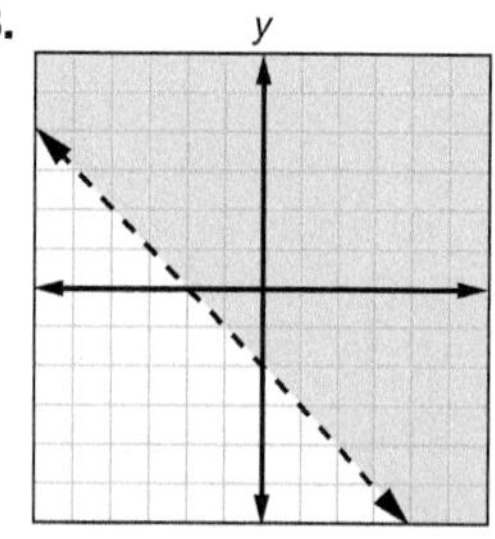

25.

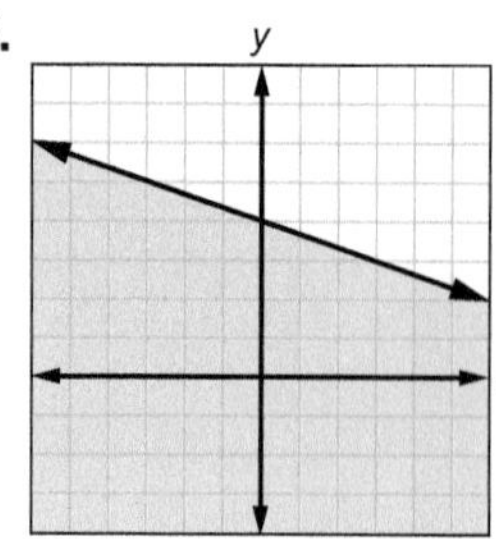

27.

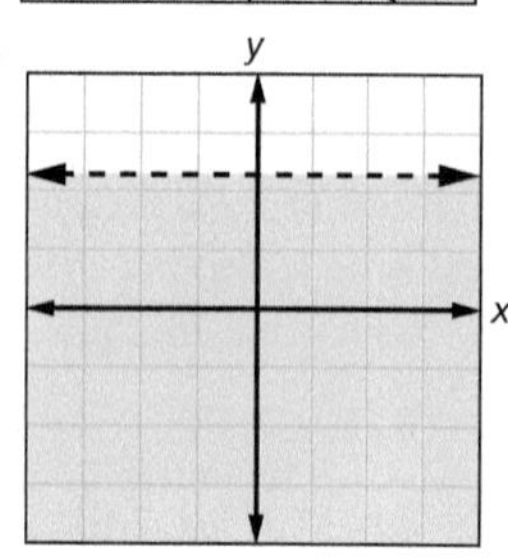

29. $y \leq \frac{2}{3}x - 3$ **31.** an infinite number

33. $y \leq 5x$

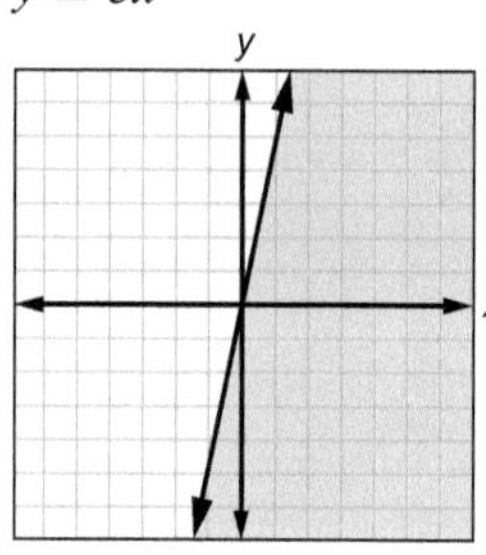

35. $4x + 3y \leq 60$

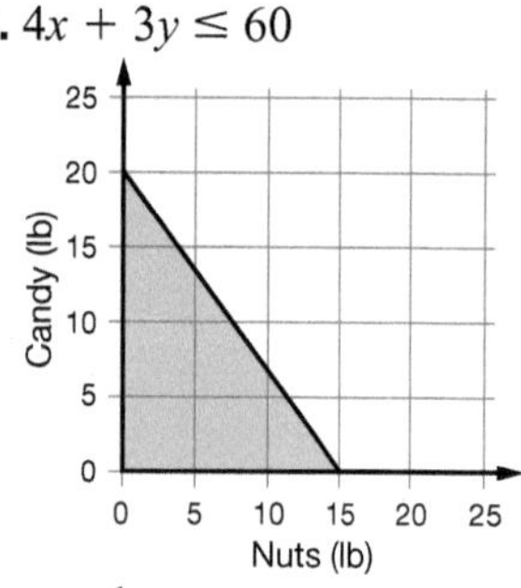

37.

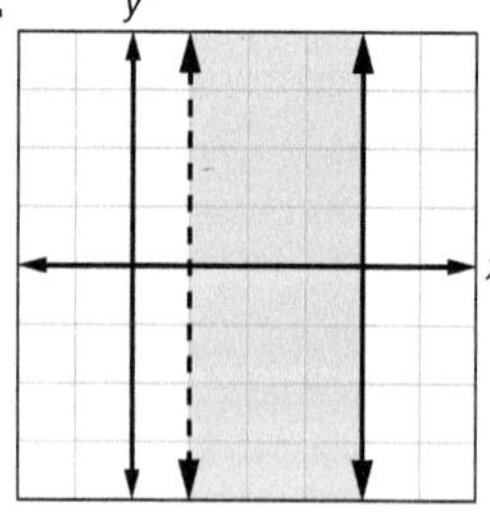

39. $y > \frac{1}{3}x$ and $y < x$

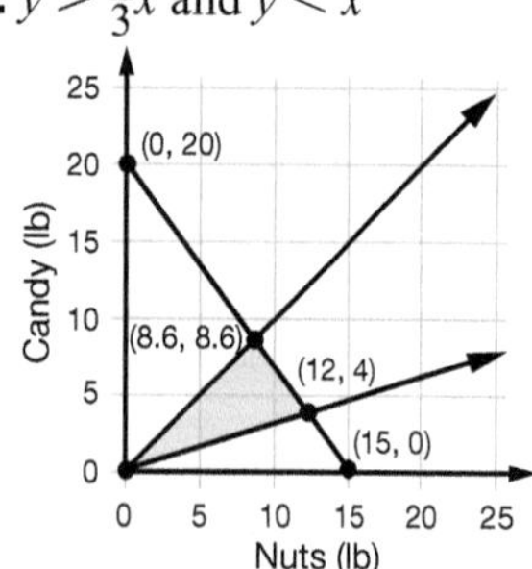

She could buy 12 lb of nuts and 4 lb of candy (a salty mix), 8 lb of each (a sweet mix), or 9 lb of nuts and 6 lb of candy.

41. $0.08\overline{3}$ **43.** 0.067; yes **45.** −196 **47.** $x = \frac{136}{27} = 5.\overline{037}$

49. $p = \frac{S - 2B}{H}$ **51.** 14,720 bu **53.** 222 ft

Chapter 6 Review

1. $y = 5x - 6$

3. Table values will vary.

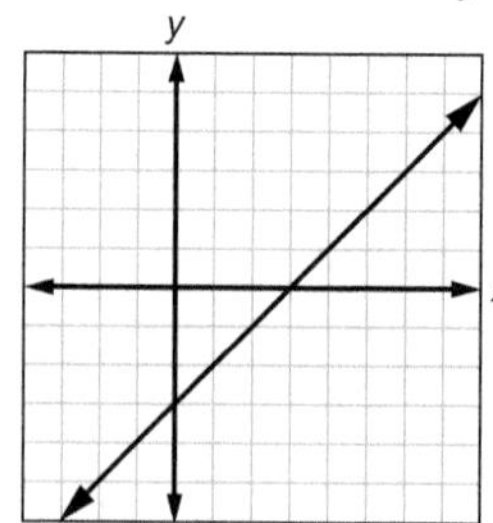

5. (4, 0), (0, −1)

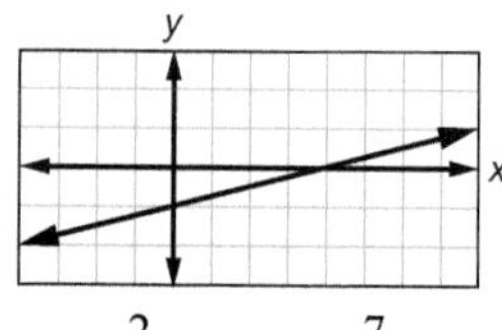

7. $m = \frac{2}{3}$ **9.** $m = \frac{7}{2}$

11. $m = 0$

13.

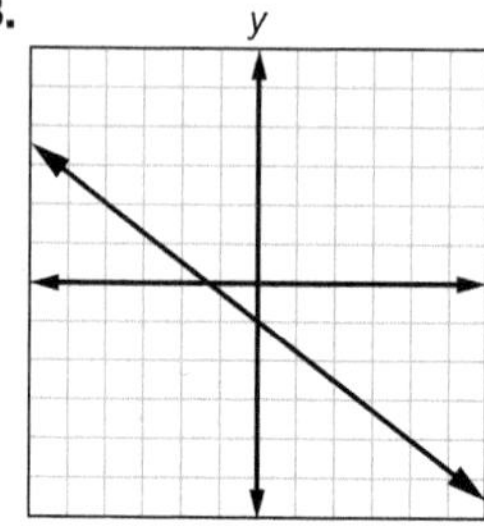

15. undefined; none

17. $y = -3x - 6$ **19.** $f(x) = \frac{2}{3}x - 3$

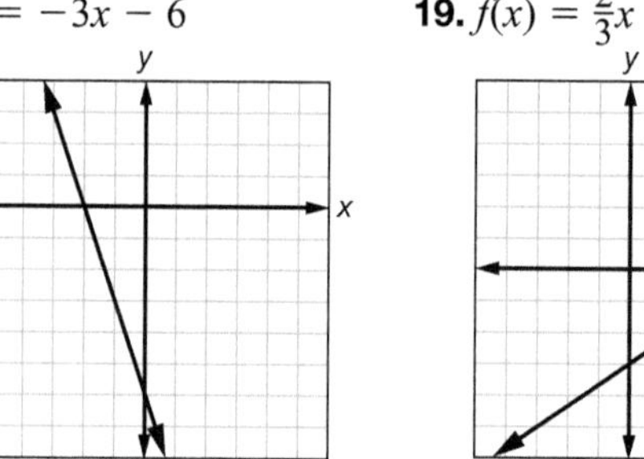

21. $y = 2x + 3$ **23.** $y = \frac{3}{2}x + \frac{5}{2}$ **25.** $y = -\frac{1}{2}x + 4$

27. $y = -3$ **29.** $y = \frac{2}{7}x + \frac{22}{7}$ **31.** $3x - 2y = 6$

33. $\frac{1}{2}$ and $-\frac{8}{3}$ **35.** neither

37. $m_c = 2$; $m_d = -\frac{1}{2}$; perpendicular **39.** dashed; above

41.

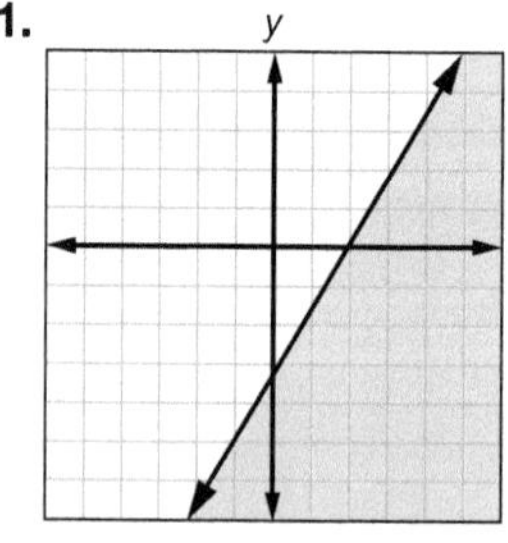

43. $f(x) = \frac{1}{4}x + 8$; 22 lb

45.

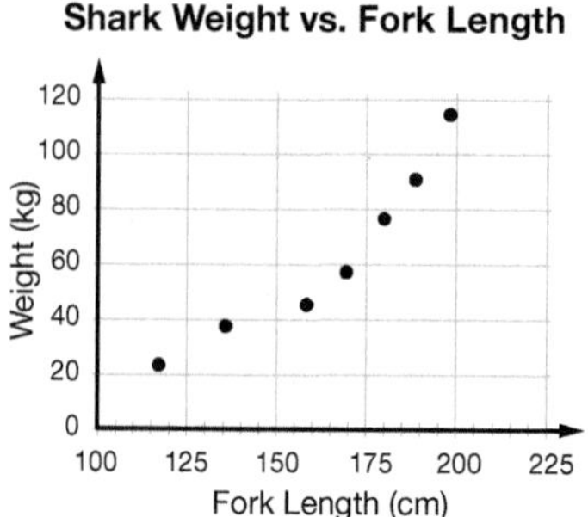

strong positive

47. 61 kg

49. c

Chapter 7—Linear Systems

7.1 Graphing Systems of Equations

1. (1, 5) **3.** ∅ **5.** yes **7.** no **9.** no **11.** no **13.** yes

15. (3, 2)

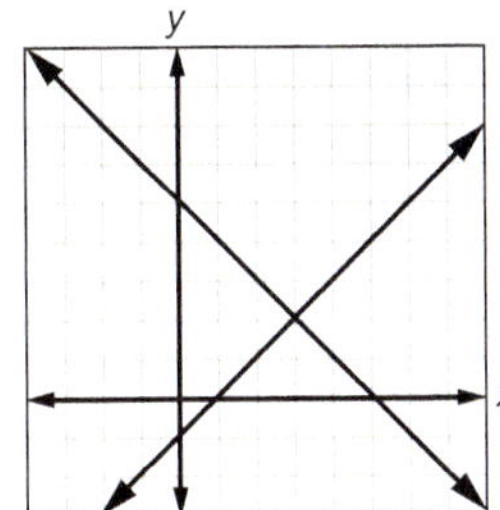

17. (1, 1)

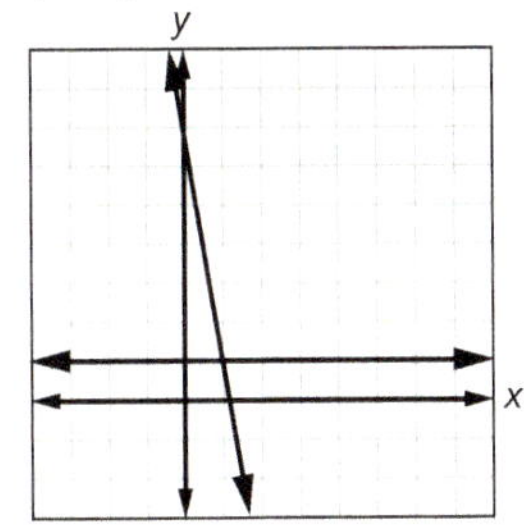

19. (3, −5)

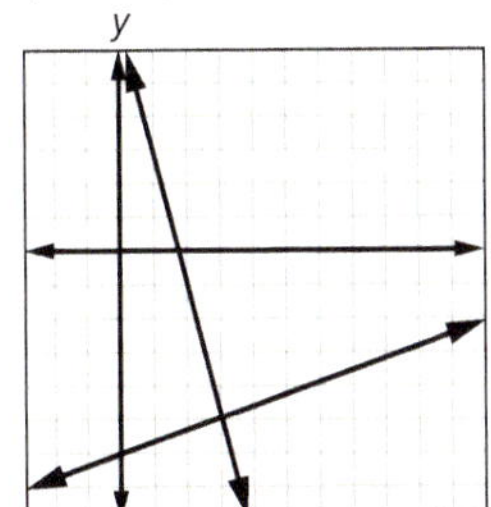

21. (−2, 6)

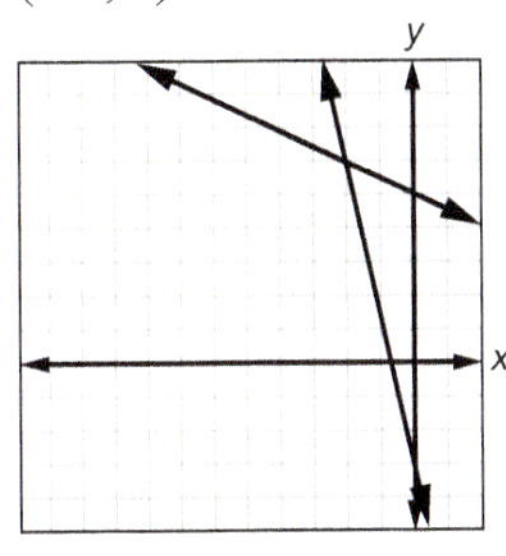

23. (5, 7)

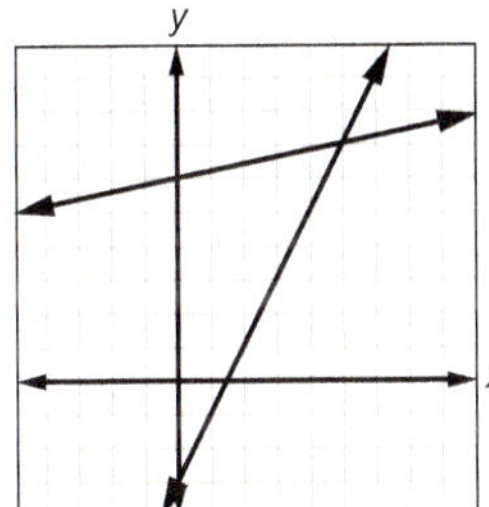

25. (2, 4)

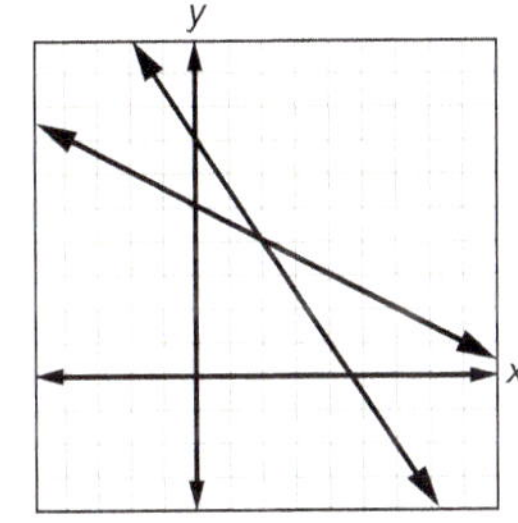

27. (−3, −1)

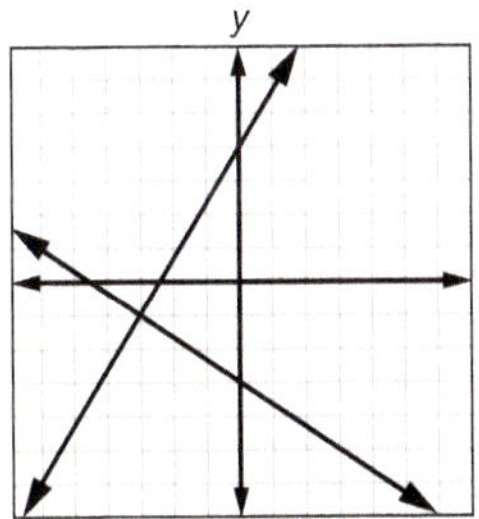

29. ∅

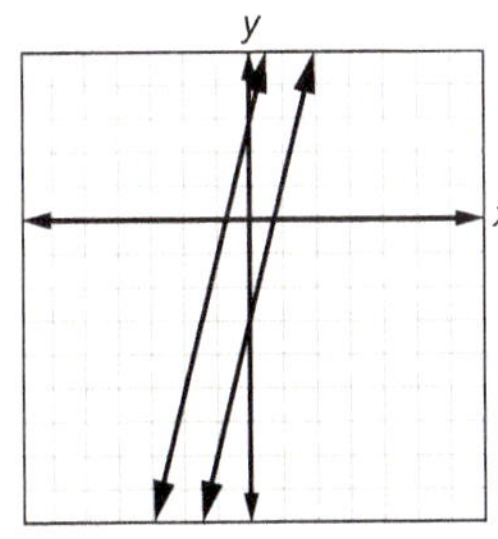

31. ∅

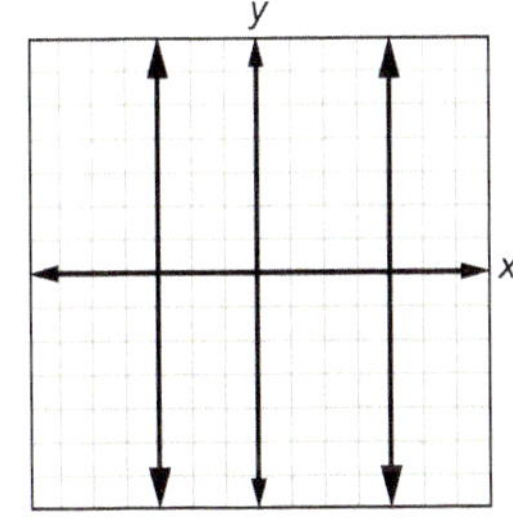

33a. $x + y = -7$ and $x - y = 3$

33b. (−2, −5)

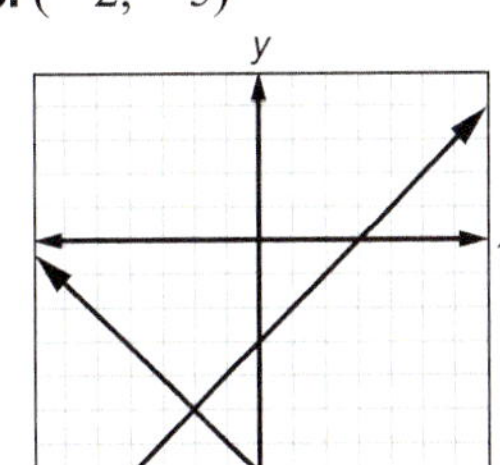

35a. $r(x) = 30x$

35b. $e(x) = 20x + 120$

35c. 12 jackets

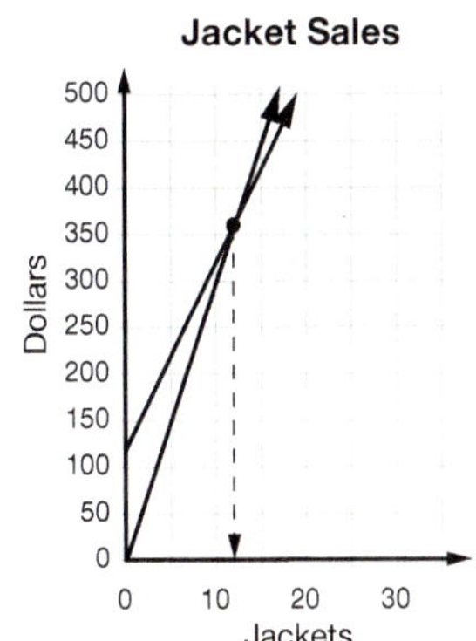

37. about 230 min

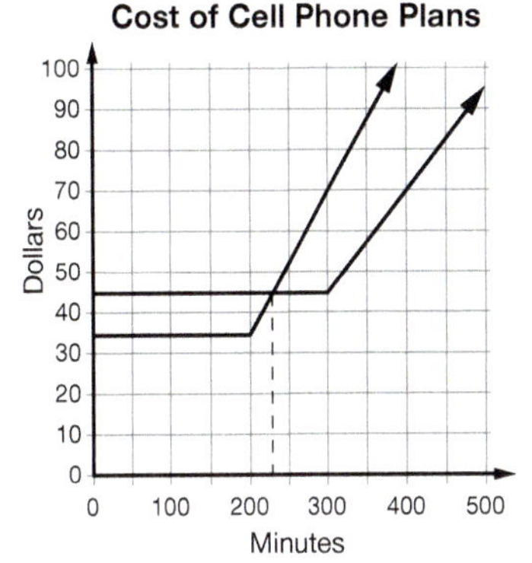

39.

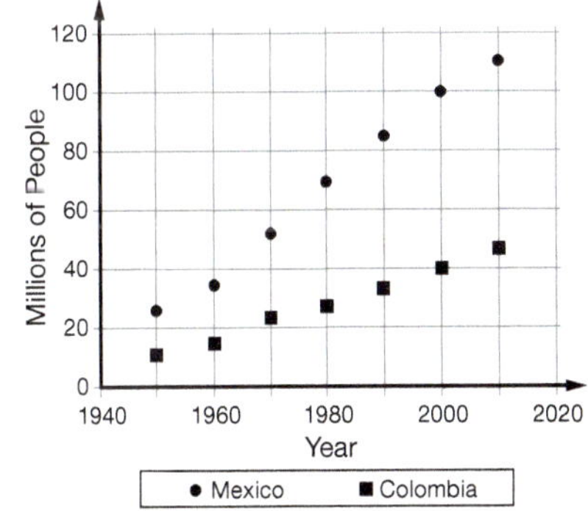

41. $\frac{7}{78}$ **43.** $x = 5$ **45.** $x = \frac{-4y + 5}{3}$ **47.** $x \geq -3$ **49.** $\mathbb{R}$

7.2 Solving Simple Systems by Substitution

1. a **3.** c **5.** (7, 21) **7.** (4, 14) **9.** (8, 3) **11.** (−2, −6) **13.** (−4, 3) **15.** (9, 2) **17.** (−2, 8) **19.** (−3, −4) **21.** (−2, 1) **23.** (−4, 5) **25.** (10, 5) **27.** $\left(\frac{28}{5}, \frac{14}{5}\right)$ **29.** 38 and 25 **31.** 19 ft × 26 ft **33.** 5 apple and 7 pecan **35.** $\left(\frac{66}{41}, \frac{26}{41}\right)$ **37.** No; the Mexican population is larger and is growing at a faster rate. **39.** b **41.** $x + 7$ **43.** $x^2 + 2x + 1$ **45.** $2x + 14$ **47.** −2

7.3 Solving Advanced Systems by Substitution

1. c **3.** d **5.** a **7.** b **9.** d **11.** $(-8, -1)$ **13.** $(9, -1)$ **15.** $(9, -1)$ **17.** $\left(3, \frac{1}{5}\right)$ **19.** $\left(-\frac{9}{26}, -\frac{25}{26}\right)$ **21.** $(-4, 3)$ **23.** $\left(\frac{5}{2}, \frac{4}{3}\right)$ **25.** $\left(\frac{1}{3}, \frac{5}{6}\right)$ **27.** $\left(\frac{63}{23}, -\frac{15}{23}\right)$ **29.** wallet: \$0.50; 5" × 7": \$5 **31.** 4% investment: ≈ 6.08 yr (6 yr 1 mo); 5% investment: ≈ 7.08 yr (7 yr 1 mo)
33. 4 cheesecakes and 5 pies **35.** $\left(-\frac{3}{2}, \frac{6}{5}\right)$ or $(-1.5, 1.2)$
37. CD: 16 yr; mutual fund: 15 yr **39.** 2500 CDs
41. $x = 3, 5$ **43.** $x = -\frac{10}{3}, 6$ **45.** ∅ **47.** $f(x) = 2x + 1$
49. $m = -1$

Sequences Limit of a Sequence

1. no **3.** yes **5.** 0 **7.** −2 **9.** 1.618

7.4 Solving Systems by Elimination

1. a) 2nd equation by −7; b) either equation by −1
3. a) 2nd equation by −2; b) either equation by −1 **5.** (1, 4)
7. $(-3, -1)$ **9.** $(-3, 4)$ **11.** a) 1st equation by −2; b) 1st equation by 5 and 2nd equation by 3 **13.** a) 2nd equation by −6; b) 1st equation by 3 and 2nd equation by 11
15. $\left(4, \frac{5}{2}\right)$ **17.** $(4, 7)$ **19.** $(-1, -3)$ **21.** $(6, -6)$
23. $\left(\frac{5}{2}, \frac{9}{2}\right)$ **25.** $\left(\frac{19}{8}, \frac{1}{16}\right)$ **27.** $\left(\frac{31}{17}, \frac{4}{17}\right)$ **29.** 365 and 256
31. 75 adult and 170 student tickets
33. 18 softballs and 25 baseballs
35. 2 oz: 23; 3 oz: 14 **37.** $(-8, 25)$
39. 200 CDs per person **41.** 10 **43.** $-2x^2 + x + 6$
45. $x = \frac{14b + c}{a}$ **47.** 43.75% **49.** 31.5 L

7.5 Special Systems

1. b **3.** b **5.** a, d **7.** c **9.** a, b **11.** c **13.** b
15. $(-3, -4)$; consistent independent **17.** ∅; inconsistent
19. (2, 4); consistent independent
21. $y = -\frac{4}{3}x + \frac{7}{3}$; consistent dependent **23.** ∅; inconsistent
25. $\left(\frac{41}{7}, -\frac{3}{7}\right)$; consistent independent
27. $\left(\frac{63}{23}, -\frac{15}{23}\right)$; consistent independent **29.** ∅; inconsistent
31. $y = \frac{3}{2}x + 4$; consistent dependent
33. They have the same slope and the same y-intercept.
35. 12 cm, 18 cm, and 18 cm **37.** Easy: $y = 0.15x + 50$; Clean: $y = 0.2x + 40$; 200 mi for \$80; over 200 mi
39. Solving for any variable results in a fractional expression.
41. $k = 2$; $y = 2x$ **43.** They both have a y-intercept of zero or they both go through the origin.
45. $p(d) = 0.445d + 14.7$ **47.** 1086 atm **49.** 73.7 atm

7.6 Motion Problems

1. $12 - w$ **3.** $4 + c$ **5.** $500 - w$
7a. b = the rate with no wind
7b. $1.125(b + 8) = 36$ or $2.25(b - 8) = 36$
7c. There is only one unknown.
9.

	r	t	d
Upstream	$b - 3$	6	$6(b - 3)$
Downstream	$b + 3$	4	$4(b + 3)$

11. x = the rate with no wind; w = the wind speed
13. $2(x + w) + 5(x - w) = 150$ or $7x - 3w = 150$
15. x = the rate with no wind; w = the wind speed
17. $2(x + w) = 3(x - w)$ **19.** 4 hr
21. plane: 275 knots; wind: 25 knots
23. southbound train: 65 mi/hr; northbound train: 55 mi/hr
25. 24 mi/hr **27.** first car: 65 mi/hr; second car: 59 mi/hr
29. express train: 59.5 mi/hr; local train: 45.5 mi/hr
31. 2 hr **33.** $400{,}000x + 4300 = 70{,}000$; $x \approx \$0.164$ per CD **35.** −32 **37.** 81 **39.** 2^2 **41.** $\frac{1}{3^6}$ **43.** $\frac{2^3}{3^5}$

7.7 Mixture Problems

1. \$40 **3.** 82 bags **5.** $30x$ **7.** 28% salt
9. $0.12x$ gal of acid **11.** 0%
13. 18 oz of nuts and 30 oz of raisins
15. 1.5 lb of caramels and 3.5 lb of butterscotch candies
17. 12 gal of the 30% salt solution and 28 gal of pure water
19. 40 mL of the 5% solution and 10 mL of the 10% solution
21. 35 lb of the original mix and 15 lb of fescue
23. solution I: 15%; solution II: 35%
25. \$4200 at 10% and \$5800 at 9.5%
27. 116.7 gal of type A and 83.3 gal of type B
29. 10.7 gal of cream and 49.3 gal of milk
31. 26.6 lb of A and 58.4 lb of B
33. dry cleaning: 8%; restaurant: 10%
35. 2.6 L of ocean water and 9.4 L of Dead Sea water
37. $16.5x \leq 2000$; $x \leq 121.\overline{21}$; 121 boxes **39.** 1700 lb
41. $x < -1$
−4 −3 −2 −1 0 1 2 3 4
43. $x \leq 3$
−4 −3 −2 −1 0 1 2 3 4
45. $y \leq -1$

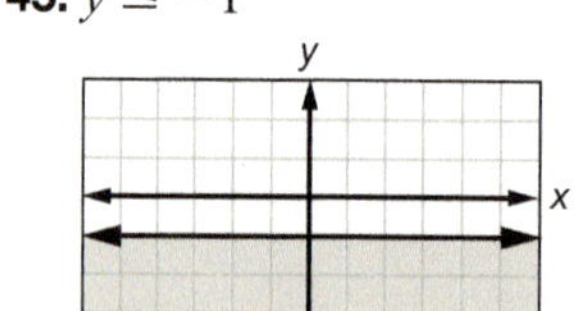

47.

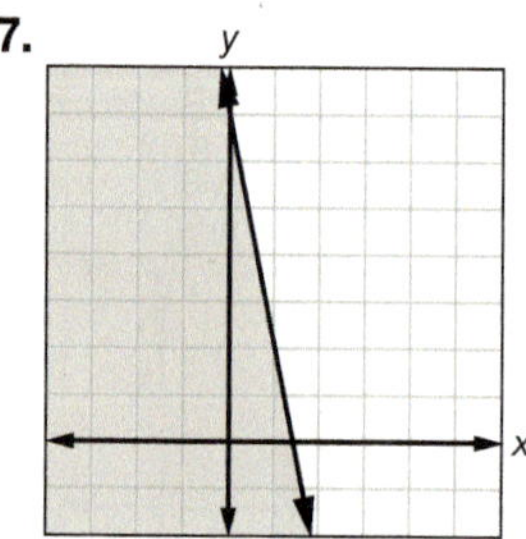

49. $(-6, 6)$

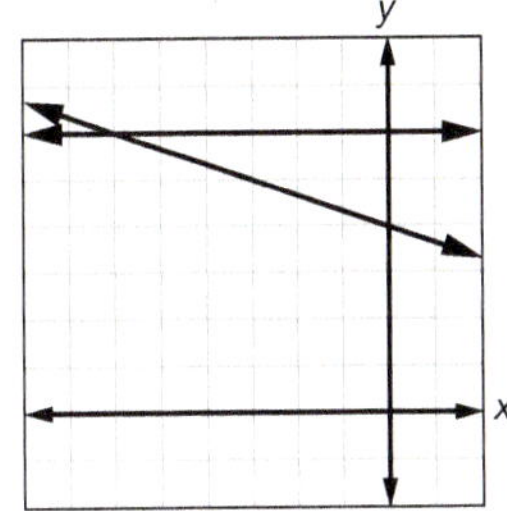

7.8 Solving Systems of Inequalities

1. I **3.** II **5.** yes **7.** no

9.

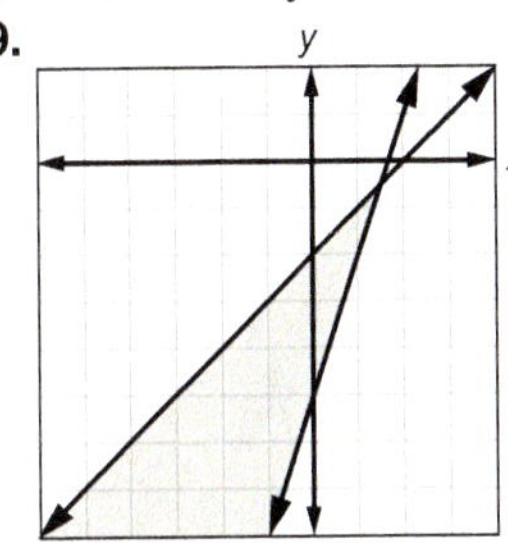

11.

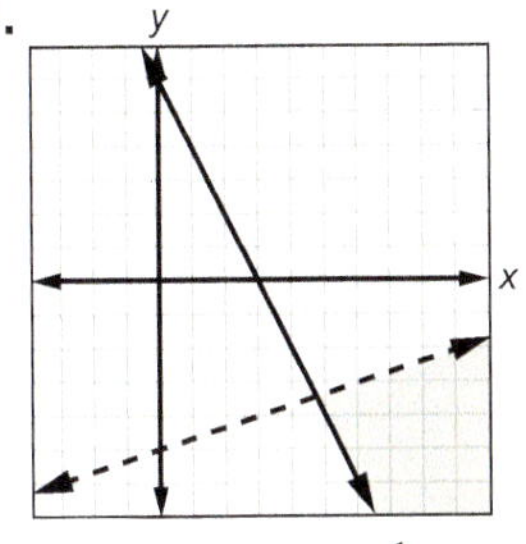

13.

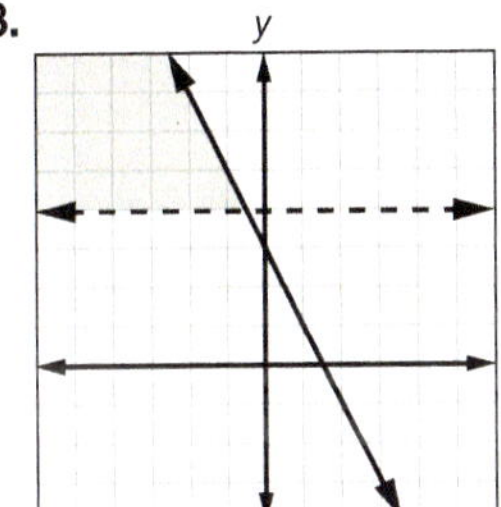

15. $y \leq 2$ and $y \leq -\frac{1}{3}x + 1$

17. $y > \frac{1}{3}x + 4$ and $y \leq 2x + 2$

19.

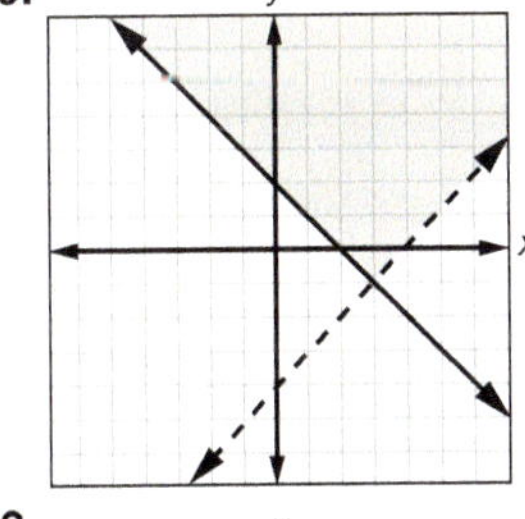

21.

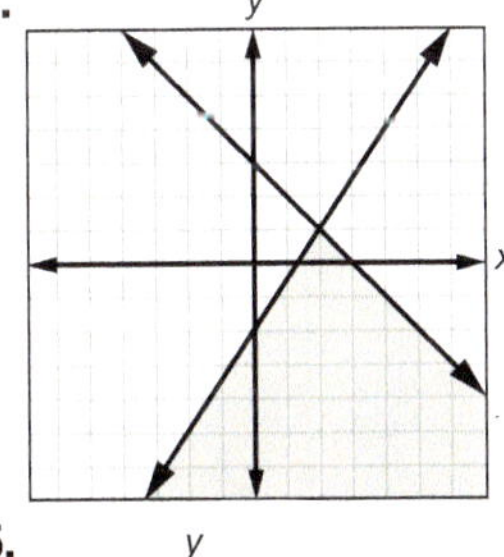

23.

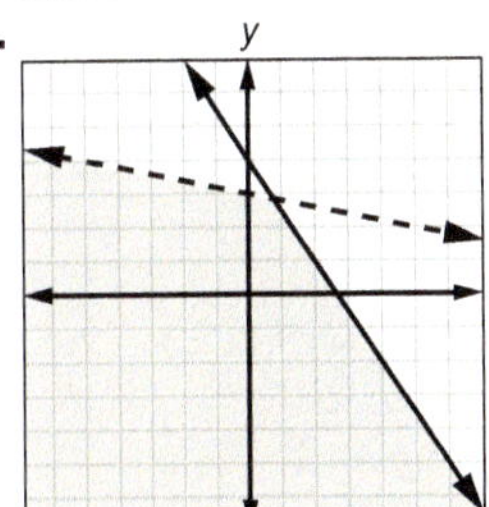

25.

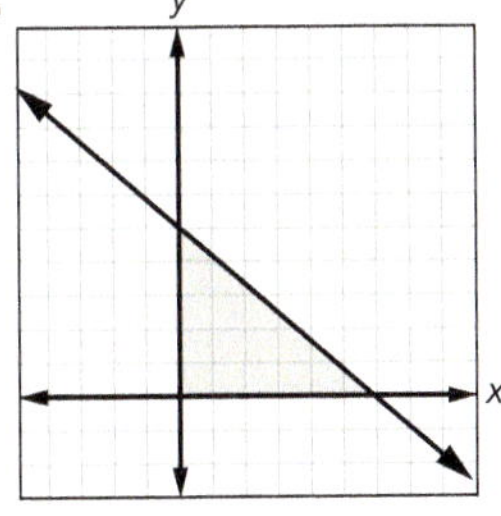

27.

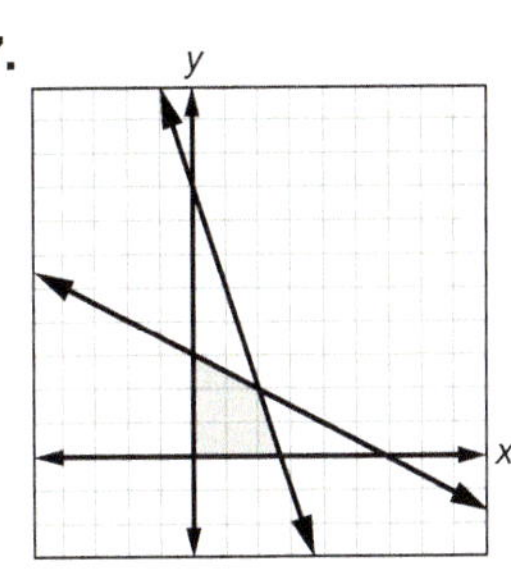

29. $y \geq -3x - 2$ and $y < -3x + 5$

31. $y \geq -3$, $x < 2$, and $y < 2x + 1$

33a. $12x + 7y \leq 70$, $8x + 12y \geq 72$, $y \geq 0$, and $x \geq 0$

33b.

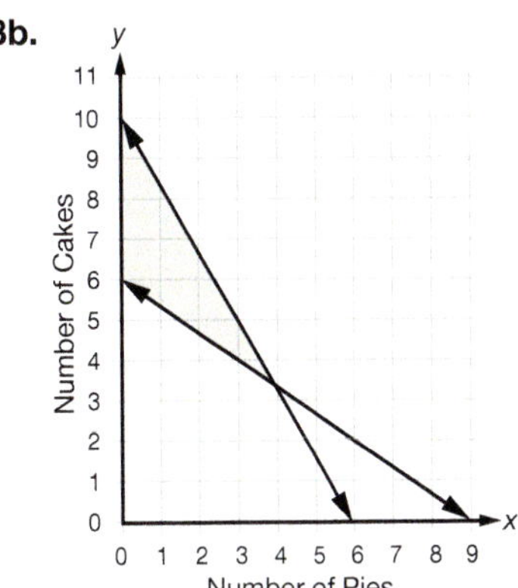

33c–d.

Chocolate Silk	Lemon Chiffon	Number of Servings	Cost
2	5	76	\$59
2	6	88	\$66
3	4	72	\$64

35. $x > 10$, $y > 12$, and $2x + 2y \leq 60$

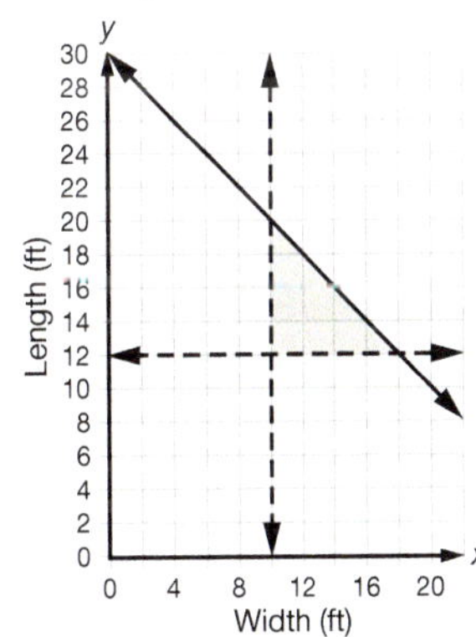

any point in the triangular region

37. $x \geq 0$, $y \geq 0$, $2.5x + 2y \leq 80$, and $2x + 3y \leq 100$

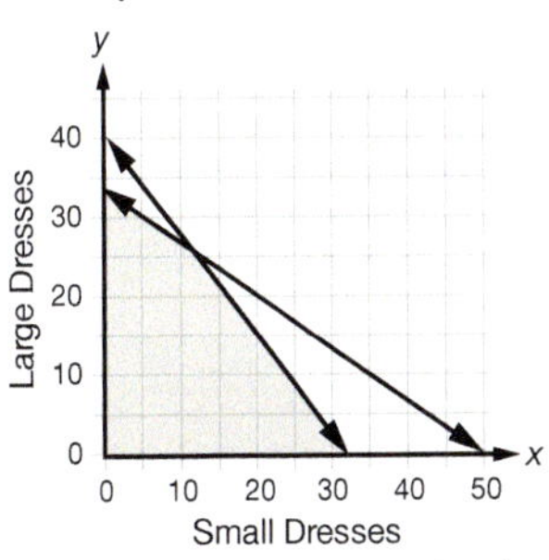

10 small and 25 large (75 hr and \$95)

39. $1.5x \leq 1224$ **41.** $(11, 19)$ **43.** $(-2, 3)$ **45.** $(4, -2)$
47. $(8, -3)$ **49.** $(-4, -2)$

Chapter 7 Review

1. yes **3.** yes
5. (3, −5)

7. (4, 0) **9.** ∅ **11.** (−2, 5) **13.** (1, 1) **15.** (2, 4) **17.** (−4, 19) **19.** (4, 0) **21.** ∅

23. They are parallel. **25.** They intersect at their y-intercept.
27. no solutions; inconsistent
29. infinite number of solutions; consistent dependent
31. no solutions; inconsistent **33.** $12x$ **35.** $r + 40$
37. −9 and −14 **39.** 33 people
41. slower boat: 8 mi/hr; faster boat: 24 mi/hr
43. 2.5 lb of chocolates and 3.5 lb of nuts
45. 60 gal of 60% sucrose and 240 gal of 25% sucrose
47.

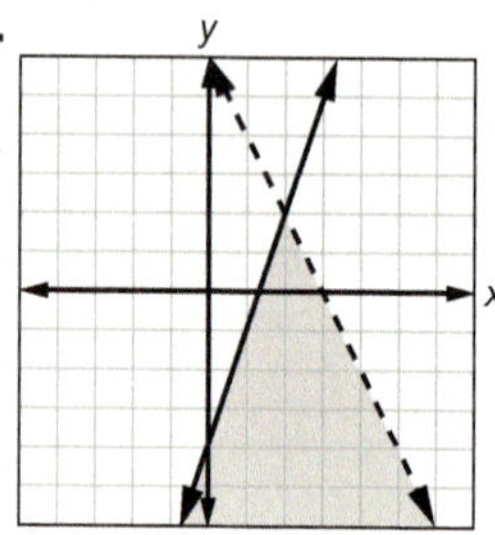

49.

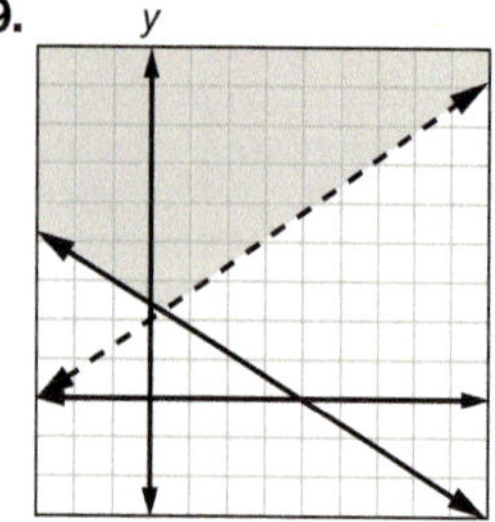

Chapter 8—Exponents

8.1 Products and Powers

1. x^8 **3.** x^7y^5 **5.** $12w^7$ **7.** a^{12} **9.** c^{13} **11.** $4b^4c^2d^{10}$
13. $-10a^2b^3$ **15.** $-12a^4b^2c^2$ **17.** $500x^{11}$ **19.** $-45x^9y^3z^{22}$
21. $16x^{16}y^{10}z^5$ **23.** $5b^2c^2 - bc^2$ **25.** $7x^5y^4 - 6x^4y^5$
27. $6x^5y^2$ **29.** $\frac{1}{6}h^6$ **31.** x^{n+3} **33.** a^{2x+2} **35.** $3x^2y^2 - 2xy^3$
37. $-4r^4s^3 + r^2s^5$ **41.** the product of x factors of b
43. $\frac{1}{-2(-2)(-2)(-2)} = \frac{1}{16}$ **45.** $-\frac{5}{4}\left(\frac{5}{4}\right) = -\frac{25}{16}$
47. The coefficient is 1. **49.** Its slope is undefined.

8.2 Quotients

1. $9x^5$ **3.** $-\frac{3ab^5}{4}$ **5.** z^2 **7.** $\frac{2}{x^3}$ **9.** x **11.** 1 **13.** $\frac{x^6}{y^{15}}$ **15.** $\frac{x^3}{y^3}$
17. $8x^5y^{-9}$ **19.** $\frac{1}{5}b^{-9}c^{-5}$ **21.** $-5a^2b^{-5}c^4$ **23.** $-\frac{6x^3z}{7}$
25. $-\frac{a^6c^5}{b^{11}}$ **27.** $\frac{1}{2a^3}$ **29.** $\frac{1}{x^4y^4}$ **31.** $-\frac{a^2bc^3}{3}$ **33.** $3a^{-3}b$ or $\frac{3b}{a^3}$
35. $6x$ **37.** a^2 **39.** x^{2a-b} **41.** $\frac{x^2}{y^4} - 1$ **43.** $\frac{1}{rs^2} - 3$
47. 0.85 sec **49.** $D = \{-2, 4, 8, 9\}$; $R = \{4, 8, 9\}$; yes
51. rate; time **53.** $t(r) = \frac{300}{r}$ **55.** $y = -2x + 6$
57. $x = -3$

8.3 Scientific Notation

1. 3.9×10^8 **3.** 5.2×10^{-5} **5.** 1.36×10^8
7. 2.356×10^{-2} **9.** 254,000,000 **11.** 0.000153
13. 6×10^7 **15.** 8.1×10^{21} **17.** 5×10^6
19. 9.5×10^3 **21.** 2.77×10^9 **23.** 3.01×10^3
25. 5.625×10^{-29} **27.** 2.73×10^7
29. capillary: 8×10^{-6} m; aorta: 2.5×10^{-2} m
31. 5×10^2 sec **33.** 2.25×10^8 m; 2.25×10^5 km
35. 7×10^5 **37.** 1 **39.** 6×10^5
41. 314×10^3; 314 thousand
43. 53.85×10^{-6}; 53.85 millionths
45. 1,000,000,000,000,000,000,000 **47.** 8×10^{32}
51. $-\frac{3}{2}$ **53.** $4x - y = -10$
55. (2, −2)

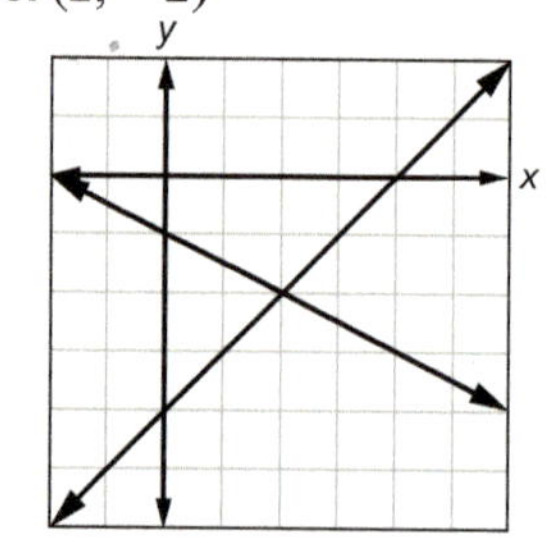

57. (2, 3) **59.** (2, 1)

8.4 Translating Power Functions

1. 5 units up **3.** 7 units left **5.** 5 units right, 6 units up
7. h units right, k units up **9.** (3, 1) **11.** (−8, 6)
13. (0, 19) **15.** (7, 0) **17.** $(h, -k)$
19. (0, 3);
Table values will vary.

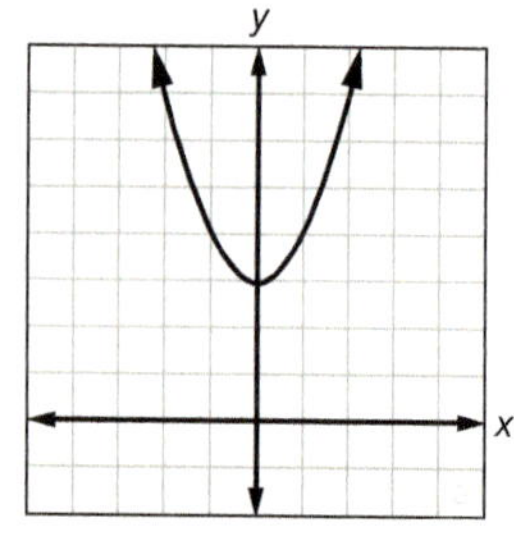

21. (−2, 0);
Table values will vary.

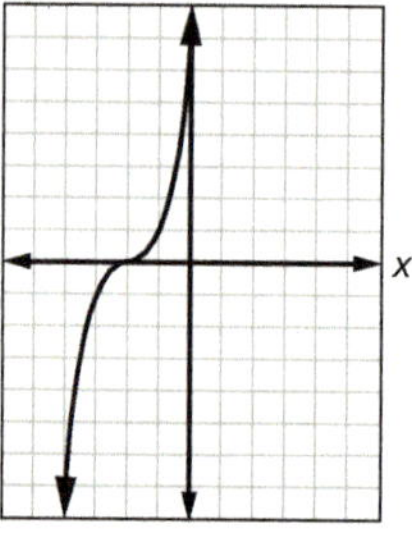

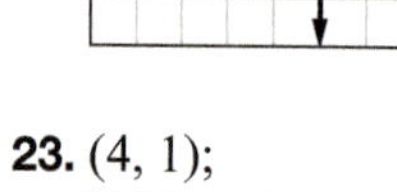

23. (4, 1);
Table values will vary.

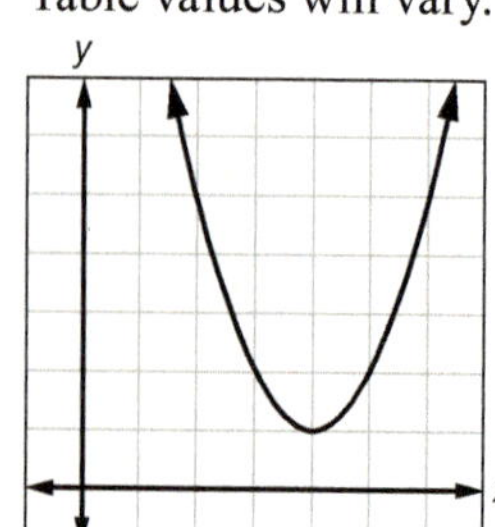

25. $f(x) = (x - 10)^2$
27. $f(x) = (x + 3)^2 - 12$
29. $f(x) = (x - 2)^2 + 1$
31. $y = x^2 + 2$
33. $y = (x - 2)^2 - 2$

35. Table values will vary.

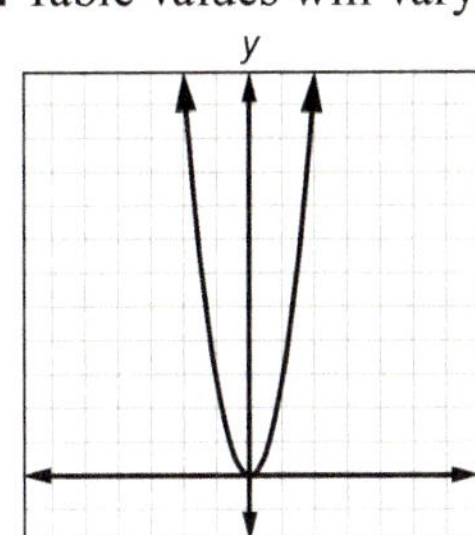

37. Table values will vary.

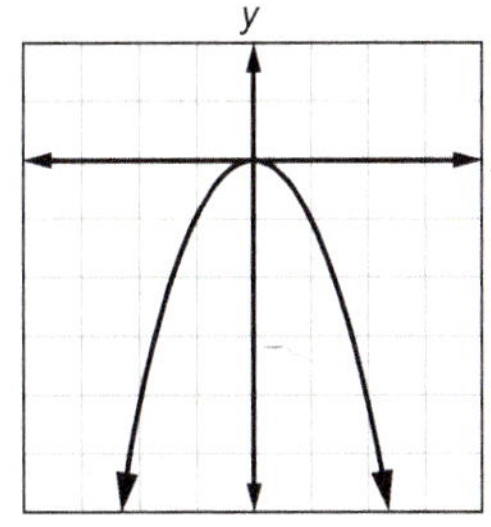

39. It gets "wider." **41.** 2 units down
43. 12 units left, 3 units up **45.** e **47.** c **49.** $(-3, -2)$

8.5 Exponential Functions

1. $\frac{1}{32}$ **3.** 192

5.

x	y
-2	$\frac{1}{9}$
-1	$\frac{1}{3}$
0	1
1	3
2	9

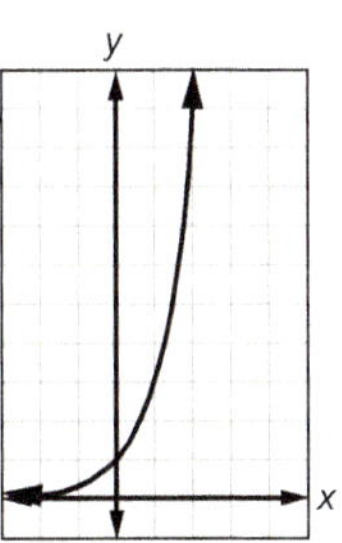

7.

x	y
-3	$\frac{27}{8}$
-2	$\frac{9}{4}$
-1	$\frac{3}{2}$
0	1
1	$\frac{2}{3}$
2	$\frac{4}{9}$
3	$\frac{8}{27}$

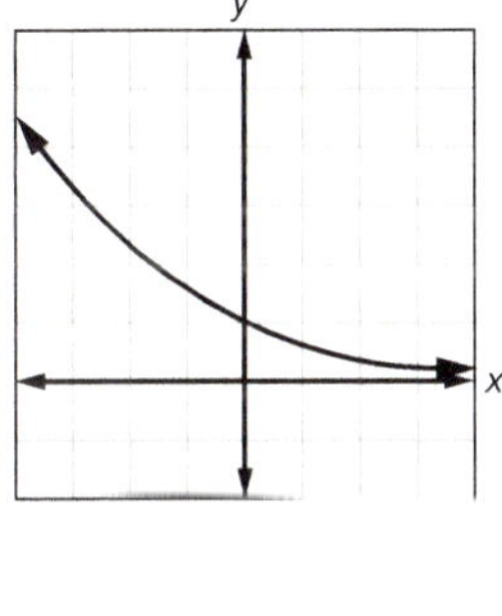

9.

x	y
-2	$\frac{2}{9}$
-1	$\frac{2}{3}$
0	2
1	6
2	18

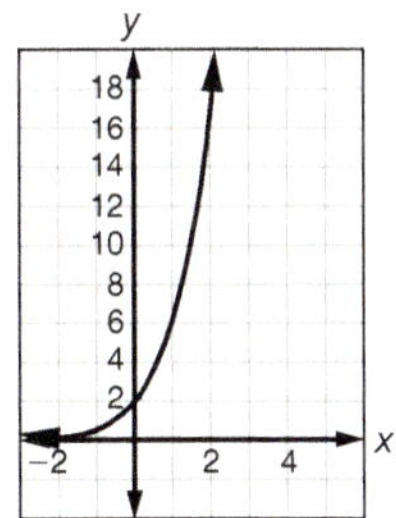

11.

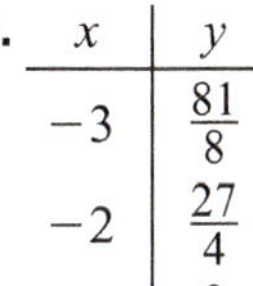
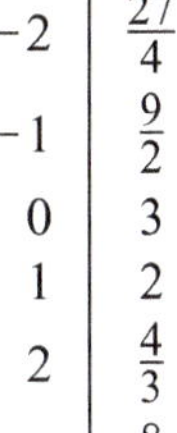

x	y
-3	$\frac{81}{8}$
-2	$\frac{27}{4}$
-1	$\frac{9}{2}$
0	3
1	2
2	$\frac{4}{3}$
3	$\frac{8}{9}$

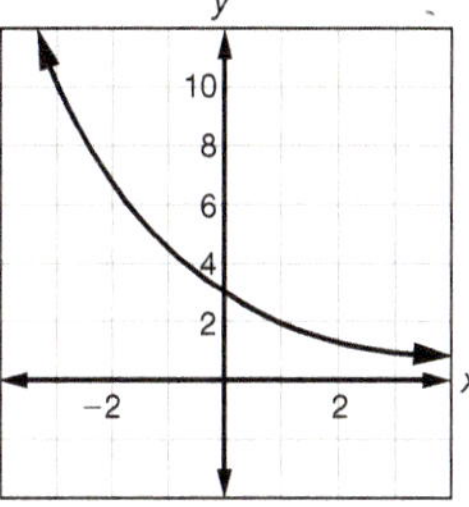

13. e **15.** As x increases, y decreases. **17.** $y = 1.25^x$; 156%
19. $y = 0.75^x$; 32% **21a.** \$1020.00 **21b.** \$1040.40
21c. \$1218.99 **21d.** \$1640.61 **21e.** \$2691.59

23.

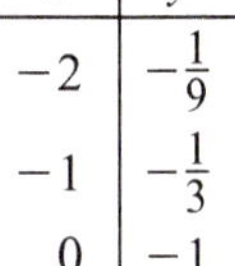

x	y
-2	$-\frac{1}{9}$
-1	$-\frac{1}{3}$
0	-1
1	-3
2	-9

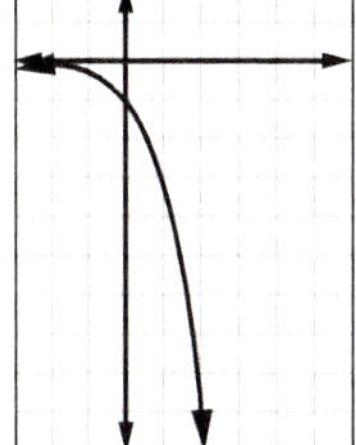

25.

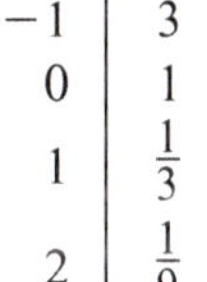

x	y
-2	9
-1	3
0	1
1	$\frac{1}{3}$
2	$\frac{1}{9}$

29.

x	y
-2	$\frac{13}{4}$
-1	$\frac{7}{2}$
0	4
1	5
2	7

31. a **33.** d **35.** 2
37. $y = a$; The graph is a horizontal line.
41. 1×10^{-6} **43.** 6×10^{12} mi **45.** 1.32×10^{19} mi

47.

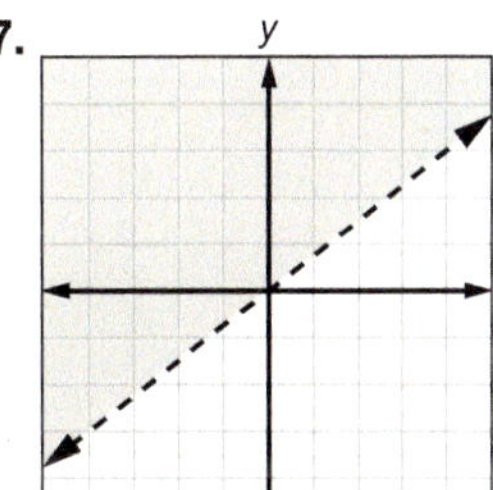

49. $\frac{3}{m^5n^3}$ **51.** $\frac{xz}{y^5}$
53. boat: 25 mi/hr; current: 5 mi/hr
55. \$7500 at 6% and \$2500 at 2%

8.6 Exponential Growth and Decay

1. 35; 5%; 1.05; growth **3.** 1; −30%; $\frac{7}{10}$; decay
5. $f(t) = 10(2.5)^t$; growth
7. $f(t) = 81\left(\frac{3}{2}\right)^t$ or $f(t) = 81(1.5)^t$; growth
9. $f(t) = 75(0.5)^t$ or $f(t) = 75\left(\frac{1}{2}\right)^t$; decay
11. $y = 800{,}000(0.9)^x$ **13.** $y = 1500(2)^x$
15. 10 periods; 6% **17.** 40 periods; 1.5%
19. 3650 periods; ≈ 0.016% **21.** $5000(1.02)^{20}$
23. $37{,}855(1.00\overline{3})^{360}$
25. $f(q) = 10{,}000(1.015)^q$; $f(48) = \$20{,}434.78$
27. $f(h) = 2000(1.03)^h$; $f(40) = \$6524.08$
29. $f(n) = 5000(1.045)^n$; $f(6.5) = \$6656.19$
31a. $f(x) = 500(0.9)^x$ **31b.** 328 mg
31c.

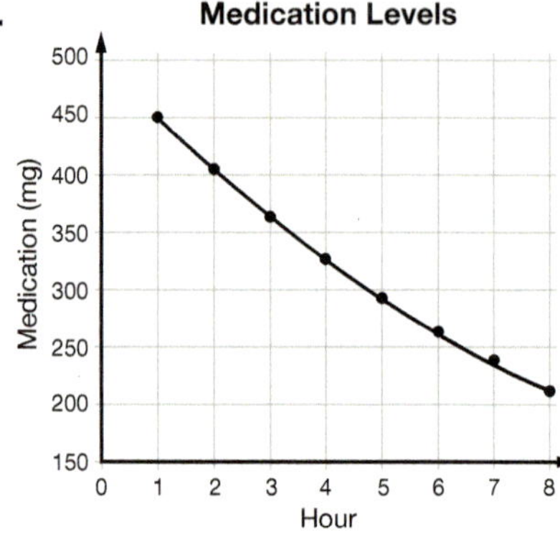

33a. $f(n) = 0.01(2)^n$ **33b.** 10.24 cm
33c.

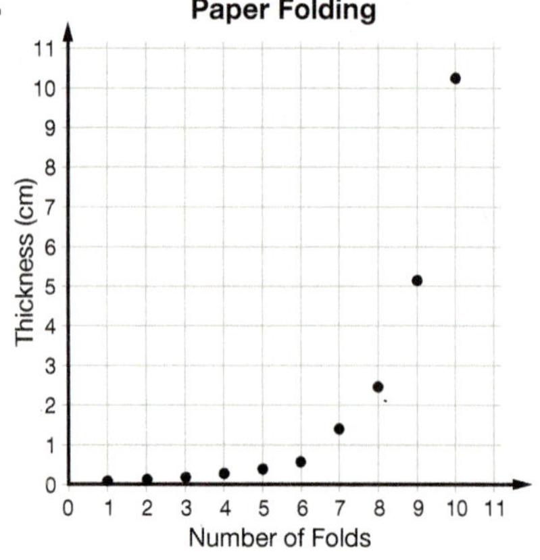

35a. $f(m) = 2(2)^{\frac{m}{3}}$ **35b.** $f(n) = 2(2)^{4n}$ or $f(n) = 2(16)^n$
35c. 8192 squirrels **37.** $0 < b < 1$ **39a.** \$153
39b. \$79 **39c.** \$54 **39d.** \$285 or \$286
41. $2.001 \times 10^9 = 2{,}001{,}000{,}000$ people
43. $1.25 \times 10^3 = 1250$ messages per second
45. no **47.** $y < -\frac{1}{3}x + 3$ and $y \le x + 4$
49. $y = -2x + 3$; consistent dependent
51. ∅; inconsistent **53.** $\frac{9x^8}{y^{11}}$

Sequences Infinite Geometric Sums

1. yes; $r = \frac{1}{5}$ **3.** no; arithmetic sequences diverge **5.** 1
7. $\frac{3}{2}$ **9.** $\frac{7}{9}$

Chapter 8 Review

1. $-18m^3n^{12}$ **3.** $9a^3b^2$ **5.** $4m^2np^4$ **7.** x^3 **9.** 2^x
11. $-3m^{-4}r^5s^{-6}$ **13.** $-\frac{3}{4}x^{-3}y^{-2}$ **15.** $-\frac{15y^3}{x^6z}$ **17.** $\frac{b^{10}}{4a^6}$
19. 3.19×10^{-4} **21.** 3×10^5 **23.** 1.25×10^{14}
25. 5.4×10^{-2} **27.** 1.08×10^9 km **29.** (0, 7) **31.** (2, −4)
33. 3 units right, 2 units up; Table values will vary. **35.** $y = (x + 1)^3 - 2$
37. $-\frac{2}{9}$

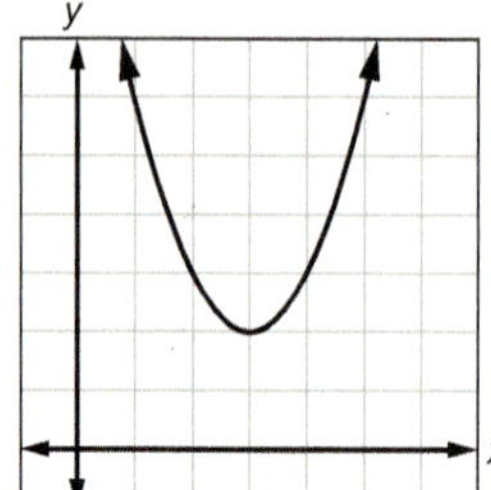

39. Table values will vary.

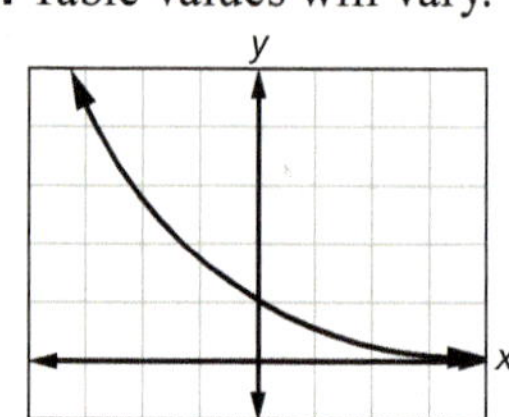

41. (0, 1)
43. 1; 50%; 1.5; growth
45. $y = 0.8^x$; 51.2%
47. $f(x) = 50{,}000(1.05)^x$; \$81,445
49. 40 periods; 0.75%; \$6741.74
51a. 200%; 3 **51b.** $f(d) = 500(3)^d$
51c. 1,093,500 mosquitoes

Chapter 9—Polynomials

9.1 Classifying and Evaluating Polynomials

1. 7 **3.** 0 **5.** binomial; 8 **7.** polynomial; 3
9. not a polynomial **11.** binomial; 3
13. trinomial; 5 **15.** $-2x^2 + 3x + 5$; 2; trinomial
17. $-6x^5 + \frac{1}{3}x^2 - 4x + 12$; 5; polynomial
19. $6x^4 - 7x^2y + 2xy$; 4; trinomial **21.** 9 **23.** −33
25. 3240 **27.** 133 **29.** −3 **31.** −23 **33.** 214 **35.** $-\frac{1}{36}$
43. a **45.** e **47.** 11, 13, and 15 **49.** 3% **51.** growth

9.2 Adding and Subtracting Polynomials

1. $19m - 6$ **3.** $7m^2 + 10m - 5$ **5.** $3a + 8$
7. $-3x + 10y - 6$ **9.** $9x + 3y$
11. $5x^3 + 9x^2 + 2x - y + 15$ **13.** $x + 2y - 14$
15. $-a^3 + 4b^2 + 9c$ **17.** $-8y^2 + 7y + 11$
19. $-4x^3 + 2x^2 - 2x + 9$
21. $-6x^3 - x^2y + xy^2 + y^3 - 2y^4$
23. $8.76a + 7.59b + 8.56$ **25.** $5x + 8y$
27. $-8x^2 - 18xy + 2y^2$ **29.** $x - 16$ **31.** $14x + 3$
33. $3x^2 + 8x + 13$ **35.** $2x + 1$ **37.** $2x^3 + 4x^2 - 8x - 10$
39. $7n + 27$ **41.** $x + (x + 1) = 2x + 1$; Doubling any number $(2x)$ results in an even number. Adding 1 to an even number always results in an odd number. **43.** ≈ 433,902
45. 36 **47.** If $a = b$, then $a + c = b + c$.
49. $8 - 3 = 5$, but $3 - 8 = -5$. **51.** $15x^5$ **53.** $15x - 10y$

9.3 Multiplying Polynomials

1. $x(x + 3) = x^2 + 3x$ **3.** $(x + 4)(2x + 1) = 2x^2 + 9x + 4$
5. $-20ab - 40b^2$ **7.** $9x^2 - 3x - 30$
9. $12x^3 + 12x^2y - 15xy^2 - 9y^3$ **11.** $-56k^3 - 63k^2 - 42k$
13. $12a^6 + 15a^3 - 54a^2$ **15.** $y^2 - 7y + 12$
17. $c^2 + 5c - 24$ **19.** $4x^3 + 5x^2 - 12x - 12$
21. $-24x^5y - 56x^4y^2 - 80x^3y$ **23.** $4x^3 + 2x^2y + 8x$
25. $-8b^4 + 24b^3 + 16b^2 - 48b$
27. $-12x^3y - 9x^2y^2 + 16xy^2 + 12y^3$
29. $6x^3 - 3x^2 - 6x + 48$ **31.** $x^3 - 27$
33. $x^4 + x^3 - 27x^2 - 18x + 54$
35. $x^3 - 15x^2 + 71x - 105$ **37.** $4y^3 - 25y^2 + 8y - 12$
39. $-4x^2 - 12x + 25$ **41.** $x^2 + 2x$ **45.** 2,220,366
47. 94,230,932 **49.** $8\frac{1}{3}$ or $\frac{25}{3}$ **51.** $x = 7$ **53.** $(2, -3)$
55. 4.24×10^{-4} **57.** 3×10^1

9.4 Multiplying Binomials Using FOIL

1. $x^2 + 7x + 10$ **3.** $a^2 + 2a - 24$ **5.** $x^2 - x - 2$
7. $v^2 - 4v - 5$ **9.** $c^2 - 6c + 9$ **11.** $x^2 - 4x - 12$
13. $s^2 - 16s + 63$ **15.** $y^2 - 25$ **17.** $a^2 - 49$
19. $x^2 - 16x + 64$ **21.** $a^2 - 2ab + b^2$ **23.** $x^2 + 6x + 9$
25. $3x^2 + 14x + 8$ **27.** $15z^2 - 8z - 12$
29. $a^2 + 12a + 36$ **31.** $49g^2 + 140g + 100$
33. $3y^2 + 17y - 28$ **35.** $6b^2 + 7b - 24$ **37.** $9h^2 - 16$
39. $6v^2 - 24$ **41.** $3x^4 + 3x^2 - 6$ **43.** $2z^6 - 3z^3 - 44$
45. $a^5 - a^4 + a^2 - a$ **47.** $8x^2 - 10x - 12$
49. $w^2 + 60w + 500$ **51.** $n^2 + 10n + 24$ **53.** $x^2 + 46$
55. $\frac{7}{2}x^2 - 2x - \frac{3}{2}$ **57.** 540,604,795 **61.** $a(bc) = (ab)c$
63. $a^nb^nc^n$ **65.** $y = 30$ **67.** $\frac{1}{2}$ **69.** $-x^3yz$

Sequences Sums of Perfect Squares and Cubes

7. inductive **9.** 44,100

9.5 Special Products

1. $2ab$ **3.** $-$ **5.** $y^2 - 32y + 256$ **7.** $36x^2 - 144x + 144$
9. $a^2 + 6ab + 9b^2$ **11.** $x^2 - 1$ **13.** $25z^2 - 9$
15. $4d^2 - 121$ **17.** $(30 - 3)(30 + 3) = 891$
19. $(30 - 7)(30 + 7) = 851$ **21.** $(60 + 3)^2 = 3969$
23. $(70 - 1)^2 = 4761$ **25.** $16x^2 - 9y^2$
27. $x^4 - 16x^2y^3 + 64y^6$ **29.** $-4b + 8$
31. $z^3 + 3z^2 - 9z - 27$ **33.** $a^3 + 3a^2b + 3ab^2 + b^3$
35. $x^3 - 3x^2 + 3x - 1$ **41.** 8.107 billion **43.** $-\frac{7}{3}$
45. $x = -25$ **47.** $y < 3$ **49.** $7x^4y^4z^7$ **51.** decreasing

9.6 Dividing Polynomials

1. $x + 5$ **3.** $4x^2 - 1$ **5.** $x^2 + 5$ **7.** $a - 6b$
9. $8a^2 - 4a + 3b$ **11.** $-\frac{15}{2}x^2 + \frac{3}{2}x - \frac{27}{2}$
13. $3n^2 + 5n - 1$ **15.** $\frac{1}{3}x^3 + 3x^2 - \frac{5}{3}$ **17.** $x - 2$
19. $3x + 1$ **21.** $a + 10 + \frac{72}{a - 6}$ **23.** $24x - 11 + \frac{2}{x + 1}$
25. $3y - 5$ **27.** $x - 4 + \frac{14}{x + 3}$ **29.** $x^2 + 2x + 11 + \frac{44}{x - 4}$
31. $x^2 + 3x + 9$ **33.** $x^2 - 12 + \frac{64}{x^2 + 4}$ **35.** $2x + 2$
37. $x^3 - 4x - 2 + \frac{16x + 14}{x^2 + 4}$ **39.** $2x + 2$ **41.** \$4.50
43. 1.54% **47.** 0 **49.** $\frac{1}{3}$ **51.** $y = -\frac{5}{2}x + 11$ **53.** c **55.** b

Chapter 9 Review

1. binomial; 1 **3.** trinomial; 3 **5.** 36 **7.** 12
9. $9x^2 + 3x - 14$ **11.** $-a^2 + 6ab - 13b^2$
13. $13x^2 - x + y$ **15.** $3x^3 - 15x^2 + 12x$
17. $-6a^7b^2c^2 - 21a^5b^2c^7$ **19.** $x^3 + 8$ **21.** $x^2 - 2x - 24$
23. $x^2 - 20x + 99$ **25.** $6x^2 + x - 35$ **27.** $y^2 - 36$
29. $x^2 - 4xy + 4y^2$ **31.** $16a^2 - 40ab + 25b^2$
33. $x^3 - 3x^2 - 9x + 27$ **35.** $5x - 7$ **37.** $4xyz + 6y^3$
39. $6x + 3$ **41.** $x + 2 + \frac{9}{x - 2}$ **43.** $x - 1$
45. $x^3 - 2x^2 + 4x - 3$ **47.** $11x^2 + 12x + 3$
49a. $-2x + 11$ **49b.** $-2x + 8.5$ **49c.** $4x^2 - 39x + 93.5$
49d. $4x^3 - 39x^2 + 93.5x$

Chapter 10—Factoring Polynomials

10.1 Factoring Common Monomials

1. 15 **3.** $8x$ **5.** x^2y^2 **7.** $3(x + y)$ **9.** $8(x - 4y)$
11. $a(a - 1)$ **13.** prime **15.** $7(x^2 - 7xy + 4y^2)$
17. $6(a^2 + 8a - 4)$ **19.** $3ab(2a + b)$ **21.** $13a^2b(7ab + 1)$
23. prime **25.** $4x(3x^2 - 4y^2)$ **27.** $14x(4x + 3y - 2)$
29. $7a(a^4 - 12a^2 + 3)$ **31.** $(x - 2)(x + 8)$
33. $(b + c)(5 + a)$ **35.** 1 in. by 42 in., 2 in. by 21 in., 3 in. by 16 in., and 6 in. by 7 in. **37.** 13 cookies
39. $2\pi(x + y)$; $x + y$
41. $17a^2b^5(2a^5b^3c^2 + 3ad^4 - 17b^4c^{10}d)$
43. 57π in.^{3} **45a.** 555.6 m **45b.** 1822 ft
45c. 405.6 m; 1330 ft **47.** $R = \{1, 2, 3, 4\}$
49. $\{(-3, 1), (-1, 3), (2, 4), (4, 2)\}$
51. $x^2 - 3x - 40$ **53.** $x^2 + 3x - 40$
55. $4x^2y - 28xy - 32y$

Sequences Factorial Sequences

1. $3 \cdot 2 \cdot 1 = 6$ **3.** 8! **5.** 2.667, 2.708, 2.717
7. 2.7182815, 2.7182818, 2.7182818

10.2 Factoring Trinomials of the Form $x^2 + bx + c$

1. $(a + 3)(a + 5)$ **3.** $(a - 3)(a - 1)$ **5.** $(b + 5)(b - 4)$
7. $(a + 4)(a - 2)$ **9.** prime **11.** $(x - 5)(x - 4)$
13. $(y + 5)(y + 2)$ **15.** prime **17.** $(x + 10y)^2$
19. $(x - 2y)(x - 48y)$ **21.** $(a - 18b)(a + 4b)$
23. $(x + 4)(x + 14)$ **25.** $(x - 3)(x + 18)$
27. $3(x - 9)(x + 1)$ **29.** $3(x + 6)(x + 2)$
31. $x^2(x + 8)(x - 2)$ **33.** $x^2 + 32x + 60$; $(x + 30)(x + 2)$
35. $4a^2b(a + 2b)(a - 5b)$

37.

39. 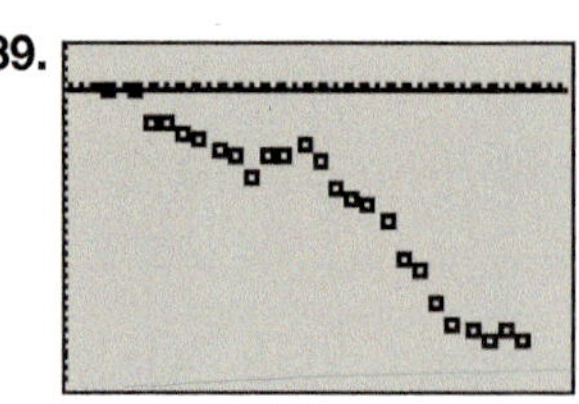

41. trinomial **43.** 6 **45.** $x^2y^3(6x - 27yz + 8)$
47. $6x^2 + x - 35$ **49.** 62 mi/hr

10.3 Factoring Trinomials of the Form $ax^2 + bx + c$

1. $(2x + 3)(x + 4)$ **3.** $(x - 4)(3x - 2)$ **5.** $(3x - 7)(x + 4)$
7. $(3x + 4)(2x - 1)$ **9.** $(3c + 2)(5c + 3)$
11. $(2a + 1)(3a - 4)$ **13.** $(5x + 1)(x - 3)$
15. $(5x - 7y)(x + 3y)$ **17.** $3(2x + 5)(x - 2)$
19. $8(2a - b)(a + 3b)$ **21.** $(m - 5n)(m + 4n)$
23. $(x - 2y)(x + 5y)$ **25.** $2(x - 8)(x + 4)$
27. $3ab(b - 1)(4b + 7)$ **29.** $3(5x + 3)(3x - 2)$
31. $2(x - 13y)(x + 4y)$ **33.** $6(8x - 3)(3x - 1)$
35. $(3x + 4)(5x + 4)$ **37.** 5 : 3 **41.** $25x^2 - 9$ **43.** (6, 2)
45. $3x - 4$ **47.** $x < 5$ **49.** 20 hr

10.4 Special Patterns

1. +, − or −, + **3.** 2, 2 **5.** $6a$, 7 **7.** yes; $3x$ doubled = $6x$
9. no; x^2y is not a perfect square
11. no; $2x(5)$ doubled = $20x$ **13.** $(a - 15b)(a + 15b)$
15. $(x - 3)^2$ **17.** $(x - 8)^2$ **19.** $(a - 5b)^2$
21. $2(a - 11)(a + 11)$ **23.** $(3x + 2)^2$
25. $x^2(a - 4)(a + 4)$ **27.** $(3x + 5z)^2$
29. $4(2x - 1)(2x + 1)$ **31.** prime **33.** $7(3x - 5)(3x + 5)$
35. $18b^4c(3a - 1)(3a + 1)$ **37.** $3xy(3x^2 - 4y)^2$
39. 17,952 days ≈ 49 yr
41a. $y \approx -3.66x^2 + 30.86x + 196.34$
41b. 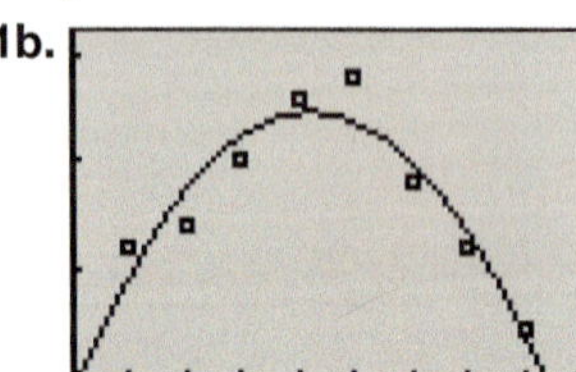
43. −15
45. $x = 2$
47. −2323

49. $\frac{-2323}{23} = -101$ or $\frac{-2323}{-101} = 23$;
$\frac{7373}{73} = 101$ or $\frac{7373}{101} = 73$
51. $-5x(8y^2 - y - 4)$; $\frac{-40xy^2 + 5xy + 20x}{-5x} = 8y^2 - y - 4$
or $\frac{-40xy^2 + 5xy + 20x}{8y^2 - y - 4} = -5x$

10.5 Factoring Completely

1. $5(a + 1)^2$ **3.** $2(10 - a)(10 + a)$ **5.** $3(x - 2)(x - 4)$
7. $5(x - 7)^2$ **9.** $(x + y)(2x - y)$ **11.** $(a + b)(t - w)$
13. $(a + b - 2)(m + n)$ **15.** $8(a - 3b^2)^2$
17. $(x^2 + 4)(x + 2)(x - 2)$ **19.** $3(3x + 4)(2x - 3)$
21. $2(4w - 3)(5w - 4)$ **23.** $(w + x)(y + z)$
25. $(x + 7)(y - 6)$ **27.** $(x + 4)(6x + y^2)$
29. $(y - 2)(y + 2)(y - 3)$ **31.** $5(y + 1)(y - 1)(y - 3)$

33. $(x - 2)(x + 3)(x - 4)$ **35.** $(y + 1)(y - 3)(2y - 5)$
37. $(x + y - 3)(x + y + 3)$ **39a.** 10,496 ft
39b. $\approx 2.3 \times 10^6$ ft^3 **41a.** 0.8 km/min **41b.** 30 mi/hr
43. $q = \frac{7n + 5}{A}$ **45.** $q = \frac{7}{x + 5}$
47. 21 quarters, 14 dimes, and 7 nickels
49. current: 0.5 mi/hr; boys: 4.5 mi/hr
51. 5 lb of the alloy and 5 lb of copper

Chapter 10 Review

1. $r(3r + 5)$ **3.** $4x^2(x - 2)$ **5.** $(y - 7)(y - 8)$
7. $(2d + 7)(d + 1)$ **9.** $(x + 7)(x - 2)$
11. $(5b + 3)(3b - 2)$ **13.** $(a - 3)(a + 3)$
15. $(a + 7)^2$ **17.** $y(x + 8)(x - 7)$ **19.** $2(9x - 5)(x - 5)$
21. $2(x - 9)(x - 6)$ **23.** prime **25.** $2(3y - 5)(3y + 5)$
27. $24(x - 2y)(x + 2y)$ **29.** $(3b - 2c)^2$
31. $(5x + 2y)(3x - y)$ **33.** prime **35.** $2(3a - 7)(a - 1)$
37. $2(3y - 5)(2y + 7)$ **39.** $(2x^2 - 3)(5x + 7)$
41. $(c^2 + 2)(c^2 + 1)$
43. $(2a - 1)(2a + 1)(4a^2 + 1)(16a^4 + 1)$
45. $(x - 1)(xy + x + 1)$ **47.** $x^2 + xy + y^2$ **49.** 35 in.^{2}

Chapter 11—Radicals

11.1 Expressing Roots

1. 7 **3.** −6 **5.** $\frac{7}{4}$ **7.** 0.1 **9.** not real **11.** 2 **13.** −3
15. $\frac{5}{4}$ **17.** 1492 **19.** 1776 **21.** 4 **23.** 17 **25.** 3.5
27. 6.557 **29.** 83.0 **31.** $3^{\frac{1}{2}}x^{\frac{3}{2}}$ **33.** $3^{\frac{1}{5}}x^{\frac{2}{5}}y^{\frac{3}{5}}$ **35.** $\sqrt{3y}$
37. $\sqrt[6]{7a^3b^2}$ **39.** $\sqrt[8]{27r^2s^4}$ **43.** 625 **45.** 1.618034
47. $x \le -1$ **49.** 6 **51.** $-10a^3b^4c^6$ **53.** $32x^8$
55. $(2x - 17)(2x + 17)$

11.2 Simplifying Radicals

1. $2\sqrt{3}$ **3.** $3\sqrt{3}$ **5.** $5\sqrt{2}$ **7.** $8\sqrt{3}$
9. cannot be simplified **11.** $3\sqrt[3]{2}$ **13.** $2\sqrt[3]{20}$
15. cannot be simplified **17.** $|x|$ **19.** $y^2\sqrt{y}$ **21.** $4|x|$
23. $3|x|\sqrt{2y}$ **25.** $5xy\sqrt{2xy}$ **27.** $3x^3y^4\sqrt{2xy}$
29. $5ac^4\sqrt{3ab}$ **31.** $\frac{|x^3y^3|}{10}$ **33.** cannot be simplified
35. $2ab^3$ **37.** 240 mi/hr **39.** 60 mi/hr
41. cannot be simplified **43.** $9x^3yz\sqrt[3]{15xy^2z^2}$
45. $66x^4y^2z^3\sqrt[4]{10y^2z}$ **47.** 1.618034 **49a.** 1.5; 1.667
49b. no **51.** 123 **53.** 9 **55.** $(a + 8)(a - 4)$
57. $(m + 14)^2$ **59.** $3b^6(2b + 1)(2b - 1)$

11.3 Multiplying Radicals

1. $\sqrt{15}$ **3.** $2\sqrt{6}$ **5.** $2\sqrt{15}$ **7.** $5\sqrt{2}$ **9.** $14\sqrt{5}$
11. $\sqrt{590}$ **13.** $\sqrt{6} - \sqrt{3}$ **15.** $7\sqrt{3} + \sqrt{14}$ **17.** $6\sqrt[3]{10}$
19. $14\sqrt[3]{2}$ **21.** $2b\sqrt{5a}$ **23.** $4w^8z^2$ **25.** $16a^4c^4\sqrt{3a}$
27. $2x^3y^2\sqrt[3]{x^2y^2}$ **29.** $5a^2bc^3\sqrt[3]{4a^2b^2c}$ **31.** 45 in.^{2}
33. x^3 units^2 **35.** 6 m^2 **37.** $57\sqrt[3]{3}$ **39.** $x^{\frac{1}{2}}$; $x^{\frac{1}{3}}$ **41.** $\sqrt[6]{x^5}$
43. $\frac{21}{13}, \frac{34}{21}, \frac{55}{34}$; 1.615, 1.619, 1.618 **45.** 1 **47.** $n = -1$
49. $-\frac{r^2s}{7}$ **51.** $\frac{x^{10}}{4}$ **53.** $n^3 + 5n - 1$ **55.** $a^{\frac{7}{8}}b^{\frac{1}{2}}c^{\frac{1}{4}}$

Sequences Summation (Sigma) Notation

1. $3 + 5 = 8$ **3.** $1 + 8 + 27 + 64 = 100$
5. $\frac{4(1 - 4^{11})}{1 - 4} = 5{,}592{,}404$ **7.** $\frac{1}{2}, \frac{1}{6}, \frac{1}{12}, \frac{1}{20}$
9. $S_n = \frac{n}{n + 1}; \frac{20}{21}$

11.4 Dividing Radicals

1. $\frac{2}{5}$ **3.** $\frac{\sqrt{5}}{2}$ **5.** $\sqrt{2}$ **7.** $\frac{\sqrt{2}}{3}$ **9.** $\frac{3\sqrt{13}}{13}$ **11.** $\frac{\sqrt{6}}{6}$ **13.** $\frac{\sqrt{6}}{3}$
15. $\sqrt{3}$ **17.** $\frac{\sqrt[3]{4}}{2}$ **19.** $\frac{\sqrt[3]{2}}{2}$ **21.** $\frac{\sqrt[3]{25}}{5}$ **23.** $\frac{2\sqrt{7x}}{7y}$ **25.** $r\sqrt{s}$
27. $\frac{z^2\sqrt{35txz}}{5t}$ **29.** $\frac{y\sqrt[3]{3x^2}}{x}$ **31.** $3\sqrt{3}$ in. **33.** $\sqrt{3} \approx 1.7$ m^2
35. $\frac{3\sqrt{5}}{5xy}$ **37.** 1 **39.** $\frac{\sqrt[4]{14x}}{3x}$ **41.** 2.618034 **43.** $m = -\frac{1}{3}$
45. 

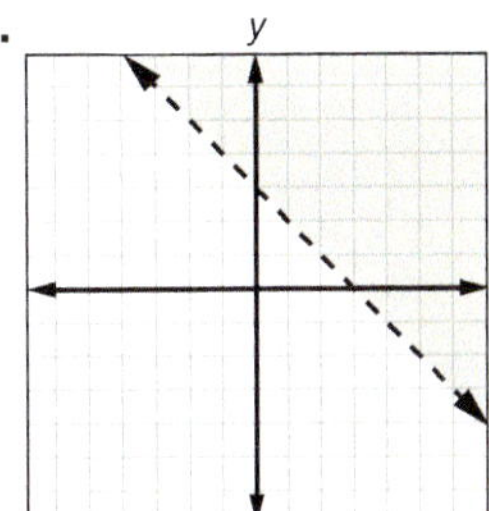

47. $4x^2 - 4xy + y$
49. $2x^3y - 11x^2y^3 - 4xy - 15$
51. $2(b - 7)^2$

11.5 Adding and Subtracting Radicals

1. $6\sqrt{3}$ **3.** $2\sqrt{3} + 4\sqrt{2}$ **5.** $5\sqrt{2}$ **7.** $4\sqrt{3}$ **9.** $\sqrt{2}$
11. $8\sqrt[3]{2}$ **13.** 5, 12, 17, 13
15. $3\sqrt{5}, 5\sqrt{3}, 3\sqrt{5} + 5\sqrt{3}, 2\sqrt{30}$ **17.** $3\sqrt{5} - 6\sqrt{2}$
19. $2\sqrt[3]{2} + \sqrt[3]{18}$ **21.** $7\sqrt[3]{2} + 3\sqrt[3]{3}$ **23.** $6\sqrt{x} + 3\sqrt{xy}$
25. $-2a\sqrt{3} + \sqrt{2}$ **27.** 0 **29.** $\frac{7x}{12}$ **31.** $16\sqrt{3}$ units
33. $\frac{3\sqrt{455}}{2}$ cm^2 **35.** $3x^2y\sqrt[3]{2y}$
37. $9x\sqrt{x} - 4\sqrt{x} + x + 2$ **39.** $2xy\sqrt[3]{x^2y}$ **41.** $\frac{13}{8}$; 1.625
43. $\frac{34}{21}$; 1.619 **45.** 1 **47.** $x = 5$ **49.** 22; 3
51. $4(x - 2y)(x + 2y)$ **53.** $2x + 7 + \frac{4}{3x - 5}$

11.6 The Pythagorean Theorem

1. 5 **3.** $\sqrt{41}$ **5.** $2\sqrt{15}$ **7.** $4\sqrt{2}$ **9.** yes **11.** no **13.** yes
15. no **17.** (4, 5) **19.** (−7.5, −7) **21.** 5 **23.** $5\sqrt{2}$
25. 13 **27.** $3\sqrt{10}$ **29.** $\sqrt{65}$ **31.** $\sqrt{71}$ **33.** $c = 4$
35. $c = 2\sqrt{6}$ **37.** $(3\sqrt{2}, -2\sqrt{5})$ **39.** $7 + \sqrt{17}$
41. $2\sqrt{10} + 2\sqrt{29}$ **43.** $\sqrt{0.45} \approx 0.67$ mi
45. $\sqrt{157} \approx 12.5$ nautical miles **47.** 15 ft **49a.** $\sqrt{l^2 + w^2}$
49b. $\sqrt{l^2 + w^2 + h^2}$ **51.** (44, 240, 244) **55.** $2\sqrt{x} - 3$
57. $2\sqrt{x - 3}$ **59.** 1.5×10^{-5} **61.** $\sqrt{x} - 2$
63. $15x^2 + x - 6$

11.7 Multiplying and Dividing Radical Expressions

1. 1 **3.** −8 **5.** $8 - 2\sqrt{15}$ **7.** $28 - 7\sqrt{11} + 4\sqrt{3} - \sqrt{33}$
9. $4\sqrt{3} + 4\sqrt{6} - 4 - 4\sqrt{2}$ **11.** $-3 - \sqrt{5}$ **13.** $2 - \sqrt{3}$
15. $\sqrt{33} - \sqrt{22}$ **17.** $\frac{\sqrt{2}}{2}$ **19.** $-4\sqrt{2} - 4$
21. $5\sqrt{105} + 10\sqrt{7} + 25\sqrt{3} + 10\sqrt{5}$
23. $x^2\sqrt{3} + 2\sqrt{3x} - 5x^2 - 10\sqrt{x}$
25. $5\sqrt{3} - 6\sqrt{15} - 5\sqrt{2} + 6\sqrt{10}$ **27.** $12\sqrt{5}$
29. $\sqrt{3} + 1$ **31.** $\frac{-5\sqrt{2} - 5\sqrt{6}}{4}$ **33.** $4\sqrt{7} - 4\sqrt{3}$
35. $42\sqrt{10} + 12\sqrt{30}$ units^2
37. $\frac{12\sqrt{6} - 3\sqrt{10} - 32\sqrt{21} + 8\sqrt{35}}{43}$
39. $\sqrt[6]{32}$ **41.** 6.8 cm **43a.** 633 pixels **43b.** 391 pixels
43c. 1.619 **45.** $x < -7$ **47.** $1 + \frac{-2x + 1}{x^2 - 3x + 2}$
49. $2(a - 11b)(a + 2b)$ **51.** $\frac{8\sqrt[3]{25}}{5}$ **53.** $3\sqrt{7}$

11.8 Radical Equations

1. $x = 16$ **3.** $x = 62$ **5.** $x = 36$ **7.** $x = -5$ **9.** $x = -1$
11. $x = 100$ **13.** no solution **15.** $y = 124$ **17.** $x = 19$
19. $x = \frac{73}{16}$ **21.** 72 **23.** 180 ft **25.** $D = \frac{S^2}{30f}$ **27.** 99.4 cm
29. 968 ft **31.** $d = \frac{484V^2}{225g}$ **33.** $x = 2$ **35.** 40 m
37. 1.618 **43.** $m = \frac{1}{2}$; (0, 2) **45.** $\left(x + \frac{1}{3}\right)\left(x - \frac{1}{3}\right)$
47. $x - 13$ **49.** $-\frac{9\sqrt{5}}{10}$ **51.** $f(x) = (x - 4)^2 - 1$

11.9 Radical Functions

1. $x \geq 0$ **3.** $x \geq 0$ **5.** $x \geq 5$ **7.** $x \geq -2$

9. Table values will vary.

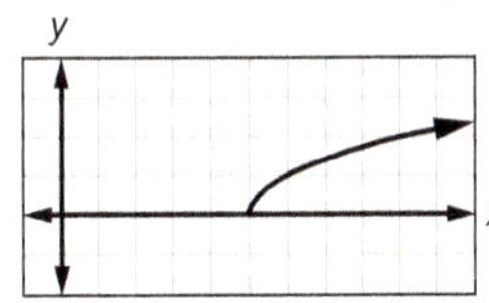

11. Table values will vary.

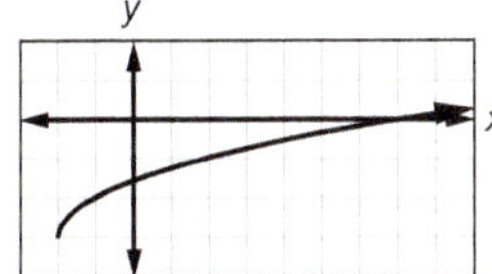

13. Table values will vary.

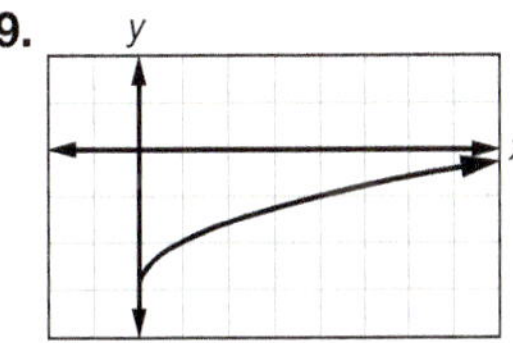

15. Table values will vary.

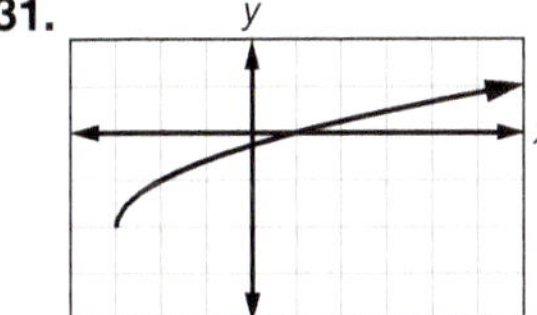

17. a **19.** d **21.** e **23.** $f(x) = \sqrt{x} + 1$
25. $f(x) = \sqrt{x - 2} - 2$ **27.** $f(x) = \sqrt{-x}$
29.
31.

33.

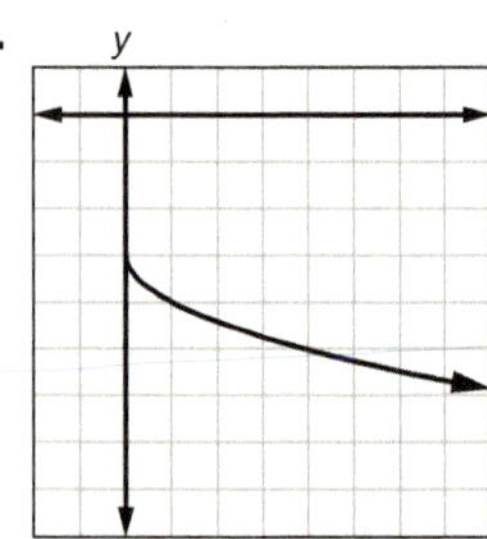

35. It is the reflection of $y = \sqrt{x}$ across both axes, which causes the graph to lie in the third quadrant.

37.

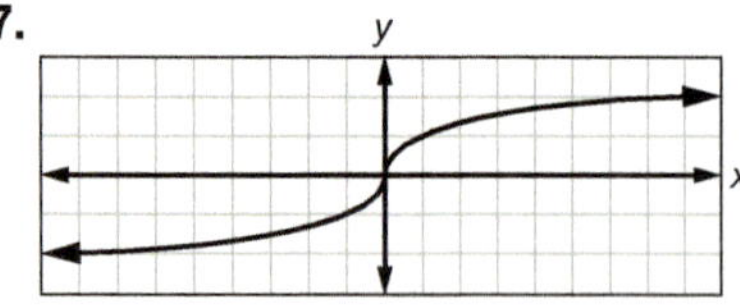

39.

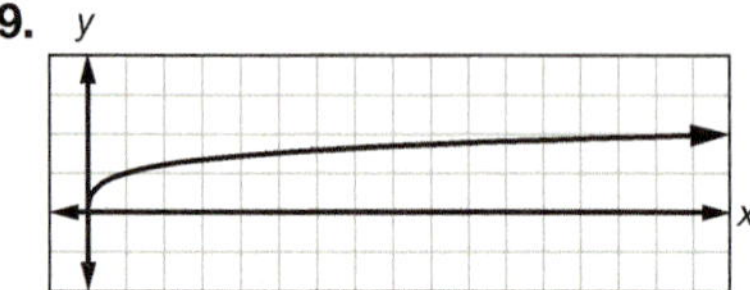

43. $\sqrt{51} \approx 7$ units **45.** $15; \left(\frac{5}{2}, 2\right)$ **47.** $y = -\frac{3}{4}x$
49. $\frac{a^3\sqrt{6}}{3b}$ **51.** $3 - 2\sqrt{2}$

Chapter 11 Review

1. 11 **3.** 0.79 **5.** $2^{\frac{1}{2}}x^{\frac{3}{2}}$ **7.** $a\sqrt[3]{9a^2}$ **9.** -2 **11.** $2\sqrt{21}$
13. $2|a|b^2\sqrt{3a}$ **15.** $3\sqrt[3]{7}$ **17.** $6ab^2\sqrt{35a}$
19. $7\sqrt{5} + 5\sqrt{7}$ **21.** $\frac{5\sqrt{6}}{6}$ **23.** $\frac{\sqrt{21xy}}{9xy}$ **25.** $4\sqrt{6}$ **27.** 0
29. -1 **31.** $\frac{18 - 6\sqrt{2}}{7}$ **33.** $6\sqrt{2} \approx 8.49$ units **35.** $2\sqrt{5}$
37. $(-1, 8)$ **39.** $x = 256$ **41.** no solution **43.** $4\sqrt[3]{25}$ cm²
45. $\frac{\pi\sqrt{110}}{2} \approx 16.5$ sec **47.** 36.6 in. × 20.6 in.
49. $x \geq 2$; Table values will vary. **51.** e

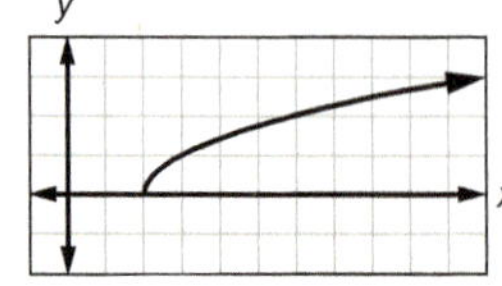

Chapter 12—Quadratic Functions

12.1 Solving Quadratic Equations by Factoring

1. $x = 2, -4$ **3.** $z = 0, -7$ **5.** $x = 0, -1, 10$
7. $b = -\frac{1}{4}, -\frac{2}{3}$ **9.** $y = -3$ **11.** $x = \pm 3$ **13.** $a = -9, 4$
15. $\{0, 1\}$ **17.** $\{0, 7\}$ **19.** $\{2, 8\}$ **21.** $\{-4, 6\}$
23. $\{\pm 13\}$ **25.** $\left\{\pm\frac{3}{2}\right\}$ **27.** $\{-3, 24\}$ **29.** $\left\{-\frac{3}{2}, \frac{4}{5}\right\}$
31. $\{-2, 0, 1\}$ **33.** 15 and 16 or -16 and -15 **35.** 3 ft
37. 14 **39.** $\left\{-\frac{6}{5}, 0, \frac{7}{3}\right\}$ **41.** ≈ 9.8 ft/sec **43.** $x = 2, 6$
45. $x = -9, -5$ **47.** $k = 7, 8, 13$ **49.** $6\sqrt{3}$ **51.** $2\sqrt{33}$

12.2 Solving Quadratic Equations by Taking Roots

1. $x = \pm 5$ **3.** $x = \pm\sqrt{6}$ **5.** $x = \pm 10$ **7.** $x = \pm 6$
9. no real solution **11.** $x = \pm\sqrt{17}$ **13.** $x = \pm 3\sqrt{3}$
15. $x = \pm\frac{5\sqrt{2}}{2}$ **17.** $x = \pm 5$ **19.** $x = \pm\sqrt{7}$
21. $x = -2 \pm 2\sqrt{3}$ **23.** $x = 8, -14$ **25.** $x = -2$
27. $x = \frac{1}{2}, -1$ **29.** $x = \frac{1}{2}, -\frac{5}{2}$ **31.** 12 ft **33.** 64 ft
35. ≈ 3.5 sec; 5 sec; The object continues to fall faster as time progresses. **37.** $x = \frac{5 \pm 4\sqrt{2}}{2}$
39a. $d = |h_c - h_i|$ or $d = |h_i - h_c|$ **39b.** $h_c = -\frac{1}{2}gt^2 + h_i$
39d. 9.8 m/sec²; $h_c = -4.9t^2 + h_i$
41a. $a \approx -4.06$; $h(x) = -4.06x^2 + 65$
41b. $a \approx -2.15$; $h(x) = -2.15x^2 + 65$
41c. $a \approx -2.88$; $h(x) = -2.88x^2 + 65$
43. Double the denominator. **45.** $\left(\frac{1}{2} \cdot \frac{3}{8}\right)^2; \frac{9}{256}$
47. $x = \pm\frac{8}{3}$ **49.** $x = -\frac{1}{2}, \pm 2$
51. 21 nickels, 24 dimes, and 19 quarters

12.3 Completing the Square

1. 16 **3.** 64 **5.** 1 **7.** $\frac{9}{4}$ **9.** $\frac{121}{4}$ **11.** $\frac{9}{16}$ **13.** $\frac{25}{121}$
15. $a = 3, -5$ **17.** $x = 1, -5$ **19.** $\{-2, 12\}$
21. $\{-9, 11\}$ **23.** $m = \frac{4}{3}, 0$ **25.** $x = 1, -2$
27. $x = -10 \pm \sqrt{85} \approx -0.78, -19.22$
29. $x = 9 \pm 6\sqrt{2} \approx 17.49, 0.51$
31. $x = 4 \pm 2\sqrt{3} \approx 7.46, 0.54$
33. $x = 2 \pm \sqrt{7} \approx 4.65, -0.65$ **35.** 10 in. on each side
37. $x = \frac{-3 \pm \sqrt{5}}{2}$ **39.** 44 ft × 54 ft **41a.** 1520 ft²
41b. 28% **43.** $-\frac{26}{49}$ **45.** $x^2 + \frac{5}{3}x - 5$ **47.** $x^2 - 6x + 2$
49. no real solution **51.** $x = 13$

12.4 Completing the Square with Leading Coefficients

1. $x = -3 \pm \sqrt{11}$ **3.** $a = 8, 0$ **5.** $y = -4 \pm \sqrt{19}$
7. no real solution **9.** $x = 4 \pm 2\sqrt{5}$ **11.** $y = 1, -\frac{3}{2}$
13. $c = \frac{5}{2}, -\frac{1}{2}$ **15.** $x = 1, -\frac{1}{2}$ **17.** $g = \frac{1 \pm \sqrt{17}}{4}$
19. no real solution **21.** $k = 2, -\frac{4}{3}$ **23.** $x = -\frac{1}{2}, -1$
25. $m = \frac{1 \pm 2\sqrt{3}}{2}$ **27.** $y = \frac{-12 \pm \sqrt{142}}{2}$ **29.** 11 and 21
31. 32 ft × 49 ft **33.** 13 and 25 **35.** opponent's 13.5 yd line
37. $x = \frac{-12 \pm \sqrt{139}}{5}$ **39.** $y \approx 0.56, -3.56$; $x \approx \pm 0.75$
41. 42 **43.** $-\frac{38}{27}$ or $-1\frac{11}{27}$ **45.** $4x^2 + 22x + 30$ **47.** $4\sqrt[4]{6}$
49. $\frac{5 + \sqrt{5}}{2}$

12.5 The Quadratic Formula

1. $a = 1, b = 6, c = 5$ **3.** $a = 1, b = 5, c = 6$ **5.** $a = 1, b = 0, c = -6$ **7.** $\{-4, -2\}$ **9.** $\{-5, -3\}$ **11.** $\{-3, 3\}$ **13.** $\left\{-1, \frac{3}{2}\right\}$ **15.** $\{3\}$ **17.** $x = 4, \frac{1}{2}$ **19.** $y = \frac{5 \pm \sqrt{17}}{4}$ **21.** $x = \pm\frac{\sqrt{21}}{3}$ **23.** $x = \frac{5 \pm \sqrt{105}}{4}$ **25.** $m = -\frac{2}{7}, -1$ **27.** $k = \frac{1 \pm \sqrt{31}}{3}$ **29.** $x = \frac{7}{2}, -\frac{1}{4}$ **31.** $x \approx 2.39, 0.28$ **33.** Multiplication Property of Equality **35.** 12 and 18 **37.** ≈ 4.6 in. **39.** $x \approx 2.30, 0.66$ **41.** $y = \frac{-s \pm \sqrt{s^2 - 4rt}}{2r}$ **43.** yes **45.** $0 = 2.2x - 22$ **47.** At 10 ft from the first fountainhead, the two fountains intersect at a height of 3 ft. **49.** irrational **51.** rational **53.** irrational **55.** 40% **57.** 47.5%

12.6 More Quadratic Equations

1. a **3.** b **5.** d **7.** 196; a **9.** 0; c **11.** 97; b **13.** 17; b **15.** $x = 3, -15$; rational **17.** $a = \frac{-3 \pm \sqrt{5}}{2}$; irrational **19.** $x = \frac{7 \pm \sqrt{33}}{4}$; irrational **21.** $x^2 + 4x - 32 = 0$ **23.** $8x^2 - 10x + 3 = 0$ **25.** $y = 2, \frac{1}{6}$ **27.** $a = -\frac{4}{3}$ **29.** $x = -4 \pm 3\sqrt{2}$ **31.** $a = -22 \pm \sqrt{482}$ **33.** $x = \frac{8}{3}, -\frac{3}{2}$ **35.** no real solution **37.** 12 in. × 15 in. **39.** $y = -\frac{7}{6}$ **41.** $x \approx 7.110, -2.110$ **43.** $9x^2 + 18x + 4 = 0$ **45.** $x = \frac{14}{5}$

47.

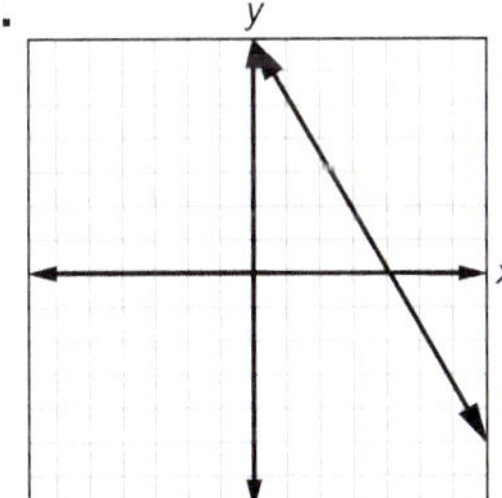

49–51. odd

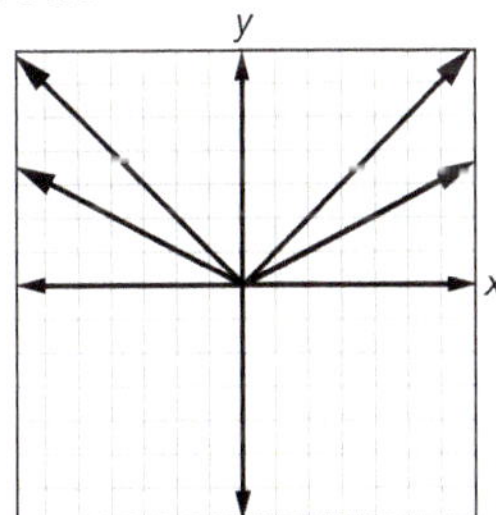

53. ≈ 8.9 ft

12.7 Quadratic Functions: $f(x) = ax^2 + c$

1. 0, 4, 1, 16 **3.** $-4, 0, -3, 12$ **5.** $0, -8, -2, -32$ **7.** $-7, 5, -4, 41$ **9.** up, (0, 0), minimum, steeper **11.** down, (0, 4), maximum, same

13.

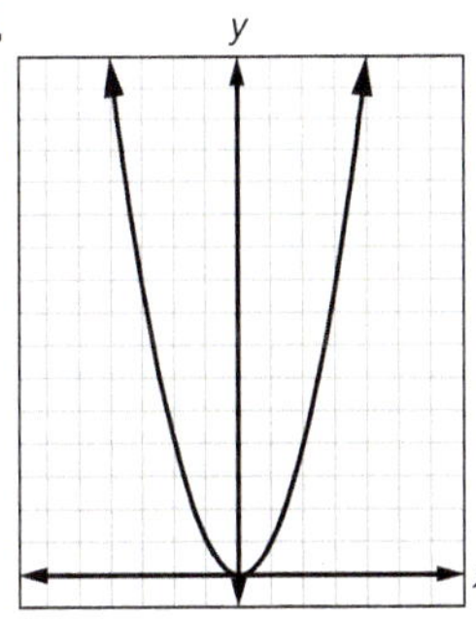

15.

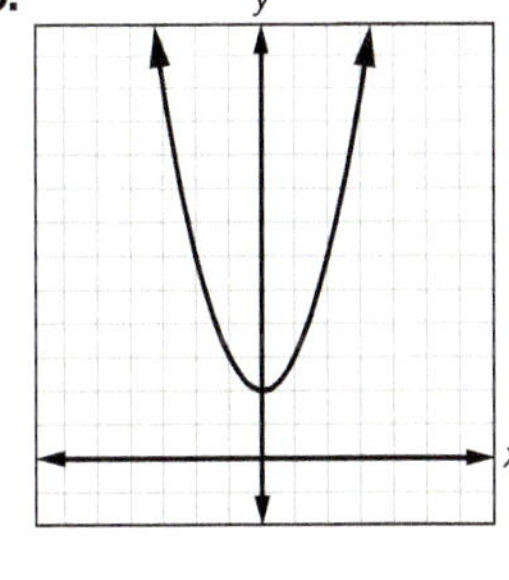

17.

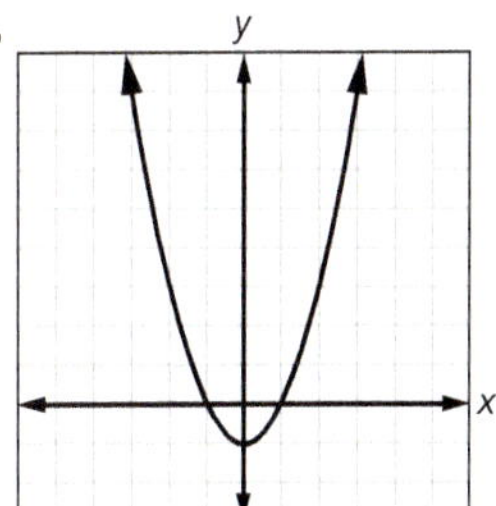

19.

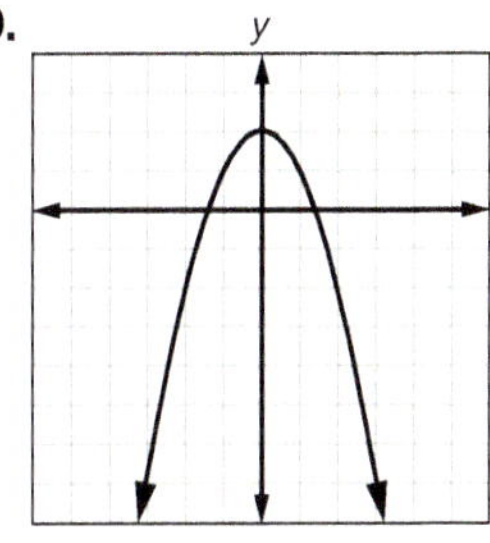

21.

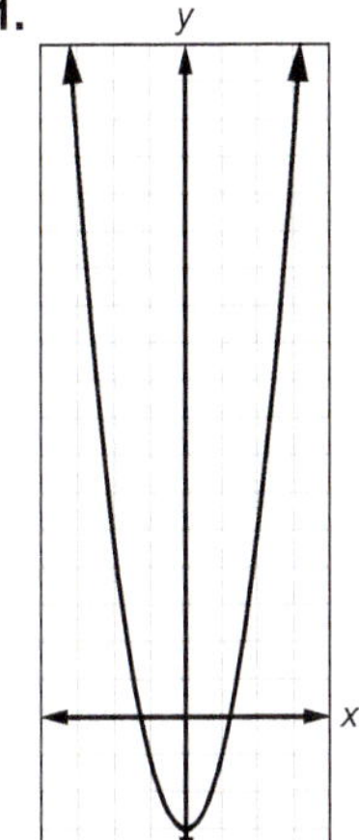

23.

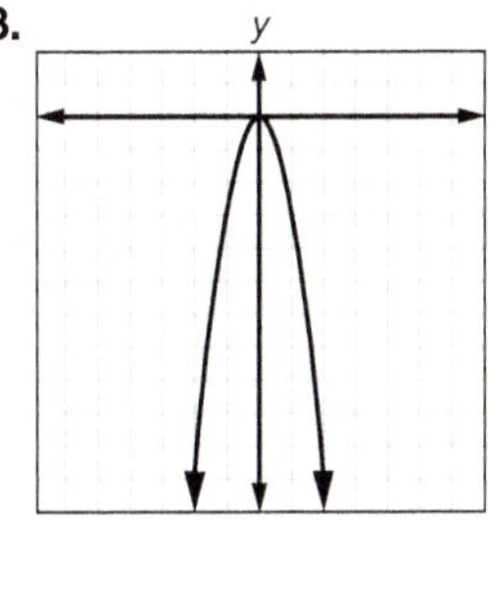

25.

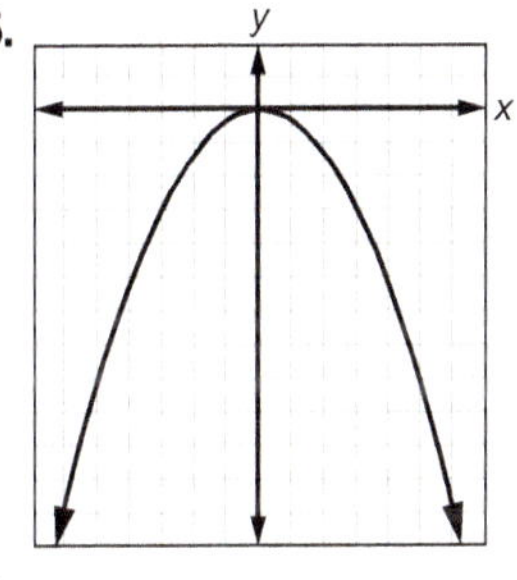

27.

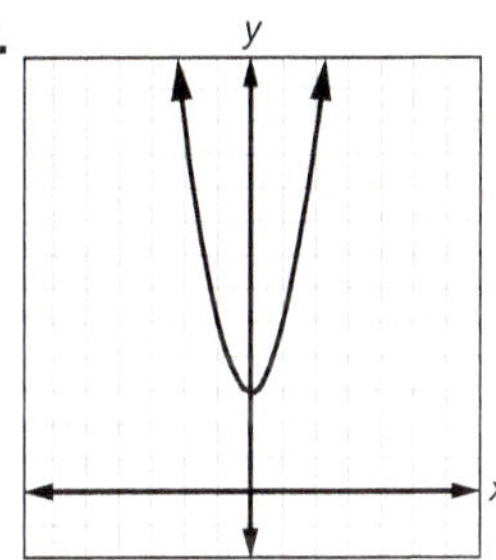

29.

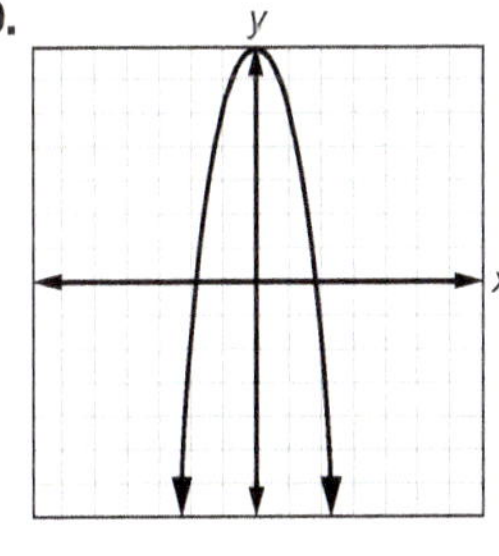

31a–b.

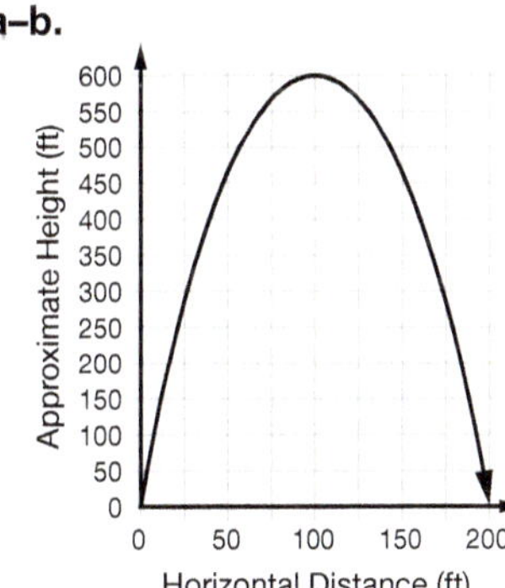

31b. 125, 150, 175, 200 **31c.** 200 ft **33.** (0, 0); (0, 0) **35.** $(0, c)$ **37.** all four **39.** $f(x) = -1.44x^2 + 7.2x$ **41.** Multiplication Property of Equality **43.** $\frac{\sqrt{13}}{26}$ **45.** 8 **47.** $x < -3$ **49.** 131 pennies and 21 dimes

12.8 Quadratic Functions: $f(x) = a(x - h)^2 + k$

1. (2, 9), $x = 2$, 1, up, same **3.** (2, -4), $x = 2$, 3, up, steeper
5. (-1, 0), $x = -1$, -5, down, steeper
7. (0, 7), $x = 0$, $-\frac{7}{4}$, down, steeper
9. (4, -7), $x = 4$, 1, up, same

11.

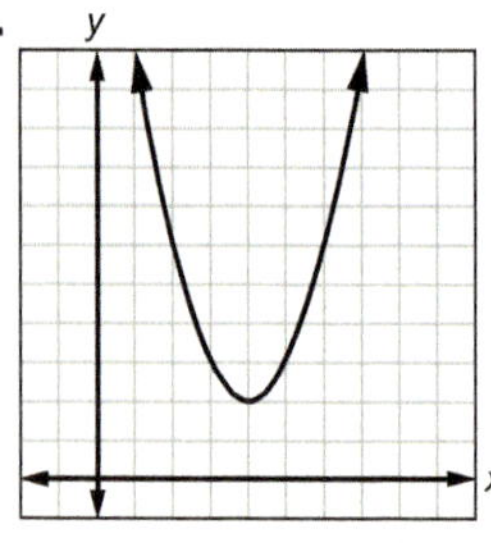

13.

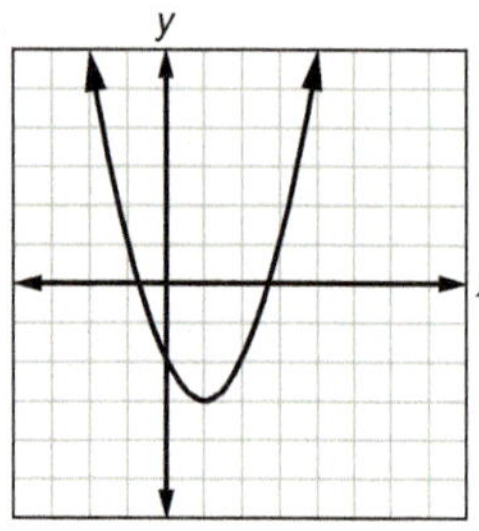

15.

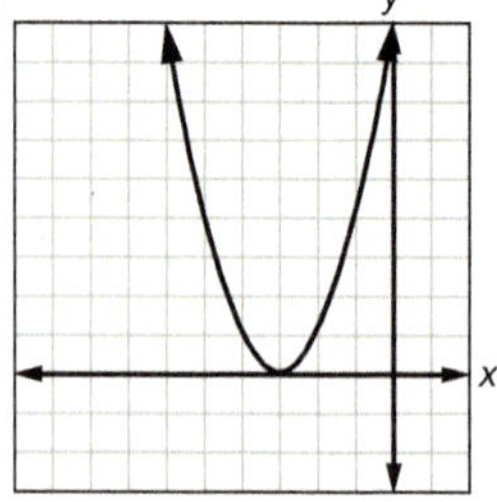

17.

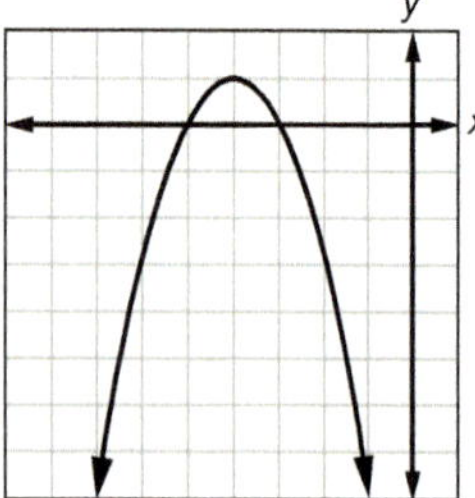

19.

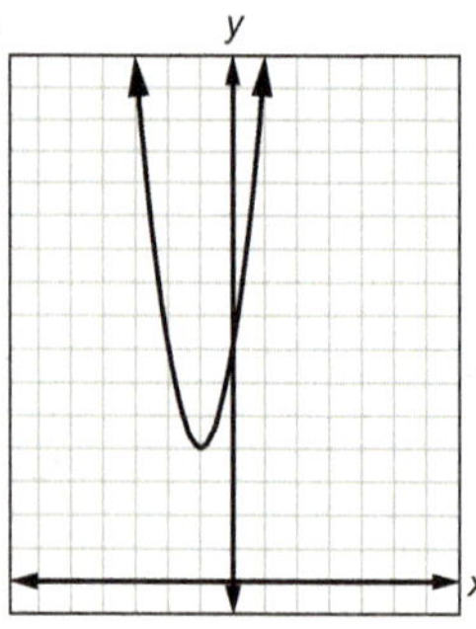

21. $f(x) = (x + 2)^2 - 16$; (-2, -16)
23. $f(x) = (x + 7)^2 - 9$; (-7, -9)
25. (-5, -1)

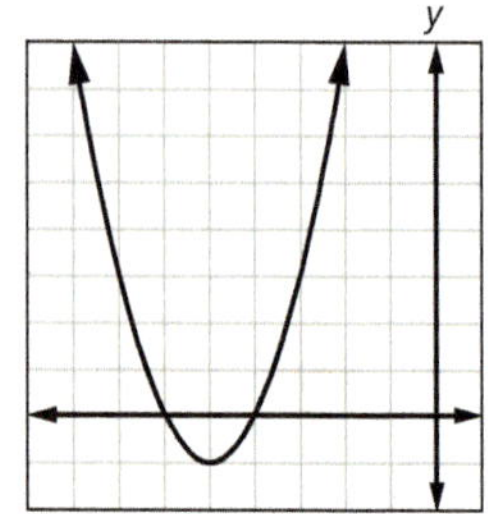

27. (-4, 4)

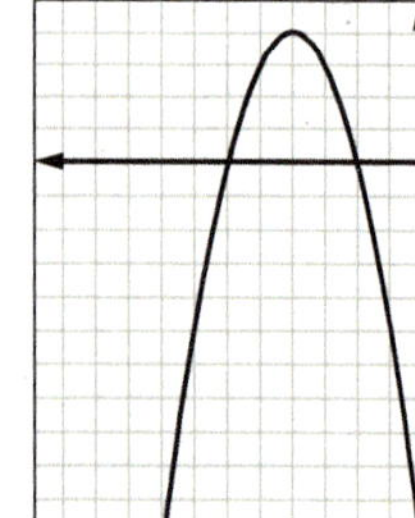

29. (3, -7)

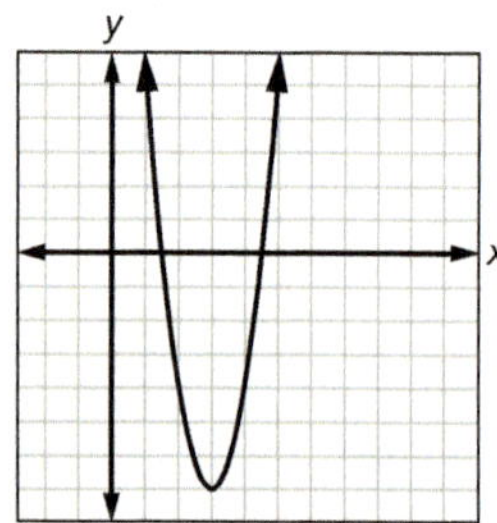

31a. \$1750 lost; \$125 gained **31b.** 200 items **31c.** \$750
33. 74.0625 ft; 2.0625 sec **35a.** $90 - 2x$
35b. $f(x) = x(90 - 2x)$ or $f(x) = -2x^2 + 90x$
35c. (22.5, 1012.5) **35d.** 22.5 ft × 45 ft **35e.** 1012.5 ft^2
37. (-1, -1.5)

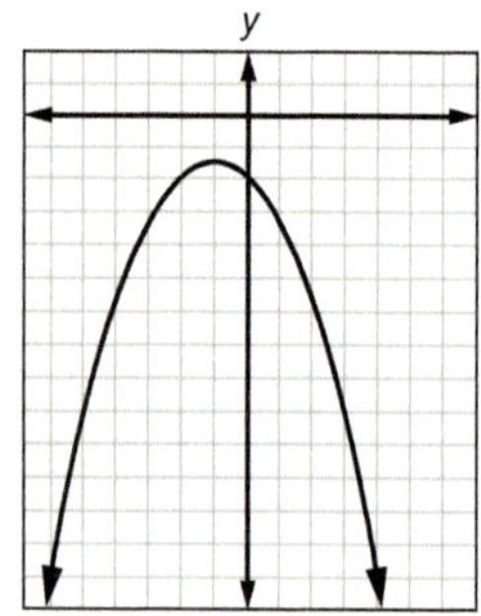

39. (0, 6); (20, 12) **41.** $f(x) = -0.015(x - 20)^2 + 12$; $f(x) = -0.015x^2 + 0.6x + 6$
43. between the opponent's 30 and 25 yd lines
45. $x + 4 - \frac{11}{x + 1}$ **47.** 71 and 174 **49.** $x = -15, 7$
51. $x = \frac{1}{2}, 1$ **53.** \$20.79

12.9 Zeros of a Function

1. (0, 0) **3.** (0, -5) **5.** (0, 1) **7.** (0, 8) **9.** (0, -5)
11. -2, 4 **13.** 3 **15.** no real zero **17.** 1, -3
19. $-3 \pm \sqrt{17}$ **21.** (0, c) **23.** the x-axis or $y = 0$
25. (0, 0); 0; (0, 0)

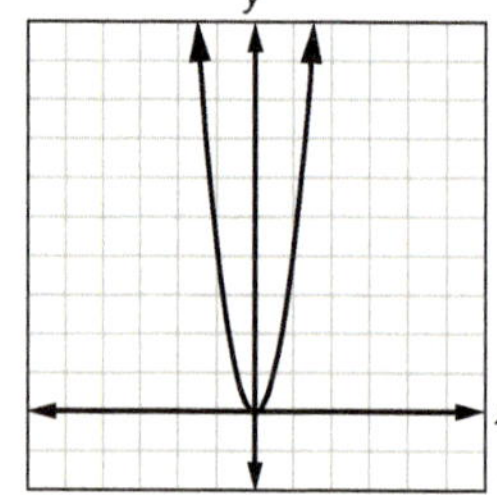

27. (0, 0); -3, 0; $\left(-\frac{3}{2}, -\frac{9}{4}\right)$

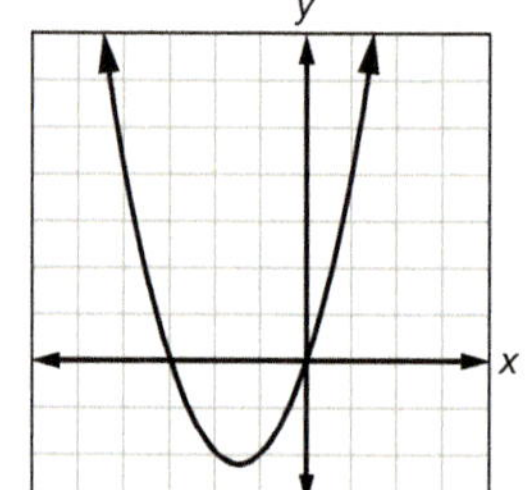

29. (0, 6); 2, 3; $\left(\frac{5}{2}, -\frac{1}{4}\right)$

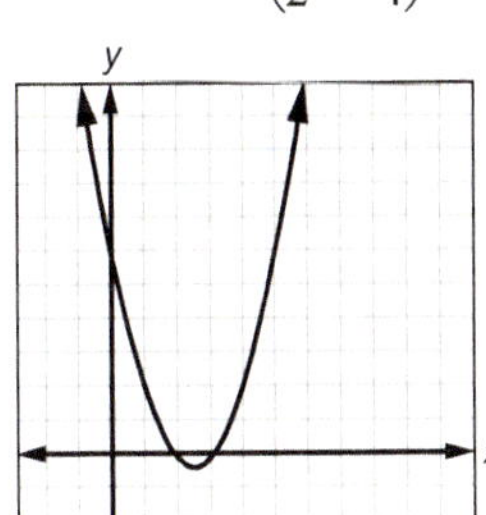

31. (0, 9); 1.5; (1.5, 0)

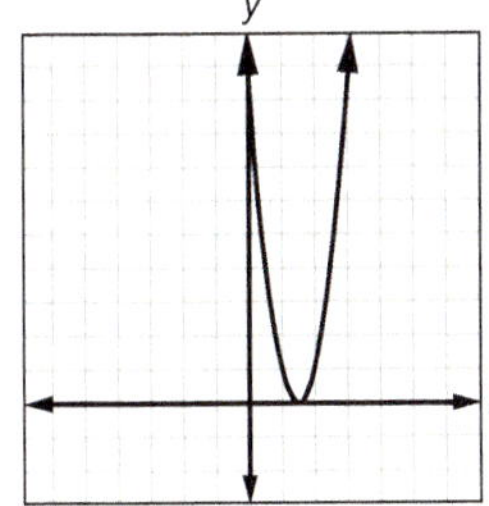

33. (0, −3); no real zero; (−1, −2)

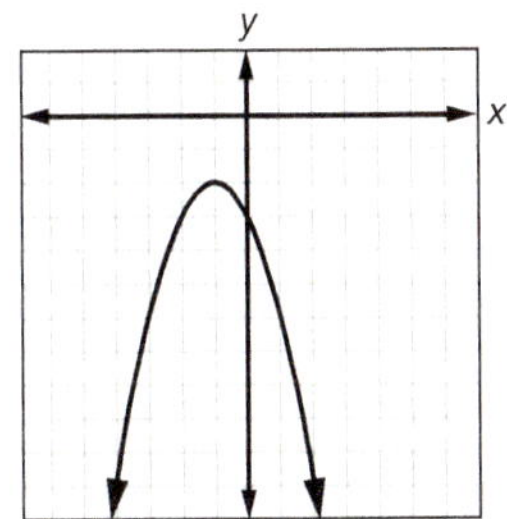

35a. 200 refrigerators **35b.** \$34,000
35c. $15 < x \le 384$ or $16 \le x \le 384$ **37.** $ah^2 + k$
39a. $\left(-\frac{7}{3}, \frac{121}{6}\right)$ or $(-2.\overline{3}, 20.1\overline{6})$ **39b.** $x = -\frac{7}{3}$ or $-2.\overline{3}$
39c. downward **39d.** steeper **39e.** $-6, \frac{4}{3}$
43. $m = 2$; $(0, -7)$; $\frac{7}{2}$ **45.** $\sqrt{29}$
47.

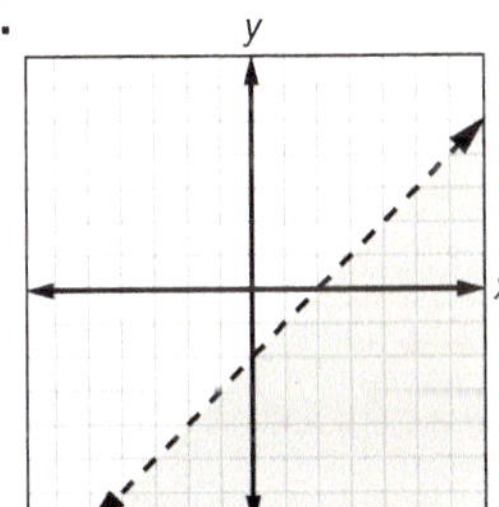

49. $x = -2$
51. $x = 4 \pm 2\sqrt{3}$

Sequences Quadratic Arithmetic Sums

1. inductive **3.** substitution **5.** Commutative Property of Addition **7.** $S_n = 3n^2 + 3n$ **9.** $A_n = 5n - 2$

Chapter 12 Review

1. $x = 2, -3$ **3.** $a = 0, \frac{1}{4}, -3$ **5.** $x = -1, 7$ **7.** $x = \pm\sqrt{6}$
9. $x = 4 \pm \sqrt{11}$ **11.** $x = -1, -7$ **13.** $x = 5 \pm \frac{\sqrt{110}}{2}$
15. $a = 1, b = 4, c = -1$ **17.** 0; one rational
19. −4; no real **21.** $x = 3 \pm \sqrt{7}$ **23.** $x = 1, -\frac{5}{3}$
25. {−2, 7} **27.** {±3} **29.** $\left\{\frac{1 \pm \sqrt{21}}{2}\right\}$ **31.** {−6, −3}
33. −7 and 12 or −12 and 7 **35.** 15 ft × 8 ft
37. −11 **39.** −3 **41.** 1; $x = 3$

43. (0, 0); 0

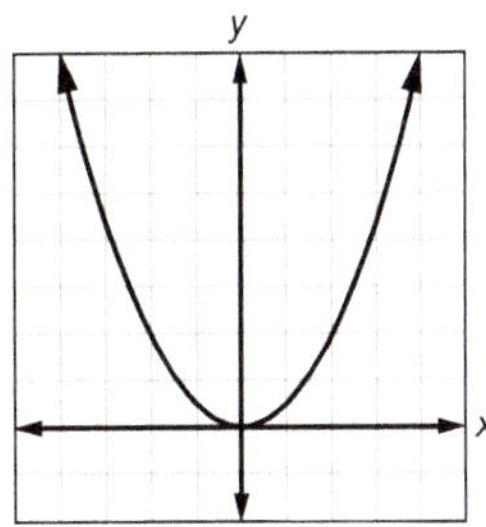

45. (2, 0); 2

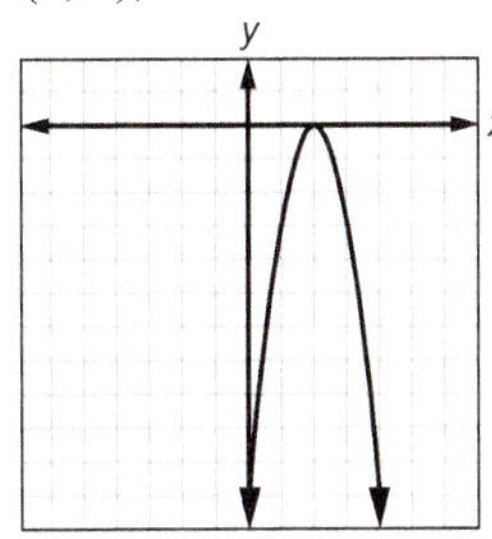

47. $(-3, -8)$; $-3 \pm 2\sqrt{2}$

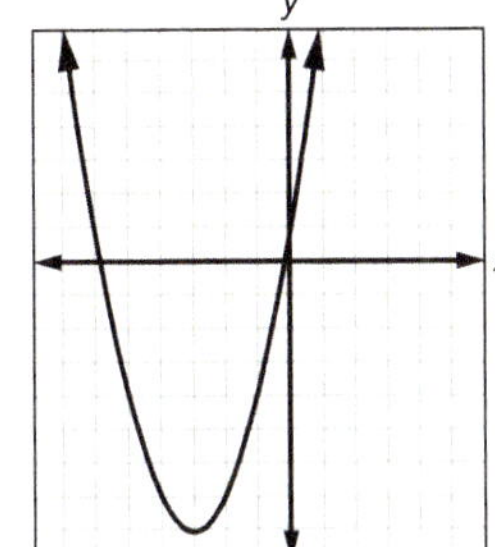

49. (0, 6)
51. (2, −2); minimum

Chapter 13—Rational Expressions

13.1 Simplifying Rational Expressions

1. $x = 0$ **3.** $a = -2$ **5.** $x = -8, 3$ **7.** $c = 4, 6$ **9.** $\frac{x}{5}$
11. $a + 7$ **13.** $x - 1$ **15.** $2a + 1$ **17.** $\frac{x+5}{x-1}$; $x \ne 1, 9$
19. $x - y$; $x \ne -y$ **21.** $\frac{3x}{x-y}$; $x \ne \pm y$
23. $\frac{x-5}{x}$; $x \ne 0, -2$ **25.** $\frac{\pi}{4} \approx 78.5\%$ **27.** $\frac{2}{\pi} \approx 63.7\%$
29. $\frac{5-x}{x}$ **31.** $\frac{x}{5}$ **33.** $\frac{z}{2(x-1)}$; $x \ne \pm 1$ **37.** $\frac{1}{4}l$ **39.** $m = \frac{r}{p}$
41. $\frac{8}{15}$ **43.** $\left(\frac{11}{3}, 0\right), \left(0, -\frac{11}{5}\right)$ **45.** inverse
47. direct; $y = 3x$

13.2 Multiplying and Dividing Rational Expressions

1. $\frac{19}{70}$ **3.** a **5.** 5 **7.** $-\frac{3}{2}$ **9.** $\frac{a^2b}{a^2 - b^2}$ **11.** $\frac{7}{2}$ **13.** $\frac{7y^2}{3}$
15. $x + y$ **17.** $\frac{x+2}{2}$ **19.** $(x + 8)(x + 3)$ **21.** $\frac{2x + y}{3x - 4y}$
23. $\frac{x + y}{(x - y)^2}$ **25.** $\frac{x}{x + 1}$ **27.** $\frac{(3x + y)(x + y)}{(x - y)^2}$ **29.** $\frac{a + 2b}{a}$
31. 1 **33.** $(x + 2)(x - 2)$ **35.** $\frac{(4x - 7y)(x + 2y)}{(x + 3y)(x + y)}$
37. 1.875 gal/min **39a.** ≈ 0.28 chapters/hr **39b.** ≈ 3.6 hr
41. $\frac{4b(a - b)(a - 2b)}{a(a + 7b)}$ **43.** $\frac{2x - 1}{x - 1}$ **45.** $6\sqrt{2} - 3\sqrt{26}$

47.

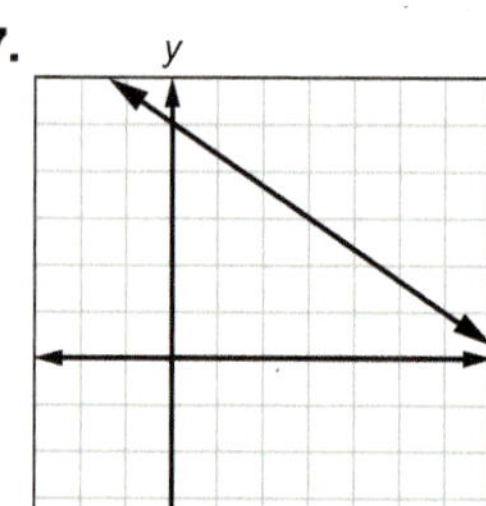

49. $x = \frac{-1 \pm \sqrt{241}}{6}$

51. $m = -\frac{7}{5}$

53. $\sqrt{65}$

Sequences Harmonic Sequences

1. yes **3.** no **5.** $A_n = \frac{1}{2n-1}$ **7.** $A_n = \frac{1}{5n-7}$ **9.** inductive

13.3 Adding and Subtracting Expressions with Common Denominators

1. $\frac{14}{x}$ **3.** $-\frac{x}{5}$ **5.** $-\frac{1}{c}$ **7.** 2 **9.** $\frac{9x}{x^2+y}$ **11.** 1 **13.** $\frac{3(3x+y)}{yz}$

15. b, d **17.** $-\frac{3}{m-4}$ **19.** 1 **21.** $\frac{3(2-a)}{a+b}$

23. $-\frac{2}{a^2+3a-2}$ **25.** $\frac{2x-7}{x^2-4x-7}$ **27.** $\frac{3m+n}{m-n}$ **29.** $\frac{x+7}{x+9}$

31. $\frac{3x^2-11x-15}{x^2-8x+15}$ **33.** $\frac{1}{a-b}$ **35.** $\frac{x^2+x-2}{x}$

37. $\frac{x^3+2x^2-4x-8}{x^2}$ **39.** $\frac{-3x+1}{x-4}$ or $\frac{1-3x}{x-4}$ **41.** $\frac{3}{4}l$

43.

Octave 5th 4th

0 $\frac{1}{2}l$ $\frac{2}{3}l$ $\frac{3}{4}l$ l

45. radical in the denominator; $\frac{20\sqrt{7}+4\sqrt{14}}{23}$

47. uncombined like terms; $5x - 4$

49. common factor in the numerator and the denominator; $3x$

51. 1 **53.** $\frac{-5x-y}{3}$ or $-\frac{5x+y}{3}$

13.4 Adding and Subtracting Expressions with Different Denominators

1. $24x^3$ **3.** $6(r+6)$ **5.** $(a+b)(a-b)^2$ **7.** $\frac{12x+9}{16}$

9. $\frac{2x+3y}{15}$ **11.** $\frac{7a^2+6a-45}{10}$ **13.** $\frac{-c+3}{c(c+1)}$ **15.** $\frac{20x^2y+5}{2x^3y^2}$

17. $\frac{2b^2-11b-15}{(b-4)(b+5)}$ **19.** $\frac{2y^2-23y-36}{(y-7)^2(y+4)}$ **21.** $\frac{-2x-6}{(x+2)(x-4)}$

23. $\frac{d(d+28)}{6(d+4)(d-4)}$ **25.** $\frac{7(x+1)}{12(x-3)(x+4)}$

27. $\frac{5x-5y+3}{(x+y)(x-y)}$ **29.** $\frac{y(2y^2+y-15)}{(y+2)(y+5)^2}$ **31.** $\frac{3(x-2)}{x+5}$

33. $\frac{-8a-25}{(a-5)(a+5)}$ **35.** $\frac{1}{x-2}$

37. second term should be negative; $\frac{-x+4}{(x-2)(x-5)}$

39. $\frac{5}{6}l$; $\frac{4}{5}l$ **41.** harmonic; arithmetic **43.** $-\frac{a^2}{16b^2}$

45. $a+b$ **47.** $3a(a+3)(a+4)$ **49.** $x = -2, 0, 3$

51. $2\sqrt{109} \approx 20.9$ ft

13.5 Complex and Mixed Expressions

1. $\frac{2x+1}{x}$ **3.** $\frac{2y^2+2y+1}{y+1}$ **5.** $\frac{2ab-a}{b}$ or $\frac{a(2b-1)}{b}$

7. $\frac{2}{5}$ **9.** $\frac{ad}{bc}$ **11.** $\frac{xz}{y}$ **13.** $\frac{ac^3}{d^2}$ **15.** $\frac{y}{c^2+cy}$

17. $\frac{x^2+2x+1}{x}$ or $\frac{(x+1)^2}{x}$

19. $\frac{2y^2+5y+2}{y+1}$ or $\frac{(2y+1)(y+2)}{y+1}$ **21.** $\frac{Ac+b}{c}$ **23.** $\frac{x^2}{x+1}$

25. $\frac{x}{x-y}$ **27.** $\frac{1}{(x-y)^2}$ **29.** $\frac{1}{4}$ **31.** $\frac{2}{(x+2)^2}$

33. $\frac{-x+6}{(x+5)(x+3)}$ **35.** 429 Hz **37.** $\frac{4-3x}{3x^3}$

41. negative exponent in the denominator; $3x^2$

43. radicand has a perfect square factor; $5\sqrt{2}$ **45.** $x = \frac{8}{9}$

47. $a = 4$ **49.** $24x^3y^2z^5$

13.6 Solving Rational Equations

1. $m = 27$ **3.** $x = \frac{1}{4}$ **5.** $x = \frac{2}{3}$ **7.** no solution **9.** $a = \frac{9}{2}$

11. $m = 22$ **13.** $a = -\frac{241}{8}$ **15.** $x = -\frac{1}{6}$ **17.** $x = -\frac{103}{2}$

19. no solution **21.** $x = -3$ **23.** $m = \frac{51}{10}$

25. 36 in. × 18 in. **27.** 13 **29.** 4 and 10

31a. 25 made free throws **31b.** 8 missed free throws

33. $x = \frac{3 \pm \sqrt{137}}{4}$ **35.** $x = -\frac{11}{5}$ **39.** $\frac{3(x+3)}{x+5}$

41. $\frac{3x+16}{x(x+2)}$ **43.** $y = \frac{nq+t}{6p}$ **45.** $y = \frac{3}{4}x - 3$

47. 3 cm, 4 cm, and 5 cm

13.7 Applying Rational Equations

1.

	r	*t*	*d*
Going	$r+3$	$\frac{200}{r+3}$	200
Returning	r	$\frac{200}{r}$	200

3.

	Time (hours)	**Rate of Work**
Pete	3	$\frac{1}{3}$
Adam	6	$\frac{1}{6}$
Together	x	$\frac{1}{x}$

5. 2.5 mi **7.** 29.8 hr **9.** 21.3 hr

11. slower plane: 750 mi; faster plane: 1500 mi

13. walking: 3 mi/hr; riding: 10 mi/hr **15.** 15 mi/hr

17. 1.2 hr **19.** Liam: 75 min; Lucas: 150 min

21. Fran: 6.8 hr; Rachel: 9.8 hr **23.** 5.8 mi/hr

25. 1.2 hr **27.** going: 40 mi/hr; returning: 50 mi/hr; 2.5 hr

29. Carla: 7.1 hr; Pamela: 9.1 hr

31.

	r	*t*	*d*
Going	20 mi/hr	$\frac{5}{2}$ hr	50 mi
Returning	30 mi/hr	$\frac{5}{3}$ hr	50 mi

33. 24 mi/hr

35.

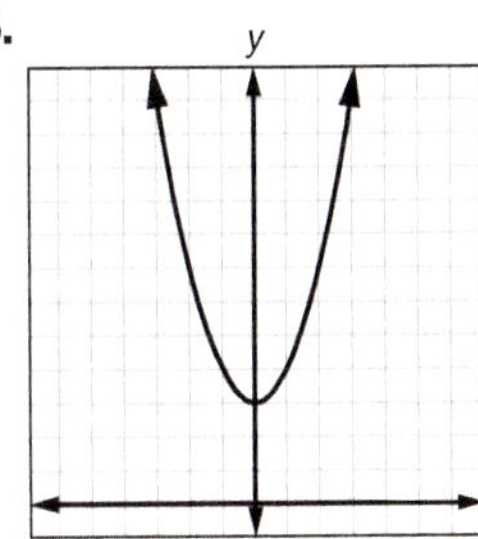

37.

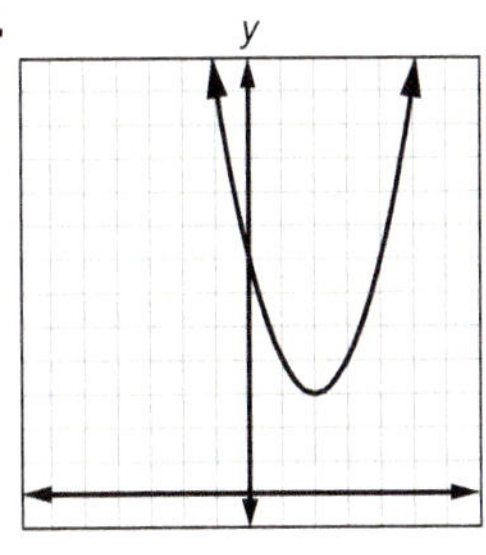

39. $(-8, -10)$ **41.** $x + 2 + \frac{6}{x - 1}$

43. $|48\% - a| \leq 3\%$; $45\% \leq a \leq 51\%$

13.8 Graphing Rational Functions

1. b **3.** a **5.** c **7.** b **9.** e **11.** $x = 5$

13. $(0, 0)$; $x = 0$ and $y = 0$ **15.** $(1, 7)$; $x = 1$ and $y = 7$

17. $\left(\frac{1}{2}, 1\right)$; $x = \frac{1}{2}$ and $y = 1$ **19.** $(0, 0)$; $x = 0$ and $y = 0$

21.

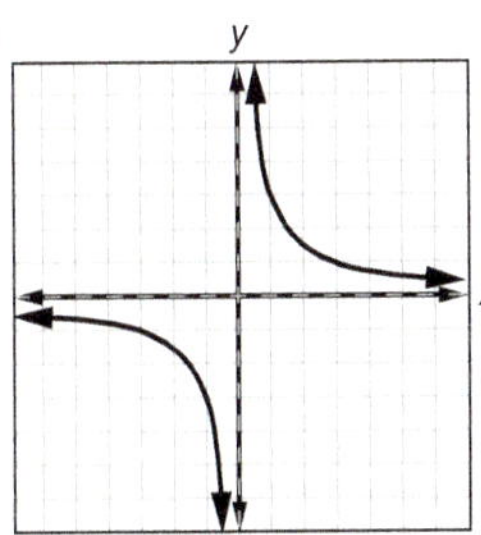

23.

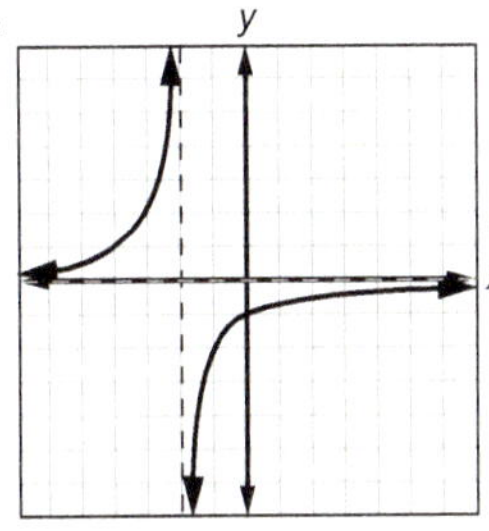

25.

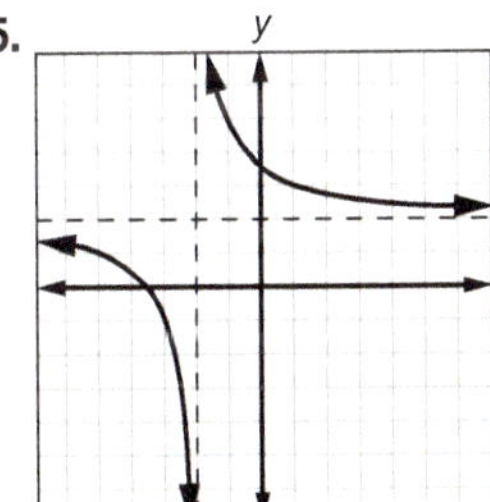

27. 10 units right

29. 7 units up

31. 4 units left and 9 units down

33.

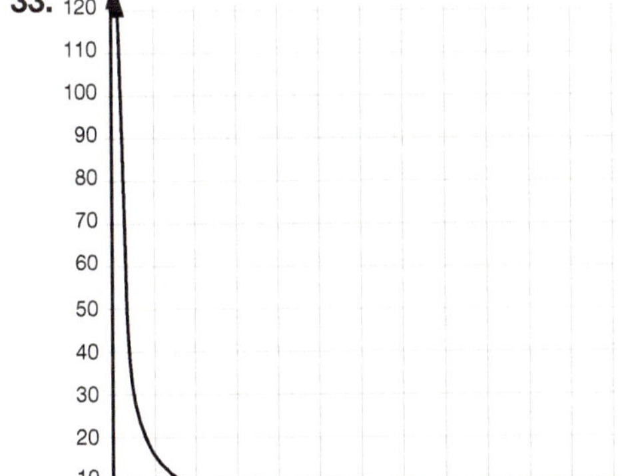

35. The current increases.

37.

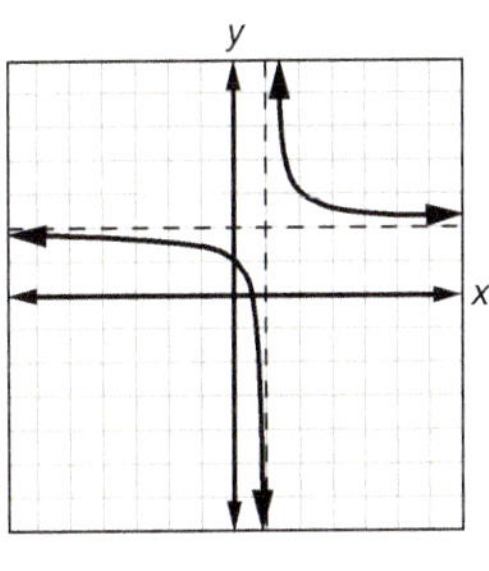

39. The equation is not defined for those values of x.

41.

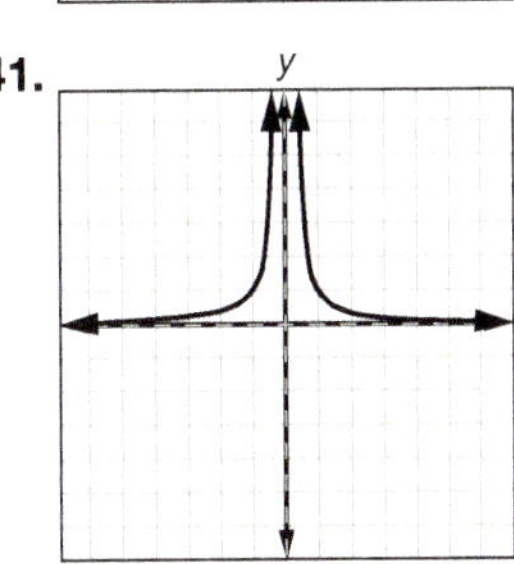

45. $x \leq \frac{2}{5} \vee x > \frac{4}{5}$

47. $x = \pm 4$

49. $(x - 3)(x + 3)(3x^2 - 4)$

51. $m = -\frac{2}{3}$

53. 20 lb

Chapter 13 Review

1. $x = 3$ **3.** $x = 0, \pm 1$ **5.** $\frac{3}{2}$ **7.** $\frac{x + 5}{x - 3}$ **9.** $56x^4y$ **11.** $\frac{ac}{b}$

13. $\frac{3x + 10}{3x + 5}$ **15.** $7x + 5$ **17.** $3(x + 11)$ **19.** $\frac{1}{x - y}$ **21.** $-\frac{x}{y}$

23. $\frac{4}{x - 1}$ **25.** 0 **27.** $\frac{(x - 1)(x - 5)}{x^2(x + 3)}$ **29.** $\frac{x - 2}{x}$

31. $\frac{x}{(3x + 2)(x + 2)}$ **33.** $\frac{1}{xy}$ **35.** $x = -\frac{3}{2}, 8$

37. $x = 26$ **39.** $x = \frac{11}{6}$ **41.** $x = 7$

43. $7(x - 5) = 4(x + 5)$ **45.** $\frac{1}{5} + \frac{1}{2} = \frac{1}{x}$ **47.** $1\frac{7}{8}$ hr

49. Micah: 24 mi/hr; Alexis: 60 mi/hr **51.** $\frac{\sqrt{3}\pi}{6} \approx 90.6\%$

53. $(-3, 0)$; $x = -3$ and $y = 0$

55.

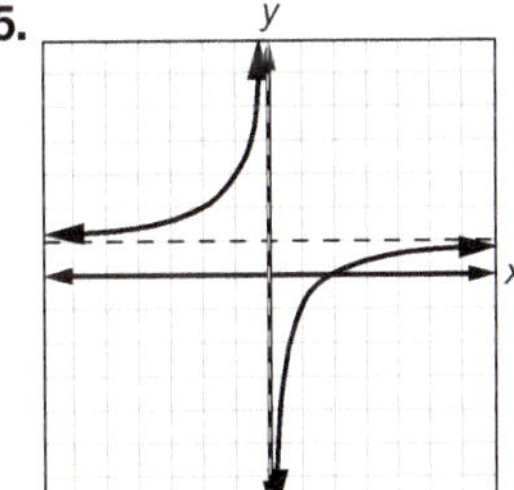

57. b

GLOSSARY

abscissa (p. 183) The first coordinate of an ordered pair.

absolute value (pp. 9, 169) The number of units (distance) between zero and a number's graph on a number line, denoted $|x|$. Formally, $|x| = x$ if $x \geq 0$ and $|x| = -x$ if $x < 0$.

Addition Property of Equality (p. 64) If a, b, and c are real numbers such that $a = b$, then $a + c = b + c$.

Addition Property of Inequality (p. 148) If a, b, and c are real numbers such that $a < b$, then $a + c < b + c$.

additive identity (p. 15) The number 0; the value that when added to another number yields the same number.

Additive Identity Property (p. 15) For any real number a, $a + 0 = a$. (Adding zero to a number does not change the value of that number.)

additive inverse (p. 15) The additive inverse of a, denoted $-a$, is the opposite of a.

Additive Inverse Property (p. 15) For any real number a, $a + (-a) = 0$. (The sum of a number and its additive inverse is always zero, the additive identity.)

algebraic expression (p. 46) A mathematical expression containing numbers, variables, and operational signs.

arithmetic mean (p. 531) The sum of n numbers divided by n. Commonly called the *mean* or *average*.

Associative Property of Addition (p. 15) For any real numbers a, b, and c, $(a + b) + c = a + (b + c)$. (Addends can be grouped in any order without affecting the sum.)

Associative Property of Multiplication (p. 23) For any real numbers a, b, and c, $a(bc) = (ab)c$. (Factors can be grouped in any order without affecting the product.)

axis (p. 183) A horizontal or vertical reference line in a Cartesian coordinate plane.

base of an exponential expression (p. 33) A number, variable, or expression that is raised to a power indicated by an exponent (e.g., in the expression a^3, a is the base).

binomial (p. 370) A polynomial with exactly two terms.

Cartesian coordinate plane (p. 182) A plane with a rectangular coordinate system that associates every point in the plane with an ordered pair of numbers. Named after René Descartes.

coefficient (p. 58) The constant factor accompanying the variables of a term. Also called a *numerical coefficient*.

common denominator (p. 13) A common multiple of the denominators of two or more fractions. The least common denominator (LCD) is the smallest multiple shared by two or more denominators.

common monomial factor (p. 404) A factor common to all of the terms of a polynomial.

Commutative Property of Addition (p. 15) For any numbers a and b, $a + b = b + a$. (Addends can be arranged in any order without affecting the sum.)

Commutative Property of Multiplication (p. 23) For any numbers a and b, $ab = ba$. (Factors can be arranged in any order without affecting the product.)

complementary angles (p. 48) Two angles with measures whose sum is 90°. Each is the complement of the other.

completing the square (p. 495) A method of solving quadratic equations by obtaining a perfect square trinomial.

complex rational expression (p. 552) An algebraic expression of the form $\frac{n}{d}$ in which the numerator, n, the denominator, d, or both are rational expressions.

compound sentence (p. 158) Combining two or more equations or inequalities using the word *and* or the word *or*.

conjugates (p. 391) Two algebraic expressions in the form $a + b$ and $a - b$.

conjunction (p. 158) A compound sentence consisting of mathematical statements connected by the word *and*, meaning intersection, and symbolized by $\wedge$.

consistent system (p. 302) A system of equations with one or more solutions.

constant (p. 46) A symbol that represents a fixed number.

constant of variation (pp. 215, 217) The constant relating two variables in a direct or inverse variation. In the expressions $y = kx$ and $yx = k$, k is the constant of variation.

coordinate (p. 7) The number corresponding to a point on a number line.

cube root (p. 432) One of a number's three equal factors.

deductive reasoning (p. 221) Reasoning from a general premise to a specific conclusion.

degree of a polynomial (p. 371) The degree of the highest-degree term in a polynomial.

degree of a term (p. 370) The sum of the exponents of the variables of a term.

dependent system (p. 302) A consistent system of equations with an infinite number of solutions. The graphs coincide because the equations are equivalent.

direct variation (p. 215) A function in which the ratio of variables is a nonzero constant, k. That is, $\frac{y}{x} = k$, where $k \neq 0$.

disjunction (p. 163) A compound sentence consisting of mathematical statements connected by the word *or*, meaning union, and symbolized by $\vee$.

distance formula (p. 459) The distance, d, between any two points (x_1, y_1) and (x_2, y_2) can be found using the formula $d = \sqrt{(x_2 - x_1)^2 + (y_2 - y_1)^2}$.

Distributive Property (p. 23) For any real numbers a, b, and c, $a(b + c) = ab + ac$. (The product of a number and a sum is equal to the sum of the products of the number and each addend.)

domain of a relation (p. 188) The set of first elements (x-coordinates) of the ordered pairs in a relation.

domain of a variable (p. 46) The set of numbers the variable may represent.

Dominion Mandate (p. 45) God's command in Genesis 1:28 for humans to fill, subdue, and rule over the earth.

element (p. 2) A distinct object or member of a set.

elimination method (p. 295) Solving a system of equations by adding or subtracting multiples of the equations to eliminate one of the variables.

empty set (p. 2) A set containing no elements. Also called a *null set*.

equation (p. 63) A mathematical sentence stating that two expressions are equal.

evaluate (p. 34) To find the number that is represented by a mathematical expression.

exponent (p. 33) The number of times a base occurs as a factor, indicated by a superscript numeral to the right of the base (e.g., in the expression a^3, 3 is the exponent).

exponential decay (p. 358) A quantity exhibits exponential decay if it decreases by the same percent during each unit of time.

exponential function (p. 350) A function of the form $f(x) = ab^x$, where $b > 0$ and $b \neq 1$.

exponential growth (p. 358) A quantity exhibits exponential growth if it increases by the same percent during each unit of time.

exponential notation (p. 33) A simplified form of writing repeated multiplication (e.g., 5^3). Also called *exponential form*.

extraneous solution (p. 470) A derived solution that is invalid because it does not make the original equation true.

factor (pp. 23, 404) One of several numbers or expressions multiplied to form a product.

factoring (p. 404) The process of determining the individual factors of a product.

finite set (p. 3) A set in which the number of elements in the set is a whole number.

FOIL method (p. 385) A method of multiplying two binomials by finding the products of the first, outer, inner, and last terms and then combining like terms.

formula (p. 51) An equation that describes a relationship between quantities.

fractals (p. 1) Rough or fragmented geometric shapes that are repeated at an increasingly smaller scale.

function (p. 188) A relation in which each element in the domain is paired with exactly one element in the range.

function notation (pp. 189, 245) A means of expressing the dependence of one variable upon another (e.g., $y = x + 5$ is expressed as $f(x) = x + 5$ in function notation).

golden ratio (pp. 104, 435–36) The irrational mathematical constant $\frac{1 + \sqrt{5}}{2} \approx 1.6180339887$, represented by the lowercase Greek letter phi (ϕ). Also called the *divine proportion*, *golden mean*, or *golden section*.

graph (p. 7) A visual representation of a mathematical solution on a number line or a coordinate plane.

greatest common factor (GCF) (pp. 6, 404) The largest positive integer that evenly divides two or more given integers. The GCF of a polynomial is the product of all factors shared by every term of the polynomial.

harmonic mean (pp. 531, 546, 573) The number of values divided by the sum of their reciprocals. The harmonic mean of n numbers can be found using the formula $H = \frac{n}{\frac{1}{x_1} + \frac{1}{x_2} + \dots + \frac{1}{x_n}}$. When there are just two values, a and b, this simplifies to $H = \frac{2ab}{a + b}$.

hypotenuse (p. 457) The side of a right triangle opposite the right angle.

inconsistent system (p. 302) A system of equations with no solution.

independent system (p. 302) A consistent system of equations with a finite number of solutions.

index (p. 432) A superscript number within the radical sign that indicates the root (e.g., in $\sqrt[3]{x}$, the index is 3). If no index is given, the index is understood to be 2.

inductive reasoning (p. 12) Reasoning from specific instances to a general conclusion.

inequality (p. 144) A mathematical sentence stating that two quantities are not always equal.

infinite set (p. 3) A set that is not finite.

integers (p. 3) The set of whole numbers and their opposites, $\{\dots, -3, -2, -1, 0, 1, 2, 3, \dots\}$.

intersection of sets (p. 2) The set of elements common to all of the given sets.

inverse operation (p. 63) A mathematical operation that reverses another operation, such as addition and subtraction or multiplication and division.

inverse variation (p. 217) A relation in which the product of the two variables is a nonzero constant, k. That is, $xy = k$, where $k \neq 0$.

irrational number (p. 4) A number that cannot be expressed as a ratio of two integers.

least common multiple (LCM) (p. 84) The smallest positive integer that is evenly divided by two or more given integers.

like terms (p. 57) Terms that contain an identical variable or variables, all raised to the same respective powers.

linear equation (p. 232) An equation whose graph is a straight line. The standard form of a linear equation is $Ax + By = C$, where A, B, and C are real numbers.

linear function (p. 232) A function described by the equation $f(x) = mx + b$, where m and b are real numbers.

literal equation (p. 94) An equation containing several variables.

mathematical property (p. 15) An equation or statement that is true for any values of the variables. Also called an *identity*.

midpoint formula (pp. 10, 460) If a and b are coordinates of the endpoints of a segment on a number line, then the coordinate of the segment's midpoint is $m = \frac{a + b}{2}$. The midpoint of a segment on a Cartesian coordinate plane with endpoints $P_1(x_1, y_1)$ and $P_2(x_2, y_2)$ is $M\left(\frac{x_1 + x_2}{2}, \frac{y_1 + y_2}{2}\right)$.

monomial (p. 370) A polynomial with exactly one term.

Multiplication Property of Equality (p. 64) If a, b, and c are real numbers such that $a = b$, then $ac = bc$.

Multiplication Property of Inequality (p. 150) If a, b, and c are real numbers such that $a < b$, then $ac < bc$ if $c > 0$ and $ac > bc$ if $c < 0$.

multiplicative identity (p. 23) The number 1; the value that when multiplied by another number yields the same number.

Multiplicative Identity Property (p. 23) For any real number a, $a \cdot 1 = a$. (Multiplying a number by one does not change the value of that number.)

multiplicative inverse (p. 29) The multiplicative inverse of a, denoted $\frac{1}{a}$, is the reciprocal of a.

Multiplicative Inverse Property (p. 29) For any real number a, where $a \neq 0$, $a \cdot \frac{1}{a} = 1$. (The product of a number and its reciprocal is always one, the multiplicative identity.)

natural numbers (p. 3) The set of counting numbers, $\{1, 2, 3, \dots\}$.

null set (p. 2) A set containing no elements. Also called an *empty set*.

opposites (p. 8) Numbers that when located on a number line are the same distance from zero but on opposite sides of zero.

ordered pair (p. 182) A pair of numbers (x, y) describing a point in the Cartesian coordinate plane.

order of operations (p. 39) The rules for determining the order in which operations should be performed in a mathematical expression: parentheses and other grouping symbols, exponents and roots, multiplication and division from left to right, and addition and subtraction from left to right.

ordinate (p. 183) The second coordinate of an ordered pair.

origin (p. 183) The point at which the axes of a coordinate plane intersect.

parabola (p. 344) The curved graph of a quadratic function.

percent (p. 112) The ratio of a number to one hundred.

perfect square trinomial (p. 421) A trinomial that is the square of a binomial.

point-slope form (p. 256) A linear equation in the form $y - y_1 = m(x - x_1)$, where m is the slope of the line and (x_1, y_1) is a given point.

polynomial (p. 370) An expression that can be written as a sum of terms consisting of constants, variables with nonnegative integral exponents, or products of a constant and one or more such variables.

Power Property of Exponents (p. 328) For any real numbers x, a, and b, $(x^a)^b = x^{ab}$. (To raise an exponential expression to a power, multiply the exponents.)

principal (p. 124) The amount of money initially invested or borrowed.

Product Property of Exponents (p. 328) For any real number x and any integers a and b, $x^a \cdot x^b = x^{a+b}$. (To multiply exponential expressions with like bases, add the exponents.)

Product Property of Radicals (p. 437) For $x \geq 0$ and $y \geq 0$, $\sqrt[n]{xy} = \sqrt[n]{x} \cdot \sqrt[n]{y}$. (The root of a product equals the product of the roots of the individual factors.)

Pythagorean theorem (p. 457) The sum of the squares of the lengths of the legs of a right triangle is equal to the square of the length of the hypotenuse. If a right triangle has legs of lengths a and b and a hypotenuse of length c, then $a^2 + b^2 = c^2$.

quadrant (p. 183) One of the four regions of the Cartesian coordinate plane determined by the x- and y-axes.

quadratic equation (p. 484) A second-degree equation that can be written in the standard form $ax^2 + bx + c = 0$, where a, b, and c are real numbers and $a \neq 0$.

quadratic formula (p. 503) The solutions to a quadratic equation of the form $ax^2 + bx + c = 0$, where a, b, and c are real numbers and $a \neq 0$, are given by the formula $x = \frac{-b \pm \sqrt{b^2 - 4ac}}{2a}$.

quadratic function (p. 511) A function that can be written in the standard form $f(x) = ax^2 + bx + c$, where a, b, and c are real numbers and $a \neq 0$.

Quotient Property of Exponents (p. 333) For any real number x, where $x \neq 0$, and any integers a and b, $\frac{x^a}{x^b} = x^{a-b}$. (To divide exponential expressions with like bases, subtract the exponents.)

Quotient Property of Radicals (p. 447) For $x \geq 0$ and $y > 0$, $\sqrt[n]{\frac{x}{y}} = \frac{\sqrt[n]{x}}{\sqrt[n]{y}}$. (Roots of numerators and denominators of a fraction under a radical can be found separately.)

radical (p. 432) An expression in the form $\sqrt[n]{x}$.

radical equation (p. 470) An equation that contains a variable in a radicand.

radical expression (p. 464) An algebraic expression containing at least one radical.

radical sign (p. 432) The symbol $\sqrt{\ }$ or $\sqrt[n]{\ }$, used to indicate a square root or nth root. When n is even, the principal (positive) root is indicated.

radicand (p. 432) The number under the radical sign (e.g., in $\sqrt[n]{3x}$, $3x$ is the radicand).

range (p. 188) The set of second elements (y-coordinates) of the ordered pairs in a relation.

rate of change (p. 241) The comparison of two changing quantities, such as time and distance or rise and run on a graph.

ratio (p. 98) The comparison of two numbers by division, often expressed as a fraction or in the form $x : y$.

rational equation (p. 556) An equation containing rational expressions.

rational expression (p. 532) A ratio of polynomials whose denominator is not zero.

rational function (p. 568) A function expressed as a ratio of polynomials whose denominator is not zero.

rationalizing the denominator (p. 447) The process of eliminating any radicals from the denominator of a fraction.

rational number (p. 4) A number that can be expressed as a ratio of two integers $\frac{a}{b}$, where $b \neq 0$.

real numbers (p. 4) The union of the sets of rational and irrational numbers.

reciprocals (p. 29) Two numbers whose product is 1 (e.g., 3 and $\frac{1}{3}$ are reciprocals). Also called *multiplicative inverses.*

relation (p. 188) Any set of ordered pairs.

relatively prime numbers (p. 404) Two numbers or terms whose greatest common factor is one.

right angle (p. 457) An angle whose measure is 90°.

right triangle (p. 457) A triangle that has a right angle.

set (p. 2) A group or collection of numbers or objects.

slope-intercept form (p. 245) A linear equation in the form $y = mx + b$, where m is the slope of the line and $(0, b)$ is the y-intercept.

slope of a line (p. 238) The ratio of its rise (vertical change) to its run (horizontal change).

solve (p. 63) To find the values that make a mathematical equation true.

square root (p. 432) One of a number's two equal factors.

subset (p. 3) A set, all of whose elements are contained in another set.

substitution (p. 284) The process of replacing one quantity with an equal quantity.

substitution method (p. 284) Solving a system of equations by first solving an equation for one variable and then replacing that variable in the other equation, resulting in a single equation in one variable.

supplementary angles (p. 48) Two angles with measures whose sum is 180°. Each is the supplement of the other.

system of equations (p. 278) Two or more equations in two or more variables to be solved simultaneously.

term (pp. 57, 370) A constant, a variable, or the product or quotient of a constant and one or more variables.

theorem (p. 457) A statement that has been proved to be always true.

trinomial (p. 370) A polynomial with exactly three terms.

union of sets (p. 2) The set of elements that appear in any of the given sets.

variable (p. 46) A symbol used to represent an unknown quantity.

vertex (pp. 222, 344) The turning point of a graph at which it reaches its highest or lowest point.

vertical line test (p. 193) The method of moving a vertical line across a graph of a relation to determine whether the relation is a function.

whole numbers (p. 3) The set of counting numbers and zero, $\{0, 1, 2, 3, \ldots\}$.

x-axis (p. 183) The horizontal reference line in a Cartesian coordinate plane.

x-coordinate (p. 183) The first coordinate of an ordered pair, indicating the horizontal distance from the y-axis. Also called the *abscissa.*

x-intercept (p. 234) The point at which the graph of an equation crosses, or intersects, the x-axis.

y-axis (p. 183) The vertical reference line in a Cartesian coordinate plane.

y-coordinate (p. 183) The second coordinate of an ordered pair, indicating the vertical distance from the x-axis. Also called the *ordinate.*

y-intercept (p. 234) The point at which the graph of an equation crosses, or intersects, the y-axis.

zero of a function (p. 522) A value of x such that $f(x) = 0$. It is the x-coordinate of any point at which a graph of the function intersects the x-axis. Also called a *root* of the function.

Zero Product Property (p. 484) If $pq = 0$, then $p = 0$ or $q = 0$. (If the product of two or more factors is zero, then at least one of the factors is zero.)

Zero Property of Multiplication (p. 23) For any number a, $a \cdot 0 = 0$. (The product of any number and zero is zero.)

INDEX

A

B

C

D

E

M

N

O

P

Q

R

S

T

PHOTOGRAPH CREDITS

The following agencies and individuals have furnished materials to meet the photographic needs of this textbook. We wish to express our gratitude to them for their important contribution.

Alamy
Associated Press
BigStockPhoto
BJU Photo Services
Bob Jones University Museum & Gallery
Centre National d'Études Spatiales (CNES)
CORBIS
COREL Corporation
David Malin Images (DMI)
Dickinson, Cindy
Diehl, Deb, A Dyno-Might Amusements Co. LLC
DigitalGlobe
DigitalSTOCK
Dreamstime.com
Electric Boat Division, General Dynamics Corp
Fotolia
Garrison, Odell
General Bathymetric Chart of the Oceans (GEBCO)
GeoEye
Getty Images
iStockphoto
JupiterImages Corporation
Kaminski, Peter
Kromer Co.
Lucas, John, South Carolina Department of Natural Resources (SCDNR)
Mitchell, Rita
National Aeronautics and Space Administration (NASA)
National Geospatial-Intelligence Agency (NGA)
National Institute of Allergy and Infectious Diseases (NIAID)
National Institutes of Health (NIH)
National Oceanic and Atmospheric Administration (NOAA)
New York National Guard (NYNG)
OceanwideImages.com
PhotoDisc, Inc.
Ring, Mark
Rocky Mountain Laboratories
SuperStock
Thinkstock
United States Air Force (USAF)
United States Department of Agriculture (USDA)
United States Geological Survey (USGS)
United States Marine Corps
United States Navy
Unusual Films
Visuals Unlimited, Inc.
Washer, Kay
Wetzel, Mark
Wikimedia Commons
Wikipedia
www.travelsd.com

Cover
©iStockphoto.com/stariot

Introduction
© iStockphoto.com/Dudarev Mikhail iii (top); Album / Alamy Stock Photo iii (bottom right); ScanPyramids Mission iii (bottom); NASA/JHUAPL/Gordan Ugarkovic viii; ©iStockphoto.com/beerkoff ix (top); Asimzb/Wikimedia Commons/CC 3.0 ix (bottom)

Chapter 1
©CORBIS/SuperStock x–1; PhotoDisc, Inc. 2 (top), 33; ©dinostock/Fotolia 2 (bottom); © 2008 JupiterImages Corporation 7; DigitalSTOCK 8 (left); COREL Corporation 8 (right), 13, 28; ©iStockphoto.com/Erkki Makkonen 17; ©iStockphoto.com/Sean Locke 18; ©Tetra Images/SuperStock 20; NOAA 21; Getty Images/iStockphoto/Thinkstock 22 (top); USDA 22 (bottom); Getty Images/Thinkstock 32; Ralph Hutchings/Visuals Unlimited, Inc. 37; Getty Images/Hemera/Thinkstock 41

Chapter 2
iStockphoto/Thinkstock 44–45, 50 (right); COREL Corporation 46, 51, 72; Brand X Pictures/Thinkstock 49; Hemera/Thinkstock 50 (left); ©iStockphoto.com/Nikada 56; www.jupiterimages.com/Thinkstock 57, 63; ©iStockphoto.com/Rich Matts 70; DigitalSTOCK 74; ©Earnest Prim/Fotolia 78; ©iStockphoto.com/nicolamargaret 80; Getty Images/Comstock Images/Thinkstock 83; Kromer Co. 85; Unusual Films 88

Chapter 3
©Blend Images/SuperStock 92–93; ©Lisa F. Young/Fotolia 94; ©Inspirestock Inc./Alamy 96; ©Phillip Minnis. Image from BigStockPhoto.com 107; ©Joe Mule/Workbook Stock/Getty Images 112; ©iStockphoto.com/Eu Touch 113; Getty Images/iStockphoto/Thinkstock 115; Thomas Doerfer/Wikimedia Commons/GNU/CC 3.0 117 (top); OSX/Wikimedia Commons/Public Domain 117 (bottom); Sarah Mason/Digital Vision via Getty Images 120; www.jupiterimages.com/PhotoObjects.net/Thinkstock 121; Digital Vision/Thinkstock 123; Getty Images/Goodshoot RF/Thinkstock 124; ©iStockphoto.com/Don Joski Photography 127; EThamPhoto / Alamy Stock Photo 128; Denis Boissavy/Taxi/Getty Images 130; iStockphoto/Thinkstock 132; ©vichie81/Fotolia 133; Tractorboy60/Wikimedia Commons/Public Domain 135

Chapter 4
©Ron Brown/SuperStock 142–43; ©iStockphoto.com/Bill Grove 144; PhotoDisc, Inc. 148, 158; iStockphoto/Thinkstock 151; Vincent van Gogh's *Portrait of Dr. Gachet*/Wikimedia Commons/Public Domain 153; Getty Images/Hemera/Thinkstock 155; Norbert Wu/Minden Pictures II/Getty Images 161 (top left); ©Rudie Kuiter/OceanwideImages.com 161 (bottom left); Wikimedia Commons/Public Domain 161 (right); Comstock/Thinkstock 163; ©Neil Julian Photography/Alamy 169; ©imagebroker.net/SuperStock 175

Chapter 5
©iStockphoto.com/staphy 180–81; ©Royalty-Free/CORBIS 182; Image © 2010 DigitalGlobe/© 2010 CNES/Spot Image/Data SIO, NOAA, U.S. Navy, NGA, GEBCO/Image © 2010 GeoEye 186; ©Mehmet can. Image from BigStockPhoto.com 187; Associated Press 193; ©iStockphoto.com/George Peters 200; BJU Photo Services 204 (all); ©Dennis MacDonald/Alamy 208; iStockphoto/Thinkstock 212; ©age footstock/SuperStock 218

Chapter 6
BJU Photo Services/Thinkstock 234–35; SSGT SUZANNE M. JENKINS, USAF/Defenseimagery.Mil 232; Getty Images/Comstock Images/Thinkstock 238; COREL Corporation 243; PhotoDisc, Inc. 245; Photo by www.travelsd.com 250; Getty Images/Hemera/Thinkstock 256; ©iStockphoto.com/Michael Svobada 267; iStockphoto/Thinkstock 272

Chapter 7
Cindy Dickinson 276–77; Getty Images/Goodshoot RF/Thinkstock 278; Getty Images/iStockphoto/Thinkstock 287, 295, 323; ©torono reaction/Fotolia 290; ©Tetra Images/Alamy 292; Courtesy of Mark Ring 299; U.S. Air Force photo/Staff Sgt. Kristi Machada 303; U.S. Navy photo courtesy of Electric Boat Div., General Dynamics Corp 306; Getty Images/Hemera/Thinkstock 307; Erik Isakson/Getty Images 308; Comstock/Thinkstock 311; COREL Corporation 313; ©Pixtal/SuperStock 316; CORBIS 320

Chapter 8
Getty Images/Nikolaevich 326–27; Kay Washer 328; ©iStockphoto.com/Mike Kiev 331; ©Oleksandr Pakhay/Dreamstime.com 333; ©tarajane/Fotolia 338; Getty Images/iStockphoto/Thinkstock 342; ©iStockphoto.com/gechutka 350; Getty Images/Hemera/Thinkstock 352; ©Akira Fujii/DMI 355; John Lucas—SCDNR 357; Rocky Mountain Laboratories, NIAID, NIH 362

Chapter 9
©Visual&Written SL/Alamy 368–69; ©redbrickstock.com/Alamy 370; Charles Thatcher/Getty Images 374; PhotoDisc/Getty Images 375; PhotoDisc, Inc. 379; ©Science and Society/SuperStock 385; Unusual Films 387; U.S. Marine Corps photo by Staff Sgt. Danielle M. Bacon 390; Postdlf/Wikipedia/GNU Free Documentation License, CC 3.0 Unported 394

Chapter 10
©Chris Warham/Alamy 402–3; PhotoDisc/Getty Images 404, 420; ©Ashley Cooper/Alamy 407; USGS 412 (all); iStockphoto/Thinkstock 415; ©iStockphoto.com/Ronna Nichter 425; H. Pelloux/Wikimedia Commons/Public Domain 427

Chapter 11
Getty Images/iStockphoto/Thinkstock 430–31; ©iStockphoto.com/bigworld 432; ©iStockphoto.com/Jbryson 437; ©iStockphoto.com/Annie Greenwood 437 (inset); NASA 440 (both); ©Kike Calvvo/Alamy 447; ©Judith Collins/Alamy 450; Jose Gil/iStockphoto/Thinkstock 453; Getty Images/Comstock Images/Thinkstock 455; Oliviero Olivieri/Getty Images 457; *Wounded Christ Between Angels*, Bartolommeo Vivarini, From the Bob Jones University Collection 463; John Foxx/Stockbyte/Thinkstock 467; ©iStockphoto.com/Steve Cole 469; iStockphoto/Thinkstock 470; ©iStockphoto.com/Sam Castro 473; Rita Mitchell 474; Arnaud 25/Wikimedia Commons 481

Chapter 12
©Psalm 113. Image from BigStockPhoto.com 482–83, 526; ©iStockphoto.com/calvio 484; ©freakart.lv/Fotolia 488; ©iStockphoto.com/mattnomad 491; ©Odell Garrison 492; ©iStockphoto.com/RBFried 500; PhotoDisc, Inc. 502, 507; Peter Kaminski/Flickr/CC 2.0 511; Getty Images/Hemera/Thinkstock 515; ©Pearson Art Photo/Fotolia 516; ©iStockphoto.com/Fredy Thuerig 521

Chapter 13
Getty Images/iStockphoto/Thinkstock 530–31, 565; PhotoDisc, Inc. 532; Mark Wetzel 536; ©iStockphoto.com/RBFried 538; Digital Vision/Thinkstock 547; ©iStockphoto.com/scorpion56 552; Unusual Films 559; Deb Diehl, A Dyno-Might Amusements Co. LLC 564; Photo by Sgt. 1st Class Steven Petibone, New York National Guard 567